✱ 58 M.C.

✱   ✱

"C" thus

The U.S. Naval Institute
Blue & Gold Professional Library

For more than 100 years, U.S. Navy professionals have counted on specialized books published by the Naval Institute Press to prepare them for their responsibilities as they advance in their careers and to serve as ready references and refreshers when needed. From the days of coal-fired battleships to the era of unmanned aerial vehicles and laser weaponry, such perennials as *The Bluejacket's Manual* and the *Watch Officer's Guide* have guided generations of Sailors through the complex challenges of naval service. As these books are updated and new ones are added to the list, they will carry the distinctive mark of the Blue and Gold Professional Library series to remind and reassure their users that they have been prepared by naval professionals and meet the exacting standards that Sailors have long expected from the U.S. Naval Institute.

EM 300 Principles of Naval Engineering: Propulsion and Auxiliary Systems

Propulsion Technology					Auxiliary Systems				
Internal Combustion			Steam		HVAC + Psychrometrics	Desalination	Fluid Power and Flow		Heat Exchangers
Recip							Hydraulics	Pneumatics	
Otto (SI)	Diesel (CI)	Brayton (Ship + Aircraft)	Conv. Steam	Nuc Steam			Pumps		
Combustion of Hydrocarbons									
Drive Train Components and Efficiencies									

Thermodynamic Basics
Units and Unit Algebra   Properties   Power   Energy   Efficiency
General Energy Equation   Ideal Gas Laws   Open and Closed Systems   First Law   Second Law

Foundation
Physics   Algebra   Geometry   Chemistry

This sketch illustrates the variety of topics covered in this book and
the foundation knowledge and skills expected of students taking the course.

PRINCIPLES OF NAVAL ENGINEERING

PROPULSION AND AUXILIARY SYSTEMS

Matthew A. Carr

NAVAL INSTITUTE PRESS
Annapolis, Maryland

Naval Institute Press
291 Wood Road
Annapolis, MD 21402

© 2012 by the United States Naval Institute
Annapolis, Maryland
All rights reserved. No part of this book may be reproduced or utilized in any form or by any means, electronic or mechanical, including photocopying and recording, or by any information storage and retrieval system, without permission in writing from the publisher.

Library of Congress Cataloging-in-Publication Data
Carr, Matthew A.
  Principles of naval engineering : propulsion and auxiliary systems / Matthew A. Carr.
     p. cm.
  Includes bibliographical references and index.
  ISBN 978-1-61251-104-7 (alk. paper)
  1. Marine engineering. 2. Ship propulsion. I. Title.
  VM600.C37 2012
  623.87—dc23
                                                                                        2012014681

Printed in the United States of America on acid-free paper: ∞

20 19 18 17 16 15 14 13 12   9 8 7 6 5 4 3 2 1
First printing

Contents

Preface ix

UNIT 0 Introduction
0.1 Course Perspectives and Description 1
0.2 Engineering and Weapons Division Core Course Attributes of Graduates 2
0.3 Course Mission 2
0.4 Course Outcomes 2
0.5 Key Concepts 3
0.6 Foundational Skill Sets for Success in This Course 3
0.7 Systematic Approach to Problem Solving 3
0.8 Reasons Why Students Don't Excel in EM300 5
0.9 The Best Ways to Mess Up Practically Any Exam 6
0.10 Unit Algebra Review 6
0.11 Historical Background of Naval Engineering 8

UNIT 1 Fundamentals of Thermodynamics
1.1 Unit Learning Objectives 9
1.2 Introduction 10
1.3 Concepts 10
1.4 Units and Conversions 11
1.5 Force, Weight, and Mass 11
1.6 Properties 12
1.7 First Law of Thermodynamics 16
1.8 Steady Flow Energy Equation (SFEE) 16
1.9 Non-Flow Energy Equation (NFEE) 21
1.10 Second Law of Thermodynamics 21
1.11 Processes 22
1.12 Cycles 22
1.13 Efficiency 23
1.14 Torque and Power 26
1.15 Ideal Gas Law 27
■■■ PRACTICE PROBLEMS 31

UNIT 2 Incompressible Fluid Flow
2.1 Unit Learning Objectives 34
2.2 Introduction 35
2.3 Fleet Applications 35
2.4 Overview of Valves and Valve Symbology 36
2.5 Theory Development and Problem Solving Methodology 39
2.6 Pumps 46
2.7 Electric Motor Review 53
2.8 Venturi Meters 55
2.9 Siphons 55
2.10 Piping Identification 56
■■■ PRACTICE PROBLEMS 60

UNIT 3 Hydraulic and Pneumatic Systems
3.1 Unit Learning Objectives 64
3.2 Introduction 64
3.3 Fleet Applications 65
3.4 Theory Development and Problem Solving Methodology 66
3.5 Control of Fluid Power 69
3.6 Summary 74
■■■ PRACTICE PROBLEMS 75

UNIT 4 Internal Combustion Engines
4.1 Unit Learning Objectives 78
4.2 Introduction 79
4.3 Fleet Applications 80
4.4 Ideal Gas Processes 80
4.5 Otto Cycle 85

4.6	Diesel Analysis	92
4.7	Otto and Diesel Comparison	95
4.8	Hydrocarbon Combustion	97
	■■■ PRACTICE PROBLEMS	101

UNIT 5 Gas Turbines

5.1	Unit Learning Objectives	105
5.2	Introduction	106
5.3	Fleet Applications	106
5.4	Theory Development and Problem Solving Methodology	106
	■■■ PRACTICE PROBLEMS	126

UNIT 6 Steam Power

6.1	Unit Learning Objectives	131
6.2	Introduction	132
6.3	U.S. Navy Current Applications of Steam Power	133
6.4	States of Water	133
6.5	Steam Tables	136
6.6	Enthalpy: Entropy (*h-s*) or Mollier Diagrams	142
6.7	The Basic Steam Plant	144
6.8	The Rankine Cycle (The Basic Steam Plant with Ideal Components)	144
6.9	Basic Steam Cycle Analysis	145
6.10	Non-Ideal Steam Plant	147
6.11	Steam Plant Analysis	149
6.12	Conventional Boilers	154
	■■■ PRACTICE PROBLEMS	157

UNIT 7 Heat Exchangers

7.1	Unit Learning Objectives	162
7.2	Introduction	163
7.3	Modes of Heat Transfer	163
7.4	Heat Exchanger Classifications	166
7.5	Indirect Contact Heat Exchanger Construction	167
7.6	Heat Exchanger Analysis	168
	■■■ PRACTICE PROBLEMS	175

UNIT 8 Nuclear Power

8.1	Unit Learning Objectives	178
8.2	Introduction	179
8.3	Current U.S. Navy Applications of Nuclear Power	179
8.4	Basics of Nuclear Fission	180
8.5	Types of Nuclear Reactors	181
8.6	Navy Nuclear Propulsion Plant	182
	■■■ PRACTICE PROBLEMS	195

UNIT 9 Heating, Ventilation, Air Conditioning & Refrigeration (HVAC&R)

9.1	Unit Learning Objectives	198
9.2	Introduction	199
9.3	Historical Overview	199
9.4	Psychrometrics	201
9.5	Psychrometric Chart	202
9.6	HVAC Processes	205
9.7	Human Comfort Zone	205
9.8	HVAC Technology	205
	■■■ PRACTICE PROBLEMS	217

UNIT 10 Desalination

10.1	Unit Learning Objectives	221
10.2	Introduction	221
10.3	Distillation Basics	222
10.4	Methods of Distillation	223
10.5	Conclusions	228
	■■■ PRACTICE PROBLEMS	229

Equation Summary	231
Nomenclature	236
Appendix: Answers to Practice Problems	239
Refrigerant 134a Data	247
Air Tables	271
Index	283

Preface

This textbook has been developed for the course *Principles of Naval Engineering: Propulsion and Auxiliary Systems*. This is a "core engineering" course at the U.S. Naval Academy that is taught to all midshipmen who are not enrolled in an engineering major. This course has been taught in some form or other to generations of midshipmen. Reflecting the traditional emphasis throughout its history, the course continues to be referred to by its nickname "*Steam*."

The overarching goals of the course are to provide a basic understanding of the technology of the U.S. Naval Forces and to develop the technical analytical skills of midshipmen in view of future assignment to the Navy and Marine Corps. This one course covers Fleet applications of what would make up four or five courses in a standard Mechanical Engineering curriculum and does it in the "English" unit system (something increasingly rare in the American educational system, but still used extensively by the U.S. Fleet). Textbooks developed for undergraduate programs at civilian colleges and universities do not cover the breadth and depth of this course, so developing text materials "in house" at the Naval Academy has been the pattern for many years.

A brief history of the text resources used over the past several decades includes: in the 1970s, course notes developed by the Naval Systems Engineering Department that supplemented *Principles of Naval Engineering*, NAVPERS 10788 series (1958, 1966, and 1970); *Introduction to Naval Engineering* (1980) edited by Edward F. Gritzen; *Introduction to Naval Engineering, 2nd Ed.* (1985) edited by Professor Emeritus Arthur E. Bock; *Thermodynamics of Marine Engineering Systems* (1995) by Hawley, Vining, and Wiggins; *Thermodynamics of Marine Engineering Systems, 2nd Ed.* (1998) by Hawley, Wiggins, Vining, and Lindler.

The Fleet underwent a significant transition in technology and shift in propulsion plant configuration over the period of use of the preceding texts that was not reflected in these earlier books. In response to an extensive program assessment in 2001 and curriculum review conducted by the superintendent during the 2005–2006 academic year, the topical coverage of this course was revised:

(1) to reflect the current state of engineering systems in the Fleet;
(2) to ensure coverage of practical engineering issues published in the *Professional Core Competencies* and from the various sources of feedback from the Fleet;
(3) to include aviation turbo-engine variants to assist the significant number of graduates selecting aviation careers; and
(4) to remove more theoretical elements in the course that did not directly apply to the operation and maintenance of current Fleet systems.

As a result, a team of officers with extensive Fleet experience participated in a multi-year project to redevelop a textbook for the Navy of the early twenty-first century. The following personnel contributed to the writing and assembly of course notes used as a developmental draft. These officers worked to produce a resource that would prove valuable to newly commissioned officers for personal review and for training others under their command. The authors' ranks and warfare specialties are listed so students may see the diversity of perspective that went into this textbook.

- Fundamentals of Thermodynamics
  LT Alexander Dutko, USN, Helicopter Pilot
- Incompressible Fluid Flow
  LT Nathan Williams, USN, CEC
- Hydraulic and Pneumatic Systems
  LT Nathan Williams, USN, CEC

- Internal Combustion Engines
  CDR Matthew Carr, USN, Submariner
- Gas Turbine Engines
  CDR Leonard Hamilton, USN, Test Pilot
- Steam Power
  CDR Matthew Carr, USN, Submariner
  LT Joseph Root, USN, Submariner
- Heat Exchangers
  CDR Matthew Carr, USN, Submariner
  CDR Kevin Leeds, USNR, Surface Warfare
- Nuclear Power
  CAPT Murray Snyder, USN, Submariner
  CDR Matthew Carr, USN, Submariner
  LT Joseph Root, USN, Submariner
- Heating, Ventilation, Air Conditioning & Refrigeration (HVAC&R)
  CDR Matthew Carr, USN, Submariner
  CDR Kevin Leeds, USNR, Surface Warfare
- Desalination
  LT Jason Zeda, USN, CEC

The course notes were used over a number of semesters and progressively edited by CDR Carr and CDR Leeds. In addition, a number of retired officers and professors have taught this material as long-term adjunct instructors. Their experience and input have been invaluable in smoothing out the scope and sequence of the topics. Noteworthy contributors during this phase of development include:

- CDR Bruce Duncan, USN (ret): CDR Duncan is a career submariner, including having commanded a nuclear submarine. His perspective in developing junior officers and the demands placed upon them has been invaluable.
- CDR Wallace Elger, USN (ret): CDR Elger was an Engineering Duty Officer (EDO) with significant shipyard experience and served on the Navy's Board of Inspection and Survey (INSURV) as well as teaching experience with The Johns Hopkins University and USNA.
- CDR Ashley Yetman, USNR (ret): CDR Yetman has taught this course material for more than five years and has developed a strong record of teaching.
- Professor Eugene Keating: Dr. Keating is a Fellow of the Society of Automotive Engineers (SAE) with teaching experience at the U.S. Merchant Marine Academy, the U.S. Naval Academy, and the University of Maryland. He is an author of engineering books on thermodynamics and combustion.

We believe that this textbook will prove to be a helpful resource and is worthy to be placed in the personal professional library of each of our graduates. Best wishes in your naval careers!

CDR Matthew A. Carr, USN
Editor-in-Chief

CDR Kevin J. Leeds, USNR
Associate Editor

0 Introduction

The USS *Winston Churchill* (DDG-81) is one of the *Arleigh Burke*–class of Navy destroyers. The *Burke*-class, as planned, will be the largest class of destroyers ever built.

This introductory unit is intended to provide an overview perspective of this course. It describes the mission and scope of topical coverage of the *Principles of Naval Engineering: Propulsion and Auxiliary Systems* course, based upon the Professional Core Competencies (PCCs) issued by the Chief of Naval Education and Training (CNET) and other local requirements satisfied by this course. Further, it promulgates the course learning outcomes. Lastly, it identifies the typical issues students face when enrolled in this course.

0.1 COURSE PERSPECTIVES AND DESCRIPTION

This is a course about how to think and, specifically, how to think about mechanical technology of which Naval Academy graduates will be placed in charge during their careers. The authors have attempted to make this course apply to the types of things division officers see in their Fleet assignments.

This course is part of the engineering portion of the "divisional requirements" courses within the U.S. Naval Academy Academic Program. It is a required course for all students enrolled in majors from Divisions II and III and may be incorporated into the course sequence for General Engineering students from Division I.

The mission statement recognizes that the role of the fleet officer is to lead people in the management of technologically sophisticated systems, regardless of warfare platform. Further, approximately 40 percent of each graduating class will be engaged in some aspect of Naval Aviation. ==This course teaches students about gas turbine engines used in aviation applications.==

COURSE CATALOG DESCRIPTION. EM300 *Principles of Naval Engineering: Propulsion and Auxiliary Systems (3-2-4)*. A study of naval engineering propulsion and auxiliary systems, including the principles of energy conversion and the basic operation of internal combustion engines, ship and aircraft gas turbine engines, and conventional and nuclear steam engines; and the basic operating principles of auxiliary systems and components, such as air conditioning, desalination, hydraulic and pneumatic power systems, and heat exchangers.

PREREQUISITES.
SP211 General Physics I
or
SP221 Physical Mechanics I

Note: Although SP211 or SP221 are the only official prerequisite courses, the fundamental skills of high-school-level Algebra and Geometry and the topics of chemical stoichiometry, chemical thermodynamics, and shipboard feed water chemistry controls, covered in SC111 and SC112, are assumed.

0.2 ENGINEERING AND WEAPONS DIVISION CORE COURSE ATTRIBUTES OF GRADUATES

The following four attributes/objectives have been defined for the Engineering and Weapons contribution to the USNA Core Program. This contribution seeks to:

1. Provide officers who posses knowledge of fundamental engineering principles that build on the knowledge provided by the basic sciences.
2. Provide officers who can apply their knowledge of engineering fundamentals to understand and analyze Navy relevant engineering systems.
3. Provide officers who can communicate effectively on technical topics related to the Navy's engineering systems.
4. Provide officers who are capable of building on this engineering foundation.

These are the things that you should be able to do better when you are commissioned because you've taken this course.

0.3 COURSE MISSION

The mission of this course is to develop the technical foundations of the enrolled midshipmen with the goal of competency in technical communication and reasoning for the newly commissioned unrestricted line officer in the naval service.

0.4 COURSE OUTCOMES

Course outcomes are things that students should be able to do at the conclusion of the course. These are more detailed than the attributes of graduates listed above and are supported by more detailed unit learning objectives.

<u>Periodic review of these course outcomes and the unit learning objectives will provide a comprehensive view of what the student is expected to demonstrate.</u>

Student self-assessment of these outcomes will be conducted as part of the end-of-course critique.

1. Communicate technical information using proper terminology, units, and ranges of values. Apply analytical problem solving skills and unit analysis to evaluate technical information.
2. Apply the first law of thermodynamics and the continuity concept to solve fluid flow problems. Apply the concepts of head loss, pump head, and Mechanical power to determine major and minor head losses, pump performance, and power requirements.
3. Apply Pascal's law, the ideal gas law, and the principle of continuity to evaluate operation of hydraulic and pneumatic systems.
4. Identify the basic components of reciprocating engines and describe the operation of both Otto and Diesel ideal cycles. Use ideal gas law relationships to determine cycle parameters, performance, and efficiency.
5. Identify the basic components ==of gas turbine engines== and describe the operation of the Brayton cycle. Use Air Tables to determine cycle parameters, performance, and efficiency.
6. Identify the basic components of current steam engines and describe the operation of a Rankine cycle. Use steam tables to determine cycle parameters, performance, and efficiency.

7. Explain how the heat from fission powers naval heat engines. Describe the main components and fundamental operation of a naval pressurized water reactor.
8. Describe the construction of common heat exchangers. Use heat transfer relations to determine the transfer of heat between systems. Explain the source and effect of fouling on heat exchanger performance.
9. Explain how air is conditioned for naval applications and use a psychrometric chart to determine air properties. Describe the operation of a vapor-compression refrigeration plant and use pressure-enthalpy (p-h) charts to determine cycle parameters and performance.
10. Discuss needs for different grades of water and methods for producing it.

0.5 KEY CONCEPTS

In general the student should have competency in selecting, analyzing, and presenting information about naval heat engines and auxiliary systems and be able to reason about, compare, and identify abnormalities in their performance. This means, for all featured power and auxiliary systems:

- Explain the thermodynamics of machinery operation—what it does and how it does it—using energy and mass transfer and balance principles.
- Compare system performance under different conditions by applying appropriate principles.
- Compare real to ideal machines.

Commit the following foundational concepts to memory. These are "Carry-Around Knowledge," which will serve you well as you analyze the systems covered in this course.

- Fluid flows in response to a pressure gradient, from high pressure to low pressure.
- Heat flows in response to a temperature gradient, from high temperature to low temperature.

0.5.1 Unit Systems

- Convert from SI to USCS, as necessary.
- Use the unique units and ranges found on gages and other instruments in U.S. Fleet systems.

0.5.2 Reasoning

The professional world is full of technical problems for which there are missing information and extraneous data. The technical thinker will recognize and discard extraneous data; make reasonable assumptions for needed data that are not readily available within the required time; and evaluate available parameters and use them to obtain a satisfactory solution. This course will help to develop this way of thinking. In addition, students are encouraged to:

- Develop expectations for, evaluate changes in, and select among system outcomes as conditions change.
- Evaluate strategies for improving efficiency or reducing losses.

0.5.3 Fleet Applications

- Think about where the equipment being studied is used in the Fleet and Marine Forces. You will see this identical or similar equipment again.

0.6 FOUNDATIONAL SKILL SETS FOR SUCCESS IN THIS COURSE

- Be able to do algebra involving variables with associated units.
  - *Apply conversion factors when and where appropriate.*
- Look up data from the appropriate table or graph or calculate them from equations of state.
  - *Correctly interpolate using tabular or graphical data sources.*
- Interrelate:
  system schematic (symbols)
      AND
  process graph(s) (lines/curves/areas)
      AND
  physical components (equipment).
- Solve word problems in a systematic/organized and neat fashion.

0.7 SYSTEMATIC APPROACH TO PROBLEM SOLVING

You may consider this a checklist, but it is really a disciplined thought process that will help you to develop competence in technical analysis. Refer back to this section as each new unit is covered.

Read the Problem Statement

- Diagnose the problem. (The 5 S's)
  - **System**—Open (flowing) or closed (non-flowing)?
  - **Substance**—What material(s) is(are) involved?
  - **State**—gas, liquid, solid, or phase mixtures?
  - **State Change**—What kind of <u>process</u> is happening between state points?
  - **Sequence**—Does a series of processes make a cycle by returning to the starting point?
- Review the given information.
  - Associate the words with the correct variable symbols and vice versa.
  - Evaluate the given units for needed conversions and to corroborate the associated variables.
  - Consider if any given information is extraneous and not needed.
- List what you are asked to find and note if the units are specified.
- Consider which equation(s) you will use <u>and their limits of applicability</u>.
  - Remember that some equations were developed with limiting restrictions, like for ideal gases only. Some require absolute temps and pressures, not relative ones.
  - Think about data sources for info needed, but not given:
    - standard conditions from the course equation sheets.
      Note: If information is given which leads you to use "non-standard" condition values, then you are expected to take that into account.
    - look-ups from data tables (typically Air or Steam).
    - thermodynamic graphs that are associated with the problem.
    - change in (Δ) one property based upon changes in other property(ies).
    - assumptions with reasonable values.
  - Make preliminary conversions and determine intermediate answers that you'll need in the final equation. It is a good idea to make the intermediate answers stand out by underlining or other technique that isn't confused with the final answers in boxes.
    Note: It is usually better to accept some round-off rather than carry the preliminary equations into the primary equation for the problem. Carry extra digits while doing this, but then round at the end to minimize too much preliminary rounding error.

Draw "The System" Schematic and Include Appropriate Process Graphs

- Use appropriate symbols and nomenclature.
- Identify state points to analyze on the schematic(s) and the process graph(s).
  - One sketch per state point if a non-flowing problem.
  - SPs should match between schematic and process graphs.
- Show appropriate data on the schematic.
  - Make the sketch communicate "the essence" of the problem.
  - Think about the fluid flows, thermodynamic processes, and basic feedback control signals.
- Draw process graphs and additional sketches to show greater detail or concepts, as necessary.
  - process graph(s) (p-v, T-s, h-s, p-h, etc.)
    - Include real and isentropic points, as necessary.
  - efficiency block diagrams
  - heat source and sink first law concept sketches
  - pressure relationship graphs

Organize and Display Data

- Lay out a state point table for cycle problems.
  - Show the units in the variable cells.
  - Circle the givens and other entering information.
  - Indicate which cells are NA, if any.
  - Adjust the table to include isentropic efficiencies, as necessary.

Write the Appropriate Equation(s)

- Eliminate terms that don't apply, are negligible, or cancel.
- Solve algebraically for the requested variable.
  - Make sure that your state point numbers are consistent in the equation.
- Substitute in the values and units for the remaining variables.
- Analyze units for any needed conversion factors.
  - All additive terms must have the same units.

Answer(s)

- Box your answer(s)
  - Variable = number and units.
    - Check that you have solved it in the prescribed units.
    - Use customary units for the answer variable if none are specified.
- Check the answer for reasonability and against the answer key.

- ☐ Check and correct or annotate if it seems out of line with expected values and you can't figure out why.
- ■ One box per item asked for in the problem statement.

0.8 REASONS WHY STUDENTS DON'T EXCEL IN EM300

We recognize that this is a challenging course for most students. However, the following is a synopsis of instructor observations and student self-assessments regarding the causes of less-than-optimum performance in this course. Make the commitment to learn from others' mistakes and not to repeat them for yourself. If you don't do well, this section is formatted so you can check the blocks identifying your issues and develop a remediation plan. **The Bottom Line—FALLING BEHIND IS A KILLER!** Keep up with the flow of this course!

POOR ATTITUDE AND MOTIVATION—"YOUR ATTITUDE DETERMINES YOUR ALTITUDE."

- ■ Loss of Big Picture—Many have convinced themselves that their major's courses are the primary reason that they chose to attend USNA and then do not apply themselves well to courses outside of their major. Remember that for most of the people enrolled, this course is part of the reason you get a Bachelor of Science degree, rather than a Bachelor of Arts.
- ■ Some think that their service selection won't use the material covered in this course. Are you really sure? Or is that wishful thinking? Don't you think your much more experienced instructors who have been out in the naval service know better?
- ■ Others are not pursuing excellence and are doing the minimum necessary to "get by."

ABSENTEEISM + INEFFECTIVE EFFORTS AT MAKING UP MISSED MATERIAL

- ■ Inattentiveness in class—Being disengaged is essentially absenteeism.

LESS-THAN-ADEQUATE ORGANIZATION SKILLS

- ■ Poor note taking skills in class.
  - ☐ Sloppy
  - ☐ Illegible
  - ☐ Disorganized
  - ☐ Copying vs. listening, thinking, synthesizing, and writing
- ■ Failure to organize notes and other printed resources.
  - ☐ Bound notebooks don't allow organizing handouts, quizzes, exams—Don't use a bound notebook.
  - ☐ Use of pockets in folders vs. a 3-ring filing system.
  - ☐ Loose items not sorted and filed.

LESS-THAN-ADEQUATE TIME MANAGEMENT

- ■ Ineffective budgeting of time during exams and quizzes.
- ■ Not allowing enough time to do the HW and lab practicals.
  - ☐ Produce sloppy and incomplete work.
- ■ Not working HW closer to the class session in which the material was covered.
  - ☐ Deadline motivator vs. effectively building capability.
  - ☐ Time-Delay Knowledge Decay Factor.

FAILURE TO TAKE ADVANTAGE OF AVAILABLE RESOURCES

- ■ EI
  - ☐ Avoid scheduling EI.
  - ☐ Don't prepare for EI.
- ■ Textbook
  - ☐ Won't obtain
  - ☐ Won't read
- ■ Course Website
  - ☐ methodology, examples, answer keys, extra credit opportunities.
- ■ Handouts
- ■ Returned Graded Work
  - ☐ Filed without (or insufficient) analysis of the feedback and commitment to incorporate lessons learned in future work.

LESS THAN ADEQUATE STUDY SKILLS

- ■ Lecture Prep—Don't preview topics before a class session.
- ■ Lecture Review—Don't review topics after a session.
- ■ Memorization vs. Understanding
  - ☐ Don't really know how to analyze a worked example or answer key.
- ■ Ineffective Group Work
  - ☐ Copying others' work vs. using it as a learning tool.

UN-REMEDIATED POOR ALGEBRA SKILLS (some are out of practice and don't review it)

0.9 THE BEST WAYS TO MESS UP PRACTICALLY ANY EXAM

This section is adapted from guidance put out by CAPT Len Hamilton of the USNA Mechanical Engineering Department to his students. It is a bit "tongue in cheek," but if it prevents you from making mistakes and losing points, then it is worth printing. It is probably also worth you reading it…

8. Use the ideal gas law for everything, even liquids and solids!
7. Don't draw system sketches or process graphs. Sure, they clear up confusion, but look at the valuable 5 or 10 seconds they waste!
6. Forget about units! They'll probably work out on their own.
5. Who needs variables? Just start plugging numbers into your calculator to save time. Who needs partial credit anyway?
4. Don't sweat extra batteries. Your calculator will never quit during an exam. That only happens to other people!
3. Don't box your answers. Make it challenging for the instructor to find them. It serves him/her right for making such a hard test anyway!
2. If you get stuck on a problem, stick with it until time runs out. If you don't get to the other problems, it is clear that the test was too long!

And the #1 answer is….

1. Don't bother to read the question thoroughly. It's better to answer what you think it said or what you wanted it to say, rather than what it actually said!

0.10 UNIT ALGEBRA REVIEW

This section is included in this preliminary unit as a refresher for those who have not done algebra for some time. If you are current in your algebra skills, then you may choose to skip this section without consequences.

A *variable* is a symbol that represents an attribute or property, such as length (L), temperature (T), time (t), or many others. A *term* is any of the quantities of variables connected by addition or subtraction in an equation, or on opposite sides of the equal sign, such as in $a = b + (c/d)$. Both "b" and "c/d" are terms, as well as the "a."

Equations used in EM300 are *dimensionally homogeneous*, which means that the net units (after cancellation) of all terms in an equation must be the same if the terms are being added or subtracted.

In order to successfully use the equations presented in this course, **YOU MUST KEEP TRACK OF ALL UNITS** and apply *unit conversions* as necessary to get the different groups of terms to have the same units, or to provide the answer in the desired units.

You will find unit conversion factors in your *Steam Tables* and on the course equation sheets. A unit conversion provides the relationship between units with the same basic dimensions, e.g., 1 in = 2.54 cm (length is the basic dimension). Dividing both sides of this unit conversion by 2.54 cm:

$$\frac{1 \text{ in}}{2.54 \text{ cm}} = \frac{2.54 \text{ cm}}{2.54 \text{ cm}} = 1$$

The "unit ratio" is 1, so we can multiply or divide any quantity by unit ratios and change the units without changing the meaning of the mathematical term. To determine the number of centimeters in the length (L) of a marathon, we proceed as follows:

$$L = 26.2 \text{ miles} \cdot \frac{5{,}280 \text{ ft}}{1 \text{ mile}} \cdot \frac{12 \text{ inches}}{1 \text{ ft}} \cdot \frac{2.54 \text{ cm}}{1 \text{ inch}}$$

$$= 4{,}216{,}481.28 \text{ cm} = 4.22 \times 10^6 \text{ cm}$$

Note that all the units except "cm" cancel and give us the number of "cm" that is equal to 26.2 miles. **Also note that we're still talking about the same variable, L, regardless of whether the units are in miles or cm.** This procedure may have been introduced to you previously using "railroad track" analysis. This is a helpful organizational technique to many students.

$$L = \frac{26.2 \text{ miles}}{} \; \bigg| \; \frac{5{,}280 \text{ ft}}{1 \text{ mile}} \; \bigg| \; \frac{12 \text{ inches}}{1 \text{ ft}} \; \bigg| \; \frac{2.54 \text{ cm}}{1 \text{ inch}}$$

The back section of your *Steam Tables* booklet has conversion factors written in the following format example: "Multiply Btu by 778.172 to obtain ft-lb$_f$." The associated variable could be either heat (Q) or work (W), but this format means that the units associated with the conversion factor's number have the desired ending units in the numerator and the starting units in the denominator of the conversion factor, e.g.,

$$Q = 500 \text{ Btu} \left\{ 778 \frac{\text{ft} \cdot \text{lb}_f}{\text{Btu}} \right\} = 389{,}000 \text{ ft} \cdot \text{lb}_f$$

Note that I have rounded the conversion factor to the more commonly used approximate value of 778. This conversion factor is also known as the *mechanical equivalent to heat*, since Btu units are more typically associated with heat, while ft-lb$_f$ units are associated with work, which is mechanical energy.

Typically you will need to convert units while solving an equation to be able to add/subtract terms. As an example, consider the frictionless form of Bernoulli's equation:

$$z_1 + \frac{p_1}{\gamma} + \frac{\vec{v}_1^2}{2g} = z_2 + \frac{p_2}{\gamma} + \frac{\vec{v}_2^2}{2g}$$

This equation relates pressure (p), velocity ($\vec{v}$), and elevation (z) at two points in a fluid with known specific weight (γ), and can be used to solve for an unknown pressure, velocity, or elevation at one of the points provided that the correct unit conversions are applied. Note that there is an arrow over the velocity variable. This is to distinguish it from ambiguous uses of the same letter—v also stands for specific volume, which is also different from total volume, V, as we'll discuss later in the course.

EXAMPLE 0.1: UNIT CONVERSIONS METHODOLOGY

If $z_1 = 100$ ft, $z_2 = 50$ ft, $p_1 = 30$ lb$_f$/in^2, $\vec{v}_1 = 25$ ft/s, $\vec{v}_2 = 1$ ft/s, and $\gamma = 62.4$ lb$_f$/ft^3, then the unknown pressure p_2 can be determined by rewriting Bernoulli's equation using algebra as follows:

$$p_2 = p_1 + \gamma(z_1 - z_2) + \gamma\left(\frac{\vec{v}_1^2 - \vec{v}_2^2}{2g}\right)$$

Note that each term has basic dimensions of force per area (length^2) and you must make sure that the units for each term are the same before adding and subtracting terms:

$$p_2 = 30\frac{\text{lb}_f}{\text{in}^2} + 62.4\frac{\text{lb}_f}{\text{ft}^3} \cdot (100\text{ft} - 50\text{ft}) \cdot \left(\frac{1\text{ ft}}{12\text{ in}}\right)^2$$

$$+ 62.4\frac{\text{lb}_f}{\text{ft}^3} \cdot \frac{\left(25\frac{\text{ft}}{\text{s}}\right)^2 - \left(1\frac{\text{ft}}{\text{s}}\right)^2}{2 \cdot 32.2\frac{\text{ft}}{\text{s}^2}} \cdot \left(\frac{1\text{ ft}}{12\text{ in}}\right)^2$$

$$\boxed{p_2 = 55.87\frac{\text{lb}_f}{\text{in}^2}}$$

Using the "railroad track" method:

$$p_2 = \frac{30\text{ lb}_f}{1\text{ in}^2} + \frac{62.4\text{ lb}_f}{1\text{ ft}^3}\bigg|\frac{(100-50)\text{ ft}}{1}\bigg|\frac{1\text{ ft}^2}{144\text{ in}^2} +$$

$$\frac{62.4\text{ lb}_f}{1\text{ ft}^3}\bigg|\frac{(625-1)\text{ ft}^2}{1\text{ s}^2}\bigg|\frac{1}{2}\bigg|\frac{1\text{ s}^2}{32.2\text{ ft}}\bigg|\frac{1\text{ ft}^2}{144\text{ in}^2}$$

$$p_2 = \frac{30\text{ lb}_f}{\text{in}^2} + \frac{21.67\text{ lb}_f}{\text{ft}^3}\bigg|\frac{\text{ft}}{}\bigg|\frac{\text{ft}^2}{\text{in}^2} +$$

$$\frac{4.20\text{ lb}_f}{\text{ft}^3}\bigg|\frac{\text{ft}^2}{\text{s}^2}\bigg|\frac{\text{s}^2}{\text{ft}}\bigg|\frac{\text{ft}^2}{\text{in}^2}$$

$$p_2 = \frac{30\text{ lb}_f}{\text{in}^2} + \frac{21.67\text{ lb}_f}{\text{in}^2} + \frac{4.20\text{ lb}_f}{\text{in}^2}$$

$$\boxed{p_2 = 55.87\text{ lb}_f / \text{in}^2 = 55.87\text{ psi}}$$

As shown in the first line, there is an "implied" 1 above or below each non-1 number.

Note that the equal signs are lined up vertically to better illustrate the relationships and substitutions. This is a good technique to emulate.

Note that the unit of "ft/s^2" (associated with 32.2) is itself a fraction and is written as the reciprocal in the "railroad tracks" since 32.2 is in the denominator. This follows the rules of complex fractions, with which students should be familiar.

Also notice that you can apply an exponent to a unit ratio conversion factor and still retain the validity of the conversion factor.

If $1\ ft = 12\ in$, then the conversion factor is $\frac{1\text{ft}}{12\text{in}}$

and $\left(\frac{1\text{ft}}{12\text{in}}\right)^2 = \frac{1\text{ft}^2}{144\text{in}^2}$.

The unit ratio conversion factor $(1\text{ft}/12\text{in})^2$ was applied to the last two terms in order to guarantee that all terms have units of lb$_f$/in^2. Without the unit ratio conversion factors, the last two terms would have units of lb$_f$/ft^2 and could therefore not be added to the first term, which has units of lb$_f$/in^2.

In order to recognize when unit conversion factor ratios are required, **YOU MUST ATTACH THE APPROPRIATE UNITS TO EVERY NUMBER YOU WRITE DOWN**.

Also note that when you apply an exponent to a variable, you apply the same exponent to the units. In the above example,

$$\bar{v}^2 = \left(25\,\frac{\text{ft}}{\text{s}}\right)^2 = 25^2\,\frac{\text{ft}^2}{\text{s}^2} = 625\,\frac{\text{ft}^2}{\text{s}^2}$$

Yes, it is a small amount of extra work. However, you will catch many mistakes by always writing down units since you will be able to see that the basic dimensions of terms don't match (because you wrote the equation down wrong, or plugged in a length for a velocity, or forgot to square something, or any number of other minor mistakes that will cost you points on exams, quizzes, and homework) or that unit ratio conversion factors are even required for the problem you are working.

The "**SOLVE**" button on your very expensive graphing calculator will not sort out the units for you. If you are unable to perform basic algebra and evaluate the units to determine whether a conversion factor must be applied, you may not pass this course—you definitely won't do well. Diagnose your situation and take appropriate remedial actions to address any weaknesses in this foundational skills area.

Extra Credit Opportunity—If you still need some extra practice, work Navy Knowledge Online Unit #045DA01 "Dimensional Analysis" and submit the completion certificate for extra credit.

0.11 HISTORICAL BACKGROUND OF NAVAL ENGINEERING

The U.S. Navy built the first steam-powered warship in 1815, the *Demologos* (later re-named the *Fulton* after Robert Fulton, its designer). The post-war (War of 1812) mission requiring ships on distant stations and the technological limitations of this period resulted in a sail-only Navy until improved commercial steam vessels and the introduction of steamships into the European navies caused the U.S. Navy to re-investigate steam propulsion in the mid-1830s. The American Civil War, with significant naval forces employed in the Mississippi and other rivers and in the typically shallow coastal environment of the Confederacy, accelerated the rate of introduction of steam warships and the need for officers trained to employ this technology. The post-war (Civil War) Navy reverted to sailing ships with steam auxiliary propulsion until the 1880s, when steam technology development reached the state that allowed abandoning sails for primary propulsion. Steam propulsion continues to this day in warships, primarily in submarines and larger ships, but has largely been replaced by diesel engines in merchant ships and gas turbines in smaller warships.

The early twentieth century saw the addition of diesel engines to the naval propulsion mix, and it is a prime mover of ships and electrical generators to this day. Gas turbine ship propulsion came into the force in the 1970s. Currently, the Navy is, once again, investigating electric-drive ships. This propulsion concept was used on ships up to battleship size starting around the WWI timeframe and on submarines until nuclear power replaced diesel-electric drive. Subsequent and ongoing improvements in electric motor and combat systems technology make electric drive an attractive feature of proposed ship classes. Regardless of what drives the ship's propulsors, electrical generators will continue to be driven by thermodynamic engines (most likely gas turbine and diesel engines) for the foreseeable future.

Ship auxiliary systems have likewise seen significant developments over the years. Fluid systems (hydraulics, pneumatics, potable water, wastewater, cooling water, lubricating oil, etc.) are common and necessary systems for ship operation. Energy, volume, weight restrictions, and maintenance needs along with stealth requirements have caused the Navy to re-design many of these necessary auxiliary systems. Air conditioning and refrigeration systems and other atmospheric control equipment make our ships and submarines habitable, if not always comfortable. However, environmental concerns that surfaced in the 1970s have forced the Navy to adopt new refrigerants. Increased concern over the use of hazardous materials and the environmental impacts of emissions has added another layer of requirements on the Navy of today. The need for technologically competent officers has never been greater.

Aircraft engines were traditionally not taught in this course. While there is overlap in the technology (specifically spark ignition and gas turbine engines) with marine propulsion systems, aircraft have unique components and drive train configurations with which our graduates need to be familiar, especially as about 40 percent of our graduates are assigned to some aspect of naval aviation.

The topical coverage taught to midshipmen has adapted with new technology, and this course will continue to adapt as the Navy continues to develop new propulsion and auxiliary systems.

Some version of this course has been taught to midshipmen since shortly after the American Civil War. If your predecessors handled it and, through better understanding of how their warfare platforms worked, they helped to create the most powerful navy in the world, then so can you.

1 Fundamentals of Thermodynamics

This sailor is calibrating a pressure gage to ensure that it indicates properly. Having accurate instrument readings is essential to proper operation of systems and equipment.

1.1 UNIT LEARNING OBJECTIVES

1.1.1 • Terminology: definitions, variables, and typical units

1.1.1.1 thermodynamics, working substance, material properties, density, specific gravity, specific volume, specific weight, gage pressure, absolute pressure, vacuum pressure, energy, potential energy, kinetic energy, internal energy, heat, work, power, state point, thermodynamic process, reversible process, ideal process, adiabatic, enthalpy, entropy, specific heat, thermodynamic cycle, efficiency

1.1.2 • Concepts: ideas and engineering expressions of them

1.1.2.1 State the difference between stored energy and transitional energy.

1.1.2.2 State the difference between thermal and mechanical energy.

1.1.2.3 State the first law of thermodynamics and explain the meaning of each term.

1.1.2.4 State the second law of thermodynamics and discuss its graphical application.

1.1.2.5 Explain the concept of efficiency. Apply the concept to thermodynamic cycles.

1.1.2.6 Distinguish between steady flow, non-flow, and steady state.

1.1.2.7 Explain reversibility and irreversibility in a thermodynamic process.

1.1.2.8 Relate the universal gas constant to specific heats.

1.1.2.9 Explain the concept of torque and relate it to a mechanical power train.

1.1.3 Skills: procedures, practices, or methods that enable reasoning

- *1.1.3.1* Convert systems and basis (specific, total, rate) of units.
- *1.1.3.2* Calculate pressure using a manometer.
- *1.1.3.3* Relate absolute, gage, vacuum, and barometric pressures.
- *1.1.3.4* Write and apply the steady flow energy equation (SFEE).
- *1.1.3.5* Write and apply the non-flow energy equation (NFEE).
- *1.1.3.6* Calculate the heat exchanged during a thermodynamic process using mathematical expressions.
- *1.1.3.7* Sketch thermodynamic processes on *T-s* and *p-v* coordinates and relate the meaning of areas on the graphs.
- *1.1.3.8* Apply the equation of state for an ideal gas to solve for an unknown property.
- *1.1.3.9* Apply the first law of thermodynamics in analysis of a cycle.
- *1.1.3.10* Select the first law expression appropriate to system conditions.
- *1.1.3.11* Calculate heat supplied in a thermodynamic cycle.
- *1.1.3.12* Calculate heat rejected in a thermodynamic cycle.
- *1.1.3.13* Calculate net-work in a cycle.
- *1.1.3.14* Calculate cycle thermal efficiency.
- *1.1.3.15* Apply the various efficiencies represented in a ship propulsion system to calculate the flow of power and conversion of types of power.

1.2 INTRODUCTION

Unit 1 reviews some concepts that are typically covered in chemistry and physics courses, both at the high school and at the post-secondary level. One important difference between the coverage of these topics in the prerequisite courses and this course is that we will use the United States Customary System (USCS) of units rather than Système International (SI). Americans, in practical terms, have been extremely resistant to shift to the SI system in spite of the fact that the American education system shifted to SI in textbooks and instruction decades ago. Similarly, the Navy continues to use USCS and related traditional units in the gages and instruments installed on naval equipment. This is because the costs of transitioning to SI would be extensive in hardware, technical manuals, personnel training, and the increased risk of mishaps while gaining familiarity with a "new" unit system. So, since the Navy continues to use USCS, the students of this course will be expected to gain proficiency in using USCS and converting the few cases of SI data to USCS units.

1.2.1 Thermodynamics

Stemming from the Greek terms *therme*, meaning heat, and *dynamis*, meaning power, *thermodynamics* is the science of transforming and transferring energy, and the accompanying change in the state of the matter that is involved. Though energy can take myriad different forms, the applicable forms for this course are *heat*, the thermal energy transferred due to a temperature difference, and *work*, energy transferred by forces acting on a body. The interaction between heat and work is the true basis for any thermodynamic analysis. While the term thermodynamics may conjure notions of thermal energy in motion, our discussion will be limited to *steady state*, or equilibrium conditions. Our focus, therefore, will be the initial and final conditions—or states—of the matter being analyzed. We will not look at transients, other than in concept.

1.3 CONCEPTS

Fundamental to the study of thermodynamics is the concept of a *system*. A thermodynamic system is a quantity of matter contained within a prescribed boundary. This boundary, though not necessarily physical, defines the outer limit of the matter under analysis. As shown in Figure 1.1, the surroundings include everything outside the system boundary.

There are three main ways to classify a system, each depending on the transfer of energy or matter between a system and its surroundings. (1) In an *open* system, both energy and matter may cross the system boundaries. (2) A *closed* system, however, is identified by a fixed mass, and, as a consequence, energy only may be exchanged between a closed system and its surroundings, but the system's matter remains within the boundaries. While both closed and open systems interact with their surroundings through energy exchange, (3) an *isolated* system can neither transfer matter nor energy across its boundaries.

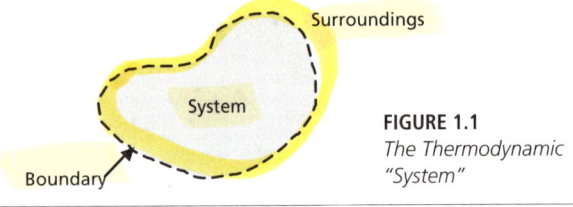

FIGURE 1.1
The Thermodynamic "System"

The matter contained within the system is termed the *working substance* or *working fluid*. It is defined as any substance (or fluid) that undergoes changes to affect a change in the system. The state—or unique condition—of the working fluid is defined by certain material *properties*. Any change in a single property will alter the state of the working fluid. Such a change in the system state—through a change in any material property of the working fluid—is termed a *process*. A system undergoes a *cycle* when the system with a given initial state experiences a series of processes and returns to the initial state. The concepts of processes and cycles will be addressed more fully in Section 1.12.

The properties that define the state of the working fluid within a system may be *intensive* or *extensive*. Intensive system properties do not depend on the system size or the amount of working fluid in the system. Examples of intensive properties include temperature, pressure, density, and boiling point. Some intensive properties are per unit mass. Specific volume and specific internal energy are examples of such intensive properties. By contrast, an extensive property depends directly on the size—or extent—of the system, or the amount of working substance contained within the system boundaries. Mass, volume, and total energy are examples of extensive properties. Our study of thermodynamics requires a firm grasp of several extensive and intensive properties that we will address in greater detail later in this unit.

1.4 UNITS AND CONVERSIONS

Prior to examining the individual properties of a working fluid, we will review the basic systems of units. Although SI (Système International) units are likely familiar, virtually every gage on board naval ships and aircraft displays USCS (U.S. Customary System) units, also known as "English" units. As such, our thermodynamic analysis will primarily employ USCS.

Fundamental units are the foundational units of any system of measurement. They do not get simpler than fundamental units. In familiar SI units, the fundamental units of length (L), mass (m), and time (t) are meters (m), kilograms (kg), and seconds (s), respectively. Combinations of units are referred to as *derived* units. The concept of force (F) originates from Newton's law where $F = ma$ and is thus a derived attribute. In SI terms, the units of F are Newtons (N) with $1\ N = 1\ kg\text{-}m/s^2$. The corresponding fundamental units in the USCS are feet (ft), pounds-mass (lb_m), and seconds (sec).

Units may be converted from one system to the other, or to desired units within the same system, by application of a conversion factor. By the nature of a factor within algebra, it is acceptable to multiply any number by a *unity conversion factor*, sometimes also referred to as a *unity ratio* to change the units. For example, if you wanted to determine the length of a mile in feet, then multiply L by the appropriate unity conversion factor:

$$L = \frac{1\ \text{mile}}{} \cdot \frac{5280\ \text{ft}}{1\ \text{mile}} = 5280\ \text{ft}$$

The numerator units cancel with "like" denominator units, in this case, miles. The variable is still length, L. The variable has not changed, just its units have changed. In the above example, the "railroad track" technique is employed to keep numerator units and denominator units organized. Any number of unity conversion factors may be multiplied in sequence. For example, to determine the number of inches in a mile, multiple conversion factors may be applied.

$$L = \frac{1\ \text{mile}}{} \cdot \frac{5280\ \text{ft}}{1\ \text{mile}} \cdot \frac{12\ \text{inches}}{1\ \text{ft}} = 63{,}360\ \text{in}$$
$$= 6.336\ \text{E4 in}$$

Successful students in this course consistently prove to be proficient in recognizing the need for and applying appropriate conversion factors.

1.5 FORCE, WEIGHT, AND MASS

Newton's second law of motion relates the force exerted on an object to the rate of change of the object's momentum through the familiar equation, $F = ma$. By definition, one Newton of force accelerates one kilogram of mass at the rate of one meter per square second, i.e., $1\ N = 1\ kg\text{-}m/s^2$. In USCS, however, one pound-force accelerates not one, but 32.2 pounds-mass (commonly known as one *slug*), at one foot per square second. That is, $1\ lb_f = 32.2\ lb_m\text{-}ft/sec^2 = 1\ slug\text{-}ft/sec^2$. Note that the units of force in USCS are pounds-force (lb_f). Since the majority of mass measurements are specified in pounds-mass, a gravitational constant, g_c, is applied to Newton's second law to obtain consistent units, or dimensional homogeneity. This conversion factor, obtained through examination of the units of the relationship $F = ma$, is given by $g_c = 32.2\ lb_m\text{-}ft/lb_f\text{-}sec^2$. Newton's second law can thus be restated as:

$$F = \frac{ma}{g_c} \tag{1.1}$$

where the mass, m, is specified in pounds-mass (lb_m) and the acceleration, a, of the object is given in ft/sec^2.

Since the weight of an object is the force due to gravity, the familiar relationship of the product of the object's mass and the local acceleration due to gravity, $W = mg$, must be similarly modified to obtain dimensional homogeneity. Weight, therefore, can be determined from:

$$W = \frac{mg}{g_c} \qquad (1.2)$$

where the object's mass is specified in pounds-mass, but the gravitational acceleration, g, is taken as a constant 32.2 ft/sec² at sea level. Note that the acceleration due to gravity, g, and the gravitational constant, g_c, have identical numerical values, but significantly different units. Restating both quantities: $g = 32.2$ ft/sec² and $g_c = 32.2$ ft-lb_m/lb_f-sec². Since the ratio of these two quantities appears regularly, it is helpful to note that $g/g_c = 1$ lb_f/lb_m. At sea level, therefore, one pound-mass weighs one pound-force.

EXAMPLE 1.1

If an astronaut weighs 200 lb_f on earth, how much does the astronaut weigh on the moon where the local acceleration due to gravity is 5.3 ft/sec²?

GIVEN: $W_{earth} = 200$ lb_f
$g_{earth} = 32.2$ ft/sec²
$g_{moon} = 5.3$ ft/sec²

FIND: W_{moon} (lb_f)

SOLUTION: From Newton's second law, $W = \frac{mg}{g_c}$.

On earth, therefore, the astronaut has a mass of

$$m_{earth} = \frac{W_{earth} g_c}{g_{earth}}$$

$$= \frac{(200 lb_f)(32.2 \, ft \cdot lb_m / lb_f \cdot sec^2)}{32.2 \, ft / sec^2} = 200 lb_m.$$

Since the mass of the astronaut remains constant, i.e., $m_{earth} = m_{moon}$, the astronaut's weight on the moon is given by

$$W_{moon} = \frac{m_{moon} g_{moon}}{g_c} = \frac{(200 lb_m)(5.3 \, ft / sec^2)}{32.2 lb_m \cdot ft / lb_f \cdot sec^2} = 32.9 lb_f$$

1.6 PROPERTIES

Properties define a system of material being analyzed. There are two categories of properties, *extensive* (also referred to as *total*), and *intensive* (also referred to as *specific*). Extensive properties depend upon the quantity of material present. The value of the property changes if there is more or less of the material present in the system. Intensive properties are independent of how much of a material is present.

Most extensive properties have corresponding intensive properties. We use a convention of indicating extensive properties with capital letters and their associated intensive properties with lowercase letters. For example, total enthalpy (H) has units of [Btu] and specific enthalpy (h) has units of [Btu/lb_m].

With a few notable exceptions, total properties are converted to specific properties by dividing the total property of the system by the mass of the system. For example, specific volume (v) is calculated by dividing the total volume by the mass in that volume: $v = V/m$. This is amplified in the discussions below.

1.6.1 Density and Specific Volume, Specific Gravity and Specific Weight

The density of a working fluid (ρ) depends on its temperature and pressure. It is defined as the mass of the substance per unit volume.

$$\rho = \frac{m}{V} \qquad (1.3)$$

Though the density of a substance is frequently utilized, a more common quantity is the specific volume (v). It is defined as the volume per unit mass and is the reciprocal of the density.

$$v = \frac{V}{m} = \frac{1}{\rho} \qquad (1.4)$$

In some instances, the density of a substance is given relative to the density of a standard substance. This ratio of densities is termed the specific gravity. The reference standard for comparison of liquids is freshwater and for gases is standard air.

$$s.g._{fluid} = \frac{\rho_{fluid}}{\rho_{reference \, fluid}} \qquad (1.5)$$

Recall from Equation 1.2 that the weight of a substance is defined as the product of its mass and the local acceleration due to gravity divided by the conversion factor g_c. Specific weight is the weight per volume.

$$\gamma = \frac{W}{V} = \frac{mg}{Vg_c} = \frac{m}{V}\frac{g}{g_c} = \rho\frac{g}{g_c} \qquad (1.6)$$

The numerical value of g, therefore, is identical to that of r. The units, however, differ: g (lb$_f$/ft^3) and r (lb$_m$/ft^3). Note that with the exception of s.g. and g, all remaining specific properties of interest can be defined on a per-unit-mass basis.

1.6.2 Temperature

The temperature of a substance is a measure of the activity of the molecules comprising the substance. Although this definition may not seem immediately useful, the units of temperature—and conversions among them—are utilized regularly. Two temperature scales—relative and absolute—are fundamental to our discussion. The familiar relative temperature scales, Fahrenheit (T_F) and Celsius (T_C), are related by:

$$T_F = \frac{9}{5}T_C + 32 \quad (1.7)$$

Each of the relative temperature scales has an associated absolute temperature scale. The absolute temperature on the Celsius scale (°C) is given in Kelvins (K) and is governed by:

$$T_K = T_C + 273.15 \quad (1.8)$$

The Celsius scale was defined by the freezing and boiling point of water at standard pressure. On the Fahrenheit scale (°F), meanwhile, the absolute temperature is given in degrees Rankine (°R) and is governed by:

$$T_R = T_F + 459.67 \quad (1.9)$$

The rounded-off values 273 and 460 are typically used in Equations 1.8 and 1.9, respectively, without loss of too much precision. Notice that two of these temperature scales are SI and the other two are USCS.

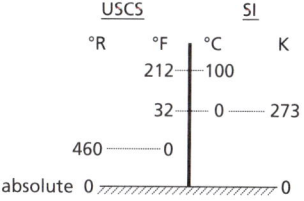

FIGURE 1.2 *Temperature Scales and Key Equivalent Values*

The fundamental concepts of temperature are reflected in the "zeroth law of thermodynamics" which is a statement concerning systems in thermal equilibrium. The zeroth law states that if two systems are in thermal equilibrium with a third system, then they must be in thermal equilibrium with each other. The ability to use a thermometer comes from the zeroth law. The reason that it is numbered "zero" is that this reality came to light after the first law of thermodynamics was well established and this principle is considered even more fundamental than the first law.

1.6.3 Pressure

For liquids and gases—our most prevalent working substances—the force acting normal to a unit area of the substance's surface is known as the pressure exerted on the substance.

$$p = \frac{F}{A} \quad (1.10)$$

This relationship will form the foundation for our discussion of hydraulics and pneumatics in Unit 3.

Various types of gages are utilized to measure and indicate the pressure of the working fluid within a particular system. As mechanical devices, gages are influenced by the pressure of the environment in which the gage sits. To clearly specify the gage value, pressure readings from gages are typically reported in psig. In most applications, however, absolute pressure must be utilized in order to look up the correct values of properties in data tables. Absolute pressure, P_{abs}, is the sum of the measured—or gage—pressure, P_{gage}, and the local atmospheric pressure, P_{atm}.

$$P_{abs} = P_{atm} + P_{gage} \quad (1.11)$$

The pressure that the atmosphere exerts, P_{atm}, essentially comes from the weight of the air above the location of measurement and, consequently, varies with elevation and weather conditions. At sea level, standard atmospheric pressure is 14.696 psia. This value is commonly rounded off to 14.7 psia.

A special purpose pressure gage, called a barometer, is used to measure the atmospheric pressure. Because atmospheric pressure is related to weather, a barometer is useful in predicting weather events, such as impending storms and weather front passages. Barometers always indicate absolute pressure, regardless of the units. Common barometer units include: in-Hg; mm-Hg (Torr); milliBars (mB); and kPa. Table 1.1 relates standard atmospheric pressure in common barometer units.

TABLE 1.1: *Standard Atmospheric Pressure Equivalents*

1 std atmosphere	= 14.7 psia
	= 29.92 in-Hg
	= 760 Torr
	= 101.3 kPa
	= 1013 mBar

As noted above, atmospheric pressure varies with elevation. Figure 1.3 illustrates the pressure vs elevation relationship based upon standard atmospheric pressure at sea level. The addition of an "a" on the units of atmospheric pressure denotes an absolute value, not that it is atmospheric. Negative elevation (below sea level) will increase barometric pressure. The surface of the Dead Sea in Israel is about 1,100 feet below sea level. The average atmospheric pressure there is about 31.5 in-Hg. An interesting proof of this concept can be conducted by students when traveling. Seal an empty plastic beverage bottle when driving across a mountain range and then examine the bottle at sea level. The bottle should have partially collapsed.

Weather conditions also affect atmospheric pressure with extreme low pressure associated with storms. The lowest sea-level barometric pressure ever recorded was 25.69 in-Hg in the eye of the storm, Typhoon Tip, west of the island of Guam in the Pacific Ocean on October 12, 1979. The U.S. record is 26.35 in-Hg, produced by the 1935 Labor Day hurricane (back then they didn't yet use names for hurricanes as we do now), the eye of which crossed the U.S. coastline at Matecumbe Key, Florida, on September 2, 1935. Hurricane Katrina, which caused extensive damage to New Orleans and neighboring Gulf Coast areas in 2005, had a reported eye pressure of 26.86 in-Hg. Even lower barometric pressures may occur in the eye of a tornado, but they are more dangerous to measure. In contrast, the highest barometric pressure ever recorded on Earth was 32.01 in-Hg, measured in northern Siberia, in 1968. The weather was clear and very cold (dense) at the time, with temperatures between -40 °F and -58 °F.[1] The typical range of barometric pressure varies between locations, but is on the order of 1 to 3 in-Hg. Weather stations, regardless of their elevation, correct their barometric pressure readings to what they would be at sea level. Lines of constant barometric pressure on a weather map are called *isobars*. The closer that isobars are together, the higher the wind velocity will be.

Now, back to gage pressures. Gage pressure can be positive or negative, with negative gage pressures called vacuum. Equation 1.11 holds for negative gage pressures, though vacuum pressures may be annotated P_{vac}.

$$p_{abs} = p_{atm} + (-p_{gage}) = p_{atm} - p_{vac} \qquad (1.12)$$

The relationships among absolute, atmospheric, gage, and vacuum pressures are shown graphically in Figure 1.4.

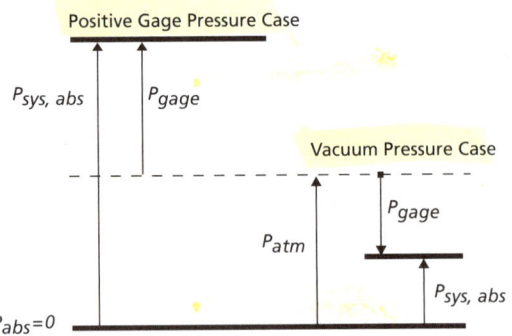

FIGURE 1.4 *Pressure Relationships*

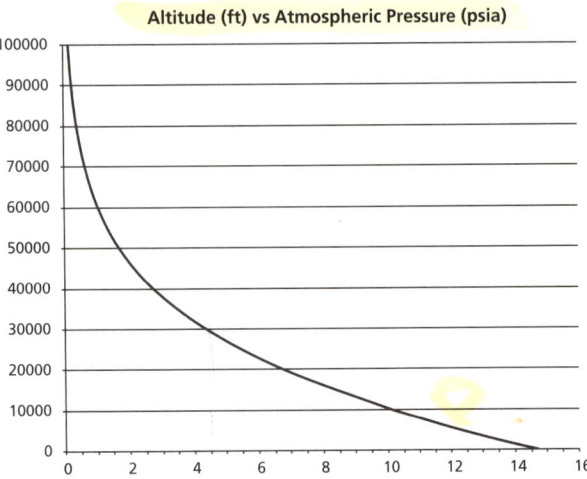

FIGURE 1.3 *Atmospheric Pressure vs Altitude (Based upon 14.7 psia at Sea Level)*

Of special note, the U.S. Navy standard is that vacuum gages indicate pressure in units of "inches of mercury" (in-Hg). The genesis of this convention is the historical use of mercury-filled manometers in the early condensers used in naval steam plants. While manometers are not currently in common use for indicating condenser pressure, the units persist. Whenever these units are used, students should be aware that the units *could* indicate a vacuum reading. Be careful to interpret whether the problem statement is referring to the gage reading or the absolute pressure.

1.6.3.1 Manometers

A pressure gage of particular interest is the U-tube manometer, an example of which is shown in Figure

1.5. It is a device utilized to measure the pressure of an enclosed fluid or gas using hydrostatics. That is, the static height of a column of liquid, h, indicates the gage pressure of the substance.

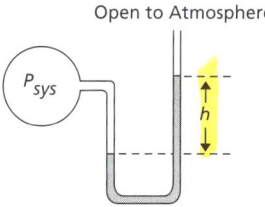

FIGURE 1.5 Simple U-Tube Manometer

The absolute pressure of the enclosed gas or fluid can be determined from Equation 1.11. The gage pressure is simply the weight of the liquid column, W, acting on the cross-sectional area, A, of the vertical section.

$$p_{gage} = \frac{F}{A} = \frac{W}{A} = \frac{1}{A}\left(\frac{mg}{g_c}\right) \quad (1.13)$$

The area of the vertical section can be written as the volume, V, divided by the height of the column such that Equation 1.13 can be simplified.

$$p_{gage} = \left(\frac{\rho g}{g_c}\right)h \quad (1.14)$$

Finally, noting that the quantity in parentheses is the specific weight of the liquid, Equation 1.14 is simplified to the hydrostatic equation. The manometer fluid must be steady for the hydrostatic equation to apply.

$$p_{gage} = \gamma h \quad (1.15)$$

The pressure calculated from 1.15 can be greater or less than atmospheric, depending on the height of the liquid column.

Manometers that are open to the atmosphere are limited in the pressures that they can measure due to the height available to install and read the fluid levels in the tubing. The pressure can be increased by using a denser manometer fluid, such as mercury instead of water. Mercury-filled manometers were used in early steam condensers to measure the vacuum pressure achieved in the condenser. This fact lives on as a legacy item in the Navy in that we continue to measure steam plant condenser vacuum pressure in the units of in-Hg, even though mercury-filled manometers have been replaced by mechanical gages and electronic pressure transducers. Recall that 14.7 psi = 29.92 in-Hg.

Yet another customary unit for several unique applications is inches of water (in-H_2O) or inches water gage (in-w.g.). Water-filled manometers used to be used to measure small pressures or pressure differences. For example, air duct pressure losses and supply pressure for fuel gases, such as natural gas, continue to be measured in in-H_2O. Recall that 14.7 psi = 29.92 in-Hg = 407 in-H_2O.

A manometer may also be used to measure differential pressure across a device as shown in Figure 1.6. If the density of the manometer fluid is much greater than the measured fluid, then the h_1 may be ignored. Regardless, since h_3 applies equally to both sides of the device with level piping, h_3 may be ignored. Note that with this configuration, it is impossible to determine the absolute pressure. Only the differential pressure may be obtained.

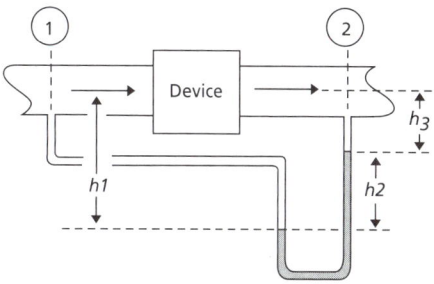

FIGURE 1.6 Differential Pressure Measuring Manometer

EXAMPLE 1.2

A simple U-tube manometer uses mercury (γ_{Hg} = 850 lb$_f$/ft^3) to measure the pressure in a water pipe (γ_{H2O} = 62.4 lb$_f$/ft^3) as shown below. If the vertical section is open to the atmosphere (p_{atm} = 14.7 psia), the mercury level difference is 4 inches, and water level in the manometer is 12 inches below the centerline of the pipe, calculate the absolute pressure in the pipe.

GIVEN: γ_{Hg} = 850 lb$_f$/ft^3
γ_{H2O} = 62.4 lb$_f$/ft^3
h_1 = 12 inches
h_2 = 4 inches

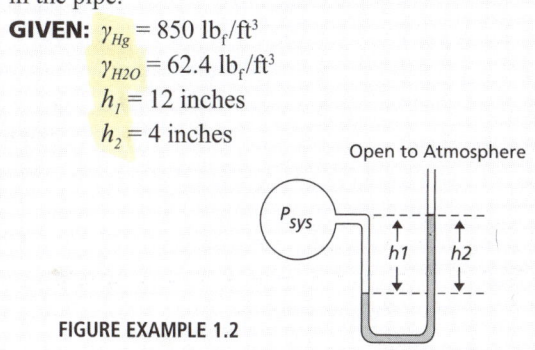

FIGURE EXAMPLE 1.2

FIND: $p_{sys,abs}$ (psia)

SOLUTION: The system being measured is the pipe. Notice that there is an extra hydrostatic component in the manometer due to a fairly dense material, water, in the manometer tube. We usually can neglect this when measuring low-density gases with most liquids. In this case the pressure in the pipe will be slightly less than the pressure at the interface of the water and mercury. From Eqns 1.11, 1.15, and including a correction for water in the tube, $p_{sys} = p_{atm} - \gamma_{H2O}h_1 + \gamma_{Hg}h_2$. We substitute the given information and include appropriate conversion factors to obtain

$$p_{sys,abs} = \left(14.7 \frac{lb_f}{in^2}\right)\left(144 \frac{in^2}{ft^2}\right) - \left(62.4 \frac{lb_f}{ft^3}\right)(12in)$$
$$\left(\frac{1}{12}\frac{ft}{in}\right) + \left(850 \frac{lb_f}{ft^3}\right)(4in)\left(\frac{1}{12}\frac{ft}{in}\right)$$

$$p_{sys,abs} = \left(2116.8 \frac{lb_f}{ft^2}\right) - \left(62.4 \frac{lb_f}{ft^2}\right) + \left(283 \frac{lb_f}{ft^2}\right)$$

$$= 2337.4 \frac{lb_f}{ft^2}$$

Applying our conversion factor (144 in²/ft²) yields

$$p_{sys,abs} = 16.23 \text{ psia.}$$

1.6.3.2 Pressure Gages

With the extended discussion of manometers, one would think that these are the primary pressure gages used. In reality, manometers are rarely used. In their place, a variety of mechanical devices have been developed. The most common gage today uses a Bourdon tube connected to a mechanical gear system that moves a needle over a graduated faceplate (with numbers marked on the faceplate). In addition, bellows and diaphragm gages will also be found in typical Fleet applications. A simplex gage reads one location. Other gages include compound, duplex, vacuum, and differential. If the gage is subjected to vibrations, a fluid such as glycerin may be added to dampen the indicating needle. These gages will be discussed in a lab experience. The important point to make here is that these gages are influenced by the environment in which they sit. Consequently, the difference between gage pressure and absolute pressure must be accounted for, just as described above for manometers.

1.7 FIRST LAW OF THERMODYNAMICS

The first law of thermodynamics requires that all energy, E, input to a system equals the energy leaving the system:

$$\sum E_{in} = \sum E_{out} \qquad (1.16)$$

Formally, the first law states that the only manner in which a system may change its total energy is through changes in heat and/or work. This relationship is commonly referred to as *conservation of energy*. Energy conservation implies that energy can neither be created nor destroyed, merely converted from one form to another. We'll see, later, that nuclear power violates this in that matter is converted to energy, but for the rest of the material covered in this course, the first law holds true.

1.8 STEADY FLOW ENERGY EQUATION (SFEE)

Most students can, at this point in their education, name several forms of energy such as potential, kinetic, electromagnetic, and nuclear. A general energy equation could be written that would include every form of energy, but when considering that some of the energy forms are inconsequential to engines and auxiliaries, they may be ignored. It is also noteworthy to recognize that nuclear fission and fusion reactions are associated with conversion of mass into energy. Nuclear energy requires special accounting to keep track of this form of energy.

We, then, can write a modified version of the general energy equation that only includes the forms that may be consequential to the study of engines and auxiliaries. The energies into and out of the system may be classified and accounted for in their various forms. The forms of interest to us are potential, kinetic, internal, and flow work energy. Additionally, the first law requires accounting for heat and/or work to cause change in the state of the working fluid.

Consider a working fluid flowing at steady state through a piece of equipment (steady state means no transients are occurring). The boundaries of the system are typically taken as the casing of the component and data measurement points in the fluid where it enters and exists the component. Fluids generally enter components via pipes. With one fluid inlet and one fluid outlet, it is convenient to designate state point (SP) 1 at the inlet and SP2 at the outlet. State points are placed at the inlet and outlet of components in steady flow systems. When considering open streams of flowing fluid, such as water falling over a dam or the stream of water from a fire hose nozzle, it is convenient to consider the boundary at the interface between the working fluid and its surroundings and designate cross-sectional locations where data are known for the state points.

The potential, kinetic, internal, and flow work energies traveling with the working fluid may be accounted for at the inlet and outlet state points. These energies are *stored energies* in that they are stored in the working fluid. Heat and work occur across system boundaries and are called *transitional energies*.

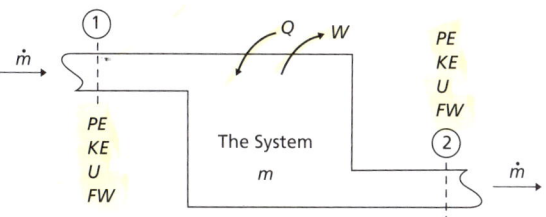

FIGURE 1.7 Steady Flowing Open System

Thus, the first law for open, flowing systems takes the form of the SFEE:

$$\Sigma(PE + KE + U + FW)_{in} + Q_{if}$$
$$= \Sigma(PE + KE + U + FW)_{out} + W_{if} \quad (1.17)$$

We make special note of the different subscripts applied to the heat and work terms. Since the working fluid possesses potential (PE), kinetic (KE), internal (U), and flow work (FW) energies, each term is labeled with subscripts indicating their values at either the inlet (*i*) or outlet (*f*) state points. Heat and work, however, are transferred by the working fluid between the initial and final states and cross the system boundaries. Consequently, the terms receive subscripts that emphasize that subtle, yet very important difference. For most of our systems, there will be one inlet and one outlet. If you have multiple inlets and/or outlets, then each flowing boundary point gets its own state point number. Sum up the inlets on the left hand side of the SFEE and the outlets on the right.

Before discussing each form of energy individually, we note that the USCS units of energy are British Thermal Units (Btu) or ft-lb$_f$. It is helpful to note that *1 Btu = 778 ft-lb$_f$*. This conversion factor is very useful in thermodynamic analyses throughout the course and is known as the *mechanical equivalent of heat*. A Btu is analogous to a calorie. It takes 1 Btu of heat to raise the temperature of 1 lb$_m$ of water (at typical room conditions) by 1 °F.

We'll now look into the various energy components of the SFEE.

1.8.1 Potential Energy

Energy stored within a system due to the position or elevation of the working fluid relative to a specified datum is termed potential energy. The amount of total potential energy, PE, contained within a substance is determined directly from its ability—or potential—to perform work against gravity.

$$PE = \frac{mgz}{g_c} \quad (1.18)$$

On a specific basis:

$$pe = \frac{PE}{m} = \frac{gz}{g_c} \quad (1.19)$$

where the elevation of the substance, z, is given in feet.

1.8.2 Kinetic Energy

The kinetic energy of a substance is stored due to the velocity, *v*, of the working fluid. The amount of total kinetic energy, KE, possessed by a substance is based on the work required to accelerate the fluid to a given velocity from rest.

$$KE = \frac{m\vec{v}^2}{2g_c} \quad (1.20)$$

Specific kinetic energy is calculated by:

$$ke = \frac{KE}{m} = \frac{\vec{v}^2}{2g_c} \quad (1.21)$$

where the USCS units of velocity are typically ft/sec.

1.8.3 Internal Energy

The total thermal energy stored within a substance is termed the internal energy. It is associated with the translation, rotation, and vibration of the molecules comprising the substance and is indicated by the variable *U* or—on a specific basis—*u*. Although internal energy is always present in a substance possessing molecular activity, we need not calculate its absolute value. Rather, we will often focus on the magnitude of the change of a working fluid's internal energy.

1.8.4 Flow Work

The amount of stored energy required to maintain the continuous steady flow of working fluid is termed the displacement energy, or flow work. This form of energy applies only to open systems. That is, no displacement energy is stored within a working substance unless it is flowing into or out of the system. The total amount of flow work, *FW*, possessed by a substance is related directly to the pressure required to move a unit volume of the substance across the system boundary.

$$FW = pV \quad (1.22)$$

Specific flow work is calculated by

$$fw = \frac{FW}{m} = pv \quad (1.23)$$

where the pressure must be converted to lb_f/ft^2 since the volume (or specific volume) is typically given in ft^3 (or ft^3/lb_m).

1.8.5 Steady Flow Energy Equation (SFEE)

Substituting Eqns 1.18, 1.20, and 1.22 into 1.17 yields the Steady Flow Energy Equation (SFEE) in total terms:

$$\left(\frac{mgz}{g_c} + \frac{m\bar{v}^2}{2g_c} + U + pV\right)_i + Q_{if}$$
$$= \left(\frac{mgz}{g_c} + \frac{m\bar{v}^2}{2g_c} + U + pV\right)_f + W_{if} \quad (1.24)$$

On a specific basis, the SFEE takes the familiar form:

$$\left(\frac{gz}{g_c} + \frac{\bar{v}^2}{2g_c} + u + pv\right)_i + q_{if}$$
$$= \left(\frac{gz}{g_c} + \frac{\bar{v}^2}{2g_c} + u + pv\right)_f + w_{if} \quad (1.25)$$

Finally, we note that it is important to understand that Eqns 1.24 and 1.25 are merely mathematical expressions of the first law of thermodynamics. The student will become very familiar with these equations.

1.8.6 Enthalpy

The sum of internal energy and flow work is termed the enthalpy of the working fluid. Enthalpy is a measure of the ability of a fluid to do work. The total enthalpy (*H*) of a substance is of fundamental importance to thermodynamic analysis of engines and thermal energy systems.

$$H = U + pV \quad (1.26)$$

The specific enthalpy of a substance is given by

$$h = u + pv \quad (1.27)$$

Either identity may be substituted into the SFEE. Take special note of the units associated with each term in Equations 1.26 and 1.27—conversion factors are required to obtain dimensional homogeneity!

1.8.7 Heat, Work, and Power

The term *work* encompasses a broad spectrum of energy transfer methods. The work, *W*, performed on or by a working fluid, however, is typically defined as the mechanical work—the product of the force, *F*, applied to or generated by the working fluid and the distance, *d*, moved in the direction of the applied force. Work is given by the familiar equation $W = F \times d$. For the engines that we will be dealing with, *W* is produced at a rotating shaft. Regardless of its form, work is energy transferred across the boundary of a system and *specific work* is $w = \frac{W}{m}$.

The rate of performing work is termed *power* and is abbreviated as $\dot{W}$ or P. The rate of performing mechanical work, therefore, is termed mechanical power (see Eqn 1.28). Rate is often indicated by placing a dot over the term. For example, mass flow rate is $\dot{m}$. The mechanical power performed by a working fluid is calculated by the product of $\dot{m}$ of the working fluid and the specific work conveyed by the substance.

$$\dot{W} = P = \dot{m}w \quad (1.28)$$

The widespread use of the unit of horsepower (hp) in engine ratings requires us to note that *1 hp = 746 W = 550 ft-lb_f/sec*. Horsepower is an energy rate.

Energy transferred across the system boundary that is not accounted for by any of the various work modes is termed *heat*. Heat (*Q*) is thermal energy transferred due to a temperature gradient—or difference—between the system and surroundings. It is important to understand that systems do not contain heat. Rather, they contain energy, and heat is energy in transition. The amount of heat transferred by a unit mass of working fluid—the *heat in specific terms*—is given by $q = \frac{Q}{m}$. The heat in specific terms, or *specific heat*, must not be confused with the *specific heat capacity*, *C*. While *q* in specific terms is the amount of heat transferred per unit mass of the working fluid, *C* is defined as the energy required to increase the temperature of the unit mass of a substance by one degree under specified conditions. Of interest are the specific heat capacities of a substance at constant volume (C_v) and constant pressure (C_p). Physically, C_v and C_p can be viewed as the energy required to raise the temperature of a unit mass of a substance by one degree while the volume is held constant, or while the pressure is held constant, respectively. We will address and utilize the specific heat capacities of ideal gases later.

Just as the rate of performing mechanical work is termed mechanical power, the rate at which heat is transferred across the system boundary is termed thermal power ($\dot{Q}$). As with mechanical power, thermal power is calculated from the product of the mass flow rate of the working fluid and the specific heat transferred by the substance.

$$\dot{Q} = \dot{m}q \quad (1.29)$$

In subsequent units, we will study additional methods of calculating thermal power. We now see, however, that the energy contained within a system may be transferred to the surroundings through the performance of work or transfer of heat. Thus, heat and work are quantitatively equivalent.

Using the concepts just developed, the SFEE may also be expressed in total rate basis format.

$$\dot{m}\left(\frac{gz}{g_c} + \frac{\bar{v}^2}{2g_c} + u + pv\right)_i + \dot{Q}_{if}$$
$$= \dot{m}\left(\frac{gz}{g_c} + \frac{\bar{v}^2}{2g_c} + u + pv\right)_f + \dot{W}_{if} \quad (1.30)$$

The energy equations have been developed to reflect the typical situation of one flow path of working fluid in and out of the system and flowing from the inlet to the outlet. If there is more than one input stream or output stream of working fluid, then Equations 1.24, 1.25, and 1.30 must be modified to reflect the summation of input energies and output energies.

1.8.7.1 Heat and Work Energy Sign Convention

Whether heat or work is positive or negative comes from the concept of a heat engine. In a heat engine, heat is supplied by burning a fuel and the heat released by combustion is converted to work. Therefore, heat is accounted for on the input side of the SFEE and if heat is "in" to the system it is considered to be positive. Similarly, work is the desired product of an engine. The work is accounted for on the output side of the SFEE and is arbitrarily designated as positive if out of the system.

Two useful mnemonics to remember this convention are:

> Heat In or Work Out – Positive (HIWOP)
> Heat Out or Work In – Negative (HOWIN)

It is interesting to note that a vapor compression refrigeration system, such as the air conditioner in an automobile or a household refrigerator, requires work input to absorb heat into the system and is opposite to the heat engine example. Refrigeration will be addressed in great detail in a later unit.

Since most of the applications of this course deal with engines that convert heat to work, it is useful to illustrate the energy flows using concept sketches.

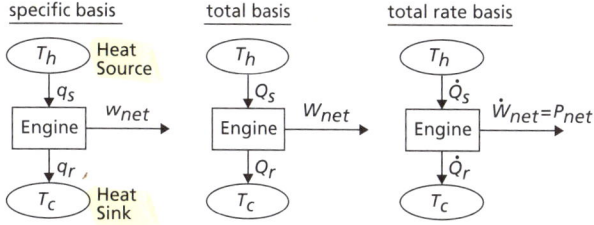

FIGURE 1.8 *Heat Engine Energy Concept Sketches*

In Figure 1.8, the left sketch deals with energy on specific terms; the middle in total terms; and, the right is total rate terms. A heat engine converts heat supplied from a high temperature source (q_s, Q_s, or $\dot{Q}$) to net work (w_{net} or W_{net}) or power ($\dot{W}$ or P). A heat engine also rejects heat, q_r or Q_r to a low temperature sink. Then the first law applied is: The energy input must equal the outputs:

$$q_s = w_{net} + q_r;\ Q_s = W_{net} + Q_r;\ \text{and},\ \dot{Q}_s = P_{net} + \dot{Q}_r.$$

EXAMPLE 1.3

A steadily flowing system containing 3 lb_m of working fluid emits 15 Btu of heat. If the following initial and final conditions are observed, calculate the work (Btu) performed on or by the system.

	INITIAL, SP1	FINAL, SP2
p (psia)	50	37
v (ft³/lb_m)	10.3	11.7
u (Btu/lb_m)	170	182
v (ft/sec)	350	50
z (ft)	900	100

GIVEN: $m = 3 \; lb_m$
$Q_{12} = -15$ Btu (Note the negative sign)
Data summarized in table above.

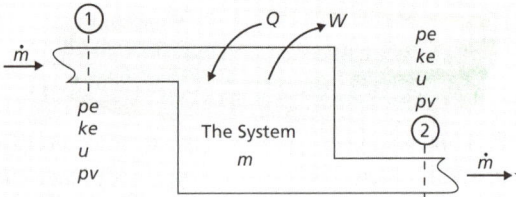

FIGURE EXAMPLE 1.3

FIND: W_{12} (Btu)

SOLUTION: We're looking for total work, W. Notice that the mass given is the mass inside the boundaries and not the mass flow rate. I've shifted to referring to the state points by numbers, with 1 representing the entering and 2 representing exiting boundaries. The given information permits a steady flow analysis using either Equation 1.24 or 1.25. Since the majority of the given information is on a specific basis, we employ the specific form of the SFEE. Our approach, therefore, will be to solve for the specific work using the following:

$$w_{12} = \left(\frac{gz}{g_c} + \frac{\bar{v}^2}{2g_c} + u + Pv\right)_1 - \left(\frac{gz}{g_c} + \frac{\bar{v}^2}{2g_c} + u + Pv\right)_2 + q_{12},$$

and then multiplying by the mass to obtain the total work, i.e., $W_{12} = (m)w_{12}$. Finally, we must correctly interpret the sign of the calculated quantity in accordance with our sign convention.

Substituting the given information and ensuring dimensional homogeneity, we obtain the following:

$$w_{12} = \left(\frac{1 lb_f}{1 lb_m}\right) 900 \, ft + \left(\frac{1}{2 \cdot 32.2 \frac{ft \cdot lb_m}{lb_f \cdot sec^2}}\right)\left(350 \frac{ft}{sec}\right)^2 +$$

$$\left(170 \frac{Btu}{lb_m}\right)\left(778 \frac{ft \cdot lb_f}{Btu}\right) + \left(50 \frac{lb_f}{in^2}\right)\left(144 \frac{in^2}{ft^2}\right)$$

$$\left(10.3 \frac{ft^3}{lb_m}\right) - \left(\frac{1 lb_f}{1 lb_m}\right)(100 \, ft) - \left(\frac{1}{2 \cdot 32.2 \frac{ft \cdot lb_m}{lb_f \cdot sec^2}}\right)$$

$$\left(50 \frac{ft}{sec}\right)^2 - \left(182 \frac{Btu}{lb_m}\right)\left(778 \frac{ft \cdot lb_f}{Btu}\right) - \left(37 \frac{lb_f}{in^2}\right)$$

$$\left(144 \frac{in^2}{ft^2}\right)\left(11.7 \frac{ft^3}{lb_m}\right) + \left(-5 \frac{Btu}{lb_m}\right)\left(778 \frac{ft \cdot lb_f}{Btu}\right)$$

$$w_{12} = 1259.8 \frac{ft \cdot lb_f}{lb_m}$$

$$W_{12} = (3 \; lb_m)\left(1259.8 \frac{ft \cdot lb_f}{lb_m}\right)$$

$$= (3779.3 \, ft \cdot lb_f)\left(\frac{1 \, Btu}{778 \, ft \cdot lb_f}\right) = 4.9 \; Btu$$

Thus, +4.9 Btu of work is generated by the working fluid (work out is positive).

Now let's apply these principles to a real engine application.

EXAMPLE 1.4

H-3 HELICOPTER

Each T58-GE-402 engine that powers the H-3 Sea King produces 1500 hp. From temperature measurements, it is known that air enters each engine with an enthalpy of 100 Btu/lb_m and exhausts with an enthalpy of 300 Btu/lb_m, with negligible changes in kinetic or potential energies. If 360 Btu/lb_m of heat are added to every pound of air in the engine's combustion chamber, determine the mass flow rate of the air (lb_m/min) through a single engine.

GIVEN: $\dot{W} = P = 1500$ hp (recall that the rate of total work is power, P)
$h_1 = 100$ Btu/lb_m
$h_2 = 300$ Btu/lb_m

$ke_1 \cong ke_2, pe_1 \cong pe_2$
$q_{12} = 360$ Btu/lb$_m$

FIND: $\dot{m}_{air}$ (lb$_m$/min)

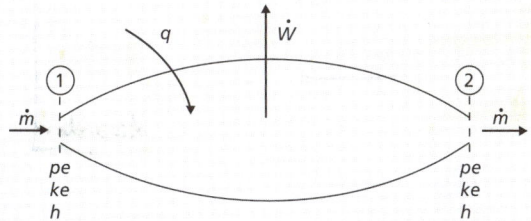

FIGURE EXAMPLE 1.4

SOLUTION: The given information suggests that solution of the problem will utilize Equation 1.30. A more straightforward approach, however, will utilize 1.25 to solve for the work performed by (or on) the air, then substitute into 1.26 to determine the mass flow rate. First, however, we must recall 1.27: $h = u + pv$. Applying this definition of enthalpy and the given information, 1.26 simplifies to $h_1 + q_{12} = h_2 + w_{12}$. Rearranging this expression for the work term and substituting given values yields $w_{12} = (100 - 300)\frac{Btu}{lb_m} + 360\frac{Btu}{lb_m} = 160\frac{Btu}{lb_m}$. Now, the definition of mechanical power provides the final result:

$$\dot{m}_{air} = \frac{\dot{W}_{12}}{w_{12}} = \frac{(1500 hp)\left(42.42\frac{Btu}{min \cdot hp}\right)}{\left(160\frac{Btu}{lb_m}\right)} = 397.7\frac{lb_m}{min}$$

1.9 NON-FLOW ENERGY EQUATION (NFEE)

The first law applies to closed systems as well as open systems. In typical closed systems, however, the changes in elevation and velocity of the working fluid are relatively negligible. Changes in potential and kinetic energies are thus often neglected in first law analyses of closed systems. Furthermore, as stated above, no displacement energy (flow work) is stored within a working substance unless it is flowing into or out of the system. With these simplifications, $[(PE_i - PE_f) + (KE_i - KE_f) + (FW_i - FW_f)] + U_i + Q'_{if} = U_f + W_{if}$ (1.17 rearranged) becomes the *Non-Flow Energy Equation* (NFEE).

$$U_i + Q'_{if} = U_f + W_{if} \quad (1.31)$$

* In a closed system, the boundary can flex

On a specific basis, the NFEE takes a similar form:

$$u_i + q_{if} = u_f + w_{if} \quad (1.32)$$

* The change of volume in a closed system is called bound work

1.10 SECOND LAW OF THERMODYNAMICS

Prior to formally stating the second law of thermodynamics, we must first discuss the concept of *entropy, s or S*. Formalized by Clausius in the mid-nineteenth century, entropy is a thermodynamic property of a working substance that is a measure of the disorder or randomness of the molecules comprising the substance. While our analysis of various thermodynamic cycles will occasionally utilize particular values of entropy at a state point, we are most often concerned with the <u>change in the entropy of the system</u>. Formally, the entropy change during a process is calculated from the sum of the differential heat transferred divided by the absolute temperature: $\Delta S = \int_1^2 \frac{\delta Q}{T}$. Rearranging this relationship, we see that the differential amount of heat transferred during a process can be quantified by the product of the temperature and the change in entropy: $\delta Q = TdS$. The total amount of heat transferred during a process, therefore, can be calculated from $Q = \int_1^2 TdS$. This can be illustrated graphically. As shown in Figure 1.9, heat in specific terms, q, is the area under the process curve on a *T-s* graph and total heat, Q, is the area under the process curve on a *T-S* graph. In this illustration, <u>the heat is $q_{supplied}$ to the system or q_{in}.</u>

* Integral = area under a curve

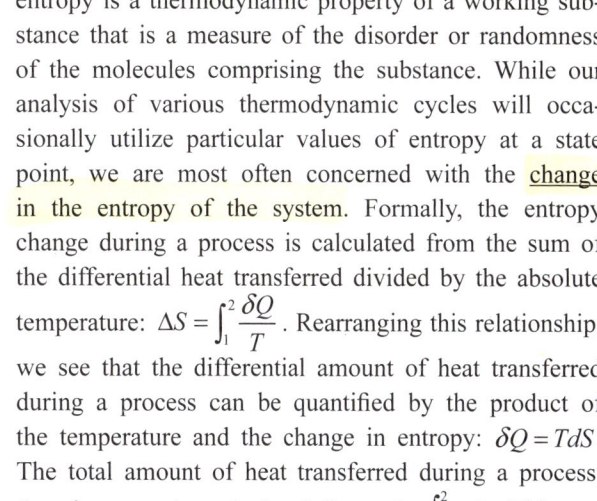

FIGURE 1.9 *T-s and T-S Graphs for a Heat Addition Process*

The second law of thermodynamics mandates the direction of processes by ensuring that the overall change of entropy of the system plus surroundings is never less than zero. The net entropy change for a reversible process is identically zero, while the entropy of the working fluid increases for all other—real—processes. A truly reversible process is one that, having occurred, leaves no change in either the system or surroundings. Real processes, however, are fraught with irreversibilities such as friction, fluid turbulence, electric resistance, and chemical reactions. These irreversibilities (as well as heat transfer into the fluid) generate entropy, ensuring that the process will alter the system or—at the very least—its surroundings at its conclusion. Real processes are thus irreversible since they have a required direction.

Common experience supports this concept. A hot cup of coffee left in a cool room eventually cools off. The reverse process—a hot cup of coffee absorbing heat and getting even hotter in a cool room—may satisfy the first law, but we know that this process will not occur. Nature thus dictates that processes occur in a certain direction and not in the reverse direction. The second law of thermodynamics helps to explain this phenomenon.

Moreover, the second law eliminates the possibility that any heat engine could operate with 100 percent efficiency. That is, it is impossible to construct a device that operates on a cycle and produces no other effect than the production of work and the transfer of heat from a single source. This form of the seond law of thermodynamics, termed the Kelvin-Planck Statement, implies that it is impossible to build an engine that extracts energy from a high temperature reservoir, performs work, and does not transfer heat to a lower temperature reservoir. In other words, all the heat supplied to an engine cannot be converted into work and q_r must be > 0.

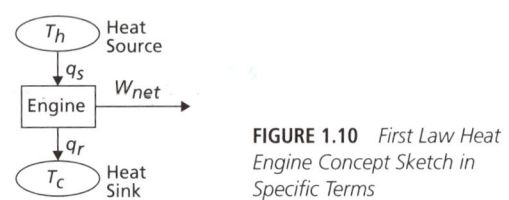

FIGURE 1.10 *First Law Heat Engine Concept Sketch in Specific Terms*

It is important to note that this form and others of the second law are expressions of experimental observations and have never been formally proven. We will address the concept of efficiency more fully below.

Some additional definitions follow. A process in which there is no heat transferred is termed *adiabatic*. This can be approached physically by insulating the system such that a negligible amount of heat crosses the boundary or by ensuring that the system and its surroundings are at the same temperature. By eliminating heat transfer and all irreversibilities, the entropy of a working fluid will not change during a process. Thus, an adiabatic and reversible process is characterized by constant entropy and is termed *isentropic*.

1.11 PROCESSES

As stated in Section 1.3, the state of a working fluid is changed by thermodynamic processes. The processes that we will discuss are:

- isobaric—no change in pressure
- isochoric (isometric)—no change in volume
- isothermal—no change in temperature
- isentropic—no change in entropy

It is often convenient to display the processes graphically. We've already covered the meaning of the area under a *T-s* graph. Another important relationship is illustrated by graphing on *p-v* coordinates. Notice that we have ambiguous use of a variable, *P*. We're talking about pressure in this application and not power. Familiarity will help the student to understand which attribute the variable represents from context. For a closed system, work is

$$w = \int_i^f p\,dv \quad \text{and} \quad W = \int_i^f p\,dV$$

These relationships are for boundary work, where the boundary moves, as with a piston moving inside a sealed cylinder. It is important to emphasize that boundary work only applies to a closed system. The meaning of the integral is the area under the curve.

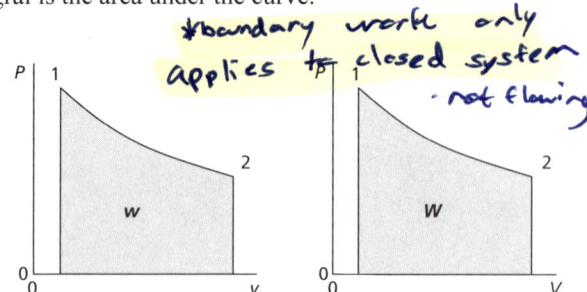

FIGURE 1.11 *p-v and p-V Graphs for an Expansion Process*

Figure 1.11 shows *p-V* and *p-v* graphs for an expansion process. The system proceeds from a high pressure, low volume state to a lower pressure, higher volume state. Notice that when integrating for specific volume, *v*, the result is specific work, *w*, and when integrating over the total volume, *V*, the area is total work, *W*. In this illustration, the work is out of the system and $w_{if} = w_{out}$ and $W_{if} = W_{out}$.

1.12 CYCLES

A cycle is a series of two or more thermodynamic processes that return the working fluid to its starting condition. The typical heat engines that will be covered in this course operate with four engine processes. A useful mnemonic for the four general heat engine processes is CAER.

- Compression (w_{in})
- Addition of heat ($q_{supplied}$)
- Expansion (w_{out})
- Rejection of heat ($q_{rejected}$)

Graphing the thermodynamic processes associated with each engine process will be a useful tool in solving engine problems in this course. For instance, in the four state point cycle graphed below, there is no heat shown in processes 1-2 and 3-4. Process 2-3 proceeds to the right (a positive heat process—heat in). The amount of heat input corresponds to the shaded area in the center sketch. Similarly, the 4-1 process also subtends an area representing heat out (proceeding left, a negative process). Applying the first law to the engine concept sketch shows that $w_{net} = q_s - q_r = q_{net}$.

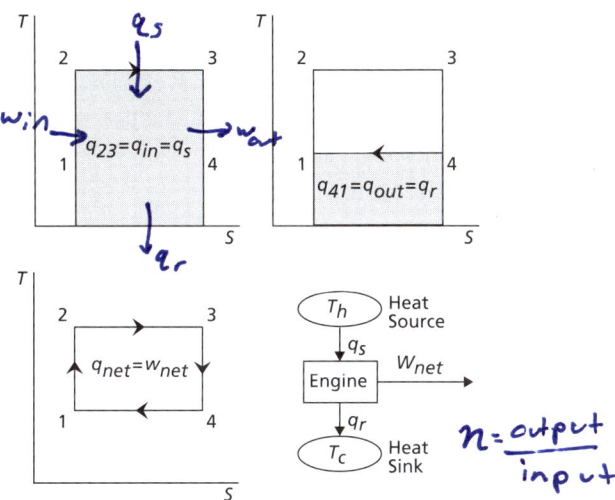

FIGURE 1.12 *T-s Process Graphs and Engine First Law Concept Sketch*

It should be noted that all engine cycles proceed clockwise on the *T-s* and/or *p-v* process graphs. Similarly, the area enclosed in a *p-v* cycle graph shows w_{net} from calculating the difference between w_{out} and w_{in}.

1.13 EFFICIENCY

Efficiency (η) relates how well a process achieves its desired output from the required input. This is shown graphically in Figure 1.13.

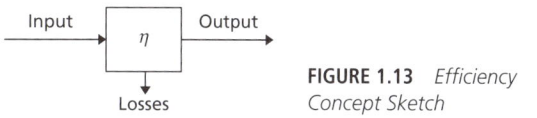

FIGURE 1.13 *Efficiency Concept Sketch*

The combination of η, input, and output are multiplicative, i.e., (Input) (η) = Output. Rearranging yields η = Output / Input. Efficiency is typically expressed as a percentage or decimal equivalent, therefore the units in the numerator must be identical to those in the denominator. Since the numerator and denominator may consist of more than one term, conversion factors may be required to obtain dimensional homogeneity. The upper limit of efficiency is unity, i.e., $\eta = 1$ for an ideal or perfect process. The second law of thermodynamics imposes this bound. Efficiencies less than unity ($\eta < 1$) are realistic values for real processes.

The first law of thermodynamics states that energy is only converted into other forms of energy. The second law forbids all of the heat supplied to an engine from being converted into useful work. Some heat must be rejected from the system to comply with the laws of thermodynamics. This heat rejection takes many forms, but all are irreversible. A basic heat engine may be modeled with Figure 1.14. Figures 1.13 and 1.14 are related to each other.

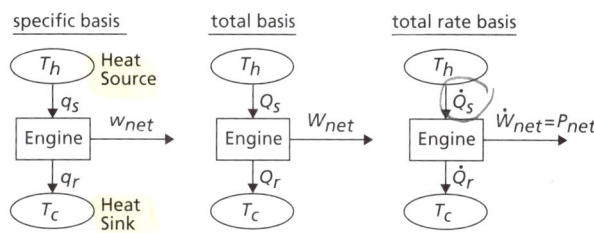

FIGURE 1.14 *Heat Engine Concept Schematics*

A first law analysis of Figure 1.14 requires that all the input energy or power equate to the energy or power leaving the engine. Second law requirements dictate that the energy lost to the surroundings (the heat sink) cannot be zero. Substituting the terms from Figure 1.14, we obtain $Q_s = W_{net} + Q_r$. Now, we can define the *thermal efficiency* of the engine as $\eta_{th} = \dfrac{desired\ output}{required\ input} = \dfrac{W_{net}}{Q_s}$. On a specific basis, the thermal efficiency is obtained from $\eta_{th} = \dfrac{w_{net}}{q_s}$. Similarly, the thermal efficiency can be calculated on a total rate basis from $\eta_{th} = \dfrac{w_{net}}{q_s} = \dfrac{w_{net}}{q_s}\left(\dfrac{\dot{m}}{\dot{m}}\right) = \dfrac{\dot{W}_{net}}{\dot{Q}_s} = \dfrac{P_{net}}{\dot{Q}_s}$. Since the net work can be written as the difference between the supplied and rejected heat, $w_{net} = q_s - |q_r|$, thermal efficiency can be rewritten as $\eta_{th} = \frac{w_{net}}{q_s} = \frac{q_s - |q_r|}{q_s} = 1 - \frac{|q_r|}{q_s}$. Note that η_{th} is focused on the thermodynamics of the working fluid in the engine.

EXAMPLE 1.5

A closed thermodynamic cycle has a cycle thermal efficiency of 25%. 100 Btu/lb$_m$ of heat is rejected by the working fluid. Determine the heat supplied to the engine (Btu/lb$_m$) and the net work of the cycle (Btu/lb$_m$).

GIVEN: $\eta_{th} = 0.25$
$q_r = 100$ Btu/lb$_m$

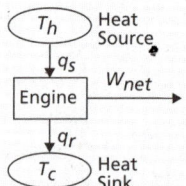

FIGURE EXAMPLE 1.5

FIND: q_s (Btu/lb$_m$)
w_{net} (Btu/lb$_m$)

SOLUTION: We first solve for the heat supplied (q_s). Once the amount of heat added is calculated, we can then utilize the first law to determine the net work. Solving for q_s and substituting given information yields

$$q_s = \frac{|q_r|}{1-\eta_{th}} = \frac{100 \frac{Btu}{lb_m}}{0.75} = 133.3 \frac{Btu}{lb_m}$$

From the efficiency relationships we see

$$w_{net} = \eta_{th} \cdot q_s = (0.25)\left(133.3 \frac{Btu}{lb_m}\right) = 33.3 \frac{Btu}{lb_m}$$

Alternatively, from the first law, the net work is calculated to be

$$w_{net} = q_s - |q_r| = (133.3 - 100)\frac{Btu}{lb_m} = 33.3 \frac{Btu}{lb_m}$$

When you think of a heat engine—the most common experience of students being the internal combustion engine that powers an automobile—you clearly don't use the fuel provided to directly supply power. Rather, the engine is an energy conversion device that burns the hydrocarbon-based fuel and uses the released heat energy to perform work inside the engine, which is then mechanically transmitted to the wheels of your vehicle. This sequence of energy transmissions and conversions within an engine system implies a sequence of efficiencies that effectively reduces the amount of energy liberated from the supplied fuel source, transmitted through each subsequent conversion and transmission process, and to the output shaft. This sequence defines the overall efficiency, η_{OA}, of the engine and will be discussed in greater detail in a later unit. The overall efficiency of the engine, however, is still obtained from comparison of the desired output and the required input. It is calculated from

$$\eta_{OA,\,engine} = \frac{output}{input} = \frac{\dot{W}_{brake}}{\dot{m}_{fuel} HV}.$$ The second term in the

denominator, HV, refers to the heating value of the fuel—that amount of energy liberated from the combustion of a unit mass of the fuel. Ultimately, the amount of real power generated by the engine is significantly less than that released during combustion due to numerous irreversibilities and parasitic loads within the engine. It is crucial to remember that the terms "supplied" and "rejected" refer to the heat supplied to and rejected from the working fluid in the engine. The working fluid in a gasoline engine is air.

Considering a ship propulsion train, Figure 1.15 illustrates the efficiency and multiple energy conversion concepts. This basic model can be adapted to other power

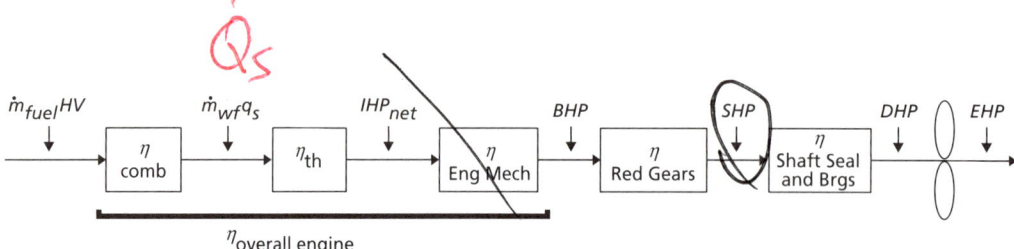

FIGURE 1.15 Propulsion Train Efficiency Block Diagram

train configurations for aircraft, locomotives, and motor vehicles that use heat engines for propulsion.

Tracing the total rate of energy from left to right, the fuel energy is converted through combustion and heat absorption processes to thermal energy (q_s) absorbed by the working fluid (subscript "wf"). The thermal efficiency governs the conversion of heat supplied to net indicated horsepower (IHP_{net})—recall that the thermal conversion is analyzed strictly on the thermodynamics of the working fluid. There are additional mechanical losses that reduce the real amount of power produced at the output coupling of the engine (brake horsepower, BHP). These include engine-driven parasitic loads, such as mechanical fuel, oil and water pumps; engine-driven auxiliaries such as electrical generation and air conditioning; and any engine strokes not included in the analysis (intake and exhaust). Note: Depending on the type of engine, not all of these categories apply to mechanical efficiency. For instance, a reciprocating piston/cylinder engine used in an automotive application typically runs the power steering pump and the alternator off the same crankshaft that provides rotational power to the transmission. However, a steam propulsion system will generally run electrical turbo-generators separately from the propulsion turbines so the propulsion turbine mechanical efficiency will not include these necessary but parasitic losses. The hydraulic plant that powers the steering system has an electric motor to pressurize the hydraulic oil that gets its power from the turbo-generator, not the propulsion turbine. Clearly, one must think about the losses that would apply to the efficiency category called "engine mechanical."

The first three blocks multiply together to yield the engine's overall efficiency. Most rotative engines require a set of reduction gears (aka transmission) that reduce the engine's output speed and increase the torque. The gears have mechanical friction losses that show up in heat that must be removed from the gearbox. Shaft horsepower (SHP) is measured downstream of the reduction gears. There are seals installed that allow the propeller shaft to penetrate the hull. There may also be additional shaft support bearings. The friction in these elements reduces the amount of power delivered (delivered horsepower, DHP) to the propeller. The power it takes to push a vessel through the water is effective horsepower (EHP). (This is also the power it would take to tow the vessel at a given speed.) The overall propulsion efficiency can be measured at either the EHP, the SHP, or DHP levels—traditionally, it is measured at the SHP level. In this case, $\eta_{overall\ propulsion} = SHP / \dot{m}_{fuel} HV$.

Note that this discussion is for a ship propulsion engine that burns fuel, such as fuel oil. A nuclear propulsion system does not "burn" fuel, so the fuel flow rate and heating value do not apply in this type of analysis.

The relationship between SHP and EHP is the *propulsive efficiency*.

Considering that every efficiency is less than unity (<1), that all of the propulsion train efficiencies are multiplicative, and including the efficiencies of a motor vehicle's differential, shaft and wheel bearings, and tires in a propulsion train similar to that shown in Figure 1.15, it is no wonder that the overall efficiency of most motor vehicles is on the order of about 18 percent.

1.13.1 Other Measures of Performance

There are many other measures of performance used in various industries and applications. Whether a higher number or a lower number is better really depends on the specific application. For example, most students should be familiar with vehicle fuel economy as measured in miles driven per gallon of fuel consumed (mpg). The concept of desired output for a required input still holds true, but unlike efficiency, the resulting value has associated units. It is clear that a high fuel economy number is preferable to low fuel economy.

Fuel economy is often used to compare vehicles, but implicit in this is that the vehicles perform the identical purpose. When considering the purpose of a vehicle, fuel economy might not be the best measure of performance. For example, passenger-carrying vehicles often use passenger-miles as the desired output. Using this measure of performance, a bus carrying 40 people and getting 5 miles per gallon will have a passenger fuel economy of 200 passenger-miles per gallon. Yet a taxi carrying three passengers but getting 20 miles per gallon will have a rating of 60 passenger-miles per gallon. Clearly, the bus is better than the taxi using this metric. This example might prompt consideration of the fuel cost as the next level in passenger transportation performance measurements. Gasoline is the typical fuel for a taxi, and buses are usually diesel-powered. While fuel costs may vary considerably over time, in many areas diesel fuel is more expensive than gasoline. So if gasoline costs $3 per gallon and diesel fuel is $3.25, the cost per passenger-mile is $0.016 for the bus and $0.05 for the taxi. Notice that when costs are considered, the lower number is better.

The next tier performance measure would then factor in operator costs and maintenance expenses. The business community uses measures of performance like these examples involving costs to make procurement, retirement, and replacement decisions all the time. In the aviation community, another common measure of performance is the number of maintenance hours per flight hour. When considering the manpower cost of maintenance personnel, it is understandable, when considering essentially comparable aircraft, why the one with a lower ratio of maintenance hours to flight hours will be preferred. This was one of the primary reasons the F/A-18 replaced the F-14 fighter aircraft.

When considering various engine configurations, a typical measure of performance is brake specific fuel consumption (*bsfc*). An engine's *bsfc* is the mass flow rate of fuel required to generate a single horsepower, $bsfc = \frac{\dot{m}_{fuel}}{\dot{W}_{brake}}$. Lower *bsfc* is better. This concept will be used in a later unit.

1.14 TORQUE AND POWER

With very few exceptions, all of our current engines produce rotational power via spinning shafts. Whether turning a propeller, helicopter rotor, or wheels, the engine provides mechanical power via a spinning shaft. Brake power is measured at the output of an engine (refer back to Figure 1.15). For instance, an automobile engine is rated for a certain amount of horsepower, but that rating only applies at a specific speed. Rotational power is related to the output torque and speed.

$$BHP = 2\pi\tau N \tag{1.33}$$

Torque (τ) is a moment, which is the product of a moment arm and a force perpendicular to the moment arm. The USCS units for τ are ft-lbs. The same form of equation applies at the shaft power and delivered power levels. Let's practice using this equation.

EXAMPLE 1.6

GIVEN: A pick-up truck's diesel engine produces 440 ft-lbs of torque at 1500 RPM.
FIND: BHP
SOLUTION: We'll use Equation 1.33 to solve this directly. Annotate τ with a subscript to indicate it is engine output, or brake, torque, τ_B.

$$BHP = \dot{W}_B = 2\pi\tau_B N$$
$$= \frac{(2\pi \text{ radians/rev})(440 \text{ ft} \cdot \text{lbs})(1500 \text{ revs/min})}{(33{,}000 \text{ ft} \cdot \text{lb}_f/\text{hp} \cdot \text{min})}$$
$$= 126 hp$$

$$\boxed{BHP = \dot{W}_B = 126 hp}$$

Take note of the unit analysis. It would seem that radians should appear in the answer. To sort this out, we need to think about the definition of a radian. A radian is a ratio of arc length of a circle to the radius length. Therefore, radians are <u>unitless</u>.

Students may be familiar with this form of the equation:

$$\dot{W}[hp] = \frac{\tau[ft \cdot lbs] N[RPM]}{5252}$$

In this case, the 5252 is the combination of the 2π and the 33,000 shown in the example.

The output characteristic of turbines and many internal combustion engines is high rotational speed, but low torque. Ship and submarine propellers are low speed, high torque components. Thus the need for the conversion device known as reduction gears. In automotive applications, the gearbox is known as the transmission. Ship reduction gears may allow disconnecting engines with a clutch, but ships do not shift gears as motor vehicles do. The reduction gears convert high speed / low torque to low speed / high torque. A common reduction gear configuration is shown in Figure 1.16. The "1st reduction pinions" are the power input from the engines. The "bull gear" is the output toward the propeller.

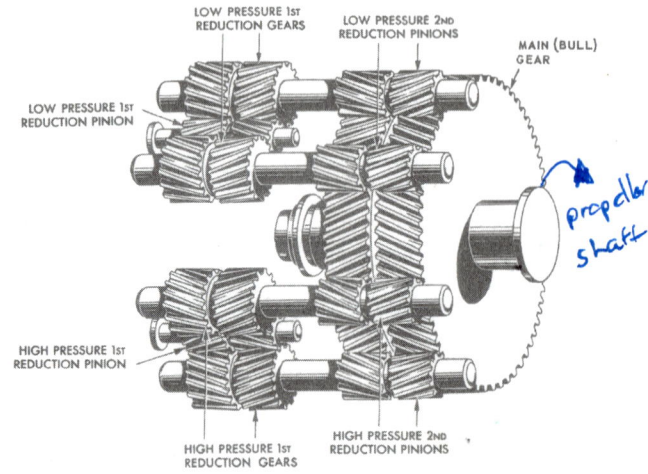

FIGURE 1.16 *Locked Train, Herringbone Pattern, Double Reduction Gears*

Figure 1.16 is labeled for a steam configuration. In a gas turbine ship, two separate gas turbine engines will connect via clutches to the first reduction pinions.

1.15 IDEAL GAS LAW

We'll close out this unit with a discussion of ideal gas laws and set the table for using them in the air cycle engines we will cover in Unit 4.

The pressure, specific volume, and temperature of a low-density vapor are related by the ideal gas equation of state. An equation of state interrelates properties and allows calculation of a property, given the others.

A cautionary note before we progress any further: In order to use ideal gas law equations, <u>pressures and temperatures must be in absolute terms.</u> That means you cannot use gage pressures or temperatures in either Celsius or Fahrenheit and expect to get the correct answer.

$$pv = RT \qquad (1.34)$$

The volume in Equation 1.34 is specific volume, v. Incorporating total volume and mass, where $v = V/m$,

$$pV = mRT \qquad (1.35)$$

The gas constant, R, is unique to the particular gas (or gas mixture) and is thus termed the *characteristic gas constant*. R is related to the *universal gas constant*, R_u, through the molar mass, M, of the substance. M is also called "molecular weight." The USCS units of M are lb_m/pmole, where pmole stands for "pound-mole" or moles on a pound basis. The molecular weight of an element or compound that students are familiar with in the SI unit system is the same numerical value in USCS. For example, the molecular weight of O_2 is 32 grams/gmole = 32 lb_m/pmole. Relating molar mass to the mass and number of moles present,

$$M = m/n \qquad (1.36)$$

The characteristic gas constant, R, is related to the universal gas constant, R_u, as follows:

$$R = \frac{R_u}{M} \qquad (1.37)$$

where R_u = 1545.33 ft-lb_f/ pmole-°R.

Using 1.36 and 1.37, 1.35 can be rewritten in the more familiar form covered in the typical chemistry course utilizing the number of moles, n, of the gas.

$$pV = nR_uT \qquad (1.38)$$

EXAMPLE 1.7

GIVEN: Air is comprised of 78% N_2, 21% O_2, and 1% CO_2 by mass.
FIND: M_{air}; R_{air}.
SOLUTION: Determine M for the mixture by taking a weighted average.

$$M_{air} = (0.78)\left(28\frac{lb_m}{pmole}\right) + (0.21)\left(32\frac{lb_m}{pmole}\right) +$$
$$(0.01)\left(44\frac{lb_m}{pmole}\right) = 29.0\frac{lb_m}{pmole}$$

$$\boxed{M_{air} = 29.0\frac{lb_m}{pmole}}$$

$$\boxed{R = \frac{R_u}{M} = \frac{1545.3\frac{ft-lb_f}{pmole-°R}}{29.0\frac{lb_m}{pmole}} = 53.3\frac{ft-lb_f}{lb_m-°R}}$$

Make sure that you can do this unit analysis!

The specific heat capacities of an ideal gas at constant pressure and constant volume, C_p and C_v, are related to the characteristic gas constant in a very simple manner.

$$C_p = C_v + R \qquad (1.39)$$

It is also convenient—as we will see in the following section—to define another ideal gas property called the *specific heat capacity ratio*, k.

$$k = \frac{C_p}{C_v} \qquad (1.40)$$

Although not typically identified as an ideal gas, air behaves as such over a wide range of temperatures and pressures. As a common working fluid, we note the characteristic gas constant, specific heat capacities, and specific heat capacity ratio for *air at standard conditions*: $R = 53.3\frac{ft \cdot lb_f}{lb_m \cdot °R}$, $C_p = 0.24\frac{Btu}{lb_m \cdot °R}$, $C_v = 0.171\frac{Btu}{lb_m \cdot °R}$, and $k = 1.4$. Standard air is 70 °F and 29.92 in-Hg. Note that a conversion factor must be employed to obtain dimensional homogeneity in Equation 1.39 with the values given above.

Changes in internal energy and enthalpy are related to C_p and C_v.

$$\Delta h = C_p \Delta T \quad \text{and} \quad \Delta H = Cm_p \Delta T \quad (1.41)$$

and

$$\Delta u = C_v \Delta T \quad \text{and} \quad \Delta U = Cm_v \Delta T \quad (1.42)$$

EXAMPLE 1.8

Determine the mass of air (lb_m) contained within a submarine whose internal volume is 141,000 ft^3. The air is at 70 °F and the barometric pressure inside the sub is 29.7 in-Hg.

GIVEN: $V = 141,000\ ft^3$
$T = 70\ °F = 530\ °R$
$p = 29.7$ in-Hg absolute (recall that barometers read in absolute terms)

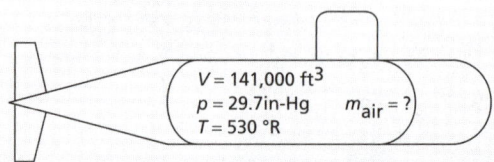

FIGURE EXAMPLE 1.8

FIND: m_{air} (lb_m)

SOLUTION: Treating air as an ideal gas, we employ 1.35 with the given information. $pV = mRT$. Solving for the mass of air, substituting the given information, and including necessary conversions:

$$m = \frac{pV}{RT}$$

$$= \frac{\left[(29.7"Hg)\left(\frac{14.7\ psia}{29.92"Hg}\right)\left(144\frac{in^2}{ft^2}\right)\right](141,000\ ft^3)}{\left(53.3\frac{ft \cdot lb_f}{lb_m \cdot °R}\right)\left[(70+460)°R\right]}$$

$$= 10479.4\ lb_m$$

EXAMPLE 1.9

SCENARIO: A sub's main air flasks are recharged to 4500 psig while the boat is deep and the seawater temp is 55 °F. The boat then comes shallow and enters the Gulf Stream where the seawater temperature is 80 °F. Because the air flasks are in the external ballast tanks, which are flooded, assume the air flasks equalize with the new ocean environmental temperature. Also assume that no air is discharged from the flasks during this time. What will the air flask pressure gage read in psig? What amount of heat was added to the air in the flasks? Classify the process that occurred.

GIVEN: $p_1 = 4500$ psig = 4515 psia (we must use absolutes when using ideal gas law)
$T_1 = 55\ °F = 515\ °R$ (absolute temperature, too)
$T_2 = 80\ °F = 540\ °R$

FIND: a) p_2 (psig) (answer units are specified)
b) q_{12} (Since we don't have air mass and can't calculate it from given info, I'll answer heat in specific terms.)
c) Process classification

SOLUTION: Assume standard atmospheric pressure. Rounding at high pressures, such as 4500 psi, does not cause much error; therefore the absolute pressure inside the air flasks at the start of this problem was rounded to 4515 psia vice 4514.7 psia. This is an example of a two state point problem as represented in the sketch.

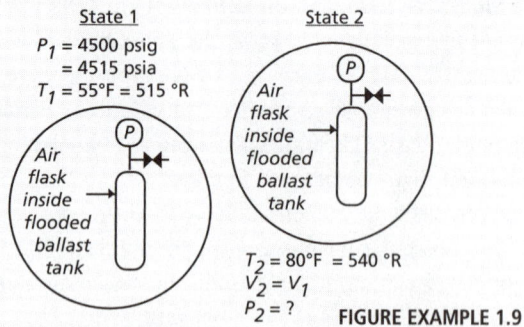

FIGURE EXAMPLE 1.9

Answer c) first. Process classification is isochoric since the air flasks are assumed to have no change in volume. In reality there is some change in volume due to the elasticity of the air flasks.

a) $p \underbrace{V}_{\text{no change}} = \underbrace{mR}_{\text{no change}} T \quad \therefore \quad \frac{p_1}{T_1} = C$

$$\frac{p_1}{T_1} = \frac{p_2}{T_2} \Rightarrow p_2 = p_1\left(\frac{T_2}{T_1}\right) = 4515\ psia\left(\frac{540}{515}\right)$$

$$= 4734\ psia - 14.7\ psia = 4719\ psig$$

$$\boxed{p_2 = 4719\ psig}$$

b) $q_{12} = c_v(T_2 - T_1) = \left(0.171\frac{Btu}{lb_m - °R}\right)(80-55)°F$

$$= 4.28 \frac{Btu}{lb_m}$$

$$\boxed{q_{12} = +4.28 \frac{Btu}{lb_m}}$$

Note that the change in T is the same for °F and °R. It is not necessary to convert the temperature to absolute when calculating differences, but it is necessary when taking ratios of the temperatures.

1.15.1 Combining Hydrostatics with Ideal Gas Law

Navy personnel spend a lot of time around and under the water and need to have an appreciation for the effects on gases exposed to pressure that is related to depth of submergence.

For example, this is particularly important for the safety of scuba divers. As divers descend to greater depths, the air that they are breathing must be supplied to them at increasing pressure in order to overcome the hydrostatic pressure of the water on the body. However, when ascending, the reduction of external hydrostatic pressure will allow the air (or other gases if using alternate gas mixtures) inside the human lungs to expand. If the diver does not allow the expanding gas to escape by exhaling, the diver's lungs may expand to the point of physiological damage or even rupture.

Alternately, snorkelers, who inhale air at the surface and then descend in the water, will feel the effects of the external pressure, typically by pain felt in their ears. The remedy for that is known as the Valsalva maneuver, in which the diver pinches the nose and attempts to exhale with the mouth closed. This action forces air into the middle ear via the Eustachian tubes that connect the middle ear to the throat. Equalizing the internal pressure of the middle ear with the external water pressure will relax the eardrum.

More on the hardware side of things, submarine main ballast tanks (MBTs) are filled with air when a submarine is on the surface. The MBTs are always open to the sea at the bottom of the tank via louvered gratings. The MBTs are located at the ends of U.S. submarines and are subdivided into various tank spaces along their length as well as port and starboard. Subdividing the tanks prevents outer hull damage in one area from compromising the functionality of all of the MBTs at that end of the boat. Figure 1.17 shows a simplified sketch of a submarine and illustrates the situation when a submarine is surfaced. In this condition, the MBTs are filled with air and an interface exists between the air inside the tanks and the sea.

EXAMPLE 1.10

GIVEN: A submarine displaces 5000 LT when submerged and 4400 LT when surfaced. The air used to blow the water out of the MBTs is stored in high pressure air flasks at a pressure of 4500 psig. The MBT openings (grates) are 30 feet below the sea surface when the sub is "surfaced."

FIND: The volume of air necessary to blow the MBTs "dry" when the submarine is at the sea surface. (Note: The real design criterion is to be able to blow the tanks completely when the submarine is operating at depth in order to surface during a flooding casualty.)

SOLUTION: This is a problem that combines hydrostatics and the ideal gas law. The "system" being tracked is the air. The high pressure air flasks are located inside the MBTs, but the valves that allow the air out of the flasks and into the MBTs are inside the boat and connected to both the MBTs and the air flasks. Since the air transitions from being in the air flasks to being in both the air flasks and the MBTs, this is a "before" and "after" problem. I'll show these two conditions as State 1 and State 2. I'm going to use the ideal gas law to solve for the required volume of the air flasks, so I'll need to convert the given pressure to absolute.

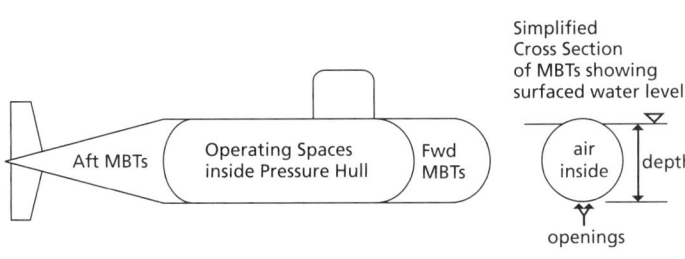

FIGURE 1.17 *Submarine Main Ballast Tanks Simplified Sketch*

$$pV = mRT$$

Since m, R, and T are all constant in this problem (mass doesn't leak out; R is the gas constant for air; and T is assumed to be the same before and after since the air flasks are physically located in the MBTs and equalized with the sea temperature), assuming no air is lost out the bottom of the MBTs, the mass of air is constant. The ideal gas law can be adapted to this problem by the following logic:

$$p_1 V_1 = \overbrace{mRT}^{Constant} = p_2 V_2$$

$$P_1 V_1 = P_2 V_2$$

Referring this to a sketch, we get the following.

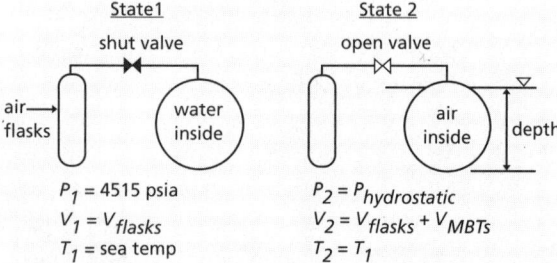

State 1: shut valve
- $P_1 = 4515\ psia$
- $V_1 = V_{flasks}$
- $T_1 = $ sea temp

State 2: open valve
- $P_2 = P_{hydrostatic}$
- $V_2 = V_{flasks} + V_{MBTs}$
- $T_2 = T_1$

FIGURE EXAMPLE 1.10

Note that there is still air in the air flasks when the MBTs have been blown (State 2). The air has merely equalized with the MBTs and stopped moving.

The difference in a submarine's displacement is the weight of water in the ballast tanks. We can solve for the MBT volume. Be aware that ship and submarine displacement is given in long tons (LT), which equal 2240 lb$_f$/LT. Starting with the definition of g and re-arranging to solve for the volume of the MBTs:

$$\gamma = \frac{W}{V} \Rightarrow V_{MBTs} = \frac{W_{seawater\ in\ MBTs}}{\gamma_{seawater}}$$

$$V = \frac{(600\,LT)\left(2240\dfrac{lb_f}{LT}\right)}{64\dfrac{lb_f}{ft^3}} = 21{,}000\,ft^3$$

The specific weight (γ) is for standard seawater, since we only operate submarines in saltwater environments.

When the sub is surfaced, the pressure at the interface between the air inside and the sea outside the MBTs must be equal in order to keep the sea from pushing into the MBTs. The sea pressure will be the hydrostatic pressure at the depth of the gratings. Assume standard atmospheric pressure exists above the sea level.

$$p_{hydrostatic} = p_{atm} + \gamma h = 14.7\,psia$$
$$+ \left(64.0\dfrac{lb_f}{ft^3}\right)(30\,ft)\left(\dfrac{1\,ft^2}{144\,in^2}\right)$$
$$= 14.7\,psia + 13.3\,psi = 28.0\,psia$$

Now substituting into $P_1 V_1 = P_2 V_2$ and solving algebraically for V_{flasks}:

$$p_{flasks} V_{flasks} = p_{hydrostatic}\left(V_{flasks} + V_{MBTs}\right)$$

$$p_{flasks} V_{flasks} = p_{hydrostatic} V_{flasks} + p_{hydrostatic} V_{MBTs}$$

$$\left(p_{flasks} - p_{hydrostatic}\right) V_{flasks} = p_{hydrostatic} V_{MBTs}$$

$$V_{flasks} = \frac{p_{hydrostatic} V_{MBTs}}{\left(p_{flasks} - p_{hydrostatic}\right)} = \frac{(28\,psia)(21{,}000\,ft^3)}{(4515 - 28)\,psia}$$

$$= 131\,ft^3$$

$$\boxed{V_{flasks} = 131\,ft^3}$$

---

1. http://www.usatoday.com/weather/resources/askjack/wfaqpres.htm, 6/22/07.

Practice Problems

"Whether you think you can, or whether you think you can't, you're right."
—Henry Ford

1.1 Ten pounds of a substance occupies a volume of 20 ft³. Calculate the specific volume (ft³/lb$_m$), density (lb$_m$/ft³), and specific weight (lb$_f$/ft³).

1.2 Determine the specific volume of air in a room assuming air is an ideal gas at standard temperature and pressure. Provide your answer in USCS units.

1.3 Common barometer readings are given in units of inches of mercury. Draw the pressure relationship graph (see Fig. 1.4), label all arrows, and indicate which arrow represents a barometer's reading.

1.4 Gage pressure is 21.4 in-Hg (inches of mercury). The corresponding barometer reading converts to 14.7 psia. Recall that pressure gages reading in inches of mercury typically read vacuum pressures. Draw the pressure relationship graph, show the values of this problem on the graph, and calculate the absolute pressure in psia and inches of mercury absolute.

1.5 The ship's barometer reads 28.95 in-Hg. A pressure gage at a turbine inlet reads 475 psig. A vacuum gage at the turbine exhaust reads 25.40 in-Hg. Draw the pressure relationship graphs for these gages. What are the turbine inlet and exhaust pressures in psia?

1.6 Determine the gas pressure p (psia) for the manometer in the sketch.
Given:
p_{atm} = 14.7 psia
L = 60 inches of fluid
fluid density = 78.8 lb$_m$/ft³

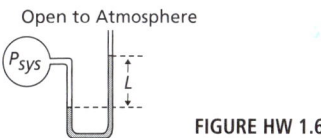

FIGURE HW 1.6

1.7 The main ballast tanks in a submarine are open to the sea at the bottom of the tanks. The tanks are filled with air when the submarine is on the surface. If the ballast tanks' bottom grates are 30 feet below the surface of the sea, what is the air pressure in the ballast tanks (psia; psig)?

1.8 Water contained by a dam exits a spill-way 685 ft above the base of the dam. Determine the velocity of the water as it strikes the base of the dam if its initial velocity as it leaves the spillway is negligible (ft/sec).

1.9 15 pounds of working fluid is moving with a velocity of 75 ft/sec through a process. As it moves from state point 1 to state point 2, its elevation increases 79 ft, it delivers 9003 ft-lb$_f$ of work, and it receives 69 Btu of heat. During this movement the pressure and density of the fluid do not change. Determine the change in internal energy of the working fluid (Btu/lb$_m$).

1.10 50 pounds of a working substance undergoes a non-flow process. The internal energy at the start of the process is 89 Btu/lb$_m$ and at the completion of the process is 150 Btu/lb$_m$. A total of 350 Btu of work is done on the working substance. How much heat was added or removed from each pound of the working substance (Btu/lb$_m$)?

1.11 At a vacuum of 22.73 inches of mercury, steam has a specific volume of 101.7 ft³/lb$_m$. The internal energy is 1058.7 Btu/lb$_m$. Calculate the specific enthalpy of the steam (Btu/lb$_m$).

1.12 A centrifugal compressor used in a gas turbine engine increases the pressure of 2000 lb$_m$/min of air from 15 psia to 150 psia. The enthalpy of the entering air is 150 Btu/lb$_m$ and of the compressed air is 300 Btu/lb$_m$, and 1.53 Btu/lb$_m$ of heat is lost to the environment. Assuming no change in the kinetic or potential energy of the air, determine how much power is required to drive the compressor (hp).

1.13 A tank is filled with air at a pressure of 42 psia and a temperature of 118 °F. What is the specific volume of the air in the tank (ft³/lb$_m$)?

1.14 An ideal gas initially at 15 psia and 200 °F is compressed from 20 ft³ to 13 ft³. If the final temperature is 407 °F, what is the final pressure (psia)?

1.15 a) The basis for a British Thermal Unit (Btu) is:

b) The basis for a "refrigeration ton" is:

c) 1 kW = x Btu/hr (determine this number using the conversion factors on the course standard equation sheet)

d) The mechanical energy equivalent of heat energy is:

1.16 A closed thermodynamic cycle has a thermal efficiency of 30%. The energy rejected to the heat sink is 800 Btu/lb_m.
What is the net work (Btu/lb_m) of the cycle?
What is the heat supplied (Btu/lb_m) to the cycle?

1.17 According to *Jane's Fighting Ships*, the propulsion plant on the *Sturgeon*-class of submarines had a reactor that produced a maximum of 78 MW of thermal power. The submarine's engine was rated at 15,000 SHP at flank speed. Draw the efficiency block diagram that represents this situation. Properly label all arrows and the block contents. What was the overall efficiency of a *Sturgeon* propulsion plant? What is the rejected heat transfer rate (Btu/hr)? Note that, ultimately, even the power put into the water by the propeller dissipates as heat, but this problem is asking about the heat put into the water by all other means.

1.18 An *Arleigh Burke*–class DDG is a twin-shaft ship with two LM-2500 gas turbine engines capable of powering each propeller shaft via two sets of reduction gears. The rated brake power of this version of LM-2500 is 29,500 hp at 3600 RPM. Each set of reduction gears has an efficiency of 92% and an overall gear ratio of 16. Draw a schematic that illustrates this configuration and an efficiency block diagram. What are the shaft speed, torque and power?

1.19 A submarine running at flank speed produces 15,000 shp with the propeller shaft spinning at 250 RPM. The reduction gears ratio is 17 with an efficiency of 92%. What is the torque on the propeller shaft? What speed are the turbines running? What is the power output by the turbines? If two turbines power the reduction gears and each has equal output, what is the torque output of one turbine? If the reduction gear losses go to the sub's lube oil system, what is the heat transfer to the lube oil (Btu/hr)?

1.20 For the hypothetical thermodynamic cycle shown below made up of processes:
area A = 250 Btu/lb_m
area B = 125 Btu/lb_m
area C = 400 Btu/lb_m
Work out = 270 Btu/lb_m

Determine:
a) Heat added
b) Heat rejected
c) Net work
d) Work in
e) Cycle thermal efficiency

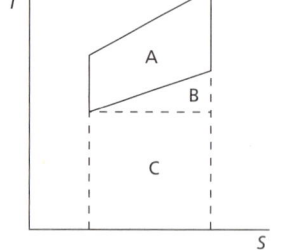

FIGURE HW 1.20

1.21 A SEAL team doing a HALO insertion from a C-130 flying at 30,000 ft has included scuba equipment in their gear bags. If a scuba tank's pressure gage read 3000 psi when on the ground, assuming no change in temperature, what should it read at altitude? If the temperature at 30,000 is really -10 °F, what should the pressure gage read at altitude? Clearly state any assumptions.

1.22 Scuba tanks are described by the volume of air at standard atmospheric pressure that can be pumped into them (by a compressor) to reach their rated pressure. A very common size tank is rated at 80 cubic feet and 3000 psi. Assuming that the temperature does not change and no tank expansion occurs, what is the air mass and volume inside this scuba tank (in^3)? State any assumptions. Make sure that your answers are reasonable!
FYI: Some interesting scuba tank explosion photographs and energy analyses are on the web. Find them by searching "scuba tank explosions" using your web brouser.

1.23 Navy divers from the Mobile Salvage Unit have been assigned to recover an F4U Corsair that ditched off Key West, Florida, during WWII. The aircraft was discovered during sonar tests in the area and found to be intact and sitting upright in 300 feet of water. The technique the divers plan to use is to attach three "air lift" bags, one to each of the wing roots and one to the fuselage using nylon slings. The air bags will then be filled with air to create lift. Once the plane is close to the surface, they will transfer the slings to a crane and lift the plane onto a barge. The air bags appear similar to parachutes—they are open at the bottom and will hold air that is added by a diver holding the nozzle of an air hose under the bottom of the lift bag. The air hose leads to the seafloor from an air compressor on the support barge at the surface. Significant statistics of this aircraft are: wingspan 41 ft; length 33'-4"; dry empty weight 8982 lb; dry loaded weight 14,000 lb. Assume the air bag system must be able to lift 12,000 lb due to some sediments and marine growth on the aircraft. Each air lift bag, when fully inflated, has a lift capacity of 4000 lb when used in seawater. Overfilling the bag only allows air to bubble out of the opening at the bottom. What is the pressure in the air in the lift bags before the plane starts its ascent? What is the mass of air contained in each lift bag? How much air spills out of each lift bag between 300 feet and 10 feet from the surface?

1.24 A submarine's main air flasks are recharged to 4500 psig while the boat is deep and the seawater temp is 55 °F. The boat then comes shallow and enters the Gulf Stream where the seawater temperature is 80 °F. Because these flasks are in the main ballast tanks, which are flooded with seawater, assume the air flasks equalize with the new environmental temperature. Also assume that no air is discharged from the flasks during this time. What will the air flask pressure gage read (psig)? What amount of heat was added to the air? Classify the thermodynamic process and explain why this is the correct answer.

1.25 On July 9, 1972, the fast attack submarine USS *Barb* (SSN-596) surfaced off Guam in the middle of Typhoon Rita to rescue the survivors of a B-52 crash. FYI: This event is shown in a painting in Alumni Hall. You can also read about the *Barb* rescue at the following web address: http://members.aol.com/brittvanm/ssn596/rescue.htm. If the *Barb*'s steam plant condenser pressure gage read 28.0 in-Hg with barometric pressure inside the boat of 15.0 psia while submerged, what did the condenser pressure gage read after the OOD opened the snorkel valve, thereby allowing the pressure in the boat to equalize with the outside atmosphere (equalizing the pressure is done before opening the conning tower hatch: (1) to prevent blowing the hatch open, which may bounce off its open latch and damage the mechanisms and (2) to prevent ejecting the personnel who open the hatch)? Recall that condensers operate at a vacuum. Assume: (1) that Typhoon Rita had the same barometric pressure as the record-setting Typhoon Tip (25.69 in-Hg) and (2) that the absolute pressure inside the condenser remained the same before and after the snorkel valve was opened. Draw the pressure relationship graph showing the "before" and "after" pressure situation.

1.26 The USS *Merrimac* (commissioned in 1856 and converted to the CSS *Virginia* during the American Civil War) had a set of double-piston rod, horizontal, condensing, reciprocating steam engines (with 72 inch diameter pistons and a stroke of 3 feet) driving a 17'-4" diameter two-bladed propeller. The engine developed 1294.4 brake horsepower. The ship averaged 5.25 knots at 36.5 shaft RPM with 12.8 psig boiler pressure and 20.4 in-Hg vacuum condenser pressure. The stokers had to shovel 3,400 lbs of anthracite coal per hour to keep the steam pressure at the level cited above. (Ref: *C.S. Ironclad Virginia and U.S. Ironclad Monitor* by Sumner B. Besse, The Mariner's Museum, 1937) The heating value of anthracite coal is about 14,000 Btu/lb$_m$. Find the overall efficiency of this engine (%).

1.27 The very first steam engine was built in 1712 by Thomas Newcomen. His machine pumped water out of a coal mine in England. This pump lifted water from the depth of 150 feet and produced 120 gallons per minute. If a gallon of water weighs 8.3 pounds, calculate the power of this pump (hp). Recall that $W = F \times d$.

2 Incompressible Fluid Flow

Many tug boats also have fire fighting capability. This tug boat is spraying water in celebration of the arrival of USS *Iwo Jima* for Fleet Week in New York City.

2.1 UNIT LEARNING OBJECTIVES

2.1.1 Terminology: definitions, variables, and typical units

    2.1.1.1 continuity equation, pressure head, velocity head, potential head, head loss, water horsepower (WHP), net positive suction head (NPSH), viscosity, laminar flow, turbulent flow, friction coefficient for pipes, loss coefficient for fittings, pump classifications, valve classifications, common pipe fittings, pump efficiency, siphon.

2.1.2 Concepts: ideas and engineering expressions of them

    2.1.2.1 Adapt the steady flow energy equation (SFEE) to incompressible fluid flow in a pipe. Compare Bernoulli's equation to the steady flow energy equation. Understand the meanings of the various terms of Bernoulli's equation.

    2.1.2.2 Describe various types of pumps and their applications.

    2.1.2.3 Discuss pump performance characteristics and similarity relations (pump laws).

- **2.1.2.4** Describe the conditions necessary for a siphon to operate.
- **2.1.2.5** Describe valve and fitting types and their operating characteristics.

2.1.3 Skills: procedures, practices, or methods that enable reasoning

- **2.1.3.1** Calculate fluid flow using a Venturi:
  - **2.1.3.1.1** Given physical dimensions.
  - **2.1.3.1.2** Given a Venturi performance graph.
- **2.1.3.2** Convert between head and pressure.
- **2.1.3.3** Calculate the head loss in a piping system.
  - **2.1.3.3.1** Calculate the NPSH available at the pump in a system.
- **2.1.3.4** Interpret a manufacturer's pump curve chart and obtain data from the chart.
- **2.1.3.5** Obtain data from a Moody Friction Factor Chart for flow in pipes.
- **2.1.3.6** Calculate water horsepower for a pump. Calculate pump efficiency.
- **2.1.3.7** Calculate pump flow characteristics for multiple speeds using pump affinity laws.
- **2.1.3.8** Calculate series and parallel pump performance for multiple speeds.
- **2.1.3.9** Evaluate the effect of pump or fluid system changes on head, flow, and power.
- **2.1.3.10** Interpret fluid system schematics to describe flow and control.

2.2 INTRODUCTION

This unit covers pumps, pipes and fittings and the fluids that are typically pumped in naval applications. The scope will be limited to *relatively incompressible fluids* or liquids that do not change appreciably in density, regardless of the amount of pressure exerted upon them. Water, fuels, and lubricating oils are examples of common incompressible fluids. Under some limitations, gases may also be considered as *relatively incompressible* and the same methodology may be employed. Typical engineering applications for liquids include storage in tanks, movement through pipes or open channels, and pressurization for a specific use. A simple and general methodology for analyzing these types of problems is presented in this chapter.

Students will do well to remember the fundamental rule that fluids move in response to a pressure gradient, from high pressure to low.

2.3 FLEET APPLICATIONS

Aboard modern ships, submarines, aircraft, land vehicles, and forward operating bases one will find a wide variety of essential applications for the fluid flow principles learned in this unit.

Jet fuel is pumped from tanks deep in the hull of the ship to aircraft on the flight deck. Freshwater is pumped throughout vessels to serve such purposes as cleaning, cooking, and drinking water, and propulsion plant needs. Saltwater is pumped through a wide variety of heat exchangers and may be used for fire fighting purposes. Another damage control function is de-watering a flooded compartment by pumps and eductors. In addition, wastewater from the galley and the heads must be pumped off the vessel.

Pumping systems may be permanently installed or portable. For example, P-100 and P-250 portable pumps may be used to directly dewater flooded compartments that have no installed or operable drainage system. They can also be used to provide water for driving portable eductors or fire fighting water.

Various water systems are found in the expeditionary environment. Sanitary conditions affect both troops and civilians. One recent military example of this was the situation after the fall of Fallujah, Iraq, during operation Phantom Fury in 2005. Areas of the city were under sewage-contaminated water after the offensive. The Seabees of NMCB-4 repaired several preexisting pumping stations and installed supplementary pumps and transmission hoses to quickly improve living conditions and reduce the risk of disease. The Seabees also restored the city's potable water treatment, storage, and distribution systems to encourage residents to return after the assault.

FIGURE 2.1: *Members of NMCB-25 Installing a Water Tank for an Elementary School in the Philippines (left) and a Water Storage Tank in a Flooded Area of Iraq (right).*

2.4 OVERVIEW OF VALVES AND VALVE SYMBOLOGY

Valves are installed in fluid systems to control and direct fluid flow, isolate portions of systems for maintenance or casualties, relieve overpressure conditions, and allow flow in one direction only. Other valves reduce pressure or perform special functions, such as allowing water but not allowing steam to pass through the valve. Most valves are manually operated, but some will be fitted with pneumatic, hydraulic, or electric actuators that allow for remote or automatic operation. For the purposes of this chapter, we'll limit our discussion to valves in piped systems. Recognize, however, that ducted air conditioning systems also have dampers in them that perform valve functions.

Figure 2.2 shows the basic parts of a valve. Take note of the nomenclature.

The basic valves are globe, gate, ball, plug, butterfly, and check. Another classification of valves is relief and safety valves.

A globe and a gate valve are shown in Figure 2.3. Note that the valve disc rises to open the valve. A needle valve is a special case of globe valve. The valve disc in a needle valve is a long tapered cone shape that improves its ability to throttle the flow.

Figure 2.4 shows butterfly, ball, and plug valves. The noteworthy common operating feature is that the valve disk rotates 90-degrees on the valve stem. The convention for handle installation is that the valve is open when the actuator handle is in line with valve body.

Figure 2.5 shows a swing check and a lift check valve. These valves serve to allow flow in one direction only. There are other check valve configurations, such as a butterfly check valve and a stop check valve. The stop check valve includes a valve handle and stem that can lock the disk in the shut position and is useful as a maintenance boundary. It should be apparent why check valves have installation orientation requirements. If one of the valves shown was installed upside down, it would not function properly if at all.

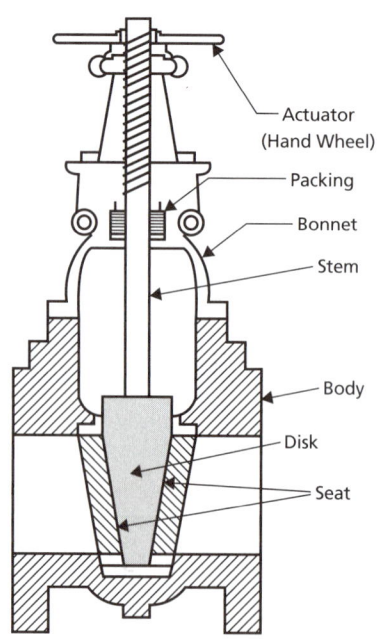

FIGURE 2.2: *Basic Parts of a Valve*

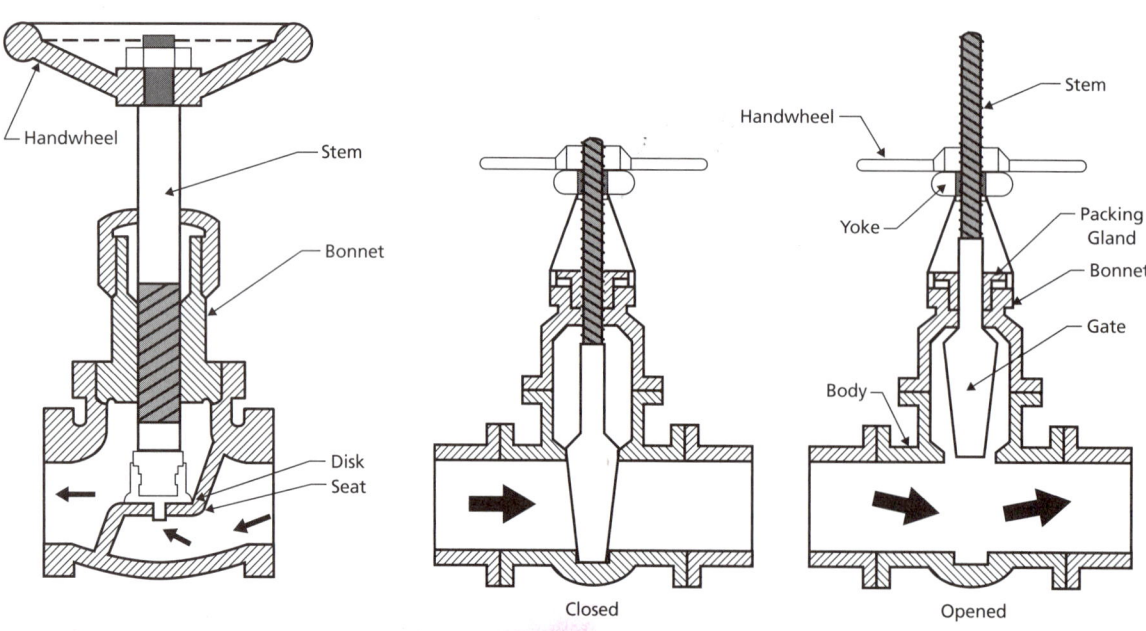

FIGURE 2.3: *Globe Valve and Gate Valve*

Relief and safety valves serve to prevent equipment damage due to accidental overpressurization. These valves open automatically to relieve system pressure. The difference between a relief valve and a safety valve is how much it opens at the setpoint and what happens as the pressure is reduced. A relief valve gradually opens above the setpoint, re-shuts at the setpoint, and is typically used in incompressible fluid systems such as water or hydraulic systems. A safety valve snaps open at the setpoint and will stay fully open until a lower pressure is achieved. The difference in opening and shutting pressures is called *blowdown* and is typically expressed as a percent of the opening setpoint pressure. Safety valves are typically used for compressible fluids, such as gases or steam systems. Safety valves typically have an external lever on them to allow manually opening the safety valve to check operability.

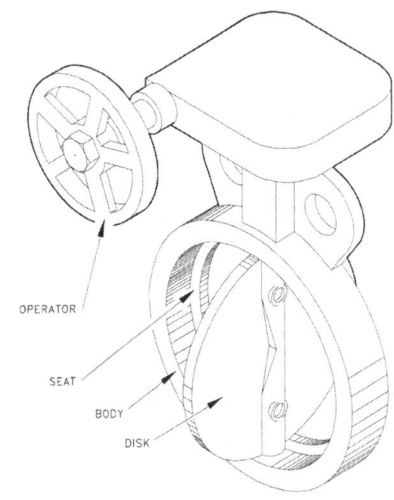

FIGURE 2.4a: *Butterfly Valve*

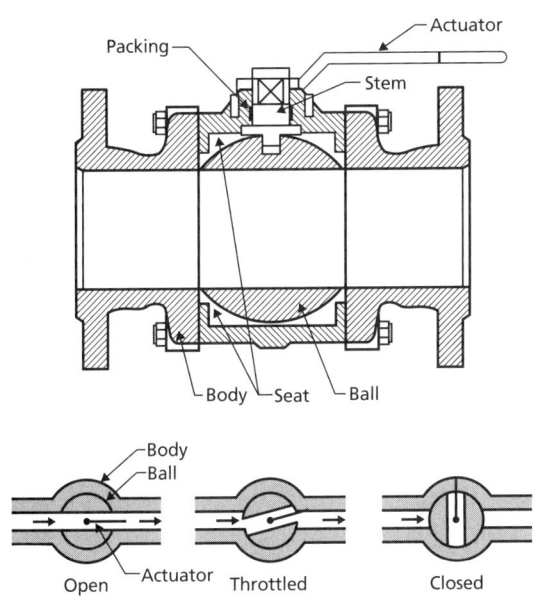

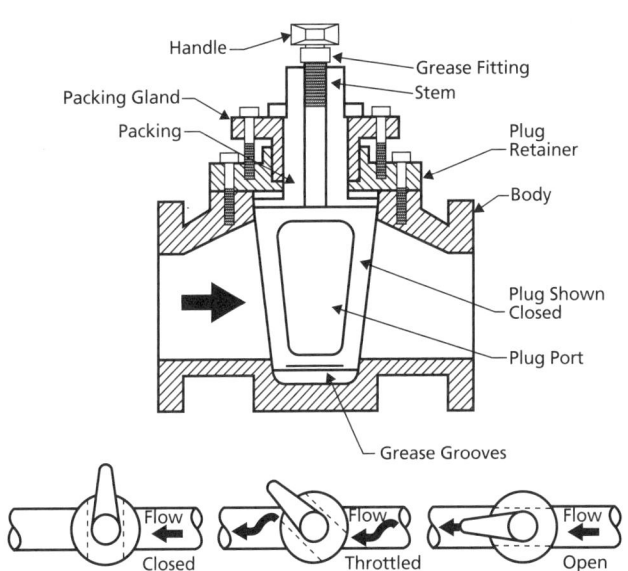

FIGURE 2.4b: *Ball Valve and Plug Valve*

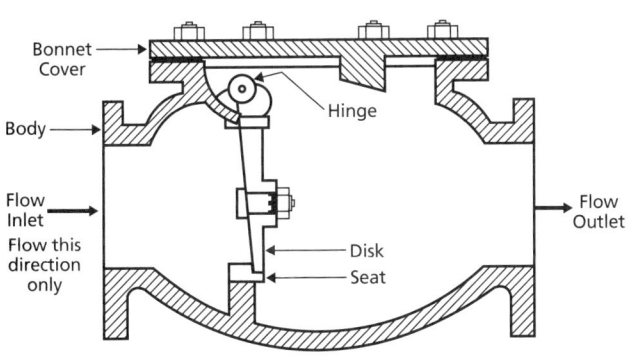

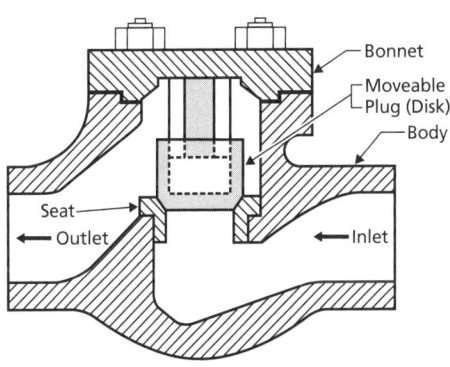

FIGURE 2.5: *Swing Check Valve and Lift Check Valve*

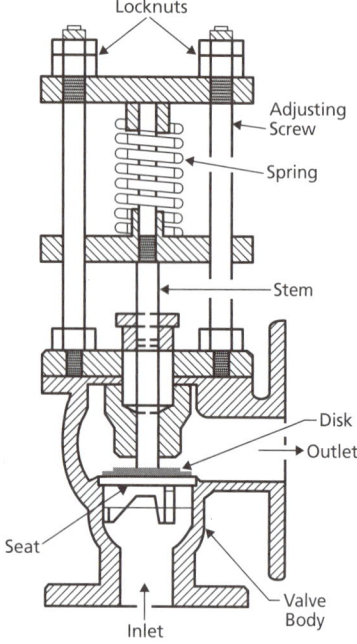
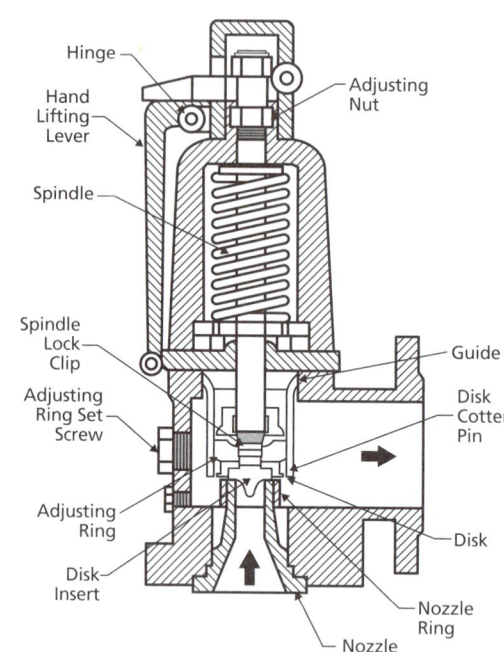

FIGURE 2.6: *Relief Valve and Safety Valve*

All valves have the same basic components and function to control flow in some fashion. However, the method of controlling the flow differs between the different valve types. In general, there are four methods of controlling flow through a valve:

1. Move a disc, or plug into or against an orifice (for example, globe or needle type valve).
2. Slide a flat, cylindrical, or spherical surface across an orifice (for example, gate and plug valves).
3. Rotate a disc or ellipse about a shaft extending across the diameter of an orifice (for example, a butterfly or ball valve).
4. Move a flexible material into the flow passage (for example, diaphragm and pinch valves).

Each method of controlling flow has characteristics that make it the best choice for a given application. The table below lists significant characteristics for some of the common valves.

TABLE 2.1: *Common Valve Comparison*

Valve	Strengths	Weaknesses
Globe	□ Good throttle □ Excellent stop valve	□ Pressure drop due to tortuous flow path
Needle	□ Excellent throttle □ Excellent stop valve	□ Large disk and seat contact area can cause binding during temperature change
Gate	□ Minimal pressure drop when fully open □ Excellent stop valve	□ Poor throttle □ Can be hard to open with large Δp or temperature changes
Ball	□ 90-degree actuation minimal motion □ Good stop valve	□ Poor throttle □ Large Δp increases friction against valve seats, makes it harder to open
Butterfly	□ 90-degree actuation minimal motion □ Lighter weight in larger applications	□ Poor throttle □ Marginal sealing when shut; typically uses a rubber-type gasket for sealing, which also has temperature limits

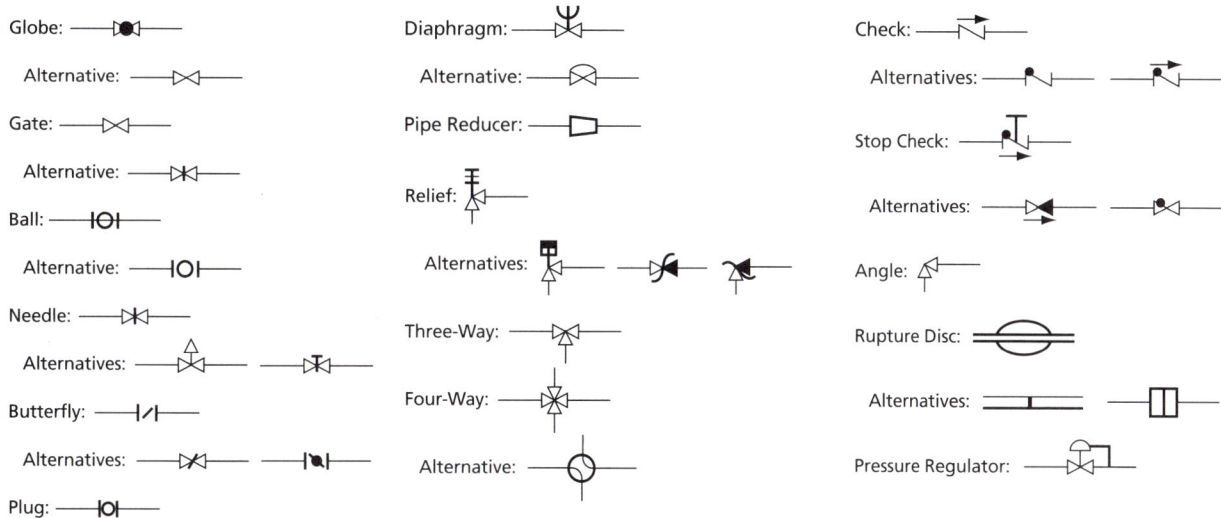

FIGURE 2.7: *Standard Symbols for Common Valves*

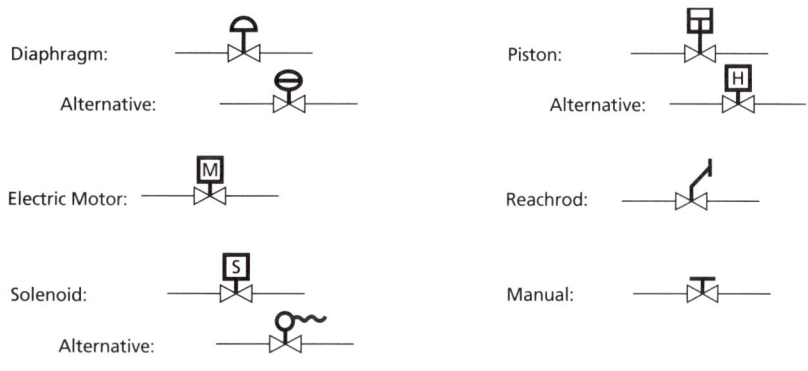

FIGURE 2.8: *Valve Actuator Symbols*

Interpreting drawings is a skill that all officers assigned to the Fleet will develop. The type of drawing that applies to this unit is known as a piping & instrument drawing (P&ID).

Only the most basic valves are discussed in this unit. Additional information on this topic is available in the DOE Fundamentals Handbooks 1016/1-93 and 1018/2-93. These documents are available for free online.

2.5 THEORY DEVELOPMENT AND PROBLEM SOLVING METHODOLOGY

2.5.1 The Continuity Equation

The basic concepts of the open system that were discussed in Chapter 1 can be applied to mass flow and are very important for nearly all calculations dealing with fluid flow. Just as energy was conserved in the first law of thermodynamics as expressed in the steady flow energy equation (SFEE), mass is conserved in the continuity equation. Under steady state conditions, the sum of mass flow rates into a system boundary or control volume must equal the mass flow rate out, or:

$$\sum \dot{m}_{in} = \sum \dot{m}_{out}$$

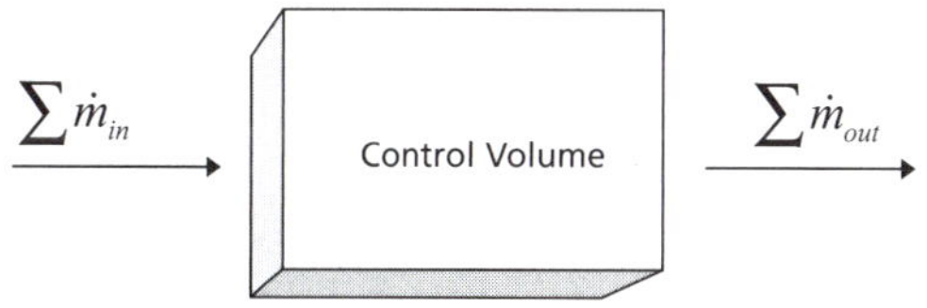

FIGURE 2.9: *Control Volume*

Typically the units of the mass flow rate will be in lb_m / (unit time). When applied to the SFEE, the most useful denominator unit will be seconds.

The mass flow rate ($\dot{m}$) of a fluid stream or flow can be decomposed into its two subcomponents: volumetric flow rate ($\dot{V}$) and density (ρ).

$$\dot{m} = \rho \dot{V} \quad (2.1)$$

Volumetric flow rate is commonly given in gallons per minute (GPM), cubic feet per minute (cfm), or ft^3/sec. The volumetric flow rate can be further decomposed into the <u>cross sectional area</u> (A) of the fluid and the average velocity ($\bar{v}$):

$$\dot{V} = A\bar{v} \quad (2.2)$$

Then, an alternate version of the continuity equation is:

$$\dot{m} = \rho A \bar{v} \quad (2.3)$$

If the fluid is incompressible, then the density of the fluid does not change ($\rho_{in} = \rho_{out}$), and several other forms of the continuity equation are then valid with this limitation:

$$\sum \dot{V}_{in} = \sum \dot{V}_{out} \quad (2.4)$$

$$\sum A_{in}\bar{v}_{in} = \sum A_{out}\bar{v}_{out} \quad (2.5)$$

A common application of this result is fluid flowing through an expansion or contraction in a piping system as shown in Figure 2.10.

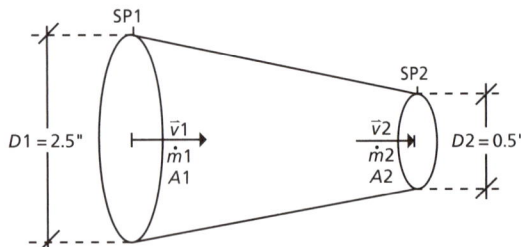

FIGURE 2.10: *Flow in a Converging Nozzle*

Piping systems use standardized, commercially available circular-section (round) pipe stock. Assuming that one knows both the inlet and outlet diameters (D_{in} and D_{out}) and is given the inlet velocity or volumetric flow rate, one can find the outlet velocity by applying the continuity equation.

$$\sum A_{in}\bar{v}_{in} = \sum A_{out}\bar{v}_{out}$$

$$A = \frac{\pi D^2}{4} \quad (2.6)$$

$$\frac{\pi D_{in}^2 \bar{v}_{in}}{4} = \frac{\pi D_{out}^2 \bar{v}_{out}}{4}$$

Canceling like terms and solving for $\bar{v}_2$:

$$\bar{v}_{out} = \frac{D_{in}^2}{D_{out}^2} \bar{v}_{in} = \left(\frac{D_{in}}{D_{out}}\right)^2 \bar{v}_{in}$$

Thus the velocity decreases in an expanding pipe by the ratio of the diameters of the cross sections squared. The opposite is true for a contracting pipe. Also note that it is not necessary to convert D to standard USCS units; as long as the D units are the same, the units will cancel.

EXAMPLE 2.1: NOZZLE CONTINUITY

GIVEN: A nozzle on a fire hose, as shown in Figure 2.10, that contracts from 2.5 inches in diameter to 0.5 inches, and the average water velocity in the 2 ½ inch fire hose is 10 ft/sec.

FIND: the stream velocity at the discharge of the nozzle.

SOLUTION: Water is assumed to be completely incompressible. The mass flow rate at state point 1 is the same as the mass flow rate at state point 2. Note that the numbering of the state points connotes that the flow occurs from SP1 to SP2.

$$\dot{m}_1 = \dot{m}_2$$

$$\rho_1 A_1 \bar{v}_1 = \rho_2 A_2 \bar{v}_2$$

The water does not change density going through the nozzle, so $r_1 = r_2$. Thus the density terms cancel. The cross section is circular, so the area, A, is $pD^2/4$. Making this substitution and performing some algebra to solve for the requested variable yields the following:

$$\bar{v}_2 = \left(\frac{D_1^2}{D_2^2}\right)\bar{v}_1 = \left(\frac{D_1}{D_2}\right)^2 \bar{v}_1$$

$$\bar{v}_2 = \left(\frac{2.5}{0.5}\right)^2 (10\,ft/sec) = (5)^2 (10\,ft/sec)$$

$$\boxed{\bar{v}_2 = 250\,ft/sec}$$

As noted above, as long as the diameter units are the same, they cancel. Therefore, it is not necessary to convert diameter to feet to match the velocity term units.

EXAMPLE 2.2: UNIT CONVERSION EXERCISE

Now for an example where unit conversions will be necessary.

GIVEN: A 5/8″ garden hose streaming water into a 5-gallon bucket. It takes 2 minutes to fill the bucket from empty.

FIND: the mass flow rate and the stream velocity at the discharge of the hose. (Note that we're asked to find two things and the units were not specified.)

SOLUTION: Water is assumed to be incompressible. Rate of flow is assumed to be steady. Calculate the volumetric flow rate, then substitute that into some form of the continuity equation. Answer the stream velocity first, since that is a subset of the continuity equation.

$$\dot{V} = V/t = 5\,\text{gal}/2\,\text{min} = 2.5\,GPM$$

Note that units in gallons are volume, and not mass. This will require at least one conversion factor. Also note that there is only one SP given for this problem, the hose discharge, so I dropped the subscripts. Standard units for velocity are ft/sec, so I'll work in that direction.

$$\dot{V} = A\bar{v} \Rightarrow \bar{v} = \frac{\dot{V}}{A} = (\dot{V})\left(\frac{4}{\pi D^2}\right)$$

$$= \left(2.5\,\frac{\text{gal}}{\text{min}}\right)\left(\frac{4}{\pi 0.625^2\,\text{in}^2}\right)\left(\frac{1\,\text{ft}^3}{7.48\,\text{gal}}\right)$$

$$\left(\frac{144\,\text{in}^2}{\text{ft}^2}\right)\left(\frac{1\,\text{min}}{60\,\text{sec}}\right)$$

$$\boxed{\bar{v} = 2.6\,\text{ft/sec}}$$

Water density will be taken as freshwater at standard conditions, since we don't know the temperature and will assume that the water coming out of a garden hose is freshwater. The density shouldn't be too far from standard, because it comes out of the water distribution system. (It might be different from standard if the hose had been sitting in the sun and the insolation heated the water to some higher temperature.)

$$\dot{m} = \rho A \bar{v} = \rho \dot{V}$$

$$= \left(62.4\,\frac{\text{lb}_m}{\text{ft}^3}\right)\left(2.5\,\frac{\text{gal}}{\text{min}}\right)\left(\frac{1\,\text{ft}^3}{7.48\,\text{gal}}\right)\left(\frac{1\,\text{min}}{60\,\text{sec}}\right)$$

$$\boxed{\dot{m} = 0.35\,\text{lb}_m/\text{sec} = 20.9\,\text{lb}_m/\text{min}}$$

Note that the first answer was less than 1, so since the units weren't specified by the problem I chose to answer it with units that gave a number greater than 1, but not requiring scientific notation.

2.5.2 The Frictionless Bernoulli Equation

Bernoulli's equation is a rearrangement of the SFEE that includes potential energy, kinetic energy, flow work, internal energy, heat, and work of a fluid. Recall that the SFEE is:

$$\left(\frac{g}{g_c}\right)z_i + \left(\frac{1}{2g_c}\right)\bar{v}_i^2 + p_i v_i + u_i + q_{if}$$

$$= \left(\frac{g}{g_c}\right)z_f + \left(\frac{1}{2g_c}\right)\bar{v}_f^2 + p_f v_f + u_f + w_{if} \quad (2.7)$$

Assuming that there is no change in temperature (therefore no change in internal energy) of the fluid and that the flow is adiabatic (no heat transfer) (both reasonable assumptions for pipe flow) and that specific volume is constant for incompressible fluids, the equation reduces to:

$$\left(\frac{g}{g_c}\right)z_i + \left(\frac{1}{2g_c}\right)\bar{v}_i^2 + p_i v_i$$

$$= \left(\frac{g}{g_c}\right)z_f + \left(\frac{1}{2g_c}\right)\bar{v}_f^2 + p_f v_i + w_{if} \quad (2.8)$$

Typical units for each of the terms are $\frac{\text{ft}\cdot \text{lb}_f}{\text{lb}_m}$. A further simplification can be made by multiplying the entire equation by g_c/g.

$$z_i + \left(\frac{\bar{v}_i^2}{2g}\right) + \frac{g_c}{g}p_i v_i$$

$$= z_f + \left(\frac{\bar{v}_f^2}{2g}\right) + \frac{g_c}{g}p_f v_i + \frac{g_c}{g}w_{if} \quad (2.9)$$

One last simplification can be made by substituting the expression for specific weight into the flow work portion of the SFEE:

$$\gamma = \rho \frac{g}{g_c} = \frac{g}{g_c v} \quad (2.10)$$

Finally, the resulting Bernoulli equation is:

$$z_i + \frac{\bar{v}_i^2}{2g} + \frac{p_i}{\gamma} = z_f + \frac{\bar{v}_f^2}{2g} + \frac{p_f}{\gamma} + \frac{g_c}{g} w_{if} \quad (2.11)$$

It should be obvious that the units of this form of Bernoulli's equation are *feet* or what is commonly referred to by engineers as *head*. Head is a surrogate term for pressure and can be thought of as the height of a column of the flowing fluid that a given pressure could support. Within the Bernoulli equation there are several different types of head due to differing energy sources. The velocity term ($\bar{v}$) contributes a *velocity head*, the height (z) above a reference point contributes an *elevation head*, and the pressure term (p) contributes a *pressure head*. Lastly, the work term (w) is the amount of *head* added by an *isentropic* or *ideal* pump. In this form of Bernoilli's equation, the sign applied to the w term follows the "work in is negative" convention. An alternate form of Bernoulli's equation solves for ideal pump work and resolves the pump's work into the fluid to be a positive number. This is called the *pump head equation without friction* and is often a more convenient starting point for analyses.

$$\underbrace{w_{p,s}\left(\frac{g_c}{g}\right)}_{\text{pump head}} = \underbrace{(z_f - z_i)}_{\text{elevation head}} + \underbrace{\left(\frac{\bar{v}_f^2 - \bar{v}_i^2}{2g}\right)}_{\text{velocity head}} + \underbrace{\left(\frac{p_f - p_i}{\gamma}\right)}_{\text{pressure head}} \quad (2.12)$$

The standard units of each term are feet. If specific work units are desired, then multiply both sides of the equation by g/g_c to resolve $w_{p,s}$ in the typical units of ft-lb$_f$/lb$_m$, which is also convertible to Btu/lb$_m$. Used in conjunction with the continuity equation, one can find pressure, velocity, and height at any point in a non-constant diameter piping system using the Bernoulli equation.

EXAMPLE 2.3: FRICTIONLESS FLOW

GIVEN: Freshwater enters a 4 inch diameter pipe at 100 psia and 10 ft/sec (SP1) and exits through a 2 inch cross section 10 ft below the entrance (SP2). The water does not pass through a pump.

FIND: What are the velocity and pressure (psia) of the water at the exit?

SOLUTION: We'll assume standard water and establish the elevation datum $z_2 = 0$.

From continuity:

$$\bar{v}_2 = \left(\frac{D_1^2}{D_2^2}\right)\bar{v}_1 = \left(\frac{4in}{2in}\right)^2 10\frac{ft}{\sec} = 40\frac{ft}{\sec}$$

Now, start with the pump head equation, then eliminate terms that either cancel or do not exist. In this case, there is no pump, so $w_{p,s} = 0$, but no like terms cancel.

$$w_{p,s}\left(\frac{g_c}{g}\right) = (z_2 - z_1) + \left(\frac{\bar{v}_2^2 - \bar{v}_1^2}{2g}\right) + \left(\frac{p_2 - p_1}{g}\right)$$

$$0 = (z_2 - z_1) + \left(\frac{\bar{v}_2^2 - \bar{v}_1^2}{2g}\right) + \left(\frac{p_2 - p_1}{\gamma}\right)$$

Then solve for the parameter of interest, in this case, p_2.

$$\frac{p_2}{\gamma} = (z_1 - z_2) + \left(\frac{\bar{v}_1^2 - \bar{v}_2^2}{2g}\right) + \left(\frac{p_1}{\gamma}\right)$$

$$p_2 = \gamma\left[(z_1 - z_2) + \left(\frac{\bar{v}_1^2 - \bar{v}_2^2}{2g}\right)\right] + p_1$$

Next, substitute values and evaluate for necessary conversion factors.

$$p_2 = 62.4\frac{lb_f}{ft^3}\left[(10-0)ft + \left(\frac{(10^2 - 40^2)ft^2/\sec^2}{2\left(32.2\frac{ft}{\sec^2}\right)}\right)\right]$$

$$\left(\frac{1ft^2}{144in^2}\right) + \left(100\frac{lb_f}{in^2}\right) absolute$$

$$p_2 = 62.4\frac{lb_f}{ft^3}[(10)ft + (-23.3)ft]\left(\frac{1ft^2}{144in^2}\right)$$

$$+ \left(100\frac{lb_f}{in^2}\right) absolute$$

$$p_2 = (-5.8 + 100)\frac{lb_f}{in^2} = 94.2\, psia$$

$$p_2 = 94.2\ psia$$

Next we'll address friction losses.

2.5.3 Pipe and Fitting Head Loss

In reality, head cannot be transferred entirely from one point in a system to another. There are always frictional losses due to interaction of the fluid with the pipe walls and components such as valves, elbows, and other pipe fittings. In general, these losses are determined empirically and can be added to the ideal Bernoulli equation to improve accuracy. As such, a variable representing the head loss (H_L) is added to Bernoulli's equation:

$$z_i + \frac{\bar{v}_i^2}{2g} + \frac{p_i}{\gamma}$$
$$= z_f + \frac{\bar{v}_f^2}{2g} + \frac{p_f}{\gamma} + \left(\frac{g_c}{g}\right) w_{if,isen} + \overset{head\ loss}{\widetilde{H_{L,if}}} \qquad (2.13)$$

Head loss (H_L) is a positive quantity added to the downstream state point. In effect, H_L takes head away from the usable velocity, elevation, and pressure head components and requires additional work from the pump. Another convenient rearrangement of the Bernoulli equation that also incorporates friction losses is referred to as the *pump head equation*. Under this form, the equation is rearranged to solve for the pump work term and the sign of the equation is taken as positive for work put into the fluid by the pump. As introduced above, the pump work term is referred to as *isentropic pump work*, or *ideal pump work* supplied to the fluid, so the subscripts for the head terms are in the order (f, final minus i, initial) to reinforce this concept.

$$w_{p,s}\left(\frac{g_c}{g}\right)$$
$$= (z_f - z_i) + \frac{(\bar{v}_f^2 - \bar{v}_i^2)}{2g} + \frac{(p_f - p_i)}{\gamma} + H_{L,if} \qquad (2.14)$$

The units of the terms in the above equation are all in feet, but may be converted to ft-lb_f/lb_m by multiplying both sides by g/g_c. This is usually best done after simplifying the right side of the equation. Once converted to ft-lb_f/lb_m, it is also easily converted to Btu/lb_m using Joule's constant (778 ft-lb_f = 1 Btu) as a conversion factor.

2.5.3.1 Major Losses—Pipes

The first component of H_L is the major loss due to pipe wall friction in a straight pipe. The losses in bends of pipes, valves, and other fittings are referred to as minor losses. Major losses are directly attributable to the frictional contact between the *real* fluid and the pipe wall. These losses are proportional to the length (L) of the pipe, are inversely proportional to the diameter (D), and increase with the square of the fluid velocity ($\bar{v}$). This relationship is known as Darcy's equation for head loss and includes a friction factor (f) for differing pipe attributes:

$$H_{L-Major} = \left(\frac{fL}{D}\right)\frac{\bar{v}_{pipe}^2}{2g} \qquad (2.15)$$

The friction factor (f) is found using the Moody Friction Factor Chart (Figure 2.11) and depends upon the Reynolds number (R) and relative roughness:

$$R = \frac{\bar{v}D}{\upsilon} \qquad (2.16)$$

$$relative\ roughness = \frac{\varepsilon}{D} \qquad (2.17)$$

It is important to note that the Moody Friction Factor Chart denotes velocity V, diameter D, and the *kinematic viscosity* (υ) of the fluid. Do not become confused by this traditional but ambiguous use of the variables. The Reynolds number must be dimensionless, so the units of viscosity, υ, often must be converted.

The surface roughness (ε) of the specific piping material is measured in ft, mm, or mils. Relative roughness is also dimensionless, so the diameter may require conversion to the same units as e. Typical values of e for several materials (White, 1999) are given in Table 2.2:

TABLE 2.2: *Typical Roughness Values*

Material	Condition	ε (ft)
Steel	Stainless, New	0.00016
	welded, seamless	0.0002
	Rusted	0.007
Iron	Cast, New	0.00085
	Galvanized, New	0.0005
Brass	Drawn tubing	0.000005
Rubber	Smoothed	0.000033
Plastic	Drawn tubing	0.000005

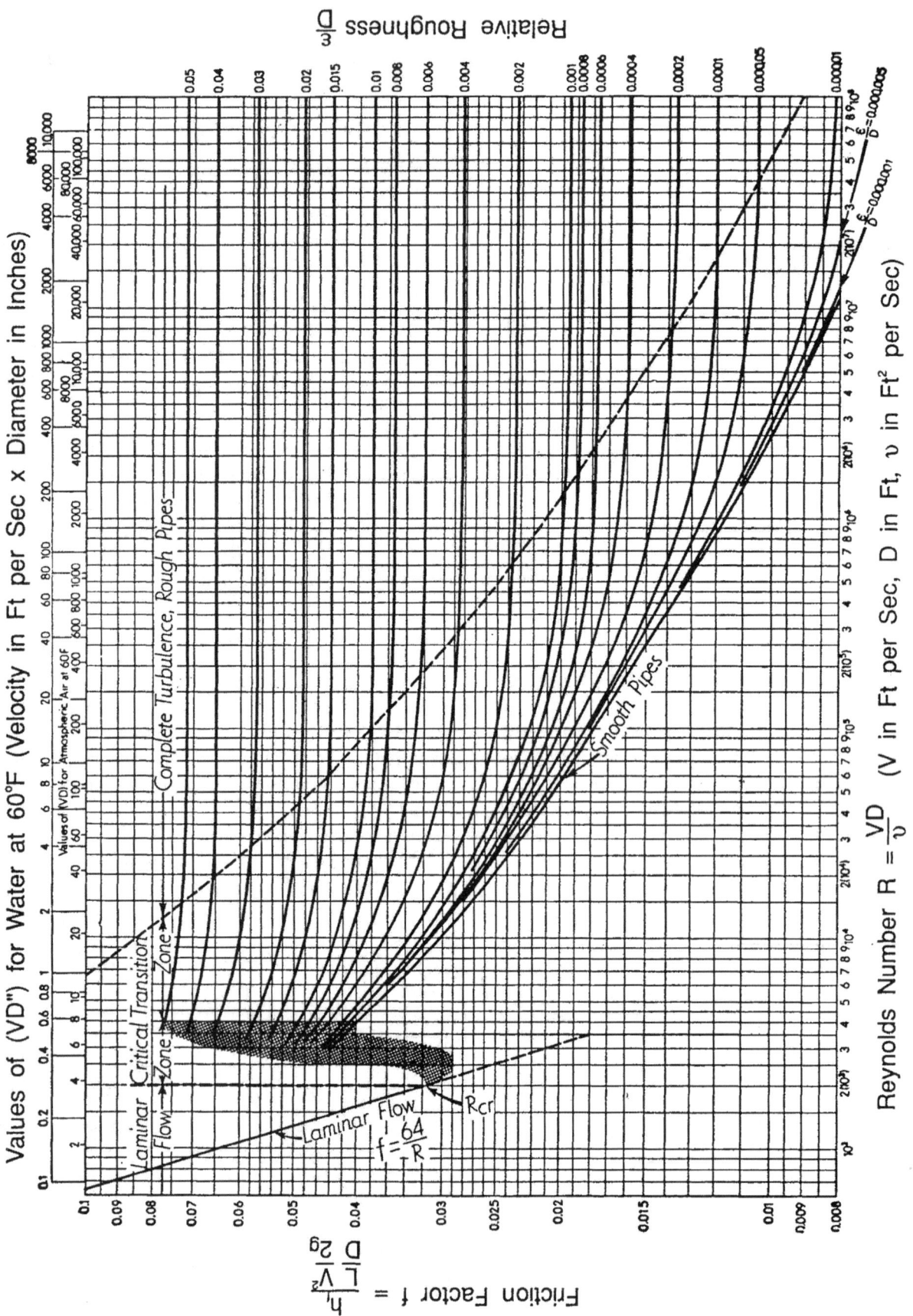

FIGURE 2.11: *Moody Friction Factor Chart*

Moody Friction Factor Chart from "Friction Factors for Pipe Flow," by Moody, L. F., Transactions, Vol 66, 1944, reprinted by permission from ASME

EXAMPLE 2.4: CALCULATING FRICTION FACTOR

GIVEN: Water is pumped from a well to a house using high density polyethylene pipe (HDPE) (plastic). The pipe is 1 inch diameter and the water is flowing with a Reynolds number (R) of 50,000.

FIND: What is the friction factor (f) of the pipe in this situation?

SOLUTION: Look up e for this material in Table 2.2 and calculate *relative roughness*:

$$\text{relative roughness} = \frac{\varepsilon}{D} = \frac{0.000005\,ft}{1in\left(\frac{1ft}{12in}\right)} = 0.00006$$

Entering the Moody Friction Factor Chart (Figure 2.11) with *relative roughness* from the right and find the intersection with the value of R. Then read the value of f on the left axis.

$$f = 0.021$$

2.5.3.2 Minor Losses—Fittings

Minor losses are attributable to piping system components such as tank inlets and outlets, pipe elbows, tees, valves, etc., which are referred to as *fittings*. The minor loss in each component is reflected in a fitting loss coefficient (k). The fitting loss coefficient can be derived from several sources or experimentally, but typically one relies on a manufacturer's data sheet for components such as valves and elbows. It is important to note that the k value is not completely constant for all conditions, but rather depends upon Reynolds number, the exact valve configuration (open, partially closed, etc.), and proximity to other fittings. The scope of this text is limited to assumed constant k values for each component. The minor losses are taken into account in the following way:

$$H_{L-Minor} = (\Sigma k)\frac{\vec{v}_{pipe}^2}{2g} \quad (2.18)$$

2.5.3.3 Combined Head Loss

The combined head loss equation that accounts for major and minor losses is as follows and may be substituted into the extended Bernoulli equation:

$$H_{L,if} = \left(\frac{fl}{D} + \Sigma k\right)\frac{\vec{v}_{pipe}^2}{2g} \quad (2.19)$$

It is important to note that this equation is valid for a constant diameter section of the piping system. If the diameter of the system changes at some point we know from continuity that the velocity will change and therefore the head loss equation will have to be broken into two analyses. This is true for subsequent diameter changes.

2.5.4 Velocity Profiles in Pipes

There are two major regimes of fluid flow—*laminar* and *turbulent*. A practical impact of fluid viscosity is that a pumped fluid adheres to the pipe walls.

Laminar flow is associated with low Reynolds numbers—low velocity or high viscosity. In laminar flow, the velocity profile from pipe wall to pipe wall will be parabolic and the streamlines are all parallel. Another implication of laminar flow is that if a dot of dye were injected into the fluid stream, the dot would retain its shape as it moved along with the fluid. Students should note the laminar region on the Moody Friction Factor Chart as it will affect the friction factor (f).

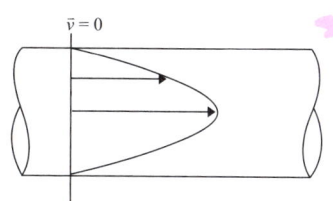

FIGURE 2.12: Laminar Flow Parabolic Velocity Profile in Pipes

Turbulent flow is associated with higher Reynolds numbers. The velocity transitions from where viscosity causes the fluid to adhere to the pipe wall to the velocity of the bulk fluid in the center. This transitional portion is called the *boundary layer*. Turbulence means there is random motion that results in mixing in the bulk section of the fluid. Hence, a dot of dye injected into turbulent flow would rapidly become mixed with the fluid. Refer back to Figure 2.11 and note that in fully turbulent flow, the friction factor is fairly constant as velocity and R changes.

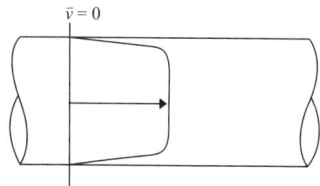

FIGURE 2.13: Turbulent Flow Velocity Profile in Pipes

Regardless of type of flow, the velocity ($\vec{v}$) used in the Bernoulli equation, and associated equations, is the mean velocity. This is the velocity calculated from volumetric flow rate ($\dot{V}$).

2.5.5 Fluid Flow Problem Solving Methodology

The fundamentals of conservation of energy and continuity still apply. The steps below set out a thought process that amplifies the steps listed in Unit 0. The additional points are best practices for solving a fluid mechanics type of problem.

1. Diagnose that you have a Bernoulli-type problem
   a. Fluid flowing—typically freshwater, seawater, or various oils
2. Explicitly list Given and Find:
   a. Be careful if units for the answer are specified. Otherwise use typical units.
   b. Be careful if you are asked to find gage pressures to give the answer in gage units.
3. Draw the system and include given information.
   a. Identify state points for analysis
      i. At end locations
         a) Interface of the liquid and air
      ii. Where you are given or can calculate a significant system parameter, like pressure
      iii. Across any component that you will be analyzing
4. Do preliminary analysis
   a. Pressures to PSF (lb_f/ft^2)
   b. Pipe cross-section areas in SF (ft^2)
   c. V-dot & D allow calculation of Velocity (ft/sec) using Continuity
5. Write the Bernoulli equation or pump head equation
   a. Simplify terms
      i. Significant assumptions and simplifications
         a) Tanks have negligible velocity due to large interface surface area.
         b) Velocity in valves and fittings are the connected pipe's velocity regardless of what happens inside the valve or fitting—e.g., a 6-inch valve is connected to 6-inch pipe unless told differently.
         c) Free liquid streams into air are at atmospheric pressure.
         d) Some of the like terms may not change and therefore cancel.
   b. Solve for the requested parameter algebraically.
   c. Substitute the values and units into the equation.
   d. Evaluate for any needed conversion factors for dimensional homogeneity.
   e. Solve for the numerical answer.
   f. Box your answer.

$$\boxed{\text{variable = value and units}}$$

2.6 PUMPS

Pumps in general add energy to a fluid in order to supply pressure head, overcome head loss, provide sufficient flow (velocity), or to raise the height of the fluid. Pumps may be classified as variable displacement or positive (fixed) displacement. Common variable displacement pumps are centrifugal and axial pumps. Positive displacement pump subclassifications are reciprocating and rotary.

2.6.1 Variable Displacement Pumps

Centrifugal pumps are the most common variable displacement pumps. They are commonly used in piping systems aboard naval vessels and many other applications. The centrifugal pump operates by creating a low pressure area at the pump entrance (the "eye") and increasing the velocity of the fluid by directing the particles along impeller guide-vanes (see Figure 2.14). The higher velocity particles enter the volute with increased kinetic energy that is then converted to pressure as the fluid interacts with the volute (casing) and subsequently moves to the discharge nozzle. The impeller can be driven electrically or mechanically through the shaft.

A variable displacement pump may be one of three types: radial flow, mixed flow, and axial flow. A radial flow pump's pressure is developed entirely by the centrifugal force of the impeller and depends upon the density of the fluid being pumped. The axial flow pump develops pressure by the lifting action of the impeller blade, which acts similar to a boat or aircraft propeller (see Figure 2.15). A mixed flow pump is a combination of the two.

2.6.2 Pump Curves and Pump Affinity Laws

Centrifugal pumps can be operated under a wide variety of conditions and generally impart between 5 and 100 to 150 psi of additional pressure to the working fluid. At a given rotational speed (N) a centrifugal pump can impart a change of pressure (Δp) that depends upon the flow rate of the system, viscosity of the fluid, density of the fluid, and many other factors. Holding all factors constant except volumetric flow rate ($\dot{V}$) and (Δp) one can develop a characteristic pump curve. Curves for speed N_1 and a higher speed N_2 are shown in Figure 2.16.

There are several key conditions that are shown on the pump characteristic curve. The maximum head developed is known as *pump shutoff head* and occurs when there is no flow due to excessive restrictions, such as a shut valve in the discharge line. Graphically, this is the value of the pump characteristic curve at the y axis. The maximum flow rate is known as *pump runout* and occurs when the pump has no backpressure. This is located along the x axis. In reality, the operating point for a system with a pump in it is the intersection of the pump characteristic curve and the system curve. Since the system curve is the total head seen by the pump and the velocity head and head loss both are dependent on the velocity squared, the system curve tends to be parabolic and curving up. The desired operating point should give the most efficient operation, if the pump is properly selected by the design engineer and the system properly operated by the crew.

Centrifugal pumps obey a simple set of rules known as the *Pump Affinity Laws*. These laws assume that the viscosity, fluid density, and other factors that affect performance are held constant, as before, but the rotational speed (N) of the pump is allowed to vary.

■ **LAW 1:** Volumetric flow rate is proportional to pump speed.

$$\dot{V} \propto N \quad \text{or} \quad \frac{\dot{V}_f}{\dot{V}_i} = \frac{N_f}{N_i} \quad (2.20)$$

■ **LAW 2:** The pressure change across the pump is proportional to the square of pump speed.

$$\Delta p \propto N^2 \quad \text{or} \quad \frac{\Delta p_f}{\Delta p_i} = \left(\frac{N_f}{N_i}\right)^2 \quad (2.21)$$

Note: Pump curves show the pressure <u>rise</u> measured across the pump, but the pressure <u>drop</u> across any component also follows the same affinity law.

Figure 2.17 illustrates the first two pump affinity laws and shows the system curve and how the affinity laws apply to the operation of the system.

■ **LAW 3:** The power required by the pump is proportional to the cube of pump speed.

$$\dot{W} \propto N^3 \quad \text{or} \quad \frac{\dot{W}_f}{\dot{W}_i} = \left(\frac{N_f}{N_i}\right)^3 \quad (2.22)$$

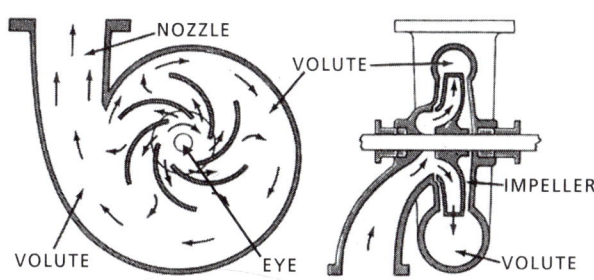

FIGURE 2.14: *Centrifugal Pump (radial)*

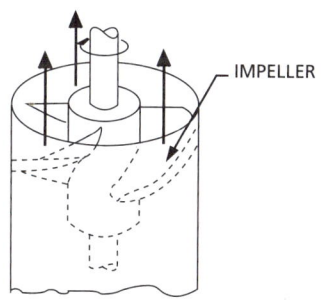

FIGURE 2.15: *Propeller Pump (axial)*

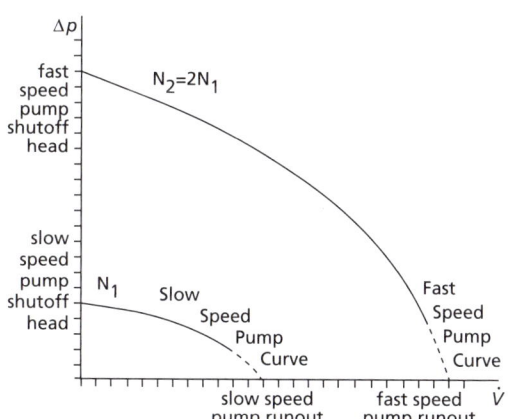

FIGURE 2.16: *Pump Curves for a Variable Displacement Multi-Speed Pump*

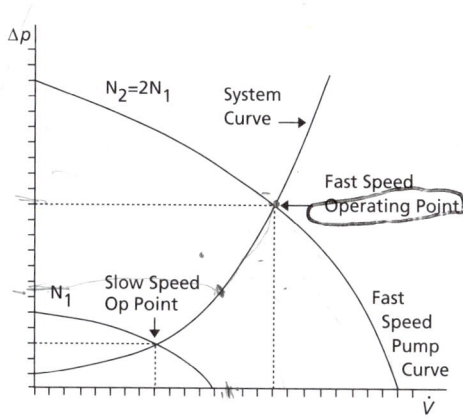

FIGURE 2.17: *Pump and System Curves for Multi-Speed Variable Displacement Pump*

> **EXAMPLE 2.5: PUMP LAWS**
>
> **GIVEN:** If a centrifugal pump raises the pressure of the working fluid by 10 psi at 500 RPM, then what will the pressure rise be if the same pump is operated at 600 RPM?
>
> **SOLUTION:**
>
> $$\Delta p \propto N^2$$
>
> $$\frac{\Delta p_2}{\Delta p_1} = \left(\frac{N_2}{N_1}\right)^2$$
>
> $$\Delta p_2 = \Delta p_1 \left(\frac{N_2}{N_1}\right)^2 = 10\,psi \left(\frac{600\,RPM}{500\,RPM}\right)^2$$
>
> $$= 10(1.2)^2 = 10(1.44) = 14.4\,psi$$

2.6.3 Cavitation and Net Positive Suction Head

Cavitation is the formation and subsequent collapse of bubbles (or cavities) of working fluid vapors. Bubbles form where local pressure is low enough that some liquid changes phase to vapor. Collapse occurs where the local pressure is higher and the vapor abruptly returns to liquid phase and the surrounding liquid rushes in to "fill" the void. Cavitation can occur anywhere in fluid flow sytems where the local pressure is low enough. Cavitation frequently occurs in the impeller of centrifugal pumps. Vapor bubbles, which are compressible and reduce the fluid density, degrade the performance of the pump while bubble collapse can physically damage the pump such as by erosion of the metal parts the bubbles collapse on. People operating pumps can change system conditions (and therefore fluid properties) so that fluid pressure is kept high enough to prevent cavitation. The concept of *net positive suction head* (NPSH) and how system conditions affect it is important to understand so the operator can avoid cavitation-related damage to the pump.

Liquid entering the suction of a centrifugal pump is accelerated by the impeller. The pressure at the impeller lowers as the velocity increases (consistent with the Bernoulli equation). If the pressure lowers to *saturation pressure* for the fluid being pumped some liquid changes to vapor (i.e., "boils") and a vapor bubble is formed. Bubbles are carried by the fluid stream to locations where the local pressure is greater than saturation pressure and there collapse. Pressure is increased anywhere that the fluid velocity is reduced, such as the outlet of the impeller or the pump casing around the impeller. These tiny implosions can erode the metals from which pump components are fabricated and ultimately reduce the pump's efficiency as the remaining metal parts are eroded into inefficient shapes. In extreme cases of fluid flashing to vapor, the impeller area may be filled with vapor. In this more severe scenario, the reduction in density reduces the ability of the impeller to function and the mass flow rate will drop significantly. This situation is referred to as "gas binding" or "vapor binding." Without the higher density liquid in the impeller, the pump speeds up and loses cooling to internal seals, which may cause damage to these components. The pump will typically re-fill and then "slug" the liquid through the system, causing large pressure surges. The gas binding process tends to repeat over and over without operator intervention.

Net positive suction head (NPSH) is a measure of how close a fluid is to saturation conditions. NPSH is the Net amount the pressure Head at the Suction of the pump is greater than (Positive) the saturation pressure of the fluid being pumped.

$$NPSH = p_{suction} - p_{saturation} \qquad (2.23)$$

There are locations in the pump where, due to design and flow rate, the pressure is lower than at the suction, so the pump manufacturer specifies a minimum pressure to be maintained at standard location of the pump suction to ensure cavitation does not occur anywhere in the pump internals. This pressure is called the $NPSH_{required}$. $NPSH_{required}$ is typically shown on the manufacturer-supplied pump curves (see Figure 2.21). The NPSH that can be achieved at the pump suction by adjusting fluid system conditions is called $NPSH_{available}$. By maintaining $NPSH_{available}$ greater than $NPSH_{required}$ cavitation can be avoided.

$$NPSH_A > NPSH_R \quad \text{(to prevent cavitation)} \qquad (2.24)$$

The operator of the fluid system is in the best position to detect cavitation and to prevent both it and the degraded performance and damage associated with it. Signs that cavitation is occurring are: 1) noise from the collapse of vapor bubbles, 2) fluctuating discharge pressure, 3) fluctuating pump flow rate, and 4) fluctuating pump motor power. The operator can prevent cavitation by raising $NPSH_A$ of the fluid being pumped. Each of the fluid properties that are factors in the Bernoulli equation can be changed to raise $NPSH_A$. Also, the saturation pressure can be reduced by lowering the temperature of the liquid being pumped. Changing some or all of these properties as system procedures or the situation allow, will increase NPSH:

- Elevation (z): lowering the pump suction elevation relative to the liquid level in the tank being pumped (or raising tank liquid level)
- Velocity ($\bar{v}$): lowering velocity through the pump such as by reducing the speed of a variable speed pump, running two parallel pumps at reduced speed instead of one pump at fast speed, or throttling the pump discharge valve
- Pressure (p): raise the pressure head at the pump suction by raising the pressure in an expansion tank (used to maintain system pressure) or pressurizing the tank being pumped from with air or inert gas.
- Head loss (H_L): reducing the head loss in the suction pipe connected to the pump suction will directly increase the pressure head available at the pump suction.
- Temperature (T): lowering the temperature of the liquid being pumped will lower the saturation pressure of the liquid

EXAMPLE 2.6: CALCULATING AVAILABLE NPSH

GIVEN: A centrifugal pump is mounted next to a tank where atmospheric conditions are standard. The pump is taking suction on (removing fluid from) the tank. The tank holds room temperature salt water and the surface of the water is 10 feet below the suction inlet on the pump. The suction pipe is 20 feet long, 2½ inch diameter, and has 4 feet of head loss at the current volumetric flow rate of 200 GPM. The manufacturer specifies the $NPSH_R$ is 8 feet for this flow rate.

FIND: Determine the $NPSH_A$ (feet).

SYSTEM SKETCH:

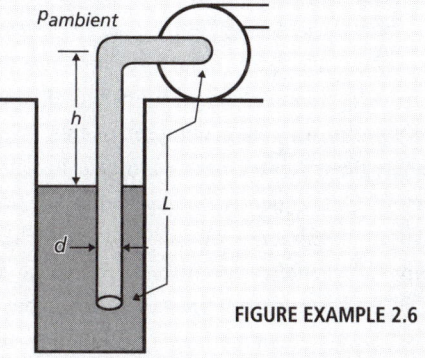

FIGURE EXAMPLE 2.6

SOLUTION: This solution will demonstrate an expanded logic using the methodology of Section 2.5.5.

Classify the problem.
1. System—water flowing in tank and suction pipe → open system
   Boundaries—water surface inside the tank, pipe wall, pump suction inlet—does not include the pump itself.
2. Substance—salt water
3. State—liquid; stays liquid
4. State Change / Process—pumping process
5. Sequence—fluid flows through a pipe to the pump inlet. This is not a power cycle.

Organize the data and associate the values with variables.

p_{amb} = 14.7 psia (standard) $H_{L\text{-}20ft}$ = 4 ft
$T_{water} = T_{amb}$ = 70 °F (standard) $\dot{V}$ = 200 GPM
h = 10 ft L = 20 ft
d = 2½ inch $NPSH_A$ = ?

List the appropriate equations.
$NPSH_A$ definition: $NPSH_A = p_{suction} - p_{saturation}$
Bernoulli equation:

$$z_i + \frac{\bar{v}_i^2}{2g} + \frac{p_i}{\gamma} = z_f + \frac{\bar{v}_f^2}{2g} + \frac{p_f}{\gamma} + \left(\frac{g_c}{g}\right)w_{if,s} + H_{L,if}$$

Select state points: Shifting to numbered state points, we'll set SP1 as the initial (i) location and SP2 as the final (f). SP2 will be at the pump suction since our desired property (pressure @ suction) is there. SP1 should be where the most properties are known or cancel like terms in the Bernoulli equation. Options include:

a. At the lowest end of suction pipe?
   p_1 varies with depth, $\bar{v}_2 = \bar{v}_1$. We know the pipe length, but not the elevation of the open pipe end (z_1) since the pipe appears to have a horizontal portion.
b. At the surface of the water in the tank?
   p_1 known (= p_{amb}), $\bar{v}_1$ can be assumed negligible (due to large tank surface area), z_1 can be established as the datum ($z_1 = 0$).
c. Inside the suction pipe at the same elevation as the water surface?
   p_1 varies with velocity and is effected by head loss, $\bar{v}_2 = \bar{v}_1$, the L between 1 and 2 is unknown.

Choose Option b. More terms are eliminated; unknowns in a or c are too much trouble to resolve.

ASSUMPTIONS:
- fittings head loss is negligible.
- water is incompressible.
- specific weight (γ) of the seawater is 64.0 lb$_f$/ft^3.
- saturation pressure (p_{sat}) of seawater is same as freshwater.

REQUIRED INFORMATION: p_{sat} of water @ 70 °F = 0.36 psia (from *Steam Tables*)

Simplify the Bernoulli equation: $w = 0$ (no pump inside the "system"), $z_1 = 0$ (baseline), $\bar{v}_1 = 0$ (large surface area in the tank); solve for p_2 which is $p_{suction}$.

$$z_1 + \frac{\bar{v}_1^2}{2g} + \frac{p_1}{\gamma} = z_2 + \frac{\bar{v}_2^2}{2g} + \frac{p_2}{\gamma} + \left(\frac{g_c}{g}\right)w_{12} + H_{L,12}$$

$$0 + 0 + \frac{p_1}{\gamma} = z_2 + \frac{\bar{v}_2^2}{2g} + \frac{p_2}{\gamma} + 0 + H_{L,12}$$

$$p_2 = \gamma\left(\frac{p_1}{\gamma} - z_2 - \frac{\bar{v}_2^2}{2g} - H_{L,12}\right)$$

We need $\bar{v}_1$; we have $\dot{V}$. Use the continuity equation relationship:

$$\dot{V} = A\bar{v} = 200\,GPM \rightarrow$$

$$\bar{v}_2 = \left(\frac{200\,gal}{min}\right)\left(\frac{1}{\pi(1.25in)^2\{1\,ft^2/144in^2\}}\right)$$

$$\left\{\frac{1\,min}{60\,sec}\right\}\left\{\frac{1\,ft^3}{7.48\,gal}\right\} = 13\,ft/sec$$

Bernoulli re-arranged to solve for p_1.

$$p_2 = \gamma\left(\frac{p_1}{\gamma} - z_2 - \frac{\bar{v}_2^2}{2g} - H_{L,12}\right)$$

$$p_2 = \left(\frac{64\,lb_f}{ft^3}\right)\left(\frac{(14.7\,lb_f/in^2)\{144\,in^2/ft^2\}}{64\,lb_f/ft^3} - 10\,ft\right.$$

$$\left. - \frac{13^2\,ft^2/s^2}{2(32.2\,ft/s^2)} - 4\,ft\right) = 1053\,\frac{lb_f}{ft^2}$$

$$H_1 = \frac{p_1}{\gamma} = \left(\frac{1053\,lb_f}{ft^2}\right)\left(\frac{1\,ft^3}{64\,lb_f}\right) = 16.46\,ft$$

Recall that the definition of *NPSH* is the difference between the pressure at the pump's suction inlet and the pressure at which the fluid will begin to boil.

$$NPSH_A = p_{suction} - p_{saturation} = [16.46\,ft - (0.36\,lb_f/in^2)$$

$$\{144\,in^2/ft^2\}]/(64\,lb_f/ft^3) = 15.7\,ft$$

$$\boxed{NPSH_A = 15.7\,ft}$$

CHECK:
This answers the specific question that was asked and is in the specified or customary units.

The numeric value is not unreasonable (a negative or a much larger value would be unreasonable).

ADDITIONAL THOUGHTS:
Note that the head can always be converted between units of feet or psi since they are related by the specific weight of the fluid ($\Delta p = \gamma h$). This was done a couple of times in this example.

Will this pump cavitate under these conditions? ($NPSH_A > NPSH_R$) This is the big "so what" of this problem.

How can an operator increase NPSH (raise $NPSH_A$ or lower $NPSH_R$)?

2.6.4 Pump Power and Pump Efficiency

The pumping of fluids is necessary for many shipboard and shore-based systems, and it should be no surprise that this function comes at a cost in terms of the power necessary to run these components. Pumps used in modern ship applications are typically driven by electric motors, or in some steam propulsion systems, by steam turbines. In addition, portable, engine-driven pumps, such as the P-100 and P-250 portable pump series, exist in multiple configurations, some of which are driven by gasoline engines and others by diesel engines. Internal combustion engines typically drive auxiliary pumps to circulate lube oil and cooling water via gears or belts driven by the engine crankshaft. These necessary functions do reduce the power that can be delivered to the drive train. This all comes at a cost.

A pump may be a component that is part of a heat engine system, but a pump is not a heat engine in and of itself. Consequently, the concepts developed discussing thermal efficiency do not fully apply here. However, pumps are energy conversion devices that convert mechanical energy into pressure energy, but do so imperfectly, so the general concept of efficiency does apply. So, let's develop the equations for looking at the energy of a pump in specific terms and then we'll shift to the total rate or power basis and apply that toward understanding a pump's efficiency.

If we concentrate on a pump as the thermodynamic "system" being analyzed, the Conservation of Energy and the Conservation of Mass govern the analysis. We have already seen how to modify the SFEE into the Bernoulli equation with head loss (Equation 2.13) and the re-arrangment that solves for the work input into the fluid, called the Pump Head Equation (Equation 2.14).

Starting with the SFEE and substituting the definition of h.

$$\left(\frac{g}{g_c}\right)z_i + \left(\frac{1}{2g_c}\right)\bar{v}_i^2 + \overbrace{p_i v_i + u_i}^{h_i} + q_{if}$$
$$= \left(\frac{g}{g_c}\right)z_f + \left(\frac{1}{2g_c}\right)\bar{v}_f^2 + \overbrace{p_f v_f + u_f}^{h_f} + w_{if}$$

Looking only at the pump, then the following simplifications/assumptions apply: (1) Δpe is negligible since the pump inlet and outlet are close to the same elevation; (2) Δke cancels if the inlet and outlet connections are the same size; and, (3) the pump is adiabatic with its environment, so $q_{if} = 0$. Solving for the work and shifting to the "delta" format yields:

$$w_{if} = \left(\frac{g}{g_c}\right)\overbrace{(z_i - z_f)}^{cancels} + \overbrace{\left(\frac{\bar{v}_i^2 - \bar{v}_f^2}{2g_c}\right)}^{cancels} + (h_i - h_f)$$

$$+ \underbrace{q_{if}}_{adiabatic = 0} = h_i - h_f$$

$$w_{if} = h_i - h_f$$

This SFEE analysis proves that the thermodynamic work of the pump (w) is the difference in enthalpy (Δh) through the pump. The answer will be negative, representing work "in" to the fluid. Looking at it from the energy standpoint, the enthalpy at the discharge will be higher than the enthalpy at the suction. That should make sense at this point in the course.

Similarly, we can resolve the pump head equation (Equation 2.14) for the conditions measured across the pump. By definition, H_L applies to pipes and fittings, but not to the interior of the pump, so $H_L = 0$.

$$w_{p,s}\left(\frac{g_c}{g}\right) = \overbrace{(z_f - z_i)}^{cancels} + \overbrace{\left(\frac{\bar{v}_f^2 - \bar{v}_i^2}{2g}\right)}^{cancels} + \frac{p_f - p_i}{\gamma} + \underbrace{\widetilde{H_{L,if}}}_{does\ not\ apply} = \frac{p_f - p_i}{\gamma}$$

Solving for w, including the relationship between density (ρ) and specific weight (γ), and recognizing ρ is the inverse of specific volume (v) results in the following:

$$w_{p,s} = \left(\frac{g}{g_c}\right)\frac{p_f - p_i}{\gamma} = \left(\frac{g}{g_c}\right)\left(\frac{g_c}{\rho g}\right)(p_f - p_i)$$
$$= \frac{(p_f - p_i)}{\rho} = v(p_f - p_i) = v\Delta p$$

Recall that the pump head equation solves for pump work as a positive number. Also, we're assuming that the fluid is incompressible, so neither ρ nor v changes. The resulting relationship only applies to an ideal pump.

Shifting the work derived from the SFEE analysis to a positive value and then setting that equal to the results of the pump head equation analysis provides the following useful identity. Note: We must annotate the enthalpy to reflect that it is based upon the ideal pump work.

$$w_{pump,ideal} = v(p_{out} - p_{in}) = h_{out,ideal} - h_{in} \quad (2.25)$$

Now, thinking about ideal pump work, the term $w_{p,s}$ represents the energy that must be put into the fluid to achieve the resulting pressure rise. However, the fluid's movement through the pump will have turbulent fluid friction and recirculation around the edges of the impeller and internal friction, all of which contribute to the pump's fluid irreversibilities. So some energy is lost, which shows up in the temperature of the fluid, which increases slightly through the pump. This means that $w_{p,real} > w_{p,ideal}$ and the work, whether ideal or real, is still related to the change in h.

$$w_{pump,real} = h_{out,real} - h_{in}$$

Therefore, $h_{out,real} > h_{out,ideal}$ even though the pressure increase is the same for the real and ideal pump.

Now, to this point, we have looked only at what happens to the fluid. Considering that a pump is a mechanical device, there are shaft and impeller seals and bearings in the mechanical portion that have some friction that shows up as energy losses. This is shown graphically in Figure 2.18, below.

Clear segregation of the mechanical and fluid pump efficiencies isn't a simple process. In practicality, the pump manufacturers measure and provide the overall pump efficiency for their products, usually in the form of graphed curves (which we'll cover in the next subsection). We then approximate the mechanical input by equating that with the Δh that really occurs across the pump. The graphic representing this is Figure 2.19.

It should be recalled that, as a power-consuming mechanical device with efficiency <100%, a pump requires more mechanical work input than the pump produces as output. This shows up in the temperature of the fluid being pumped, whereas the perfect pump would be isothermal. The ideal pump work input to the fluid results in the same measured pressure increase as the real input with losses.

Power is a total rate basis term. We can shift from the specific basis to the total rate basis by multiplying the Δh by the mass flow rate ($\dot{m}$). When we focus on the pressure increase in the fluid, we call the resulting power term water horsepower (*WHP*). *WHP is the amount of power required under isentropic or ideal conditions to impart the specified change in pressure for a given flow rate.* Equation 2.26 provides some equivalent expressions for *WHP*.

$$WHP = \dot{m} w_{p,s} = (\dot{m} v)\Delta p = \dot{V} \Delta p \quad (2.26)$$

Like every other component in thermodynamics, the pump is not perfectly efficient and draws more power from the prime mover (electric motor, turbine, diesel engine, etc.) than is ideally required. The real power drawn from the prime mover is known as the pump's *brake horsepower* or *BHP*. *BHP* is the power measured at the shaft coupling of the machine. The input and output power are related to one another by the concept of pump efficiency (η_{pump}) as shown in Figure 2.20.

Equation 2.27 summarizes the relationships developed above.

$$\eta_{pump} = \overbrace{\frac{w_{p,ideal}}{w_{p,real}} = \frac{h_{out,ideal} - h_{in}}{h_{out,real} - h_{in}}}^{\text{specific basis}} = \overbrace{\frac{\dot{m}\Delta h_{ideal}}{\dot{m}\Delta h_{real}} = \frac{WHP}{BHP}}^{\text{total rate basis}} \quad (2.27)$$

EXAMPLE 2.7: PUMP POWER/BERNOULLI PROBLEM

A 98% efficient pump raises freshwater by 10 ft as it is pumped from one vented tank to another at a rate of 10 ft³/sec. Assuming a head loss of 10 ft, calculate the *BHP* (hp) required to maintain these conditions:

$$w_{pump,isen} = \frac{g_c}{g}\left[(z_2 - z_1) + \left(\frac{\vec{v}_2^2 - \vec{v}_1^2}{2g}\right) + \frac{p_2 - p_1}{\gamma} + H_{L,12}\right]$$

$$w_{pump,isen} = \frac{g_c}{g}\left[(10\,ft - 0\,ft) + \left(\frac{0-0}{2g}\right) + \frac{14.7\,psia - 14.7\,psia}{\gamma} + 10\,ft\right]$$

$$w_{pump,isen} = \frac{32.2\,\frac{ft \cdot lb_m}{lb_f \cdot s^2}}{32.2\,\frac{ft}{s^2}}(20\,ft) = 1\frac{lb_f}{lb_m}20\,ft$$

$$= 20\,\frac{ft \cdot lb_f}{lb_m}$$

$$\dot{m} = \rho\dot{V} = 62.4\,\frac{lb_m}{ft^3} \cdot 10\,\frac{ft^3}{s} = 624\,lb_m/s$$

$$WHP = \dot{m}w_{pump} = 624\,\frac{lb_m}{s} \cdot 20\,\frac{ft \cdot lb_f}{lb_m}\frac{1\,hp}{550\,\frac{ft \cdot lb_f}{s}}$$

$$= 22.69\,hp$$

$$BHP = \frac{WHP}{\eta_{pump}} = \frac{22.69\,hp}{.98} = 23.15\,hp$$

$$\boxed{BHP = 23.15\,hp}$$

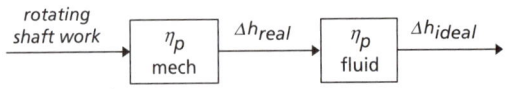

FIGURE 2.18: *Pump Efficiencies*

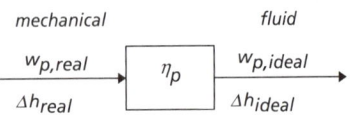

FIGURE 2.19: *Pump Efficiency*

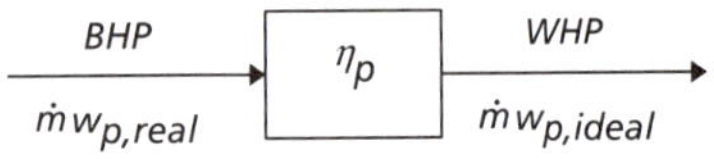

FIGURE 2.20: *Pump Efficiency Concept Block Diagram*

2.6.5 Combined Pump Characteristics Curves

Manufacturers produce pump curves that include a variety of information. This includes the head versus flow information as discussed above. The information can also include pump efficiency, power, and *NPSH* required. (See Figure 2.21, next page.)

2.6.6 Multiple Variable Displacement Pumps

Pumps can be used in series or parallel configurations to achieve certain operating characteristics. In order to increase the volumetric flow rate of a pump, a second pump can be added in parallel. The volumetric flow rate of the system is now the sum of the two pumps or roughly double if they are two identical pumps (see Figure 2.22):

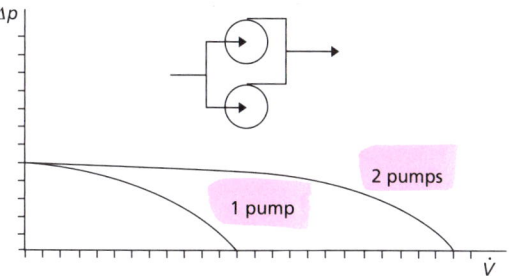

FIGURE 2.22: *Pumps in Parallel*

Similarly, two centrifugal pumps can be operated in series in order to overcome a large pressure head. The pressure developed by the combined pumps is now the sum of the two individual pumps while the volumetric flow rate stays the same (Figure 2.23).

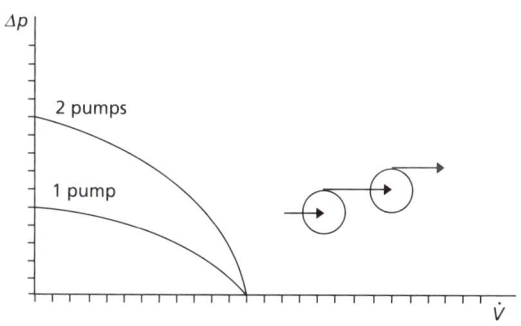

FIGURE 2.23: *Pumps in Series*

2.6.7 Positive Displacement Pumps

Another category of pumps are positive displacement pumps. Instead of a centrifugal impeller that creates pressure by flinging the fluid against the pump casing or a propeller pump that has similar variable displacement characteristics, these pumps essentially isolate small portions of the incoming fluid through (1) reciprocating piston/cylinder, (2) screws, or (3) intermeshing gear impellers and push the incompressible fluid into the discharge. The characteristic curve of a positive displacement pump is a vertical line on Δp vs $\dot{V}$ coordinates. For every revolution of the pump shaft, these pumps push the same volume of fluid, regardless of the backpressure until extreme pressures exceed the ability of the pump seals to prevent recirculation. These pumps are typically used for hydraulic systems that function at high pressure, to pump lube oil through engines, and to meter chemicals into various processes at precise rates. Hydraulic systems will be discussed in more detail in Unit 3. Note that the volumetric flow rate can be adjusted by changing the speed of the pump.

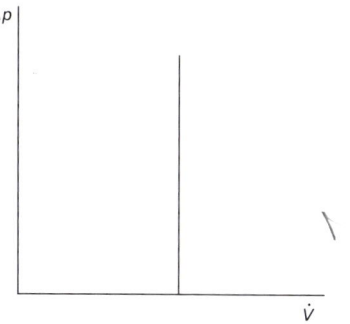

FIGURE 2.24: *Positive Displacement Pump Ideal Characteristic "Curve"*

2.7 ELECTRIC MOTOR REVIEW

While some pumps are driven by gasoline or diesel engines, as in the P100 and P250 portable pumps used for dewatering and fire fighting, and a small number are driven by steam engines, most pumps today are driven by electric motors.

Motors are energy conversion devices that convert electrical power to mechanical rotational power. As with other energy conversions, the efficiency is less than unity. In the case of a motor, the input power is usually expressed in Watts (or kW or MW). Motors are rated based upon their output power (*BHP*). The motor input power may be converted to horsepower and considered to be "electrical horsepower" (*EHP*). Pumps, however, are rated based upon their input power (*BHP*).

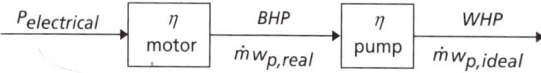

FIGURE 2.25: *Motor and Pump Efficiency Concept Block Diagram*

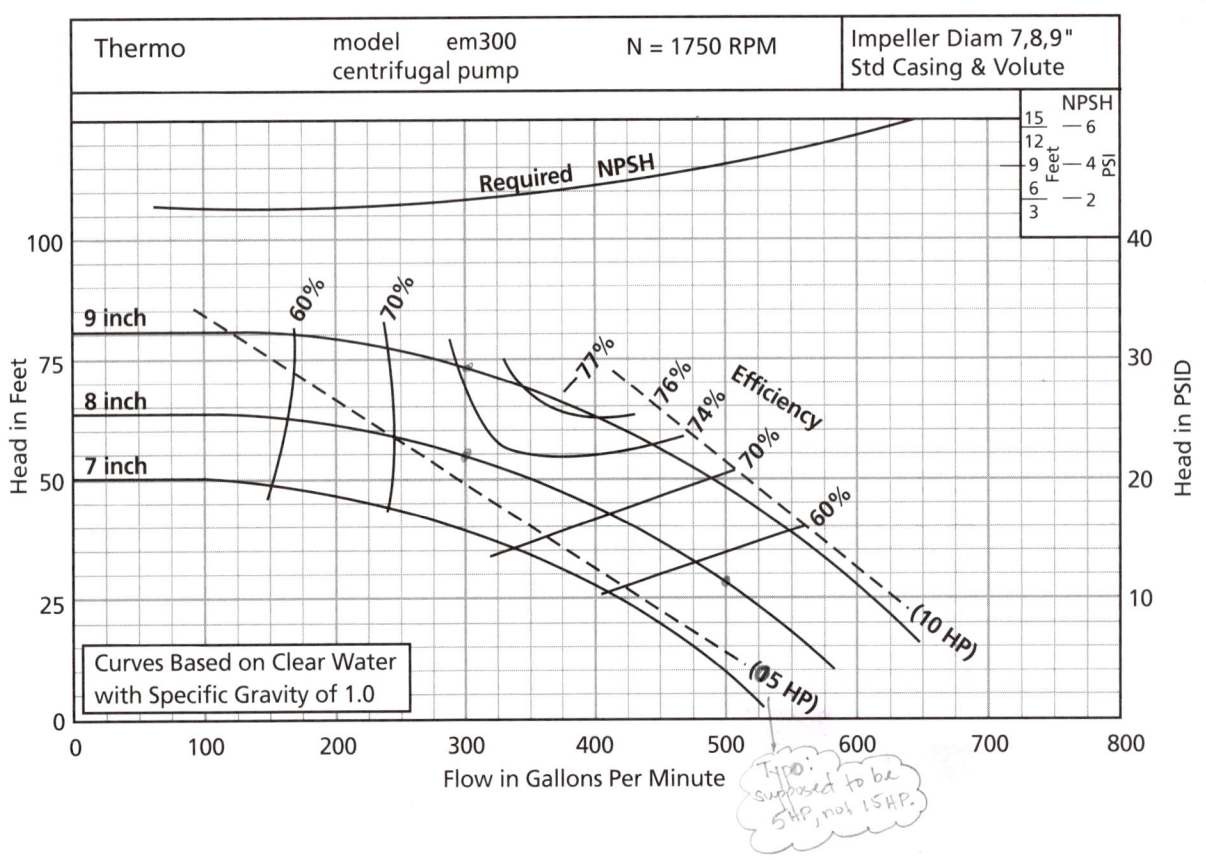
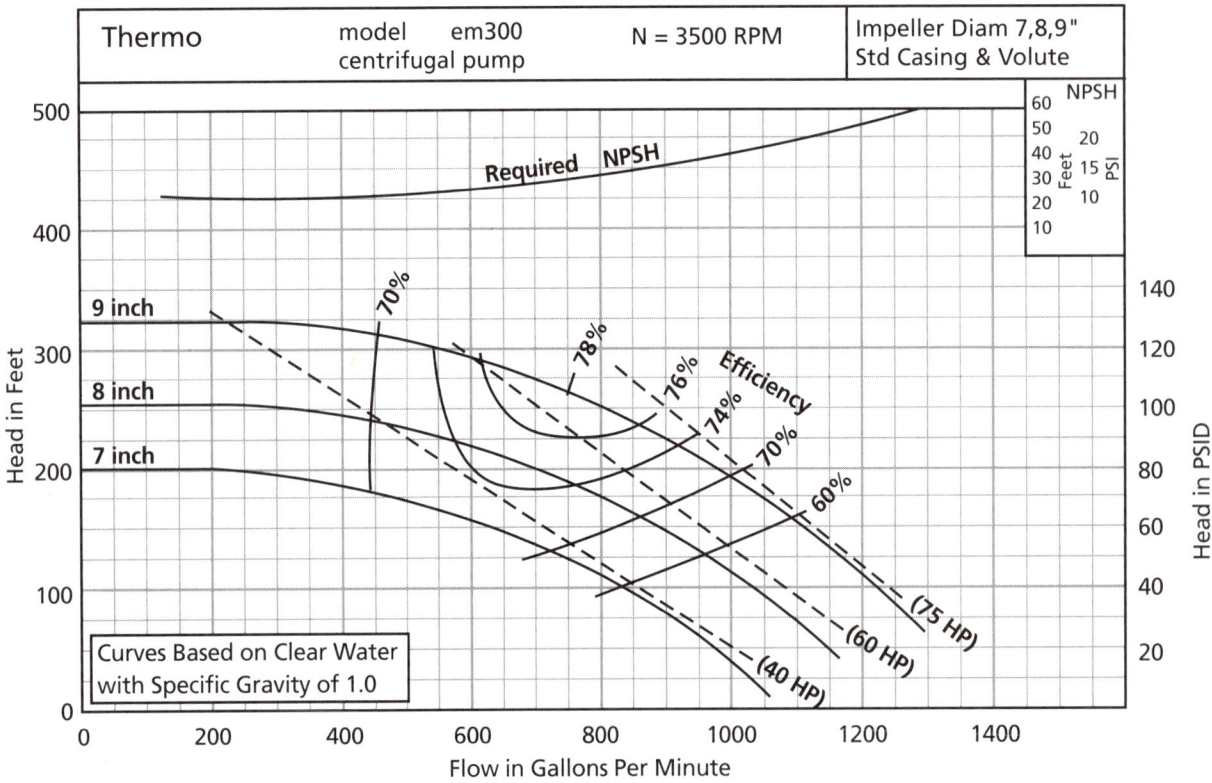

FIGURE 2.21: *Manufacturer's Pump Curves for 1750 and 3500 RPM*

Large pumps require powerful motors. U.S. Navy ships use 3-phase (3-ϕ) alternating current (AC) motors to drive their larger pumps. Synchronous motor speed is related to the number of magnetic poles in the rotor by the following equation:

$$NP = 120f \qquad (2.28)$$

where N is the rotational speed in RPM; P is the number of poles in the motor (either 2 or 4); f is the electrical frequency in Hz; and 120 is a constant that accounts for the various units in this relationship. The United States standard frequency for AC power is 60 Hz. So, an electrical motor connected to a 60 Hz electrical system and wired with 2 poles will have a synchronous rotational speed of 3600 RPM. Multi-speed motors typically have only two speeds, and changing speeds is accomplished by electrically changing the number of poles to 4. The synchronous speed will then be 1800 RPM.

Some motors are fitted with a variable frequency drive (VFD) unit that receives 60 Hz electrical input, but can electronically change the output frequency sent to a motor. If the electrical frequency is changed, the motors will change speed proportionally.

Another factor to consider is "motor slip." Loads on a motor will cause it not to rotate at synchronous speed. Slip represents the departure from synchronous and is the reason that typical full-load motor speeds are around 3450 to 3500 RPM in fast speed and around 1750 RPM in slow speed. For the purposes of this course, if a two-speed pump's rotational speeds are not known, then assume the speed ratio is exactly 2.

2.8 VENTURI METERS

It is possible to calculate the flow rate in a pipe using the continuity equation, Bernoulli's equation, and the geometry of a device known as a *venturi meter*. Consider the diagram of the venturi meter in Figure 2.26:

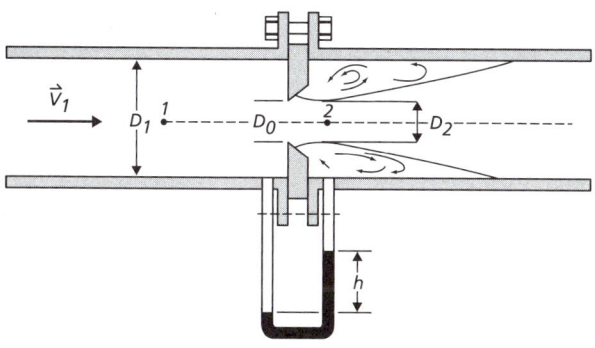

FIGURE 2.26: *Venturi Meter*

Fluid flows through state point 1 with a relatively low velocity and high pressure and passes through state point 2 with a higher velocity and lower pressure. The pressure difference (Δp) between the points can be measured using a conventional manometer. From continuity we can write $\vec{v}_1$ in terms of $\vec{v}_2$ and the known geometry of the venturi:

$$\dot{m}_1 = \rho A_1 \vec{v}_1 = \dot{m}_2 = \rho A_2 \vec{v}_2$$
$$\vec{v}_1 = \frac{A_2}{A_1} \vec{v}_2 \qquad (2.29)$$

Assuming the elevation change of the system is negligible, no head loss internal to the device, and no pump work, Bernoulli's equation simplifies to:

$$\frac{\vec{v}_1^2}{2g} + \frac{p_1}{\gamma} = \frac{\vec{v}_2^2}{2g} + \frac{p_2}{\gamma} \qquad (2.30)$$

Combining this result with the continuity equation yields:

$$\frac{\left(\frac{A_2 \vec{v}_2}{A_1}\right)^2}{2g} + \frac{p_1}{\gamma} = \frac{\vec{v}_2^2}{2g} + \frac{p_2}{\gamma}$$

$$\frac{\left(\frac{A_2^2}{A_1^2} - 1\right)}{2g} \vec{v}_2^2 = \frac{p_2 - p_1}{\gamma}$$

$$\vec{v}_2 = \sqrt{\frac{\frac{p_2 - p_1}{\gamma} 2g}{\left(\frac{A_2^2}{A_1^2} - 1\right)}}$$

The student should be able to determine the venturi velocity, the volumetric flow rate, and the fluid velocity in the pipe from the venturi's geometry and pressure difference. The net units inside the radical should be ft²/sec², so use this knowledge to insert any necessary conversion factors.

2.9 SIPHONS

A siphon is an inverted tube that can drain a liquid from a higher reservoir to a lower one. Naval professionals need to be aware of how a siphon works because they can be inadvertently started and can lead to undesired drainage of a fluid. Figure 2.27 shows a representative schematic of a siphon draining a tank. No pump is necessary, but the loop must be full of the fluid in order to flow. The height of the loop crest is limited by atmospheric pressure, fluid density, and its vapor pressure (the pressure at which the fluid will begin to boil at a given temperature). And the outlet must be below the source level. Once established, the flow will continue until the level at the discharge equals the level in the source reservoir or until the siphon is "broken" by sucking in air. An example of a device designed to function as a syphon is a standard toilet, which operates by establishing a siphon to draw the contents of the toilet bowl out and into the drain.

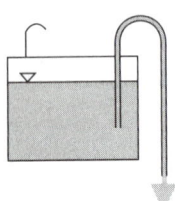

FIGURE 2.27: *A Siphon Draining a Tank*

2.10 PIPING IDENTIFICATION

Identification of piping systems is aided in the Fleet by painting and marking the pipes. From the Naval Ships' Technical Manual (NSTM):

> 505-7.8.1 GENERAL. Piping system designations and markings assist in training and troubleshooting and permit quick identification and proper system operation during casualty control. The designations and markings used on ships should conform to the descriptions provided in the following paragraphs. However, consult ship drawings for specific designations and markings.

Piping systems aboard ship are identified by color codes designated in NSTM Table 505-7-1, reproduced here:

TABLE 2.3: *NSTM Table 505-7-1*

Fluid	Color	FED STD 595 Color Number and Chip	Extent
Steam and Steam Drains	White	17886	Note A
Potable Water	Dark Blue	15044	Note A
Nitrogen	Light Gray	16376	Note A
Bleed Air	Green-gray	16555	Note A
Bleed Air Anti-Icing	Striped green-gray/light blue	16555/15200	Note A
Bleed Air Masker	Striped green-gray/light yellow	16555/13655	Note A
Bleed Air Prairie	Striped green-gray/dark blue	16555/15044	Note A
HP Air (>1000 psig)	Dark Gray	16081	Note A
MP Air (>150 psig & <1000 psig)	Striped dark-gray/tan	16081/10324	Note A
LP Air and Salvage Air	Tan	10324	Note A
Deballast Air	Striped Tan/Black	10324/17038	Note A
Oxygen	Green	14449	Note B
Seawater (other than fire main and sprinkling). Includes Main and Secondary Drainage, Waste Drainage, Distilling Plant Feed, Distilling Plant Brine Overboard, and Countermeasure Wash Down.	Dark Green	14062	Note A
JP-5	Light Purple	17142	Note B
Fuel	Yellow	13538	Note A
Lube Oil (including PolyAlphaOlefin (PAO) electronic cooling)	Striped Black/Yellow	17038/13538	Note A
Foam Discharge Plugs (AFFF)	Striped Red/Dark Green	11105/14062	Note A
Gasoline	Yellow	13538	Note B
Freshwater, Condensate, Feed, and Distillate (Submarines only)	Dark Blue	15044	Note A
Freshwater, Condensate, Feed, and Distillate (Surface Ships only)	Light Blue	15200	Note A
Primary Coolant and Charging Water (Submarines only)	Light Blue	15200	Note A

TABLE 2.3: *NSTM Table 505-7-1 (Continued)*

Fluid	Color	FED STD 595 Color Number and Chip	Extent
Hydraulic	Orange	12246	Note A
Refrigerant	Dark Purple	17100	Note B
Hydrogen	Chartreuse	23814	Note A
Amine Dry Cleaning Fluid	Brown	10080	Note B
Helium	Buff	10371	Note A
Helium/Oxygen	Striped Buff/Green	10371/14449	Note A
Sewage	Gold	17043	Note A
Halon	Striped Gray/white	16187/17886	Note A
Fire Main (including root valves)	Red	11105	Note C
Chilled Water (Submarines only)	Dark Blue	15044	Note A
Chilled Water (Surface Ships only)	Striped Light Blue/Dark Green	15200/14062	Note A
Demineralized Electronic Cooling Water (Submarines only)	Dark Blue	15044	Note A
Demineralized Electronic Cooling Water (Surface Ships only)	Striped Light Blue/Dark Purple	15200/17100	Note A
AFFF Concentrate	Striped Light Blue/Red	15200/11105	Note A
Access fittings (Submarines only)	Black	17038	
Carbon Dioxide (Submarines only)	Violet		Note A
Oil Pollution Abatement (OPA – Surface Ships only)	Black	17038	Note B and Note E
Jacket Water/Waste Heat	Stripe Light Blue/Black	15200/17038	Note A
Divers life support system	Various	Various	Note F
AFFF Solution (concentrate/saltwater mix)	Striped Red/Dark Green	11105/14062	Note D

SYMBOLS LIST:

Note A – Color code only valve handwheels and levers on valves not exposed to the weather. Valves and handwheels exposed to the weather (ship board connection) shall have label plates or plain language markings clearly delineate the service for each connection.

Note B – Color code valve bodies and handwheels exposed to the weather and all interior piping. Piping in tanks, voids, cofferdams, and bilges shall not be color-coded.

Note C – All fire plugs and handwheels including associated components (strainer, wyegate, applicators, wrenches, and hose racks) shall be color coded.

Note D – Color code all handwheels.

Note E – OPA piping in the bilge area shall be painted terra-cotta red (approximately chip 20152).

Note F – See NAVSEA 0994-LP-001-9010, Vol. 1 and 2, USN Diving Manual.

REFERENCES:

DOE Fundamentals Handbook, *Thermodynamics, Heat Transfer and Fluid Flow,* Vol. 3 of 3, DOE-HDBK-1012/3-92, U.S. Department of Energy, 1992.

DOE Fundamentals Handbook, *Engineering Symbology, Prints, and Drawings*, Vol. 1 of 2, DOE-HDBK-1016/1-93, U.S. Department of Energy, 1993.

DOE Fundamentals Handbook, *Mechanical Science*, Vol. 2 of 2, DOE-HDBK-1018/2-93, U.S. Department of Energy, 1993.

Naval Ships' Technical Manual, S9086-RK-STM-010, Chapter 505 *Piping Systems.*

F. M. White, *Fluid Mechanics*, 4th ed. (Boston: McGraw Hill, 1999).

D. W. Wolansky, J. Nagohosian, and R. W. Henke, *Fundamentals of Fluid Power* (Boston: Houghton Mifflin, 1986).

FIGURE 2.28: *Standard Symbology*

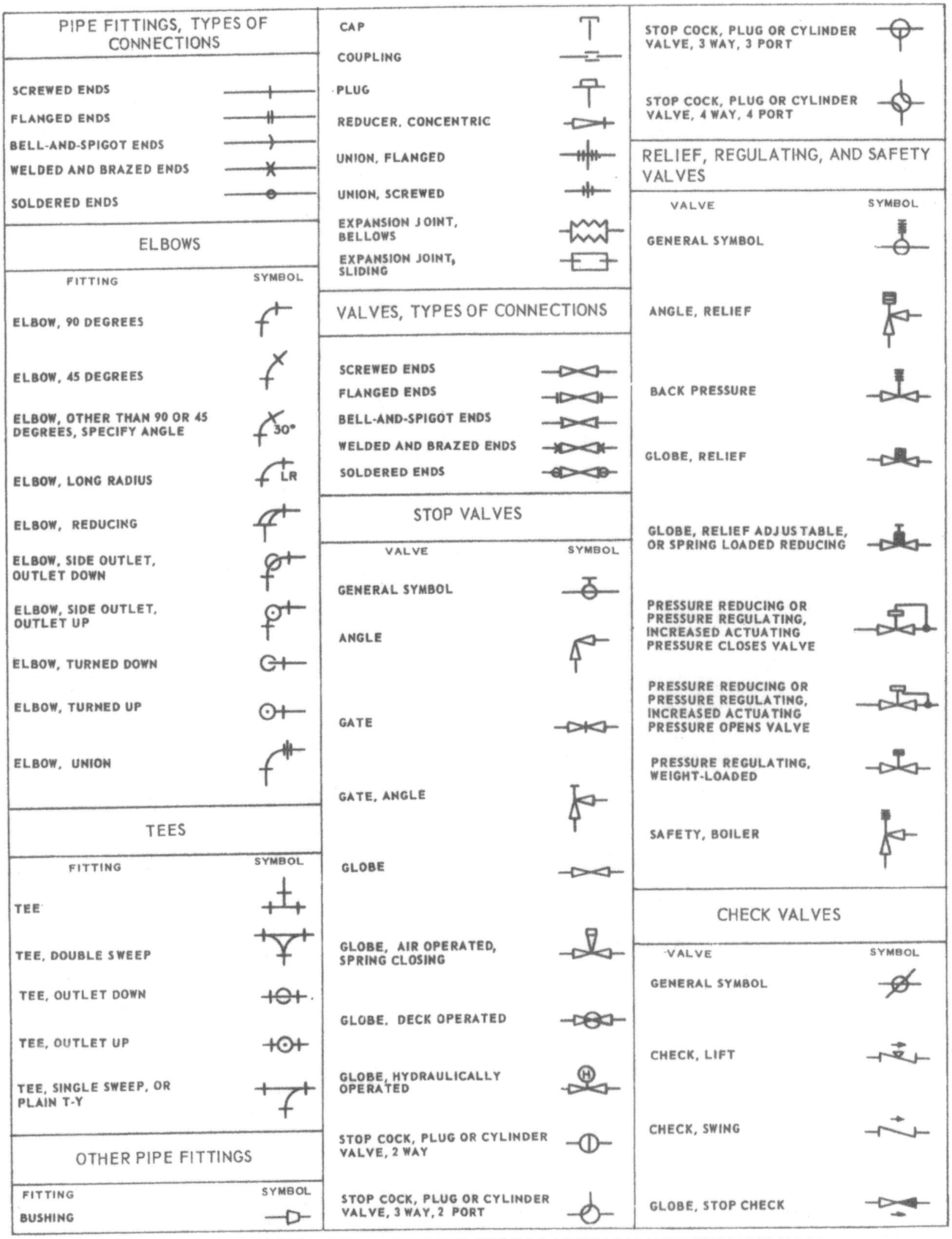

PIPE FITTINGS, TYPES OF CONNECTIONS	
SCREWED ENDS	—+—
FLANGED ENDS	—++—
BELL-AND-SPIGOT ENDS	—)—
WELDED AND BRAZED ENDS	—✕—
SOLDERED ENDS	—⊙—

ELBOWS	
FITTING	SYMBOL
ELBOW, 90 DEGREES	
ELBOW, 45 DEGREES	
ELBOW, OTHER THAN 90 OR 45 DEGREES, SPECIFY ANGLE	30°
ELBOW, LONG RADIUS	LR
ELBOW, REDUCING	
ELBOW, SIDE OUTLET, OUTLET DOWN	
ELBOW, SIDE OUTLET, OUTLET UP	
ELBOW, TURNED DOWN	
ELBOW, TURNED UP	
ELBOW, UNION	

TEES	
FITTING	SYMBOL
TEE	
TEE, DOUBLE SWEEP	
TEE, OUTLET DOWN	
TEE, OUTLET UP	
TEE, SINGLE SWEEP, OR PLAIN T-Y	

OTHER PIPE FITTINGS	
FITTING	SYMBOL
BUSHING	

CAP	
COUPLING	
PLUG	
REDUCER, CONCENTRIC	
UNION, FLANGED	
UNION, SCREWED	
EXPANSION JOINT, BELLOWS	
EXPANSION JOINT, SLIDING	

VALVES, TYPES OF CONNECTIONS	
SCREWED ENDS	
FLANGED ENDS	
BELL-AND-SPIGOT ENDS	
WELDED AND BRAZED ENDS	
SOLDERED ENDS	

STOP VALVES	
VALVE	SYMBOL
GENERAL SYMBOL	
ANGLE	
GATE	
GATE, ANGLE	
GLOBE	
GLOBE, AIR OPERATED, SPRING CLOSING	
GLOBE, DECK OPERATED	
GLOBE, HYDRAULICALLY OPERATED	
STOP COCK, PLUG OR CYLINDER VALVE, 2 WAY	
STOP COCK, PLUG OR CYLINDER VALVE, 3 WAY, 2 PORT	
STOP COCK, PLUG OR CYLINDER VALVE, 3 WAY, 3 PORT	
STOP COCK, PLUG OR CYLINDER VALVE, 4 WAY, 4 PORT	

RELIEF, REGULATING, AND SAFETY VALVES	
VALVE	SYMBOL
GENERAL SYMBOL	
ANGLE, RELIEF	
BACK PRESSURE	
GLOBE, RELIEF	
GLOBE, RELIEF ADJUSTABLE, OR SPRING LOADED REDUCING	
PRESSURE REDUCING OR PRESSURE REGULATING, INCREASED ACTUATING PRESSURE CLOSES VALVE	
PRESSURE REDUCING OR PRESSURE REGULATING, INCREASED ACTUATING PRESSURE OPENS VALVE	
PRESSURE REGULATING, WEIGHT-LOADED	
SAFETY, BOILER	

CHECK VALVES	
VALVE	SYMBOL
GENERAL SYMBOL	
CHECK, LIFT	
CHECK, SWING	
GLOBE, STOP CHECK	

FIGURE 2.28: *Standard Symbology (Continued)*

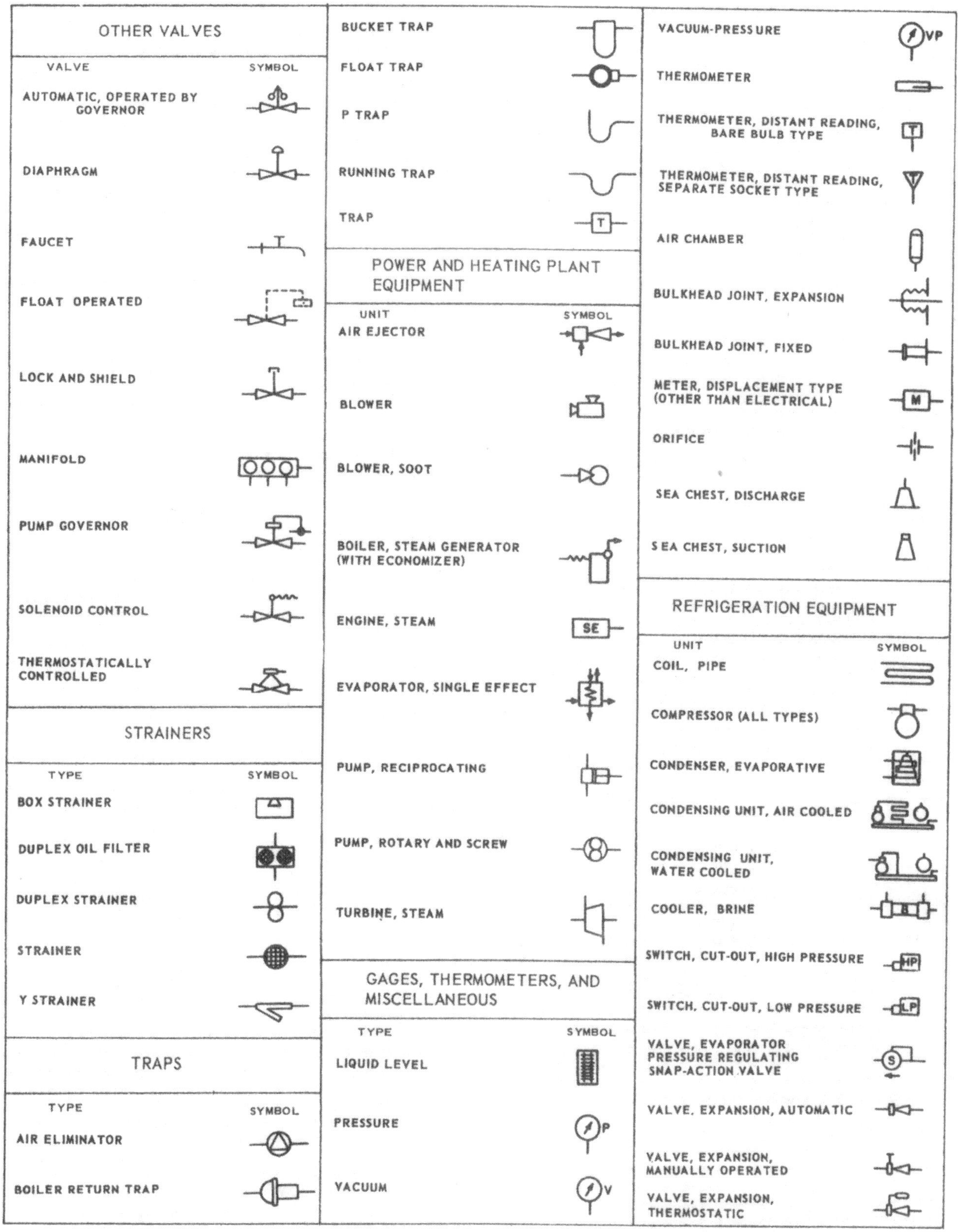

OTHER VALVES	
VALVE	SYMBOL
AUTOMATIC, OPERATED BY GOVERNOR	
DIAPHRAGM	
FAUCET	
FLOAT OPERATED	
LOCK AND SHIELD	
MANIFOLD	
PUMP GOVERNOR	
SOLENOID CONTROL	
THERMOSTATICALLY CONTROLLED	

STRAINERS	
TYPE	SYMBOL
BOX STRAINER	
DUPLEX OIL FILTER	
DUPLEX STRAINER	
STRAINER	
Y STRAINER	

TRAPS	
TYPE	SYMBOL
AIR ELIMINATOR	
BOILER RETURN TRAP	
BUCKET TRAP	
FLOAT TRAP	
P TRAP	
RUNNING TRAP	
TRAP	

POWER AND HEATING PLANT EQUIPMENT	
UNIT	SYMBOL
AIR EJECTOR	
BLOWER	
BLOWER, SOOT	
BOILER, STEAM GENERATOR (WITH ECONOMIZER)	
ENGINE, STEAM	SE
EVAPORATOR, SINGLE EFFECT	
PUMP, RECIPROCATING	
PUMP, ROTARY AND SCREW	
TURBINE, STEAM	

GAGES, THERMOMETERS, AND MISCELLANEOUS	
TYPE	SYMBOL
LIQUID LEVEL	
PRESSURE	P
VACUUM	V
VACUUM-PRESSURE	VP
THERMOMETER	
THERMOMETER, DISTANT READING, BARE BULB TYPE	T
THERMOMETER, DISTANT READING, SEPARATE SOCKET TYPE	
AIR CHAMBER	
BULKHEAD JOINT, EXPANSION	
BULKHEAD JOINT, FIXED	
METER, DISPLACEMENT TYPE (OTHER THAN ELECTRICAL)	M
ORIFICE	
SEA CHEST, DISCHARGE	
SEA CHEST, SUCTION	

REFRIGERATION EQUIPMENT	
UNIT	SYMBOL
COIL, PIPE	
COMPRESSOR (ALL TYPES)	
CONDENSER, EVAPORATIVE	
CONDENSING UNIT, AIR COOLED	
CONDENSING UNIT, WATER COOLED	
COOLER, BRINE	
SWITCH, CUT-OUT, HIGH PRESSURE	HP
SWITCH, CUT-OUT, LOW PRESSURE	LP
VALVE, EVAPORATOR PRESSURE REGULATING SNAP-ACTION VALVE	S
VALVE, EXPANSION, AUTOMATIC	
VALVE, EXPANSION, MANUALLY OPERATED	
VALVE, EXPANSION, THERMOSTATIC	

Practice Problems

> "I never worked on anything by accident, and none of my inventions ever happened by accident. They came from work."
> – Thomas Edison, holder of 1,093 patents

2.1 Water is flowing through a long pipe that has a progressively increasing diameter. At state point 1 in the pipe the internal diameter is 1 ft and the fluid velocity is 20 ft/sec. At state point 2 the internal diameter is 2 ft. Assume that the water temperature does not change between state points. Determine:

   a) the mass flow rate (lb_m/sec)
   b) the fluid velocity (ft/sec) at position 2

2.2 A well pump supplies water to an elevated storage tank that is 150 feet high. The pump draws water from the ground where the water level is 75 feet below the surface. If the pump delivers 500 GPM, what is the *WHP* of this pump system?

2.3 A straight, level run of steel pipe, nominally 8 inches in diameter, with a friction factor of 0.015 is filled with freshwater flowing at 6 ft/sec. The inside diameter of the 8-inch pipe is 7.981 inches; the outside diameter is 8.625 inches; and the weight of the pipe is 28.55 lb_f/ft. What is the pressure drop in each 100 foot length of pipe? What is the weight of the pipe and contents (ignoring the presence of flanges and insulation)? What is the density of the steel used to make this pipe? Sketch the schematic for the *Dp* calculation and the geometry for the weight and density calculations.

2.4 Use the Moody Friction Factor Chart to determine the pipe friction factor (*f*) for 50 °F water being pumped through 4-inch, schedule-40 welded steel pipe at 250 GPM. The ID of this pipe is 0.3355 ft and u (water @ 50 °F) = 1.41E-5 ft²/sec.

2.5 An 8" feedwater regulating valve (FRV) has water flowing through it. When the FRV is throttled such that the flow rate is 1800 GPM, the differential pressure across this valve is 22.5 psi. Note: An FRV functions to control the amount of feedwater entering the boiler. It does this by automatically throttling to provide the programmed flow rate for the boiler conditions.

   a) What is the mass flow rate of the feedwater?
   b) What is the average velocity in the pipe?
   c) What is the value of the fitting loss coefficient (k) in this condition?

2.6 The Main Feed Pump (MFP) in a steam plant has a flow rate of 650 GPM with a pressure rise across the pump of 775 psi. If the pump brake horsepower is 432 horsepower, what is the pump efficiency (%)? Be sure to include a schematic and the efficiency concept sketch, both properly labeled. Start with the pump head equation and derive the two versions of the water horsepower relationships shown in the course equation sheets.

2.7 See the schematic below. Fuel oil with a density of 50 lb_m/ft³ is to be transferred from a storage tank. The fuel oil transfer pump is driven by a 30 horsepower motor. The pump has an efficiency of 75% at a flow rate of 500 GPM. The piping system between the tank and the pump consists of 200 feet of 9 inch diameter pipe with a friction factor of 0.015. If the tank is vented to the atmosphere with barometric pressure = 30.53 in-Hg and the pump discharge pressure is 100 psia, what is the height of the oil in the tank at the start of the transfer operation (ft)? Write your solution using the correct state points from the sketch to the right.

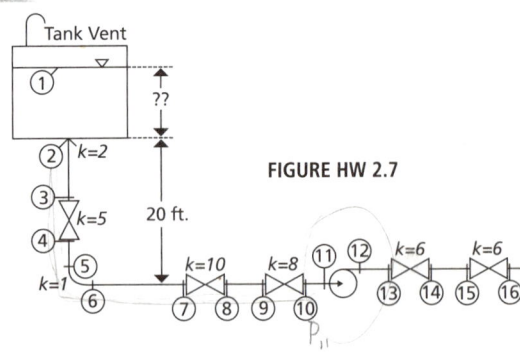

FIGURE HW 2.7

2.8 You are in an operating space on a ship and see pipes painted solid green. What fluid should be inside that pipe? What colors of pipes would you expect to see connected to the ship's diesel engine? List the fluids and their associated color scheme. Would a relief valve or a safety valve be installed in each of these systems? Explain.

2.9 For a closed piping system with one centrifugal pump, it is desired to increase the flow rate by a factor of 5. By how much will the power consumption of the pump have to increase? If only 3 times the power is available, how much can the flow be increased?

2.10 It is known that a given centrifugal pump operating at speed "N" has a shut-off head of 25 psi and a maximum flow rate of 50 GPM. Sketch and label the pump characteristic curve (psi vs GPM). On the same graph, sketch the characteristic curve when two of these pumps are operating in series, in parallel, and when it is operated alone at speed 2N.

2.11 Sketch the pump curve for a positive displacement pump and the pump curve for a variable displacement pump on separate axes. Add a line segment illustrating the shape of the system curve on both graphs. Label the intersection. On the variable displacement graph, show what happens to the system curve if a valve on the discharge side of the pump is throttled in the "shut" direction, but not completely closed.

2.12 The Main Seawater (MSW) Pump pumps 1250 GPM to the main condensers in slow speed and develops 30 psid across the pump. The pump run-out value for the slow speed pump is 1600 GPM, and the pump shutoff head for the slow speed pump is 50 psid. Draw the pump curves for this pump in slow and fast speed and the system curve, labeling all significant points with their values. What are the expected head and flow rate in fast speed if fast speed is twice slow speed?

2.13 Given the system in the schematic shown below, find the following:

a) $\dot{V}$ (ft³/s) to 3 decimal places

b) A_{pipe} (ft²) to 3 decimal places

c) velocity in the pipe (ft/s) to two decimal places

d) mass flow rate (lb_m/s) to one decimal place

e) kinetic energy of the pipe flow (ft)

f) total head loss (ft)

g) $w_{p,s}$ (ft-lb_f/lb_m)

h) WHP (HP)

i) BHP (HP)

j) Dp_{pump} (psid)

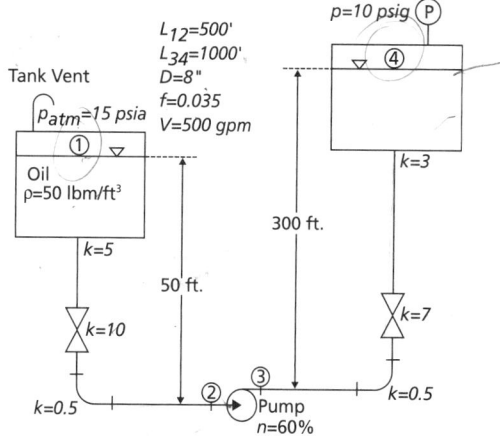

FIGURE HW 2.13

Hint: Note that there are 4 state points drawn on this schematic. Be careful to adjust the pump head equation to reflect the system between the state points that you are analyzing—adjust the state point numbers of the equation and only count the pipe length and k factors between the SPs. You'll need to write the equation from SP1 to SP4 to determine part (g). Then to determine part (j), either write the equation from SP2 to SP3 using $w_{p,s}$ determined in (g), or solve the pump head equation from SP1 to SP2 (solving for p_2) and then from SP3 to SP4 (solving for p_3) and take the difference between p_2 and p_3.

2.14 Show by calculation whether a motor rated at 10 horsepower is adequate to drive a model em300 centrifugal pump with an 9-inch impeller at slow speed in each of the following system conditions:

a) Flow greatly reduced by valve throttling (200 GPM, 33 psid)

b) Normal flow with throttle valve in mid-position (415 GPM, 26 psid)

c) High flow with throttle valve open wide (550 GPM, 16 psid)

2.15 During fuel oil transfer operations, 24.5 GPM of diesel fuel ($r = 58$ lb_m/ft^3) is pumped from the fuel oil storage tank (FOST) to the clean fuel oil tank (CFOT). The FOST is vented to atmosphere. The pressure rise across the pump (differential pressure) is 60.4 psi. Piping is 1-inch diameter ($f = 0.03$), the total piping length between tanks is 50 ft. The total loss coefficient (k) for all of the fittings in this system is 23. Determine the velocity of the fluid in the pipe (ft/s); the pump work (ft-lb_f/lb_m); and the pressure in the CFOT (psig). What color should these pipes be painted?

2.16 A Thermo model em300 centrifugal pump with a 9-inch impeller is pumping 300 GPM with the motor running in slow speed (see the manufacturer's pump curve figure in the unit). Calculate the mass flow rate and the water horsepower (*WHP*) and determine the power input to the pump by the motor (hp). What is the pump's efficiency? What is the required *NPSH* in this operating condition (psia)? The pump is shifted to fast speed. What is the new efficiency?

2.17 A two-speed pump is running in a system in slow speed. The slow speed power demand of the pump is 45 kW. If the ship's service turbine generator (SSTG) output power (pump plus other loads) was initially 1400 kW, what is the expected final SSTG output power after the pump is shifted to fast speed (twice slow speed)? What is the power consumed by all of the other loads when the pump is running in fast speed? SSTGs are typically rated at 1500 kW with a 10% overcurrent before the breaker trips. Should you start and bring online a second SSTG before shifting the pump speed?

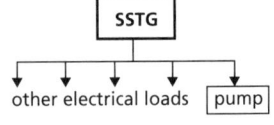

FIGURE HW 2.17

2.18 A model em300 centrifugal pump with a 9-inch impeller (see pump curves in text) is operating in a cooling system at slow speed, at a flow rate of 375 GPM, and a pressure across the pump of 28 psid. Determine: a) The water horsepower (*WHP*) developed by the pump (hp), b) the brake horsepower (*BHP*) required at the pump drive shaft to operate this pump (using the appropriate pump mechanical efficiency), and c) using the pump affinity laws, the volumetric flow rate (GPM) and pump differential pressure (psid) expected when this pump is shifted to fast speed in the same system.

2.19 The pump in the schematic is 50% efficient. It is driven by an electrical motor that is 90% efficient. Recall that pumps are rated by their input power and motors are rated by their output power. The cost of electricity is $0.15 per kilowatt-hour. The pump is a two-speed pump that pumps 500 GPM in fast speed. Determine the following:

a) The differential pressure across the valve between state points 2 and 3 when the pump is running in fast speed.

b) The absolute pressure at state point 5 when the pump is running in fast speed.

c) The cost to run the pump continuously for one hour when the pump is running in fast speed.

d) The rated power of the motor.

e) The expected flow rate and differential pressure across the pump in slow speed.

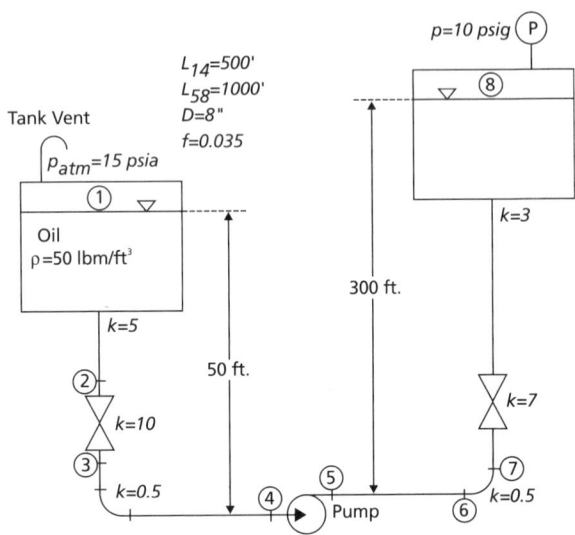

FIGURE HW 2.19

2.20 The venturi used in the pipe flow lab has the dimensions shown on the sketch below. Recall that the venturi was used to measure the volumetric flow of freshwater in a system using the difference in pressure as the incompressible fluid accelerates through the throat of the venturi. Water-filled manometers were connected at state points 1 and 2, and the difference in their height was recorded. Hint: Use the pump head version of Bernoulli's equation and the continuity equation to <u>derive</u> an equation that relates (1) the pressure difference at the two state points measured in inches of water to (2) volumetric flow rate of water in ft³/sec. Program the equation into an Excel spreadsheet and graph the flowrate for the range of 0 to 20 inches of water in 0.25 inch increments. Produce a smoothed-line graph of the results. Submit your derivation, the spreadsheet, and the graph. Compare your results to the graph provided with the lab handout.

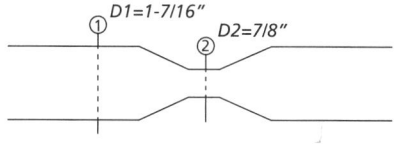

FIGURE HW 2.20

2.21 Fuel Oil (S.G. = 0.9) service tank and fuel cutoff valve for the ship's engines are connected by a new 6″ steel pipe 3000 feet long. The cutoff valve is 60 feet above the service tank outlet. Pressure gage at the cutoff valve reads 50 psig, and flow rate is 750 GPM through the cutoff valve to the engines. If the friction factor is 0.03, what is the pressure (psig) at the outlet of the service tank?

2.22 Brine (S.G. = 1.2) flows through a 200 GPM pump. The pump outlet is 6″ pipe and is 4 feet above the 8″ pipe inlet. The inlet vacuum is 6″ Hg. Outlet pressure is 20 psig. What is the water horsepower (*WHP*) of the pump? What color should the pipes be painted that are connected to this pump?

2.23 A model em300 centrifugal pump with an 8-inch impeller (see pump curves in text) is operating in a cooling system at slow speed, at a flow rate of 250 GPM. What motor torque is required to run this pump?

3 Hydraulic and Pneumatic Systems

Gas turbine–powered ships do not reverse the direction of rotation of the propellers to go astern. They change the pitch of the blades. Both the rudders and the pitch of the propeller blades of this destroyer are controlled by hydraulic power systems.

3.1 UNIT LEARNING OBJECTIVES

3.1.1 Terminology: definitions, variables, and typical units

3.1.1.1 Pascal's law, ram, actuator, positive displacement pump, accumulator, dual acting, single acting, control valve

3.1.2 Concepts: ideas and engineering expression of them

3.1.2.1 Sketch and describe the characteristics of a positive displacement pump.

3.1.2.2 Describe the operation of a hydraulic control valve.

3.1.2.3 List the similarities and differences between hydraulic power systems and pneumatic power systems. *[high press / low press.]*

3.1.2.4 Calculate the work effect of the energy stored in a hydraulic accumulator.

3.1.3 Skills: procedures, practices, or methods that enable reasoning

3.1.3.1 Calculate the force generated by a hydraulic ram including accounting for back-pressure effects.

3.1.3.2 Apply conservation of mass and Pascal's law to determine actuator displacement.

3.1.3.3 Trace the flow of hydraulic fluid through a hydraulic piping system.

3.1.3.4 Determine the effect on an actuator for changes in control valve position.

3.2 INTRODUCTION

Fluid power systems store, collect, and transmit power with great efficiency using a gaseous or liquid medium. *Pneumatics* is the term used for the study of power transmission with gases. Similarly, *hydraulics* is the study of power transmission with liquids. Fluid power has a deceptively short history, as the laws of fluid power were not fully understood until about 1650 with the discovery of what is called *Pascal's law*. This law asserts that a liquid at rest transmits pressure equally in all directions and that an increase in pressure on the surface of a confined fluid is transmitted undiminished throughout the fluid. Exploitation of these facts to develop mechanical advantage will be the primary thrust of this chapter.

Hydraulic and pneumatic systems have distinct advantages over purely mechanical (gears, levers, etc.) and electrical systems in many regards, but are also not without their drawbacks. Some of the salient distinguishing characteristics and performance advantages/disadvantages are listed in Table 3.1.

3.3 FLEET APPLICATIONS

Naval applications for hydraulic systems are numerous. Most notably a ship's rudder is controlled by hydraulics. The ailerons and flaps of a fixed-wing aircraft are deployed with hydraulic actuators. The pitch of a controllable pitch propeller and the angle of attack of a rotating wing (helicopter blade) are also controlled with hydraulic systems. Examples on the ground abound as well. Modern tactical vehicles used by the Marine Corps and Navy Seabees use hydraulics extensively. The High Mobility Multi-Purpose

TABLE 3.1: *Pros and Cons of Various Power Systems*

	Advantages	Disadvantages
Electrical Systems	☐ Many available power sources ☐ Flexibility in locating equipment and transmission lines	☐ Difficult to store large amounts of energy in batteries ☐ Shock hazard, fire hazard ☐ Wear of brushes on electrical generators and motors ☐ Corrosion of relays, contacts, etc. ☐ Short circuits in wet environment prevent proper operation
Mechanical Systems (gears, levers, etc.)	☐ Simplicity ☐ Instantaneous transmission of force ☐ Low maintenance	☐ Wearing of parts ☐ Difficult to implement safety controls
Hydraulic systems (liquids)	☐ Ease of locating parts with the use of transmission tubes, hoses, etc. ☐ Transmission of tremendous forces with little or no wear ☐ Infinitely variable speed, direction, and rotation of actuator depending on design ☐ No electrocution hazard ☐ Ability to include safety relief valves	☐ Equipment seal failure ☐ Fluid subject to contamination—requires filtering or change out ☐ Weight of fluid for certain applications can be costly ☐ Environmental hazards of fluids ☐ Fire retardant, not fireproof
Pneumatic systems (gases)	☐ Ease of locating parts with the use of transmission tubes, hoses, etc. ☐ Transmission of tremendous forces (less than hydraulics though) ☐ Little or no wear of parts with the addition of some lubricant in air stream ☐ Infinitely variable speed, direction, and rotation of actuator depending on design ☐ Storage of energy in simple tanks ☐ Electrical controls not required ☐ No electrocution hazard ☐ Ability to include safety relief valves ☐ Unnecessary to change fluids aside from lubricant ☐ Unlimited supply of power transmission medium (air)	☐ Equipment seals vulnerable to failure ☐ Lower system pressures relative to hydraulics ☐ Elastic nature means leaks continue to flow until depressurized

Wheeled Vehicle (HMMWV) and the Medium Tactical Replacement Vehicle (MTRV) use hydraulic braking, steering, and lift/jack systems. All modern motor vehicles have hydraulic braking systems.

Pneumatic systems are also found aboard modern ships and aircraft. Applications include power sources for air tools and control mechanisms for heating, ventilating, and air conditioning systems (HVAC). Due to the elimination of the shock hazard associated with electrical tools, pneumatic tools are ideal for working in and around the wet environments typically encountered in naval service.

3.4 THEORY DEVELOPMENT AND PROBLEM SOLVING METHODOLOGY

3.4.1 The Lever Rule

At this point it is useful to review Newton's laws of static equilibrium for the purpose of analyzing leverage. We know from Newton's third law that for every action there is an equal and opposite reaction. This can be extrapolated to the realization that the sum of the moments about any point is zero for a system in static equilibrium. Recall that a moment is equal to force times distance (FxD). For any lever one can find an unknown reaction force with the knowledge of lever arm lengths and a force input:

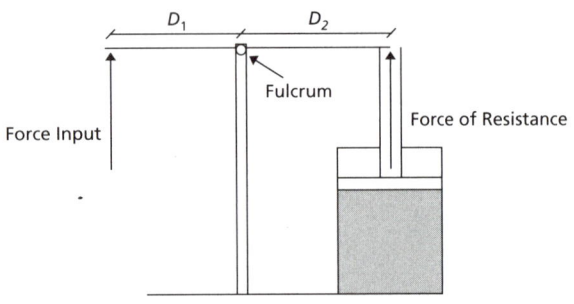

FIGURE 3.1: *Lever Principle*

Assuming positive moments are clockwise we can set the sum of the moments about the fulcrum equal to zero and determine the unknown force:

$$\sum M = 0$$
$$F_{input} \cdot D_1 - F_{resist} \cdot D_2 = 0$$

$$F_{resist} = \frac{F_{input} \cdot D_1}{D_2} \tag{3.1}$$

This form of mechanical advantage will be used in conjunction with mechanical advantage developed by constrained fluids throughout this chapter.

3.4.2 The Simple Hydraulic Press

Pascal's law, while interesting, was difficult to exploit until the advent of acceptable seals for fluid power systems. The key to this law is that force is transmitted undiminished through a "confined fluid." Power is transmitted through movement, which means mechanical seals must be capable of "confining" the fluid or handling the pressure while allowing parts to slide and move. One of the first applications for fluid power and working hydraulic seals was the invention of the hydraulic press in 1795 by an Englishman, John Brahmah. The hydraulic press can be designed to transmit a load equivalent to, greater than or less than the input force over long distances or through spaces that are too awkward for earlier systems consisting of conventional gears and levers. Both the input and output devices are piston/cylinder combinations. The piston is sealed to the cylinder so as to prevent (or at least minimize) leakage of the hydraulic fluid, but not so tightly as to prevent the piston from moving.

In the hydraulic press below, the input force is equal to the output force assuming that the fulcrum is at the center of the lever and the two hydraulic actuators are the same diameter. The length of hydraulic line between the input cylinder on the left and the output on the right is nearly limitless.

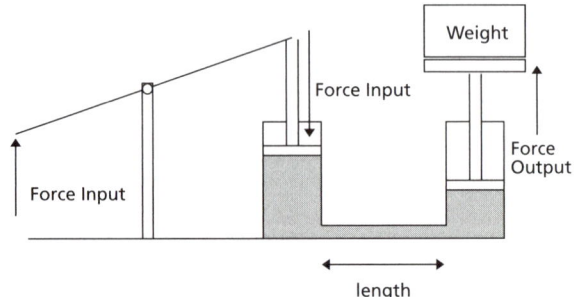

FIGURE 3.2: *Pressure and Force Transmission with Lever and Fluid*

As introduced in Unit 1, Pascal's law in algebraic form is:

$$p = \frac{F}{A} \tag{3.2}$$

where p is pressure (psi), F is force (lb$_f$), and A is area (in^2). A circular-section piston and cylinder combination is used in virtually all applications of this technology in order to facilitate the task of sealing the piston to the cylinder. So, the area of the piston is the area of a circle, over which the pressure of the hydraulic fluid is converted to force. Note that the cylinder is also subjected to very high radial forces, as the pressure of the fluid also pushes on the cylinder walls. For this reason, the cylinder must also be sized to handle the stresses placed upon it and not flex to where the piston seals fail or the cylinder blows apart. Also note that the end plate sees the same pressure and must also be sized to handle the stress without separating from the cylinder.

For the system in Figure 3.2, one can calculate the output force by setting the pressure on the face of piston #1 to the pressure on the face of piston #2 as dictated by Pascal's law of undiminished transmission of force:

$$p_{input} = \frac{F_{input}}{A_{input}}$$

$$p_{output} = \frac{F_{output}}{A_{output}}$$

$$p_{output} = p_{input}$$

$$\frac{F_{output}}{A_{output}} = \frac{F_{input}}{A_{input}}$$

$$A_{output} = A_{input}$$

$$\text{Therefore} \quad F_{output} = F_{input}$$

More important, hydraulics can be used to amplify force using an arrangement similar to the one shown in Figure 3.3 by using a larger output actuator than the input.

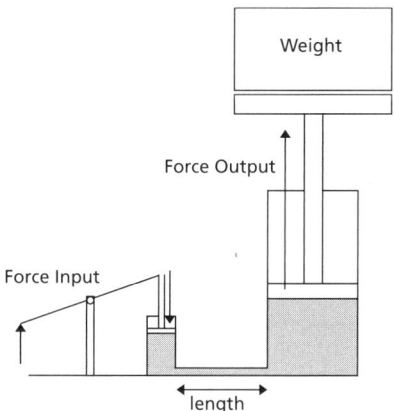

FIGURE 3.3: *Hydraulic Press*

In Figure 3.3, for the same input force as shown in Figure 3.2 one is able to lift a much larger weight. One can calculate the output force in a similar fashion as the previous example:

$$p_{input} = \frac{F_{input}}{A_{input}}$$

$$p_{output} = \frac{F_{output}}{A_2}$$

$$p_{output} = p_{input}$$

$$\frac{F_{output}}{A_{output}} = \frac{F_{input}}{A_{input}}$$

$$F_{output} = F_{input} \frac{A_{output}}{A_{input}} \quad (3.3)$$

Clearly, if A_{output} is greater than A_{input} the force F_{output} will be greater than F_{input}.

Figure 3.3 also illustrates the configuration of hydraulic jacks used to lift motor vehicles and other large loads. This is the arrangement used by an MTVR dump truck's hydraulic actuator to raise the bed to drop its load, or a hydraulic lift on a Maritime Pre-positioning Force Ship in the offloading of Marine equipment (M1 Abrams tanks, armored personnel carriers, etc.). A ship's steering system also requires a large mechanical advantage that is provided by the hydraulic system and actuator mechanics.

The mechanical advantage of the system does not come without a cost. It is important to note that the volume displaced in the movement of the input cylinder is equal to the displacement of the output cylinder. Consider Figure 3.4 where the input stroke length L_{input} logically will be longer than the output stroke length L_{output}.

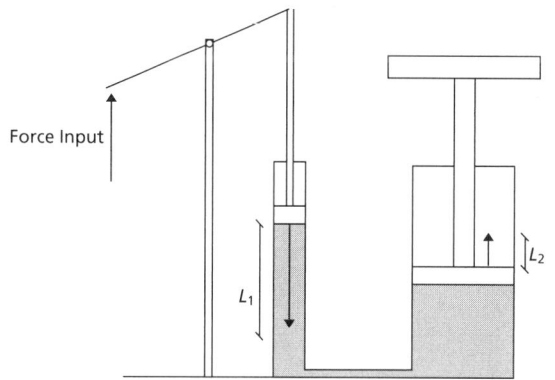

FIGURE 3.4: *Simplified Hydraulic Jack*

Since the two volumes of displaced fluid must equal each other the following relationships exist:

$$V = \frac{\pi \cdot D^2 \cdot L}{4}$$

$$V_1 = V_2$$

$$\frac{\pi \cdot D_{input}^2 \cdot L_{input}}{4} = \frac{\pi \cdot D_{output}^2 \cdot L_{ouput}}{4}$$

V = Displaced Volume
D = Diameter

$$L_{ouput} = L_{input} \frac{D_{input}^2}{D_{ouput}^2} \qquad (3.4)$$

To overcome this volumetric limitation of the simple hydraulic press the input force can be replaced with a high pressure supplied by a pump and a large reservoir of fluid.

3.4.3 Actuators with More than One Pressure and Calculating Resultant Forces

An actuator is *single acting* if it is designed to push in one direction only. The actuator is reset by springs or gravity forces by letting the hydraulic fluid flow out of the actuator. The actuator on a dump truck is an example of gravity return. The weight of the bed, even when empty, will push the actuator back to its starting position when the hydraulic fluid is released from the actuator.

Another classification of actuators is *double-acting*. Double-acting actuators can push or pull. Since actuators work on the basis of positive pressure applied to the piston, and not at vacuum pressure, then double-acting actuators are fitted with two ports, one on each end of the cylinder, to allow pressurizing either face of the piston.

This requires allowing the hydraulic fluid on the low pressure side to be squeezed out of the cylinder via the second cylinder port by the movement of the piston that is being pushed by high pressure fluid on the "supply" side of the piston. The resistance to "return flow" creates a *back-pressure* that reduces the actuator force.

Consider the double connecting rod actuator shown in Figure 3.5, where the pressure in the left side p_1 is higher than that of the right p_2 and therefore the resulting force is to the right. We'll consider forces to the right as positive (+) and forces to the left as negative (-).

The resulting force can be calculated from the pressures and effective surface areas of the piston in the following manner:

$$A_{face\,1} = A_{piston} - A_{rod}$$
$$A_{face\,2} = A_{piston} - A_{rod}$$

$$F_1 = A_{face\,1} p_1$$
$$F_2 = A_{face\,2} p_2$$

Summing the two forces yields the net force and direction. Since the areas are the same on the left and right, the equation can be rearranged to group the difference in pressure and yield:

$$F_{resultant} = F_{net} = F_1 - F_2 = A_{net}(p_1 - p_2)$$

Note that the area of the face is the piston's cross-sectional area minus the cross-sectional area of the connecting rod.

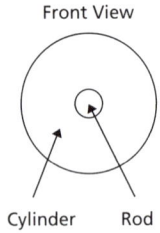
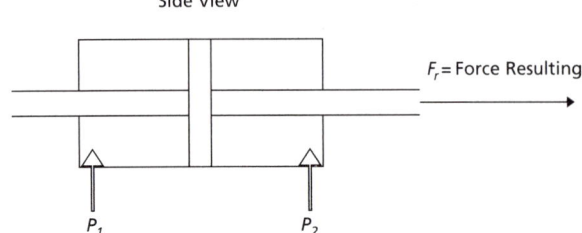

FIGURE 3.5: *Hydraulic or Pneumatic Actuator*

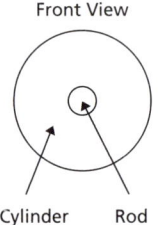
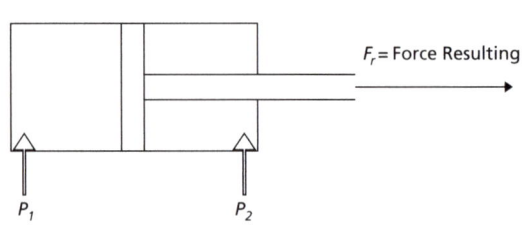

FIGURE 3.6: *Actuator with Unequal Piston Effective Areas*

When the actuator has a single connecting rod (Figure 3.6) the different effective surface areas on the right versus the left face of the actuator's piston must be accounted.

For this actuator:

$$A_{face\,1} = A_{piston}$$
$$A_{face\,2} = A_{piston} - A_{rod}$$

$$F_1 = A_{face\,1} p_1 = A_{piston} p_1$$
$$F_2 = A_{face\,2} p_2 = (A_{piston} - A_{rod}) p_2$$

And summing the two opposing forces yields the net force and direction. Because the areas are not equal, the forces are best calculated separately and then summed.

$$F_{resultant} = F_{net} = F_1 - F_2$$

3.5 CONTROL OF FLUID POWER

There are two special purpose valves that make hydraulics possible and safe: Directional control valves and pressure relief valves.

3.5.1 Pressure Relief Valves

It is important to protect against overpressurization of hydraulic and pneumatic systems. This can be accomplished by placing a pressure relief valve on the high pressure side of the fluid power system. The simple pressure relief valve, shown in Figure 3.7, is adjustable using the screw adjustment and spring to seat the ball in the relief line. If the hydraulic pressure gets too high, the hydraulic force applied to the ball will overcome the spring force and cause the ball to move, creating a path for the hydraulic fluid to flow to a low pressure area of the system and thereby reduce the pressure in the high pressure portion of the system.

FIGURE 3.8: *Seabees in Afghanistan Using Up-Armored Heavy Construction Equipment to Build Protective Berms at a Forward Operating Base (2009). Hydraulic Actuators Are Significant Components of These Machines.*

There are several causes of possible overpressurization. The most likely is the fact that a load transmitted to the actuator can easily exceed the design limits of the system. This could happen to a hydraulic actuator such as the one controlling the bucket on the front-end loader shown below (Figure 3.8).

The fluid flow of Figure 3.7 would change to the configuration shown in Figure 3.9 if too large of a back pressure on the system were to occur. As the pressure rises the force on the bottom of the ball seal rises to the point where the spring is compressed and averts a rupturing of the system by allowing the fluid to escape to the reservoir.

Since hydraulic fluid is relatively incompressible, any net loss of fluid from the high pressure side results in very rapid depressurization. The exception is if the

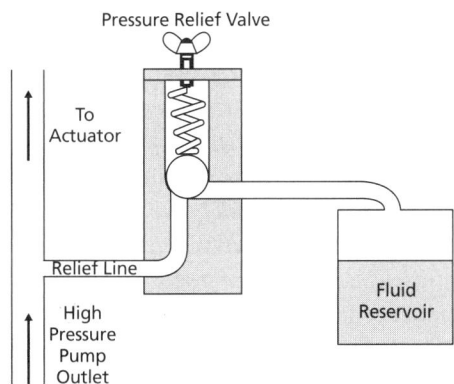

FIGURE 3.7: *Illustrated Concept Pressure Relief Valve*

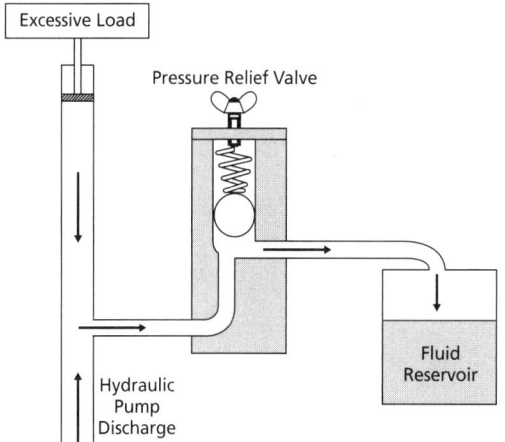

FIGURE 3.9: *Pressure Relief Valve Relieving Pressure*

capacity of the pressure source is larger than the relief path (or leak). For this reason, relief valves must be sized to let more fluid pass than the system's pump can supply. Pneumatics, however, take longer to depressurize due to the elastic nature of compressed gases, but the relief capacity must still be larger than the pressure source to protect the system.

3.5.2 Directional Control Valves

A large number of directional control valves are available with many variations manufactured for industrial applications. Directional control valves allow a system to operate many actuators in several directions from the same pump and reservoir, or to operate multidirectional elements such as the double-acting actuators discussed in the previous section.

The simplest directional control valve is a two-way, two-position directional control valve. The control valve shown in Figure 3.10 is displayed in both possible positions along with the American National Standards Institute (ANSI) symbol that would appear on system schematics and drawings.

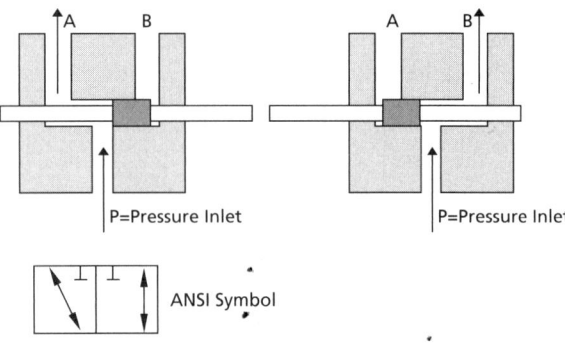

FIGURE 3.10: *Two-Way, Two-Position Sliding Control Valve*

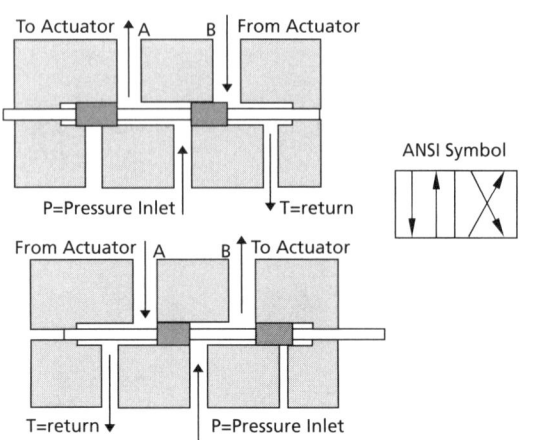

FIGURE 3.11: *Four-Way, Two-Position Control Valve*

Another type of control valve typically found in fluid power systems is the two-position, four-way sliding valve, which can be used to extend or retract double-acting actuators: (See Figure 3.11.)

3.5.3 Prime Movers: Pumps and Compressors

Pumps are used as the power inputs to hydraulic systems while compressors are used for pneumatic systems.

3.5.3.1 *Pumps*

The most general classification of pumps is whether they are positive displacement or variable displacement. A positive displacement pump delivers a fixed volume of fluid to the high pressure side of the pump every cycle. A seal exists between the low pressure side of the system and the high pressure side. Positive displacement can be accomplished through the use of a number of different designs, but common ones include screw pumps, gear pumps, lobe pumps, and rotary piston pumps. The fact that a seal exists between the low and high side allows the positive displacement pump to produce the high pressures necessary to take full advantage of hydraulic systems, such as the Navy's typical 3000 psi hydraulic systems. One potential drawback is that the positive displacement pump has to deliver a fixed volume each cycle and thus experiences catastrophic failure if the maximum system back pressure is reached. Thus, the importance of having an operable pressure relief valve in a hydraulic system is immediately evident.

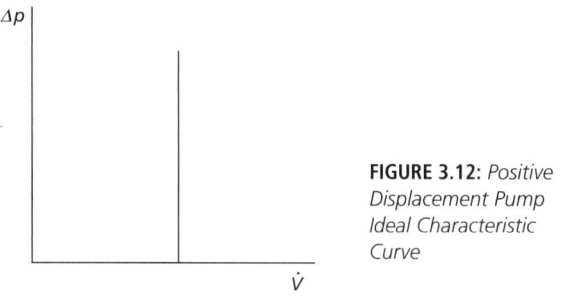

FIGURE 3.12: *Positive Displacement Pump Ideal Characteristic Curve*

A variable displacement pump, by contrast, delivers a continuous stream of fluid to the high pressure side of the system and is frequently used to raise the pressure prior to entering a positive displacement pump. This rise in pressure is essentially accomplished through the acceleration of fluid particles and is appropriate for low pressure/high volume applications. A centrifugal pump is an example of a variable displacement pump. This class of pumps continues to move fluid so long as the back pressure on the high pressure side does not reach the "pump shutoff" value for the pump. As the backpressure on a variable displacement pump increases, the pump will experience fluid slippage and recirculation inside the impeller and around the clearances between the impeller and the casing. Upon reaching pump shutoff, a variable displacement pump will churn in the fluid with no flow and the fluid's temperature will rise. If the temperature continues to go up, it could reach the boiling point for the fluid and flash to a vapor. And the internals of these pumps are lubricated and cooled by the pumped fluid. So, running a pump at shutoff conditions can also result in damage to components due to loss of lubrication and cooling.

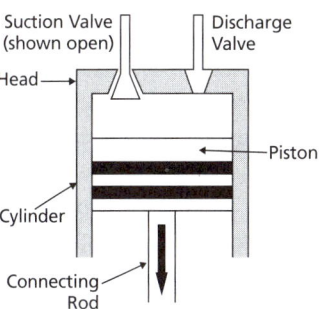

FIGURE 3.14: *Basic Reciprocating Air Compressor (Intake Stroke Shown)*

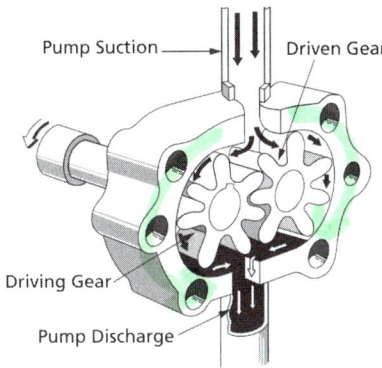

FIGURE 3.13: *Typical Gear Pump*

3.5.3.2 Compressors

Similar to pumps, there are a number of mechanical configurations that can be used to increase the pressure of air, such as reciprocating pistons (Figure 3.14) or rotating vanes, and compressors can be classified as positive or variable displacement. The most common air compressor used for pneumatic power tools uses a positive displacement reciprocating piston arrangement, with a high pressure safety valve to protect the system from overpressure if the compressor fails to shut down.

Portable pneumatic tools use low pressure air, nominally 100 psig. Ships and submarines use higher pressure air for gas turbine or diesel engine starting, emptying and transferring water out of ballast tanks, valve actuation, launching torpedoes, and other operations at pressures up to nominal 4500 psig. Getting air to that elevated pressure is accomplished by multiple stage compressors, where the discharge of one stage is sent on to the next stage. This is shown schematically in Figure 3.15. Four-stage compressors are fairly common in naval applications. Note that putting work into the fluid raises its temperature. This is often referred to as the "heat of compression" but thermodynamically is <u>work</u> energy put into the fluid. Regardless, for this reason air compressors must incorporate some heat removal mechanisms. Low pressure air compressors typically incorporate fins around the cylinder to transfer heat to the environment. Higher capacity air compressors will use water cooling to both the cylinders and the air. Cooling the air between stages increases the density and efficiency of the follow-on stages.

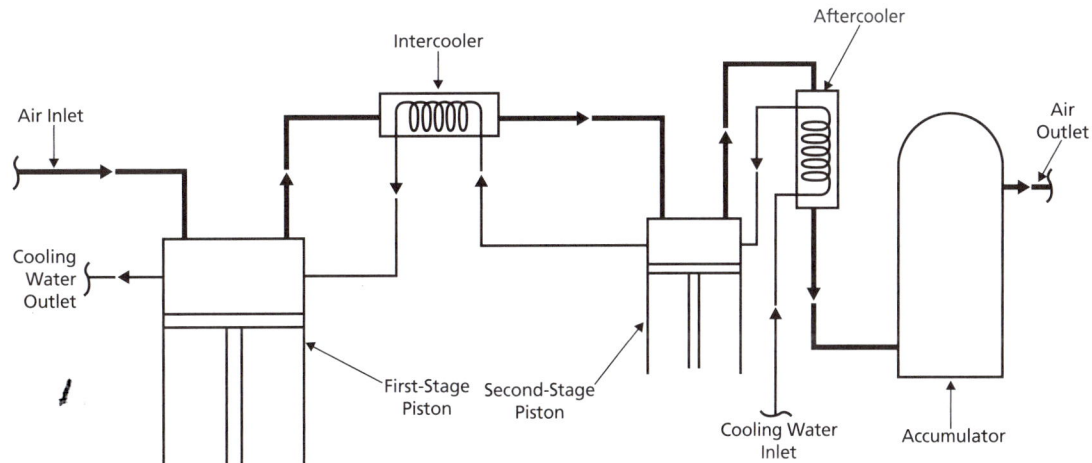

FIGURE 3.15: *General Arrangement of Multi-Stage Compressed Air Plant*

3.5.4 Accumulators, Receivers, and Storage Tanks

As mentioned previously, demand on the hydraulic or pneumatic supply system, whether going to actuators, tools, relief valve paths, or leaks, reduces the system pressure. This pressure drop is particularly high in the relatively incompressible hydraulic systems. In order to keep the hydraulic pumps from cycling on and off at rapid intervals, an *accumulator* is installed in many systems that serves two primary purposes: (1) to dampen the pressure surges caused by the positive displacement pump; and (2) to provide a pressurized stored source of hydraulic oil that minimizes the rapid cycling of the hydraulic pump on and off. In some systems, the accumulator also provides a source of pressurized hydraulic oil for some period of time when the pump has lost power due to a casualty.

A hydraulic accumulator is typically a piston-cylinder arrangement with gas on one side of the piston and hydraulic oil on the other. The gas is usually nitrogen and not air, to minimize any combinations of oil and oxygen, which might cause flammable or explosive combinations. As the hydraulic pump adds oil to the system, the piston is forced against the gas and pressurizes the gas. Hydraulic pumps are controlled by system pressure. When the pressure reaches the pump shutoff pressure, the pump's motor shuts off and the hydraulic accumulator provides the hydraulic oil to meet the system demands. When the demand reduces the pressure to the low pressure set-point, the hydraulic pump's motor starts and raises the system pressure.

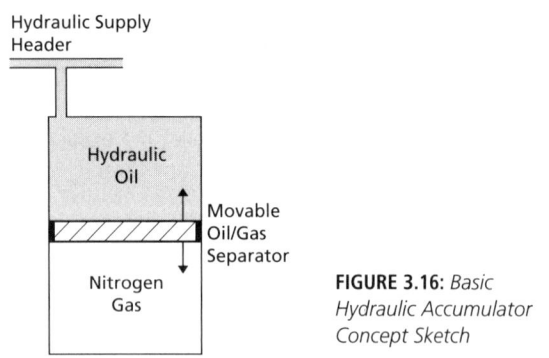

FIGURE 3.16: *Basic Hydraulic Accumulator Concept Sketch*

Similarly, air compressors are connected to a *receiver* that serves the same main purposes as a hydraulic accumulator and is also called an accumulator. For mobile air compressors, the receiver is a tank on which the compressor is mounted.

However, on simple, small hydraulic systems, such as an hydraulic log splitter, there is no accumulator and the relief valve relieves oil back to the storage tank when the demand is less than the volume that the pump supplies and the pump driver (usually in this application a small gasoline-powered engine) runs constantly. The storage tank in this case is vented to the atmosphere.

3.5.5 Fluid Power Systems: Combining Elements

A fluid power system includes pumps or compressors, air storage tanks or hydraulic accumulators, pressure relief valves and control valves. The combination of these elements allows a fluid to operate in a useful manner. One such system that could be used to control the rudder of a ship or aileron of a fixed-wing aircraft is shown in Figure 3.17. In normal operation, a pump controlled by an input signal from the helm provides a high pressure to one side of the actuator depending upon the position of the control valve. If, for example, this were the hydraulic system for a ship's steering the hydraulic cylinder would be forced left in order to move the rudder right through the application of fluid power and mechanical leverage as in Figure 3.17. If the control valve is switched to the other position the system flow will change to Figure 3.18.

Lastly, if the force input from the aileron or rudder were to exceed the design pressure of the system the pressure relief valve would open as in Figure 3.19.

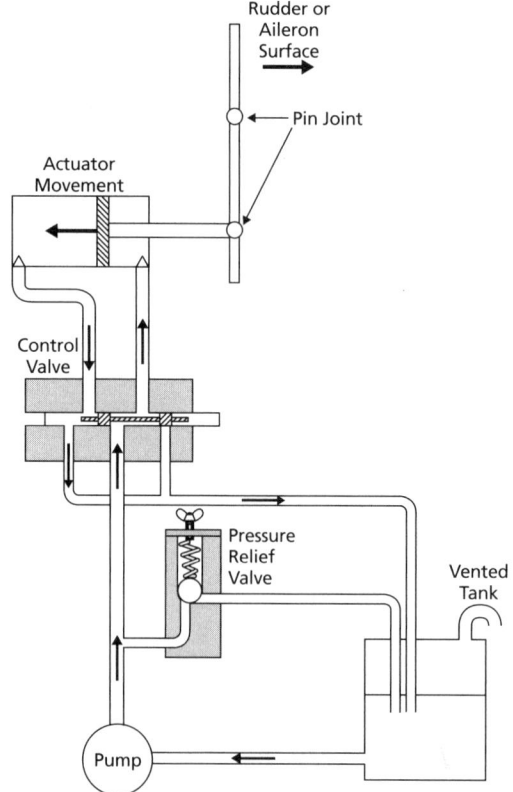

FIGURE 3.17: *Simplified Hydraulic Control Surface Positioning System Showing Right Control Surface Actuation*

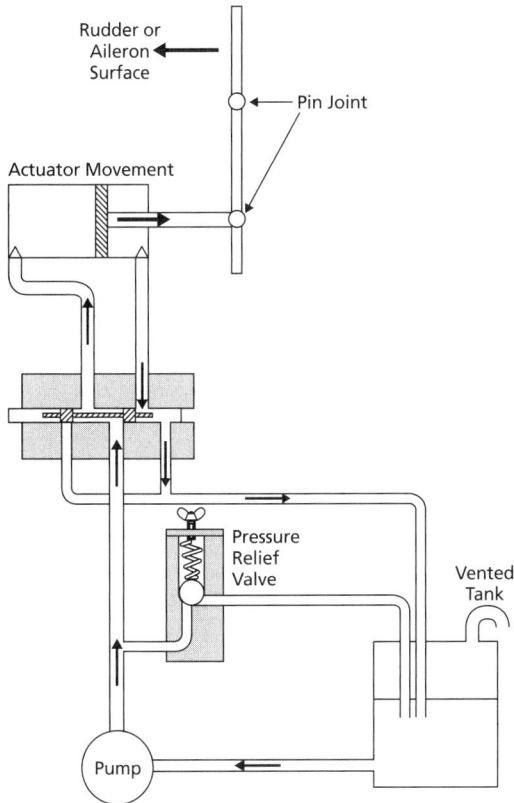

FIGURE 3.18: *Simplified Hydraulic Controls Showing Left Control Surface Actuation*

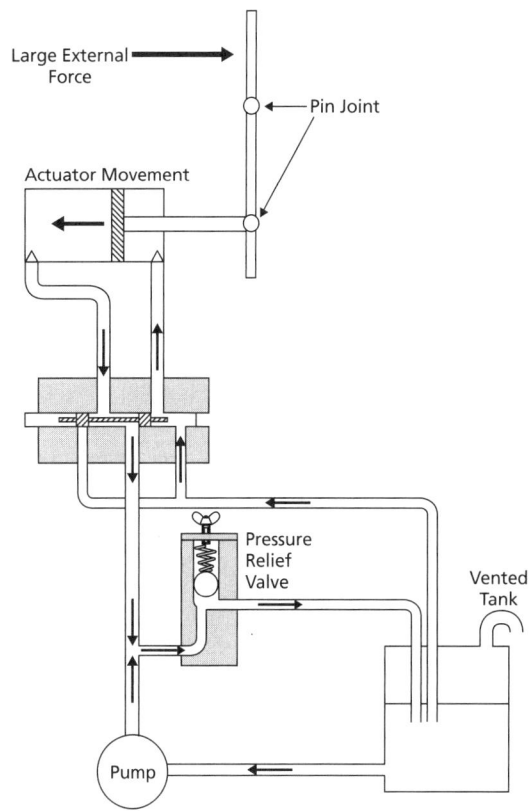

FIGURE 3.19: *Simplified Hydraulic Controls Showing Over-Pressure Protection Functioning*

Pneumatic circuits are slightly different in that generally the prime mover can be separated from the circuit given a large enough tank. The tank provides air to the system at a constant pressure using a regulating valve. Controls and actuators for pneumatic systems can be used similar to the previous hydraulic example. Figure 3.20 is an example of pneumatic control of a damper. This is a typical application for the low pressure air lines found throughout commercial buildings and ships. The linear actuator rotates a pulley or gear that closes the damper when the thermostat calls for such a setting.

3.5.5.1 Miscellaneous Air Compressor Safety Issues

A phenomenon that occurs in air compressor systems is humidity in the air will reach saturation conditions in the air receiver tank and will condense out of solution when the tank contents cool off. The result is a supply of liquid water inside a, typically, carbon steel tank. This moisture must be drained regularly in order to prevent creating significant corrosion problems in the tank's bottom.

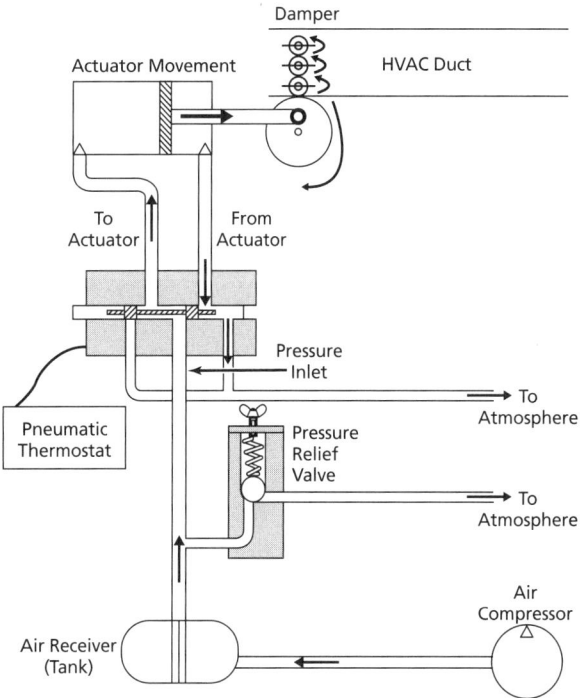

FIGURE 3.20: *Pneumatic Thermostat HVAC Control System*

Corrosion of the air tank thins the tank wall and could lead to an explosive failure of the tank.

Another reality element in air compressor systems is that the compressor is usually lubricated by oil. Compressed air that is used by any forced air breathing apparatus, such as scuba tanks or "Scott air packs" in fire fighting, must be oil-free in order to prevent human health effects. Oil-free air compressors are specialty equipment. Otherwise, special filters are required to ensure no oil is entrained in the air that will be inhaled by divers or fire fighters wearing this equipment.

Related to the oil used for lubrication, it is possible to achieve the auto-ignition temperature of the lube oil which can cause the oil to combust. This phenomenon is known as "dieseling" and is typically associated with reduced cooling due to fouling or improper cooling water valve closure. Dieseling can result in explosion of the cylinder.

3.6 SUMMARY

Hydraulic and pneumatic systems have several advantages over electrical and purely mechanical systems, including safety, flexibility, durability, amplification of force, and many others. Taking these advantages into consideration with the limited number of disadvantages it is little surprise that fluid power systems are ubiquitous in Navy and Marine Corps weapons, platforms, and support systems. A fundamental knowledge of these systems has been presented in this chapter with a primary emphasis on the major components that allow a fluid power system to perform amazing tasks in a safe manner.

REFERENCES:

L. Stewart and T. Philbin, *Pneumatics and Hydraulics* (New York: Macmillan, 1966).
D. W. Wolansky, J. Nagohosian, and R. W. Henke, *Fundamentals of Fluid Power* (Boston: Houghton Mifflin, 1986).
NAVEDTRA 14105, *Fluid Power.*

Practice Problems

"If you tried to do something and failed, you are vastly better off than if you had tried to do <u>nothing</u> and succeeded."

—Richard Martin Stern

3.1 The connecting rod has a diameter of 3" and the cylinder has diameter of 1 ft. If the resulting output force of the actuator is 10,000 lb_f to the right and "backpressure" of p_2 is 100 psia, what is p_1 (psia)? If the control valve allows high-pressure hydraulic oil to be sent to either port p1 or port p2, is the actuator single acting or double acting? Explain your answer.

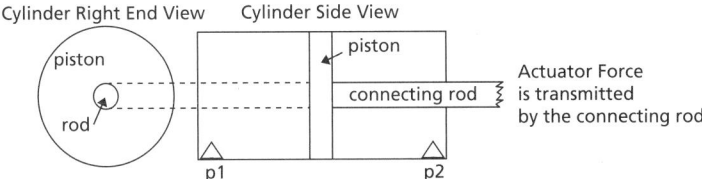

p1 and p2 are ports into the cylinder where tubing is connected.
The piston is sealed to prevent leakage from either side of the piston.
The connecting rod is sealed to prevent leakage from the cylinder to the outside.

FIGURE HW 3.1

3.2 For the differential area piston system below determine the force (F) exerted by the piston rod if the fluid is pressurized to 2000 psi. The piston is 3 inches in diameter and the rod is 1 inch in diameter.

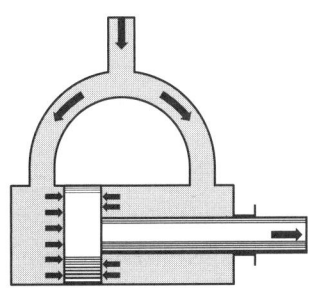

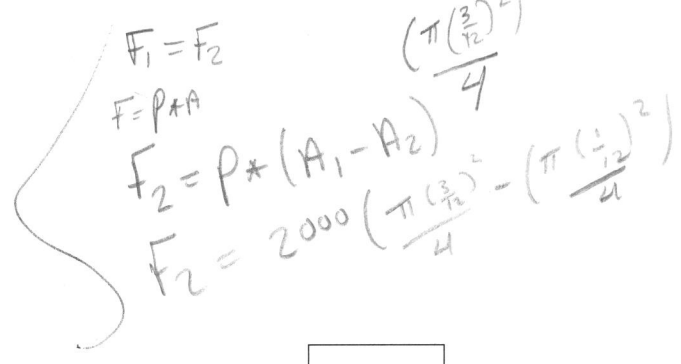

FIGURE HW 3.2

3.3 A mechanism known as a "hydraulic press" is used to move very heavy objects and has the configuration shown below. A force input of 100 lb_f is applied to the lever below and an equivalent force presses down on the smaller piston. If the smaller piston has a diameter of 1" and the larger piston has a diameter of 12", what is the force output? If the smaller piston travels ten inches downward, how far does the large piston move? (See the conversion factors in the back of your *Steam Tables* book if you don't already know what a "mil" is.)

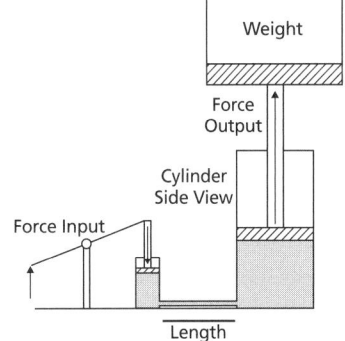

FIGURE HW 3.3

3.4 A DDG has a twin rudder system with each rudder being 15 ft long and with the rudder post (a pivot point) at the leading edge of the rudder. The proportional servo control system shifts the control valve to send hydraulic oil to the actuator and cause the rudder to rotate as ordered by the helm. Once the rudder is in the ordered position, the servo shifts the control valve to the "shut" position to hold the rudder at the ordered position. This isolates the actuator from both the supply and return piping and is referred to as "locked hydraulics." However, that doesn't mean that there aren't pressures from reaction forces in the hydraulic actuator. At the ship's full speed and with the rudder at the "right full rudder" position, the reaction force due to the flow of water across each rudder, transmitted through the linkage mechanism, results in a force applied to the actuator of 100,000 lb_f. The actuator has a cylinder with a diameter of 16 inches and the connecting rod is 4 inches in diameter. Determine the pressure of the hydraulic oil on the left side of the actuator piston under these conditions and with the control valve in the shut position.

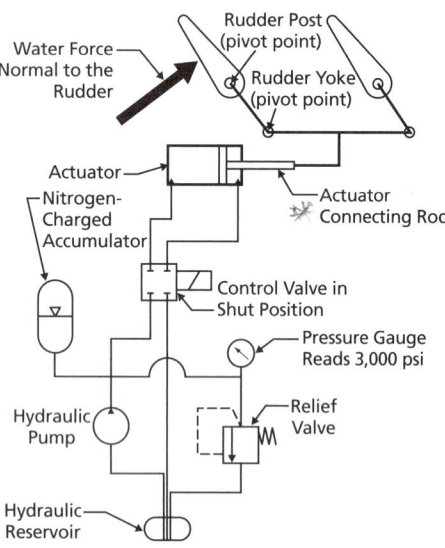

FIGURE HW 3.4

3.5 A sentinel valve with a 2-inch diameter disc, similar to the one shown to all students in the Pipes and Valves lab, is designed to warn plant operators of an overpressure situation in a steam plant. If the desired alarm set point is 875 psi, what is the force that the valve's spring must exert to hold the actuator disc shut when the steam pressure is less than the alarm set point? Draw "the system" and show your analysis. Referring to your sketch, describe what opens the valve and what holds it shut until the setpoint. To where does a sentinel valve discharge?

3.6 Print a copy of PR-02, page 28, of DOE-HDBK-1016/1-93 (available at http://www.hss.energy.gov/NuclearSafety/ns/techstds/standard/hdbk1016/h1016v1.pdf and labeled "Figure 25 Valve Symbol Development," page 80 of 120 page pdf) and determine whether the rudder moves to satisfy a left or right ordered position from the helm, when the control valve is in Position "C." On the sketch shown above (Figure HW 3.4), highlight the piping exposed to high pressure hydraulic oil with one color and low pressure hydraulic oil with another color with the control valve in Position "C."

3.7 Early generation naval steam reciprocating engines (piston and cylinder) boiled saltwater in their boilers because desalination technology was not available and they could not store enough freshwater to make up for system leaks. As a consequence of saltwater-induced corrosion in the boiler materials, early U.S. Navy engineers were forced to use very low steam pressure in the engines. One example: the USS *Merrimac* (commissioned in 1856 and converted to the CSS *Virginia* during the American Civil War) had a reciprocating steam engine consisting of two cylinders, each with 72-inch diameter pistons and a stroke of 3 feet. The connecting rods of the pistons were connected to a crankshaft that turned a 17′4″-diameter two-bladed propeller. During an in-service test, the engine developed 1294.4 brake horsepower and the ship averaged 5.25 knots at 36.5 shaft RPM with 12.8 psig boiler pressure and 20.4 in-Hg condenser vacuum pressure. (Ref: *C.S. Ironclad Virginia and U.S. Ironclad Monitor* by Sumner B. Besse, The Mariner's Museum, 1937) An arrangement known as a "sliding valve" simultaneously ported the steam from the boiler to one end of the cylinder and ported the opposite end of the cylinder to the condenser. When the piston had traveled full stroke, the sliding valve shifted to reverse the cylinder connections, thereby creating what is referred to as a "double-acting" piston. This means that one side of the piston was exposed to steam pressure while the opposite side of the piston was simultaneously exposed to condenser vacuum pressure and these conditions shifted at the end of each stroke. Ignoring the size of the connecting rods, what net force was applied by a single piston to its connecting rod? How much larger was the force of the piston because the opposite side was exposed to the condenser vacuum, rather than being exhausted to the atmosphere (as was done in steam train locomotives)? Hint: You should draw the piston/cylinder for each pressure scenario and analyze the resultant forces on the piston.

3.8 In the classic 1968 submarine movie, *Ice Station Zebra*, an attempted sabotage of the nuclear submarine USS *Tigerfish*, is overcome by a single person using brute force to shut an open torpedo tube inner door while seawater was rushing in. Is this scenario feasible or even possible? If not, then why not? For the sake of this analysis, assume the submarine was surfaced and the torpedo tube was only at a depth of 15 feet. U.S. submarine torpedo tubes are 21 inches in diameter and the inner door swings into the boat to allow re-loading the tube. Hint: Estimate the force this human being had to exert to stop the flooding. Don't neglect that the human will likely grab the handle on the edge of the door and not push from the center—there is a 2:1 mechanical advantage in doing so.

Ice Station Zebra *plot summary: Commander James Ferraday, USN, has new orders: get David Jones, a British civilian, Captain Anders, a tough Marine with a platoon of troops, Boris Vasilov, a friendly Russian, and the crew of the nuclear sub USS* Tigerfish *to the North Pole to rescue the crew of Drift Ice Station Zebra, a weather station at the top of the world. The mission takes on new and dangerous twists as the crew finds out that all is not as it seems at Zebra, and that someone will stop at nothing to prevent the mission from being completed. (http://us.imdb.com/Plot?0063121, 9/11/02)*

3.9 A hydraulic accumulator using nitrogen gas is connected to a 3000 psi hydraulic system. The accumulator cylinder is 20 inches inside diameter and 36 inches inside height. The floating piston between the oil and nitrogen is 2 inches thick. The hydraulic pump shutoff setpoint is 3000 psig and the accumulator gas is ¼ of the accumulator volume when shutoff occurs. In an emergency, the accumulator provides oil at pressure if the hydraulic pump fails. Assuming pump failure with a full accumulator, what is the pressure in the accumulator and how much oil was delivered to the hydraulic header at the point that the accumulator empties?

4 Internal Combustion Engines

The bow of the USS *New York* includes steel salvaged from the World Trade Center buildings that were destroyed in the terror attacks of September 11, 2001. The ship is propelled by four turbocharged marine diesel engines driving two propellers and capable of producing a total of 41,600 shaft horsepower.

4.1 UNIT LEARNING OBJECTIVES

4.1.1 Terminology: definitions, variables, and typical units

4.1.1.1 polytropism, isobaric, isometric, isentropic, isothermal, cylinder, piston, compression ratio, displacement, swept volume, clearance, volumetric efficiency, crankshaft, valves, rocker arms, push rods, flywheel, Otto cycle, spark ignition, Diesel cycle, compression ignition, stoichiometric air, combustion efficiency, rich fuel mixture, lean fuel mixture, heating values, natural induction, forced induction, measures of performance

4.1.2 Concepts: ideas and engineering expression of them

4.1.2.1 Sketch ideal gas processes (isobaric, isometric, isentropic, and isothermal) on *p-v* and *T-s* coordinates. Relate each graph to a physical process (expansion, heating, etc.).

4.1.2.2 Sketch and describe the differences between a stroke and a process.

4.1.2.3 Sketch and describe the differences between a 2-stroke and a 4-stroke reciprocating engine.

4.1.2.4 Sketch the ideal Otto and Diesel cycles on *T-s* and *p-v* or *p-V* coordinates.

4.1.2.5 Describe the chemical combustion process for hydrocarbon fuels in air and

list the typical products of combustion in both ideal and real combustion.
- *4.1.2.6* Choose the proper heating value (*HHV/LHV*) for given conditions.
- *4.1.2.7* Describe the effects of running rich compared to those of running lean.
- *4.1.2.8* Understand the effects of charging (turbo-, super-) and scavenging.
- *4.1.2.9* List preferred applications of diesel engines and gasoline engines based on engine characteristics.
- *4.1.2.10* Show numerically or graphically the effects of parasitic and auxiliary loads on engine performance.

4.1.3 Skills: procedures, practices, or methods that enable reasoning

- *4.1.3.1* Calculate the change in ideal gas properties in a reversible process using the general polytropic equation.
- *4.1.3.2* Calculate the work done in a reversible process with ideal gases using the *p-v* diagram and the general polytropic equation.
- *4.1.3.3* Calculate the efficiency of an ideal, air standard Diesel and an Otto cycle.
- *4.1.3.4* Calculate the net-work and power of an ideal Diesel and an Otto cycle.
- *4.1.3.5* Relate single-cycle performance parameters to engine speed, power output, and fuel and air needs.
- *4.1.3.6* Write and balance the chemical equation for ideal combustion of a given hydrocarbon in air.
- *4.1.3.7* Calculate the effects of rich and lean conditions on ideal combustion stoichiometry.
- *4.1.3.8* Explain real limitations on the use of internal combustion engines in naval applications.

4.2 INTRODUCTION

This unit is written assuming the use of companion material. The companion material is available as articles at www.HowStuffWorks.com. The graphics on that website are superb and often include motion. Consequently, this unit material will focus more on the calculation aspects of the subject. Students will need to learn the vocabulary and interaction of the engine components from other sources, specifically, HowStuffWorks.

Internal combustion engines have been crucial to our society for more than one hundred years. These engines use pistons that reciprocate inside cylinders and are therefore referred to as "reciprocating" internal combustion engines. The typical orientation of the reciprocating action is mostly up and down, although many engines have the cylinders mounted on angles ("slant-six" V-8, etc.), and some are horizontal. Many piston-engined aircraft throughout history used radial engines in which the cylinders were oriented like spokes on a wheel. An example of a radial engine-driven airplane is the Wright Double Cyclone engine, four of which powered the B-29 Superfortress of WWII. That engine had fourteen cylinders in two staggered banks of seven. However, for the illustrations used in this text, the reciprocating action will be shown as vertical and only one cylinder will be shown.

The major categories of reciprocating engines are: (1) *spark ignition* (SI) engines, which use a spark plug to initiate the combustion process; and (2) *compression ignition* (CI) engines, which inject fuel into the cylinder when the conditions are suitable for auto-ignition of the fuel. Diesel engines are CI engines. Both categories also exist in 4-stroke and 2-stroke varieties. Spark ignition engines mostly use gasoline for fuel, and compression ignition engines run on fuel oil, otherwise known as diesel fuel.

4.2.1 IC Engine Historical Overview

Internal combustion reciprocating engines power essentially all motor vehicles on the road today. Based upon their impact on society, internal combustion engines were some of the most significant inventions in history. Over most of the twentieth century, SI engines operating on the Otto cycle and CI engines operating on the Diesel cycle, using petroleum-derived fuels, were dominant. The combination of environmental regulations and rising petroleum prices has encouraged consideration of alternate fuels. Engineers have recently re-investigated using hydrogen, natural gas, and numerous other alternate fuels to run these engines.

A web search on internal combustion engine history will find many sites with historical notes on engine development. Some highlights follow.

The first person to experiment with an internal combustion engine was the Dutch physicist **Christian Huygens**, about 1680. Huygens experimented with gunpowder to make the piston move inside a cylinder. He achieved this, but experienced difficulty repeating the process for a cycle. Piston and cylinder engines were used in steam engines from the year 1712. However, no effective spark ignition engine was developed until 1859, when the Belgian-born French engineer **J. J. Étienne Lenoir** built a double-acting spark ignition engine that could be operated continuously. Lenoir adapted an existing steam engine and used coal gas as the fuel. In 1863 Lenoir constructed a vehicle propelled by his engine and completed a fifty-mile trip.

In 1862 **Alphonse Beau de Rochas**, a French scientist, patented but did not build a 4-stroke engine. Sixteen years later, when **Nikolaus A. Otto** (1832–1891) built a successful 4-stroke engine, it became known as the "Otto cycle."

The first successful 2-stroke engine was completed in the same year by Sir **Dugald Clerk**, in a form that (simplified somewhat by **Joseph Day** in 1891) remains in use today.

In 1885 **Gottlieb Daimler** constructed what is generally recognized as the prototype of the modern gas engine: small and fast, with a vertical cylinder, it used gasoline injected through a carburetor. In 1889 Daimler introduced a 4-stroke engine with mushroom-shaped valves and two cylinders arranged in a V, having a much higher power-to-weight ratio. With the exception of electric starting, which would not be introduced until 1924, most modern gasoline engines are descended from Daimler's engines.

Rudolph Diesel applied for a patent for his internal combustion engine concept in 1892. The fundamental feature of this new engine was that the compression ratio was so high that the temperature achieved exceeded the auto-ignition temperature of the fuel. The desired fuel was initially coal dust, which existed as a waste product. The focus shifted to liquid fuels, which instigated the development of the fuel injector. Various configurations were attempted until **Robert Bosch** developed the fuel injector design in the 1930s that is the forerunner of the fuel injectors used on most diesel engines today. The auto parts company that he started still bears his name.

4.3 FLEET APPLICATIONS

Gasoline engines are very common in automotive applications, but the fire hazards associated with this volatile fuel limits its use in military vessels. Outboard boat motors are the primary naval use of gasoline engines in the marine environment. However, special precautions must be taken to minimize the risks associated with storing and handling gasoline.

Due largely to its reduced flammability and stability in storage, diesel fuel has proven preferable in naval applications. Diesel engines are used for propulsion on a number of ship classes within the U.S. Fleet, mostly medium-sized and smaller amphibious ships (the largest amphibious ships tend to be steam-powered). Further, diesel engines powering electrical generators provide emergency electrical power for steam-powered ships and submarines, as well as the primary source of electrical power on diesel-powered amphibious ships, auxiliary ships, and service vessels.

On the landside of the naval service, diesel engines power wheeled combat vehicles such as the high-mobility multipurpose wheeled vehicle (HMMWV), the light armored vehicle (LAV-25), and the construction equipment used by the construction battalions (the Sea Bees). The expeditionary fighting vehicle (EFV) has a diesel engine for propulsion and a second diesel for electrical generation. Note, however, the M1 Abrams tank is powered by a gas turbine engine.

4.4 IDEAL GAS PROCESSES

In Unit 1, we reviewed the basic ideal gas law, aka ideal gas equation of state, which has two forms based upon the treatment of volume (specific or total).

$$pv = RT \tag{4.1}$$

$$pV = mRT \tag{4.2}$$

In addition, we went over how the ideal gas law is derived into various alternate equations that are known as Charles' law, Boyle's law, and others. In these variations of ideal gas equations, the equations related two state points, or the "initial" and "final" situations.

$$p_i V_i / T_i = p_f V_f / T_f \qquad p_i V_i = p_f V_f$$
$$p_i / T_i = p_f / T_f \qquad V_i / T_i = V_f / T_f$$

The processes that defined the path in between the state points were related to the three main variables in the ideal gas law: pressure constant ($\bar{p}$)—*isobaric*; volume constant ($\bar{v}$ or $\bar{V}$)—*isometric* or *isochoric*; and temperature constant ($\bar{T}$)—*isothermal*. Note that in this use the bar over the variable indicates that the property does not change during the process. There is another parameter that has not been covered to this point, which affects the way an ideal gas receives or delivers work, and that is entropy (s). An *isentropic* process ($\bar{s}$) is yet another of the series of ideal gas processes.

Further, we covered the concept that for closed systems the area under a *p-v* graph indicates work done ("by" or "on" the working fluid depends on the direction of the process). This work is referred to as "boundary work." Regardless of whether the system is open or closed, the area under a *T-s* graph indicates heat ("in" or "out" of the working fluid depends on the direction of the process). An appreciation of these tenets will aid the student in grasping the importance of the following subject.

4.4.1 Polytropic Processes

Pressurized air has many uses industrially and shipboard. Air compressors "pump" or pressurize this gas mixture to use in various pneumatic tools and devices. Early engineers recognized that individual air compressors differed in their performance. They were able to quantify the performance with an equation, $pV^n = constant$. This phenomenon is called *polytropism* and the value of the exponent depends on the equipment. For efficient air compressors, n is typically between 1.25 and 1.30. $pv^n = c$ also holds true.

$$p_i v_i^n = p_f v_f^n \qquad (4.3)$$

Taking Equation 4.3 together with the ideal gas equation of state (4.1 and 4.2), we can define other cases where $n = 0, 1, k,$ and ∞.

- $n = 0$ substituted into $pv^n = c$, yields $p = c$. Constant pressure processes are termed *isobaric* ($\bar{p}$).
- $n = 1$ substituted into $pv^n = c$, yields $pv^1 = c$ and noting that $pv = RT$, means that $RT = c$. Since R is constant, then $n = 1$ describes a constant temperature —or *isothermal* ($\bar{T}$)—process.
- $n = k$ substituted into $pv^n = c$, yields $pv^k = c$ which holds for a constant entropy, or *isentropic* ($\bar{s}$) process.
- $n = \infty$ substituted into $pv^n = c$, yields $pv^\infty = c$. This corresponds to a constant volume process, aka *isochoric* or *isometric* ($\bar{v}$ or $\bar{V}$).
- While not showing rigorous proofs of these concepts, in the following sections we will examine each type of process individually and graphically compare the *p-v* and *T-s* graphs. Recall:
- The area under a *p-v* graph shows "boundary" work, if the system is a closed system.
- The area under a *T-s* graph will show heat, regardless of whether the system is open or closed.

The equations that govern ideal gas behaviors for the various cases are listed on the course standard equation sheets. Students should familiarize themselves with the layout and contents of these course equation sheets and the limitations on applicability.

4.4.2 Isobaric Processes (n = 0)

Isobaric—constant pressure—processes are governed by $pv^n = c$ with $n = 0$. Diagrams of isobaric cooling are shown in Figure 4.1. It should be clear that the reverse process (heating) will appear similar yet follow the opposite path.

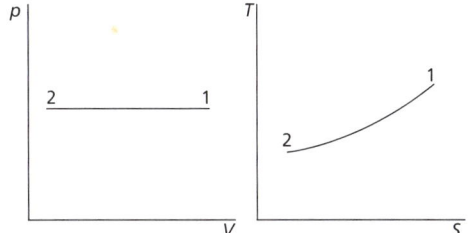

FIGURE 4.1: *Isobaric Cooling/Compression*

To gain physical insight into these diagrams, think of a fully inflated balloon on a warm summer day. The balloon has a certain pressure, temperature, and specific volume, let's say p_i, T_i, and v_i. As night falls, the temperature of the air surrounding the balloon decreases. A small temperature gradient thus exists across the boundary of the system—the surface of the balloon. Heat is therefore transferred from the air within the balloon, lowering its temperature to that of the surroundings. To keep the pressure of the air within the balloon constant, the ideal gas equation of state tells us that the volume of the air must decrease. As such, isobaric cooling is also termed compression, while constant pressure heating corresponds to expansion of the working fluid. As heat is transferred from the air within the balloon, the entropy of the air decreases. Our understanding of ideal gas behavior thus enables depiction and, more important, understanding of the isobaric cooling—or compression—process in *p-v* and *T-s* coordinates.

The work performed on or by the air in our balloon can therefore be graphically depicted as the area below the process line in the *p-v* diagram. The heat transferred to or from the air can similarly be represented by the area beneath the process line in the *T-s* diagram.

If the initial and final properties of the air are known, we can calculate the amount of work performed or heat transferred during an isobaric process from:

$$w_{i-f} = R(T_f - T_i) \qquad (4.4)$$

$$q_{i-f} = C_p(T_f - T_i) \qquad (4.5)$$

The signs for both of these are negative, since heat is "out" and work is "in." Another indication is of the signs of both heat and work are the directions of the process lines from right to left.

4.4.3 Isothermal Processes (n = 1)

The polytropic processes in which the temperature remains constant are governed by $pv^n = c$, where $n = 1$. Process graphs of isothermal compression are shown in Figure 4.2. Isothermal expansion would follow path lines in the opposite direction.

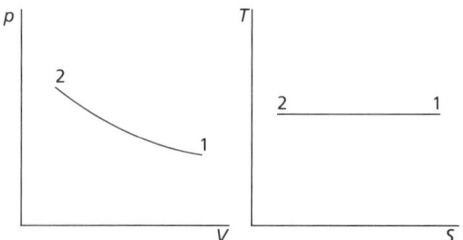

FIGURE 4.2: *Isothermal Compression*

Such a physical process is identical to the compression of air in a piston-cylinder device shown in Figure 4.2. Again, the work performed on or by the working fluid can be depicted as the area below the path line in the *p-v* diagram. The heat transferred to or from the ideal gas can similarly be represented by the area beneath the process line in the *T-s* diagram. For isothermal processes, these areas are identical. Thus, the amounts of work performed and heat transferred are equivalent and can be calculated from the initial and final properties of the working fluid.

$$w_{i-f} = q_{i-f} = p_i v_i \ln\left(\frac{v_f}{v_i}\right) = RT \ln\left(\frac{p_i}{p_f}\right) \quad (4.6)$$

4.4.4 Isentropic Processes (n = k)

As stated at the beginning of this section, polytropic processes in which the entropy remains constant are governed by $pv^n = c$, where $n = k$. The k in this equation is C_p/C_v. The *p-v* and *T-s* diagrams of isentropic compression are depicted in Figure 4.3. We note that isentropic expansion would follow similar yet reversed path lines.

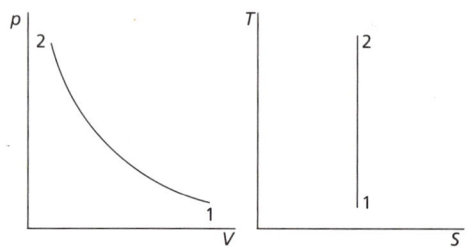

FIGURE 4.3: *Isentropic Compression*

The work performed on or by the working fluid can again be displayed as the area below the process line in the *p-v* diagram. For isentropic processes, the amount of work performed can be calculated from:

$$w_{i-f} = C_v(T_i - T_f) \quad (4.7)$$

Although the heat transferred to or from the working substance can be depicted by the area beneath the process line in *T-s* coordinate space, examination of Figure 4.3 shows that $q = 0$ for isentropic processes. If we recall the definition of isentropic (adiabatic and reversible), we see that the *T-s* diagram merely displays the fact that no heat is transferred during isentropic processes.

4.4.5 Isometric/Isochoric Processes (n = ∞)

Reversible processes conducted with an ideal gas held at constant volume are governed by $pv^n = c$ where $n = \infty$. Isometric—aka isochoric—heating is depicted in *p-v* and *T-s* coordinates in Figure 4.4. Isometric cooling would follow opposite process path lines.

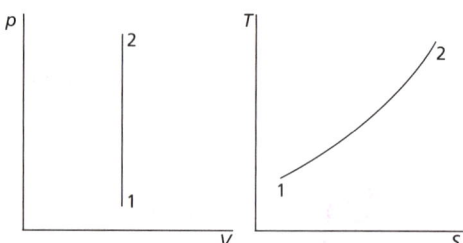

FIGURE 4.4: *Isometric Heating*

To visualize this process, think of a fully pressurized scuba tank left out on a pier. In its initial state, the air in the tank has a certain pressure, temperature, and specific volume: p_i, T_i, and v_i. As the sun sets on our warm tank, the temperature of the air surrounding the tank decreases. A small temperature difference thus develops between the system—the air in the tank—and the surroundings. Heat is therefore transferred from the air within the tank until thermal equilibrium is reached with the outside air. Since the scuba tank is a rigid device, the volume that the air can occupy cannot decrease. With the temperature of the air decreasing to T_f and a constant volume, the equation of state mandates that the pressure of the air in the tank must decrease to some new lower value p_f. As heat is transferred from the air within the tank, Equation 4.1 requires the entropy of the air to decrease.

Again, the work performed on or by the working fluid can be depicted as the area below the process line in the p-v diagram. Examination of Figure 4.4, however, reveals no area beneath an isometric process line in p-v space. As such, no boundary work is performed during isometric/isochoric processes, i.e., $w_{i-f} = 0$.

The heat transferred to or from the air can be represented by the area beneath the path line in the T-s diagram. For isometric/isochoric processes, the amount of heat transferred can be calculated from the initial and final properties of the working fluid with

$$q_{i-f} = C_v(T_f - T_i) \tag{4.8}$$

4.4.6 Consolidated p-v and T-s Process Graphs

Figure 4.5 shows the relationships of isothermal, isentropic, isobaric, and isometric processes on the same coordinate axes. Notice: (1) that the straight lines on one graph are curves on the other; (2) the direction and curvature of a process on the opposite graph; and (3) which curved process has the higher slope.

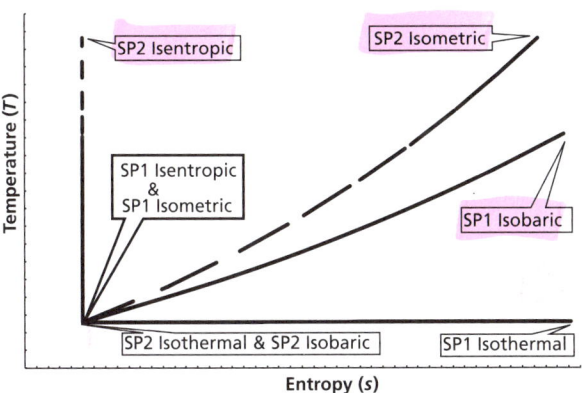

FIGURE 4.5: *Consolidated p-v and T-s Concept Graphs for Various Ideal Gas Processes*

Each of the formulas put forth to determine the amount of work performed or heat transferred during a particular polytropic process has required knowledge of the initial and final properties of the working fluid. What if one of the properties is unknown? We address this situation in the following section.

4.4.7 Ideal Gas Property Calculation

Combining an identity of Equation 4.2

$$\frac{p_i v_i}{T_i} = R = \frac{p_f v_f}{T_f} \tag{4.9}$$

with an identity of Equation 4.3

$$p_i v_i^n = c = p_f v_f^n \tag{4.10}$$

leads to three p-v-T expressions that relate the pressure of the gas to the specific volume it occupies; the temperature to its pressure; and its temperature to the specific volume of the gas.

$$\frac{T_f}{T_i} = \left(\frac{p_f}{p_i}\right)^{\frac{n-1}{n}} \tag{4.11}$$

$$\frac{T_f}{T_i} = \left(\frac{v_i}{v_f}\right)^{n-1} \tag{4.12}$$

$$\frac{p_f}{p_i} = \left(\frac{v_i}{v_f}\right)^n \tag{4.13}$$

Other expressions may be developed. The most useful of these are on the course standard equation sheets.

As mentioned above, work can be supplied to or extracted from a working fluid. Similarly, heat can be transferred to or from the substance. These processes—each of which necessarily modifies the properties of the working fluid—can be carried out in an isobaric, isothermal, isentropic, or isometric/isochoric manner. Equations 4.11, 4.12, and 4.13 can be used to calculate the final properties of a working fluid from known initial conditions—provided the proper value of the polytropic exponent, n, is employed!

EXAMPLE 4.1: ISENTROPIC COMPRESSION OF AIR

Air at 200 °F and 30 psia is compressed isentropically to 1/20th its original volume. Sketch the process in p-v and T-s coordinate space. Calculate the final temperature (°F) and pressure (psia) of the air.

GIVEN: $T_i = 200$ °F
$p_i = 30$ psia
$n = k$ (isentropic), $k = 1.4$ (air)

$$V_f = \frac{1}{20} V_i$$

FIND: p-v and T-s diagrams
T_f (°F) and p_f (psia)

SOLUTION: From known initial conditions, we'll use Eqn 4.13 to obtain the final pressure and 4.12 to calculate the final temperature of the air since $\frac{V_f}{V_i} = \frac{v_f}{v_i} = \frac{1}{20}$. I'll shift to using state point (SP) numbers in which the process proceeds from SP1 to SP2 in solving the rest of this problem.

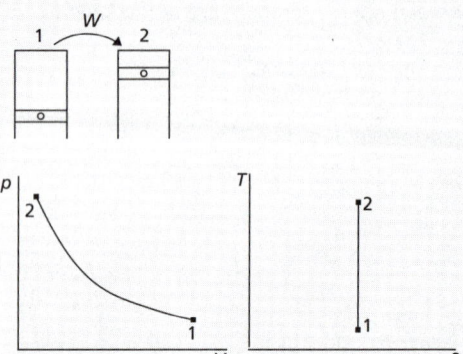

FIGURE EXAMPLE 4.1

Employing $n = k = 1.4$ for air at standard temperature conditions and substituting given information into Equation 4.13 yields

$$p_f = p_i \left(\frac{v_i}{v_f}\right)^k = 30\, psia\, (20)^{1.4} = 1989\, psia$$

Similarly, substituting into Equation 4.12 gives

$$T_f = T_i \left(\frac{v_i}{v_f}\right)^{k-1} = 660°R\, (20)^{0.4} = 2188°R = 1728°F.$$

Note that absolute scales <u>must</u> be utilized in these formulas.

EXAMPLE 4.2: SEQUENTIAL COMBINATION OF PROCESSES

Air in a piston-cylinder device is cooled from 300 °F to 50 °F at a constant volume. It is then isentropically compressed by a factor of 5. Sketch the processes on p-v and T-s diagrams. Calculate the net heat transferred (Btu/lb$_m$) and net work performed (ft-lb$_f$/lb$_m$) on or by the working fluid.

GIVEN: $T_1 = 300\,°F = 760\,°R$
$T_2 = 50\,°F = 510\,°R$
process 1-2 = isometric
process 2-3 = isentropic
$V_3 = \frac{1}{5}V_2$

FIND: p-v and T-s diagrams
w_{net} (Btu/lb$_m$)
q_{net} (Btu/lb$_m$)

SOLUTION: We note that $w_{net} = w_{12} + w_{23}$ and $q_{net} = q_{12} + q_{23}$. Knowledge of the particular type of polytropic processes occurring enables us to choose the correct expression to substitute for w_{12}, w_{23}, q_{12}, and q_{23}. Since the air is cooled at constant volume from state point 1 to state point 2, $w_{12} = 0$ and $q_{12} = c_v (T_2 - T_1)$. Similarly, isentropic compression from state point 2 to state point 3 determines that $w_{23} = c_v (T_2 - T_3)$ and $q_{23} = 0$. Thus, $w_{net} = w_{23} = c_v (T_2 - T_3)$ and $q_{net} = q_{12} = c_v (T_2 - T_1)$. Given information, particularly T_1 and T_2 would permit solution of the net heat transferred. We see, however, that the temperature at state point 3 must be determined to obtain the net work. We must employ the appropriate p-v-T relationship, therefore, to calculate T_3. With $n = k$, we refer to Eqn 4.12 with the given volume ratio between state points 2 and 3 to obtain

$$T_3 = T_2 \left(\frac{v_2}{v_3}\right)^{k-1} = 510°R\,(5)^{0.4} = 971°R = 511°F$$

With all three temperatures, we now substitute into our expressions for the net heat and work to obtain the solution:

$$w_{net} = w_{2-3} = C_v (T_2 - T_3)$$
$$= 0.171 \frac{Btu}{lb_m \cdot °R}(510°R - 971°R) = -78.8 \frac{Btu}{lb_m}$$

and

$$q_{net} = q_{1-2} = C_v (T_2 - T_1)$$
$$= 0.171 \frac{Btu}{lb_m \cdot °R}(510°R - 760°R) = -42.8 \frac{Btu}{lb_m}$$

Note that work is performed on the working fluid (negative value) to compress the air by a factor of 5. Similarly, to cool the air by 250 degrees, heat must be removed (another negative value) from the working substance. We now refer to Figure 4.5 to sketch the p-v and T-s diagrams.

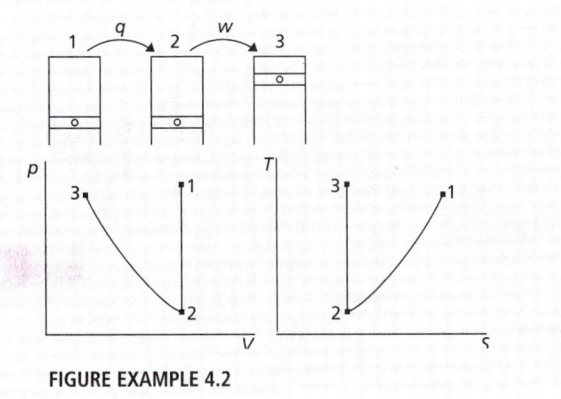

FIGURE EXAMPLE 4.2

Note that the direction in which the process proceeds on the p-v and T-s diagrams correlates directly to our sign convention. Specifically, processes that proceed from left to right are positive while those proceeding from right to left are negative. To illustrate this concept, examine the isothermal compression process shown in Figure 4.2. In the p-v diagram, note that the compression process proceeds from right to left. As we saw in the previous example, compression work is calculated as a negative number. This computation is in accordance with our sign convention and experience. That is, you must perform work on the working fluid—a negative value by convention—to compress air. Similarly, examination of the T-s diagram reveals that the process proceeds from right to left. Indeed, the only way to compress air without changing its temperature is to remove heat—again negative by convention—from the working fluid. Thus, we see that just as work performed on the working fluid and simultaneous heat removal from the working fluid are negative quantities by convention, processes that proceed from right to left on p-v and T-s diagrams are negative quantities. In contrast, processes that proceed from left to right on p-v and T-s diagrams are positive quantities. Remembering this small detail can prove helpful in understanding p-v and T-s diagrams for unfamiliar processes and cycles.

4.5 OTTO CYCLE

4.5.1 Air Standard Otto Cycle Analysis

We will start by discussing the Otto cycle engine. As previously mentioned, the spark ignition (SI) engine is modeled after the Otto cycle. Figure 4.6 shows the parts and the strokes of a 4-stroke SI engine (see page 86).

The fundamental part of this engine is a piston inside of a cylinder as shown in Figures 4.6, 4.7, and 4.8. The piston reciprocates in the cylinder between bottom dead center (BDC) and top dead center (TDC). A **stroke** is a one-way movement of the piston from TDC to BDC or from BDC to TDC. A **connecting rod** runs between the piston and a **crankshaft** and transmits force to or from the crank. The **crank** produces rotational motion, and it takes 180° of rotation to produce one stroke.

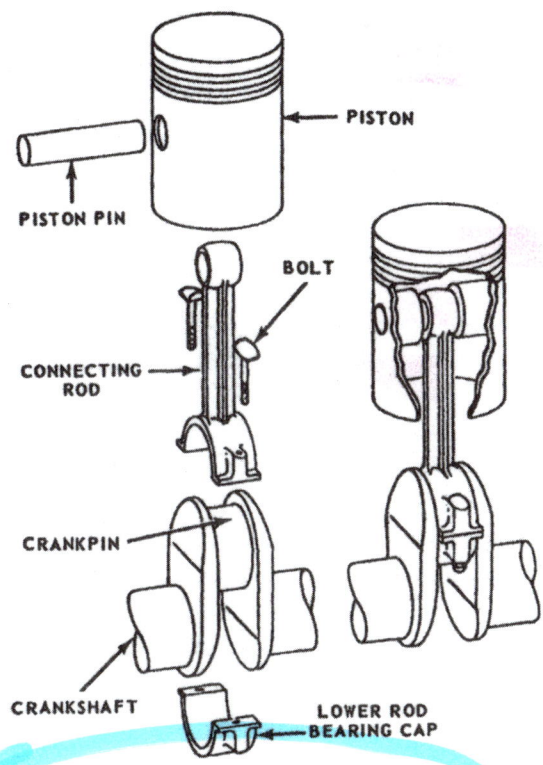

FIGURE 4.7: Piston to Crankshaft Assembly

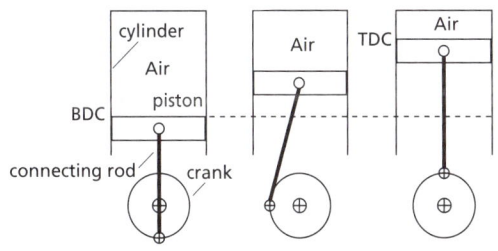

FIGURE 4.8: Piston Motion inside the Cylinder, Illustrating One Stroke

Smaller engines, such as those used for lawn mowers, chain saws, and other yard care applications, typically have one cylinder. Engines that are designed to produce larger amounts of power have multiple pistons connected to the same crank and typically phased, also referred to as "timed," so that there are always one or more pistons putting power into the crank while other pistons are going through other portions of the cycle. Lawn tractors and

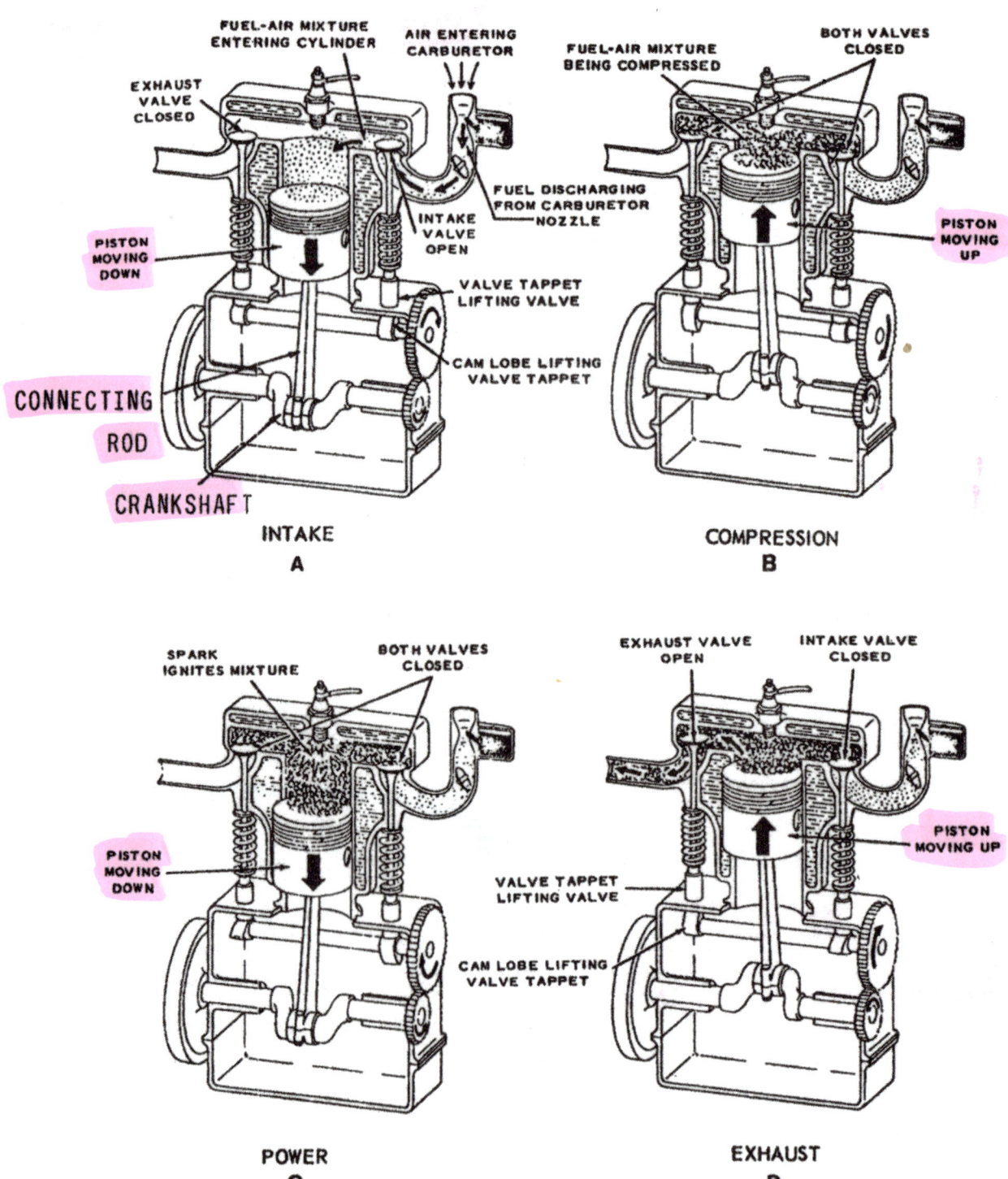

FIGURE 4.6: *4-Stroke Cycle of a Spark Ignition Engine*

some motorcycles have 2-cylinder engines. Most motor cycles have 4-cylinder engines. Most automobiles and pick-up trucks use either a 4-, 6- or 8-cylinder engine, but there are a few models that use 3- or 5-cylinder engines and a number of models use 10- or 12-cylinder engines.

Spark ignition engines may be constructed to operate either as a 2-stroke or a 4-stroke engine. Four-stroke engines tend to be less polluting than 2-stroke engines, so larger SI engines are required to be 4-stroke engines by regulation. (2-stroke SI engines are limited by regulation to smaller engines, such as are found on chain saws, weed-eaters, and similar size tools.) These engines typically have two valves—an intake valve that opens to admit a fresh charge of air and an exhaust valve that opens to allow the exhaust to exit the cylinder. Intake and exhaust are aided by a piston stroke—the exhaust stroke by pushing the burned fuel-air mixture out of the cylinder via the exhaust valve and the intake stroke by drawing fresh fuel-air mixture into the cylinder via the open intake valve. Figure 4.9 shows a configuration of the mechanism that times the opening and closing of the intake and exhaust valves. Note that most SI engines mix the fuel with the air before it is drawn into the cylinder, so that the intake process draws in the mixture.

We'll cover combustion later in this unit, but most students will realize that the energy to run an internal combustion engine comes from converting the chemical energy that was bound up in the fuel into thermal energy that is heat. The engine then converts the heat energy to mechanical energy in the form of specific work (w), total work (W), or the time-dependent quantity, power (P or $\dot{W}$). In air standard analysis, the chemical changes in the air due to burning the fuel—which is an oxidation reaction that consumes oxygen from the air—are ignored. We assume that "the system" remains air. While this is not completely accurate, the analysis technique is still a valid predictor of engine performance.

Spark ignition engines operate as closed systems for several processes in the cycle. The basic rule for closed systems is to sketch "the system" at each state point in the cycle. This is similar in concept to "freeze frame photography" where each sketch captures the state of the system when it transitions from one process to another. Processes occur between the state points. Recall that in "air standard" analysis we assume that the system is air and that it behaves as an ideal gas. The location of the piston determines the volume of the system.

By definition, a "thermodynamic cycle" returns the working fluid to its original state whereupon the cycle can be repeated. As stated in Unit 1, the state of the working fluid is changed through thermodynamic processes. As also introduced in Unit 1, it is often useful to graphically display processes. The state points on process graphs correspond to conditions in the system at the point of change from one process to another. The two concept graphs that we will use for Otto and diesel engines are p-v and T-s. The

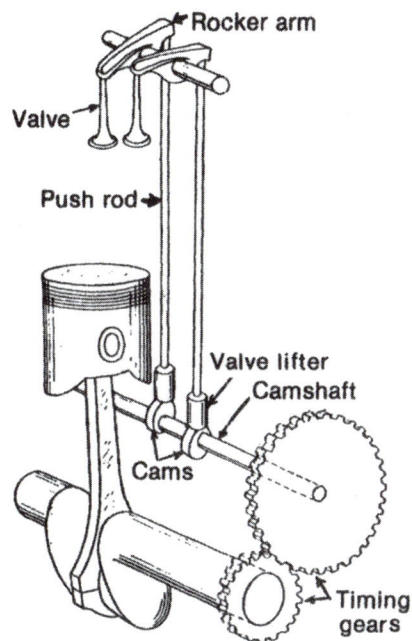

FIGURE 4.9: Example Valve Timing and Actuation Mechanism

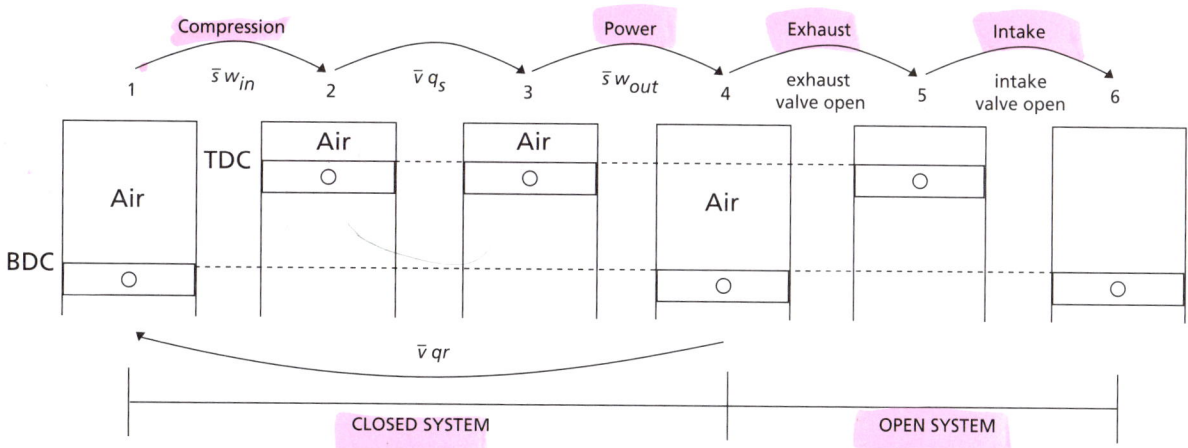

FIGURE 4.10: 4-Stroke Otto Cycle Simplified Schematic

area under a curve on a *p-v* graph indicates work (known as boundary work) and the area under the *T-s* graph indicates heat. Students are reminded that the enclosed area of both graphs represents w_{net}, since w_{net} may be calculated as the difference between q_s and q_r or w_{out} and w_{in}.

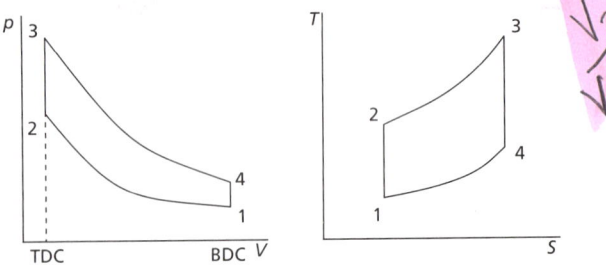

FIGURE 4.11: *Ideal, Air Standard Otto Cycle p-v and T-s Process Graphs*

Notice that Figure 4.10 illustrates 4-strokes, and yet the Figure 4.11 ideal process graphs only illustrate two-strokes. The open processes of exhaust and intake are not shown in the air standard analysis, but the restoration part of the cycle, the SP4 to SP1 process, is modeled as isometric heat rejection. So keeping track of these 2-strokes will be accounted for later. In reality, the amount of power consumed by these 2-strokes reduces the power output of the engine by about 5 percent. In air standard analysis, the exhaust and intake strokes will be assumed to be "free" strokes and their power consumption ignored. When we transition to looking at real engine performance, these free strokes will be accounted for in the mechanical efficiency of the engine.

The *p-V* graph is often referred to as an "indicator graph," because early engineers developed a device that mechanically tracked piston position vs pressure through the cycle. This device produced a graphical output called an "indicator card." As mentioned above, the area inside the cycle tracing on the indicator card indicated the amount of work that the engine should be capable of producing. Today's indicator cards are produced electronically.

The **compression ratio** is the ratio of the volume when the system becomes closed. (When the intake valve shuts before compression. In some engines, the intake valve does not close at BDC. However, unless additional information is given, assume the valve closure event occurs at BDC.) Therefore

$$r_v = v_{BDC}/v_{TDC} = V_{BDC}/V_{TDC} \qquad (4.14)$$

4.5.2 Ideal, Air Standard Otto Analysis, Constant C Approach, State Point Table Technique

Ideal connotes that the compression and expansion processes will be analyzed as isentropic. In reality, the turbulence inside the cylinder will not allow the entropy to remain the same during these processes.

Air standard means that the working fluid is analyzed as only air and that it that does not change its chemical composition. We know that, in reality, the engine draws a fuel/air mixture into the cylinder and the combustion of the fuel consumes oxygen and produces carbon dioxide and water vapor.

Constant C Approach means that the specific heat capacities (C_p and C_v) of the air and the related terms, R and k, do not change. This means that the mathematics remain in the realm of algebra rather than becoming a differential equation. However, it is important to at least recognize that these terms are temperature dependent.

These three simplifications come at a cost of accuracy in the analysis. Nonetheless, the usefulness of the ideal, air standard constant specific heat methodology is that it predicts trends and can be used to compare engines. And it can be adapted to the analysis of real engines without too much of a stretch.

A *state point table* is an excellent way to keep track of the properties of the working fluid (air) throughout the cycle. In the state point table, each column represents a state point in the cycle and each row is reserved for a critical property. For the ideal Otto cycle, four columns are required for the four state points and minimum number of rows to use in are *p*, *T* and *v*. Notice that these are the principal variables in the ideal gas law, $pv = RT$. Notice also that total volume, V, may be substituted for v (or included as one more row), if the $pV = mRT$ form of the ideal gas law is used.

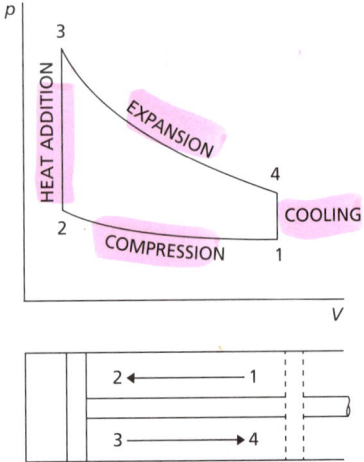

FIGURE 4.12: *Ideal Otto Cycle p-V Graph Relating Processes and Piston Locations*

A blank state point table is shown as Table 4.1. Note that additional staggered rows are shown to indicate processes or heat/work interactions that occur between state points. These can be useful in completing the state point table or using it to analyze cycle performance. Not all cells need to be completed and some do not apply. However, the *expanded* state point table is very valuable to students studying this material for the first time in helping to organize data and illustrate essential concepts. The cycle is completed by realizing that the restoration process occurs between state point 4 and state point 1.

TABLE 4.1: *Ideal Otto Cycle Expanded State Point Table*

	1	2	3	4
p (psia)				
T (°R)				
v (ft³/lb$_m$)				
r_v				
process				
q (Btu/lb$_m$)				
w (Btu/lb$_m$)				

EXAMPLE 4.3: IDEAL, AIR STANDARD OTTO ANALYSIS, CONSTANT SPECIFIC HEAT APPROACH

GIVEN: An ideal, air standard Otto cycle engine receives air at 12 psia and 70 °F (if this information isn't given then assume standard atmospheric conditions—14.7 psia and 70 °F). The engine's compression ratio is 8:1 and 200 Btu/lb$_m$ of heat is added to the air during the combustion process.

FIND: The thermal efficiency (h_{th})

1. Read the problem and associate variables with the given information.
2. Identify the cycle and whether it is using air or some other gas.
3. Draw the schematics (see Figure 4.10). Since we are assuming a closed system with a closed system heat removal process postulated from SP4 to SP1, you can leave out SP5 and SP6 in this solution.
4. Draw the p-v and T-s process graphs (see Figure 4.11).
5. Fill in the "givens" into the state point table and circle the starting point information.

Note: I've bolded the givens in the table.
Note also that I use an expanded state point table to help organize the data and results.

The process identification comes from the p-v and T-s graphs. All processes are either isentropic or isometric. Use this intelligence to select the appropriate equations from the ideal gas relationships shown on the course standard equation sheets.

List the starting equations used under the state point table using the appropriate state point numbers in the equation; algebraically solve for the desired variable; substitute values and units; consider the need for conversion factors based upon the units; and then solve for the numerical answer and units.

At room temperature, $R = 53.3$ ft-lb$_f$/lb$_m$-°R, $C_v = 0.171$ Btu/lb$_m$-°R, and $k = 1.4$.

State Point 1
$p_1 v_1 = RT_1 \rightarrow v_1 = RT_1/p_1 = (53.3 \text{ ft-lb}_f/\text{lb}_m\text{-°R})$
$(530 \text{ °R}) / (12.0 \text{ lb}_f/\text{in}^2)(144 \text{ in}^2/\text{ft}^2) = 16.34 \text{ ft}^3/\text{lb}_m$

State Point 2
$r_v = v_1/v_2 \rightarrow v_2 = v_1/r_v = (16.34 \text{ ft}^3/\text{lb}_m)/8 = 2.04 \text{ ft}^3/\text{lb}_m$
$T_2/T_1 = (v_1/v_2)^{k-1} \rightarrow T_2 = T_1 (r_v)^{k-1} = 530 \text{ °R} (8)^{0.4} = 1218 \text{ °R}$
$p_2 v_2 = RT_2 \rightarrow p_2 = RT_2/v_2 = (53.3 \text{ ft-lb}_f/\text{lb}_m\text{-°R})$
$(1218 \text{ °R})/(2.04 \text{ ft}^3/\text{lb}_m)(144 \text{ in}^2/\text{ft}^2) = 221.0 \text{ lb}_f/\text{in}^2$

	1	2	3	4	
p (psia)	**12.0**	221.0	433.2	23.5	
T (°R)	**530**	1218	2388	1039	
v (ft³/lb$_m$)	16.34	2.04	2.04	16.34	
r_v		**8**	N/A	1/8	NA
process		Isentropic	Isometric	Isentropic	Isometric	$w_{net}\downarrow$
q (Btu/lb$_m$)		NA	**+200**	NA	-87.0	$\Sigma = 113.0$
w (Btu/lb$_m$)		-117.6	NA	+230.7	NA	$\Sigma = 113.1$

State Point 3

$v_3 = v_2 = 2.04 \text{ ft}^3/\text{lb}_m$

$q_{23} = C_v (T_3 - T_2) \rightarrow T_3 = (q_{23}/C_v) + T_2 = (200 \text{ Btu/lb}_m / 0.171 \text{ Btu/lb}_m\text{-°R}) + 1218\text{°R} = 2387.6\text{°R}$

$p_3 v_3 = RT_3 \rightarrow p_3 = RT_3/v_3 = (53.3 \text{ ft-lb}_f/\text{lb}_m\text{-°R})(2387.6\text{°R}) / (2.04 \text{ ft}^3/\text{lb}_m)(144 \text{ in}^2/\text{ft}^2) = 433.2 \text{ lb}_f/\text{in}^2$

State Point 4

$v_4 = v_1 = 16.34 \text{ ft}^3/\text{lb}_m$

$T_2/T_1 = (v_1/v_2)^{k-1} \rightarrow T_4 = T_3 (v_3/v_4)^{k-1} = T_3 (1/r_v)^{k-1} = 2387.6\text{°R} (1/8)^{0.4} = 1039.3\text{°R}$

$p_4 v_4 = RT_4 \rightarrow p_4 = RT_4/v_4 = (53.3 \text{ ft-lb}_f/\text{lb}_m\text{-°R})(1039.3\text{°R}) / (16.34 \text{ ft}^3/\text{lb}_m)(144 \text{ in}^2/\text{ft}^2) = 23.5 \text{ lb}_f/\text{in}^2$

Now for the cells in the rest of the expanded rows. The order of the temperatures is determined by the NFEE ($u_1 - u_2 = w_{12}$ and $u_2 - u_1 = q_{12}$) adapted to the two state points being analyzed. The "NA" cells were determined based upon the vertical processes on the *p-v* and *T-s* concept graphs. Recall that the area under a *p-v* graph process for a closed system indicates boundary work. No area, then no work for processes 2-3 and 4-1. Also, the area under a *T-s* graph process (regardless of whether it is an open or closed system) indicates heat.

$w_{12} = C_v (T_1 - T_2) = 0.171 \text{ Btu/lb}_m\text{-°R} (530 - 1218)\text{°R}$
$= -117.6 \text{ Btu/lb}_m$

$w_{34} = C_v (T_3 - T_4) = 0.171 \text{ Btu/lb}_m\text{-°R} (2388 - 1039)\text{°R} = +230.7 \text{ Btu/lb}_m$

$q_{41} = C_v (T_1 - T_4) = 0.171 \text{ Btu/lb}_m\text{-°R} (530 - 1039)\text{°R}$
$= -87.0 \text{ Btu/lb}_m$

I added the extra column to illustrate the w_{net} calculation.

$w_{net} = \Sigma q \text{ or } \Sigma w$

$w_{net} = \Sigma q = +200.0 - 87.0 = 113.0 \text{ Btu/lb}_m$

$w_{net} = \Sigma w = +230.7 - 117.6 = 113.1 \text{ Btu/lb}_m$.

The difference is round-off error and is insignificant.

$$\boxed{\eta_{th} = w_{net} / q_s = 113 / 200 = 56.5\%}$$

4.5.3 Real Otto Cycle Engines

Displacement is the volume that is swept by a piston between BDC and TDC.

$$V_{displacement} = V_{BDC} - V_{TDC} \qquad (4.15)$$

When displacement is listed for an engine, it is for the entire engine with all its cylinders. It is often convenient to solve for performance based upon a single cylinder. To determine the displacement for one cylinder take the engine displacement and divide it by the number of cylinders ($n_{cylinders}$) in the engine.

$$V_{cylinder\ displacement} = V_{engine\ displacement} / n_{cylinders} \qquad (4.16)$$

Clearance volume ($V_{clearance}$) is the volume when the piston is at TDC (V_{TDC}). Given the engine displacement, number of cylinders, and recalling that the compression ratio is $r_v = V_{BDC} / V_{TDC}$, students should be able to solve for the V_{BDC} and V_{TDC}.

The displacement of an engine is determined by the volume that is swept by the piston times the number of cylinders in the engine, $V_{engine\ displacement} = (V_{BDC} - V_{TDC}) n_{cylinders}$. Thus, a 302 cubic inch displacement (cid) V-8 engine has 37.75 cid per cylinder. The cylinder displacement can be measured by using the cylinder bore (diameter) to determine the cross-sectional area and multiplying it by the stroke length of the piston.

$$V_{cylinder\ displacement} = (\pi D^2_{bore} / 4) L_{stroke} \qquad (4.17)$$

Recalling that power is related to the mass flow rate of the fluid, we can determine the mass flow rate of air by using the ideal gas law in the $pV = mRT$ form and determining the mass of air that enters the cylinder on the intake stroke. The best estimate of the system pressure will be the intake manifold pressure, which is the pressure available just outside the intake valve when it opens. The real cylinder pressure will be slightly lower than this value, but the in-cylinder pressure is not typically measured while the intake manifold pressure is. An alternate term for the intake manifold pressure is manifold absolute pressure (MAP), which is why the pressure sensor that provides a signal to the engine control system's computer is called a MAP sensor.

$$m = p_{intake\ manifold} V_{displacement} / RT \qquad (4.18)$$

This will yield the ideal air mass charged per intake stroke. Now considering that every other revolution on a 4-stroke engine yields an intake stroke, the mass flow rate can be determined. For example, if the 302 V-8 engine, above, is turning at 3,600 RPM, then there are 1,800 intake strokes per minute for each cylinder, or thirty intake strokes per second for each cylinder of this engine. If the MAP sensor reads 10 psia and the air temperature is 70 °F, then adapting the ideal gas equation to yield the air mass flow rate ($\dot{m}_{air}$) is:

$$\dot{m}_{air} = \frac{p_{intake}\dot{V}_{displacement}}{RT} = \frac{\left(10\frac{lb_f}{in^2}\right)\left(37.75\frac{in^3}{cylinder\,stroke}\right)\left(30\frac{cylinder\,intake\,strokes}{second}\right)\left(\frac{1\,ft}{12\,in}\right)}{\left(53.3\frac{ft-lb_f}{lb_m-°R}\right)(530°R)}$$

$$\dot{m}_{air} = 0.033\frac{lb_{m\,air}}{cylinder\text{-}sec} = 0.264\frac{lb_{m\,air}}{engine\text{-}sec}$$

This truly is an estimate of air flow rate but takes into account the pressure loss across the air intake system. Note that the measurement of air pressure in this scenario was outside the cylinder and that the air has to flow through the intake manifold and the intake valve opening to get into the cylinder. It is the mass that actually makes it into the cylinder that matters. Recalling that a fluid moving through or around physical shapes involves friction, friction is associated with head loss, and head loss is a pressure loss, the actual air charge into the cylinder will be less than predicted above. The pressure in the cylinder would be lower and the air would be heated as it traveled through the hot intake port and around the cylinder walls that are hot from the preceding combustion events.

Actual mass flow of air into the engine can be corrected by including the *volumetric efficiency* (η_ϖ) into the equation. Volumetric efficiency is a mass ratio of what air actually made it into the engine versus what could have made it in under static ambient conditions. This denominator term can be measured (1) based upon ambient atmospheric conditions or (2) based upon the air available at the intake manifold conditions. The η_ϖ of the scenario above is about 68% (10 psia/14.7 psia). If we look at the follow-on flow, the typical valve port restrictions will give an additional volumetric efficiency factor typically of about 85% to 90% (a reduction in mass flow of 10% to 15%).

Volumetric efficiency can be boosted by reducing obstructions (increasing the intake valve opening stroke with a larger cam, increasing the intake valve port diameter or putting in two intake valves, machining the flow path to be smoother, etc.); forced induction (turbo-charg-

ing, supercharging, ram air induction); or by tuning the intake or exhaust to take advantage of pressure fluctuations (opening the intake valve near the peak of a pressure fluctuation or opening the exhaust valve near the bottom of the pressure wave). The exhaust tuning may not at first be very obvious, but pressure fluctuations occur similarly to an organ pipe. The low pressure dip can actually be below atmospheric pressure and help to draw out the exhaust from the cylinder. This then allows more fresh air to be admitted on the intake stroke.

Combustion efficiency (η_{comb}), graphically shown in Figure 4.13, relates the heat released from burning fuel that is transferred to the air inside the cylinder.

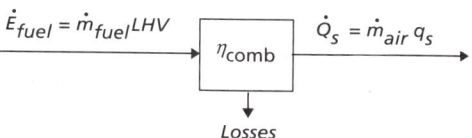

FIGURE 4.13: *Combustion Efficiency*

Typical combustion efficiency for IC engines is between 95 percent and 99 percent. We'll use 97 percent for this illustration. Now thinking about engine fuel consumption and how this factors into this analysis, consider the 302 V-8 engine already introduced above that pushes a vehicle at 80 mph while spinning at 3,600 RPM, but getting only 16 mpg in the process. In one hour, the vehicle traveled 80 miles and took 5 gallons to do it. The lower heating value of gasoline is about 18,600 Btu/lb_m. Gasoline has a specific gravity of about 0.75. This converts to a heating value of about 116,000 Btu/gallon. The energy consumption for this run was 161 Btu/sec, but also including the combustion efficiency of 97 percent means that 156 Btu/sec was transferred from the fuel to the air in the engine. Recall that there are 30 intake strokes per second for each of the eight cylinders of this engine at 3600 RPM, so each cylinder receives 19.5 Btu/sec. Dividing this by the mass flow rate of air per cylinder produces a heat supplied to the air (q_s) of 591 Btu/lb_m. The compression ratio is 8 for this engine. Doing the ideal, air standard method produces the following state point table:

	1	2	3	4
p (psia)	10.0	183.8	704.6	38.3
T (°R)	530	1218	4668	2032
v (ft³/lb_m)	19.62	2.45	2.45	19.62

The thermal efficiency is 56.5 percent and the net work is 333.2 Btu/$lb_{m\,air}$ per cylinder. Calculating the indicated horsepower (*IHP*):

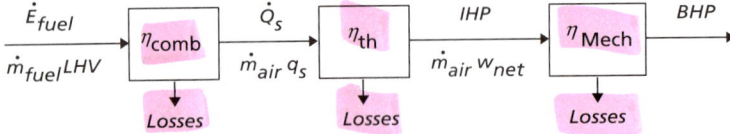

FIGURE 4.14: *Engine Overall Efficiency*

$P_{indicated}$ = IHP = (0.033 lb$_{m\,air}$/sec-cyl)
   (333.2 Btu/lb$_{m\,air}$ per cylinder)
   = 11 Btu/sec-cyl

IHP = (11 Btu/sec-cyl) (1 HP-sec/0.707Btu)
   (8 cylinders) = 125 HP

The overall efficiency for the engine is conceptually shown in the Figure 4.14 conversion sequence:

$$\eta_{engine.overall} = (\eta_{combustion})(\eta_{thermal})(\eta_{engine\,mechanical}) \quad (4.19)$$

This equation applies to the <u>engine</u> and must be modified if additional "downstream" components, such as transmission/reduction gears or electrical generator are connected to the engine. It is important to note that engine overall efficiency is measured at the output coupling of an engine. This is the location where engine brake power (*BHP*) is defined.

Besides sending mechanical power to the propulsion train or to spin an electrical generator, engines have many parasitic and auxiliary loads that consume power and therefore reduce their overall efficiency. The valves that allow intake air into the cylinders, and the exhaust valves that allow exhaust products out, are typically indirectly driven by the engine's crankshaft. Typical engine-driven auxiliaries include lube oil pump, cooling water pump, mechanical fuel pump, and diesel engine fuel injectors. (Most fuel injectors for motor vehicle gasoline engines are electrically opened valves that spray gasoline into a throttle body or the intake manifold. The gasoline is pressurized by an electrically powered fuel pump in the fuel tank. Electrical loads are then transferred indirectly to the alternator.) For motor vehicle and some marine vessels, the engine also runs an alternator for electric generation and a compressor for air conditioning or refrigeration. Motor vehicles may also have power steering and brake pumps that place mechanical loads on the engine. (Note: A trend in some engine configurations today is to remove these parasitic loads and have them driven by separate engines, electrical motors, or hydraulic means.) The engine's mechanical efficiency is reduced by friction and the parasitic/auxiliary loads, as well as the energy required to perform the exhaust and intake strokes.

Remember that the compression stroke is included in the thermal efficiency analysis. And when the transmission efficiency, differential efficiency, and tire efficiency are multiplied by the engine's overall efficiency, the typical result for overall motor vehicle propulsion efficiency is less than 20 percent.

Another useful measure of engine performance is **brake specific fuel consumption** (*bsfc*), which is the mass flow rate of fuel divided by the brake horsepower.

$$bsfc = \frac{\dot{m}_{fuel}}{BHP} \quad (4.20)$$

This can be visualized in Figure 4.14 and represents the rate of fuel needed to develop 1 horsepower measured at the engine's output coupling, which is where brake horsepower is measured. Note that that measurement location is "upstream" of the transmission (aka reduction gears).

4.5.4 Two-stroke Otto Engines

Students should be aware of the implications of 2-stroke gasoline engines. HowStuffWorks has an excellent write-up and graphics for this class of engines. The significant difference in a 2-stroke cycle is that a power stroke occurs in every down-stroke, but power is only produced during a portion of the down-stroke. The remainder of the down-stroke is also used for exhaust and then part of the intake process. Likewise, each upstroke includes both the remainder of the intake process and the entire compression process. Due to the higher pollution levels of these engines, they are restricted to smaller displacements in the United States. Other countries use a much higher proportion of 2-stroke engines due to their simplicity and therefore being less expensive. These countries do not have the environmental regulations that exist in the United States.

4.6 DIESEL ANALYSIS

Students should study the "How Car Engines Work" article on HowStuffWorks to see some excellent graphics in motion that help to illustrate what happens in one of these engines.

4.6.1 Ideal, Air Standard Diesel Constant C Method

Diesel engines are compression ignition (CI) engines. The most significant difference between the SI and CI engines is that fuel and air are drawn into the cylinder together on a SI engine. Diesel engines draw in air only during the intake stroke. They have a higher compression ratio and achieve a temperature in the cylinder at the end of compression that is above the auto-ignition temperature of the fuel. Fuel is then sprayed into the cylinder and immediately ignites. Since the timing of the fuel injection lasts longer than the relatively brief explosion that takes place in the SI engine, the piston has time to move. Consequently, the p-v and T-s concept graphs for a diesel appear as shown in Figure 4.15.

Notice that there is area under the 2-3 process on both graphs, indicating that work and heat are both occurring in this portion of the cycle. The 1-2 process is analyzed as with the Otto engine. Likewise, the 4-1 process is handled as with the Otto analysis. The 2-3 process is isobaric, so there are equations shown on the course standard equation sheets for both heat and work. Also, the 3-4 process does not use the inverse of the compression ratio in calculating the temperature at state point 4.

For Diesel cycles there is the fuel cutoff ratio (β), which is the ratio of the volume at fuel cutoff to the volume at top dead center. Note that the fuel cutoff ratio for an Otto cycle would be unity since the heat addition step (2-3) is isometric.

$$\beta = \frac{v_{FCO}}{v_{TDC}} = \frac{v_3}{v_2} \quad (4.21)$$

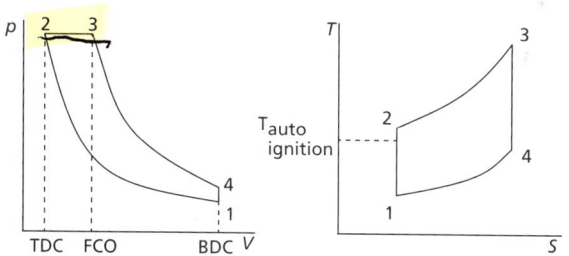

FIGURE 4.15: *Ideal, Air Standard Diesel Process Graphs*

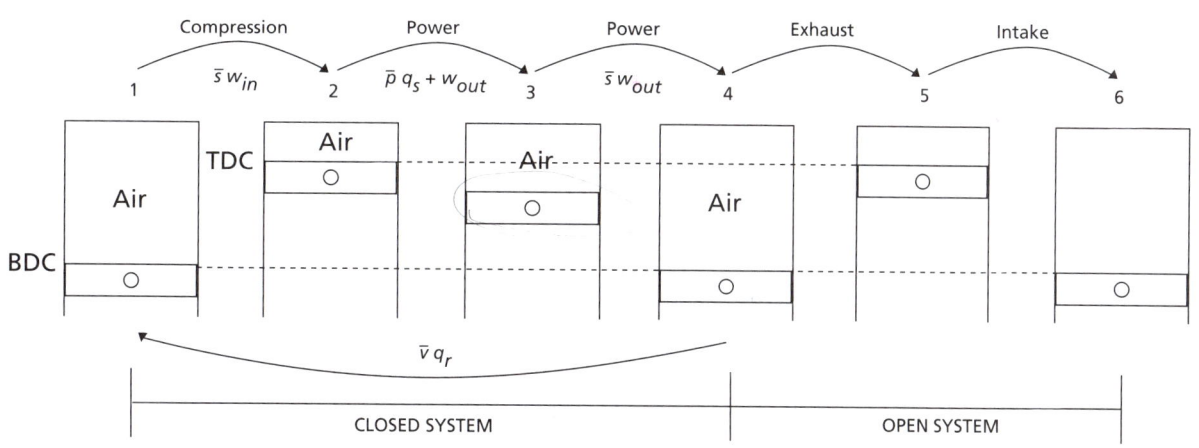

FIGURE 4.16: *Diesel Analysis Schematics*

EXAMPLE 4.4: IDEAL, AIR STANDARD DIESEL ANALYSIS USING CONSTANT C METHODOLOGY

GIVEN: An ideal, air standard Diesel cycle engine receives air at 19 psia and 100 °F (if this information isn't given then assume standard atmospheric conditions—14.7 psia and 70 °F). The engine's compression ratio is 13:1 and 425 Btu/lb$_m$ of heat is added to the air during the combustion process.

FIND: The thermal efficiency (h_{th})

1. Read the problem and associate variables with the given information.
2. Identify the cycle and whether it is using air or some other gas.
3. Draw the schematic (see the figures above for examples).

	1	2	3	4	
p (psia)	**19.0**	688.4	= 688.4	54.9	
T (°R)	**560**	1562	3333	1618	
v (ft³/lb$_m$)	10.9	0.84	1.79	10.9 =	
r_v		13	b = 2.13	NA*	NA
process		Isentropic	Isobaric	Isentropic	Isometric	$w_{net}\downarrow$
q (Btu/lb$_m$)		NA	**+425**	NA	-186.0	Σ = 239.0
w (Btu/lb$_m$)		-176.5	+121.3	+293.3	NA	Σ = 237.8

* Note that r_{v34} is NOT the inverse of the compression ratio (r_{v12}).

4. Draw the p-v and T-s process graphs.
5. Fill in the "givens" into the state point table and circle the starting point information. Note: I've bolded the givens in the table.

Note: I use an expanded state point table to help organize the data and results.

Process identification comes from the ideal Diesel p-v and T-s graphs

List the starting equations used under the state point table using the appropriate state point numbers in the equation; solve for the desired variable; substitute values and units; consider the need for conversion factors based upon the units.

SOLUTION: Air Standard & Constant "c" means R = 53.3 ft-lb$_f$/lb$_m$-°R, k = 1.4, C_p = 0.24 Btu/lb$_m$-°R, C_v = 0.171 Btu/lb$_m$-°R

Here is the sequence of calculations used to complete the state point table. The process related equations are shown on the course standard equation sheet and are selected by looking at the ideal gas relationships under the appropriate process.

State Point 1
$p_1 v_1 = RT_1 \rightarrow v_1 = RT_1/p_1$ = (53.3 ft-lb$_f$/lb$_m$-°R) (560 °R) / (19.0 lb$_f$/in²) (144 in²/ft²) = 10.9 ft³/lb$_m$

State Point 2
$r_v = v_1/v_2 \rightarrow v_2 = v_1/r_v$ = (10.9 ft³/lb$_m$) / 13 = 0.84 ft³/lb$_m$
$T_2/T_1 = (v_1/v_2)^{k-1} \rightarrow T_2 = T_1 (r_v)^{k-1}$ = 560 °R (13)$^{0.4}$ = 1562 °R
$p_2 v_2 = RT_2 \rightarrow p_2 = RT_2/v_2$ = (53.3 ft-lb$_f$/lb$_m$-°R) (1562 °R) / (0.84 ft³/lb$_m$) (144 in²/ft²) = 688.4 lb$_f$/in²

State Point 3
$q_{23} = C_p (T_3 - T_2) \rightarrow T_3 = (q_{23}/C_p) + T_2$ = (425 Btu/lb$_m$ / 0.24 Btu/lb$_m$-°R) + 1562 °R = 3332.8 °R
$p_3 = p_2$ = 688.4 psia
$p_3 v_3 = RT_3 \rightarrow v_3 = RT_3/p_3$ = (53.3 ft-lb$_f$/lb$_m$-°R) (3332.8 °R) / (688.4 lb$_f$/in²) (144 in²/ft²) = 1.79 ft³/lb$_m$

State Point 4
$v_4 = v_1$ = 10.9 ft³/lb$_m$
$T_2/T_1 = (v_1/v_2)^{k-1} \rightarrow T_4 = T_3 (v_3/v_4)^{k-1}$ = 3332.8 °R (1.79 / 10.9)$^{0.4}$ = 1618.0 °R
$p_4 v_4 = RT_4 \rightarrow p_4 = RT_4/v_4$ = (53.3 ft-lb$_f$/lb$_m$-°R) (1618.0 °R) / (10.9 ft³/lb$_m$) (144 in²/ft²) = 54.9 lb$_f$/in²

Now the secondary information can be filled in.
$b = v_3/v_2$ = 1.79 / 0.84 = 2.13
$w_{12} = C_v (T_1 - T_2)$ = 0.171 Btu/lb$_m$-°R (530 − 1562) °R = −176.5 Btu/lb$_m$
$w_{23} = R (T_3 - T_2)$ = 53.3 ft-lb$_f$/lb$_m$-°R (3333 − 1562) °R / (1 Btu / 778 ft-lb$_f$) = +121.3 Btu/lb$_m$
$w_{34} = C_v (T_3 - T_4)$ = 0.171 Btu/lb$_m$-°R (3333 − 1618) °R = +293.3 Btu/lb$_m$
$q_{41} = C_v (T_1 - T_4)$ = 0.171 Btu/lb$_m$-°R (530 − 1618) °R = −186.0 Btu/lb$_m$

I added the extra column to illustrate the w_{net} calculation.
$w_{net} = \Sigma q$ or Σw
$w_{net} = \Sigma q$ = +425.0 − 186.0 = 239.0 Btu/lb$_m$
$w_{net} = \Sigma w$ = − 176.5 + 121.3 + 293.0 = 237.8 Btu/lb$_m$. The difference is round-off error and is insignificant.

$$h_h = w_{net} / q_s = 238.4 / 425 = 56.1\%$$

4.6.2 Two-stroke Diesels

Most of the large diesels used for ship propulsion or as the prime mover of electric generators are 2-stroke diesels. Figure 4.17 illustrates the 4-stroke cycle for comparison.

Two-stroke engines must accomplish the four engine thermodynamic processes in one revolution of the crankshaft. Note that some of the events that we have assumed happen at TDC or BDC are advanced to allow for the

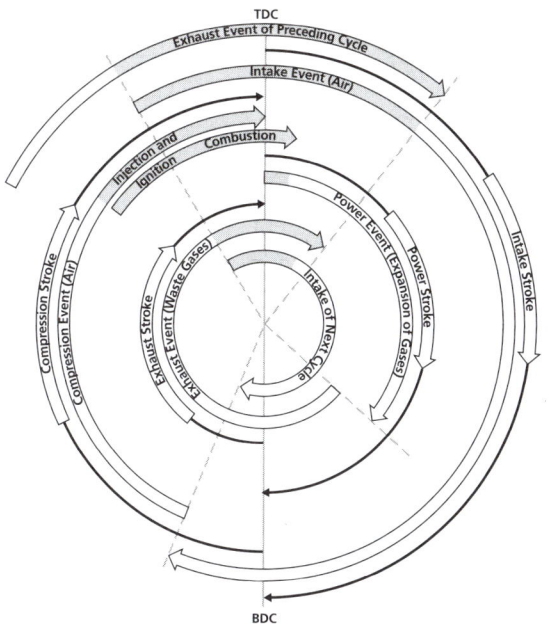

FIGURE 4.17: *The Real 4-Stroke Diesel Cycle*

finite amount of time actually needed for the effects to take place.

Figure 4.18 illustrates the real 2-stroke Diesel cycle. Highlighting the words will help you to see what is occurring.

Figure 4.19 is an alternate way of illustrating the 2-stroke and 4-stroke engine cycles.

Scavenging is the process of blowing air through the cylinder to help push out the exhaust gases. This usually occurs toward the end of the exhaust stroke when both the intake and exhaust ports are open at the same time. The air is forced in by a blower. When the exhaust ports are closed, the blower continues to charge air into the cylinder in a process called *charging*. If the blower is engine driven, then the process is called *supercharging*. If the blower is exhaust driven, then the process is called *turbocharging*.

One configuration of which students should be aware is the opposed piston diesel engine, shown in Figure 4.20, which has two crankshafts and two pistons sharing each cylinder. The pistons are timed to converge towards the center of the cylinder to create the compression process. The stroke of each piston can then be half of what it would take a single piston and crankshaft to create the same compression ratio. Opposed piston engines have inlet and exhaust ports cut into the cylinder walls, because there is no cylinder head for valves. These ports are uncovered at the appropriate point in the cycle by movement of the piston up or down the cylinder. For the same reason, the fuel

injectors are mounted in the cylinder walls in an opposed piston engine.

4.7 OTTO AND DIESEL COMPARISON

There are significant reasons why we use diesels rather than spark ignition engines in ships and other larger vehicles. The most significant are:

- Diesel fuel is much less volatile and therefore less flammable/explosive than gasoline.
- Diesel engines produce better torque output characteristics.
- Diesel engines tend to be more durable and rugged (but they are heavier too).

Diesels are also more efficient due to:

- Higher compression ratio: The higher compression ratio results in higher thermal efficiency.
- Higher volumetric efficiency: The air/fuel system in a spark ignition engine includes a throttle plate that restricts the air flow to the engine to regulate power output. Fuel and air are typically mixed in the air intake system. Diesels do not need a throttle plate. They control power by the amount of fuel injected.
- Higher energy content of the fuel: The higher octane rating of gasoline does not mean more energy content. And the presence of alcohol additives in gasoline (E10 gasoline is 10 percent ethanol) actually reduces the energy content of gasoline.

Figure 4.21 provides a side-by-side comparison of the p-V diagrams for 4-stroke diesel (on the left) and Otto (on the right). These are not to the same scale. Both illustrate naturally aspirated engines (no super- or turbocharging or scavenging). The volume is shown as "normalized" volume (the clearance at TDC is valued at 1. This technique shows the compression ratio on the V axis).

Figure 4.22 shows the power and torque comparison for a diesel and a gasoline engine of identical displacement. These 5.9L engines are used in automotive applications.

Note that the diesel engine produces higher output torque and power at lower rotational speeds.

4.7.1 Forced Induction: Supercharging, Turbocharging, and Scavenging

In a 4-stroke naturally aspirated (non-supercharged or turbocharged) engine, air is pulled into the cylinder in the intake stroke at slightly below atmospheric pressure. Two-stroke engines use scavenging pumps, blowers, or crank-

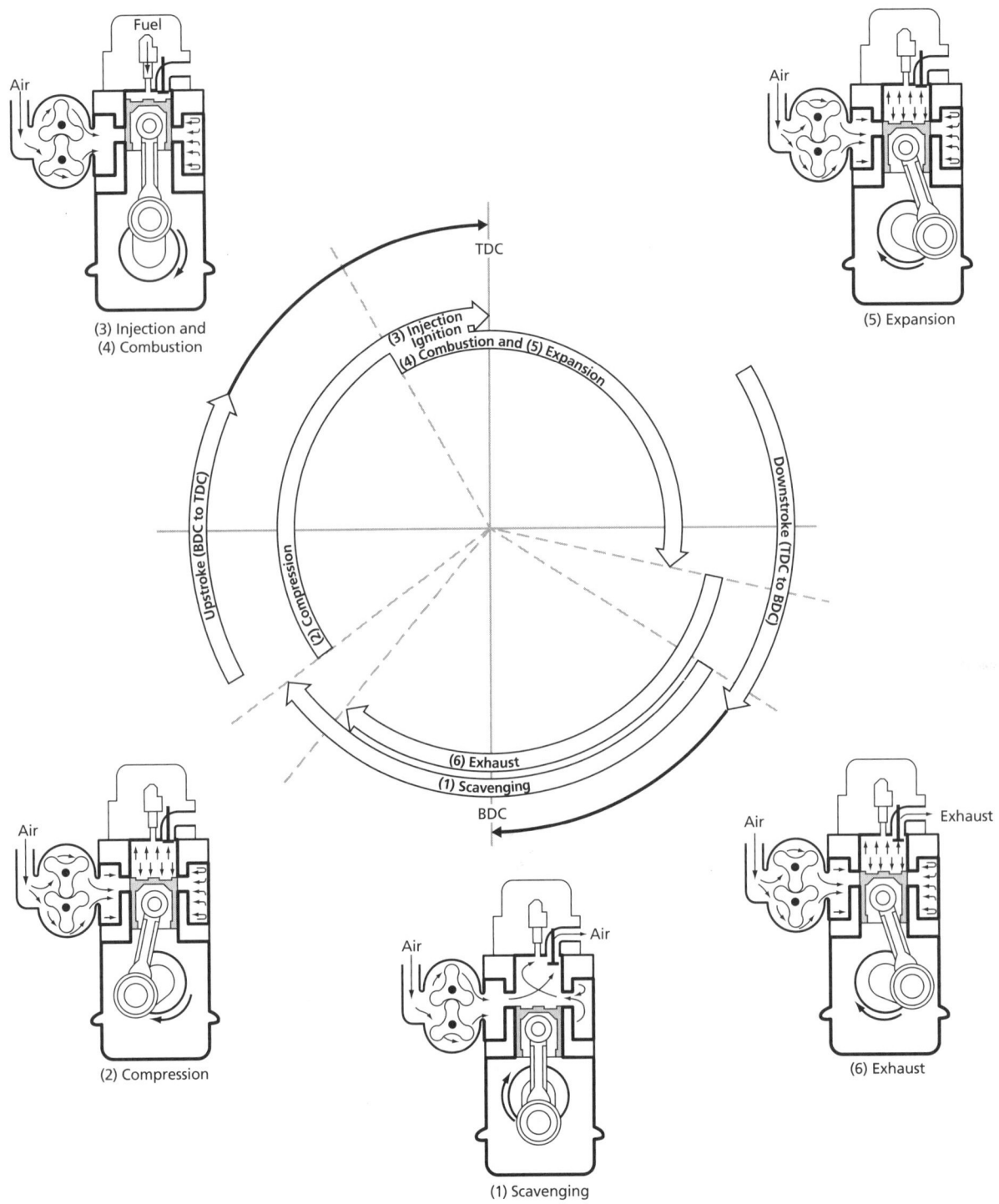

FIGURE 4.18: *The Real 2-Stroke Diesel Cycle*

case pressurization to push air into the cylinder at slightly above atmospheric pressure. In both cases, inlet pressure is typically within about 1 psid of atmospheric pressure.

Superchargers and turbochargers both substantially increase the pressure of incoming air into the cylinder (3-8 psid). Superchargers are driven by a belt off the engine while turbochargers use the flow of exhaust gases through a turbine that drives a compressor. Both use compressors to raise the pressure of the incoming air. The work "in" to run these compressors will be addressed in the next unit.

This increase in pressure is referred to as *boost*. Excessive boost (>8 psid) can cause maximum cylinder pressure to exceed design specifications of the engine. This could lead to catastrophic failure of the engine. In SI (spark ignition) engines, excessive boost can lead to pre-ignition (fuel auto-ignition prior to spark timing), which tends to push the piston down during the compression stroke and can lead to engine failure.

Boost raises the pressure of incoming air into the cylinder, which increases the mass flow rate through the engine (see volumetric efficiency). Increasing the mass flow rate increases the power output of the engine. It also increases the fuel flow rate to maintain fuel to air ratio for proper burning as discussed in the next section. In summary, turbochargers and superchargers have the following impact on engine performance:

- Raised inlet pressure of the engine above atmospheric pressure
- Raised engine power output (and raised fuel usage rate)
- Sometimes improve thermal efficiency
- Sometimes reduce thermal efficiency

Important observations include that forced induction helps to create more power, but will not increase efficiency.

4.8 HYDROCARBON COMBUSTION

The typical fuels of today's engines are hydrocarbons produced from refining petroleum. Petroleum is a veritable soup of hydrocarbons and some other contaminants, most notably sulfur. Other additives are also blended with the petroleum products to modify their storage or performance characteristics.

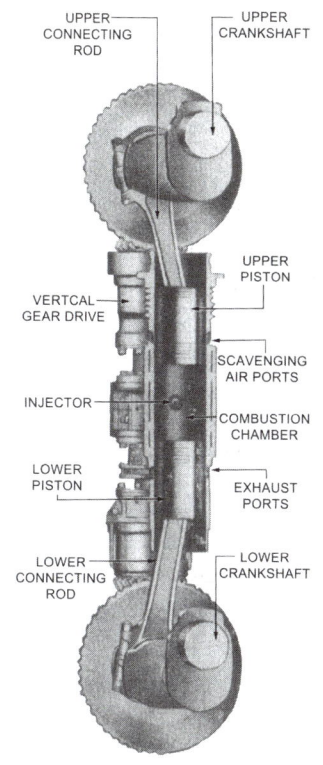

FIGURE 4.20: *Opposed Piston Diesel*

A carbon atom has four sites for other elements to associate to form compounds. These are typically linked to hydrogen atoms or to other carbons. The general chemical equation for a hydrocarbon is C_nH_{2n+2}. Hydrocarbons may form chains or loops or layers of hydrocarbons. Single bond hydrocarbon chains are referred to as alkanes, and the first twelve are listed below.

Name	Formula
Methane	CH_4
Ethane	C_2H_6
Propane	C_3H_8
Butane	C_4H_{10}
Pentane	C_5H_{12}
Hexane	C_6H_{14}
Heptane	C_7H_{16}
Octane	C_8H_{18}
Nonane	C_9H_{20}
Decane	$C_{10}H_{22}$
Undecane	$C_{11}H_{24}$
Dodecane	$C_{12}H_{26}$
...	
Cetane	$C_{16}H_{34}$
...	

Compounds with the same numbers of elements in the same pro-

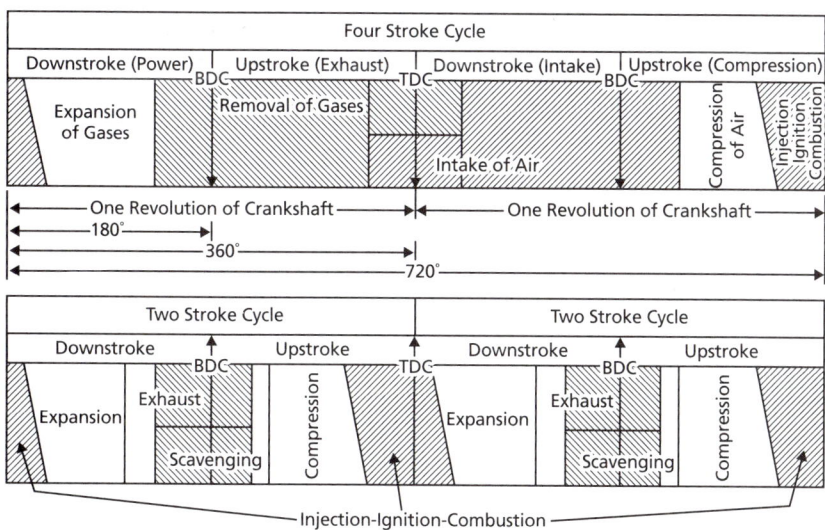

FIGURE 4.19: *Comparison of 4-Stroke and 2-Stroke Cycles*

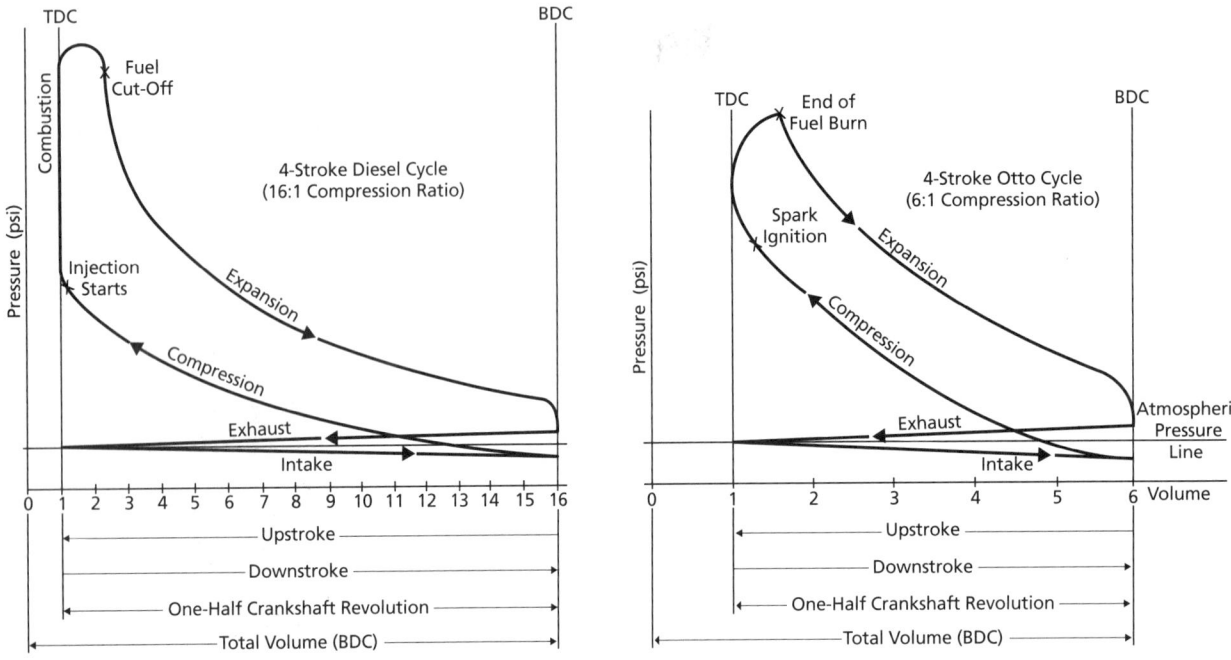

FIGURE 4.21: *Real 4-Stroke Diesel and 4-Stroke Otto p-V Indicator Diagrams*

portions are referred to as isomers. Many hydrocarbons have isomers. The difference between butane (C_4H_{10}) and isobutane (C_4H_{10}) is that isobutane has one of its carbons attached to three other carbons instead of just two. Butane, which is a chain, may also be written as $CH_3CH_2CH_2CH_3$ or $CH(CH_3)_3$. Isobutane is $(CH_3)_2CHCH_3$. This notation indicates that one of the central carbons is linked to three, not two, other carbons. Some excellent graphics on this topic are available at Answers.com.

We will limit the following discussion to the more prevalent linear hydrocarbon (alkane) chains.

C_1 through C_4 are gases at standard room temperature and pressure conditions. Hydrocarbons from C_5 up through C_{18} exist as liquids at standard room temperature and pressure, but C_5 through C_7 are napthas that easily evaporate. Gasoline is made up mostly of hydrocarbons numbering from C_7 to C_{11}. Kerosene exists between gasoline and diesel fuel with hydrocarbons C_{12} through C_{15}. Diesel fuel is predominantly hydrocarbons C_{16} through C_{19}. Chains above C_{19} transition to solid form at standard conditions with paraffin waxes, tars, and lastly asphaltic bitumens (from which asphalt road surfaces are made). The refining process results in blends that typically overlap, so it is possible to find C_{12} in gasoline and C_{20} in diesel fuel.

Data Point: In today's current debate over global climate change, it is worth noting that CH_4 is about twenty-three times more effective as a greenhouse gas than CO_2. By that standard, it is better to capture and combust

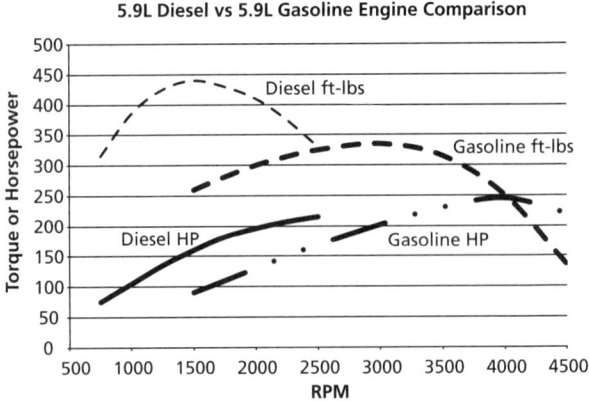

FIGURE 4.22: *Diesel and Gasoline 5.9L Engine Torque and Power Comparison*

methane than to let it escape to the atmosphere. Many students will realize that methane is a product of decomposing organic matter in an oxygen-deprived (anaerobic) environment.

Stoichiometric combustion, aka "ideal" combustion of a hydrocarbon using air, produces carbon dioxide (CO_2) and water (H_2O) and completely uses the oxygen (O_2) available in the air. Nitrogen (N_2) is passed through the engine un-reacted in ideal combustion.

Air is primarily comprised of oxygen and nitrogen with a small amount of carbon dioxide and trace amounts of other gases, such as argon and others. In thermodynamics, air is assumed to be 21 percent O_2 and 79 percent N_2 by volume. This means that there are $79/21 = 3.76$ moles of N_2 per mole of O_2.

The basic approach to balancing the chemical equation is to determine the coefficients for all of the reactants and products based upon the coefficient of the fuel being one (1). Unlike the typical convention in chemistry courses, the coefficients of all other reactants and products do not need to be integers—fractional coefficients are acceptable. The best sequence in evaluating the coefficients is to look at carbon first, then balance the hydrogen, and then determine the oxygen. Finally, the nitrogen is ratioed from the oxygen.

For combustion of butane (C_4H_{10}) the following sequence is used in balancing the equation. Note that the new coefficient is **bolded** in each line.

$C_4H_{10} + O_2 + (3.76) N_2 \rightarrow CO_2 + H_2O + (3.76) N_2$
$C_4H_{10} + O_2 + (3.76) N_2 \rightarrow \mathbf{4}CO_2 + H_2O + (3.76) N_2$
$C_4H_{10} + O_2 + (3.76) N_2 \rightarrow 4CO_2 + \mathbf{5}H_2O + (3.76) N_2$
$C_4H_{10} + \mathbf{6.5}O_2 + (3.76) N_2 \rightarrow 4CO_2 + 5H_2O + (3.76) N_2$
$C_4H_{10} + 6.5O_2 + \mathbf{6.5}(3.76) N_2 \rightarrow 4CO_2 + 5H_2O + \mathbf{6.5}(3.76) N_2$

Notice that the equation is balanced for 1 unit of the fuel and the factors for the various constituents may be left as factors.

Here's how to calculate air-to-fuel mass ratio based upon stoichiometric combustion.

$$\underbrace{C_4H_{10}}_{fuel} + \underbrace{6.5O_2 + 6.5(3.76)N_2}_{air} \rightarrow$$
$$4CO_2 + 5H_2O + 6.5(3.76)N_2$$

$$A/F = \frac{(6.5)(16)(2) + (6.5)(3.76)(14)(2)}{(12)(4) + (1)(10)}$$
$$= \frac{208 + 684.3}{48 + 10} = \frac{892.3}{58} = 15.4$$

Now let's look at what happens to the combustion reaction when we run either air deficient or with excess air. When we run an engine with excess air, we call that running "lean." There will be oxygen left over when all of the hydrocarbons are consumed. Consider combustion of decane with 110 percent of stoichiometric air (10 percent excess air).

The balanced stoichiometry of ideal combustion using the balancing sequence illustrated above is:

$C_{10}H_{22} + 15.5O_2 + 15.5(3.76) N_2 \rightarrow$
$\quad 10CO_2 + 11H_2O + 15.5(3.76) N_2$

Now with 110 percent of air required as compared to the stoichiometric case:

$C_{10}H_{22} + 1.1(15.5) O_2 + 1.1(15.5) (3.76) N_2 \rightarrow$
$\quad 10CO_2 + 11H_2O + 1.1(15.5) (3.76) N_2$
$\quad + 0.1(15.5) O_2$

Notice that the fuel is all consumed, but there was not enough C or H to react with all of the O_2, so 10 percent of the O_2 carries over to the products side of the reaction.

Now for the 10 percent rich case. In this case, we consider the reaction as 10 percent air deficient, not that there is excess fuel. This allows us to continue to use the convention that the coefficient for the fuel is 1.

$C_{10}H_{22} + 0.9(15.5) O_2 + 0.9(15.5) (3.76) N_2 \rightarrow$
$\quad 0.9(10) CO_2 + 0.9(11) H_2O + 0.9(15.5) (3.76) N_2$
$\quad + 0.1C_{10}H_{22}$

Notice that the rich case yields 90 percent of the stoichiometric CO_2 and H_2O and includes unburned fuel in the exhaust.

At this point, the student should be able to compute the air-to-fuel ratio (AFR) for the rich and lean ideal combustion cases.

In reality, additional products of combustion include small amounts of CO and various oxides of nitrogen (NO_x). If sulfur is present in the fuel, as is typically the case in diesel fuel, then various oxides of sulfur (SO_x) are produced. NO_x is associated with smog formation, and SO_x is associated with acid rain. Blood absorbs CO preferentially to O_2, so breathing in an environment with a concentration of CO for any significant time can lead to CO poisoning and death. Lean burning (higher AFR) minimizes CO production as the extra oxygen tends to fully oxidize carbon to CO_2. Lean burning also tends to create more NO_x as the extra air causes nitrogen to oxidize at the temperatures in the cylinder. This is why catalytic converters to remove some of the NO_x and metal oxides in the tailpipe emissions are mandated for automobiles.

The *octane rating* of gasolines is associated with the fuel's resistance to autoignition. Higher octane is more resistant to pre-detonation or "knocking." The *cetane rating* for diesel fuels represents the fuel's ignition delay. Higher cetane is quicker burning, and that is better.

And in addition to the petroleum originated hydrocarbons, fuels can also include additives to preserve the fuel in storage or to facilitate a cleaner combustion. Remember that fuels are refined from a range hydrocarbons and miscellaneous other materials that come out of the ground in crude oil. While we model each of the fuels as predominantly one hydrocarbon, these fuels exist as a blend of hydrocarbons around that number of carbons (C_x). And the refining process can to some degree crack longer chains and recombine shorter ones to skew the refining to the product that the refiners wish to emphasize. This ability to tune the chemical refining process also allows other hydrocarbons to be processed into vehicle fuels. The feedstocks for "biofuels" and other synthetic fuels can be vegetable and seed oils, coal that is processed into liquids, and even cellulose from plant waste.

4.8.1 Heating Values of Fuels

For any hydrocarbon fuel, two heating values exist:

HHV – Higher Heating Value

LHV – Lower Heating Value

FACT: When hydrocarbons are combusted with oxygen, water (H_2O) is produced.

FACT: Since combustion happens at high temperature, the water is produced in vapor form.

For example: burning (combusting) natural gas (methane, CH_4) in oxygen:

$$CH_4 + 2O_2 \rightarrow CO_2 + 2H_2O$$

$$HHV_{Methane} = 23{,}890 \text{ Btu/lb}_m$$

$$LHV_{Methane} = 21{,}518 \text{ Btu/lb}_m$$

The difference in **heating value** is what happens to the **water in the combustion products**.

> Note: All hydrocarbon fuels produce water during combustion and therefore have both *HHV* and *LHV*. However, if you were burning *pure carbon*, then there would be only one heating value listed for the fuel since no water is produced from burning pure carbon.

So, when do you use *HHV* vs *LHV*?

If the engine can operate at a *low enough temperature* to allow the water vapor in the combustion products to condense to a liquid, **then**, in the process of condensing, that water gives up latent heat to the working fluid in the engine where it can be converted by the engine into mechanical work. (Recall that latent heat is associated with phase transformations—in this case gas-to-liquid.) **If the engine receives that extra (latent) heat, then we use the *HHV* of the fuel.**

If the engine operates at such a high temperature that the water vapor in the combustion products is exhausted as a vapor, **then** that water condenses out in the environment and gives up its latent heat to the environment as thermal pollution and this heat is not available to the engine. **If the engine does not receive the extra (latent) heat, then we use the *LHV* of the fuel.**

> Note: This is independent of the *efficiency of combustion* ($\eta_{combustion}$). You might have 100 percent combustion efficiency, but you still need to know which heating value to use when analyzing an engine.

> Note: The focus is on the **water *in the combustion products*** and <u>not</u> the **working fluid** (which in some systems may also be water).

AN EXAMPLE OF WHERE WE DO USE **HHV**. The steam plant component that often allows us to use the *HHV* is the economizer. In a conventional steam engine, combustion happens inside a boiler, but the working fluid (water) is separated from the fire of combustion and the combustion products (which include water formed by the combustion chemical reaction) by the walls of the heat exchanger surfaces (usually tubes). The working fluid (water) enters the boiler (it is called "feedwater" at this point in the system) by way of a heat exchanger (waste energy recovery device) called an economizer. The economizer is located at the base of the exhaust stack, so the exhaust gases pass in contact with the economizer tubes. The incoming feedwater inside those tubes may be cold enough to cool down the boiler exhaust gases to the point where the *combustion product water* can condense and give off its latent heat. Since this latent heat is transferred to the *working fluid water*, *HHV* is used if the combustion product water vapor condenses where it can give up its latent heat to the working fluid, which then converts some of the extra heat input to work.

FOLLOW-UP. The logical question is, "At what temperature in the exhaust does condensation happen?" Recall from the earlier discussions of combustion that we are using air for combustion of the hydrocarbon fuels. Therefore the exhaust is a mixture of H_2O with CO_2, N_2, and other gases, such as CO, NO_x, SO_x, excess O_2, and other pollutants. The calculation for determining the dewpoint temperature involves determining partial pressures. The quick answer for common air-fuel ratios is about 165 °F.

Practice Problems

"Nothing builds self-esteem and self-confidence like accomplishment."
—Thomas Carlyle

4.1 Air in a piston/cylinder device is compressed isentropically to one-fourth its original volume. It is then cooled at constant volume from 400 °F to 175 °F. Sketch both processes on the same T-s and p-v graphs. Calculate the heat (Btu/lb_m) transferred and the work done (ft-lb_f/lb_m). Indicate whether the work and heat are added to or removed from the working fluid. Assume constant specific heats apply.

4.2 Draw the p-v and T-s graphs for an ideal Otto cycle. Identify what happens during each process as follows: (Use only those labels that apply)

1) <u>C</u>ompression, Heat <u>A</u>ddition, <u>E</u>xpansion and Heat <u>R</u>ejection

2) Isentropic, Isometric, Isobaric, Isothermal (shorthand is acceptable)

3) Work in, Work out, Heat in, Heat out (in proper variable form)

Note to students: For subsequent problems, you are expected to sketch the appropriate schematics; draw the p-v and T-s concept graphs; and complete the state point table.

4.3 Given the following partially completed state point table for an ideal, air standard Otto cycle, determine:

a) The compression ratio

b) The work during compression (Btu/lb_m)

c) The heat added during combustion (Btu/lb_m)

d) The work during expansion (Btu/lb_m)

e) The heat rejected into the atmosphere (Btu/lb_m)

f) Net work (Btu/lb_m)

g) Thermodynamic efficiency of the cycle (%)

	1	2	3	4
p (psia)	15	251.9	500.1	29.8
T (°R)	530	1187	2356	1052
v (ft³/lb_m)	13.08	1.74	1.74	13.08
r_v				
process				
q (Btu/lb_m)				
w (Btu/lb_m)				

4.4 Given the following partially completed state point table for an ideal, air standard Otto cycle, determine:

a) The compression ratio

b) The heat added during combustion (Btu/lb_m)

c) The heat rejected into the atmosphere (Btu/lb_m)

d) Thermodynamic efficiency of the cycle (%)

e) Net work (Btu/lb_m)

f) If the compression process uses 121.5 Btu/lb_m, what is the work done during the expansion of the piston? (Btu/lb_m)

	1	2	3	4
p (psia)		294.1	752.1	
T (°R)		1236	3160	
v (ft³/lb_m)	13.22	1.56		
r_v				
process				
q (Btu/lb_m)				
w (Btu/lb_m)				

4.5 An ideal, air standard Otto cycle engine receives air at 12 psia and 70 °F. The maximum cycle temperature is 3460 °F and the compression ratio of the engine is 7.5 to 1. Assume constant specific heats.

a) Complete the state point table.

b) Determine how much heat is added to each pound of air (Btu/lb_m).

c) Determine the net work of the cycle.

4.6 An ideal, air standard Otto cycle initiates the compression process with air at 14.7 psia and 75 °F. 125 Btu/lb_m of work is performed on the air in order to raise the pressure at the end of the compression process to 300 psia. 275 Btu/lb_m of heat is added to the air during the combustion process.

a) Complete the state point table.

b) Determine the compression ratio of the engine.

c) Determine the cycle thermal efficiency (%).

4.7 If an ideal, air standard Otto cycle intake is at standard atmospheric pressure and 75 °F, the temperature at the end of the expansion process is 1354 °R, the work done during compression is 128.8 Btu/lb_m, and net work is 197 Btu/lb_m, find:

a) Complete the state point table.

b) The compression ratio

c) The amount of heat supplied

d) The thermodynamic efficiency of the cycle

e) The power of the cycle (hp) if the mass flow rate is 1.5 lb_m/s

4.8 The working substance in an ideal Otto cycle engine is an ideal gas for which $Cp = 0.219$ Btu/lb_m-°R and $Cv = 0.156$ Btu/lb_m-°R. At the beginning of the compression process the gas pressure is 20 psia and its temperature is 60 °F, and at the end of the compression process the gas pressure is 200 psia. Sufficient heat is injected into the gas to raise the temperature of the gas to 2500 °F at the beginning of the expansion process.

a) What is the compression ratio for this engine?

b) What is the gas temperature at each state point (°R)?

c) What is the net work of this cycle (Btu/lb_m)?

d) What is the thermal efficiency of this cycle (%)?

4.9 Draw the *p-v* and *T-s* graphs for an ideal, air standard Diesel cycle. Identify what happens during each process as follows: (Use only those labels that apply)

1) <u>C</u>ompression, Heat <u>A</u>ddition, <u>E</u>xpansion and Heat <u>R</u>ejection

2) Isentropic, Isometric, Isobaric, Isothermal (shorthand is acceptable)

3) Work in, Work out, Heat in, Heat out (in proper variable form)

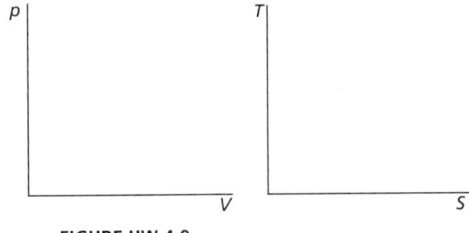

FIGURE HW 4.9

4.10 Given the following partially completed state point table for an ideal, air standard Diesel cycle, determine:

a) The compression ratio

b) The work during compression (Btu/lb_m)

c) The heat added during combustion (Btu/lb_m)

d) The work during expansion (Btu/lb_m)

e) The heat rejected into the atmosphere (Btu/lb_m)

f) Net work (Btu/lb_m)

g) Thermodynamic efficiency of the cycle (%)

h) The fuel cutoff ratio (*b*)

	1	2	3	4
p (psia)	14.5	703.3	703.3	31.4
T (°R)	510	1546	2685	1105
v (ft³/lb_m)	13.02	0.81	1.41	13.02
r_v				
process				
q (Btu/lb_m)				
w (Btu/lb_m)				

4.11 Given the following partially completed state point table for an ideal, air standard Diesel cycle determine:

a) The compression ratio

b) The heat added during combustion (Btu/lb_m)

c) The heat rejected into the atmosphere (Btu/lb_m)

d) Thermodynamic efficiency of the cycle (%)

e) Net work (Btu/lb_m)

f) If the compression process uses 186.0 Btu/lb_m, what is the work done during the expansion of the piston? (Btu/lb_m)

g) The fuel cutoff ratio (b)

	1	2	3	4
p (psia)	14.7		713	
T (°R)				
v (ft³/lb_m)			1.38	13.47
r_v				
process				
q (Btu/lb_m)				
w (Btu/lb_m)				

4.12 An ideal, air standard Diesel cycle engine operating on an air standard cycle receives air at 17 psia and 100 °F. The maximum cycle temperature is 2680 °F and the compression ratio is 16:1.

a) Complete the state point table, draw the schematics, and the p-v and T-s graphs.

b) How much heat is added to the air (Btu/lb_m)?

c) How much work was extracted while the fuel was being injected (Btu/lb_m)?

d) What is the net work of the cycle (Btu/lb_m)?

e) What is the thermal efficiency?

f) What is the fuel cutoff ratio?

4.13 At the beginning of the compression process in an ideal, air standard Diesel cycle the pressure is 15 psia, the temperature is 70 °F, and the air volume is 7 ft³. The compression ratio for this engine is 20:1, the heat supplied is 300 Btu/lb_m and the air mass flow rate is 0.5 lb_m/sec. The measured brake horsepower is 125 hp. Calculate the:

a) Maximum temperature in the cycle (°F)

b) Temperature at the end of the expansion process (°F)

c) Heat rejected (Btu/lb_m)

d) Net work of the cycle (Btu/lb_m)

e) Cycle thermal efficiency (%)

f) Indicated horsepower of the engine

g) Mechanical efficiency of the engine

4.14 Analysis of a Diesel engine operating on an ideal, air standard cycle has produced the following table of properties:

Determine the:

a) Compression ratio

b) Work done during the compression process (Btu/lb_m)

c) Work done during the heat addition process (Btu/lb_m)

d) Work done during the expansion process (Btu/lb_m)

e) Heat added (Btu/lb_m)

f) Heat rejected (Btu/lb_m)

State Point	1	2	3	4
p (psia)	14.7	651	651	47.4
T (°R)	540	1595	3679	1740

4.15 If an ideal, air standard Diesel cycle intake is at atmospheric standard, the compression ratio is 17:1, fuel cutoff ratio (b) is 1.98, and the heat added during combustion is 388.5 Btu/lb$_m$, find:

a) The maximum temperature (°F)

b) The net work

c) The thermodynamic efficiency of the cycle

4.16 A 5.9L V-8 gasoline engine produces 245 hp at 4,000 RPM. The compression ratio is 9:1. Calculate the engine output torque; the total displacement in cubic inches; and volume at TDC and BDC for each cylinder.

4.17 A 350 cid V-8 4-stroke engine with a compression ratio of 9 is operating at 2,400 RPM. The engine is being fueled by gasoline with SG = 0.7 & HV = 18,750 Btu/lb$_m$ and is consuming fuel at the rate of 4 gallons per hour. The manifold absolute pressure (MAP: the pressure available just outside the intake valves) is 4.5 psia. The combustion efficiency is 97 percent. Note: The given values are realistic for this type of engine.

a) What is the mass flow rate of fuel to each cylinder (lb$_m$/sec-cyl)?

b) Estimate the mass flow rate of air to each cylinder (lb$_m$/sec/cyl). Hint: Solve for the displaced volume of one cylinder and use the MAP in the ideal gas equation to solve for mass of air drawn into each cylinder on the intake stroke. Then factor in the engine's total number of intake strokes per second.

c) What is the A/F ratio?

d) What is the energy addition rate as measured by the fuel <u>for each cylinder</u> (Btu/sec)?

e) What is the value of q_s to use in the Otto cycle air standard analysis for a single cylinder (Btu/lb$_m$)?

4.18 Solve problem 4.17 using a volumetric efficiency based upon intake manifold conditions of 85%.

4.19 Determine the stoichiometric A/F ratio for the following fuels:

(assume that one unit of air includes 3.76 moles of N$_2$ per one mole of O$_2$)

a) C_4H_{10}

b) C_8H_{18}

c) $C_{20}H_{42}$

For each of the fuels, if the A/F ratio is 15:1, is the mixture running "rich," "lean," or at the stoichiometric ratio?

4.20 Gasoline is a mixture of hydrocarbons centered on octane (C_8H_{18}) and includes some other naturally occurring trace elements and compounds. It also has some additives from the refining process. There are 3.76 moles of nitrogen per mole of oxygen in standard air. Write the ideal equation for combustion of octane in standard air. Show what happens to the reaction if the combustion is 10% rich and 10% lean. What other products of combustion would you expect to find for real combustion of hydrocarbon fuels?

4.21 Fuels typically have two heating values—higher heating value (*HHV*) and lower heating value (*LHV*). What is the difference between *HHV* and *LHV*? Which common fuel only has one heating value? What conditions are required in order to use *HHV* in the combustion efficiency equation?

5 Gas Turbines

A landing craft air cushion (LCAC), carrying a Marine Corps amphibious assault vehicle, approaches the well deck of a dock landing ship. LCACs are powered by four gas turbine engines.

5.1 UNIT LEARNING OBJECTIVES

5.1.1 Terminology: definitions, variables, and typical units

5.1.1.1 pressure ratio, combustion chamber, rotor, compressor, turbine, nozzle, diffuser, bypass air, bleed air, thrust

5.1.2 Concepts: ideas and engineering expression of them

5.1.2.1 Describe the operation of a gas turbine engine.

5.1.2.2 Discuss the energy conversions in a gas turbine engine.

5.1.2.3 Discuss the advantages and disadvantages of a split-shaft versus a single-shaft gas turbine.

5.1.2.4 Discuss the applications where a gas turbine engine is preferable to other engine configurations.

5.1.2.5 Sketch the ideal and real Brayton cycle on h-s and T-s coordinates.

5.1.3 Skills: procedures, practices, or methods that enable reasoning

5.1.3.1 Evaluate an isentropic process using Air Tables.

5.1.3.2 Calculate gas turbine engine performance using the Variable Specific Heat method and the Air Tables.

5.1.3.3 Calculate ideal and real compressor work.

5.1.3.4 Calculate compressor efficiency.
5.1.3.5 Calculate heat supplied by a combustion chamber.
5.1.3.6 Calculate combustion chamber efficiency.
5.1.3.7 Calculate ideal and real turbine work.
5.1.3.8 Calculate turbine efficiency.
5.1.3.9 Calculate internal/indicated horsepower.
5.1.3.10 Calculate brake horsepower.
5.1.3.11 Calculate gas turbine cycle thermal efficiency.
5.1.3.12 Calculate specific fuel consumption.
5.1.3.13 Calculate inlet and exhaust duct pressure losses.
5.1.3.14 Calculate mechanical efficiency.
5.1.3.15 Calculate overall engine efficiency.
5.1.3.16 Calculate thrust.
5.1.3.17 Calculate aircraft engine diffuser pressure gain.
5.1.3.18 Analyze the effect of a subsonic nozzle or diffuser.
5.1.3.19 Evaluate changes in engine performance as operating conditions change.

5.2 INTRODUCTION

Over the past half century, gas turbine engines have revolutionized the aviation world and have also found wide utility in ship and land vehicle applications. The thermodynamic cycle upon which gas turbines are based is named after George Brayton, who filed for a patent on the principles of this engine in 1872. As an interesting aside, John Ericsson, the designer of the USS *Monitor*, invented an engine working on the same principles years earlier, the Ericsson cycle. However, it took some years before the components could be made efficient enough for the Brayton cycle to actually work. Successful gas turbine engines were developed in the 1930s by two European engineers working independently of each other. Since the first jet-powered aircraft flew in 1939, the performance advantages of gas turbines over reciprocating engines for aviation have been realized in terms of higher speeds and altitudes as well as smoother operation. Continued development of advanced gas turbine concepts will open the door to the next generation of military ships and aircraft, as well as hypersonic transport vehicles.

5.3 FLEET APPLICATIONS

Gas turbines are at the heart of almost all military aircraft engines in current use, whether traditional turbojet for direct thrust, the more recent turbofan thrust engines, or turbomachinery powering propeller craft (turboprop) or helicopters (turboshaft). Additionally, gas turbine engines are used throughout the Navy for many applications beyond aviation. Their high power-to-weight ratio and quick start capability make them well suited for shipboard operation. LM2500 gas turbine engines are used in *Perry*-class frigates, *Arleigh Burke*–class destroyers, and *Ticonderoga*-class guided missile cruisers. Electrical generation on these ships is also accomplished with gas turbine engines. And gas turbines are used to power land vehicles such as the Marine Corps M1A1 Abrams battle tank.

5.4 THEORY DEVELOPMENT AND PROBLEM SOLVING METHODOLOGY

5.4.1 Ideal Brayton Cycle Basics

The core of any gas turbine engine is the *gas generator*. The sole purpose of the gas generator is to produce high energy gases that can be used to do work such as driving a shaft or propelling an aircraft. As shown in Figure 5.2, the gas generator has three basics components, the *compressor*, *combustion chamber*, and *turbine*. Air drawn into the compressor is compressed to higher pressure and

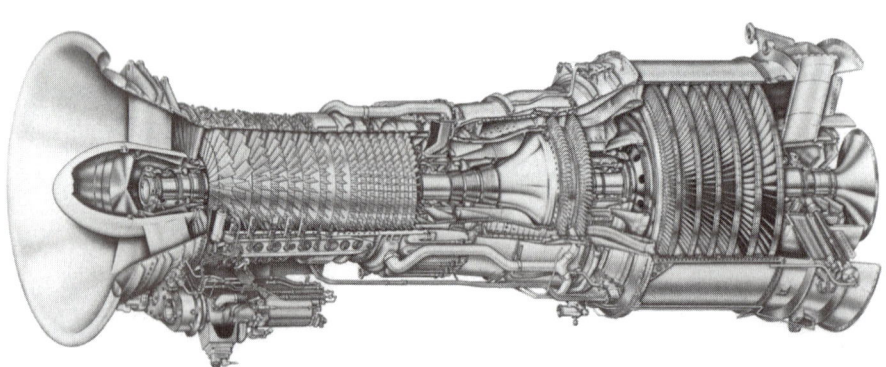

FIGURE 5.1: *The General Electric LM2500 Gas Turbine Engine Is Used in Ship Propulsion for Ships in More Than Twenty-Five Navies of the World.*

forced into the combustion chamber. Here, fuel is injected into the air and burned in continuous combustion, which increases the energy of the gases. Following combustion, the high energy gases expand across multiple turbine stages to produce work, before exiting to the atmosphere in this open cycle. Some of the energy is extracted from the turbine and is used to drive the compressor. The remaining energy can be used for useful work and is the *net work* of the cycle. The ratio of compression work to the expansion work is called the *back work ratio*.

The following schematic shows the components of a typical gas generator. The numbers in the schematic refer to specific SPs of the air and are standard throughout this text. The "s" in "2s" and "4s" indicate processes that are modeled as isentropic. For real processes, the "s" is omitted.

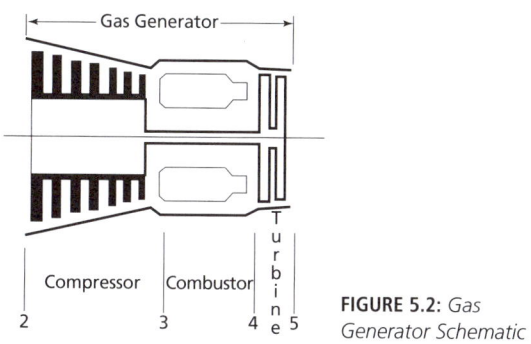

FIGURE 5.2: Gas Generator Schematic

The gas generator cycle can be analyzed more easily with the following simplifying assumptions. (1) Using the "air standard methodology" the working fluid is assumed to be air, which behaves as an ideal gas. As with the other engines that we have thus far studied, air standard methodology essentially ignores the changes in chemical composition that actually occur due to the combustion of the fuel in the working fluid, air. (2) We'll start the analysis by assuming all "work" processes to be isentropic (adiabatic and internally reversible—all components operate at 100 percent efficiency and there are no frictional losses or heat losses). (3) The combustion and exhaust processes are modeled as simple heat addition or heat rejection processes to or from the working fluid. (4) The changes in kinetic and potential energy throughout the cycle are considered negligible. Since we will restrict our analysis of the components to steady state conditions, the entire Brayton cycle can be analyzed using the SFEE.

$$pe_i + ke_i + u_i + p_i v_i + q_{if}$$
$$= pe_f + ke_f + u_f + p_f v_f + w_{if} \quad (5.1)$$

Cancelling the potential and kinetic energy terms and recalling that $h = u + pv$, the simplified version of the SFEE becomes:

$$h_i + q_{if} = h_f + w_{if} \quad (5.2)$$

An appropriate schematic is shown in Figure 5.3.

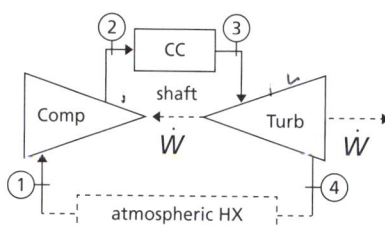

FIGURE 5.3: *Simplified Closed Cycle Gas Turbine Schematic*

State Point 1 to 2s: Process—Isentropic compression. Component—compressor.

The purpose of the compressor is to deliver high pressure air to the combustion chamber. The compressor raises the pressure of the incoming air, p_1, to the higher final pressure, p_2. The ratio of these pressures is called the compressor pressure ratio, r_p.

$$r_p = \frac{p_2}{p_1} \quad (5.3)$$

Since this process is isentropic, no heat is transferred. Thus, modelled as Equation 5.2 becomes

$$h_1 = h_2 + w_{12}$$

or

$$w_{12} = h_1 - h_2 \quad (5.4)$$

where w_{12} is the work done by the compressor. Since $h_2 > h_1$, w_{12} is negative. This is consistent because compressor work is done on the system, the working fluid, air. The enthalpy at SP2 is greater than that of SP1 by the amount of compressor work added to the system. The temperature will be higher even though only work has been added to the air.

State Point 2s to 3: Process—Isobaric Heat Addition. Component—combustion chamber.

Heat is added in the combustion chamber to increase the air's energy before it reaches the turbine. Initial light off is accomplished by igniters, and then a continuous flame is supported by the incoming fuel and air. No work is done in the combustion chamber, even though there is area under the *p-v* graph for this process. Recall that boundary work only applies for closed system analysis. Thus, after replacing the subscripts, Equation 5.2 becomes

$$h_2 + q_{23} = h_3$$

or

$$q_{23} = h_3 - h_2 \quad (5.5)$$

where q_{23} is the heat added to the air, q_s, as we have referred to it previously.

State Point 3 to 4s: Process—Isentropic expansion. Component—turbine.

In this process, the turbine extracts energy from the air leaving the combustion chamber. Like the ideal compressor, the ideal turbine is adiabatic and Equation 5.2 becomes

$$h_3 = h_4 + w_{34}$$

or

$$w_{34} = h_3 - h_4 \quad (5.6)$$

where w_{34} is the work done <u>by</u> the turbine and has a positive sign associated with it.

In a gas generator, all of the turbine work is used to drive the compressor as the two devices are connected by a common shaft. Any excess energy remaining in the working fluid can be used to drive another turbine and thereby produce additional shaft work, or it can be further expanded through a nozzle to produce jet propulsion. Jet thrust will be covered in a subsequent section.

Recall that the ratio of compressor work to expansion work is the back work ratio. This is given by the following relationship:

$$BWR = \left| \frac{w_{in}}{w_{out}} \right| = \left| \frac{h_1 - h_2}{h_3 - h_4} \right| \quad (5.7)$$

State Point 4s to 1: Process—Isobaric heat rejection. Component—a "pretend" heat exchanger.

In the final process, heat is rejected to the environment at constant pressure. No work is done in this process so Equation 5.2 becomes

$$h_4 + q_{41} = h_1$$

or

$$q_{41} = h_4 - h_1 \quad (5.8)$$

where q_{41} is the heat rejected, q_r, of the cycle. Since the engine dumps hot air into the environment, we can postulate a "pretend" heat exchanger that removes the heat from the air and restores it to the entering conditions, thus completing the "cycle." This is reasonable so long as the engine does not suck in its own or another aircraft's exhaust. The thermal efficiency of the Brayton cycle is given by the following relationship:

$$\eta_{th} = \frac{w_{net}}{q_{in}} = \frac{w_{12} + w_{34}}{q_{in}} = \frac{(h_1 - h_2) + (h_3 - h_4)}{h_3 - h_2} \quad (5.9)$$

Figure 5.4 shows the Brayton cycle plotted on *T-s* and *p-v* coordinates. Note that an *h-s* diagram would resemble the *T-s* diagram in shape since enthalpy of a gas is a direct function of temperature. Also note that I'm modeling the best way to draw the *T-s* diagram—start by sketching the lines of constant pressure and then add the "work" lines.

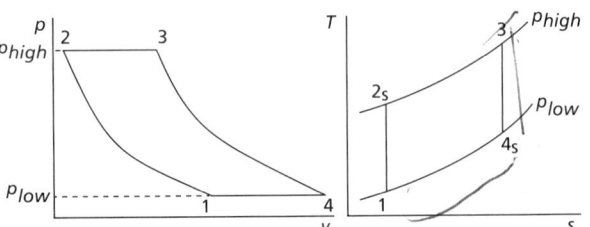

FIGURE 5.4: *The Ideal Brayton Cycle*

5.4.2 Constant C Methodology

Using the "constant *C* methodology," the *w* or *q* of each component in the cycle, including the "pretend" heat exchanger of the environment heat sink, is quantified by a difference in enthalpy, Δh, which may be calculated from the difference in temperature, $\Delta h = C_p \Delta T$. Then the constant *C* ideal gas relationships that have been discussed in the previous unit may be used to estimate the cycle's parameters. For instance, the temperature at the end of isentropic compression may be calculated using the pressure ratio (r_p):

$$T_2 = T_1 (r_p)^{\frac{k-1}{k}}$$

And the isentropic expansion will use the same form of this equation. The isobaric heat addition process will use:

$$q_s = q_{23} = C_p (\Delta T)$$

to determine q_s if T_{max} is known, or T_{max} if q_s is known (or if q_s may be determined from other data). Then once the cycle's temperatures are known, the appropriate *w* or *q* may be calculated using $\Delta h = C_p \Delta T$.

Also, be aware that while the area under the *T-s* curve still represents heat, *q*, the area under the *p-v* curve that represented boundary work when we were studying the closed system (reciprocating) engines, $p\Delta v$, does not apply to the flowing case. To illustrate this, if the combustion chamber is experiencing boundary work, it won't operate for long, because it just exploded.

5.4.3 Air Table Familiarization (Variable C Approach)

In the previous section it was shown that the Brayton cycle could be treated as an air standard cycle and that work and heat interactions can be calculated from state point enthalpies. The enthalpy at various states can be found using either constant specific heat or variable specific heat values and the resulting air temperatures. Assuming constant specific heat provides only a rough estimate of true enthalpy values since specific heat varies with air temperature. A better approach is to take this variation into account by referencing published Air Tables that provide enthalpy values for a given temperature. Problem statements will specify whether to assume constant specific heat or variable specific heat (use Air Tables).

Air tables are organized by temperature and provide air properties such as enthalpy, internal energy and relative pressure (Pr), and specific heat capacity. Table 5.1 shows an Air Table excerpt. The table can be used to determine enthalpy or temperature depending on which data are given. Additionally, the Air Tables are useful for determining final state point properties after an isentropic process is performed.

Isentropic processes can be easily analyzed through relative pressure by using the following relationship:

$$\left.\frac{p_2}{p_1}\right|_s = \left.\frac{Pr_2}{Pr_1}\right|_s \quad (5.10)$$

Note that the ratio of relative pressure values is equal to the ratio of real pressures. For a compressor, this is given by the pressure ratio.

TABLE 5.1: *Air Table Layout*

T (°R)	T (°F)	h (Btu/lb$_m$)	Pr	u (Btu/lb$_m$)	v_r
530	70	126.7	1.299	90.4	151.189
531	71	126.9	1.307	90.5	150.479
532	72	127.2	1.316	90.7	149.773
⋮					
1154	694	279.6	20.777	200.5	20.575
1155	695	279.9	20.843	200.7	20.527

The following example illustrates the use of relative pressure to determine the final state of air after an isentropic compression process:

EXAMPLE 5.1: COMPRESSOR ANALYSIS USING THE VARIABLE SPECIFIC HEAT APPROACH

Air enters the compressor of a gas turbine engine at 70 °F. The ideal compressor has a pressure ratio, $r_p = 16$. Determine the temperature and enthalpy of the air leaving the compressor.

ASSUMPTIONS: Because the compressor is ideal, the compression process is assumed to be isentropic and relative pressure ratio can be used to determine the final state.

SOLUTION STEPS:

1. Draw the schematic with SP numbers.

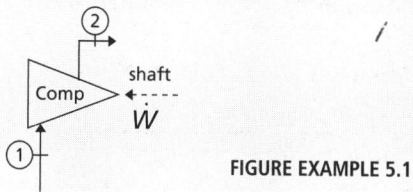

FIGURE EXAMPLE 5.1

2. Enter the Air Table using the given temperature of 70 °F
3. Read the corresponding Pr for 70 °F; $Pr_1 = 1.299$.
4. Recognize that the compression ratio is the ratio of p_2 to p_1. Using Equation 5.10

$$\frac{p_2}{p_1} = \frac{Pr_2}{Pr_1} = r_p$$

and rearranging,

$$Pr_2 = r_p\, Pr_1 = (16)(1.299) = 20.78$$

A note about relative pressure, Pr. Do not round off the values. Carry the extra decimal places until selecting the final answers from the Air Tables.

5. Enter the Air Table using $Pr_2 = 20.784$.
6. Note that this Pr_2 falls between two values in the table.

Selecting the closest value, Pr = 20.777 provides the final properties.

$$T_2 = 694°F$$

$$h_2 = 279.6\ Btu/lb_m$$

5.4.4 Gas Turbine State Point Table

As was the case with the other cycles studied to this point, a state point table is an excellent way to keep track of air properties throughout the Brayton cycle. In the state point table, each column represents a state point in the cycle and each row is reserved for a critical property. For the ideal Brayton cycle, four columns are required for the four state points and four rows are used for pressure, temperature, enthalpy, and relative pressure. A blank state point table is shown as Table 5.2. Note that three additional staggered rows are shown to indicate processes or heat/work interactions that occur between state points. These can be useful in completing the state point table or using it to analyze cycle performance. Not all cells need to be completed, and some do not apply. However, the expanded state point table is very valuable to students studying this material for the first time in helping to organize data and illustrate essential concepts.

TABLE 5.2: *Ideal Brayton Cycle Expanded State Point Table for the Variable C Approach*

	1	2	3	4
p (psia)	14.7			
T (°R)	520		2000	
h (ft³/lb$_m$)				
Pr				
r_p		10		1/10
q (Btu/lb$_m$)				
w (Btu/lb$_m$)				

EXAMPLE 5.2: SIMPLE IDEAL BRAYTON CYCLE

GIVEN: A simple ideal Brayton cycle has a pressure ratio of 10. Air enters the compressor at 14.7 psia and 520 °R. Turbine inlet temperature is 2000 °R.

FIND: Using the variable specific heat methodology, fill in the state point table and determine:
  a) heat and work transfers throughout the cycle
  b) thermal efficiency
  c) back work ratio

ASSUMPTIONS:
1. Because the cycle is ideal, the compression and expansion processes are assumed to be isentropic.
2. Additionally, since the specific heat of air varies with temperature, the Air Tables must be used to determine state points.

SOLUTION: Draw a schematic. Sketch the h-s diagram. Lay out a SP table.

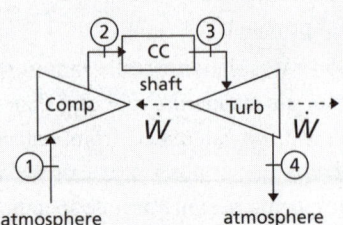

FIGURE EXAMPLE 5.2

Note that the schematic omits the atmospheric heat exchanger shown in Figure 5.3. This is an acceptable alternative schematic. This engine shows that shaft power generated by the turbine runs the compressor and there is additional power available to run something else, like a generator.

The h-s diagram will be shaped like the T-s in Figure 5.4.

SP1: The properties of the initial state point are easily found in the Air Tables since the temperature is given.

Entering the tables with T_1 = 520 °R:
h_1 = 124.3 Btu/lb$_m$ and Pr_1 = 1.215
The pressure is given as p_1 = 14.7 psia.

SP2s: The air is isentropically compressed from SP1 to SP2s. The pressure ratio, r_p, is used to determine the pressure at SP2s. Thus p_2 = 147 psia.

Additionally, the relative pressure at SP1 is multiplied by the pressure ratio to obtain the value of the relative pressure at SP2s. This relationship is used because the process from 1 to 2s is isentropic.

$$\frac{Pr_{2s}}{Pr_1} = \frac{p_{2s}}{p_1} = r_p \rightarrow Pr_{2s} = Pr_1 r_p = (1.215)(10) = 12.15$$

Pr_{2s} = 12.15 is used to enter the Air Tables to determine the remaining properties at SP2s:
T_{2s} = 997 R and h_{2s} = 240.3 Btu/lb$_m$

SP3: The properties of SP3 are found by entering the Air Tables with the given turbine inlet temperature. Note that in some problems T_3 may not be provided; instead, the amount of heat added in the combustion chamber will be given. If heat supplied is given, the enthalpy of SP3 can be determined by Equation 5.5.

$T_3 = 2000\,°R \rightarrow h_3 = 504.6\,Btu/lb_m$ and $Pr_3 = 173.796$

No pressure is lost in the combustion chamber so

$p_3 = p_2 = 147$ psia.

SP4s: The final state point can be resolved by recognizing that the air expands isentropically from 3 to 4. Thus, the relative pressure relationship can be used. Pr_3 is divided by the pressure ratio to determine Pr_4. With Pr_4, the other state point properties can be obtained from the Air Tables.

$$\frac{Pr_{4s}}{Pr_3} = \frac{p_{4s}}{p_3} = \frac{p_1}{p_{2s}} = \frac{1}{r_p} \rightarrow$$

$$Pr_{4s} = \frac{Pr_3}{r_p} = \frac{173.796}{10} = 17.38$$

Entering the Air Tables with $Pr_4 = 17.38$:
$T_{4s} = 1099\,R$ and $h_{4s} = 265.8\,Btu/lb_m$.
Finally, $p_{4s} = p_1 = 14.7$ psia

Now that the basic state point table (p, T, h, Pr) is complete, the remaining parts of the problem are easily found.

a) Heat and work transfers:

Compressor work = $w_{12} = h_1 - h_{2s}$
$= (124.3 - 240.3)\,Btu/lb_m$

$\boxed{w_{12} = -116.0\,Btu/lb_m}$

Heat added = $q_{23} = h_3 - h_{2s}$
$= (504.6 - 240.3)\,Btu/lb_m$

$\boxed{q_{23} = 264.3\,Btu/lb_m}$

Turbine work = $w_{34} = h_3 - h_{4s}$
$= (504.6 - 265.8)\,Btu/lb_m$

$\boxed{w_{34} = 238.8\,Btu/lb_m}$

Heat rejected = $q_{41} = h_1 - h_{4s}$
$= (124.3 - 265.8)\,Btu/lb_m$

$\boxed{q_{41} = -141.5\,Btu/lb_m}$

The completed state point table is shown below with "givens" **bolded**. The SP numbers reflect that the compressor and turbine are both ideal and adiabatic (isentropic).

	1	2s	3	4s
p (psia)	**14.7**	147	147	14.7
T (°R)	**520**	997	**2000**	1099
h (ft³/lb$_m$)	124.3	240.3	504.6	265.8
Pr	1.215	12.15	173.796	17.38
r_p	10	NR	1/10	NR
q (Btu/lb$_m$)	NR	264.3	NR	-141.5
w (Btu/lb$_m$)	-116.0	NR	238.8	NR

b) Thermal efficiency

$$\eta_{th} = \frac{w_{net}}{q_{in}} = \frac{w_{12} + w_{34}}{q_{23}} = \frac{(-116.0 + 238.8)\,Btu/lb_m}{264.3\,Btu/lb_m}$$

$\boxed{\eta_{th} = 0.465 = 46.5\%}$

c) Back work ratio

$$BWR = \left|\frac{w_{12}}{w_{34}}\right| = \left|\frac{-116.0\,Btu/lb_m}{238.8\,Btu/lb_m}\right| = \boxed{0.486}$$

$\boxed{BWR = 48.6\%}$

5.4.5 Real Brayton Cycle

The ideal Brayton cycle is useful for learning gas turbine analysis because of its simplicity. However, the price for simplicity is neglect of several real world losses. For example, real compressors and turbines do not operate isentropically as there are losses due to friction and heat transfer. Similarly, there are pressure losses in the compressor inlet, through the combustion chamber and in the exhaust duct. These losses will be considered as they apply to the Brayton cycle processes.

COMPRESSION PROCESS

Figure 5.5 shows both ideal and real compression processes plotted on h-s coordinates. Note that both processes result in identical pressure increases. However, the real compressor requires more work than the ideal one to overcome frictional losses and heat transferred to the environment. The ratio of ideal work to real work represents the isentropic efficiency of the compressor and is given by the following relationship:

$$\eta_{comp} = \frac{w_{comp,ideal}}{w_{comp,actual}} = \frac{w_{12s}}{w_{12}} = \frac{h_1 - h_{2s}}{h_1 - h_2} \quad (5.11)$$

The LM2500 compressor has an isentropic efficiency of approximately 92 percent.

Also, notice that in both cases the pressure at SP1 must be sub-atmospheric due to frictional losses in the intake ducting. This makes sense because the atmospheric pressure outside the duct must be greater than the pressure inside the duct in order to overcome friction and "push" the air into the compressor section. The pressure drop across the duct is known as Δp_{inlet} and is determined by:

$$\Delta p_{inlet} = p_1 - p_{atm} \quad (5.12)$$

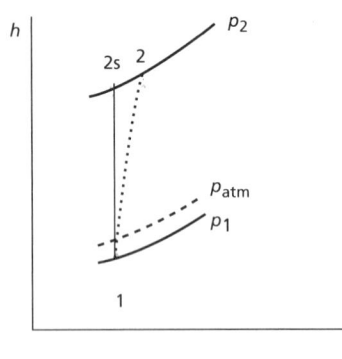

FIGURE 5.5: Ideal and Real Compressor Processes with Inlet Duct Pressure Drop

COMBUSTION PROCESS

There are two sources of losses in the combustion process, pressure losses and energy conversion losses. Figure 5.6 shows the effect of pressure loss that occurs in the real combustion process. As in the case of the intake, all real flows encounter friction losses that result in pressure drops. The pressure drop in the combustion chamber, Δp_{cc}, is given by:

$$\Delta p_{cc} = p_2 - p_3 \quad (5.13)$$

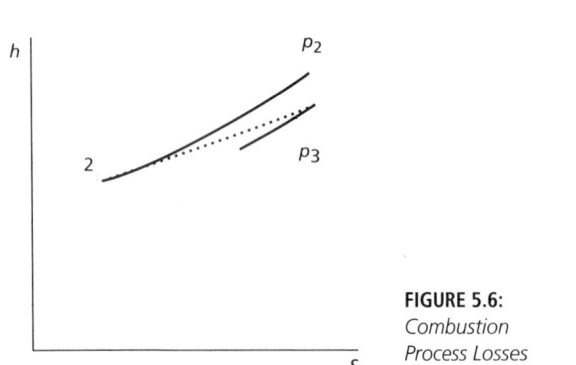

FIGURE 5.6: Combustion Process Losses

Combustion efficiency is a measure of how much available fuel energy is actually transferred to the air and is given by:

$$\eta_{cc} = \frac{energy\ to\ air}{fuel\ energy} = \frac{\dot{m}_{air}(h_3 - h_2)}{\dot{m}_{fuel} LHV} = AFR \frac{(h_3 - h_2)}{LHV} \quad (5.14)$$

where AFR is the air fuel ratio, $\frac{\dot{m}_{air}}{\dot{m}_{fuel}}$, and LHV is the fuel's lower heating value in Btu/lb$_m$.

EXPANSION PROCESS

Figure 5.7 shows both the ideal and real expansion processes plotted on h-s coordinates. As with the compression process, both cases result in identical pressure changes. However, less work is extracted from the real turbine than can be extracted from the ideal turbine due to irreversible losses such a friction and heat transfer to the environment. The ratio of real work to ideal work is a measure of the isentropic efficiency of the turbine and is given by the following relationship:

$$\eta_{turb} = \frac{actual\ turbine\ work}{ideal\ turbine\ work} = \frac{w_{34}}{w_{34s}} = \frac{h_3 - h_4}{h_3 - h_{4s}} \quad (5.15)$$

In addition to turbine inefficiencies, there are pressure losses in the exhaust ducting. To overcome friction in the exhaust duct, the pressure at the turbine exit must be greater than atmospheric pressure. The pressure drop in the exhaust duct is given by:

$$\Delta p_{exhaust} = p_4 - p_{atm} \quad (5.16)$$

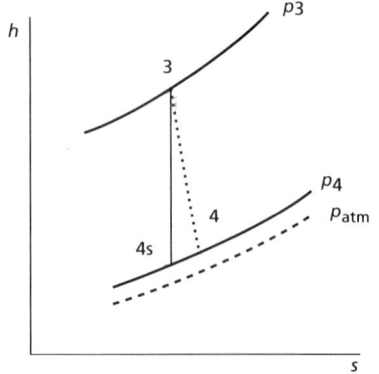

FIGURE 5.7: Ideal and Real Expansion Processes with Exhaust Duct Pressure Drop

EXAMPLE 5.3: REAL BRAYTON CYCLE

GIVEN: A real Brayton cycle using air as the working fluid has the following characteristics:
- $P_{atm} = 14.7$ psia, $T_{atm} = 520\ °R$
- Inlet duct loss, $\Delta p_{inlet} = 0.4$ psia
- Compressor pressure ratio, $r_p = 10$
- Isentropic compressor efficiency, $\eta_c = 0.85$
- Combustion chamber duct loss, $\Delta p_{cc} = 5$ psia
- Turbine inlet temp = 2000 °R
- Isentropic turbine efficiency, $\eta_T = 0.8$
- Exit duct loss, $\Delta p_{exit} = 0.2$ psia
- $\dot{m}_{air} = 13.5\ lb_m/sec$
- $\dot{m}_{fuel} = 0.2\ lb_m/sec$
- $LHV = 19{,}900\ Btu/lb_m$

FIND: Sketch the h-s diagram for this cycle and, assuming specific heat variation with temperature, determine:
a) values in the state point table
b) heat and work transfers throughout the cycle
c) combustion chamber efficiency
d) thermal efficiency
e) back work ratio

ASSUMPTION: Since the specific heat varies with temperature, the Air Tables must be used to determine state points.

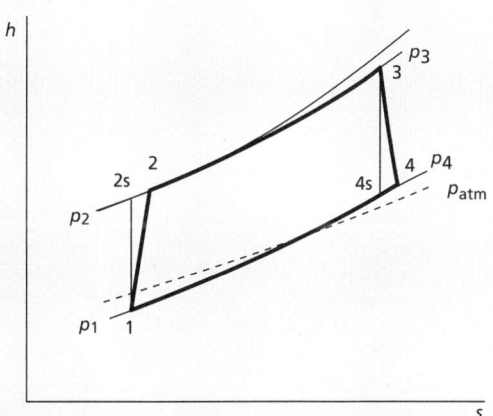

FIGURE EXAMPLE 5.3: *Sketch of actual Brayton cycle with losses*

SOLUTION: Schematic. SP Table. Process Graph.

SP1: The properties of the initial state point are easily found in the Air Tables since the temperature is given.

Entering the tables with $T_1 = 520\ °R$:

$$h_1 = 124.3\ Btu/lb_m \text{ and } Pr_1 = 1.215$$

The pressure must be calculated using the pressure loss in the duct. From Equation 5.12,

$$p_1 = p_{atm} - \Delta p_{inlet} = (14.7 - 0.4)\,psia = 14.3\,psia$$

SP2s: The air is isentropically compressed from SP1 to 2s. The pressure ratio, r_p, is used to determine the pressure at SP2s.

$$p_{2s} = r_p p_1 = (10)(14.3\,psia) = 143\,psia$$

Additionally, the relative pressure at SP1 is multiplied by the pressure ratio to obtain the value of the relative pressure at SP2s. This relationship is used because the process from 1 to 2s is isentropic.

$$\frac{Pr_{2s}}{Pr_1} = \frac{p_2}{p_1} = r_p \ \rightarrow\ Pr_2 = Pr_1\, r_p = (1.215)(10) = 12.15$$

$Pr_{2s} = 12.15$ is used to enter the Air Tables to determine the remaining properties at SP2s:

$$T_{2s} = 997\ R \text{ and } h_{2s} = 240.3\ Btu/lb_m$$

SP2: Use the compressor efficiency to determine the enthalpy at SP2.

$$\eta_C = \frac{h_1 - h_{2s}}{h_1 - h_2} \ \rightarrow\ h_2 = h_1 - \frac{h_1 - h_{2s}}{\eta_C}$$

$$= 124.3\,Btu/lb_m - \frac{(124.3 - 240.3)\,Btu/lb_m}{0.85}$$

$$h_2 = 260.8\,Btu/lb_m$$

From the Air Tables, $T_2 = 1079\ °R$. Note that there is no point in recording Pr_2 since the process from 2 to 3 is not isentropic.

Finally, $p_2 = p_{2s} = 143$ psia

SP3: The properties of SP3 are found by entering the Air Tables with the given turbine inlet temperature.

$$T_3 = 2000\ °R \rightarrow h_3 = 504.6\ Btu/lb_m$$

$$\text{and } Pr_3 = 173.796$$

The turbine inlet pressure is obtained by rearranging Equation 5.13:

$$p_3 = p_2 - \Delta p_{cc} = (143-5)\,psia = 138\,psia$$

SP4s: This state point can be resolved by recognizing that the air expands isentropically from 3 to 4s. Thus, the relative pressure relationship can be used. Pr_3 is divided by the pressure ratio to determine Pr_{4s}. With Pr_{4s}, the other properties can be obtained from the Air Tables.

First, p_{4s} can be determined by rearranging Equation 5.16:

$$p_{4s} = p_{atm} + \Delta p_{exhaust} = (14.7+0.2)\,psia = 14.9\,psia$$

$$\frac{Pr_{4s}}{Pr_3} = \frac{p_{4s}}{p_3} \rightarrow Pr_{4s} = Pr_3\left(\frac{p_{4s}}{p_3}\right)$$

$$= (173.796)\left(\frac{14.9\,psia}{138\,psia}\right) = 18.765$$

Entering the Air Tables with $Pr_{4s} = 18.765$, $T_{4s} = 1123\,°R$ and $h_{4s} = 271.8\,Btu/lb_m$.

SP4: Use turbine efficiency to determine the enthalpy at SP4.

$$\eta_T = \frac{h_3 - h_4}{h_3 - h_{4s}} \rightarrow$$

$$h_4 = h_3 - \eta_T(h_3 - h_{4s})$$

$$= 504.6\,Btu/lb_m - (0.80)(504.6 - 271.8)\,Btu/lb_m$$

$$h_4 = 318.4\,Btu/lb_m \rightarrow T_4 = 1306\,°R \text{ from the Air Tables}$$

Finally, $p_4 = p_{4s} = 14.9$ psia

a) The completed basic state point table is shown below. At this point, you may choose to expand the state point table with the staggered r_p, q and w rows.

	1	2s	2	3	4s	4
p (psia)	14.3	143	143	138	14.9	14.9
T (°R)	520	997	1079	2000	1123	1306
h (Btu/lb_m)	124.3	240.3	260.8	504.6	271.8	318.4
Pr	1.215	12.15	NR	173.796	18.765	NR

Now that the state point table is complete, the remaining parts of the problem are easily found.

b) Heat and work transfers:

Compressor work $= w_{12} = h_1 - h_2$

$$= (124.3 - 260.8)\,Btu/lb_m$$

$$\boxed{w_{12} = -136.5\,Btu/lb_m}$$

Heat added $= q_{23} = h_3 - h_2$

$$= (504.6 - 260.8)\,Btu/lb_m$$

$$\boxed{q_{23} = 243.8\,Btu/lb_m}$$

Turbine work $= w_{34} = h_3 - h_4$

$$= (504.6 - 318.4)\,Btu/lb_m$$

$$\boxed{w_{34} = 186.2\,Btu/lb_m}$$

Heat rejected $= q_{41} = h_1 - h_4$

$$= (124.3 - 318.4)\,Btu/lb_m$$

$$\boxed{q_{41} = -194.1\,Btu/lb_m}$$

c) Combustion chamber efficiency

$$\eta_{cc} = \frac{\dot{m}_{air}(h_3 - h_2)}{\dot{m}_{fuel}LHV}$$

$$= \frac{(13.5\,lb_m/s)(504.6-240.3)\,Btu/lb_m}{(0.2\,lb_m/s)(19{,}900\,Btu/lb_m)}$$

$$\boxed{\eta_{cc} = 0.827 = 82.7\%}$$

d) Thermal efficiency

$$\eta_{th} = \frac{w_{net}}{q_{in}} = \frac{w_{12}+w_{34}}{q_{23}} = \frac{(-136.5+186.2)\,Btu/lb_m}{243.8\,Btu/lb_m}$$

$$\boxed{\eta_{th} = 0.204 = 20.4\%}$$

e) Back work ratio

$$BWR = \left|\frac{w_{12}}{w_{34}}\right| = \left|\frac{-136.5\,Btu/lb_m}{186.2\,Btu/lb_m}\right| = 0.733 =$$

$$\boxed{BWR = 73.3\%}$$

Note: This is the same Brayton cycle worked in example 5.2 except that losses and inefficiencies are included. Table 5.3 shows a comparison of results from the ideal and real cycles. Compared to the ideal cycle the real cycle requires more compressor work and produces less turbine work. Most notably the real thermal efficiency is less than half of the ideal thermal efficiency and back work ratio is increased dramatically.

TABLE EXAMPLE 5.3: *Ideal/Real Brayton Cycle Comparison*

	Ideal	Real
w_c (Btu/lb$_m$)	116.0	136.5
w_T (Btu/lb$_m$)	238.8	186.2
q_{in} (Btu/lb$_m$)	264.3	243.8
η_{th} (%)	46.5	20.4
BWR (%)	48.6	73.3

5.4.6 SPLIT-SHAFT GAS TURBINE ENGINE (LM2500)

Up to this point, only single shaft gas turbine engines have been discussed. This means that a single shaft extends throughout the engine and the compressor is mechanically connected to a single turbine section. This simple configuration is useful for applications with relatively constant speed such as electrical power generation. However, for applications where load and speed vary greatly, such as ship propulsion, a split-shaft configuration is more appropriate. The LM2500 is a split-shaft gas turbine engine.

In a split-shaft design, a second turbine section is located downstream of the gas generator turbine. The two turbines are on the same axis but the shafts are not mechanically connected. Instead, the second turbine, commonly referred to as the power turbine, is driven by hot gases exiting the gas generator turbine. Figure 5.8 shows a diagram of an LM2500 split-shaft gas turbine. The dashed line highlights the "split" between the gas generator and the power turbine shafts. The power turbine shaft is mechanically linked to the reduction gearing leading to the propeller shaft. Note the numbering scheme used to identify state points within the engine. These numbers are consistent with those previously introduced for the gas generator.

As mentioned in the earlier part of this chapter, the LM2500 is widely used for shipboard propulsion. According to the Military Analysis Network LM2500 website sponsored by the Federation of American Scientists, there are nearly 900 LM2500 gas turbines in service on a variety of cruisers, frigates, destroyers, and patrol boats for more than 24 international navies. There are several versions of the LM2500 with power ranging from 21,500 BHP up to 39,000 BHP. These brake horsepower values indicate the power produced by the 6-stage low pressure power turbine. This is the net mechanical power of the engine and is equivalent to approximately 35 percent of the energy input as measured at the combustor. The other 65 percent is extracted by the 2-stage high pressure turbine (gas generator turbine) that is used to drive the 16-stage axial compressor and engine auxiliaries.

The split-shaft gas turbine has important advantages over the single shaft turbine for ship propulsion applications. First, because the gas generator is not mechanically connected to the power turbine, it is more responsive to load demands. Second, both the gas generator and power turbine can operate within their most efficient RPM ranges for a wide variety of load conditions. Typically

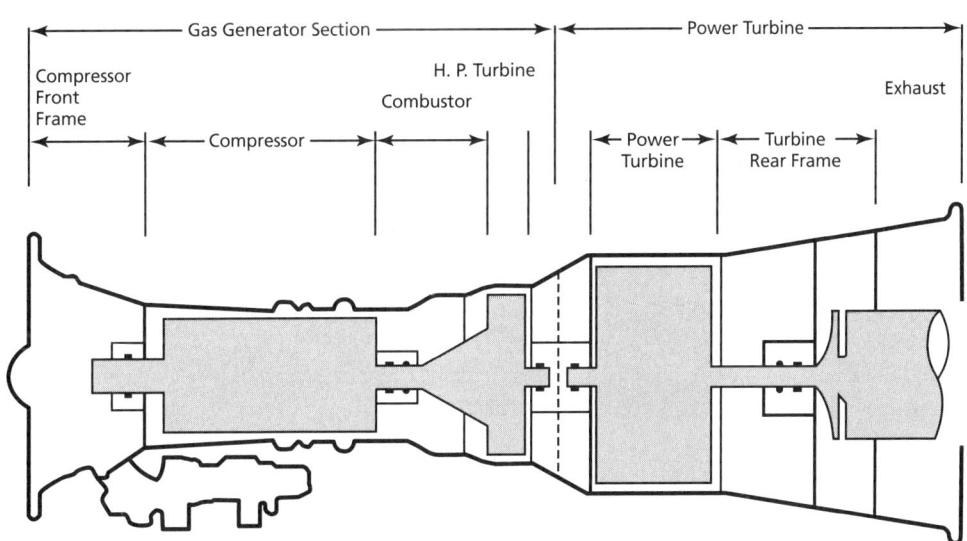

FIGURE 5.8: *LM2500 Gas Turbine Engine*

the gas generator rotates at much higher speeds than the power turbine.

Figure 5.9 shows a real split-shaft gas turbine engine cycle plotted on *h-s* coordinates. Note the pressure drops in the inlet, combustion chamber, and exhaust duct. Additionally, the isentropic compression and turbine processes are plotted with dotted lines. When analyzing the split-shaft turbine, it is important to remember that all of the gas generator turbine work is used to drive the compressor.

Thus, $w_{ggt} = -w_c$

This can be written in terms of enthalpy,

$$h_3 - h_4 = h_2 - h_1 \qquad (5.17)$$

Note: This relationship applies to real (aka actual) work, not ideal.

This relationship can be very useful when trying to fill in the state point table!

Also, since the gas generator turbine is used solely to power the compressor, all of the work produced by the power turbine can be used for propulsion and is considered net work.

Thus, $w_{pt} = w_{net}$ and

$$\eta_{th} = \frac{w_{net}}{q_{in}} = \frac{h_4 - h_5}{h_3 - h_2} \qquad (5.18)$$

The real mechanical power transmitted via the engine's output coupling to the reduction gearing depends on the mechanical efficiency of the gas turbine. The *BHP* is related to the indicated horsepower (*IHP*) by the engine's mechanical efficiency.

Since $\eta_{mech} = \dfrac{BHP}{IHP} = \dfrac{BHP}{\dot{m}_{air} w_{net}}$

$$BHP = \eta_{mech} \dot{m}_{air} w_{net} = \eta_{mech} \dot{m}_{air}(h_4 - h_5) \qquad (5.19)$$

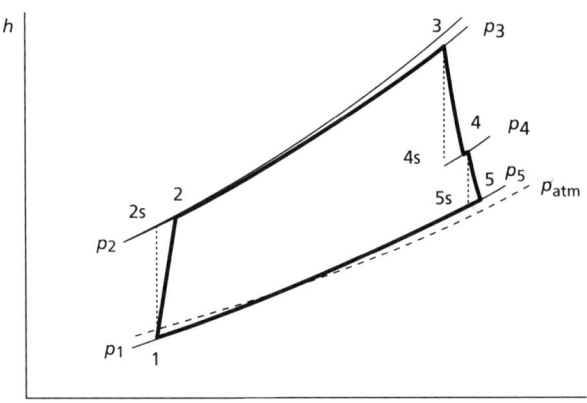

FIGURE 5.9: Split-Shaft Gas Turbine Cycle on h-s Coordinates

EXAMPLE 5.4: SPLIT-SHAFT GAS TURBINE

GIVEN: A split-shaft gas turbine inducts atmospheric air at 59 °F into a compressor operating at 84% efficiency at a rate of 140 lb_m/sec. The air loses 0.5 psi in the intake. After being compressed to 18 times the inlet pressure, the air mixes with fuel (*LHV* = 19,100 Btu/lb_m) and is burned in the combustion chamber. Due to the geometry of the combustion chamber, the pressure is reduced by 1 psi as the air passes through it. The 2000 °F combustion chamber exhaust is ducted to the gas generator turbine, which operates at 90% efficiency. The hot gases then expand through the power turbine, exiting at 988 °F. Assume variable specific heat capacities.

SKETCH the *h-s* diagram and **FIND:**
  a) work required by the compressor (Btu/lb_m)
  b) net work produced by the engine (Btu/lb_m)
  c) power turbine efficiency (%)
  d) cycle thermal efficiency (%)
  e) brake horsepower produced if the mechanical efficiency is 80%

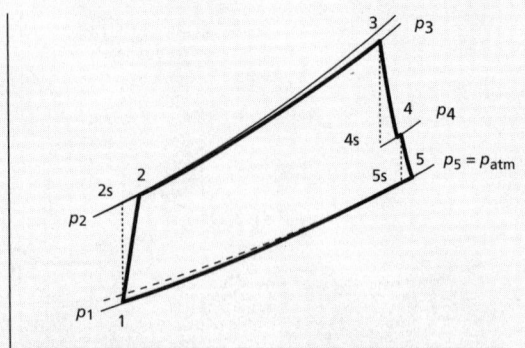

FIGURE EXAMPLE 5.4A

Note that the *h-s* diagram, Figure Example 5.4A, is very similar to the one shown in Figure 5.9 with the exception of point 5. Since no outlet losses are given, it is assumed that the outlet pressure is at atmospheric pressure. Although atmospheric pressure is not given, it can be assumed to be 14.7 psia.

The following table is used to keep track of parameters throughout the cycle. The **bold** values are "givens" and the values in italics are calculated as shown in the steps below the table. Blocks left blank indicate values that are not necessary for the problem solution.
SOLUTION: Don't forget the schematic! Here is a simplified schematic for a split-shaft GTE.

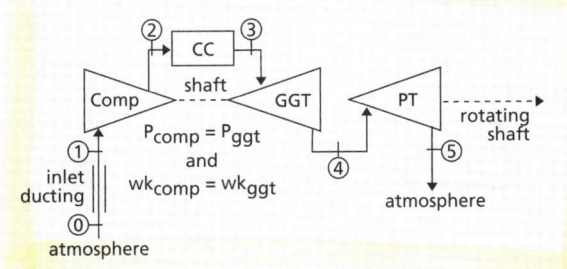

FIGURE EXAMPLE 5.4B

SP1: p_1 can be found by subtracting inlet losses from atmospheric pressure. Note that the pressure at the compressor inlet must be less than atmospheric pressure to provide the pressure gradient necessary to force the fluid through the inlet ductwork.

$$p_1 = p_{atm} - \Delta p_{in} = 14.7 \, psia - 0.5 \, psi = 14.2 \, psia$$

Since T_1 is given (59 °F) and heat capacity is assumed variable, h_1 and Pr_1 are taken from the Air Tables.
h_1 = 124.1 Btu/lb$_m$ and Pr_1 = 1.207

SP2s: This is the fictitious ideal state point following the compressor that allows the use of the Pr relationships. Since this state point is achieved isentropically, the relative pressure is used to define its parameters.

$$Pr_2 = Pr_1\left(\frac{p_2}{p_1}\right) = Pr_1 \, r_p = 1.207 * 18 = 21.726$$

Entering the Air Tables with Pr_2 = 21.726, h_{2s} = 283.2 Btu/lb$_m$

Finally, p_2 can be found using the compressor pressure ratio:

$$p_2 = r_p p_1 = 18 * 14.2 \, psia = 255.6 \, psia$$

SP2: This is the real state point following the compressor and is resolved using the compressor isentropic efficiency:

$$\eta_C = \frac{h_1 - h_{2s}}{h_1 - h_2} \rightarrow h_2 = h_1 - \frac{h_1 - h_{2s}}{\eta_C}$$
$$= 124.1 \, Btu/lb_m - \frac{(124.1 - 283.2) Btu/lb_m}{0.84}$$

So, h_2 = 313.5 Btu/lb$_m$
Note that $p_2 = p_{2s}$ = 255.6 psia.

SP3: This is the state point following heat addition in the combustion chamber.
Since T_3 is given (2000 °F), h_3 and Pr_3 are taken from the Air Tables.
h_3 = 633.9 Btu/lb$_m$ and Pr_3 = 405.967
p_3 is obtained by subtracting the combustion chamber pressure drop from p_2:

$$p_3 = p_2 - \Delta p_{cc} = 255.6 \, psia - 1 \, psi = 254.6 \, psia$$

SP4: The real enthalpy at SP4, the gas generator exit, can be found by recalling that the compressor work required for a split-shaft turbine is equivalent to the work produced by the gas generator turbine work. That is: $-w_{comp} = w_{ggt}$ or $-(h_1 - h_2) = h_3 - h_4$. So,

$$h_4 = h_3 + h_1 - h_2 = 633.9 + 124.1 - 313.5 \, Btu/lb_m$$
$$= 444.5 \, Btu/lb_m$$

SP4s: This ideal state point must be solved in order to obtain p_4.

	1	2s	2	3	4s	4	5s	5
p (psia)	*14.2*	*255.6*	*255.6*	*254.6*	*57.4*	*57.4*	**14.7**	**14.7**
T (°F)	**59**			**2000**				*988*
h (Btu/lb$_m$)	*124.1*	*283.2*	*313.5*	*633.9*	*423.4*	*444.5*		*355.4*
Pr	*1.207*	*21.726*		*405.967*	*91.537*			

$$\eta_{GGT} = \frac{h_3 - h_4}{h_3 - h_{4s}} \rightarrow h_{4s} = h_3 - \frac{h_3 - h_4}{\eta_{GGT}}$$

$$= 633.9 - \frac{633.9 - 444.5}{0.9} = 423.45\, Btu/lb_m$$

From the Air Tables, $Pr_{4s} = 91.537$. Thus,

$$p_4 = p_{4s} = p_3 \frac{Pr_{4s}}{Pr_3} = (254.6\, psia)\frac{91.537}{405.967} = 57.4\, psia$$

SP5s: There is no need to solve for h_{5s} since h_5 is known. However, if h_5 were needed, it could be obtained by finding Pr_{5s} since Pr_4, p_4 and p_5 are known or could easily be found.

a) Finding compressor work, w_c

$$w_c = h_1 - h_2$$
$$= 124.1 - 313.5\, Btu/lb_m =$$

$$\boxed{w_c = -189.4\, Btu/lb_m}$$

b) Finding net work, w_{net}

$$w_{net} = \cancel{w_{12}} + \cancel{w_{34}} + w_{45} = w_{45} = h_4 - h_5$$
$$= 444.5 - 355.4\, Btu/lb_m =$$

$$\boxed{w_{net} = 89.1\, Btu/lb_m}$$

Note: w_{12} and w_{34} cancel because they are equal and opposite in sign.

c) Finding power turbine efficiency, η_{PT}

$$\eta_{PT} = \frac{w_{PT,real}}{w_{PT,ideal}} = \frac{h_4 - h_5}{h_4 - h_{5s}}$$

$$= \frac{444.5 - 355.4}{444.5 - 304.1} = 0.635 =$$

$$\boxed{\eta_{PT} = 63.5\%}$$

d) Finding thermal efficiency, η_{th}

$$\eta_{th} = \frac{w_{net}}{q_s} = \frac{h_4 - h_5}{h_3 - h_2} = \frac{444.5 - 355.4}{633.9 - 313.5} = 0.278 =$$

$$\boxed{\eta_{th} = 27.8\%}$$

e) Finding brake horsepower if mechanical efficiency is 80%

$$\eta_{mech} = \frac{bhp}{\dot{m}_{air} w_{net}} \rightarrow$$

$$BHP = \eta_{mech}\, \dot{m}_{air}\, w_{net} = \eta_{mech}\, \dot{m}_{air}\, (h_4 - h_5)$$

$$BHP = (0.8)(140\, lb_m/s)(444.5 - 355.4)$$

$$\frac{Btu}{lb_m}\left(\frac{hp \cdot s}{0.707\, Btu}\right) =$$

$$\boxed{BHP = 14,100\, hp}$$

5.4.7 Diffusers and Nozzles

In the previous section on split-shaft turbines, the high energy exhaust from the gas generator was used to drive the power turbine and produce shaft work. Another desirable option would be to convert the energy provided by the gas generator into kinetic energy to provide jet propulsion for aircraft applications. Diffusers and nozzles provide suitable energy conversion mechanisms to make gas generators adaptable to the flight regime. Figure 5.10 shows a gas generator with a diffuser and nozzle attached. This configuration is called a turbojet and will be discussed in a later section. Note that the station numbering is selected to maintain consistency with the numbering scheme previously established for the gas generator.

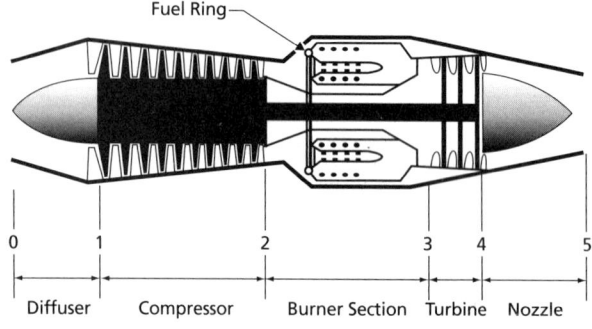

FIGURE 5.10: *Gas Generator Combined with Diffuser and Nozzle*

5.4.7.1 Diffusers

A diffuser is an open flow device that converts kinetic energy into a pressure and enthalpy increase. When installed on the front of an aircraft engine, this device provides a boost in air pressure prior to the compressor. Figure 5.11 shows diffuser shapes for both subsonic and supersonic flows. In both cases the flow velocity decreases as it moves through the diffuser.

Since the diffuser is an open flow device, the SFEE can be used to analyze fluid properties.

$$pe_i + ke_i + u_i + p_i v_i + q_{if} = pe_f + ke_f + u_f + p_f v_f + w_{if}$$

The SFEE can be simplified assuming that no potential energy changes, heat transfers, and work interactions occur in a diffuser. As a result, the SFEE can be reduced to the following form:

$$ke_i + h_i = ke_f + h_f \quad (5.20)$$

Replacing the subscripts for the diffuser, Equation 5.20 becomes

$$ke_0 + h_0 = ke_1 + h_1 \quad (5.21)$$

and

$$\frac{\vec{v}_0^2}{2g_c} + h_0 = \frac{\vec{v}_1^2}{2g_c} + h_1 \quad (5.22)$$

In Equation 5.22, v_0 equals the aircraft velocity. Also, unless otherwise told, it is assumed that the velocity of air throughout the engine is negligible compared to the velocity of the incoming air. Thus $v_1 = 0$ is a reasonable assumption.

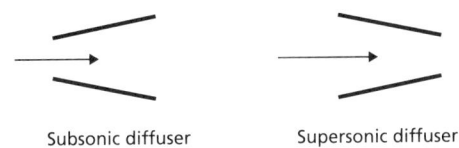

FIGURE 5.11: *Diffuser Configurations*

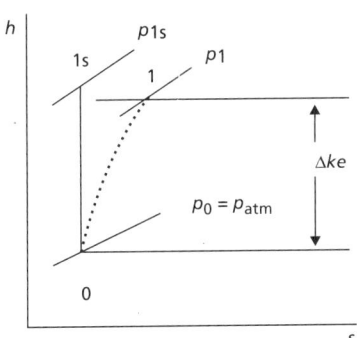

FIGURE 5.12: *Diffuser Process*

Like compressors, diffusers can either be treated as real or ideal. Figure 5.12 shows the diffuser process plotted on h-s coordinates. Note the similarity with the compression process introduced earlier. Ideal diffusers have no friction losses so all the kinetic energy is converted to an increase in pressure. However, losses must be considered in the real case. For real diffusers, these losses result in lower pressure (and enthalpy) at the compressor face for a given inlet velocity. The efficiency of a diffuser is commonly expressed as the ratio of pressure increase to the pressure increase that could be realized in an ideal diffuser for the same inlet conditions. Thus:

$$\eta_D = \frac{\Delta p_{actual}}{\Delta p_{ideal}} \quad (5.23)$$

5.4.7.2 Nozzles

A nozzle is an open flow device that converts pressure and enthalpy into an increase in kinetic energy. When installed on the back of an aircraft engine, this device provides high velocity exit gases to propel the aircraft. Figure 5.13 shows nozzle shapes for both subsonic and supersonic flows. In both cases the flow velocity increase as it moves through the diffuser. Note that the same piece of equipment can operate as either a diffuser or nozzle depending on the incoming flow speed. Therefore it is better to define diffusers and nozzles by their function rather than shape. Diffusers slow the flow, nozzles accelerate the flow.

Starting with the SFEE and assuming negligible changes in potential energy, heat transfer, and work, Equation 5.1 is directly applied to nozzles. Applying the proper subscripts for this schematic, Equation 5.20 becomes

$$ke_4 + h_4 = ke_5 + h_5 \quad (5.24)$$

and

$$\frac{\vec{v}_4^2}{2g_c} + h_4 = \frac{\vec{v}_5^2}{2g_c} + h_5 \quad (5.25)$$

Figure 5.14 shows the nozzle process plotted on h-s coordinates. This process resembles the expansion pro-

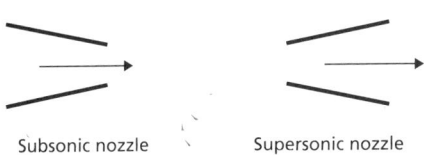

FIGURE 5.13: *Nozzle Configurations*

cess through a turbine. Just like the diffuser, losses must be considered when analyzing flow through a nozzle. Nozzle efficiency is the ratio of kinetic energy gained to the kinetic energy that could be gained with an isentropic nozzle with the same inlet state point and exit pressure and is given by:

$$\eta_N = \frac{\Delta ke_{actual}}{\Delta ke_{ideal}} = \frac{\vec{v}_5^2 - \vec{v}_4^2}{\vec{v}_{5s}^2 - \vec{v}_4^2} = \frac{h_4 - h_5}{h_4 - h_{5s}} \quad (5.26)$$

If the velocity of air through the engine is considered negligible compared to the velocity after the nozzle then it can be assumed that $v_4 = 0$.

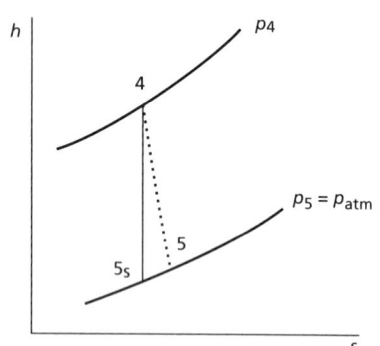

FIGURE 5.14: Nozzle Process

EXAMPLE 5.5: NOZZLE PROBLEM

GIVEN: Air enters a nozzle at 2000 °F and 22 psia at 150 ft/sec. The nozzle has an isentropic efficiency of 97% and the exit pressure is 10 psia.

FIND: Using the Air Tables, determine the velocity of the air exiting the nozzle.

SOLUTION: This could be done with a state point table with three data columns. Most students at this point will benefit from laying out a table to organize the data. Also, draw a sketch representing the nozzle and draw an h-s diagram.

1. First, because both inlet and outlet pressures are known, the relative pressure relationship can be used to determine the isentropic outlet condition. Establish SP1 at the nozzle inlet and SP2 at the nozzle discharge.
   Entering the Air Tables with $T_1 = 2000$ °F, $Pr_1 = 405.967$

$$Pr_2 = Pr_1 \frac{p_2}{p_1} = 405.967 \frac{10 \, psia}{22 \, psia} = 184.53$$

The closest Air Table entry for $Pr_2 = 184.53$ yields

$$h_{2s} = 512.9 \, Btu/lb_m$$

2. Next h_2 can be found using the nozzle efficiency equation:

$$\eta_N = \frac{h_1 - h_2}{h_1 - h_{2s}} \quad \text{rearranging:}$$

$$h_2 = h_1 - \eta_N (h_1 - h_{2s})$$
$$= 633.9 - 0.97(633.9 - 512.9) Btu/lb_m$$
$$= 516.5 \, Btu/lb_m$$

3. Finally, $\vec{v}_2$ can be solved using the nozzle equation:

$$\frac{\vec{v}_1^2}{2g_c} + h_1 = \frac{\vec{v}_2^2}{2g_c} + h_2 \rightarrow \vec{v}_2 = \sqrt{(h_1 - h_2)2g_c + \vec{v}_1^2}$$

$$\vec{v}_2 = \sqrt{(633.9 - 516.5)\frac{Btu}{lb_m}(2)\left(32.2 \frac{ft \cdot lb_m}{s^2 \cdot lb_f}\right)} \times$$

$$\sqrt{\left(\frac{778 \, ft \cdot lb_f}{Btu}\right) + \left(150 \frac{ft}{s}\right)^2} =$$

$$\boxed{\vec{v}_2 = 2429.6 \frac{ft}{s}}$$

5.4.8 AIRCRAFT APPLICATIONS

Gas turbine engines are used in aircraft applications in a variety of configurations. These configurations are all based on the basic gas generator and are designed for specific flight regimes and aircraft operating parameters. Figure 5.15 provides an overview of useful speed and alti-

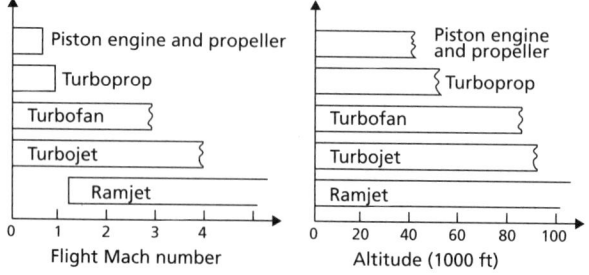

FIGURE 5.15: Aircraft Engine Applications

tude ranges for aircraft engines. In general, these engines either turn shafts or blow air. Included in the first category, turboshaft engines are used to drive helicopter rotor blades and turboprop engine power propellers. In the second category, turbojet and turbofan engines provide propulsion for jet aircraft. A ramjet is a special category of thrust engine that has no rotating compressor and turbine. In a ramjet engine, the diffuser provides the total pressurization and, similarly, the entire expansion is done in a nozzle. The German V-1 rocket was an example of a ramjet. A scramjet is a supersonic version of the ramjet and is being researched by NASA for potential applications. Both of these engines must be accelerated to their flight regimes by auxiliary rockets or other engines.

5.4.8.1 Turboshaft Engines

Figure 5.16 shows a typical turboshaft engine layout. Like the LM2500, this is a split-shaft gas turbine engine. The power shaft can be connected to a helicopter transmission to drive the main and tail rotors.

5.4.8.2 Turboprop Engines

Similarly, Figure 5.17 shows a typical turboprop configuration such as that found on the P-3C Orion maritime patrol aircraft. Notice the rotor shaft drives the prop through a reduction gearbox mounted in front of the engine. The gearbox is required to enable both the compressor and the propeller to operate in their most efficient speed ranges. The propeller produces thrust by imparting a small acceleration to a large mass of air. Some residual thrust is derived from the engine's exhaust. In some installations, this residual exhaust contributes up to 15 percent of the overall thrust produced by the engine and prop combination.

Turboprops are most efficient at speeds below 400 knots and at low altitude flight regimes. High-end speed is limited by the sonic flow characteristics and altitude is limited by the adverse effects of decreasing air density on propeller performance. For these reasons, turboprops are often selected for use in large slower aircraft that will be operated at lower altitudes.

5.4.8.3 Turbojet Engines

Figure 5.18 shows a turbojet configuration. Notice that this engine is essentially a gas generator with a diffuser on the front and a nozzle attached to the back. The station numbering scheme used is consistent with the one previously introduced for the gas generator. This basic design was used in early jet aircraft and saw continued

use in the A-4 Skyhawk, F-4 Phantom, and A-6 Intruder. Unlike turboprops, the turbojet provides a large acceleration to a small air mass. This provides greater high-end performance than the prop, but less fuel efficiency. These engines are ideal for smaller aircraft that need to go higher and faster than turboprop-equipped aircraft.

The thrust produced by a turbojet is a function of mass of air passing through the engine and the speed of that air relative to the aircraft speed. That is:

$$\bar{T} = \frac{\dot{m}_{air}\left(\vec{v}_{exhaust} - \vec{v}_{aircraft}\right)}{g_c} \qquad (5.27)$$

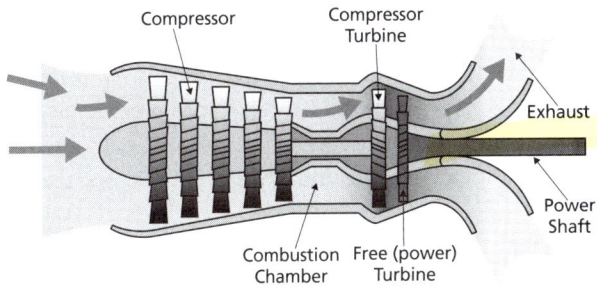

FIGURE 5.16: *Turboshaft Engine*

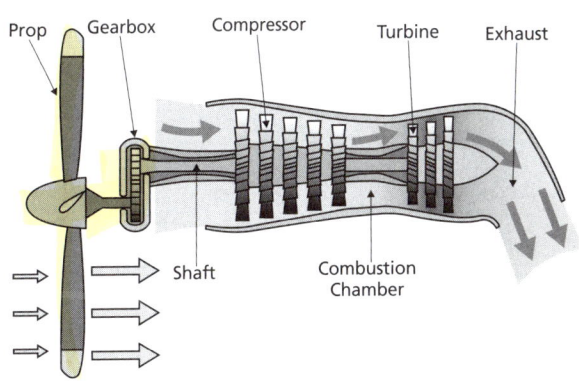

FIGURE 5.17: *Turboprop Engine*

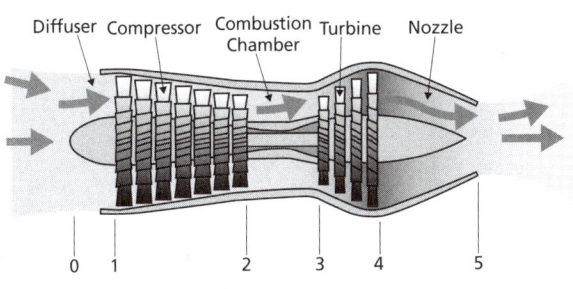

FIGURE 5.18: *Turbojet Engine*

As the aircraft speed increases, the thrust actually decreases. In the limit case, if the aircraft was able to achieve a speed equal to its exhaust velocity, thrust would equal zero.

Using the numbering scheme, this equation becomes:

$$\bar{T} = \frac{\dot{m}_{air}(\bar{v}_5 - \bar{v}_0)}{g_c}$$

Jet power, the rate of work produced by the engine, depends on the thrust and aircraft speed and is given by the following equation:

$$P_{jet} = \bar{T} \cdot (\bar{v}_{aircraft}) \qquad (5.28)$$

Note that an aircraft at full throttles sitting on the runway is producing no power until it starts moving.

EXAMPLE 5.6: TURBOJET PROBLEM

GIVEN: An F-4 Phantom is cruising at 900 ft/sec where the local atmospheric pressure is approximately 10.8 psia and the air temp is 60 °F. The engine operates with an 8:1 compression pressure ratio. Air flowing at 210 lb_m/sec enters each turbine at 1900 °F. The nozzle exhaust pressure is 11 psia. Assume that the diffuser, compressor, turbine, and nozzle operate isentropically and that no pressure drop occurs across the combustion chamber.

FIND: Draw the cycle on h-s coordinates and determine (using the Air Tables):
  a) The velocity of the air leaving the nozzle (ft/sec)
  b) The pressure at each state point during the cycle
  c) The thrust produced by the nozzle (lb_f)

RELATIONSHIPS:

$$\frac{\bar{v}_i^2}{2g_c} + h_i = \frac{\bar{v}_f^2}{2g_c} + h_f \qquad T = \frac{\dot{m}_{air}(\bar{v}_5 - \bar{v}_1)}{g_c}$$

SOLUTION: The F-4 used two turbojet engines. We'll analyze one.

Draw a representative schematic. Lay out a state point table.

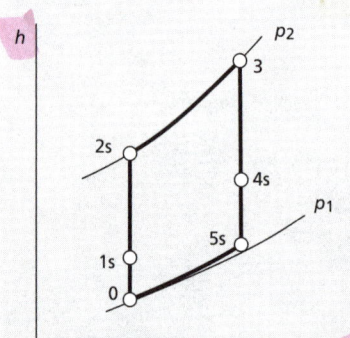

FIGURE EXAMPLE 5.6

In order to find the nozzle exit velocity, $\bar{v}_5$, the enthalpy at SP4 and SP5 and $\bar{v}_4$ must be known. Since air velocity *through* the engine is negligible compared to nozzle exit velocity, $\bar{v}_4$ will be taken as zero. The overall solution strategy will be to walk through the cycle and calculate enthalpy and pressure at each SP in order to calculate exhaust velocity and thrust.

The SP point table shown below is useful in analyzing this cycle. The **bold** numbers indicate given quantities and the *italicized* numbers indicate calculated quantities.

SP0, diffuser inlet: Enthalpy and relative pressure ratio can be directly read from the Air Tables with $T_0 = 60$ °F. From the Air Tables: $h_0 = 124.3$ Btu/lb_m and $Pr_0 = 1.215$

SP1, diffuser exit: The nozzle/diffuser equation is used to determine the condition of SP1. Since $\bar{v}_1$ is not given, it will be assumed to be zero. This is a good assumption since the velocity of air moving through the engine is much less than aircraft or nozzle exit velocity.

Rearranging the nozzle/diffuser equation:

$$h_1 = h_0 + \frac{(\bar{v}_0^2 - \bar{v}_1^2)}{2g_c} =$$

$$124.3 \frac{Btu}{lb_m} + \frac{(900\ ft/s)^2}{2\left(32.2 \frac{ft \cdot lb_m}{lb_f \cdot s^2}\right)}\left(\frac{Btu}{778\ ft \cdot lb_f}\right) = 140.5 \frac{Btu}{lb_m}$$

Entering the Air Tables with $h_1 = 140.5$ Btu/lb_m, $Pr_1 = 1.857$.

	0	1s	2s	3	4s	5s
p (psia)	**10.8**	*16.5*	*132.0*	*132.0*	*61.0*	**11.0**
T (°F)	**60**			**1900**		
h (Btu/lb_m)	124.3	*140.5*	*254.2*	*605.4*	*491.7*	*305.6*
Pr	1.215	*1.857*	*14.856*	*341.780*	*157.915*	*28.476*

This is be used to get p_1,

$$p_1 = p_0 \frac{Pr_1}{Pr_0} = 10.8\, psia \left(\frac{1.857}{1.215}\right) =$$

$$\boxed{p_1 = 26.5\, psia}$$

As expected, the pressure increases through the diffuser.

SP2, compressor discharge: Since the compressor pressure ratio is known, it is a simple matter to obtain p_2.

$$p_2 = p_1 r_p = 16.5\, psia\,(8) =$$

$$\boxed{p_2 = 132\, psia}$$

Then Pr_2 can be found:

$$Pr_2 = Pr_1 \frac{p_2}{p_1} = Pr_1\, r_p = 1.857\,(8) = 14.856$$

Entering the Air Tables with $Pr_2 = 14.856$, we find $h_{2s} = 254.2\, Btu/lb_m$

SP3, Combustion chamber exit:
Since there is no pressure drop in the combustion chamber,

$$\boxed{p_3 = p_2 = 132\, psia}$$

Entering the Air Tables with $T_3 = 1900\,°F$, we find $h_3 = 605.4\, Btu/lb_m$ and $Pr_3 = 341.780$

SP4, Turbine exit: No pressure or temperature information is provided for this point. In order to solve for SP4, it must be recalled that the turbine work is equivalent to the compressor work in a turbojet. This makes sense since no other form of power is taken from the turbine. The turbine drives only the compressor! So,

$$h_3 - h_4 = -(h_1 - h_2) \text{ or}$$

$$h_4 = h_3 + h_1 - h_2 = 605.4 + 140.5 - 254.2\, Btu/lb_m$$

$$= 491.7\, Btu/lb_m$$

From the Air Tables, $Pr_4 = 157.915$

Thus $p_4 = p_3 \dfrac{Pr_4}{Pr_3} = 132\, psia \left(\dfrac{157.915}{341.780}\right) =$

$$\boxed{p_4 = 61.0\, psia}$$

SP5, nozzle exit: First, the pressure ratio will be used to determine the relative pressure at SP5. Then the enthalpy can be found in the Air Tables for use in the nozzle equation.

$$Pr_5 = Pr_4 \frac{p_5}{p_4} = 157.915 \left(\frac{11\, psia}{61\, psia}\right) =$$

$$\boxed{Pr_5 = 28.476}$$

From the Air Tables, $h_5 = 305.6\, Btu/lb_m$
Rearranging the nozzle equation and assuming $\bar{v}_4 = 0$ yields:

$$h_5 + \frac{\bar{v}_5^2}{2g_c} = h_4 + \cancel{\frac{\bar{v}_4^2}{2g_c}} \rightarrow$$

$$\bar{v}_5 = \sqrt{2g_c(h_4 - h_5)}$$

$$= \sqrt{2\left(32.2 \frac{ft \cdot lb_m}{lb_f \cdot s^2}\right)(491.7 - 305.6)\frac{Btu}{lb_m}\left(\frac{778\, ft \cdot lb_f}{Btu}\right)}$$

$$= 3053.5 \frac{ft}{s}$$

To find the thrust, use:

$$\bar{T} = \frac{\dot{m}_{air}(\bar{v}_5 - \bar{v}_0)}{g_c}$$

$$= \frac{210\, lb_m/s\,(3053.5 - 900)\, ft/s}{32.2 \frac{ft \cdot lb_m}{lb_f \cdot s^2}}$$

$$\boxed{\bar{T} = 14{,}045\, lb_f}$$

5.4.8.4 Turbofan Engines

The turbofan is similar to the turbojet, with one important difference. The turbofan has a large fan placed in front of the compressor section. Part of the air from this fan is forced into the engine core and part of the air bypasses the core. Figure 5.19 shows how this bypass air is ducted around the core and rejoins the exhaust. This figure shows a split-shaft design since the fan is driven by the low pressure turbine with a shaft located inside the hollow gas generator shaft. Although the bypass air is slower than the core exhaust, it provides a significant increase in thrust without much of an increase in fuel consumption. This is because a large mass of slow-moving air can produce more thrust than a small mass of fast moving air. As a result, turbofans are more efficient than turbojet engines. This is why turbojet engines are being phased out of modern aircraft designs.

The ratio of bypass air to core air is called the bypass ratio and is defined:

$$BPR = \frac{\dot{m}_{bypass\,air}}{\dot{m}_{core\,air}} \quad (5.29)$$

Small and fast aircraft have bypass ratios between 0.3 and 2.0. In larger and slower aircraft, such as cargo or passenger jets, the bypass ratio can be as high as 9.0!

Turbofan thrust is calculated by combining the core thrust and the bypass thrust.

$$\vec{T} = \vec{T}_{core} + \vec{T}_{bypass} \quad (5.30)$$

where

$$\vec{T}_{core} = \frac{\dot{m}_{core}\left(\vec{v}_{core\,exit} - \vec{v}_{aircraft}\right)}{g_c} \quad \text{and}$$

$$\vec{T}_{bypass} = \frac{\dot{m}_{bypass}\left(\vec{v}_{bypass\,exit} - \vec{v}_{aircraft}\right)}{g_c}$$

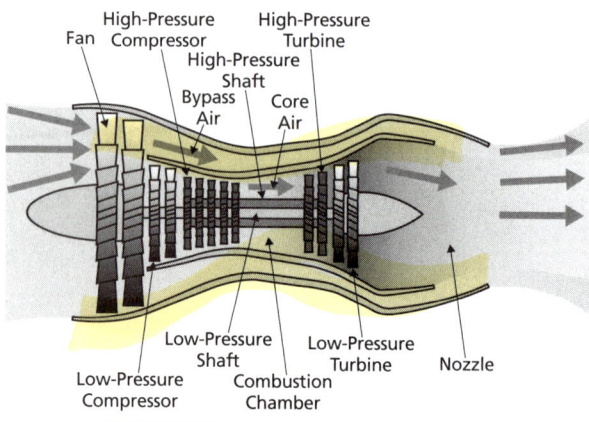

FIGURE 5.19: *Turbofan Engine*

EXAMPLE 5.7: TURBOFAN PROBLEM

GIVEN: A turbofan aircraft engine ingests air at 340 lb_m/sec while cruising at a speed of 650 ft/sec. The bypass ratio is 3:1 and the bypass air is accelerated by 1200 ft/sec as it passes through the fan. The temperature between the turbine and the nozzle is 850 °F. The exhaust temperature is 420 °F.

FIND:
a) The nozzle exit velocity (ft/sec)
b) The overall thrust developed by the engine (lb_f)
c) the power generated by the engine (hp)

RELATIONSHIPS:

$$\frac{\vec{v}_1^2}{2g_c} + h_1 = \frac{\vec{v}_2^2}{2g_c} + h_2 \qquad \vec{T} = \frac{\dot{m}_{air}\left(\vec{v}_{exit} - \vec{v}_{aircraft}\right)}{g_c}$$

$$BPR = \frac{\dot{m}_{bypass\,air}}{\dot{m}_{core\,air}} \qquad Power = \vec{T} \cdot \left(\vec{v}_{aircraft}\right)$$

SOLUTION:

a) Draw a representative schematic. Lay out a state point table. Sketch a representative h-s graph.

b) The nozzle equation is used to determine the velocity increase in the core exhaust. As before, air velocity in the engine will be considered negligible compared to the exit velocity. Also, the turbine exit will be considered SP4 and the nozzle exit will be SP5.

From the Air Tables:

$T_4 = 850\,°F \rightarrow h_4 = 319.5\,Btu/lb_m$
$T_5 = 420\,°F \rightarrow h_5 = 211.4\,Btu/lb_m$

$$\frac{\vec{v}_4^2}{2g_c} + h_4 = \frac{\vec{v}_5^2}{2g_c} + h_5 \rightarrow$$

$$\vec{v}_5 = \sqrt{2g_c(h_4 - h_5)}$$

$$= \sqrt{2\left(32.2\,\frac{ft \cdot lb_m}{lb_f \cdot s^2}\right)(319.5 - 211.4)\frac{Btu}{lb_m}\left(\frac{778\,ft \cdot lb_f}{Btu}\right)}$$

$$\boxed{\vec{v}_5 = 2,327.3\,ft/s}$$

c) To find the overall thrust, both the core and bypass thrust must be found. Bypass ratio must be used to determine bypass and core mass flow rate.

$$BPR = \frac{\dot{m}_{bypass\,air}}{\dot{m}_{core\,air}} = 3.0 \quad \text{and}$$

$$\dot{m}_{core} + \dot{m}_{bypass} = \dot{m}_{total} = 340\,lb_m/s$$

solving these two equations simultaneously yields:

$$\dot{m}_{core} = 85.0\,lb_m/s \quad \text{and} \quad \dot{m}_{bypass} = 255.0\,lb_m/s$$

$$\vec{T}_{core} = \frac{\dot{m}_{core}(\vec{v}_5 - \vec{v}_{aircraft})}{g_c} = \frac{85\frac{lb_m}{s}(2{,}327 - 650)\frac{ft}{s}}{32.2\frac{ft \cdot lb_m}{lb_f \cdot s^2}}$$

$$= 4{,}426.8\,lb_f$$

$$\vec{T}_{bypass} = \frac{\dot{m}_{bypass}(\vec{v}_{bypass} - \vec{v}_{aircraft})}{g_c} = \frac{255\frac{lb_m}{s} \cdot 1{,}200\frac{ft}{s}}{32.2\frac{ft \cdot lb_m}{lb_f \cdot s^2}}$$

$$= 9503\,lb_f$$

$$\vec{T}_{total} = \vec{T}_{core} + \vec{T}_{bypass} = (4{,}426.8 + 9503.1)\,lb_f$$

$$= 13{,}929.9\,lb_f$$

$$\boxed{\vec{T}_{total} = 13{,}930\,lb_f}$$

d) Finally, the power is calculated by:

$$Power = \vec{T} \cdot (\vec{v}_{aircraft})$$

$$= 13{,}929.9\,lb_f \left(650\frac{ft}{s}\right)\left(\frac{hp \cdot s}{550\,ft \cdot lb_f}\right)$$

$$= 16{,}462\,hp$$

$$\boxed{\dot{W} = 16{,}462\,hp}$$

Practice Problems

5.1 Complete the following table using the Air Tables. Do not interpolate values in the Air Tables. Use the nearest entry in Air Tables instead.

	T (°R)	T (°F)	h (Btu/lb$_m$)	P_r	h (Btu/lb$_m$)
1	2560				
2		75			
3			165.2		
4				50.2	
5					437.4

5.2 The following data pertain to an ideal gas turbine engine: Inlet temperature is 80 °F, maximum cycle temperature is 1800 °F, inlet pressure is 15 psia, compressor pressure ratio is 10:1, and the mass flow rate of the working substance is 135 lb$_m$/sec. Sketch the cycle on an h-s concept graph and determine:

a) The compressor discharge pressure (psia)

b) The turbine exhaust pressure (psia)

c) The net work of the cycle (Btu/lb$_m$)

d) The power produced by the gas turbine engine (hp)

e) The cycle thermal efficiency (%)

	1	2s	3	4s	
p (psia)					
T (°F)					
h (Btu/lb$_m$)					
P_r					
r_p					
q (Btu/lb$_m$)					$\Delta q =$
w (Btu/lb$_m$)					$\Delta w =$

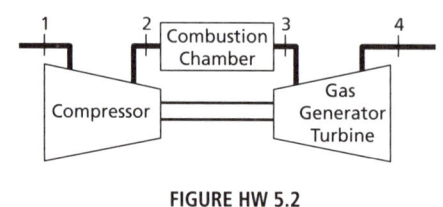

FIGURE HW 5.2

5.3 Repeat problem 5.2 with a pressure ratio 17:1.

5.4 An ideal gas turbine engine operating in an air standard Brayton cycle is served by a compressor that receives air at 14.5 psia and 70 °F and has a pressure ratio of 8:1. 340 Btu/lb$_m$ of heat is added in the combustion chamber. Find:

a) The compressor work (Btu/lb$_m$)

b) The turbine work (Btu/lb$_m$)

c) The net work of the cycle (Btu/lb$_m$)

d) The cycle thermal efficiency (%)

e) The back work ratio (%)

	1	2	3	4
p (psia)	14.5			
T (°F)	70			
h (Btu/lb$_m$)				
P_r				

5.5 Repeat problem 5.4 with isentropic efficiencies of 80% and 90% for the compressor and turbine, respectively.

	1	2s	2	3	4s	4
p (psia)	14.5					
T (°F)	70					
h (Btu/lb$_m$)						
P_r						

5.6 A single shaft gas turbine engine receives air at atmospheric conditions of 14.7 psia and 80 °F. The compressor discharge pressure is 103 psia and the compressor efficiency is 87%. The turbine inlet temperature is 1980 °F and the exhaust temperature is 1173 °F. Assume a 3 psi pressure drop in the combustion chamber and that inlet and exhaust duct losses are both 27.7 inches of water.

a) Draw the concept process graph on h-s coordinates.

b) Complete the <u>entire</u> state point table and circle all "givens."

c) Determine the compressor pressure ratio.

d) Determine the isentropic efficiency of the turbine.

e) Determine the thermal efficiency of the engine.

	1	2s	2	3	4s	4
p (psia)						
T (°F)						
h (Btu/lb$_m$)						
P_r						

5.7 For the split-shaft gas turbine engine shown below (with $\eta_c = 89\%$, $\eta_{pt} = 91\%$), find:

a) The work required by the compressor (Btu/lb$_m$)

b) The work produced by the gas generator turbine (Btu/lb$_m$)

c) The enthalpy leaving the gas generator turbine (Btu/lb$_m$)

d) The work produced by the power turbine (Btu/lb$_m$)

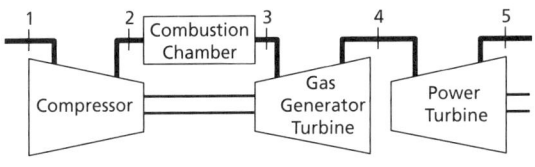

FIGURE HW 5.7

	1	2s	2	3	4s	4	5s	5
p (psia)	14.2	220	220	210	60	60	15.0	15.0
T (°F)	90			2190				
h (Btu/lb$_m$)								
P_r								

5.8 The T-34C is powered by a single Pratt and Whitney PT6A-25 turbo-prop engine. The PT6A uses a 7:1 pressure ratio. The engine delivers a max of 425 hp. Assume no pressure losses in the gas turbine engine and assume that the compressor, GGT, and PT all have isentropic efficiencies of 90%. The ITT (the temperature between the GGT and PT) for the PT6A is typically 1280 °F for continuous operations. By procedure, the pilot will usually use 100% power to climb to a higher altitude and 30% power to descend to a lower altitude. Neglecting the use of a diffuser or nozzle, if the T-34C is operating on a typical Pensacola day (90 °F, 15 psia at sea level) determine the following:

a) The enthalpy at each state point (SP1 is the inlet to the compressor and SP5 is the exhaust from the power turbine)

b) The net work of the cycle (Btu/lb$_m$) delivered to the prop shaft

c) The mass flow rate of air through the PT6A in a climb (lb$_m$/s)

d) The mass flow rate of air through the PT6A in a decent (lb$_m$/s)

	1	2s	2	3	4s	4	5s	5
p (psia)	15	105	105	105			15	15
T (°F)	90					1280		
h (Btu/lb$_m$)								
Pr								

5.9 Given an LM2500 split-shaft gas turbine engine running at 50% of rated power in a DDG-51 application (rated power for this model of LM2500 is 26,250 BHP) with the following parameters:

$p_{barometer}$ = 29.53 in-Hg $\quad\quad$ Δp_{inlet} = 0.3 psid $\quad\quad$ η_c = 85% $\quad\quad$ LHV = 18,750 Btu/lb$_m$

$T_{ambient}$ = 55 °F $\quad\quad$ Δp_{cc} = 3.0 psid $\quad\quad$ η_{cc} = 97% $\quad\quad$ HHV = 20,250 Btu/lb$_m$

T_{peak} = 2300 °F $\quad\quad$ $\Delta p_{exhaust}$ = 0.5 psid $\quad\quad$ η_{GGT} = 92% $\quad\quad$ $\dot{m}_{fuel}$ = 4 lb$_m$/s

r_p = 11 $\quad\quad\quad\quad\quad\quad\quad\quad\quad\quad\quad\quad\quad\quad\quad\quad\quad\quad\quad$ $\eta_{mech,engine}$ = 90%

a) Sketch the schematic for this system and the h-s diagram and then resolve all system pressures and enthalpies, except for the real enthalpy at the power turbine exhaust.

b) Determine the heat supplied (q_s in Btu/lb$_m$).

c) Determine the air flow rate (lb$_m$/s).

d) Find w_{PT} in Btu/lb$_m$.

e) Determine the η_{PT} (%).

f) Solve for the brake specific fuel consumption (bsfc in lb$_m$/hp-hr).

g) Determine η_{th} (%).

	1	2s	2	3	4s	4	5s	5
p (psia)								
T (°F)								
h (Btu/lb$_m$)								
Pr								
r_p								
q (Btu/lb$_m$)								
w (Btu/lb$_m$)								

5.10 The H-60 utilizes 2 GE-T700-401C turbo-shaft engines. Each engine is rated at 1662 hp continuous power. The normal operating TGT (gas generator turbine exit temperature) for each engine is 851°C (1564 °F). When the ECS (air conditioning) system is turned on, the H-60 NATOPS manual states that power output of each engine will decrease by 7% (of full power), due to the loss of air flow though the engine as the H-60's ECS system runs on bleed air from the compressor (see diagram). The temperature of the compressor discharge air is known to be 710 °F. The HV of the fuel used is 18,500 Btu/lb$_m$. Given an H-60 hovering at 80% power at sea level (15 psia) and 60 °F, a combustion chamber efficiency of 73%, no pressure losses and all isentropic components, determine the following:

a) The pressure ratio

b) The net work (Btu/lb$_m$) of the H-60 engine in this configuration

c) The air flow rate with the ECS off (lb_m/s)

d) The fuel flow rate to each engine with the ECS off (lb_m/s)

e) The fuel flow rate to each engine with the ECS **on** while maintaining the hover (lb_m/s)

  (Hint: To run the engine with the ECS on, without losing altitude in the hover, the 7% power loss must be made up for by increasing the air flow through the engine. This increase in air flow and the resulting increase in fuel flow will be **in addition** to the nominal fuel requirements with the ECS off.)

f) What is the fuel flow consumption of the H-60 (in lb_m/hr) in this hover with the ECS on? (Remember the H-60 has two engines.)

g) Does the fuel flow estimation of 1000 lb_m/hr used by H-60 pilots make sense based on your analysis?

	1	2	3	4	5
p (psia)	15				15
T (°F)	60	710		1564	
h (Btu/lb_m)					
P_r					

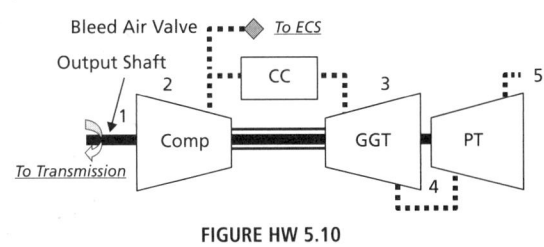

FIGURE HW 5.10

5.11 The now retired T-2 Buckeye training jet was powered by a GE J85 turbojet engine. The engine was rated at 2925 lb_f of thrust at sea level on a standard day (59 °F, 14.7 psia) and had an air mass flow rate of 45 lb_m/sec while cruising at 250 knots.

	0	1	2	3	4	5
p (psia)	14.7	18.1	126.7	126.7	60.2	14.7
T (°F)	59	80	476	1600	1250	770
h (Btu/lb_m)	124.1	129.1	225.2	521.3	425.2	299.0
P_r	1.120	1.386	9.702	195.9	93.03	26.32

a) Draw the h-s diagram for a turbojet engine assuming that all components, including the nozzle and diffuser, are ideal.

b) What is the pressure ratio of the J85?

c) What is the compressor work of this engine (Btu/lb_m)?

d) How is the turbine work related to the compressor work in a turbojet engine?

e) Determine the velocity of the exhaust gases leaving the nozzle of the J85 in this configuration (ft/s).

5.12 Most GTE engines used in naval applications typically run on JP-5. The *LHV* for JP-5 is approximately 18,500 Btu/lb_m and the *HHV* is 21,500. JP-5 may be modeled as $C_{12}H_{26}$ (170.34 lb_m/lbmole). A GTE is consuming JP-5 at a rate of 950 lb_m/hr, with a combustion chamber efficiency of 85%. Given the following data for the inlet (SPa) and outlet (SPb) conditions for a combustion chamber, complete the SP table and answer the following questions:

a) What is pressure drop across the combustion chamber and why does this occur in real GTE engines?

b) What is the specific heat supplied to this combustion chamber (q_s in Btu/lb_m)?

c) What is the stoichiometric air to fuel ratio for JP-5? (Recall that there are 4.76 moles of air per mole of Oxygen in air and 28.97 lb_m/lbmole air.)

d) Should you use the *HHV* or *LHV* for JP-5 in this engine?

	a	b
p (psia)	375	372
T (°F)	745	1750
h (Btu/lb_m)		

e) Given your answer in part d), what is the mass flow rate of air through this combustion chamber in lb_m/s? Does this GTE run rich or lean from stoichiometric?

f) Can you determine if this combustion chamber is an afterburner? Does that distinction matter in your analysis?

5.13 A turbojet aircraft engine intakes air at a rate 340 lb_m/sec while cruising at a speed of 650 ft/sec. The temperature measured between the turbine and the nozzle is 850 °F. The exhaust temperature is recorded as 420 °F. Find:

a) The increase in velocity that occurs through the nozzle (ft/sec)

b) The overall thrust developed by the engine (lb_f)

c) The power generated by the engine (hp)

5.14 A turbojet aircraft with an un-pressurized cockpit is flying with a velocity of 400 kts at an altitude where the atmospheric pressure is 10 psia and the atmospheric temperature is 10 °F. What is the airspeed in ft/s? The compressor's discharge pressure gage reads 107.9 psi. The tech manual for the engine lists the compressor's isentropic efficiency as 75% and the turbine's isentropic efficiency as 85%. The pressure difference across the combustion chamber is 5 psid. Assume that both the diffuser and the nozzle are ideal. The engine is consuming JP-5 at the rate of 14,400 pounds per hour. The *LHV* for JP-5 is approximately 18,500 Btu/lb_m. JP-5 may be modeled as $C_{12}H_{26}$. What is the stoichiometric A/F ratio? (Recall that there are 3.76 moles of Nitrogen per mole of Oxygen in air.) The actual A/F ratio is 50:1 and you may assume 100% combustion efficiency. Does this engine run rich or lean? Justify your answer. What are the mass flow rate of air, the engine thrust, and engine power in this scenario? Hint for finding p_4: Match the real work of the turbine with the real work of the compressor. Then find the isentropic work of the turbine and use relative pressures (*Pr*) to determine the turbine's pressure ratio.

6 Steam Power

The USS *Iwo Jima* (LHD-7) is powered by a steam propulsion plant capable of generating 70,000 shaft horsepower.

6.1 UNIT LEARNING OBJECTIVES

6.1.1 Terminology: definitions, variables, and typical units

6.1.1.1 saturated steam, wet vapor, saturated vapor, saturated liquid, sub-cooled liquid, superheated vapor, condensate depression, quality, moisture content, boiler / steam generator, turbine, condenser, feed pump, economizer, higher heating value

6.1.2 Concepts: ideas and engineering expressions of them

6.1.2.1 Relate water phases to given properties.

6.1.2.2 Sketch the properties of water on T-s and h-s coordinates.

    6.1.2.2.1 Identify and state the meaning of the saturated vapor line.

    6.1.2.2.2 Identify and state the meaning of the saturated liquid line.

    6.1.2.2.3 Identify the critical point.

6.1.2.3 Sketch the Rankine cycle processes on both T-s and h-s coordinates.

6.1.2.4 Define Higher Heating Value (*HHV*) and explain why it is used in this application.

6.1.3 Skills: procedures, practices, or methods that enable reasoning

6.1.3.1 Determine the properties of saturated and superheated steam using the steam tables.

6.1.3.2 Determine the properties of liquid water using the steam tables.

6.1.3.3 Calculate the moisture content and quality of "wet steam."

6.1.3.4 Interpret the Mollier diagram to determine state point properties and processes.

6.1.3.5 Sketch and evaluate cycle schematics and determine the value of enthalpy at each state point.

6.1.3.6 Calculate the energy transfer and phase transformations in a condenser, a boiler, a steam turbine, and a centrifugal pump.
6.1.3.7 Calculate the thermal efficiency of a Rankine cycle.
6.1.3.8 Calculate the efficiency of a boiler.
6.1.3.9 Calculate ideal work, real work, and isentropic and mechanical efficiency of a steam turbine.
6.1.3.10 Calculate the fuel flow rate required for a steam plant.
6.1.3.11 Calculate the overall efficiency of a steam propulsion system.

6.2 INTRODUCTION

Although conventional steam propulsion is regarded as something of a modern phenomenon, the first steam engines appeared before the American Revolution. English engineer Thomas Savery patented a crude steam engine in 1698 that pumped water. Improvements soon followed. In 1712 English blacksmith Thomas Newcomen constructed an improved steam engine designed to pump water from mines. Newcomen's engine used steam pressure to lift a piston in a cylinder and then quenched the steam with water spray creating a partial vacuum. Atmospheric pressure on the top side of the piston then provided the force to move the piston down. The piston was connected to a rocking beam. The piston being pushed down operated a reciprocating pump that was attached to the opposite side of a pivoting beam.

The leap in technology that would propel the industrial revolution was provided by the Scottish inventor James Watt. Watt (1) applied steam pressure to the piston in the engine to provide the motive force, (2) added a separate condenser to draw exhaust steam out of the cylinder, and (3) added an insulated steam chamber around the cylinder so it would remain hot. Watt's improvements, the first of which was patented in 1769, led to higher efficiency and greatly enhanced the practicality of the steam engine. The separate condenser remains a common feature on all modern steam plants.

The first application of the steam engine to marine propulsion occurred on the Delaware River in 1787 when John Fitch successfully tested a forty-five-foot steamboat. Despite the obvious promise of the steam-driven vessel, John Fitch's business ultimately failed. It was the American inventor Robert Fulton who fitted a Boulton & Watt engine to drive the steam-powered vessel *Clermont* to commercial success in 1807. The *Clermont*'s performance ignited interest in steam-powered naval vessels and earned Fulton the nickname "Father of Steam Navigation." During the War of 1812, the U.S. Navy's first steam-powered warship was built. Originally named the *Demologos*, after the death of its famous designer in 1815 it was renamed the *Fulton*. The *Fulton* was driven by a single-cylinder steam engine driving a paddle wheel, which was mounted between two hulls—a catamaran. The *Fulton*, although steam-powered, still retained sails for reserve power. The *Fulton* never deployed, but remained in New York to protect that port from invasion.

During the nineteenth century, the major navies of the world began incorporating steam propulsion into their fleets. Extremely low efficiency early on meant that steam was limited to auxiliary propulsion because the ships could not carry enough fuel for long voyages under steam power. By the end of the nineteenth century, the technology had advanced to the point that steam replaced wind as the dominant means of ship propulsion in the world. Reciprocating steam engines dominated from the time of Fulton into the early twentieth century. The high-speed steam turbine, coupled to reduction gears, was the next wave of improved steam propulsion.

Initially, wood provided the fuel for marine steam engines. However, wood was quickly replaced by coal, which contained higher energy density than wood. Coal fueled European and American imperial expansion in the latter part of the nineteenth century and early twentieth century. Part of the drive for imperial expansion by European powers and the United States involved creating a network of coaling stations around the world to supply their battleship fleets. Yet even as the coal-fired Great White Fleet of the United States was making its way around the world from 1907 to 1909, fuel oil was in the process of supplanting coal as the energy source to naval steam engines. Fuel oil could be pumped directly into boilers, rather than hand-shoveled; it could be stored in tanks distributed around the hull and transferred easily from other vessels; it was not susceptible to exploding (as coal dust was) and possessed roughly the same energy density as coal. All these factors made it more appealing than coal as the preferred fuel for vessels.

Fuel oil–fired steam propulsion was the dominant propulsion system for ships of every type up through World War II. While steam continued to dominate naval propulsion for many years following World War II, the advent of the diesel engine started to make inroads into ship propulsion for commercial vessels. Today, most commercial ships are powered by diesel engines. Some, notably a number of cruise ships, have been built with gas turbines for power. The main advantage of diesel and gas turbine engines over steam is reduced manpower for operation and maintenance.

On land in the United States, there are about 440 large coal-fired steam electric generating plants and about 100 nuclear steam power plants. These utility power plants, using steam power, generate the largest percentage of electricity of all available electric generating technologies.

6.3 U.S. NAVY CURRENT APPLICATIONS OF STEAM POWER

Relatively few surface vessels in the U.S. Navy continue to operate with *conventional* steam boilers, i.e., those where steam is created by burning fossil fuels. Those ships continuing to operate with conventional steam propulsion include various auxiliaries and some of the larger amphibious vessels, including the four ships of the *Tarawa* class (LHA-1) and the eight ships of the *Wasp* class (LHD-1). Notably, the eighth ship of the *Wasp* class, the *Makin Island* (LHD-8), is gas turbine powered. The older ten ships of the *Austin* class (LPD-4) are steam-powered, but their newer replacements, the planned twelve ships of the *San Antonio* class (LPD-17), are diesel powered. The three amphibious command ships are steam powered. The Military Sealift Command (MSC) operates more than 110 ships bearing the designation U.S. Naval Ship (USNS). Many of these ships are steam powered, but none are commissioned naval vessels, and all are crewed by civilians.

Although use of conventional steam plants is steadily being phased out of the Navy, steam propulsion remains vitally important to nuclear-powered vessels. In fact, all of the world's nuclear-powered vessels are steam powered. Nuclear steam plants, although requiring more maintenance and operator oversight relative to the other engines in use today, do not require frequent refueling and do not require air for combustion of the fuel. The nuclear reactor replaces the heat source supplied by combustion of conventional fuels, and the heat is transferred from the reactor to the boiler, in this application known as the steam generator, by hot pressurized water. The rest of the components of the nuclear steam plant are similar in form and function to their conventional counterparts. Current U.S. naval vessels employing nuclear steam plants include the *Los Angeles* class (SSN-688), the *Ohio* class (SSBN-726) and associated SSGN variants, the *Seawolf* class (SSN-21), the *Virginia* class (SSN-774), the *Nimitz* class (CVN-68), and the USS *Enterprise* (CVN-65).

Commercially, only a small number of nuclear-powered vessels were ever constructed in the world. The United States built one, the NSS (Nuclear Steam Ship) *Savannah,* which proved to be uneconomical and was retired very early in the projected life of the ship.

The steam plant on the USS *Virginia* is rated at approximately 40,000 shaft horsepower, while the *Nimitz-*class propulsion plant can produce 280,000 shaft horsepower. More specifics about nuclear technology will be covered in Unit 8.

6.4 STATES OF WATER

A firm understanding of water, specifically in its liquid and gaseous states, is essential to any discussion of steam power. The conditions at which water can exist as it pertains to steam power are more complex than what is conventionally understood. Water exists in five states. Before these states are discussed in detail, we must first discuss a concept know as *saturation*.

6.4.1 Saturation Conditions

The term *saturation* defines a condition in which a mixture of vapor and liquid can exist at a given temperature and pressure. The temperature at which vaporization (boiling) starts to occur for a given pressure is called the *saturation temperature* or *boiling point*. The pressure at which vaporization (boiling) starts to occur for a given temperature is called the *saturation pressure*. For water at 212°F the saturation pressure is 14.7 psia, and for water at 14.7 psia the saturation temperature is 212°F.

For a pure substance there is a definite relationship between saturation pressure and saturation temperature. The higher the pressure, the higher the saturation temperature. The graphical representation of this relationship between temperature and pressure at saturated conditions is called the *vapor pressure curve*. A typical vapor pressure curve is shown in Figure 6.1. The vapor/liquid mixture is at saturation when the conditions of pressure and temperature fall on the curve.

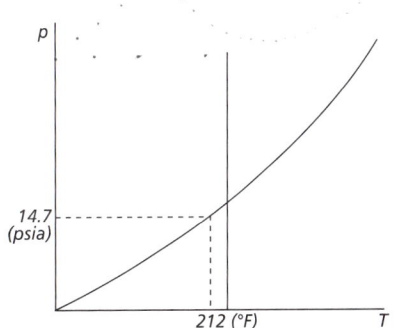

FIGURE 6.1: *Vapor Pressure Curve for Water*

6.4.2 Quality and Moisture

When water is at saturation conditions, i.e., exactly at saturation temperature and saturation pressure, a mixture of water vapor and liquid may be present. This mixture of liquid and vapor is illustrated in Figure 6.2. The percentage by mass of liquid and vapor can be defined by one of two terms known as quality (x) and moisture content (m). *Quality* is defined as the ratio of the mass of the vapor to the total mass of both vapor and liquid. Thus, if the mass of vapor is 0.2 lb_m and the mass of the liquid is 0.8 lb_m, the quality is 0.2 or 20 percent. Quality has meaning when the substance is in a saturated state only, at saturation pressure and temperature.

$$x = \frac{m_{vapor}}{m_{liquid} + m_{vapor}} \quad (6.1)$$

The *moisture content (m)* of a substance is the opposite of its quality. Moisture is defined as the ratio of the mass of the liquid to the total mass of both vapor and liquid. The moisture of the mixture in the previous paragraph would be 0.8 or 80 percent. The following equations show the relationships between quality and moisture.

$$m = \frac{m_{liquid}}{m_{liquid} + m_{vapor}} \quad (6.2)$$

and

$$m = 1 - x \quad (6.3)$$

6.4.3 T-s Diagram

The easiest way to visualize the various states of water within the region of interest is on water's temperature—entropy *(T-s)* graph. Although all students will be familiar with water's solid state (ice) and the transitions directly to/from a vapor (sublimation) and to/from a liquid (melting), we are not concerned with these phase transitions. As a reminder from earlier cycles, a T-s graph is frequently used to analyze energy transfer cycles. This is because the work done by or on the system and the heat added to or removed from the system can be visualized on the *T-s* graph. By the definition of entropy, the heat transferred to or from a system equals the area under the process curve on the *T-s* graph.

Figure 6.2 is the *T-s* diagram for pure water. A *T-s* graph can be constructed for any pure substance and will have a similar shape. The graph is not to scale and certain areas are exaggerated for clarity. As indicated in Figure 6.2, the graph will have a similar shape for *T-h* and *T-v*,

but the *T-s* version will be the one that will provide the best utility for our purposes.

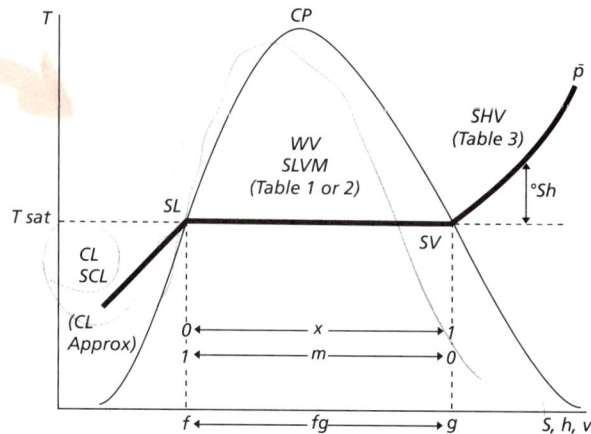

FIGURE 6. 2: *The "Steam Dome"*

The dark line is a line of constant pressure, $\bar{p}$. The abbreviations on Figure 6.2 indicate the different states of water and where the states transition from one to another at a given pressure. Note that all points lie on the line of constant p. There are an infinite number of lines of constant p on the *T-s* graph. Only one is shown. As shown in Figure 6.1, T_{sat} increases with increasing p, so higher pressures will have higher saturation temperatures, too.

Note: "x" in the Table 6.1 denotes steam quality (%)

Let's look at Figure 6.2 and examine the states of water using the example of a piston and cylinder filled with water on a stove burner, shown in Figure 6.3 (this is the concept of the actual apparatus originally used to measure steam properties). Assume the cylinder is filled to about ¼ full with pure water and the piston's weight is negligible so that atmospheric pressure above the piston controls the pressure on the water. Assume the atmospheric pressure is standard (14.7 psia).

SUB-COOLED LIQUID (SCL). The temperature of SCL water is less than T_{sat} at a given pressure. Cold tap water is typically around 60 °F.

$$T < T_{sat}(@\,p)$$

Now let's consider our cylinder of water. Initially, our cylinder of water is at 60 °F and has 14.7 psia exerted on it from the atmosphere with a negligible hydrostatic pressure component, since the depth of the water is only several inches. As a result, the temperature is less than the saturation temperature (i.e., boiling point for a given p). Thus the water in the cylinder is an *SCL*. Now, turn the burner on and add heat to the water. As "sensible heat" is added to the water, the heat shows up as a temperature

$T < TSat$

TABLE 6.1: *The Five States of Water on the "Steam Dome"*

State	Notes	Description
Sub-cooled liquid (SCL) aka Compressed Liquid (CL)	$T < T_{sat}$ (for given press)	Liquid water. Temperature of the water is below the saturation temperature for water at a given pressure.
Saturated Liquid (SL) subscript "f"	$T = T_{sat}$ (for given press) $x = 0\%$	Liquid water. Temperature of the water is equal to the saturation temperature of water at a given pressure, but none of the water has been converted to vapor.
Wet Vapor (WV) aka Saturated Liquid & Vapor Mixture (SLVM)	$T = T_{sat}$ (for given press) $0\% < x < 100\%$	Mixture of liquid and vapor (steam). Both steam and water exist together with both of them being at saturation temperature and pressure.
Saturated Vapor (SV) subscript "g"	$T = T_{sat}$ (for given press) $x = 100\%$	Vapor (steam). All of the water exists as steam at the saturation temperature for a given pressure.
Superheated Vapor (SHV)	$T > T_{sat}$ (for given press)	Vapor (steam). All of the water exists as steam. The temperature of the vapor is above the saturation temperature for a given pressure.

increase. The water will increase in temperature until it gets to T_{sat}, where its classification becomes a saturated liquid (SL).

SATURATED LIQUID (SL). At this point, the water is at saturation conditions, meaning that the temperature of the water equals T_{sat} for the given p and additional heat added will not result in a temperature increase. At saturation conditions, water and vapor can exist together. However, no vapor has yet formed, meaning that the percent steam is zero ($x = 0$).

$$T = T_{sat} (@\, p) \quad \text{and} \quad x = 0\%$$

So, given the definition of an SL, the water in our cylinder is just at the point where it is ready to boil, but it hasn't yet. The water is sitting in the cylinder at exactly 212 °F at standard atmospheric pressure. Note that the water will have become less dense as the heat is added and therefore take up additional volume as compared to the SCL.

WET VAPOR (WV). Naturally, the water doesn't stay an SL for long, because as we continue adding heat, the water begins to boil, which puts it in the next region, wet vapor

(WV). T remains constant due to the heat being applied to the latent heat of vaporization. The WV region lies between the SL and SV points. In this region, water is at saturation, meaning that water liquid and vapor exist together at the T and p. As Figure 6.2 shows, no matter what point you pick along the pressure line, the p_{sat} and T_{sat} do not change. The location of a specific state point along the T_{sat} line is defined by (1) specific entropy (s), (2) specific enthalpy (h), (3) specific volume (v), or (4) quality (x). At the SL point, $x = 0\%$. At the SV point, $x = 100\%$. So, between the SL and SV points, x ranges from 0% to 100%.

$$T = T_{sat} (@\, p) \quad \text{and} \quad 0\% < x < 100\%$$

As we watch our cylinder, the water boils and vapor fills the space above the water. Now, there's a mixture of water and vapor in the cylinder, and it is all at the same temperature. Over time, a higher and higher percentage of the water becomes vapor, meaning that the longer the cylinder is on the stove, the higher the quality that water in the cylinder achieves. While there is a mixture of water and vapor, we call it a wet vapor. Eventually, all the water turns to vapor, which creates the SV conditions.

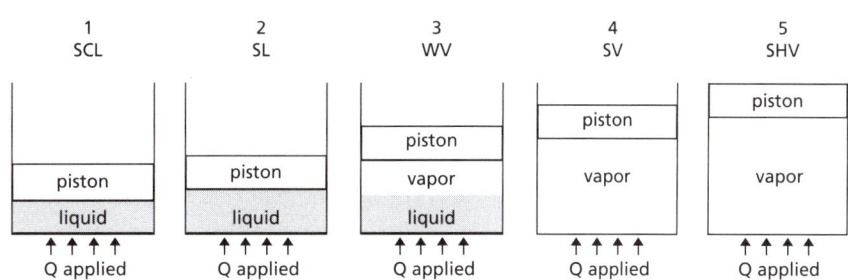

FIGURE 6.3: *Liquid and Vapor Phases of Water*

SATURATED VAPOR. At the *SV* point, water is still at saturation conditions, thus the *T* and *p* are the same as they are for the *SL* point and the WV region. The principal difference between the WV region and the SV point is that at the *SV* point, ALL of the water is vapor, meaning that the quality $x = 100\%$.

$$T = T_{sat}(@\,p) \quad \text{and} \quad x = 100\%$$

At saturated vapor conditions, our cylinder still has the same temperature and pressure that it did when it was saturated liquid and wet vapor, but now, looking at the cylinder, all of the water has turned into steam, meaning that the quality of the saturated mixture is 100% and the temperature is still 212 °F at standard atmospheric pressure. The steam doesn't stay saturated vapor for long. As we continue adding heat via the burner, we quickly turn the steam into superheated vapor.

SUPERHEATED VAPOR (SHV). In the *SHV* region, the steam is 100% vapor and the temperature of the steam is higher than T_{sat} for a given *p*. The heat applied is sensible heat, since it manifests as a temperature increase.

$$T > T_{sat}(@\,p)$$

As we continue adding heat to the steam, the temperature of the steam will rise above T_{sat}, and will continue rising as long as we add heat. Figure 6.2 illustrates this point. Remember that our boiling cylinder scenario assumes pressure remains constant at *p*. The temperature difference between T_{sat} and the real *T* in this region is °*Sh*.

Quality (*x*) loses its meaning outside of the saturated state. Likewise, °*Sh* does not make sense outside of the SHV region.

6.5 STEAM TABLES

Steam tables consist of two sets of tables of the energy transfer properties of water and steam: (1) saturated steam tables and (2) superheated steam tables. Both sets of tables are tabulations of pressure (*p*), temperature (*T*), specific volume (*v*), specific enthalpy (*h*), and specific entropy (*s*). The following notation is used in the steam tables.

- T = temperature (°F)
- p = pressure (psi)
- v = specific volume (ft³/lb$_m$)
- v_f = specific volume of saturated liquid (ft³/lb$_m$)
- v_g = specific volume of saturated vapor (ft³/lb$_m$)
- v_{fg} = specific volume change of vaporization (ft³/lb$_m$)
- h = specific enthalpy (Btu/lb$_m$)
- h_f = specific enthalpy of saturated liquid (Btu/lb$_m$)
- h_g = specific enthalpy of saturated vapor (Btu/lb$_m$)
- h_{fg} = specific enthalpy change of vaporization (Btu/lb$_m$)
- s = specific entropy (Btu/lb$_m$-°R)
- s_f = specific entropy of saturated liquid (Btu/lb$_m$-°R)
- s_g = specific entropy of saturated vapor (Btu/lb$_m$-°R)
- s_{fg} = specific entropy change of vaporization (Btu/lb$_m$-°R)
- °*Sh* = number of degrees of superheat (°F)

*Values with the subscript "*f*" are the values of the steam if at SL conditions
*Values with the subscript "*g*" are the values of the steam if at SV conditions
*Values with the subscript "*fg*" are just the mathematical difference between "*f*" and "*g*" values. These are mathematical tools useful in equations later but have no other substantial meaning.

A note about the Steam Tables *data: There are two common versions of the* Steam Tables *in current circulation. The American Society of Mechanical Engineers (ASME) produced a version dated 1967. Updated computational power in the intervening years ultimately resulted in recalculating the values in a more precise manner in the year 2000. The numbers are not significantly different, but the examples of this text are based upon the 2000 version. So, if you are using the older edition, the techniques are the same, but there may be some difference in results.*

The saturated steam tables (Tables 1 and 2) give the energy transfer properties of saturated water and saturated steam for temperatures from 32 °F to 705.47 °F (the critical temperature) and for the corresponding pressure from 0.08865 to 3200.1 psia. Normally, the saturated steam tables are divided into two parts: temperature tables, which list the properties according to saturation temperature (T_{sat}); and pressure tables, which list them according to saturation pressure (P_{sat}). The values of enthalpy and entropy given in these tables are measured relative to the properties of saturated liquid at 32 °F. Hence, the enthalpy (h_f) of saturated liquid and the entropy (s_f) of saturated liquid have values of approximately zero at 32 °F.

Remember: Tables 1 and 2 have the same information. It is just indexed differently!

Most practical applications using the saturated steam tables involve steam-water mixtures. The key property of such mixtures is steam quality (*x*), defined as the mass of steam present per unit mass of steam-water mixture, or steam moisture content (m), defined as the mass of water present per unit mass of steam-water mixture. The following relationships exist between the quality of a liquid-vapor mixture and the specific volumes, enthalpies, or entropies of both phases and of the mixture itself. These relationships are used with the saturated steam tables, where s_x, h_x, and v_x are real values at a given state.

$$h_x = h_f + x_x h_{fg} \quad v_x = v_f + x_x v_{fg} \quad s_x = s_f + x_x s_{fg} \quad (6.4)$$

These expressions can be rearranged to solve for quality at a given state.

$$x_x = \frac{h_x - h_f}{h_{fg}} \quad x_x = \frac{v_x - v_f}{v_{fg}} \quad x_x = \frac{s_x - s_f}{s_{fg}} \quad (6.5)$$

6.5.1 Determining the State and Properties of Steam

The most difficulty students studying steam encounter is how to use the *Steam Tables* to determine all of the properties of steam under different conditions. To determine the state of steam, you must know two separate and independent pieces of information regarding the steam. So, for example, in order to be able to determine all properties of steam at a given state, you must know any two independent intensive properties. Intensive properties, in contrast to extensive properties, do not depend upon how much of the substance exists. All of the following listed below are independent intensive properties.

1. Temperature (°F)
2. Pressure (psia)   Note: T and p are <u>not</u> independent if the system is saturated.
3. Quality (%)
4. Whether sub-cooled, saturated liquid, or saturated vapor
5. Specific enthalpy (Btu/lb$_m$)
6. Specific entropy (Btu/lb$_m$-F)
7. Specific volume (ft^3/lb$_m$)

6.5.2 How to Determine the State of Water

To find all properties, you must first determine the state. Use the following techniques with the steam tables.

1. Given temperature and pressure (T & p). Go to Steam Table 1 or Steam Table 2 and look up your value of pressure. Note: <u>At lower pressures</u>, Steam Table 1 tends to have more resolution for pressure values than Steam Table 2. If a problem gives a pressure that is not specifically listed in Steam Table 2, try to find a very close pressure in Steam Table 1 instead. Use the second column of Steam Table 1 for the pressure. Look at T_{sat} as listed in the table and compare it with the given value of T. Do not interpolate on the low pressures in Table 2. The resulting error is excessive.

EXAMPLE 6.1: SCL

LOGIC: If $T < T_{sat}$ then the water is a Sub-cooled liquid (SCL).
GIVEN: Pressure is 220 psia and temperature is 300 °F

Look in Steam Table 2 (saturated steam: pressure table) and find the row with 220 psia.

The saturation temperature for 220 psia is

$$T_{sat} = 389.886 \text{ °F}.$$

The given temperature for the water is 300 °F, so because $T < T_{sat}$, the water is sub-cooled. Note the location marked on the *T-s* graph.

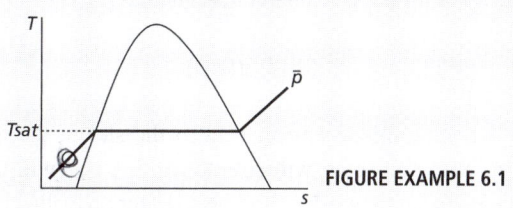

FIGURE EXAMPLE 6.1

EXAMPLE 6.2: SL, WV, SV

LOGIC: If $T = T_{sat}$ then the water is at saturation conditions, and must be saturated liquid, saturated vapor, or wet vapor.
GIVEN: $p = 250$ psia and $T = 400.983$ °F.

Look in Steam Table 2 (saturated steam: pressure table) and find the row with 250 psia.

The saturation temperature for 250 psia is $T_{sat} = 400.983$ °F.

The given temperature for the water is also 400.983 °F, so because $T = T_{sat}$, the water is a SL, WV, or SV. To determine any more information about the state in this example, we must either know the quality of the steam or be able to calculate it using h, s, or v, if known. Note the location marked on the *T-s* graph and that this condition describes a region and not a discrete point on the line of pressure.

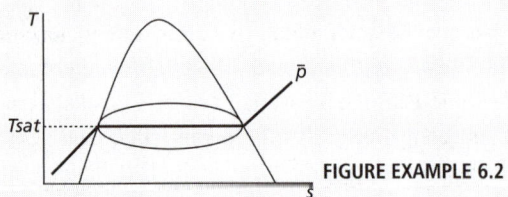

FIGURE EXAMPLE 6.2

EXAMPLE 6.3: SHV

LOGIC: If $T > T_{sat}$ then the water is *SHV*.
GIVEN: $p = 270$ psia and $T = 550\ °F$.

Look in Steam Table 2 (saturated steam: pressure table) and find the row with 270 psia.

The saturation temperature for 270 psia is
$$T_{sat} = 407.817\ °F.$$

The given temperature for the water is 550 °F, so because $T > T_{sat}$, the steam is a SHV.

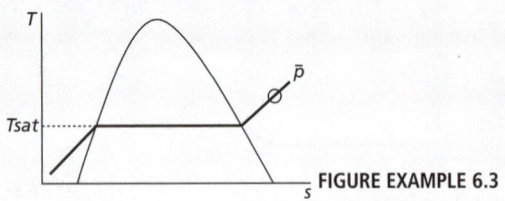

FIGURE EXAMPLE 6.3

EXAMPLE 6.4: WV

LOGIC: If x is defined and between 0% and 100%, then the water is a *WV*.
GIVEN: $p = 300$ psia; $x = 50\%$

Because a quality has been defined, we know that the steam is at saturation conditions at 300 psia and is a WV.

Look in Steam Table 2 (saturated steam: pressure table) and find the row with 300 psia. Using this row, the temperature of the steam is equal to the saturation temperature $T_{sat} = 417.35\ °F$. Note that the indicated location is 50% of the way across the horizontal portion of the line of constant p.

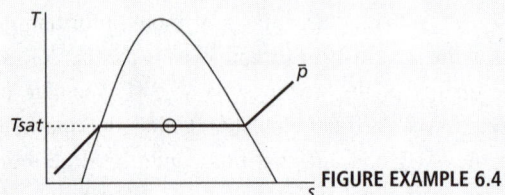

FIGURE EXAMPLE 6.4

Important Note: At low pressures (below 10 psia), Table 2 is too coarse for good accuracy and straight line interpolation yields unnecessary errors due to the data actually being on a curve. In this case, use Table 1 and enter it using the second column. Using this technique will reduce the interpolation needed to obtain data.

2. GIVEN THE STATE. Frequently, the state will be defined, i.e., a vapor will be described simply as an *SV* at 150 psia. Other common ways of defining state include saying some water is "sub-cooled" by some number of degrees or has some number of degrees of "condensate depression." These terms will be described in detail later, but if seen, these terms mean that the water is sub-cooled (SCL). Additionally, you may see a statement that steam is "superheated" by some number of degrees. Once again, this term will be described in detail shortly, but suffice to say, a statement like this means that the vapor is superheated (SHV).

3. GIVEN A TEMPERATURE (OR PRESSURE) AND THEN ONE OF THE FOLLOWING: *h, v,* OR *s*. Say, for example, that the given value is a value for *h*. Find the row containing the *T* or *p* of the steam in either Table 1 or Table 2. Try to find the closest temperature (or pressure) that is possible in either Table 1 or Table 2. That row will contain three values for *h, n,* and *s*. For example, specific enthalpy will have h_f, h_g, and h_{fg}. To determine the state of the water, determine if the given value of h falls between h_f and h_g because, as you may recall, h_f is the enthalpy of the water when it is saturated liquid and h_g is the enthalpy of the water when it is saturated vapor.

- If $h < h_f$ then the water is *SCL*.
- If $h = h_f$ then the water is *SL*.
- If $h_f < h < h_g$ then the water is a *WV* whose quality can be determined using the following expression from Equation 6.5:

$$x_x = \frac{h_x - h_f}{h_{fg}}$$

Note: If *s* or *v* were the given values, then the equation would be identical except that *h* would be replaced with *s* or *v*.

- If $h = h_g$ then the water is *SV*.
- If $h > h_g$ then the water is *SHV*.

Note: The above methods would work regardless of whether the *h, s,* or *v* is the given value.

EXAMPLE: 6.5:

GIVEN: $T = 400\ °F$ and $h = 1000\ Btu/lb_m$.

Look in Table 1 (saturated steam: temperature table) and find the row with 400 °F. Using this row, then we can determine the SL specific enthalpy (h_f) and the SV specific enthalpy (h_g).

At 400 °F, $h_f = 375.10\ Btu/lb_m$ and $h_g = 1201.5\ Btu/lb_m$.

Because h is between h_f and h_g, the steam is a WV. Notice that the value of h is closer to h_g than h_f and the circle on the line of pressure is a bit bigger. We'll calculate the exact location using x as illustrated in the next section.

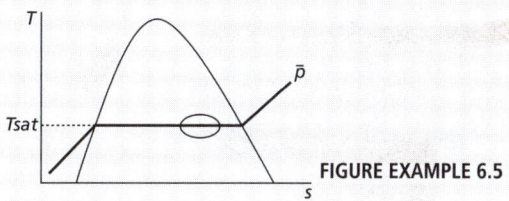

FIGURE EXAMPLE 6.5

6.5.3 Determine the Remaining Properties of Water and Steam

Now that the state of the water has been determined, finding the other properties is much easier.

1. SUB-COOLED LIQUID. When water is an SCL, properties depend primarily on temperature. So, we approximate the properties of an SCL as identical to the SL properties for the given T. Because SCLs are approximated as incompressible, p has negligible effect on an SCL's properties for pressures below about 1000 psia. This is called the "Compressed Liquid Approximation Technique."

- T = sub-cooled liquid temp
- p = given pressure (not saturation pressure)
- $h = h_{sat\ liquid}(@\ T)$
- $v = v_{sat\ liquid}(@\ T)$
- $s = s_{sat\ liquid}(@\ T)$

The influence of pressure on the enthalpy of water becomes more substantial at higher pressures. Using the CL Approximation Technique for temperatures around 100 °F will have about 2 percent error at 500 psia and about 3 percent at 800 psia for enthalpy. If compressed liquid tables are available, these data sources will reduce the error. If compressed liquid tables are not available, then at high pressures (>1000 psia), enthalpy may be determined more accurately by the following equation:

$$h(T,p) = h_f(T) + v_f(T)[p - p_{sat}(T)]$$

The CL Approximation Technique works well for specific volume (and density) throughout the range of data that will be used in this course.

EXAMPLE 6.6: CL APPROXIMATION TECHNIQUE

GIVEN: Pressure is 220 psia and temperature is 300 °F

$T_{sat}(220\ psia) = 389.886\ °F$; $T < T_{sat}$. We find the remaining properties by looking up 300 °F in Table 1 (saturated steam: temperature table), and use the SL values at 300 °F.

$h = h_{sat\ liquid}(at\ 300\ °F) = 269.76\ Btu/lb_m$

$v = v_{sat\ liquid}(at\ 300\ °F) = 0.017449\ ft^3/lb_m$

$s = s_{sat\ liquid}(at\ 300\ °F) = 0.4372\ Btu/lb_m\text{-}°R$

Note that p is still 220 psia even though p_{sat} for 300 °F is listed as 67.021 psia because pressure has a negligible effect on sub-cooled liquid properties.

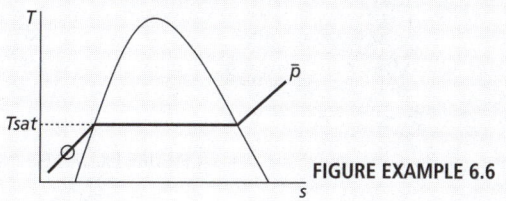

FIGURE EXAMPLE 6.6

2. SATURATED LIQUID. Find T_{sat} or p_{sat} in Table 1 or 2. The SL values listed for the given T or p are the correct properties.

EXAMPLE 6.7: SL

GIVEN: $p = 250\ psia$ and the system is an SL.

Knowing that the water is a saturated liquid, look in Table 2 (saturated steam: pressure table) and find the row with 250 psia. The properties of the water are those in the saturated liquid column.

$h = h_f(250\ psia) = 376.16\ Btu/lb_m$

$v = v_f(250\ psia) = 0.018653\ ft^3/lb_m$

$s = s_f(250\ psia) = 0.5679\ Btu/lb_m\text{-}°R$

$T = T_{sat}(250\ psia) = 400.983\ °F$

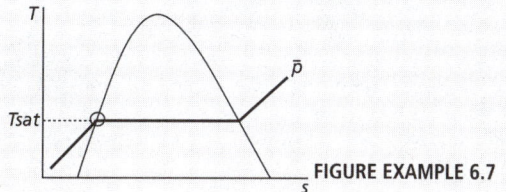

FIGURE EXAMPLE 6.7

3. WET VAPOR. A WV is part of a saturated system, and the saturation pressure and temperature are located in the same row of Table 1 or 2. Once you know T_{sat} (or p_{sat}) you may use Equation 6.5 to determine a quality, if necessary.

Then Equation 6.4 may be used to calculate other properties. The quality (x) as calculated by any of Equation 6.5 can be used in Equation 6.4 to calculate the other property values.

EXAMPLE 6.8: WV

GIVEN: $p = 300$ psia and $x = 50\%$

Because quality has been defined, we know that the steam is at saturation conditions at 300 psia and is a WV.

Look in Table 2 (saturated steam: pressure table) and find the row with 300 psia. Using this row, the temperature of the steam is equal to the saturation temperature, $T_{sat} = 417.366$ °F.

The remaining properties are calculated using the following:

$$h = h_f + x h_{fg} \quad v = v_f + x v_{fg} \quad s = s_f + x s_{fg}$$

$$h_f(300\,psia) = 394.0 \frac{Btu}{lb_m}$$

$$h_{fg}(300\,psia) = 809.38 \frac{Btu}{lb_m}$$

$$v_f(300\,psia) = 0.018897 \frac{ft^3}{lb_m}$$

$$v_{fg}(300\,psia) = 1.5434 \frac{ft^3}{lb_m}$$

$$s_f(300\,psia) = 0.5883 \frac{Btu}{lb_m \cdot °R}$$

$$s_{fg}(300\,psia) = 0.9229 \frac{Btu}{lb_m \cdot °R}$$

$$h = 394.0 \frac{Btu}{lb_m} + (0.50)\left(809.38 \frac{Btu}{lb_m}\right) = 798.69 \frac{Btu}{lb_m}$$

$$v = 0.018897 \frac{ft^3}{lb_m} + (0.50)\left(1.5434 \frac{ft^3}{lb_m}\right) = 0.790597 \frac{ft^3}{lb_m}$$

$$s = 0.5883 \frac{Btu}{lb_m \cdot °R} + (0.50)\left(0.9229 \frac{Btu}{lb_m \cdot °R}\right)$$

$$= 1.04975 \frac{Btu}{lb_m \cdot °R}$$

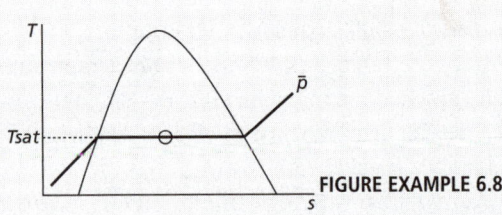

FIGURE EXAMPLE 6.8

A word about rounding: Do not round off the values looked up from the tables until your final answer. Then p, T, and h may be rounded to the nearest tenth; s and v should include a similar number of digits as looked up from the tables.

EXAMPLE 6.9: QUALITY (X)

Now, let's look at a more complicated case using quality.

GIVEN: Steam is operating at 420 psia and $h = 1000$ Btu/lb$_m$,

FIND: the v of the steam.

First, determine the state of the steam. Find 420 psia in Table 2 (saturated steam: pressure table). From this we get the following:

$$h_f = 429.56 \text{ Btu/lb}_m \quad h_{fg} = 775.58 \text{ Btu/lb}_m$$

and

$$h_g = 1205.1 \text{ Btu/lb}_m$$

The given value for h is between h_f and h_g, which means that the steam in this case is a WV. Now we must determine x of the WV using the following:

$$x = \frac{h - h_f}{h_{fg}} = \frac{1000 \frac{Btu}{lb_m} - 429.56 \frac{Btu}{lb_m}}{775.58 \frac{Btu}{lb_m}}$$

$$= 0.7355 \Rightarrow 73.6\%$$

Using x, v can be calculated as shown below. Incidentally, s and h of the steam could also be calculated using the following method (see Equation 6.5).

$$v = v_f + (x)v_{fg} \quad \text{where} \quad v_f(420\,psia) = 0.019427 \frac{ft^3}{lb_m}$$

$$\text{and} \quad v_{fg}(420\,psia) = 1.0869 \frac{ft^3}{lb_m}$$

$$v = 0.019427 \frac{ft^3}{lb_m} + (0.7355)1.0869 \frac{ft^3}{lb_m} = 0.81884 \frac{ft^3}{lb_m}$$

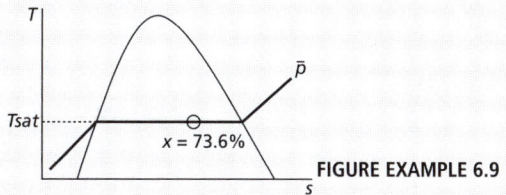

FIGURE EXAMPLE 6.9

4. SATURATED VAPOR. Find the T_{sat} or p_{sat} in Table 1 or 2. The SV values listed for the given T or p are the steam properties.

EXAMPLE 6.10: SV

GIVEN: Steam is a saturated vapor with a pressure of 500 psia.
FIND: the other steam properties from the steam table.

Knowing that the water is a SV vapor, look in Table 2 (saturated steam: pressure table) and find the row with 500 psia. The properties of the water are those in the saturated vapor column.

$T = T_{sat}$ (500 psia) = 467.047 °F

$h = h_{sat\ vapor}$ (500 psia) = 1205.0 Btu/lb$_m$

$v = v_{sat\ vapor}$ (500 psia) = 0.9282 ft^3/lb$_m$

$s = s_{sat\ vapor}$ (500 psia) = 1.4643 Btu/lb$_m$-°R

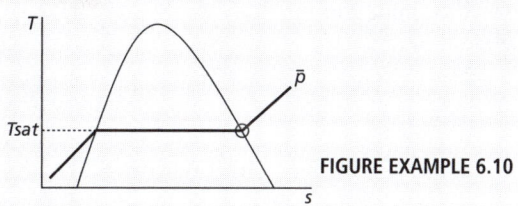

FIGURE EXAMPLE 6.10

5. SUPERHEATED VAPOR. The values for superheated vapor are found in Table 3. Enter with the two given values and select the property values. Some interpolation may be necessary

EXAMPLE 6.11: SHV

GIVEN: $p = 500$ psia; $T = 700$ °F.
FIND: the remaining properties for the steam

First, to verify the state of the steam, check Table 2 (saturated steam: pressure table). At 500 psia, we find that Table 2 lists T_{sat} = 467.047 °F. Notice also that Table 3 lists T_{sat} in parentheses (467.05).

Because $T > T_{sat}$ the steam is a SHV.

SHV properties are in Table 3, from which the following values can be drawn:

$h = h$ (500 psia, 700 °F) = 1357.0 Btu/lb$_m$

$v = v$ (500 psia, 700 °F) = 1.3044 ft^3/lb$_m$

$s = s$ (500 psia, 700 °F) = 1.6117 Btu/lb$_m$-°R

°Sh = °Sh (500 psia, 700 °F) = T − T_{sat} = 700 − 467.05 = 232.95 °F

(°Sh stands for *degrees of superheat*, which will be discussed shortly)

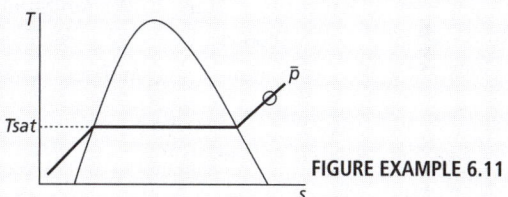

FIGURE EXAMPLE 6.11

6.5.4 Sub-cooling and Condensate Depression

The terms *sub-cooling* and *condensate depression* mean the same thing. They are terms used to describe the temperature of water relative to T_{sat} of the water at a given pressure.

When water is sub-cooled by a number of degrees it means that the water temperature is that number of degrees below the saturation temperature for a given pressure. In other words:

$$T = T_{sat} \text{ (at given pressure)} - \text{Degrees sub-cooled} \quad (6.6)$$

Condensate depression means the same thing, except that this term is typically associated with the discharge of condensers in steam plants.

$$T_{condensate} = T_{sat} \text{ (at given pressure)} - \text{Condensate Depression} \quad (6.7)$$

EXAMPLE 6.12: CONDENSATE DEPRESSION

GIVEN: A main condenser is discharging water at pressure of 2 psia that is sub-cooled by 6 degrees (or it could be stated that condensate is experiencing 6 degrees of condensate depression).

FIND: the properties of the water

First look in Table 2 (saturated steam: pressure table) and try to find 2 psia. When you look in Table 2, note that the table lists values of 1.0 and 10 psia, but nothing in between. DO NOT INTERPOLATE TABLE 2 AT LOW PRESSURES! Notice that Table 1 provides better resolution because when you look in the saturation pressure column in Steam Table 1, you find that there is a row with P_{sat} = 1.9985 psia corresponding to a T_{sat} = 126 °F.

Using the following equation:

$T = T_{sat}$ (at given pressure) – Degrees sub-cooled

$T = 126\ °F - 6\ °F = 120\ °F$

The state of the water is SCL. Therefore, we use the CL Approx Technique to determine the other values.

$h = h_f (120\ °F) = 88.002\ Btu/lb_m$

$v = v_f (120\ °F) = 0.016205\ ft^3/lb_m$

$s = s_f (120\ °F) = 0.1647\ Btu/lb_m\text{-}°R$

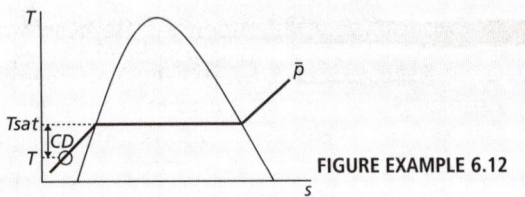

FIGURE EXAMPLE 6.12

6.5.5 Degrees of Superheat

Superheated steam is sometimes described by the phrase "degrees of superheat." This term is applied when the steam is known to have a temperature that is some number of degrees higher than the saturation temperature (for a given pressure), and that number is the number of degrees that the steam is superheated by.

$T = T_{sat} + °Sh$ (degrees of superheat) (6.8)

EXAMPLE 6.13: °SH

GIVEN: Steam at 500 psia is superheated by 132.95 degrees.

FIND: the temperature of the steam in degrees Fahrenheit (°F)

$T = T_{sat}$ (500 psia) = 467.05 °F

$T = 467.05\ °F + 132.95\ °F = 600\ °F$

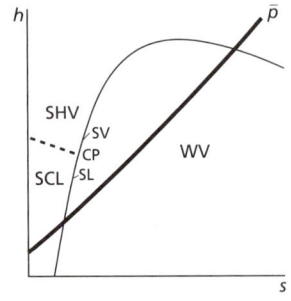

FIGURE EXAMPLE 6.13

6.6 ENTHALPY: ENTROPY (h-s) OR MOLLIER DIAGRAMS

The Mollier diagram is a chart on which specific enthalpy (h) versus specific entropy (s) is plotted. All the information available in the steam tables is also available on the Mollier diagram because the Mollier diagram is just a graphical representation of data in the steam tables.

It is sometimes known as the h-s diagram and has an entirely different shape from the T-s diagrams. The chart contains a series of constant temperature lines, a series of constant pressure lines, a series of constant moisture or quality lines, and a series of constant superheat lines. The Mollier diagram is used only when quality is greater than about 50 percent and for superheated steam.

FIGURE 6.4: *Mollier (h-s) Concept Diagram Showing Fluid States*

6.6.1 Using the Mollier Chart (Figure 6.5)

Using the Mollier chart is just a matter of practice. Make sure you know what all of the lines on the Mollier chart mean. Once you're familiar with the layout, if you enter with any two independent values and figure out where they intersect on the Mollier chart, other properties can be determined. You are encouraged to highlight the words on the chart to better orient yourself to what the various lines mean—the words line up with the lines to which they refer.

One key skill that is not too obvious is to determine the saturation temperature for some condition inside the vapor dome—WV conditions. Recognize the saturation temperature and saturation pressure are linked—they are

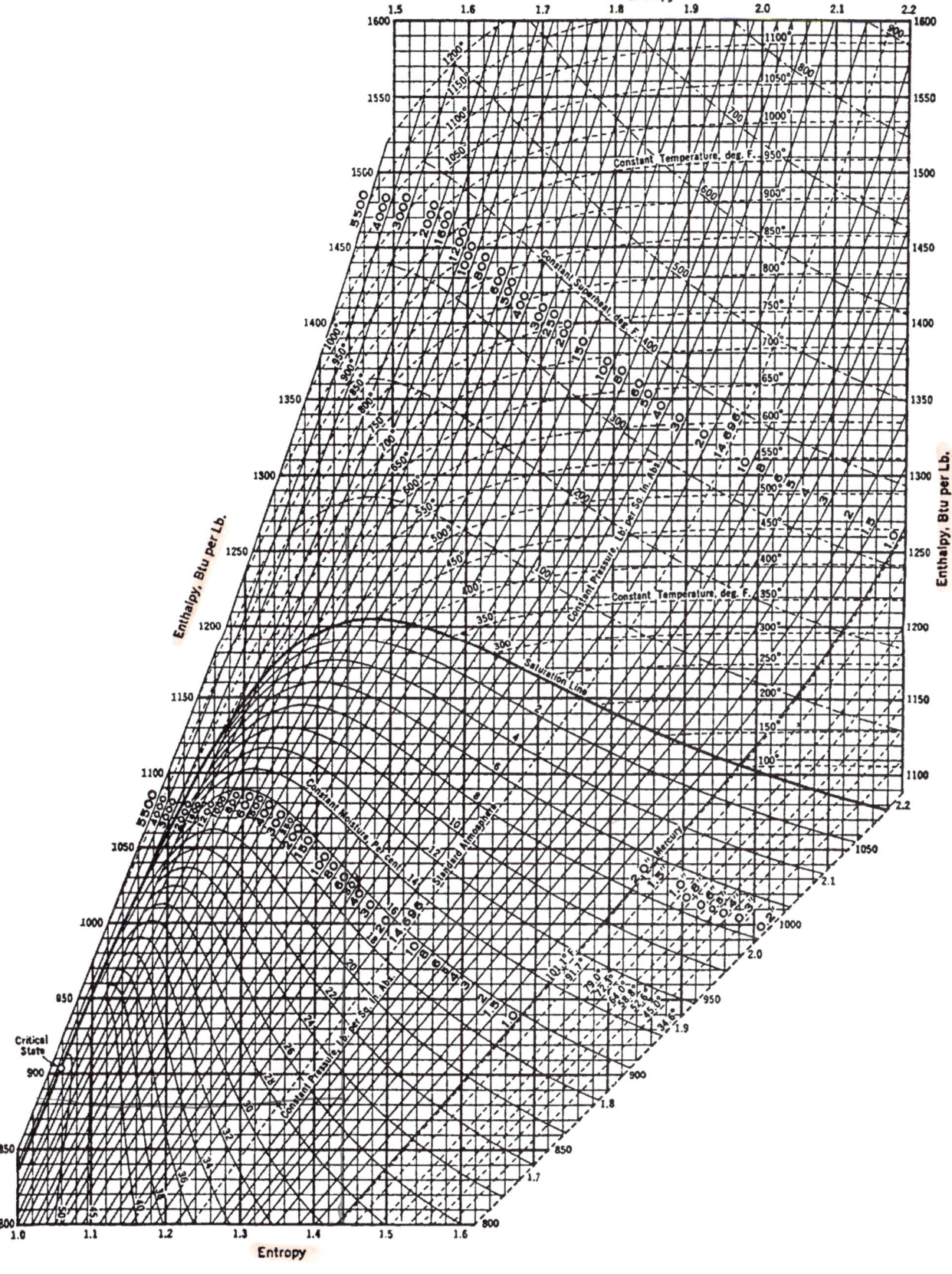

FIGURE 6.5: *Mollier (h-s) Chart*

Mollier chart from "1967 ASME Steam Tables," by IAPWS, reprinted by permission from ASME

not independent. Therefore, follow the pressure line up from the state point to the saturation line and read the constant temperature at that intersection. This may require some graphical interpolation. That is not a difficult skill, but does require some attention to detail.

The Mollier chart allows for quicker solution of turbine expansion problems. Notice that the graph does not include CL data, so an entire steam plant problem can not be solved with a Mollier chart due to the lack of low pressure and low enthalpy data on the graph. The Mollier chart is also quicker for isenthalpic problems such as throttling of steam through a throttle or governor valve. Through careful use of the chart in the back of the *Steam Tables* booklet, students should be able arrive at answers within the closest unit of enthalpy (not tenths as may be done through using the tables).

6.7 THE BASIC STEAM PLANT

With the essential skills in determining the properties of the working fluid, we now work to apply those skills toward solving problems with steam power plants.

The sequence we will follow, starting at the most basic:

a. Orientation to the four components and sequence of a basic steam plant.
b. The Rankine cycle—the basic steam plant with ideal components.
c. The basic steam plant with real (non-isentropic) work components
d. Real steam plants with additional compents and split flows

The basic steam plant consists of four components connected in the following sequence. This follows the C-A-E-R sequence that has been discussed in earlier units on heat engines.

1 → 2 Pump: The purpose of the pump is to take condensate produced by the main condenser and raise it to a high pressure, after which it is called feedwater. This high pressure feedwater then flows to the boiler where it can be boiled. The water at this point is called feedwater, hence the feed pump. The feed pump normally only consumes a small fraction of power produced by the steam plant.

2 → 3 Boiler: The boiler is where feedwater is boiled into steam. "Steam generator" is the name used for the boiler in a nuclear power plant. On conventional steam plants, boiler water is heated by burning fuel. The fuel is normally fuel oil in ships and coal, natural gas, or fuel oil on land-based power plants. Conventional steam plant boilers typically produce superheated steam while nuclear plant steam generators only produce a saturated vapor ("dry saturated steam").

3 → 4 Turbine: The purpose of the turbine is to convert the stored energy of the steam to rotating shaft work. Turbines are usually comprised of a casing surrounding a rotor, or shaft, with rows of turbine blades. Turbine blades extract energy as the steam expands and flows over them. This causes the shaft to turn. There are usually multiple rows of turbine blades, and each row is called a "stage." Steam enters the turbine at a high pressure and exits the turbine at a low pressure, usually less than atmospheric pressure.

4 → 1 Condenser: The function of the condenser is to condense the exhaust steam from the turbine back into liquid water. The condenser is a heat exchanger that uses cooling water from another source like a river or the ocean to remove the energy from steam plant water without mixing the two together. Unfortunately, in this process, the majority of the energy put into the steam by the boiler is rejected by the condenser as waste heat. Rejection of this heat is necessary to turn the turbine exhaust steam back into liquid water so that it can be pumped back up to the high pressure of the steam generator.

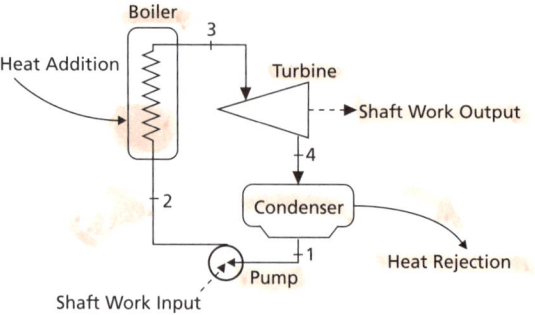

FIGURE 6.6: *The Basic Steam Plant*

6.8 THE RANKINE CYCLE (THE BASIC STEAM PLANT WITH IDEAL COMPONENTS)

The Rankine cycle is the theoretical cycle used to represent the basic steam cycle. The Rankine cycle assumes the following:

1. Only the four basic steam plant components are present.

2. The turbine and pump are 100 percent efficient (isentropic and adiabatic).
3. Steam from the boiler can be either dry saturated steam or superheated vapor. If conditions are not specified that describe a SHV, then assume SV conditions.
4. Condensate out of the condenser is saturated liquid at condenser pressure.
5. Isobaric heat exchangers.
6. No pressure drops or heat losses in the system piping.

The Rankine cycle is comprised of the four processes discussed below.

1 → 2 <u>Isentropic Compression</u> (w_{in}). The feed pump isentropically (i.e., at constant entropy) raises pressure of the condensate from the pressure of the condenser, $P_{condenser}$, to the pressure of the steam generator, P_{boiler}.

2 → 3 <u>Isobaric Heat Addition</u> (q_s). Heat is added to the water in the steam generator causing it to boil at a constant pressure P_{boiler}.

3 → 4 <u>Isentropic Expansion</u> (w_{out}). In the turbine, energy is removed from the steam, causing the pressure to drop. As pressure drops, the steam expands isentropically (i.e., at constant entropy).

4 → 1 <u>Isobaric Heat Rejection</u> (q_r). In the condenser, cooling water removes heat from the turbine exhaust, causing it to condense. While the steam is condensing, it stays at a constant pressure, $P_{condenser}$. The cooling water flows through the inside of tubes while the steam flows around the outside of the tubes. The cooling water and steam do not come into contact with each other.

The various numbered states in Figure 6.7 have specific names and properties:

    State 1 – Condensate. SL.
    State 2 – Feedwater. SCL.
    State 3 – Steam. SV
    State 4 – Turbine Exhaust. WV.

$p_1 = p_4 = p_{condenser}$ $p_2 = p_3 = p_{boiler}$

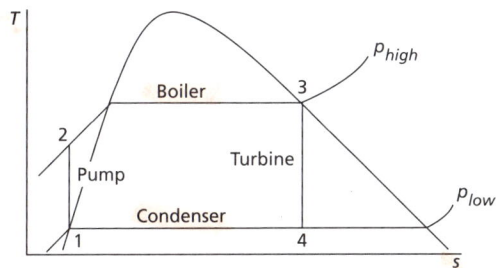

FIGURE 6.7: *Ideal Basic Rankine Cycle T-s Diagram*

6.9 BASIC STEAM CYCLE ANALYSIS

The primary emphasis of steam plant analysis is determining the enthalpies, h, at each state.

$$w_{in} = w_{pump} = h_1 - h_2 \; (Btu/lb_m) \quad (6.9)$$
$$\text{value should be negative (-)}$$

$$q_s = q_{boiler} = h_3 - h_2 \; (Btu/lb_m) \quad (6.10)$$
$$\text{value should be positive (+)}$$

$$w_{out} = w_{turbine} = h_3 - h_4 \; (Btu/lb_m) \quad (6.11)$$
$$\text{value should be positive (+)}$$

$$q_r = q_{condenser} = h_1 - h_4 \; (Btu/lb_m) \quad (6.12)$$
$$\text{value should be negative (-)}$$

At this point, students should be able to determine the properties of SP1 and SP3. SP2 and SP4 will require working with the isentropic process through the pump and turbine.

<u>SP1</u>: Condensate: "Condensate exits condenser (or enters feed pump)"
1. SP1 is a saturated liquid
   $h_1 = h_f$ at $p_{condenser}$ → go to Steam Table 1 or 2 to get enthalpy
2. Look up v_1 (ft$_3$/lb$_m$) for use in the next step.

<u>SP2s</u>: Ideal feedwater will be fed into the boiler.
This is the state of feedwater exiting the 100% efficient feed pump. SP2 is isentropic with SP1 and is therefore designated SP2s. The energy input to a pump has been covered previously in Unit 2. The appropriate equations are 6.13 and 6.14. Note why we captured v_1 in the step above.

$$w_{pump\,ideal} = -v_1(p_2 - p_1) \left(\frac{144\,in^2}{1\,ft^2} \right) \left(\frac{1\,Btu}{778\,ft-lb_f} \right) \quad (6.13)$$

$$h_{2s} = h_1 - w_{pump\,ideal} \quad \text{and don't} \quad (6.14)$$
forget that $w_{pump\,ideal}$ is negative

<u>SP3</u>: Steam from the boiler
1. Determine if SP3 is SV or SHV.
   If T_3 is not given (or some other property at SP3 that would allow determination of the state as SHV) then assume SV.
   Compare given steam temp, T_3, to T_{sat} for the given pressure in the boiler.
      If $T_3 = T_{sat}$ it is a saturated vapor

If $T_3 > T_{sat}$ it is a superheated vapor

2. If SV then:

   $h_3 = h_g @ p_{boiler}$ go to Steam Table 2

   OR

   $h_3 = h_g @ T_{sat}$ go to Steam Table 1

3. If SHV then:

   $h_3 = h(p_{boiler}, T_3)$ go to Steam Table 3

SP4s: Steam exiting the ideal (isentropic) turbine. SP4 is isentropic to SP3, therefore designated SP4s.

First, realize that $s_3 = s_{4s}$, then

1. $s_{4s} = s_f + x_{4s}(s_{fg})$ → solve for x_{4s} (steam quality)
   s_f and s_{fg} are found in Steam Table 2 (at $p_{condenser}$)

2. $h_{4s} = h_f + x_{4s}(h_{fg})$ → solve for h_{4s}

EXAMPLE 6.14: BASIC IDEAL STEAM PLANT ANALYSIS

GIVEN: An ideal Rankine cycle operating between 600 psia and 5 psia.

FIND: η_{th}

SOLUTION: Start by sketching the schematic and the processes on a T-s graph. Note that the control points are SP1 and SP3. These occur at the intersection of the pressure line and the SL curve (SP1) and the SV curve (SP3). All we were given were the pressures. We would need additional information to adjust the SP locations on this T-s graph.

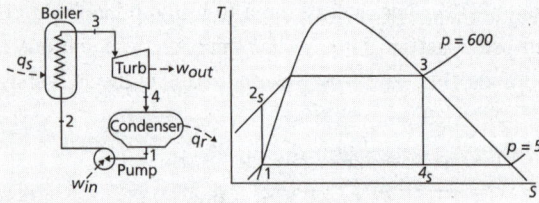

FIGURE EXAMPLE 6.14

We'll use a state point table to organize data. Given data are shown in **BOLD**. Assumptions are noted with an asterisk (*). NR means not required for the analysis. Once sufficient info to determine the h is available, the other cells are optional. It is good practice to grab some other info, just in case. Experience will help you to know the difference. The 1967 Steam Tables were used for the answers to this example. The order in which the table is filled out is shown in the solution sequence.

	1	2s	3	4s
p (psia)	**5**	**600**	**600**	**5**
T (°F)	162.24	164.0	486.20	162.24
h (Btu/lb$_m$)	130.2	132.0	1203.7	883.5
s (Btu/lb$_m$-°F)	NR	NR	1.4461	=1.4461
v (ft^3/lb$_m$)	0.016407	⇐	NR	NR
x (%)	0	NR	100	75.3
State	SL*	CL	SV*	WV

SP1: Feed pump inlet. Assumed SL at 5 psia since no other data are given for this state point. The pressure and SL condition make two independent intensive properties so we can fix the state.

$$h_1 = h_f(5\ psia) = 130.2\ Btu/lb_m$$
$$v_1 = v_f(5\ psia) = 0.016407\ ft^3/lb_m$$

SP2s: Feed pump discharge. Use the isentropic pump process to determine the h value in the CL region. Water is relatively incompressible so $v_2 = v_1$. The state will be CL by going vertically up from an SL condition on the T-s graph.

$$w_{pump,ideal} = -v\Delta p = -\left(0.016407\frac{ft^3}{lb_m}\right)$$
$$(600-5)\frac{lb_f}{in^2}\left\{144\frac{in^2}{ft^2}\right\}\left\{\frac{1 Btu}{778\ ft\cdot lb_f}\right\}$$
$$= -1.81\frac{Btu}{lb_m}$$

$$h_{2s} = h_1 - w_{pump,ideal} = \left(130.2\frac{Btu}{lb_m}\right) - \left(-1.8\frac{Btu}{lb_m}\right)$$
$$= 132.0\frac{Btu}{lb_m}$$

T_{2s} comes from the $\Delta h = C_p \Delta T$. Since C_p of liquid water is 1.0 Btu/lb$_m$-°R in this range, the change in enthalpy (Btu/lb$_m$) is the change in temperature.

SP3: Boiler outlet. Assumed SV at 600 psia since no other data were given for this state point. These make two independent items so we can fix the state.

$$h_3 = h_g(600\ psia) = 1203.7\ Btu/lb_m$$

We will need s for the next step.

$$s_3 = s_g(600\ psia) = 1.4461\ Btu/lb_m\text{-}°R$$

SP4s: Turbine outlet. First, realize that the cycle was described as ideal, so the turbine is isentropic.

Therefore, $s_{4s} = s_3$.

Then, step 1, solve for x_{4s} (steam quality).

$s_{4s} = s_f + x_{4s}(s_{fg}) \rightarrow$

$$x_{4s} = \frac{s_{4s} - s_f(5\,psia)}{s_{fg}(5\,psia)} = \frac{1.4461 - 0.2349}{1.6094} = 0.753$$

Now, step 2, use x_{4s} to calculate h_{4s}.

$$h_{4s} = h_f(5\,psia) + x_{4s} h_{fg}(5\,psia)$$
$$= 130.2 + (0.753)(1000.9) = 883.5\,Btu/lb_m$$

Graphically, it can be seen that the state is a WV, so the $T_{4s} = T_{sat}(5\,psia)$.

Now that the enthalpy row is completed, we can go on to answering the question. Start with the definition of h_{th}.

$$\eta_{th} = \frac{w_{net}}{q_s} = \frac{\sum w}{q_s} = \frac{\sum q}{q_s}$$

We'll solve this using the work, so let's calculate the turbine and pump work. With KE and PE terms canceling, the SFEE simplifies to $h_i + q_{i\text{-}f} = h_f + w_{i\text{-}f}$. The turbine and pump are both assumed to be adiabatic as part of being ideal. Therefore w is just a difference in h across the machine.

$$w_{turbine} = h_3 - h_{4s} = 1203.7 - 883.5$$
$$= +320.2\,Btu/lb_m$$

$w_{pump} = -1.8\,Btu/lb_m$ (calculated previously)

$$w_{net} = w_{turbine} + w_{pump} = 320.2 + (-1.8)$$
$$= 318.4\,Btu/lb_m$$

Similarly,

$$q_s = h_3 - h_{2s} = 1203.7 - 132.0 = 1071.7\,Btu/lb_m$$

$$\eta_{th} = \frac{w_{net}}{q_s} = \frac{318.4}{1071.7} = 0.297 = 29.7\%$$

$\boxed{\eta_{th} = 29.7\%}$

6.10 NON-IDEAL STEAM PLANT

A real steam plant uses the same basic components as the Rankine cycle and is very similar to the Rankine cycle in all respects, with the following exceptions:

1. There can be more than the four basic components, but unless otherwise stated, assume only the four basic steam components are being used. Other components that may be inserted into the system include throttle and governor valves, parallel flows to different turbines and heaters.
2. Steam out of the boiler can be either saturated (SV) or superheated (SHV).
3. The turbine and pumps can be less than 100 percent efficient, but if not otherwise stated, assume the pump and turbine are 100 percent efficient.
4. Condensate can be either SL or SCL. If SCL, then "condensate depression" exists.
5. Pressure changes can occur across the boiler or condenser, or through the connecting pipes.
6. Heat losses to the operating spaces can occur from the components.

Let's look at the logic for solving a number of these extra complications that reality introduces.

6.10.1 Condensate Depression

Condensate depression in a main condenser is the intentional sub-cooling of water, or condensate, that has been condensed on the heat exchanger tubes. Initially, after the steam is condensed on the tubes, it becomes liquid water at the saturation temperature, or saturated liquid (SL). Additional cooling of the saturated liquid drops the temperature of the water, which is now called condensate, such that the condensate has become sub-cooled (SCL), or depressed. The terms condensate depression and sub-cooling can effectively be used interchangeably. Condensate depression is necessary to protect the pumps downstream of the main condenser, but too much condensate depression can reduce the efficiency of the steam plant.

Pumps move liquid condensate from the main condenser back to the boiler or steam generator. At the suction inlet to all pumps, a low pressure region forms, and the resulting pressure drop at this low pressure point could cause the condensate to flash to steam if the temperature of the condensate is saturated leaving the condenser. Depressing the condensate, or sub-cooling it slightly, stops the condensate from reaching saturation conditions. Steam formed in a pump is known as *cavitation*, and in extreme cases *vapor binding* may occur. Cavitation wears the impeller, and vapor binding can lead to mechanical failure of the pump seals and bearings. Approximately 2 to 4 °F of condensate depression is usually sufficient to prevent these problems in naval propulsion plants. This is related to the net positive suction head (*NPSH*) discussion in Unit 2.

The downside to condensate depression is the loss of thermal efficiency because additional heat removed from the condensate must then be added again in the boiler or steam generator in order to re-boil the water.

SP1: Condensate: "Condensate exits condenser (or enters feed pump)"

a. Determine if SP1 is saturated liquid or sub-cooled liquid. See examples earlier in the chapter.

If saturated liquid, then:

- $h_1 = h_f$ at $p_{condenser}$ → go to Steam Table 1 or 2 to get enthalpy

If sub-cooled liquid, then use the CL Approximation technique:

$T_1 = T_{sat}$ (at $p_{condenser}$) – Degrees Sub-cooled

$h_1 = h_f$ at T_1 → go to Steam Table 1 or 2 to get enthalpy

b. v_1 (ft³/lb$_m$) = v_f (at T_1) regardless of whether state 1 is saturated or sub-cooled liquid

Refer back to Section 6.5.4 to see how to adjust the graphed location of the pump suction state point.

6.10.2 Pump and Turbine Efficiency <1

The *T-s* diagram for a saturated steam plant with real work components is shown in Figure 6.8 below. The primary difference in the *T-s* diagram for a real steam cycle and the ideal Rankine cycle are the dotted lines between states 1 and 2 and states 3 and 4. Note that SPs 2s and 4s represent the ideal (isentropic—the pump and turbine are 100 percent efficient) exit conditions of the pump and turbine, respectively, and SPs 2 and 4 represent the real discharge conditions.

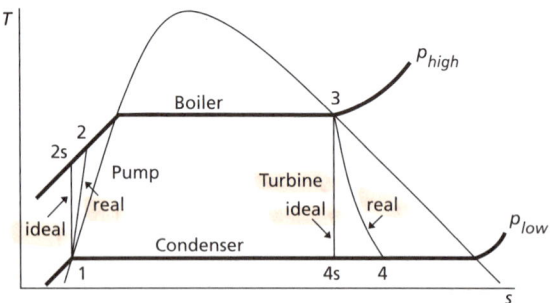

FIGURE 6.8: *Real Saturated Steam Cycle (No Superheat)*

NON-IDEAL PUMP. The line for the pump between SP1 and SP2 indicates that there have been losses due to turbulence and friction, causing a rise in entropy (*s*).

SP2s: Ideal feedwater
This is the state of feedwater exiting the feed pump if the pump is 100 percent efficient.

If the pump is 100 percent efficient, like in the case of a pump in a Rankine cycle, then state 2s and state 2 are the same.

$$w_{pump\ ideal} = -v_1(p_2 - p_1)\left(\frac{144\ in^2}{1\ ft^2}\right)\left(\frac{1\ Btu}{778\ ft-lb_f}\right)$$

$h_{2s} = h_1 - w_{pump\ ideal}$ and don't forget that $w_{pump\ ideal}$ is negative

SP2: Real feedwater
This is the state of feedwater exiting the feed pump if the pump is *less than* 100 percent efficient.

$$\eta_{PUMP} = \frac{\text{Water Horse Power}}{\text{Brake Horse Power}} = \frac{WHP}{BHP}$$

$$= \frac{\dot{m}_{STEAM}(w_{pump\ ideal})}{\dot{m}_{STEAM}(w_{pump\ real})} \quad (6.15)$$

Thus $\quad \eta_{pump} = \dfrac{w_{pump\ ideal}}{w_{pump\ real}} \quad (6.16)$

$$w_{pump\ real} = \frac{w_{pump\ ideal}}{\eta_{pump}}$$

$$= \frac{-v_1(P_2 - P_1)\left(\dfrac{144\ in^2}{1\ ft^2}\right)\left(\dfrac{1\ Btu}{778\ ft-lb_f}\right)}{\eta_{pump}} \quad (6.17)$$

$$h_2 = h_1 - w_{pump\ real} \quad (6.18)$$

Note that $h_2 > h_1$ and pump work is into the fluid.

In real plants, a single pump can not achieve the desired pressure increase. Consequently, a cascade of pumps is used, with one pump discharging to the suction of the next pump. This becomes a sequence of vertical lines in the CL region. The inlet to the second pump is the real discharge of the first pump.

NON-IDEAL TURBINE. In the turbine, the process between SP3 and SP4, we see the same situation. Friction-related losses cause a rise in entropy for the "real" turbine. Note that for both of these work components, the increase in entropy is due to friction and turbulence and not heat (*q*) being added.

SP4s: Steam exiting the turbine (or entering the condenser)
This is the **isentropic (ideal)** turbine exhaust state point
   a) First, realize that $s_3 = s_{4s}$
   b) Then, step 1,
      $s_{4s} = s_f + x_{4s}(s_{fg})$ → solve for x_{4s} (steam quality)
      s_f and s_{fg} are found in Steam Table 2 (at $p_{condenser}$)
   c) Last, step 2.
      $h_{4s} = h_f + x_{4s}(h_{fg})$ → solve for h_{4s}
      Note that h_f and h_{fg} are from Steam Table 2 (at $p_{condenser}$)

SP4: Steam exiting the turbine (or entering the condenser)
This is the **non-isentropic (real)** turbine exhaust state point.

$$\text{Turbine Efficiency } \eta_T = \frac{w_{T\,REAL}}{w_{T\,IDEAL}} = \frac{h_3 - h_4}{h_3 - h_{4s}} \quad (6.19)$$

⇒ Solve for h_4

Similar to pumps, sometimes turbines are cascaded in a sequence. This also results in a series of vertical lines on the right side of the *T-s* process graphs. The inlet to the second turbine is the real discharge of the first turbine.

6.10.3 Superheated or Saturated Boiler Output

SP3: Steam from steam generator or boiler
Determine if SP3 is saturated vapor or superheated vapor.
   Compare given steam temp, T_3, to T_{sat} for the given pressure in the steam generator or boiler.

   If $T_3 = T_{sat}$ is it saturated vapor

   If $T_3 > T_{sat}$ is it superheated vapor

   If saturated vapor then:
      $h_3 = h_g$ @ p_{boiler} go to Steam Table 2
         OR
      $h_3 = h_g$ @ T_{sat} go to Steam Table 1

   If superheated vapor, then:
      $h_3 = h_{(S/H,\,PBOIL,\,T3)}$ go to Table 3

6.11 STEAM PLANT ANALYSIS

The following state point table is useful in organizing information, but you will never need to fill out the entire table. So, don't try to complete every cell! Only fill in the cells that you need for the analysis. Number the columns for the state points involved in the problem and split the exit state points for all non-isentropic pumps and turbines.

TABLE 6.2: *Steam Plant State Point Table*

p (psia)				
T (°F)				
h (Btu/lb$_m$)				
s (Btu/lb$_m$-°F)				
v (ft³/lb$_m$)				
x (%)				
State				

6.11.1 Cycle Thermal Efficiency

The thermal efficiency is:

$$\eta_{TH} = \frac{w_{NET}}{q_S} = \frac{w_T + w_P}{q_S}$$

$$\text{OR} \quad \eta_{TH} = \frac{q_{NET}}{q_S} = \frac{q_S + q_R}{q_S} \quad (6.20)$$

Keep in mind that w_P and q_R are negative values.
   This also could be solved on the power basis. Note: When not all of the components see 100 percent of the flow, the thermal efficiency *must* be calculated on the total rate basis.

$$\eta_{th} = \frac{\dot{W}_{net}}{\dot{Q}_s} = \frac{\Sigma \dot{W}_{Turbines} + \Sigma \dot{W}_{Pumps}}{\Sigma \dot{Q}_s}$$

$$= \frac{\Sigma \dot{Q}_s + \Sigma \dot{Q}_r}{\Sigma \dot{Q}_s} \quad (6.21)$$

6.11.2 Boiler Efficiency

$$\eta_{boiler} = \frac{\text{Energy entering steam}}{\text{Energy produced by fuel}}$$

$$= \frac{\dot{m}_{STM}(h_3 - h_2)}{\dot{m}_{FUEL}(HHV)} \quad (6.22)$$

where $(h_3 - h_2) = q_S$
and *HHV* is the higher heating value of the fuel

Note: *HHV* is used if the entering temperature of the feedwater to the boiler is below about 160-165 °F. This is the temperature to which the exhaust gases must cool down to in order to get the water in the combustion products to condense and transfer its latent heat to the working fluid. If the feedwater temperature to the boiler is above this value, then use of *LHV* is more appropriate. The component that allows this to happen is the economizer. The economizer sits in the boiler stack with the hot exhaust gases flowing through and around a tube bundle. The feedwater runs through the tubes of the economizer and is pre-heated before entering the boiler. The ideal output of the economizer is an SL at the boiler pressure. If the temperature of the feedwater entering the economizer is above 165 °F, then it is more appropriate to use *LHV* in the boiler efficiency equation.

6.11.3 Power

$$POWER = Rate\ of\ Work$$

so $\quad POWER = \dot{W} = \dot{m}(w) \quad$ (6.23)

Net power of a steam plant:

$$\dot{W}_{NET} = \dot{m}_{STM}(w_{NET}) \quad where\ w_{NET} = w_T + w_P \quad (6.24)$$

**Don't forget that pump work is negative*

Power of the turbine:

$$\dot{W}_T = \dot{m}_{STM}(w_T) \quad usually\ in\ hp \quad (6.25)$$

Power required to run the pump (aka BHP):

$$\dot{W}_P = \dot{m}_{STM}(w_P) \quad usually\ in\ hp \quad (6.26)$$

EXAMPLE 6.15: RANKINE CYCLE WITH CONDENSATE DEPRESSION, SUPERHEATED STEAM, AND IDEAL WORK COMPONENTS

GIVEN: A Rankine cycle steam power plant produces steam at a pressure of 1200 psia and a temperature of 900 °F. The steam is isentropically expanded through a turbine and exhausted to a condenser operating with a vacuum of 27.894 in-Hg. The resultant condensate is sub-cooled to a temperature of 96 °F.

FIND:
a) pump work (Btu/lb$_m$)
b) heat supplied (Btu/lb$_m$)
c) turbine work (Btu/lb$_m$)
d) heat rejected (Btu/lb$_m$)
e) cycle thermal efficiency (%)

	1	2s	3	4s
p (psia)	1	1200	**1200**	1
T (°F)	96	NR	**900**	101.694
h (Btu/lb$_m$)	64.044	67.62	1440.9	886.8
s (Btu/lb$_m$-°F)	NR	NR	1.5882	1.5882
v (ft³/lb$_m$)	0.016118	0.016118	NR	NR
State	CL	CL	SHV	WV
x (%)	NA	NA	NA	78.89%

SOLUTION: You should (1) draw a schematic with appropriate state points referenced; (2) draw a *T-s* diagram with the state points identified at their proper state conditions; and then (3) fill out the state point table. The SP table, above, is started with the "givens" in **BOLD**. The condenser pressure should be converted to psia and shown as a "given." This is somewhat of an iterative process, as you might not know that the boiler output is an SHV until you look it up. Likewise, you might not realize that the condenser output is an SCL until it is confirmed.

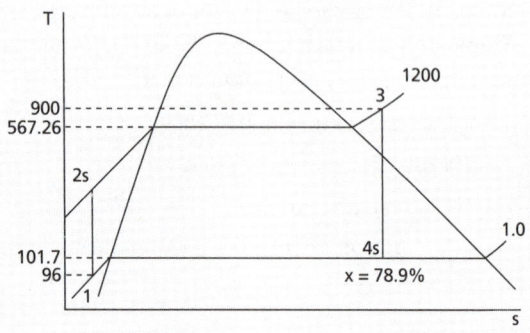

FIGURE EXAMPLE 6.15B

$p_1 = p_{atm} - p_{vac\ gage}$

$= 14.7\ psia - 27.894\ in\ Hg\left(\dfrac{14.7\ psi}{29.92\ in\ Hg}\right)$

$= 14.7\ psia - 13.7\ psi = 1\ psia$

$h_1 = h_f(96\ °F) = 64.044\ Btu/lb_m$ (Note **CLAPPROX TECHNIQUE**)

We use the ideal pump work to calculate the enthalpy at the pump discharge (h_{2s}).

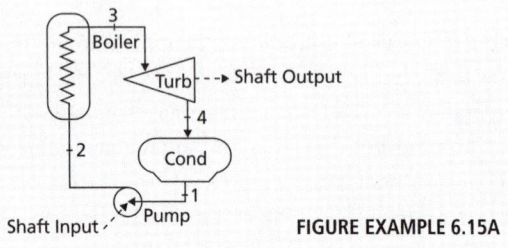

FIGURE EXAMPLE 6.15A

$$w_{p,s} = v_f \Delta p = h_{2s} - h_1 \Rightarrow h_{2s} = h_1 + v_f \Delta p$$

$$h_{2s} = 64.044 \frac{Btu}{lb_m} + \left(0.016118 \frac{ft^3}{lb_m}\right)\left(1199 \frac{lb_f}{in^2}\right)$$

$$\left(\frac{144 in^2}{1 ft^2}\right)\left(\frac{1 Btu}{778 ft - lb_f}\right) = 64.044 + 3.577 \frac{Btu}{lb_m}$$

$$h_{2s} = 67.62 \frac{Btu}{lb_m}$$

$h_3 \Rightarrow$ use superheated steam tables $\Rightarrow h_3 = 1440.9 \frac{Btu}{lb_m}$

h_4 is determined by two steps: use entropy (s_4) to calculate quality (x_4); then use x_4 to calculate enthalpy (h_4).

$$s_{4s} = s_3 \Rightarrow s_{4s} = 1.5882 \frac{Btu}{lb_m \cdot °F}$$

at 1 psia $\quad s_f = 0.1326 \frac{Btu}{lb_m \cdot °F} \quad s_{fg} = 1.8450 \frac{Btu}{lb_m \cdot °F}$

$$h_f = 69.728 \frac{Btu}{lb_m} \quad h_{fg} = 1035.7 \frac{Btu}{lb_m}$$

$$x_{4s} = \frac{s_{4s} - s_f}{s_{fg}} = \frac{1.5882 \frac{Btu}{lb_m \cdot °F} - 0.1326 \frac{Btu}{lb_m \cdot °F}}{1.8450 \frac{Btu}{lb_m \cdot °F}}$$

$$= 0.7889 \Rightarrow 78.9\%$$

$$h_{4s} = h_f + x h_{fg} = 69.728 \frac{Btu}{lb_m} + (0.7889)1035.7 \frac{Btu}{lb_m}$$

$$= 886.8 \frac{Btu}{lb_m}$$

$$\boxed{w_P = -3.577 \frac{Btu}{lb_m}}$$

$$\boxed{q_s = h_3 - h_2 = 1440.9 \frac{Btu}{lb_m} - 67.62 \frac{Btu}{lb_m} = 1373.3 \frac{Btu}{lb_m}}$$

$$\boxed{w_T = h_3 - h_4 = 1440.9 \frac{Btu}{lb_m} - 886.8 \frac{Btu}{lb_m} = 554.1 \frac{Btu}{lb_m}}$$

$$\boxed{q_r = h_1 - h_4 = 64.044 \frac{Btu}{lb_m} - 886.8 \frac{Btu}{lb_m} = -822.8 \frac{Btu}{lb_m}}$$

$$\eta_{th} = \frac{w_T - w_P}{q_s} = \frac{554.1 \frac{Btu}{lb_m} - 3.577 \frac{Btu}{lb_m}}{1373.3 \frac{Btu}{lb_m}} = 0.401$$

and alternatively,

$$\eta_{th} = \frac{q_s - q_r}{q_s} = \frac{1373.3 \frac{Btu}{lb_m} - 822.8 \frac{Btu}{lb_m}}{1373.3 \frac{Btu}{lb_m}} = 0.401$$

$$\boxed{\eta_{TH} = 40.1\%}$$

Note that the all components see the same fluid flow rate, so we can solve for η_{th} in specific terms.

EXAMPLE 6.16: REAL STEAM PLANT WITH REAL TURBINE, IDEAL PUMP, AND SATURATED STEAM

GIVEN: The secondary loop of a *Los Angeles*–class submarine propulsion plant operates as a real steam cycle. Saturated vapor at 385 psig enters the turbine from the steam generator. The condenser pressure gage reads 27.9 in-Hg vacuum and condensate exits the condenser at 90 °F. The atmospheric pressure is 14.7 psia. The pump is assumed to be isentropic and the turbine has an efficiency of 85%.

FIND: Answer questions a) through f) below.

	1	2	3	4s	4
p (psia)					
T (°F)					
h (Btu/lb$_m$)					
s (Btu/lb$_m$-°F)					
v (ft³/lb$_m$)					
x (%), State					

SOLUTION: You should draw a schematic (similar to example 6.14) with appropriate state points referenced, a T-s diagram with the state points identified at their proper state conditions, and then fill out the state point table. As you analyze my steps, go ahead and fill out the SP Table above.

Notice that the pump is listed as isentropic. This would have to be assumed unless the pump's discharge T and p were known. And the turbine has an isentropic efficiency listed. Therefore the SP Table must be adjusted to include SP4s and SP4.

a) Calculate pressure in boiler and condenser (psia)

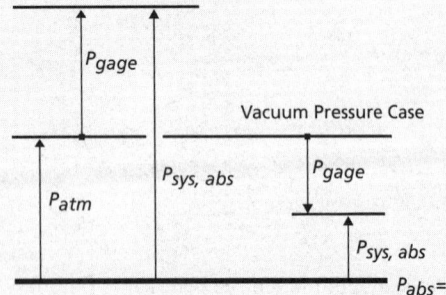

FIGURE EXAMPLE 6.16A

$p_{boiler} = p_{atm} + p_{gage} = 14.7\,psia + 385\,psig$

$\quad = 399.7\,psia \Rightarrow 400\,psia$

Note: It is OK to round at high pressures.

$p_2 = p_3 = p_{boiler}$

$p_{condenser} = p_{atm} - p_{gage}$

$\quad = 14.7\,psia - 27.9\,in-Hg\left(\dfrac{14.7\,psi}{29.92\,in-Hg}\right)$

$p_{condenser} = 14.7\,psia - 13.71\,psi = 0.99\,psia$

$p_{condenser} = 1.0\,psia$

Be careful rounding at low pressures. Round to the closest hundredth or tenth.

$p_1 = p_4 = p_{condenser}$

b) Calculate the enthalpy for each state point

$T_{sat}\,(at\,1\,psia) = 101.74°F \Rightarrow$ state 1 is sub-cooled

$h_1 = h_f(90°F) = 58.054\dfrac{Btu}{lb_m}$

$v_1 = v_f(90°F) = 0.01610\dfrac{ft^3}{lb_m}$

$w_{pump,ideal} = w_{p,s} = h_1 - h_{2s} = v_f\Delta p = v_1\Delta p_{12}$

$w_{pump,ideal} = 58.054\dfrac{Btu}{lb_m} - h_{2s}$

$\quad = \left(0.01610\dfrac{ft^3}{lb_m}\right)\left(399\dfrac{lb_f}{in^2}\right)\left(\dfrac{144\,in^2}{ft^2}\right)\left(\dfrac{1\,Btu}{778\,ft\cdot lb_f}\right)$

$\quad = -1.190\dfrac{Btu}{lb_m}$

$h_{2s} = h_1 + w_{pump,ideal} = 58.054\dfrac{Btu}{lb_m} + 1.190\dfrac{Btu}{lb_m}$

$\quad = 59.24\dfrac{Btu}{lb_m}$

$h_3 = h_g(400\,psia) = 1205.0\dfrac{Btu}{lb_m}$

Note that Table 2 doesn't have 400 psia, but Table 3 does and includes the SL and SV values. No need to interpolate!

$s_3 = s_g(400\,psia) = 1.4853\dfrac{Btu}{lb_m\cdot°F}$

$s_{4s} = s_3 = 1.4853\dfrac{Btu}{lb_m\cdot°F}$

$s_f(1\,psia) = 0.1326\dfrac{Btu}{lb_m\cdot°F}$

$s_{fg}(1\,psia) = 1.8450\dfrac{Btu}{lb_m\cdot°F}$

$x_{4s} = \dfrac{s-s_f}{s_{fg}} = \dfrac{1.4853-0.1326}{1.8450} = 0.733 \Rightarrow 73.3\%$

$h_{4s} = h_f(1\,psia) + x_{4s}h_{fg}(1\,psia)$

$\quad = 69.728\dfrac{Btu}{lb_m} + 0.733\left(1035.7\dfrac{Btu}{lb_m}\right) = 828.90\dfrac{Btu}{lb_m}$

$\eta_T = \dfrac{w_{T,real}}{w_{T,ideal}} = \dfrac{h_3-h_4}{h_3-h_{4s}} \Rightarrow$

$h_4 = h_3 - \eta_T(h_3-h_{4s})$

$\quad = 1205.0\dfrac{Btu}{lb_m} - 0.85(1205.0-828.9)\dfrac{Btu}{lb_m}$

$h_4 = 885.3\dfrac{Btu}{lb_m}$

c) Calculate the work of the turbine (Btu/lb$_m$)

$w_T = h_3 - h_4 = 1205.0 - 885.3\dfrac{Btu}{lb_m} = 319.7\dfrac{Btu}{lb_m}$

d) Calculate the cycle thermal efficiency (%)

$$\eta_{th} = \frac{w_{net}}{q_s} = \frac{w_T - |w_P|}{q_s} = \frac{319.7 - 1.19 \frac{Btu}{lb_m}}{h_3 - h_2}$$

$$\eta_{th} = \frac{318.5 \frac{Btu}{lb_m}}{(1205 - 59.24)\frac{Btu}{lb_m}} = 0.278 \Rightarrow 27.8\%$$

e) Calculate the moisture at turbine exit (%)
The turbine exit is the condenser inlet. That is SP4. Therefore we need x_4. Use the real h_4 to calculate the real quality, x_4.

$$x_4 = \frac{h_4 - h_f}{h_{fg}} = \frac{(885.3 - 69.728)\frac{Btu}{lb_m}}{1035.7 \frac{Btu}{lb_m}} = 0.787 \Rightarrow 78.7\%$$

$m = 1 - x = 1 - 0.787 = 0.213 \Rightarrow 21.3\%$

f) Calculate the condensate depression (°F)

$CD = T_{sat} - T = 101.694°F - 90°F = 11.7°F$

The T-s diagram should look like this with additional info annotated on it such as state points, pressures, and temperatures.

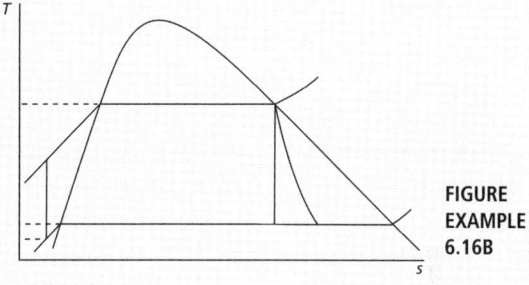

FIGURE EXAMPLE 6.16B

EXAMPLE 6.17: REAL SUBMARINE STEAM PLANT WITH THROTTLE VALVE

GIVEN: A submarine's secondary steam plant consists of an MFP, S/G, throttle valve, propulsion turbine, and condenser. S/G pressure is 600 psia, and the condenser pressure gage reads 25.4 in-Hg; condensate depression is 4 °F; MFP discharge temperature is 130 °F; the throttle valve is partially closed, dropping the downstream pressure to 300 psia. Turbine efficiency is 90%. Barometric pressure in the boat is one standard atmosphere. Recall that a submarine plant's steam generator can only produce dry saturated steam.

FIND: η_p, η_{th}, and the quality of the steam entering the condenser

SOLUTION: If the state points are not identified for you, then you put them in between the components that do something. We're not analyzing the reactor loop yet. That comes in the next unit.

The big picture concept necessary to step through this problem is that throttling is an isenthalpic process ($\Delta h = 0$). Throttle valves create a pressure drop to regulate flow, so p changes, but h doesn't.

Notice that condensate depression exists in the condenser discharge (SP1).

$\eta_{pump} = 45\%$

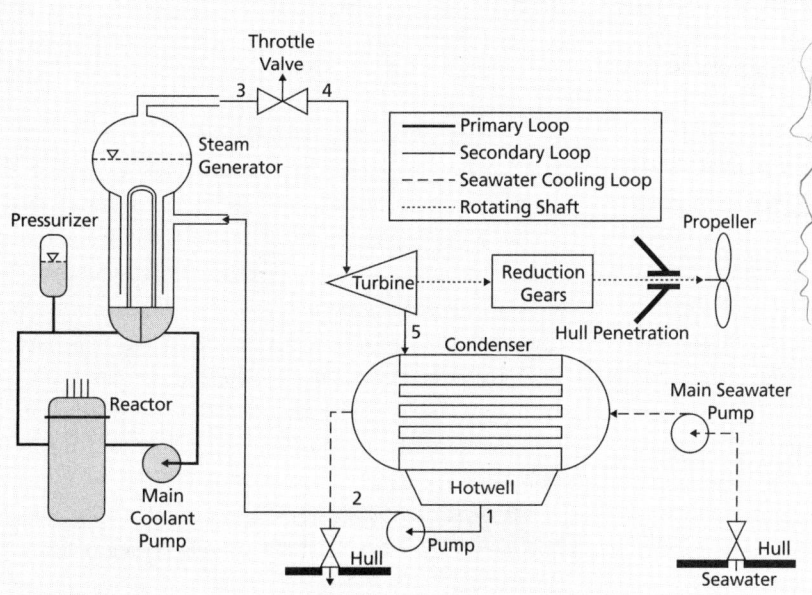

FIGURE EXAMPLE 6.17A

SP4 conditions are interpolated on Table 3 using $p = 300$, $h = 1203.9$. This is between the SV values for $p = 300$ and $T = 440\,°F$. We really only need s_4 from this interpolation. I'll show it:

$$\underbrace{\frac{s_4 - 1.5111}{1.5295 - 1.5111}}_{entropy} = \frac{1203.9 - 1203.4}{1219.7 - 1203.4} \quad s_4 = 1.51166$$

$\eta_{th} = [(1203.9 - 914.4) - (97.987 - 93.992)] / (1203.9 - 97.987) = 25.8\%$

$x = 80.1\%$

Now some explanation: This *T-s* diagram shows that the entire graph is <u>not to scale</u> and especially so in the SCL region. I posed this problem to point out that the *p* lines are exaggerated in the SCL region in order to show the pump process lines. In reality, the *p* lines are very close to the SL line, but that makes it impossible to show the pumping process. So, we exaggerate the *p* lines.

Also note that a throttling process can produce an SHV in a submarine power plant. In fact, this is the only way that these conditions can be created. h_g(300 psia) = 1203.4, so at 1203.9 it is barely superheated, but SHV nonetheless.

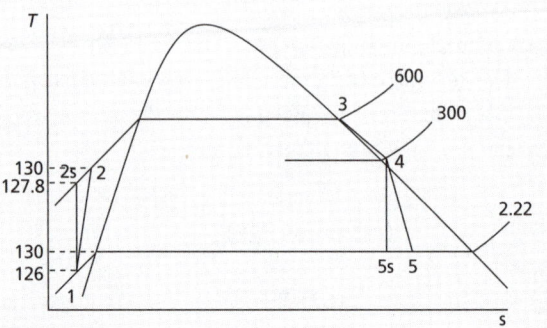

FIGURE EXAMPLE 6.17B

Secondary	1	2s	2	3	4	5s	5
p (psia)	2.22	=→	=→	600	300	=→	=→
T (°F)	126	127.8	130	486.25	418.5	130	130
h (Btu/lb$_m$)	93.992	95.79	97.987	1203.9	←=	882.2	914.4
s (Btu/lb$_m$-°R)	NR	NR	NR	NR	1.5117	←=	NR
v (ft³/lb$_m$)	0.016230	←=	NR	NR	NR	NR	NR
State; Quality	CL	CL	CL	SV	SHV	76.93%	80.1%

Givens are shown in **BOLD**. All other values are look-ups or calculated.

6.12 CONVENTIONAL BOILERS

A conventional boiler is equipped with an economizer that pre-heats the feedwater. The feedwater is injected into the "steam drum" of the boiler, ideally as a saturated liquid. From there, the water flows down under gravity to the water drum at the bottom of the boiler through tubes called "downcomers." A large number of tubes connect the water drum to the steam drum. The main group of tubes is called the generating tube bank. These tubes are exposed to the high temperature flame and combustion products of the burning boiler fuel. The water inside the generating tubes boils and this reduces the density, so the steam and water mixture rises up and enters the steam drum. Inside the steam drum, there are moisture separators and scrubbers, which serve to retain the liquid water and return it to the downcomers for another pass down and back up through the generating tubes. The dry saturated steam that passes out from the moisture separators then can be superheated in a separate heat exchanger that is also within the boundary of the boiler. The superheater tubes are also exposed to the high temperature combustion gases. Recognize that it is impossible to heat steam into the superheated region until the liquid content has been removed or boiled to a vapor. Auxiliary steam is used for auxiliary systems that use steam, but can not withstand the higher energy of the superheated main steam.

The major boiler subcomponent heat exchangers are represented schematically in Figure 6.11. Pressurized water nuclear reactor steam generators shown in Figure 6.12 can not superheat steam. They can only produce dry saturated steam. Nuclear power plants are discussed in Unit 8.

Real ship and submarine power plants must provide for both propulsion and electrical generation. Figure 6.12

shows a representative naval nuclear steam plant. Note the multiple turbines both in series and parallel. Also, note the presence of the reactor, or primary, coolant loop on the left hand side of the diagram. Figure 6.11 shows a conventional steam plant. **Note that the primary difference between the conventional and nuclear steam plant is**

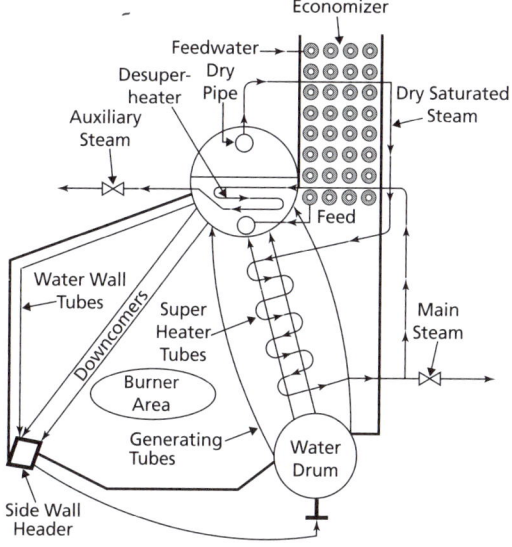
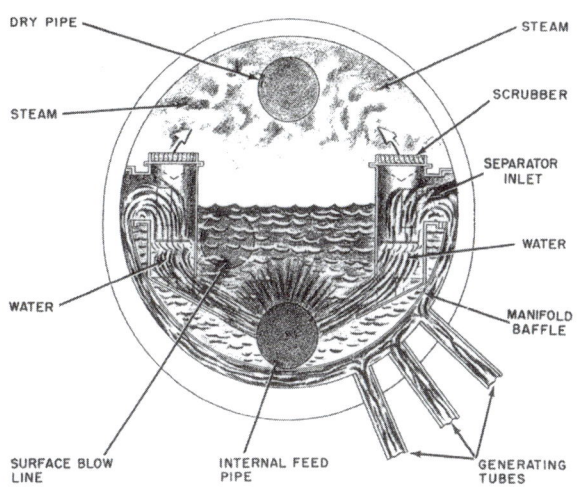

FIGURE 6.10: *Basic Boiler Configuration and Steam Drum Detail*

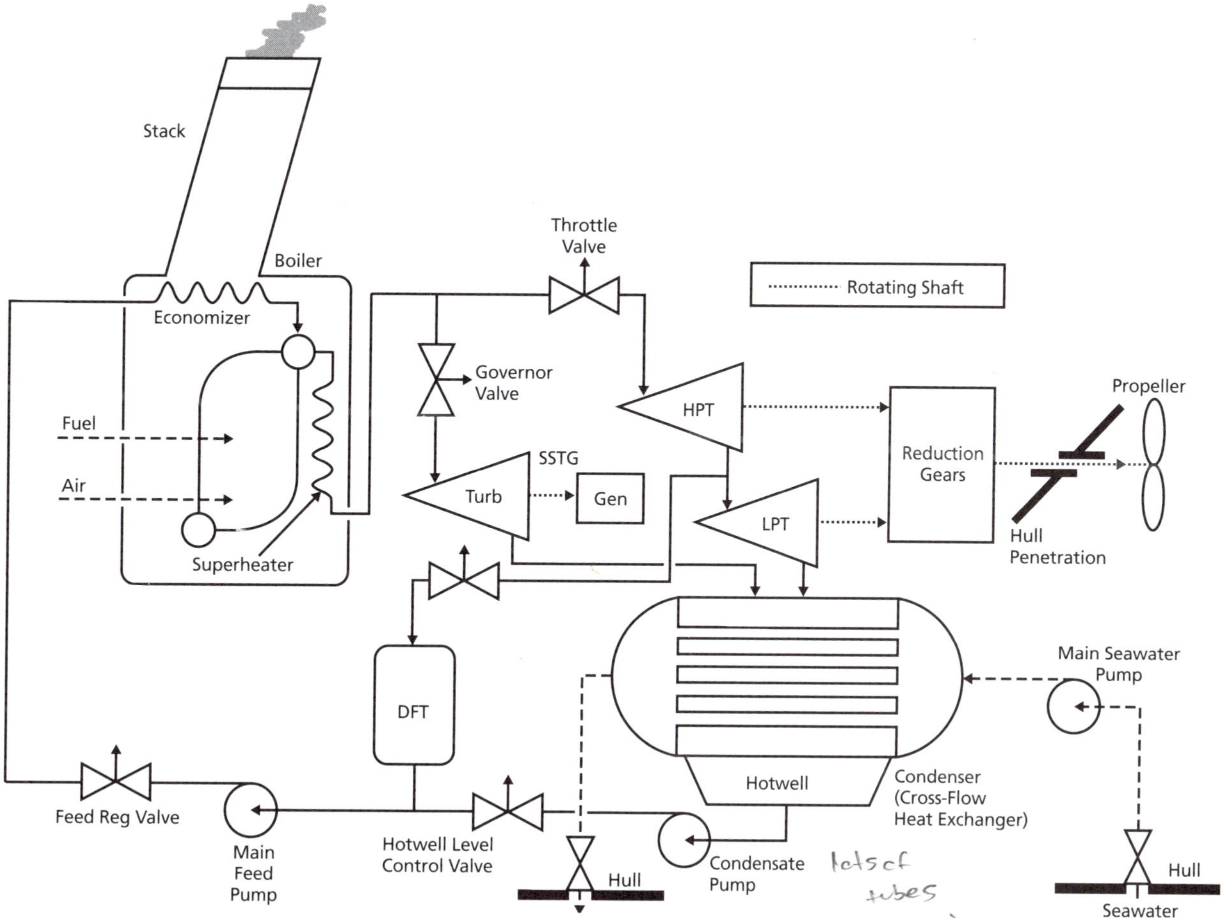

FIGURE 6.11: *Conventional Ship Steam Plant*

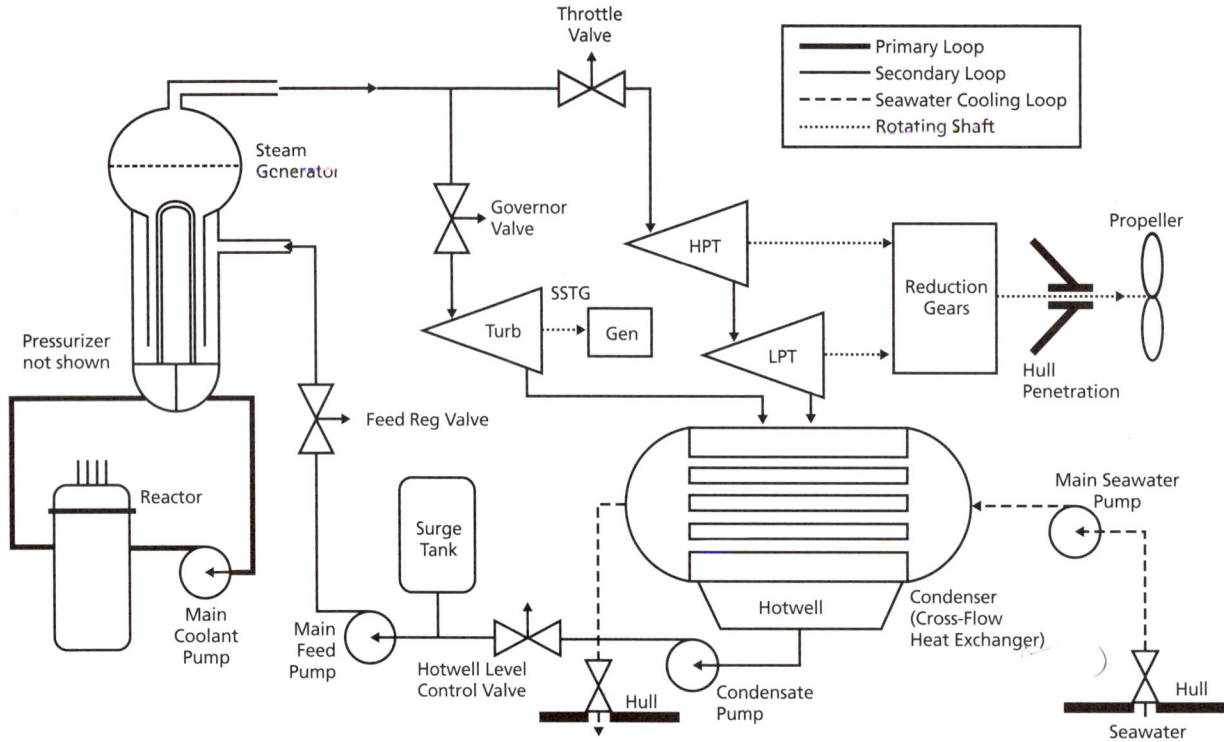

FIGURE 6.12: *Nuclear Ship/Sub Steam Plant*

the heat source. All other components are essentially the same. Surface ships typically have a DFT.

The control valves that are included in Figures 6.11 and 6.12 are for throttling purposes.

The hotwell level control valve senses the liquid condensate level in the condenser and throttles open or shut to maintain the control band in the condenser. Recall that some level of condensate is required to submerge some of the condenser tubes to produce condensate depression. The feed regulating valve maintains the boiler water level. Any imbalance between these two control valves is stored or drawn from the surge tank shown in the schematics.

A governor valve is fitted to the steam system supplying a turbine-generator to maintain steam flow sufficient to keep the generator electrical output close to the standard 60 Hz. Steam ships have multiple SSTG sets for electrical generation. Only one is shown here for concept illustration purposes.

The propulsion turbine throttle valve is to maintain the speed of the turbines, and therefore the speed of the ship as ordered from the ship's bridge or submarine's control room. Steam ships go in reverse by having a smaller, special purpose turbine attached to the turbine shaft. The backing turbine is configured to rotate the propeller in the astern direction. The throttleman must shut the ahead turbine throttle valve and then open the astern turbine throttle valve. The astern turbine is not shown on these schematics for simplicity.

REFERENCES

DOE Fundamentals Handbook: Thermodynamics, Heat Transfer, and Fluid Flow. Vol 1. (DOE-HDBK-1012/1-92).
Principles of Naval Engineering, NAVPERS 10788, Rev. B, 1970.

Practice Problems

"If you aim at nothing, you'll hit it every time."
—Zig Ziglar

6.1 Complete the following table using the *Steam Tables* to determine the property values. Draw a *T-s* diagram showing the "vapor dome" and a single line of constant pressure. Place an "x" on the line and the problem number next to the "x" for each problem below, showing where the state point is graphically. Givens are listed **BOLD**.

	p (psia)	T (°F)	v (ft³/lb$_m$)	h (Btu/lb$_m$)	s (Btu/lb$_m$-°R)	x (%)	°Sh	State
1	.95	**100**	.01613	66.8	.1295	0	N/A	SL
2	**80**		4.381			**80**		WV
3	**150**	2.8	1.49	**755.6**		49.2		
4	**600**	**1000**						
5	**150**	**340**	0.01787	311.30	.4903	—	—	SCL
6	**30**		**13.722**					
7	**50**	**320**						
8	**40**	**110**						
9		**300**				**20**		
10	**80**	312.0	5.473	1183.3		100	—	SV
11	**1000**	**1000**						
12	**80**			**1380.5**				
13		**200**			**1.762**			
14	**50**	500	11.306	1284.1	1.789		**219**	SHV
15	**20**			**1156.3**		49.2		OSC

6.2 Complete the following table using the Mollier diagram from the back of your *Steam Tables* to determine the property values. Draw an *h-s* diagram showing the "vapor dome" and a single line of constant pressure. Place an "x" on the line and the problem number next to the "x" for each problem below showing where the state point is graphically. Givens are listed **BOLD**.

	p	T	h	s	x	m	°Sh	state
1	**4**	155	950	**1.74**	92.6	7.4		WV
2	**80**	318	1020	1.41	62	**18**		WV
3	**100**	**600**	1330	1.76	—	—	270	SH
4	70	**400**	1237	1.69	—	—	**100**	SH
5				**1.90**			**400**	
6	**600**		**1100**					
7		**550**		**1.60**				
8		**400**			**90**			
9			**1150**		**0**			
10				**1.64**		**16**		
11	**300**							SV
12			**1300**				**200**	

* Note: 9 has two correct answers. Pick one.

6.3 Using the *Steam Tables*, determine the temperature above which freshwater density is more than 2% different from the standard value of 62.4 lb_m/ft^3. Describe your methodology.

6.4 What is the volume of 1 pound of water at standard atmospheric pressure? What is the volume of 1 pound of steam at standard atmospheric pressure? What is the volume ratio?

6.5 Show on *T-s* and *h-s* coordinates the following processes for steam, in continuum and in sequence:

1 → 2: isentropic work from a saturated liquid state as occurs ideally in the pump of a Rankine cycle

2 → 3: isobaric heating of water initially existing as a compressed liquid and completing the heating process as a superheated vapor as occurs ideally in the boiler of a Rankine cycle

3 → 4: isentropic expansion until the state of the steam is such that it exits as a wet vapor as occurs ideally in the turbine of a Rankine cycle

4 → 1: isobaric cooling of the wet vapor, rejecting all of the latent heat contained within the steam as occurs ideally in the condenser of a Rankine cycle

Note: For all subsequent problems, provide a schematic, a *T-s* concept graph, and a state point table with your solution to each problem. A state point table template is provided below. Not all cells are required to be filled in. You may cut and paste the table into your submitted HW. Add additional columns or delete unnecessary columns.

p (psia)							
T (°F)							
h (Btu/lb_m)							
s (Btu/lb_m-°R)							
v (ft³/lb_m)							
x (%)							
State							

6.6 An ideal Rankine cycle has steam properties at the turbine inlet of 600 psia and 800 °F and at the exit a pressure of 25.9 in-Hg. Find the enthalpies, h, (Btu/lb_m) at each of the usual state points.

6.7 A steam plant generates saturated steam at 600 psia. The condensate exits the 5 psia condenser at 158 °F. The feed pump and main turbine are isentropic.

a) Is the condensate sub-cooled? If so, how much condensate depression is there (°F)?

b) Is the generated steam superheated? If so, how much superheat is there (°F)?

c) Determine all four state point enthalpy, h, values (Btu/lb_m).

6.8 A Rankine cycle engine has a maximum pressure of 1000 psia and a maximum temperature of 950 °F. Heat is rejected under a vacuum of 27.9 in-Hg. The feed pump is 85 percent efficient. Find:

a) the isentropic and real work of the feed pump (Btu/lb_m)

b) the ideal work of the turbine (Btu/lb_m)

6.9 Saturated steam at 600 psia is expanded through a turbine to a condenser at 2 psia. The quality of the steam at the turbine exit is 80%. Determine the turbine efficiency (%).

6.10 A Rankine cycle with ideal components has steam properties at the turbine inlet of 800 psia and 1000 °F and a pressure of 1 psia at the turbine exit. Find the:

a) pump work (Btu/lb_m)

b) heat added to the cycle (Btu/lb_m)

c) heat rejected from the cycle (Btu/lb$_m$)

d) cycle thermal efficiency (%)

e) moisture content of the exhaust steam (%)

6.11 A Rankine cycle has a pump discharge pressure of 1000 psia and a condenser pressure of 1 psia. The turbine's isentropic efficiency is 87%. If the superheater raises the steam temperature 355.42 °F above the boiling point, find:

a) pump work (Btu/lb$_m$)

b) heat added to the cycle (Btu/lb$_m$)

c) turbine work (Btu/lb$_m$)

d) heat rejected from the cycle (Btu/lb$_m$)

e) cycle thermal efficiency (%)

f) mass flow rate necessary to achieve 10,000 ihp from the turbine (lb$_m$/hr)

6.12 Steam at a pressure of 1000 psia and a temperature of 800 °F is expanded through a turbine, exhausting to the condenser at a pressure of 1 psia. At the turbine exit the quality of the steam is 92%.

Determine: a) the turbine isentropic efficiency (%); and b) the turbine specific work. c) If this turbine exhausted to atmospheric pressure and the isentropic efficiency determined in part a) does not change, find the new turbine work. d) Discuss why steam plants operate with the turbine exhaust at a "deep vacuum" (as close to an absolute vacuum as the equipment can achieve). Print a copy of the Mollier chart from DOE Fundamentals Handbook 1012/1, page A-1 (PDF page 131 of 138) and show this problem on the chart by marking the points and use different colored highlighters to show the points and the two processes (turbine exhausting to vacuum; turbine exhausting to atmospheric pressure). You should check your answer on the large Mollier chart in the back of your *Steam Tables*, but don't submit that page.

6.13 In a steam plant, saturated steam enters the turbine at 585 psig. The isentropic pump work is 1.8 Btu/lb$_m$. Condensate exits the condenser at 144 °F. The pump is 80% efficient and the turbine is 90% efficient. The fuel flow rate is 2.5 lb$_m$/min and the fuel *HHV* is 19,500 Btu/lb$_m$. Engine mechanical efficiency is 72%. Assume atmospheric pressure is 15 psia.

a) Calculate the pressure (psia) in the condenser.

b) Calculate the amount of condensate depression (°F).

c) Calculate the steam temperature (°F) generated by the boiler.

d) Which should be greater: isentropic or real pump work?

e) Which should be greater: isentropic or real turbine work?

f) Calculate all state point enthalpies (Btu/lb$_m$).

g) Determine the boiler efficiency (%) if the steam flow rate is 42 lb$_m$/min.

6.14 Saturated steam at a pressure of 580 psia is provided to a turbine. The condenser pressure gage reads 27.9 in-Hg. The condensate leaving the condenser has been sub-cooled by 8.7 °F. The turbine's isentropic efficiency is 85%; the pump's isentropic efficiency is 75%. Sketch an *h-s* graph of this engine and show "a)" through "f)" on the graph. Find:

a) condenser absolute pressure (psia)

b) enthalpy of the condensate (Btu/lb$_m$)

c) real turbine work (Btu/lb$_m$)

d) ideal pump work (Btu/lb$_m$)

e) real pump work (Btu/lb$_m$)

f) heat rejected in the condenser (Btu/lb$_m$)

g) the mass flow rate of steam (lb$_m$/min) necessary to produce 15,000 ihp

6.15 The typical steam plants have a series of pumps in the feed and condensate system. Recall that the water leaving the condenser is referred to as condensate, whereas the water is referred to as feedwater when it leaves the feed pump. The feed and condensate pumps are not identical—they will produce different Δp_{pump} while pumping the same mass flow rate. This sequence of pumping is done to meet NPSH requirements of the individual pumps in a system that receives water at a vacuum and sends it to a boiler with pressure at hundreds of psi. For a two-pump feed and condensate system as shown in Figure HW 6.15, the condenser pressure and hotwell temperature gages read 25.8 in-Hg and 121 °F, respectively. The condensate system's pumping rate is 500 GPM.

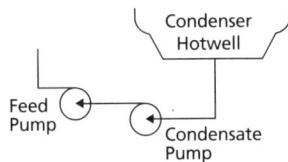

FIGURE HW 6.15

a) What is the condensate depression in this operating condition? The condensate pump discharge pressure gage reads 52.3 psi and the pump's discharge thermometer reads 121.5 °F.

b) What are the density of the condensate and the mass flow rate of the condensate?

c) What is the isentropic efficiency of the condensate pump?

d) If the pump is driven by a motor with 95% efficiency, what electrical power is required to run the condensate pump (kW)?

e) What is the rated power of the pump's motor (hp)?

Be sure to include the schematic with properly identified state points, a state point table, and a *T-s* concept graph appropriate to this problem.

6.16 A steam plant has a boiler that produces 200,000 lb$_m$/hr of steam at 1200 psia and 800 °F. The steam expands in a turbine that is 90% efficient to a condenser with a pressure gage reading 26.08 in-Hg. The barometer in the maneuvering room reads 30.15 in-Hg. The condensate has 4 degrees of condensate depression. The pump is 85% efficient and the boiler is 92% efficient. The feed pump discharge pressure is 1250 psia. The turbine's mechanical efficiency is 85%. The losses associated with the mechanical inefficiency of the turbine may be assumed to go to the lube oil system. *LHV* of the fuel is 18,400 Btu/lb$_m$ and the *HHV* of the fuel is 20,000 Btu/lb$_m$. Sketch the *T-s* and *h-s* concept graphs and find:

a) the condenser pressure (psia)

b) real turbine work (Btu/lb$_m$)

c) turbine power (bhp)

d) pump work (Btu/lb$_m$)

e) pump motor rating (bhp)

f) mass flow rate of fuel (lb$_m$/min)

g) thermal efficiency (%)

h) the rate of heat loss to the lube oil system (Btu/hr)

6.17 Refer to the marine power plant schematic, Figure HW 6.17, provided below. Steam at 850 °F and 600 psia enters a propulsion turbine producing 15,000 bhp that has an isentropic efficiency of 95% and a mechanical efficiency of 93%. The conditions downstream of the SSTG governor valve are 750 °F and 550 psia (for this part of the problem, disregard that the governor valve is isenthalpic). The SSTG turbine has the same isentropic and mechanical efficiencies as the propulsion turbine, and the generator efficiency is 96%. The generator is producing 2.5 MW of electrical power. The steam is expanded through the turbines to a condenser pressure of 2 psia. Clearly state all assumptions.

a) Determine the ihp of the propulsion and SSTG turbines.

b) What does the condenser pressure gage read (in-Hg)?

c) What is the mass flow rate of the steam to the propulsion turbines?

d) What is the mass flow rate of the steam to the SSTG turbine?

e) What is η_{th} of the overall power plant?

f) What mass flow rate of fuel oil ($LHV = 18{,}720$ Btu/lb$_m$; $HHV = 20{,}020$ Btu/lb$_m$) is necessary if the boiler efficiency is 80%? (Assume the economizer functions ideally.)

g) If this level of power output pushes the vessel at 20 knots, how much fuel will be consumed every thousand miles in gallons ($SG_{oil} = 0.9$)?

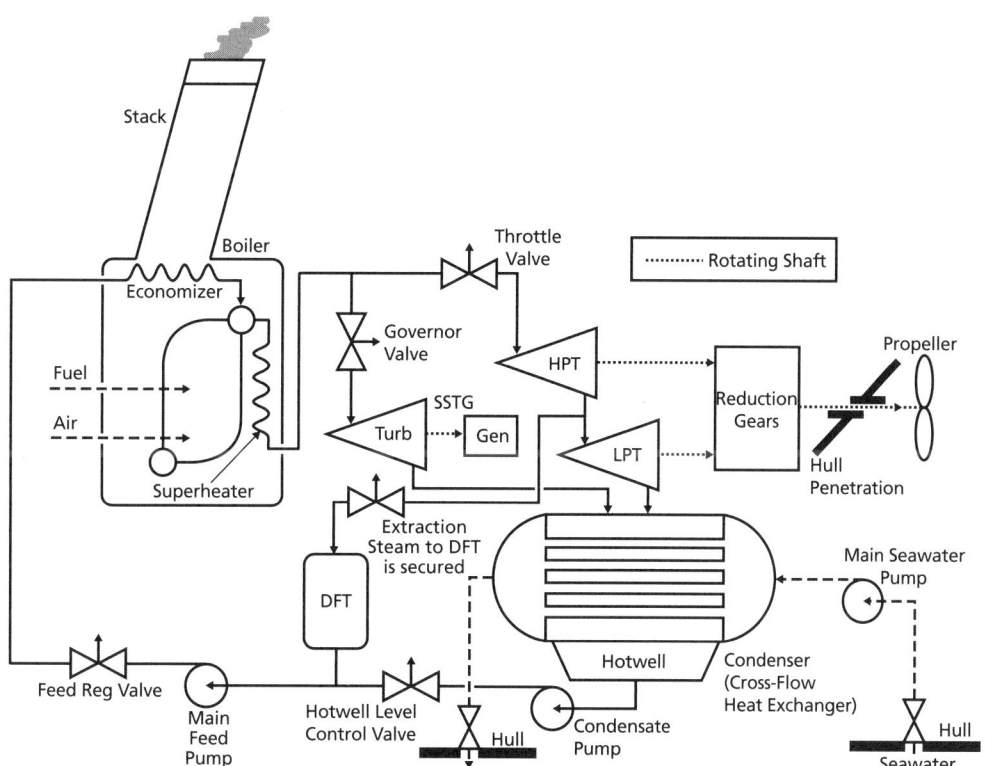

FIGURE HW 6.17

7 Heat Exchangers

Seawater-cooled heat exchangers are a significant component in naval applications and require regular maintenance to ensure proper performance. This sailor is preparing to clean a steam condenser that was fouled when numerous squid were sucked into the main seawater system.

7.1 UNIT LEARNING OBJECTIVES

7.1.1 Terminology: definitions, variables, and typical units

happens through solids

7.1.1.1 conduction, convection, radiation, logarithmic mean temperature difference ($LMTD$), cross flow, parallel flow, counter flow, surface heat exchanger, mixing heat exchanger, tube, shell, tube-sheet, baffles, thermal conductivity, convective heat transfer coefficient, overall heat transfer coefficient, fouling

7.1.2 Concepts: ideas and engineering expressions of them

7.1.2.1 Sketch the flow paths for parallel, counter, and cross-flow heat exchangers.

7.1.2.2 Graph the temperature profiles for a parallel-flow and counter-flow heat exchanger for fluids involving sensible and latent heat transfer.

7.1.2.3 Distinguish between surface-type (aka closed or indirect) and open (aka mixed-flow, spray, or direct-contact) heat exchangers.

7.1.2.4 Discuss the sources, and analyze the effects, of fouling on heat exchanger performance.

7.1.3 Skills: procedures, practices, or methods that enable reasoning

7.1.3.1 Choose relationships appropriate to each heat exchanger type.

7.1.3.2 Perform a two fluid energy balance on a heat exchanger to calculate the heat transferred and the resultant effects in each fluid.

7.1.3.3 Calculate overall heat transfer coefficient (*U*).

7.1.3.4 Use *U* to calculate heat exchanger performance.

7.1.3.5 Calculate log mean temperature difference (*LMTD*).

7.1.3.6 Use *LMTD* to calculate heat exchanger performance.

7.1.3.7 Perform a multi-fluid energy balance on a mixing tank involving combinations of sensible and latent heat transfer.

7.1.3.8 Evaluate heat exchanger performance with provided data.

7.2 INTRODUCTION

Heat transfer happens in response to a temperature difference, from the high temperature "source" material to the low temperature "sink" material. This is fundamental "carry around" knowledge that all students of this course should have memorized by this point. Whether we're talking about heat addition or rejection, the basic principle is that one fluid is giving up energy and the other fluid is receiving it—so whether we think about heat given up or absorbed, it is a matter of perspective (Which fluid are you analyzing?). Recall that heat addition or rejection (HIWOP/HOWIN) concepts for a power cycle focus on the working fluid inside the cycle.

Notice that we have been covering heat transfer processes since Unit 1 and how these processes relate to the various energy cycles covered thus far in this course. Up to this point, we have concentrated on calculating the change in properties, typically the change in enthalpy (*Δh*) of a single working fluid as it moved through a heat exchanger. We looked at the effects of the heat transfer on the other fluid in a couple of example and homework problems, but now it is time to delve more deeply into the real components that accomplish heat transfer in these applications and include the equations that allow us to better analyze the component on a fluid-to-fluid basis.

In covering gas turbine engines, we modeled the chemical combustion process in the working fluid as heat added to the fluid (air). Then the heat rejection process was the heat given up by the exhausted air to the environment. In the Steam, Nuclear Power, and HVAC&R units, we focus on the heat transfer processes across the tubes of the boiler/steam generator, steam condenser, and refrigerant evaporator and condenser. In each of these examples, there were two fluids involved, but they were separated from contact with each other by the tubes. Therefore, the heat transfer is classified as *indirect*.

Two notable exceptions to indirect heat exchangers involving two fluids are the nuclear reactor and the electric hot water heater. The reactor generates heat in the solid fuel and transfers that heat to the liquid "primary coolant." An electric hot water heater converts electrical energy to heat in a solid element and, being immersed in water in the heater's tank, the heat is transferred to the water. In both of these examples the heat is generated in a solid element and conducted out to the liquid that absorbs the generated heat.

We cover heat exchangers as a special component in greater detail because the performance of heat exchangers is influenced by corrosion, sludge originating elsewhere in the systems, biological growth, and other factors. As a result, the maintenance of heat exchangers is a significant activity in the naval service in order to keep the power systems functioning within desired operating parameters—virtually all of our graduates will be faced with maintaining heat exchangers in their futures.

We will also expand our skill set by covering mixing heat exchange processes and how to calculate their parameters.

7.3 MODES OF HEAT TRANSFER

The three modes of heat transfer are *conduction* through solids; *convection* through fluid motion (either "forced" or "natural"); and *radiation* (which occurs from surface to surface through space, including a vacuum).

For heat conduction through solids:

$$\dot{Q} = \frac{kA \cdot \Delta T}{L} \quad (7.1)$$

where k = thermal conductivity (Btu/hr-ft-R°), L = thickness of material (ft), A = heat transfer surface area (ft^2), and ΔT = temperature difference (R°).

For heat convection from the surface film into a bulk fluid:

$$\dot{Q} = hA\left(T_{surf} - T_{bulk}\right) \quad (7.2)$$

where h = convective heat transfer coefficient (Btu/hr-ft^2-R°), A = heat transfer surface area (ft^2), T_{surf} = surface temperature (such as the pipe wall), and T_{bulk} = temperature of the bulk fluid.

Natural convection involves density-driven fluid currents, whereas forced convection involves pumped fluid motion. Also recall the difference between laminar flow and turbulent flow as we covered it in Unit 2. We typically operate naval systems in a forced convection, turbulent flow regime, so we'll concentrate on this combination. Radiative heat transfer:

$$\dot{Q} = \varepsilon \sigma A \left(T_{hot}^4 - T_{cold}^4 \right) \quad (7.3)$$

where ε = emissivity (a unit-less value from 0 to 1), σ Stefan-Boltzmann constant (1.714×10^{-9} Btu/hr-ft^2- °R^4).

Note the convention that °R or °F is the temperature of a substance while °R (or °F) is a temperature difference.

Knowing radiative heat transfer and appreciating where it occurs, in concept, is sufficient for this course. All students have felt the thermal power that radiates from the sun or from a fire. Other applications for this mode of heat transfer include the heat radiating from a piece of hot metal, such as engine exhaust piping. As will be discussed in Unit 9, heat stress in personnel has a radiative component. A significant portion of the heat transfer to the tubes in a conventional steam boiler occurs via radiation.

7.3.1 Combined Convection and Conduction Heat Transfer

Indirect heat exchangers in naval applications involve flowing fluids, where the fluids are separated by heat exchanger tubes. Therefore, both convection and conduction will occur simultaneously. So, we will now develop our understanding of the combination of these two modes of heat transfer.

The temperature difference drives heat transfer like a voltage drives electrical current. In fact, thermal resistance is defined as the temperature difference divided by the heat transfer rate ($\dot{Q}$) or the heat flux ($\dot{Q}''$), where heat flux is the heat transfer rate perpendicular to an area ($\dot{Q}'' = \dot{Q}/A$). We don't typically calculate the flux, but could by dividing the thermal power by the heat transfer area.

$$R_{th} = \frac{\Delta T}{\dot{Q}} \quad (7.4)$$

Rearranging 7.1 and 7.2 we get:

$$R_{th_{conduction}} = \frac{L}{kA} \quad (7.5)$$

$$R_{th_{convection}} = \frac{1}{hA} \quad (7.6)$$

Heat flux is constant through all layers. Therefore, the thermal resistance is additive from bulk fluid inside the tube through the wall to the bulk fluid on the outside of the tube as shown in Figure 7.1. Some other things to note in Figure 7.1: (1) The curved temperature profile is associated with the boundary layer film and convective heat transfer; (2) conduction will have a straight line temperature profile through the solids (the sludges, biological fouling, corrosion layers, and the tube wall itself); (3) a high slope in the temperature profile means low thermal conductivity. (The tube material is chosen to provide high conduction rates, the required strength, and resistance to corrosion; but the corrosion and fouling layers always work against thermal performance—whether the fouling or the corrosion has the lower k, and steeper T slope, depends on the materials of the heat exchange process.)

$$R_{th} = \frac{1}{h_A A} + \frac{L_B}{k_B A} + \frac{L_C}{k_C A} + \frac{L_D}{k_D A} + \frac{L_E}{k_E A}$$
$$+ \frac{L_F}{k_F A} + \frac{1}{h_G A} = \left(\Sigma \frac{1}{h_i} + \Sigma \frac{L_i}{k_i} \right) \frac{1}{A} \quad (7.7)$$

Note that the letter subscripts in Figure 7.1 refer to the material while the numbers refer to locations. Then $T_1 = T_{bulk, Fluid A}$ and $T_2 = T_{surf, Fluid A/ Solid B}$ and similar for Fluid G on the right side of the figure.

Rearranging 7.4 to solve for $\dot{Q}$ and incorporating 7.7, the heat transfer rate, aka thermal power, becomes:

$$\dot{Q} = \frac{1}{\left(\Sigma \frac{1}{h_i} + \Sigma \frac{L_i}{k_i} \right)} A \Delta T \quad (7.8)$$

Note that when we included the ΔT in this equation, we were looking only at one specific location along the length on a heat exchanger tube.

ΔT varies at different points along the length of a heat exchanger. We typically do not have the ability to measure the temperatures internal to a heat exchanger along the length of a tube or through the layers of fouling and corrosion. So, we look at the temperatures of the fluids as they enter and exit the heat exchanger. This allows us to consider the overall heat transfer performance of the unit. To do this, we combine the inverse of the total thermal resistance and define the *overall heat transfer coefficient* (*U*) as:

$$U \equiv \frac{1}{\left(\Sigma \frac{1}{h_i} + \Sigma \frac{L_i}{k_i} \right)} \quad (7.9)$$

The units of U are Btu/hr- ft^2-R°.

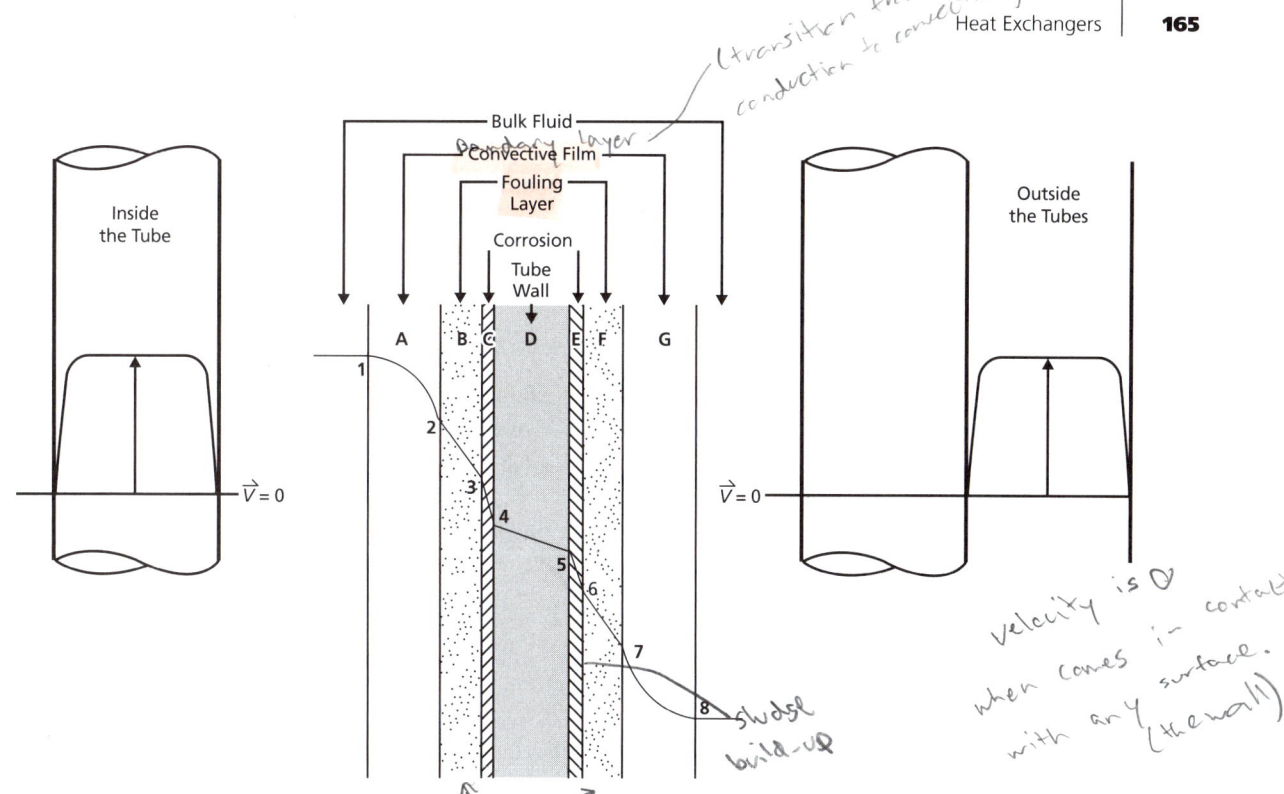

FIGURE 7.1: Fluid-to-Fluid Temperature Profile across a Heat Exchanger Tube (at a Specific Location along the Tube Length and Parallel Turbulent Flow Shown)

Typical values for the overall heat transfer coefficient (U) in common naval applications are shown in Table 7.1.

Corrosion (chemical), sludge (oil and sediments), and bio-fouling (growth of algae, attachment of barnacles or other mollusks, etc., on the tubes) increase over time and tend to insulate the heat transfer tubes (as seen in Figure 7.1). Management of these factors requires ongoing maintenance.

We now need to clarify our use of the variables. In 7.8 we used ΔT to for the difference in temperatures between the fluids at any internal location. From here on, we'll use ΔT to indicate the change in temperature of a single fluid. We'll use a Greek symbol (θ) to indicate the difference in T between the fluids. Combining 7.8 and 7.9 and adopting this new convention produces:

$$\dot{Q} = \frac{1}{\left(\sum \frac{1}{h_i} + \sum \frac{L_i}{k_i} \right)} A\theta \qquad (7.10)$$

Consider a *shell and tube heat exchanger* with the two fluids flowing as indicated in Figure 7.2. Only a single tube is shown for illustrative purposes. However, in reality there would be multiple tubes. The hot fluid is *shell-side* and the cold fluid is *tube-side* in this figure. As the two fluids move through the heat exchanger, the hot

TABLE 7.1: *Typical Overall Heat Transfer Coefficient for Common Naval Applications*

Hot Side	Cold Side	Clean Surface	Fouled Surface
Steam	Watery Solution	300-550	150-275
Steam	Air or Gases	5-10	4-8
Watery Solution	Water	195-245	105-155
Lube Oil	Water	20-30	10-20
Air or Gases	Water	5-10	4-8
Watery Solution	Refrigerant	60-90	40-60
U units shown in Btu/hr- ft²-R°			

fluid gives up energy via heat transfer to the cold fluid, which gains energy. In sensible heat transfer (not involving phase transformations), the change in energy shows up as change in enthalpy (h) and temperature (T).

Measuring the T difference between the two fluids (θ), one can see that θ changes as the fluids travel through the heat exchanger θ_x, with $x = 0 \rightarrow L$. In order to evaluate the overall heat exchanger, we need a relationship that provides an average, or mean value of θ in this range. The temperature change process can describe a curve, which would make the mathematics a bit trickier. This is simply resolved by adopting the conventional technique known as the *log mean temperature difference* (*LMTD*).

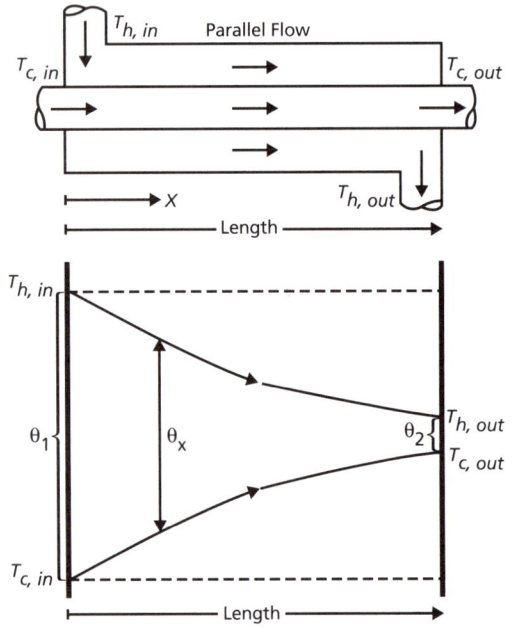

FIGURE 7.2: Simple Shell and Tube Heat Exchanger in Parallel Flow

LMTD (θ_m) is developed by taking the temperature difference between the two fluids at one end of the heat exchanger (θ_1) and the temperature difference between the two fluids at the other end (θ_2). θ_1 and θ_2 are illustrated in Figure 7.2.

$$\theta_m = \frac{\theta_1 - \theta_2}{\ln\left(\dfrac{\theta_1}{\theta_2}\right)} \quad (7.11)$$

Combining 7.10 and 7.11 then becomes:

$$\dot{Q} = UA\theta_m \quad (7.12)$$

When considering heat exchangers, we can look at the performance from the perspective of the individual fluids or between the two fluids. We can evaluate what happens to a single fluid, and then apply that to the overall heat exchanger. This is how it is done in the Fleet, and is used to determine the trend of performance and when the heat exchanger must be maintained (cleaned).

Summarizing, we have a number of relationships that allow us to calculate the performance of a heat exchanger. Considering two fluids, A and B, and the overall heat exchanger, we have the following relationships:

$$\underbrace{\dot{Q}_A = \dot{m}_A \Delta h_A = \overbrace{\dot{m}_A c_{p_A} \Delta T_A}^{if-no-phase-change}}_{A-fluid}$$

$$= \underbrace{\dot{Q}_B = \dot{m}_B \Delta h_B = \overbrace{\dot{m}_B c_{p_B} \Delta T_B}^{if-no-phase-change}}_{B-fluid} \quad (7.13)$$

$$= \underbrace{\dot{Q}_{hx} = U_{hx} A_{hx} \theta_m}_{overall-heat-exchanger}$$

Caution: Since we are talking about multiple fluids and fluid-to-fluid relationships, <u>use subscripts</u> to keep track of the parameters. A common error by students learning this material is to substitute the mass flow rate of one fluid with the temperatures or enthalpies of the other fluid. This always results in getting the wrong answer, loss of points, lower grades, and feelings of intense disappointment. These can be prevented by paying attention to this cautionary note.

7.4 HEAT EXCHANGER CLASSIFICATIONS

Heat exchangers may be classified by a number of factors. The natures of the fluids that are involved in the heat transfer process determine the optimum choices among these factors.

FLUID IDENTIFICATION. The most basic classification of a heat exchanger is by the two fluids that are involved. The direction of the heat transfer is implied by the order in which the fluids are listed. A motor vehicle radiator is a water-to-air heat exchanger, with heat transfer from the water to the air. A shipboard lube oil cooler is an oil-to-water heat exchanger.

DIRECT VS INDIRECT CONTACT. Direct contact or mixing heat exchangers have multiple fluid streams entering and exiting a mixing chamber at different energy levels. Indirect heat exchangers separate the fluids by a physical boundary, which also allows the fluids to exist at different pressures. For example, hot lube oil is cooled by seawater. Another indirect example: highly pressurized primary coolant in a nuclear steam generator will not boil even though it is at a higher temperature than the secondary (boiling) side of the steam generator.

HEAT EXCHANGE SURFACE GEOMETRY. Tubes, plates, and extended surfaces (fins) are all used in various heat transfer applications. An example of a finned surface is a motor vehicle radiator, which has metal shapes protruding beyond the water-filled portion. These fins help to carry the heat away from the water and provide additional surface area with the air, which must absorb the heat. Other surface shapes are sometimes used in order to initiate turbulent flow in one of the fluids for better mixing within that fluid.

HEAT TRANSFER MECHANISMS. Sensible/sensible or sensible/latent combinations of fluids involving boiling, condensing, or temperature change in the fluids. A shipboard vapor compression chiller's evaporator is designed for two-phase flow on the refrigerant-side and single-phase flow on the chill water side of the heat exchanger.

FLOW ARRANGEMENTS. Considering the flow paths of the two fluids, parallel-, counter-, and cross-flow configurations are the most basic. This is modified by the number of passes that either of the fluids makes through the heat transfer region of the heat exchanger. Baffles that support the tubes and redirect flow back and forth across the outside of the tubes increase the contact of that fluid with the tubes. One nuclear steam generator configuration is a vertical u-tube design. The primary coolant makes two passes as it flows up and then down the tubes.

7.5 INDIRECT CONTACT HEAT EXCHANGER CONSTRUCTION

In indirect contact heat exchangers, the two fluids are separated by a physical boundary. Common heat exchangers in marine applications fall into the shell and tube geometry classification. These heat exchangers are made by inserting round tubes into matching holes in a flat plate, called a tubesheet. The tubes are then welded or otherwise attached to the tubesheet to provide a seal. The other ends of the heat exchanger tubes are attached to the same tubesheet if the tubes are u-shaped, or attached to a second tube sheet. The tube sheets are then welded or otherwise attached to a larger, cylindrical shell. The cylindrical shell is capped by a "head" and sometimes referred to as a "waterbox." The fluid that flows through the tubes is referred to as the "tube-side" fluid. The fluid that flows around the outside of the tubes is referred to as the "shell-side" fluid. Typical shell and tube heat exchangers have hundreds or even thousands of tubes.

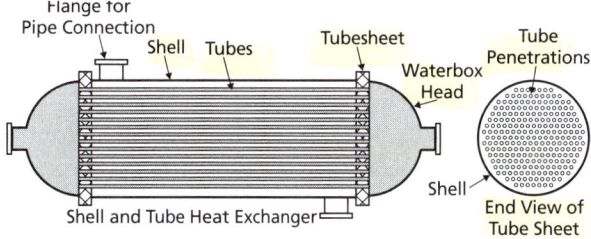

FIGURE 7.3: *Shell and Tube Heat Exchanger*

The desired properties for the material separating the fluids in a heat exchanger are strength, high thermal conductivity, and low flow resistance, and the material must not be adversely affected by either fluid. The designer must balance out competing demands for smaller and lighter heat exchangers, while at the same time not creating geometry conditions that overly restrict the fluid flow through the heat exchanger passages.

The material and geometry factors of heat exchanger design are determined by the operating characteristics of the fluids and their related systems. For example, a submarine's steam condenser must be able to handle the hydrostatic pressures of the seawater to the submarine's design depth capability. The corrosive effects of seawater mean that the parts that come in contact with seawater must be protected from corrosion, or built from materials that are not susceptible to corrosion. Seawater also includes biologics that tend to grow on warm hard surfaces, so designing the heat exchanger for easier cleaning is important. Water has higher heat transfer characteristics than steam, so some condensers include metal finned surfaces on the outside of the tubes to aid in heat transfer.

The designer's decision about which fluid flows tube-side and which flows on the shell-side is determined by:

(1) The pressures of the two fluids. Small diameter tubes can have thinner walls and still hold higher pressures. Large diameter shells require much thicker walls to contain high pressure.

(2) Cleaning requirements. It is easier to clean the inside of round tubes than the outside. Fouling due to biological growth tends to happen at higher rates than sludge deposited on the tubes. It then follows that, unless overruled by the pressure requirement, whatever will foul more quickly should flow tube-side.

Figure 7.4 shows a number of shell and tube heat exchanger configurations. The typical proportions are shortened to compact the graphics. Note the removeable flange on the heat exchanger shown in the upper right. Recall that access for cleaning is particularly important for heat exchangers using seawater. The biological growth in seawater on warm heat exchanger surfaces reduces their performance, and they must be periodically brushed or scraped to clean them.

Heat exchangers that go through large changes in temperature must account for thermal growth in the tubes. Either one of the tubesheets must be capable of shifting laterally in its seat, or U-tubes must be used.

Baffles serve not only to direct flow around the tubes, but also support the tubes and limit the amount of sagging that can occur.

There are other types of heat exchangers in common applications. Water-to-air heat exhangers typically used for motor vehicle radiators, oil coolers and air conditioning and heating applications employ fins on the air-side to aid in the heat transfer processes. These are tube-type heat exchangers that are often formed by stamping shapes into flat metal sheets and then brazing or soldering the sheets together to make the overall shape. There also are flat plate heat exchangers, which are becoming more common in marine applications.

7.6 HEAT EXCHANGER ANALYSIS

We assume that the heat transfer from the heat exchanger to the local environment is negligible. High temperature or very low temperature heat exchangers have insulation (aka "lagging") installed around them and the heat transfer to or from the operating space is minimized. In contrast to the metals with high thermal conductivity that are used to promote high heat transfer, insulation has low thermal conductivity. In reality, the engine room tends to be a warm place because insulation only slows down, but does not stop, this heat transfer. For our purposes, however, we assume that all of the heat given off by the hot fluid is absorbed by the cold fluid. Likewise, we do not credit the tube sheets with any significant heat transfer area, so we restrict our analysis to the tube area in the heat exchanger.

For direct contact heat exchange, two or more fluid streams enter but typically only one leaves. From continuity we get:

$$\Sigma \dot{m}_{in_i} = \dot{m}_{out} \tag{7.14}$$

Performing an energy balance we get:

$$\Sigma \left(\dot{m}_{in_i} h_{in_i} \right) = \dot{m}_{out} h_{out} \tag{7.15}$$

For indirect contact heat exchangers, the temperature profile helps us determine performance.

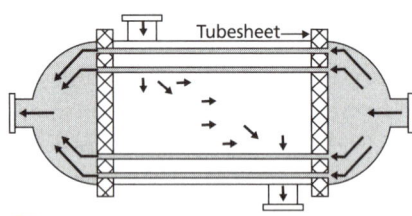

Single Pass Tube Side; Single Pass Shell Side HX

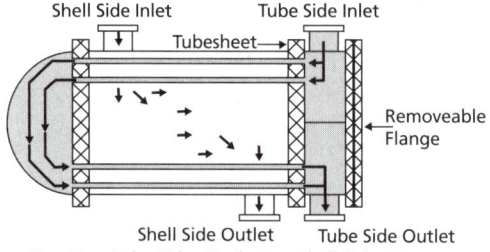

Two Pass Tube Side; Single Pass Shell Side HX

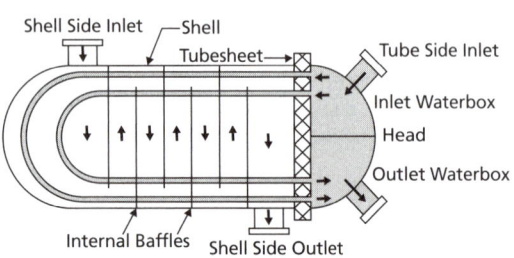

Two Pass Tube Side; Multi-Pass Shell Side HX

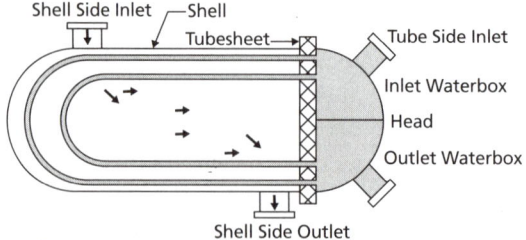

Two Pass Tube Side; Single Pass Shell Side HX

FIGURE 7.4: *Various Shell and Tube Heat Exchanger Configurations*

Even though we know there are many tubes, we sketch only one tube. This single tube graphically represents the multiple tubes that actually exist in a heat exchanger.

In a parallel-flow heat exchanger, the hot fluid enters at the same end as the cold fluid and the flows are the same direction.

In a counter-flow heat exchanger, the hot and cold fluids enter at opposite ends of the heat exchanger and θ_x for the heat exchanger is more uniform throughout yielding a more uniform heat flux (see Figure 7.5) than parallel flow.

Cross-flow heat exchangers are typically used in applications involving a phase change in one of the fluids. When the fluid undergoing the phase change is at saturation, the temperature is constant until the phase change is complete (see Figure 7.6).

For cross-flow heat exchangers the temperature profile direction should be drawn as counter-flow. In addition, as illustrated in Figure 7.7, the shape of the temperature profile of one of the fluids is available on the *T-s* graph for the cycle.

Notice that the shape of the steam line in the temperature profile graph mimics the shape in the *T-s* graph. Notice that the cooling water does not go through a phase change, so the heat gain is sensible while the steam goes through latent heat transfer followed by sensible heat transfer in creating the condensate depression.

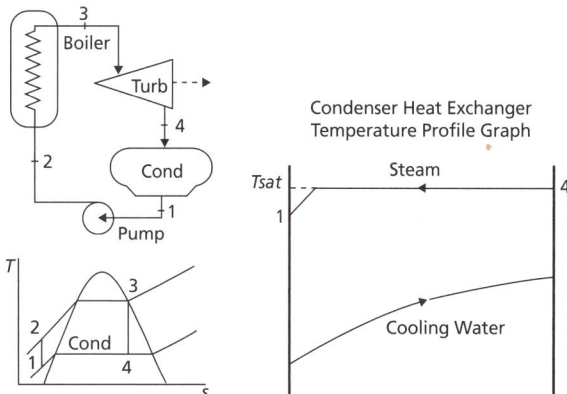

FIGURE 7.7: *Basic Saturated Steam Plant with Condensate Depression T-s and Condenser Temperature Profile Concept Graphs*

7.6.1 Problem Solving Methodology

- Read and evaluate the problem statement.
- Draw a schematic or evaluate the given schematic.
- Put state points on the schematic if not provided. State points should be placed between every component that does something consequential to the working fluid in the process.
- Look for the fluid that has mass flow rate ($\dot{m}$) given or the ability to determine it from a power ($\dot{Q}$ or P) term or volumetric flow rate ($\dot{V}$).
- If the heat exchanger is an indirect type, then think about the temperature profile graph for this heat exchanger and your other sources of data for the applicable fluids (*Steam Tables*, psychrometric chart, refrigerant *p-h* graph, etc.). The temperature profile graph will not be helpful if the problem is a direct contact (mixing) heat exchanger.
- List the "givens" on the schematic or the state point table, if used, <u>and</u> the temperature profile graph, as applicable.

<u>Caution: Common Student Errors in this Topic</u>

- Do <u>not</u> use the $C_p \Delta T$ method if the fluid is going through a <u>phase change</u>.
- Also, the *Steam Tables* will be a better source of enthalpy (h) data at higher water temperatures than assuming a constant specific heat (C_p) approach.
- Do <u>not</u> use the *Air Tables* for thermodynamic properties of humid air—use the psychrometric chart.
- List subscripts next to each term in an equation to identify the fluids in the equation being used. A com-

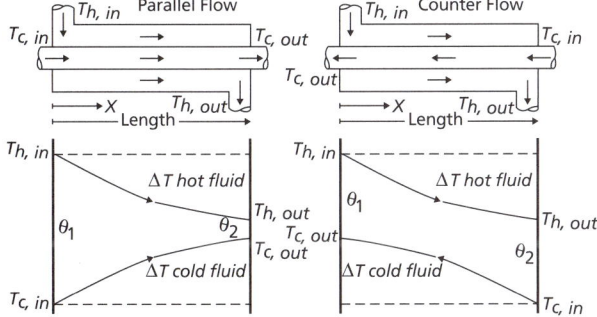

FIGURE 7.5: *Heat Exchanger Schematic and Temperature Profile Concept Graphs for Parallel and Counter-Flow Heat Exchangers*

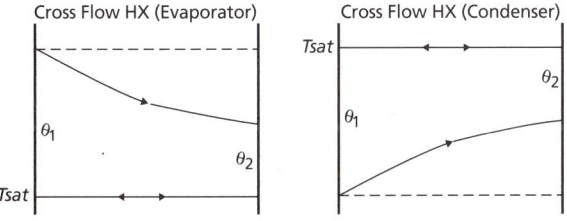

FIGURE 7.6: *Temperature Profile Concept Graphs for Phase Change Applications*

mon error is to use the $\dot{m}$ of one fluid multiplied by the Δh of the other fluid. Alternately, using the C_p for the wrong fluid in the $C_p \Delta T$ equation is a similar error. Keep your fluids straight by being organized. Use subscripts that identify which fluid you are analyzing.

EXAMPLE 7.1: LUBE OIL COOLER ANALYSIS

GIVEN: A single-shaft propulsion plant producing 15,000 SHP has main reduction gears that are 89% efficient. Lube oil is supplied to the reduction gears at 100 °F and drains out at 130 °F. The specific heat capacity of the lube oil is 0.55 Btu/lb$_m$-°F. The reduction gear lube oil is cooled by a dedicated lube oil cooler. Seawater is supplied to this counter-flow heat exchanger by an ASW pump that is pumping 1000 GPM. The injection temperature today is 75 °F.

FIND:
a) Find the required mass flow rate of the lube oil to the reduction gears (lb$_m$/min).
b) What is the ASW overboard discharge temperature?
c) What is the *LMTD* of the lube oil cooler?

SOLUTION: In all of these heat transfer problems, look for thermal power or a way to determine it. The heat put into the lube oil in the reduction gears is the difference between the power supplied to the reduction gears (*BHP*) and the output of the reduction gears (*SHP*). While *BHP* and *SHP* are both mechanical rotational power, the friction created in the reduction gears shows up as heat input to the lube oil. This is shown schematically below in a portion of the graphic that was presented in Unit 1. Note: We're only interested in the losses from the reduction gears and not the other boxes of the schematic. These are only shown for context. A conservation of energy (first law) analysis of the reduction gears will give us the heat load of the reduction gear lube oil system.

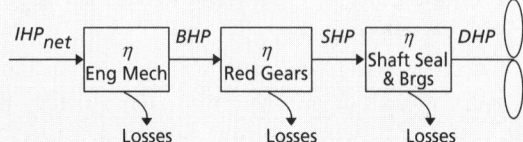

FIGURE EXAMPLE 7.1A

$$\eta_{red\,gears} = \frac{output}{input} = \frac{SHP}{BHP} \Rightarrow BHP = \frac{SHP}{\eta_{red\,gears}}$$

$$= \frac{15,000\,hp}{0.89} = 16,854\,hp$$

$BHP = SHP + losses \rightarrow$ reduction gear losses

reduction gear losses $= \dot{Q}_{lube\,oil} = BHP - SHP$

$$= 16854\,hp - 15000\,hp = 1854\,hp$$

$$\dot{Q}_{lube\,oil} = (1854\,hp)\left(2545\,\frac{Btu}{hp-hr}\right) = 4.72E6\,\frac{Btu}{hr}$$

$$\dot{Q}_{lube\,oil} = 7.86E4\,\frac{Btu}{min} = 1311\,\frac{Btu}{sec}$$

a) Find the required mass flow rate of the lube oil to the reduction gears (lb$_m$/min).

Use ample subscripts in your analysis so that you don't mix up the fluids. Here is the schematic.

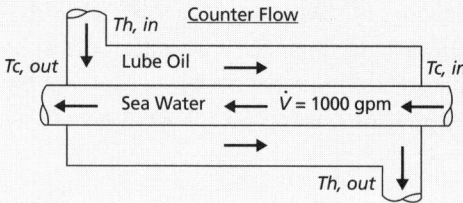

FIGURE EXAMPLE 7.1B

This schematic could also include the temperatures and C_p for the two fluids.

$$\dot{Q}_{lube\,oil} = \dot{m}_{lube\,oil}\,c_{p,lube\,oil}\,\Delta T_{lube\,oil} \Rightarrow \dot{m}_{lube\,oil}$$

$$\dot{Q}_{lube\,oil} = \frac{\dot{Q}_{lube\,oil}}{c_{p,lube\,oil}\,\Delta T_{lube\,oil}}$$

$$\dot{m}_{lube\,oil} = \frac{\dot{Q}_{lube\,oil}}{c_{p,lube\,oil}\,\Delta T_{lube\,oil}} = \frac{7.86E4\,\frac{Btu}{min}}{\left(0.5\,\frac{Btu}{lb_m - F°}\right)(30\,°F)}$$

$$\dot{m}_{lube\,oil} = 4764\,\frac{lb_m}{min}$$

$$\boxed{\dot{m}_{lube\,oil} = 4764\,\frac{lb_m}{min}}$$

b) What is the ASW overboard discharge temperature?

$$\dot{Q}_{HX} = \dot{Q}_{oil} = \dot{m}_{oil} c_{p\,oil} \Delta T_{oil} =$$
$$\dot{Q}_{seawater} = \dot{m}_{seawater} c_{p\,seawater} \Delta T_{seawater}$$

$$\dot{Q}_{HX} = \dot{Q}_{seawater} = \dot{m}_{seawater} c_{p\,seawater} \Delta T_{seawater} \Rightarrow$$

$$\Delta T_{seawater} = \frac{\dot{Q}_{seawater}}{\dot{m}_{seawater} c_{p\,seawater}}$$

$$\dot{m}_{seawater} = \rho_{seawater} \dot{V}_{seawater}$$
$$= \left(64.0 \frac{lb_m}{ft^3}\right)\left(1000 \frac{gal}{min}\right)\left(\frac{1 ft^3}{7.48 gal}\right) = 8556 \frac{lb_m}{min}$$

$$\Delta T_{seawater} = T_{sw,out} - T_{sw,in} = \frac{\dot{Q}_{seawater}}{\dot{m}_{seawater} c_{p\,seawater}} \Rightarrow$$

$$T_{sw,out} = \frac{\dot{Q}_{seawater}}{\dot{m}_{seawater} c_{p\,seawater}} + T_{sw,in}$$

$$T_{sw,out} = \frac{\dot{Q}_{seawater}}{\dot{m}_{seawater} c_{p\,seawater}} + T_{sw,in}$$

$$= \frac{7.86E4 \frac{Btu}{min}}{\left(8556 \frac{lb_m}{min}\right)\left(0.94 \frac{Btu}{lb_m - F°}\right)} + 75\,°F$$

$$\boxed{T_{sw,out} = 9.8\,°F + 75\,°F = 84.8\,°F}$$

c) What is the *LMTD* of the lube oil cooler?

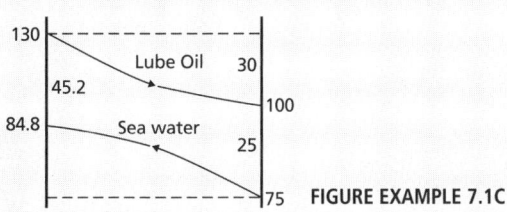

FIGURE EXAMPLE 7.1C

The temperature profile graph with given and calculated data on it is shown to the right. Note that we did not need the $\Delta T_{seawater}$, so that parameter is indicated as "NR."

$$\boxed{\theta_m = \frac{\theta_1 - \theta_2}{\ln\left(\frac{\theta_1}{\theta_2}\right)} = \frac{45.2 - 25}{\ln\left(\frac{45.2}{25}\right)} = 34.1 F°}$$

EXAMPLE 7.2: PROPULSION SYSTEM'S STEAM CONDENSER

GIVEN: A ship's main condenser has a heat exchange area of 2200 ft². Steam enters the condenser at a rate of 14,810 lb$_m$/hr and with a quality of 82%. The condenser pressure gage reads 25.4 in-Hg. The seawater injection temperature is 75 °F and the overboard discharge temperature is 102 °F.

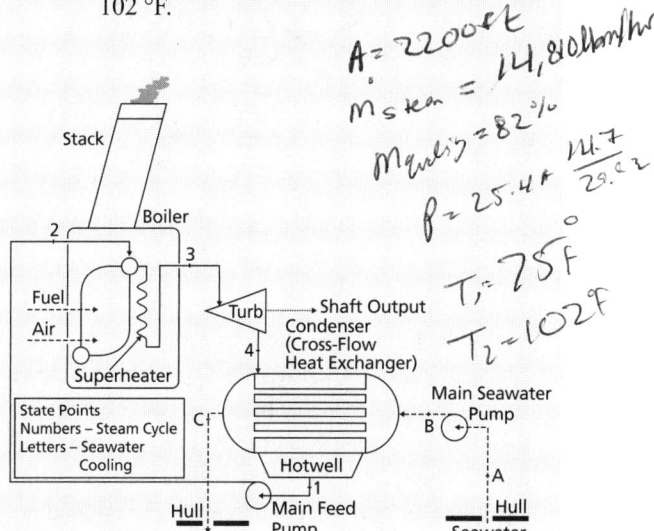

FIGURE EXAMPLE 7.2A

FIND:
a) How much heat is transferred?
b) What is the seawater flow rate (lb$_m$/hr and GPM)?
c) What is the *LMTD* of the condenser?
d) What is the overall heat transfer coefficient (U – Btu/hr-ft²-°F)?
e) If the tubes are 1 inch in diameter and limited in length to 4 feet, how many tubes are in this condenser?

SOLUTION: Evaluating the problem statement, we have the following parameters:

p_{cond} = 25.4 in-Hg (vac) {Note: Recall that steam condensers operate at a vacuum.}

$A_{hx} = 2200 \text{ ft}^2$

$T_{sw,in} = T_A = T_B = 75 \text{ °F}$ {Note: Assumes negligible temperature rise in the MSW pump.}

$T_{sw,out} = T_C = 102 \text{ °F}$

$\dot{m}_{steam} = 14{,}810 \, lb_m/hr$

$P_{condenser, absolute} = p_{atm} - p_{gage}$

$= 14.7 \, psia - \left[25.4 \, in-Hg \left(\dfrac{14.7 \, psi}{29.92 \, in-Hg} \right) \right]$

$= 2.2 \ psia$

a) How much heat is transferred?

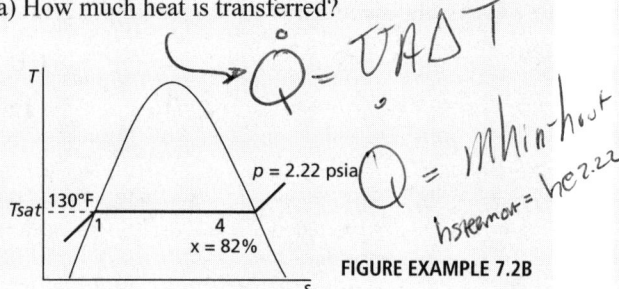

FIGURE EXAMPLE 7.2B

Notice that we have mass flow rate of the steam ($\dot{m}_{steam}$). That info along with the steam enthalpies will allow us to solve the problem. However, we are asked to find the seawater mass ($\dot{m}_{seawater}$) and volume flow rates ($\dot{V}_{seawater}$). We'll start with analyzing the steam side of the condenser. We must assume: (1) p_{atm} is standard, $p_{atm} = 14.7$ psia; and (2) the condensate out of the condenser has no condensate depression.

From the 1967 *Steam Tables* (Table 1):

$T_{sat}(@2.22 \text{ psia}) = 130 \text{ °F}$

$\dot{Q}_{steam} = \dot{m}_{steam}(h_{in} - h_{out})_{steam}$

$h_{stm,out} = h_1 = h_f(@2.22 \text{psia}) = \underline{97.96 \ Btu/lb_m}$

$h_{stm,in} = h_4 = h_f(@ \ 2.2 \ psia) + x h_{fg}(@ \ 2.2 \ psia)$

$h_{stm,in} = h_4 = 97.96 + 0.82(1019.8) = 934.2 \, Btu/lb_m$

$\dot{Q}_{steam} = \dot{m}_{steam}(h_{in} - h_{out})_{steam}$

$= 14{,}810 \dfrac{lb_m}{hr}(934.2 - 97.96)\dfrac{Btu}{lb_m} = 1.24E7 \dfrac{Btu}{hr}$

$\boxed{\dot{Q}_{steam} = 1.24E7 \dfrac{Btu}{hr}}$

b) What is the seawater flow rate (lb_m/hr and GPM)? Now consider that the heat given off by the steam is totally absorbed by the seawater. Seawater does not go through a phase change (sensible heat only) and the temperature range is not excessive, so we can use the single phase heat transfer equation (constant specific heat, C_p, approach).

$\dot{Q}_{steam} = \dot{Q}_{seawater} = \dot{m}_{seawater} c_{p \, seawater} \Delta T_{seawater}$

$\dot{m}_{seawater} = \dfrac{\dot{Q}_{seawater}}{c_{p \, seawater} \Delta T_{seawater}}$

$= \dfrac{1.24E7 \, Btu/hr}{\left(0.94 \, Btu/lb_m - hr - °F\right)(102 - 75) \, °F}$

$= 4.89E5 \, lb_m/hr$

$\boxed{\dot{m}_{seawater} = 4.89E5 \, lb_m/hr}$

Now for the volumetric flow rate in the requested units (GPM).

$\dot{V}_{seawater} = \dfrac{\dot{m}_{seawater}}{\rho_{seawater}} = \dfrac{4.89E5 \, lb_m/hr}{64.0 \, lb_m/ft^3}$

$= \left(7.64E3 \dfrac{ft^3}{hr} \right)\left(7.48 \dfrac{gal}{ft^3} \right)\left(\dfrac{1 \, hr}{60 \, min} \right)$

$= 952 \, GPM$

$\boxed{\dot{V}_{seawater} = 952 \, GPM}$

c) What is the *LMTD* (θ_m) of the condenser? The temperature profile graph is as follows. Note that although the condenser is a cross-flow heat exchanger, we draw the temperature profile graph as a counter-flow heat exchanger.

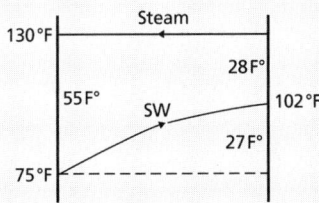

FIGURE EXAMPLE 7.2C

$$\theta_m = \frac{\theta_{left} - \theta_{right}}{\ln\left(\theta_{left}/\theta_{right}\right)} = \frac{55-28}{\ln(55/28)} = 40\ °F$$

$$\boxed{LMTD = \theta_m = 40\ °F}$$

d) What is the overall heat transfer coefficient, U?

$$\dot{Q}_{hx} = \dot{Q}_{steam} = \dot{Q}_{seawater} = UA\theta_m$$

$$U = \frac{\dot{Q}_{hx}}{A\theta_m} = \frac{1.24E7\ Btu/hr}{(2200\ ft^2)(40\ °F)} = 141\frac{Btu}{hr-ft^2-°F}$$

$$\boxed{U = 141\frac{Btu}{hr-ft^2-°F}}$$

e) If the tubes are 1 inch in diameter and limited in length to 4 feet, how many tubes are in this condenser? Realize that the heat transfer area of a tube is the surface of the tube and not the cross-sectional area that we used to calculate flow parameters in Unit 3. We adapt the equation of the surface area of a cylinder to incorporate the tube length restriction of 4 feet.

$$A_{hx} = (circumference)(L_{total}) = (\pi D_{tube})(L_{1\ tube}N_{tubes})$$

$$N_{tubes} = \frac{A_{hx}}{\pi D_{tube}\ L_{1\ tube}} = \frac{2200\ ft^2}{\pi(1/12\ ft)\left(4\ ft/tube\right)}$$

$$= 2101\ tubes$$

$$\boxed{N_{tubes} = 2101\ tubes}$$

Note that this is an example of where you round up regardless—if you need 2100.1 tubes, that means 2101.

Table 7.2 summarizes all of the heat exchangers covered by this course.

TABLE 7.2: *Summary of Heat Exchangers, Their Fluids and Data Sources*

System	Component	Fluid 1 (hot)	Fluid 2 (cold)	h, r, T Data Source(s)
Steam Propulsion	Condenser	Turbine Exhaust Steam		Steam Tables (WV)
			Seawater	$Dh = C_p DT; C_p - 0.94; r = 64$
	Nuclear Steam Generator	Primary Coolant Side		Steam Tables (CL) $\rho \neq 62.4$
			Secondary Side	Steam Tables: CL in; SV out
	Boiler	Combustion Gases		**Not calculated in EM300**
			Steam System	Steam Tables: CL in; SV or SHV out
Vapor Compression Refrigeration	Evaporator	Air in DX Systems		Psychrometric Chart
		Chill Water		Steam Tables (CL) or $Dh = C_p DT; C_p = 1.0; r = 62.4$
			Refrigerant	Refrigerant p-h chart or Tables
	Condenser	Refrigerant		Refrigerant p-h chart or Tables
			Air	Psychrometric Chart
			Condenser Water	If FW: Steam Tables (CL) or $Dh = C_p DT; C_p = 1.0; r = 62.4$
				If SW: $Dh = C_p DT; C_p = 0.94; r = 64$
HVAC	AHU coils	Air		Psychrometric Chart
			Chill Water	Steam Tables (CL) or $Dh = C_p DT; C_p = 1.0; r = 62.4$
			Refrigerant in DX Systems	Refrigerant p-h chart or Tables
Lube Oil	Lube Oil Cooler	Lube Oil		$Dh = C_p DT; C_p \neq 1; r = SG(62.4)$
			Water	If FW: Steam Tables (CL) or $Dh = C_p DT; C_p = 1.0; r = 62.4$
				If SW: $Dh = C_p DT; C_p = 0.94; r = 64$
Water (Mixing)	Hot Water Heater	Multiple Streams Possible		Steam Tables or, if temps are near to room temperature conditions, $Dh = C_p DT; C_p = 1.0; r = 62.4$
Air (Mixing)	Ductwork Tee	Outside Air	Recirc Air	Psychrometric Chart

Practice Problems

"If you don't go after what you want, you'll never have it. If you don't ask, the answer is always no. If you don't step forward, you're always in the same place."
—Nora Roberts

7.1 A 1 pint can containing a beverage (1 lb_m, assume $C_p = C_p$ of sea water) at 70 °F is submerged into one gallon of cold freshwater (initially 33 °F) held in a well insulated bucket. The beverage is cooled uniformly (by heat transfer to the cold water) until it reaches 40 °F. Calculate the temperature (°F) of the resulting uniformly distributed freshwater bath. Can the beverage be cooled more? Estimate to within 1 °F the final temperature of the beverage and water bath at equilibrium.

7.2 A closed tank contains 500 gallons of sea water at 170 °F. Determine the total energy as heat (Q) transferred from the sea water to change the temperature to 100 °F. Assume the specific heat capacity (C_p) and density of the seawater are constant at standard values.

7.3 A closed chamber contains 2 lb_m of saturated steam at a constant vacuum of 20 in-Hg. Determine the total energy transfer as heat (Q, Btu) to or from the steam to condense it to saturated liquid and then sub-cool the condensate to 120 °F.

7.4 Lubricating oil (s.g. = 0.8, C_p = 0.6 Btu/lb_m-°F) flows through a cooler at 100 lb_m/min entering at 120 °F. Determine the cooler heat removal rate ($\dot{Q}$, Btu/min) needed to reduce the oil temperature to 90 °F (at the cooler exit).

7.5 Main sea water (MSW) flows through condenser tubes at 400 gal/min and enters inlet waterbox at 50 °F. 2x10^5 Btu/min of heat energy is transferred to the MSW (from steam condensing on the "hot" side of the condenser). Determine the temperature (°F) of the MSW exiting the condenser tubes. Use standard property values.

7.6 The following flows into a drain tank that is vented to the atmosphere (assume p_{atm} = 15 psia):

Flow	Flow Rate (lb_m/hr)	Temperature (°F)
galley drains	2500	180
evaporator drains	750	190
main air ejector drains	350	200
aux air ejector drains	100	210

Assume the tank is well-insulated. Find the mass flow rate (lb_m/hr) and enthalpy (Btu/lb_m) of the flow leaving the tank.

7.7 A counter-flow lube oil cooler uses seawater as the cooling medium. The lube oil (Cp = 0.45 Btu/lb_m-°F) is flowing at a rate of 5,000 lb_m/hr and enters the cooler at a temperature of 160 °F and leaves at a temperature of 120 °F.

a) Sketch the temperature profile graph under the simplified schematic.

b) Find the flow rate (lb_m/hr) of cooling water required if the seawater inlet temperature is 50 °F and the discharge temperature is 100 °F.

c) If the overall heat transfer coefficient for the cooler is 80 Btu/hr-ft^2-°F, what is the required heat exchanger surface area (ft^2)?

7.8 For the main condenser shown in the accompanying figure, the following parameters are known:

- Mass flow rate of steam = 170,000 lb_m/hr
- Seawater injection temperature = 60 °F

- Overboard discharge temperature = 80 °F
- Condenser pressure = 1 psia
- Quality at turbine exits = 89%
- Condenser discharge may be considered ideal
- Overall heat transfer coefficient (U) = 600 Btu/hr·ft^2·°F

Find:

a) the rate of heat rejection from the steam (Btu/hr)

b) the cooling (seawater) water flow rate (lb$_m$/hr)

c) $LMTD$ (°F)

d) the cooling surface area (ft^2)

e) If the condenser tubes are 1 inch in diameter and limited to 6 feet in length, how many tubes are in the condenser?

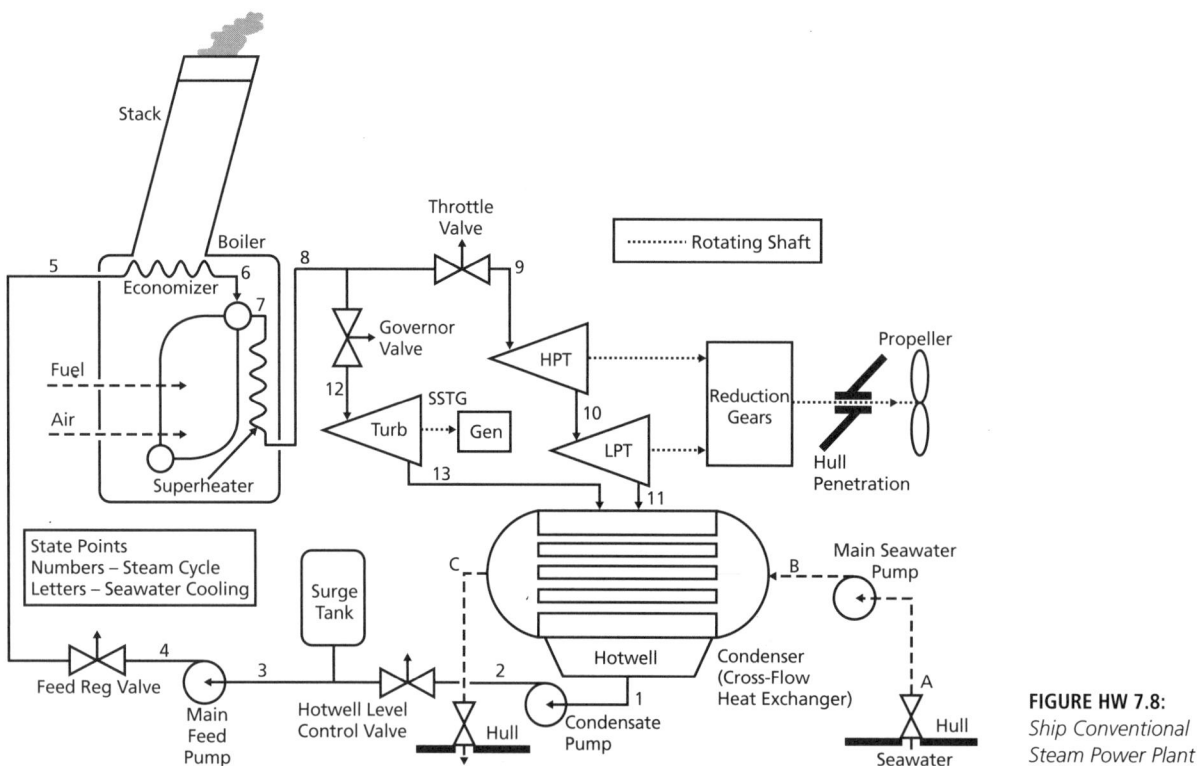

FIGURE HW 7.8: *Ship Conventional Steam Power Plant*

7.9 A counter-flow lubricating oil cooler with a net heat transfer area of 258 ft^2 cools 60,000 lb$_m$ of oil per hour from a temperature of 145 °F to 120 °F. The temperatures of the cooling water entering and leaving are 75 °F and 90 °F. The specific heat capacity of the oil is 0.5 Btu/lb$_m$-°F.

a) What is the overall heat transfer coefficient under these operating conditions?

b) What is the required area for a parallel-flow device having the same heat transfer capacity under identical operating conditions?

7.10 An auxiliary condenser (heat exchange area of 2200 ft²) operates with a vacuum of 25.4 in Hg. Steam enters the condenser at a rate of 14,810 lb$_m$/hr with a quality of 82%. The cooling medium is seawater (specific heat capacity of 0.94 Btu/lb$_m$-°F and a density of 64 lb$_m$/ft³) with an inlet temperature of 75 °F and an outlet temperature of 102 °F.

   a) How much heat is exchanged in the condenser (Btu/hr)?

   b) What is the seawater flow rate (GPM)?

   c) What is the log mean temperature difference (°F)?

   d) What is the overall heat transfer coefficient (Btu/ hr·ft²·°F) for this condenser?

7.11 A counter-flow lube oil cooler utilizes sea water as the cooling medium. The lube oil (Cp = 0.564 Btu/lb$_m$ °F) is flowing at a rate of 20,000 lb$_m$/hr and enters the cooler at a temperature of 175 °F and leaves at a temperature of 120 °F. The cooling medium temperatures are 50 °F and 70 °F.

   a) Find the mass flow rate (lb$_m$/hr) of the cooling medium.

   b) Determine the value of the $LMTD$ (°F).

   c) If the overall heat transfer coefficient of the cooler is 75 Btu/ hr·ft²·°F, what is the required heat exchange surface area (ft²)?

   d) If the cooling medium temperatures change to 90 °F and to 110 °F, how much larger would the heat exchanger have to be in order to remove the same amount of heat from the oil (ft²) (assume the other parameters are unchanged)?

7.12 Each maintenance period the overall heat transfer coefficient of the main condenser (which is configured as a cross-flow heat exchanger) in a steam power plant is calculated to determine the extent of the marine growth that has occurred since last inspection. The required heat transfer rate of the condenser is 8.82E6 Btu/hr. The area of the condenser is 2,100 ft². During the current maintenance period, the steam entering and water exiting the condenser are saturated at 1 psia. The seawater injection temperature is 52 °F and the seawater discharge temperature is 67 °F. During the last maintenance period the $LMTD$ of the condenser was recorded as being 35 °F. Determine the change in the value of the overall heat exchanger coefficient of the condenser between maintenance periods.

7.13 Some of the Navy's diesel-powered amphibious ships use the hot diesel exhaust to pre-heat seawater for a distillation-based desalination plant. The diesel engines receive air at 15 psia and 60 °F. They operate with a compression ratio of 16:1, a max cycle temperature of 2,700 °F, and the ship cruises while the engines produce a total of 10,000 SHP. Engine mechanical efficiency is 90% and reduction gear efficiency is 95%.

   a) What is the mass flow rate and temperature of the exhaust (assume air standard methodology and no other heat losses)?

   b) If the seawater is heated by the exhaust from 50 °F to 180 °F in a counter-flow heat exchanger while the diesel exhaust leaves the stack at 350 °F, what is the mass flow rate of hot seawater that can be sent to the desalination plant at these operating conditions?

   c) What is the $LMTD$ of the heat transfer process?

   d) If 50% of the seawater is converted to distillate, how many gallons at normal room temperature could be produced in a day at these conditions?

8 Nuclear Power

An artist's rendering of a *Virginia*-class fast attack submarine shooting a torpedo. United States Navy submarines and aircraft carriers can operate for many years on the nuclear fuel loaded in a reactor core.

8.1 UNIT LEARNING OBJECTIVES

8.1.1 Terminology: definitions, variables, and typical units

8.1.1.1 pressurized-water reactor (PWR), boiling water reactor (BWR), gas-cooled reactor, moderator, light water, fission, fusion, steam generator, reactor or main reactor coolant pump, reactor coolant, pressurizer, control rods, control rod drive motors, primary loop, secondary loop, scram

8.1.2 Concepts: ideas and engineering expressions of them

8.1.2.1 Understand the basics of nuclear fission.

8.1.2.2 List and describe the major categories of nuclear power reactors.

8.1.2.3 Be able to sketch and describe typical PWR system arrangement and component functions.

8.1.2.4 Understand the applications where a nuclear power plant is preferable to other propulsion configurations.

8.1.2.5 Complete simple Rankine cycle problems typical for a naval nuclear power plant.

8.1.3 Skills: procedures, practices, or methods that enable reasoning

8.1.3.1 Apply Steam Plant analysis skills to analyze the primary and secondary systems of a PWR nuclear power plant.

8.2 INTRODUCTION

On January 17, 1955, the USS *Nautilus* (SSN-571) signaled "Underway on nuclear power." This started the U.S. Navy's operations with nuclear-powered vessels. The *Nautilus* initiated many changes in our Navy, especially in the ways submarines operate. Higher speed, ability to maintain that speed almost indefinitely, and the ability to remain submerged for periods limited only by the food carried for the crew made transits to operating areas shorter and opened up areas previously impossible for any ships to conduct sustained operations, such as under the polar ice. *Nautilus* was the first vessel to reach the geographic North Pole—90 degrees north—on August 3, 1958. *Nautilus* was decommissioned on March 3, 1980, after a career spanning twenty-five years and having steamed almost half a million miles.

Nuclear power was adapted to aircraft carriers and cruisers. U.S. Navy reactors use a designation number system, the first letter being for the intended platform (S—submarine, A—aircraft carrier, C—cruiser, D—destroyer leaders); the second being the number of the design; and the third being the prime contractor (G—General Electric, W—Westinghouse, C—Combustion Engineering, B—Bechtel). The *Nautilus* was fitted with the S2W reactor design (submarine, second design, by Westinghouse; S1W was a land-based prototype reactor).

The USS *Enterprise* (CVN-65) was commissioned on November 25, 1961. This nuclear-powered platform had eight A2W nuclear reactors providing 280,000 SHP total to four shafts, plus all of the electrical and steam-operated catapult needs of the ship. By the time the USS *Nimitz* (CVN-68) was built, improvements in reactor design allowed the ships of this class to have the same shaft power output provided by two A4W nuclear reactors.

The Navy built a number of surface combatants with nuclear propulsion. The *Bainbridge*, *Truxtun*, *California*, and *Virginia* classes of guided missile cruisers were built with the twin D2G reactors. Historically, the U.S. Navy developed a class of ships referred to as "destroyer leaders" that were intended to be sized between a destroyer and a cruiser. These ships were all ultimately redesignated as cruisers. The USS *Long Beach* (CGN-9) was built specifically as a cruiser and used two reactors of the C2W design.

All of the U.S. Navy's nuclear-powered vessels employed pressurized water reactors, except one, the USS *Seawolf* (SSN-575), the second nuclear-powered submarine, which used a liquid sodium cooled reactor (S2G). After two year of service, *Seawolf's* original S2G reactor was replaced with a modified *Nautilus* reactor (S2Wa).

Other countries have adapted nuclear power to their naval vessels. The Soviet Union (now Russia), Great Britain, France, and China all have nuclear-powered submarines. India has leased a number of nuclear submarines from Russia and is now working on indigenous designs. Nuclear-powered surface vessels in other countries have been limited to a handful. France's *Charles de Gaulle* is a nuclear-powered aircraft carrier. As of this writing, a number of other countries including Brazil are developing nuclear power plants for naval submarines and ships, but these have yet to prove operational.

Commercially, only a small number of nuclear-powered vessels were ever constructed in the world. The United States built one, the NSS (Nuclear Steam Ship) *Savannah,* which proved to be uneconomical and was retired very early in the projected life of the ship. Russia continues to operate a number of nuclear-powered ice breakers that were originally built by the USSR.

8.3 CURRENT U.S. NAVY APPLICATIONS OF NUCLEAR POWER

For submarines, nuclear power is the only feasible propulsion option to achieve sustained submerged endurance. Of note, during World War II, most submarine losses were due to detection while surfaced or snorkeling or due to inability to evade at high speed while submerged (the United States lost 52 submarines in World War II, while Nazi Germany lost 662).

Nuclear power is the preferred power system for aircraft carriers as it allows more fuel to be carried for aircraft and surface escorts and dramatically reduces the number of tankers required to support a battle group. Current aircraft carrier reactor designs can operate for approximately twenty years before the nuclear fuel (or "core") must be replaced.

The primary disadvantages of nuclear-powered compared to conventionally powered surface vessels are: much higher initial cost, much higher maintenance costs, and significantly more personnel required for operation and maintenance. It is also politically unpopular with some groups and countries, and consequently, there may be restrictive port visitation requirements.

Current U.S. naval vessels employing nuclear steam plants make up about 40 percent of the major naval combatant vessels. The overall propulsion efficiency of nuclear power plants is about 15 percent.

8.4 BASICS OF NUCLEAR FISSION

Recall that a chemical element is defined by the number of protons in its nucleus and that neutrons are also present in the nucleus of most elements. An element can also have various *isotopes* due to differing numbers of neutrons accompanying the same number of protons. The term *fissionable* refers to the ability of an element to undergo a fission reaction. There are very few elements that are fissionable, and not all isotopes of a fissionable element are readily fissionable. This is the case with the principal nuclear fuel used in fission reactors—uranium. Uranium has ninety-two protons in its nucleus.

U.S Navy nuclear propulsion plants, and most commercial reactor designs, use fission of the rare uranium isotope U-235 ($^{235}_{92}U$). Only approximately 0.7 percent of natural uranium, as it is mined from the earth, is U-235; the balance is predominately U-238, which is more stable and much more difficult to fission than U-235. Commercial reactors use uranium that has been *enriched* to approximately 3 percent U-235; Navy reactors, because a higher power density is desired, use uranium with much higher U-235 enrichment.

Depleted uranium, used in projectiles, is the waste product U-238 from the enrichment process. There are two enrichment processes in current use in the world: gaseous diffusion and gaseous centrifuging. *Gaseous diffusion* uses a process of micro-filtering uranium that has been gasified into uranium hexafluoride (UF_6). *Centrifuge enrichment* spins UF_6 at high speeds, and the UF_6 with the heavier isotope (U-238) tends to migrate to the outer region of the centrifuge while the lighter isotope tends to stay closer to the centrifuge axis. Many multiple centrifuges in series are required to achieve significant enrichment by this technique.

Once the UF_6 gas reaches the required degree of enrichment, it is reformulated to solid uranium dioxide, UO_2. The UO_2 is shaped and clad with a metal *cladding* to protect the UO_2. The cladding metal is selected based upon having high thermal conductivity, low susceptibility to corrosion, and hardness to minimize erosion effects from water that is pumped through the reactor. The fuel is assembled into fuel elements that are inserted into the reactor in geometric shapes referred to as *fuel element assemblies* that make up the *reactor core*.

The U-235 fission reaction is:

$$^{235}_{92}U + ^{1}_{0}n \rightarrow FP_1 + FP_2 + 2.43\ ^{1}_{0}n + Energy \quad (8.1)$$

where *n* is a neutron and FP_1 and FP_2 are fission products (e.g., iodine, barium, cesium, etc.). Note that, on average, 2.43 neutrons are released from each fission of U-235. These neutrons are available to cause subsequent fissions in what is referred to as a *chain reaction*. In a nuclear power reactor this chain reaction is controlled to maintain the desired power level. In an atomic bomb, with approximately 80 percent U-235 enrichment, the chain reaction runs unchecked, which results in a massive release of energy in a fraction of a second.

The amount of energy released in fission, or U-235 heating value, is approximately 3.51×10^{10} Btu/lb_m of U-235, which is many orders of magnitude greater than that released in a typical chemical combustion reaction (e.g. heating value of aircraft fuel JP-5 is approximately 18,300 Btu/lb_m).

The student may be familiar with the terms *light water reactor*, *heavy water reactor*, or *graphite-moderated reactor*. These terms relate to the way in which the neutrons that are produced by fission are slowed down

TABLE 8.1: *U.S. Navy Nuclear Power Plant Summary*

Class	Reactor Design	# of Reactors	Reactor Power Rating	Propulsion Plant Shaft Horsepower[1]
Los Angeles Class (SSN-688)	S6G	1	165 MW[2]	35,000
Ohio Class (SSBN/SSGN-726)	S8G	1		60,000
Seawolf Class (SSN-21)	S6W	1		45,000
Virginia Class (SSN-774)	S9G	1		40,000
Enterprise (CVN-65)	A2W	8		280,000
Nimitz Class (CVN-68)	A4W	2		280,000

[1] Unclassified values from *Jane's Fighting Ships*
[2] Unclassified value from http://en.wikipedia.org/wiki/S6G_reactor

(moderated) to the appropriate energy level to cause another fission. As noted in the equation above, there is an average of 2.43 neutrons produced per fission event. Neutrons are produced at an energy level that is too high to cause a fission and must be slowed down to the proper energy level without losing too many of them in the process. Neutron moderation slows the neutrons and re-directs them to the core in order to sustain the fission chain reaction. Navy nuclear power uses light water reactors. Light water reactors use water, H_2O. This is in contract to heavy water reactors, which use water containing deutreium, an isotope of hydrogen, in the compound.

Control rods are devices made from materials that absorb neutrons. Control rods are inserted into spaces inside the core to shut it down. This may be done either by inserting the rods slowly with their drive motors (control rod drive mechanisms) or dropping them in quickly in what is called a *scram*. A scram may be manually initiated by plant operators or automatically initiated by the reactor safety control systems. Either way, the inserted control rods absorb neutrons and prevent these neutrons from causing fission in the fuel. Once a reactor has been run, there are always neutrons available to cause fission. Some do cause fission in a shut down reactor, but not enough to raise the reactor power to significant levels. To start up a reactor, the rods are withdrawn to allow some of the neutrons that would otherwise have been absorbed to cause additional fissions. This power level is raised until the chain reaction is self-sustaining and the reactor is referred to as *critical*.

Each time a reactor core is operated to make steam, fission products, many of which themselves are radioactive, accumulate in the core. Radioactive fission products decay by emission of particles and radiation and, therefore, continue to produce heat after the fission reaction has been shutdown. This *decay heat* must be removed in order to prevent heating the water in the reactor to boiling conditions, which may result in overheating the nuclear fuel in the core. The rate of decay heat generation is about 10 percent of the reactor power just prior to shutdown, and it decreases exponentially from the moment of shutdown. Providing for decay heat removal is a nuclear safety issue for many weeks or even months after the reactor is shutdown. It is an all-hands responsibility to ensure that decay heat is always removed from the core and to be aware of critical systems for removing decay heat from the reactor.

In contrast to splitting an atom in fission to produce lighter fission products, *nuclear fusion* employs combining elements to make a heavier element. This technology is not practical for power generation in the near term but holds great promise in the future.

8.5 TYPES OF NUCLEAR REACTORS

Nuclear reactors are broadly characterized as power reactors and breeder reactors. Power reactors are used to produce power for propulsion and/or generation of electricity. Breeder reactors are used to produce nuclear materials for nuclear medicine, nuclear weapons production, and related areas of research. Naval and electrical utility applications use nuclear power reactors.

Current naval applications of nuclear power all use the *pressurized water reactor* (PWR) configuration. In a PWR, high-pressure water is pumped through the reactor to carry the heat from the reactor core to a steam generator. The water that flows around the fuel elements and against the fuel cladding is referred to as *primary water*. A consequence of using high-pressure water is that all components exposed to this high pressure must be stronger and thicker, hence heavier. The steam generator is an "indirect" heat exchanger that separates the high-pressure primary water from *secondary water*, which is at a lower pressure and allowed to boil. The steam generator is analogous to the boiler in a conventional (oil-fired) steam plant. The steam produced by a PWR's steam generator is dry, saturated steam (not superheated). The steam portion of the overall system is otherwise similar to a conventional steam plant. However, since the turbine inlet conditions are not superheated, the steam expansion to the condenser vacuum pressure will then produce steam of lower quality at the turbine discharge. The turbines must be capable of handling this additional moisture content.

The U.S. and Soviet navies built several *liquid metal cooled reactors* (LMCR). Instead of pumping water through the reactor to absorb the heat from the core, LMCRs used metals that melt at low temperatures. Metals with this characteristic are called *eutectics*. Sodium (Na), with tremendous heat absorption capabilities and a reasonable melting temperature of about 208 °F, was the chosen eutectic. Since metallic sodium will not vaporize in the temperature range of core operation, it does not require pressurization to prevent boiling and may be operated at low pressures. The nuclear components can then be thinner (less weight). LMCRs also incorporated superheaters for producing superheated steam. The principal disadvantage of LMCR technology is that sodium burns when exposed to the oxygen in air and explodes when in contact with water. Hence, any system leaks created serious additional problems. This technology was abandoned by the

U.S. Navy early in the history of nuclear power (S2G). The Soviets built a small number of LMCR-powered submarines that were operational in the 1970s and 1980s.

Commercial nuclear power plants produce about 20 percent of the electricity in the United States. PWR technology similar to that used in naval applications is common. Another configuration used in commercial nuclear power is the *boiling water reactor* (BWR). This configuration eliminates the steam generator but must have a reactor that is designed to allow boiling to take place within the reactor core. The steam that is produced is piped to turbines that, in turn, spin electrical generators. BWRs, while eliminating the efficiency-reducing heat transfer that takes place in the PWR's steam generator, have the disadvantage of being too large for naval applications. Furthermore, radiological activation and contamination produced in the reactor is distributed to the steam system, thereby increasing maintenance costs.

High temperature gas-cooled reactors (HTGCRs) are the nuclear version of the closed-loop Brayton cycle. Gaseous helium (He) is the typical coolant that is pressurized by a compressor, heat added in the reactor, expansion through a turbine, heat rejection in a heat exchanger, and then back to the compressor to repeat the cycle. Historically, GCRs were more common in other countries than in the United States. However, this technology is being reinvestigated in the form of the *pebble bed reactor* (PBR), which is currently under development as possibly one of the next generation power reactors.

8.6 NAVY NUCLEAR PROPULSION PLANT

Figure 8.1 shows a simplified nuclear propulsion system. Real naval plants include redundant elements, isolation valves to allow for continued operation with battle damage or other casualties, and other control features. Figure 8.1 illustrates the three "loops" of working fluids and the essential components: (1) the high-pressure primary loop; (2) the steam cycle secondary loop; and (3) the seawater cooling loop to the ocean environment.

The *primary system's* working fluid is water that is pressurized to such high levels (typically in the range of 1500 to 2500 psi, depending upon the specific design) that it will not boil in the primary loop, even at the fairly high temperatures in the reactor. The state of the *primary coolant* (water) is thus a compressed liquid (CL) in the flowing portion of the system. The reactor is basically a heat exchanger in which thermal power is generated in the nuclear fission taking place in the solid fuel, the *reactor core*. The core is comprised of multiple fuel elements with the elements shaped and spaced to provide optimum heat transfer and coolant flow conditions. Primary coolant is pumped through the reactor, where it receives heat produced by the nuclear fission reaction in the core. The primary coolant then moves to the *steam generator* (S/G) via pipes called the *hot leg*. The S/G is an indirect heat exchanger, meaning the primary coolant water does not come into contact with the secondary system's water. The S/G's heat transfer takes place across numerous small-diameter tubes. The configuration of the S/G in Figure 8.1

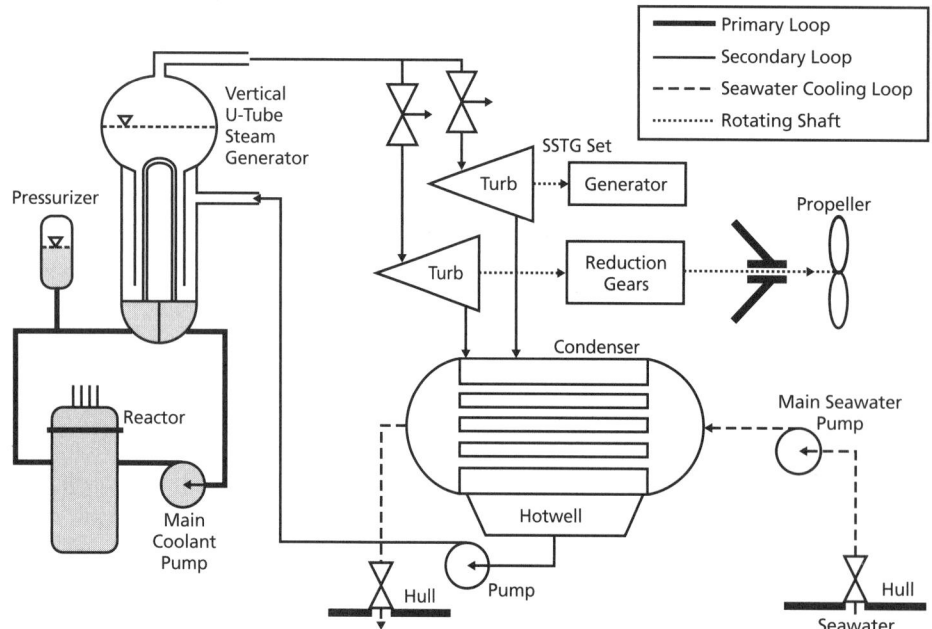

FIGURE 8.1: *Simplified Nuclear Propulsion Plant Schematic*

is a "vertical U-tube steam generator." There are other S/G configurations used in both naval and commercial applications. Regardless, the primary coolant flows through the inside of the tubes, while the secondary system's water boils on the outside of the tubes. After giving up some of its heat, the primary coolant is returned to the reactor via the *cold leg* with the main coolant pumps (MCPs) providing the pressure rise necessary to overcome the frictional head losses of the pipes, valves, reactor, and S/G. The pressurizer is a special component that is discussed in greater detail below.

The secondary side of the steam generator receives water at CL conditions and, since there is no superheater, produces SV conditions at reasonably high steam pressures. The rest of the secondary loop is similar to a conventional steam plant. Steam is expanded in parallel through propulsion turbines and turbines driving the ship's service turbine generators (SSTGs). The propulsion and SSTG turbines may exhaust to the same condenser or have their own condenser. As with the conventional steam plant, the condenser is operated at a deep vacuum. Since the turbine inlet conditions are SV (not SHV as with a conventional steam plant, since a PWR has no superheater) the quality entering the condenser is also lower. As with the conventional steam plants, the condenser is an indirect heat exchanger with multiple small-diameter tubes. The condenser's function is to reject the heat necessary to turn the secondary water from WV conditions to SL conditions (for the ideal steam plant) or CL conditions (in a real plant to prevent cavitation and consistent with the *NPSHr* for the condensate pump). Recall that in order to produce condensate depression, some number of the condenser tubes are submerged in condensate.

The seawater cooling loop removes the heat given off by the turbine exhaust steam in the process of condensing to a liquid state. The seawater side of the condenser is exposed to sea pressure and the pump head necessary to overcome the frictional head losses in the seawater system. The seawater intakes are typically on the bottom of the hull, and the pressure available at the intakes is the hydrostatic pressure. This is fairly low pressure for a surface ship, but can be substantial for a deep-diving submarine.

Figure 8.2 illustrates the redundancy provided by having two primary loops connected to the same reactor and also shows redundancy with multiple MCPs that can be isolated in the event of system damage. This is generally characteristic of submarine systems. Aircraft carriers have four S/Gs per reactor. Not only is steam needed for propulsion and electrical generation, but steam is also needed to operate the aircraft catapults. Approximately 100 to 250 gallons of water are lost from the secondary system per catapult launch. The catapult steam is not recovered, so aircraft carriers include significant desalination capacity to make up the water lost during flight ops.

The *pressurizer* is a special primary system component. The pressurizer is connected to the hot leg via the *surge line*. Hence the pressure in the hot leg is essentially the pressurizer pressure. The pressurizer is fitted with electric heaters that serve to boil the water in the pressurizer and create a steam bubble. This means that the conditions inside the pressurizer may be considered to be saturated (overall WV conditions with the SV steam bubble at the top and the SL water at the bottom). Since steam is elastic, any volume changes in the primary loops (due to temperature-induced density changes) either allow the steam to compress and condense or steam to expand and liquid to flash to vapor as needed to keep the primary system pressure within specified bands. If there were no pressurizer, since the primary loops are at CL conditions (this scenario would be referred to as "solid" even though we are talking about a liquid—the meaning of the term is that the system is completely filled with liquid with no expansion volume available), and if the average temperature of the primary were to go up, there would be a huge pressure increase that could rupture the components. Conversely, if the primary average temperature goes down, the density and pressure would also drop precipitously and the highest temperature area of the primary system (in the core) could reach saturation and begin to boil. Since steam is not as good at absorbing heat as liquid water, the area of the core with steam could overheat. So, the pressurizer is an essential piece of equipment for safe plant operation.

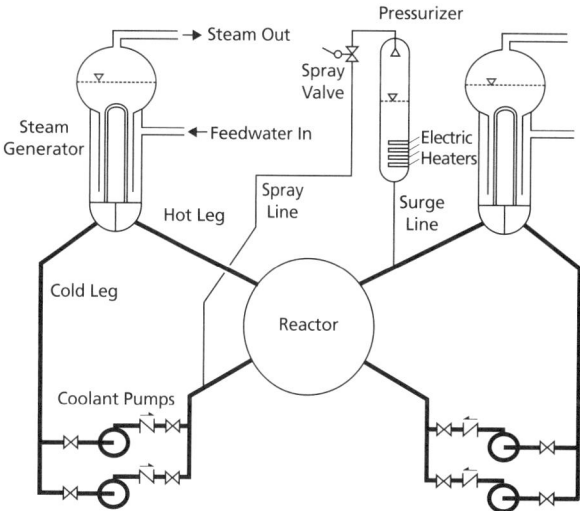

FIGURE 8.2: *Two-Loop Primary System with Pressurizer Shown*

Beyond the normal small pressure variations due to operational transients, if the primary pressure is lower than desired, the electric heaters are energized to raise the saturation temperature (and hence saturation pressure). If pressure is too high, the spray valve is remotely opened to allow water to spray in via the spray nozzle. Notice that the MCPs raise the pressure while the pipes, reactor, and S/G all have frictional head losses. For convenience, the pressure losses in the piping are lumped in with the reactor and S/G. This means that the highest primary pressure is between the MCP discharge and the reactor (the cold leg of the reactor); that the pressure in the hot leg is higher than the S/G cold leg pressure; and the cold leg of the S/G has lower pressure than the cold leg of the reactor. So, if the spray valve opens, the difference in reactor cold leg and hot leg pressures provides the pressure gradient to force some flow into the pressurizer steam space via the spray nozzle. Since the spray water is relatively colder, it will condense some of the steam bubble and reduce primary system pressure.

In military applications, the MCPs are typically multiple speed pumps, as was discussed in the fluid flow unit (commercial reactors rarely operate at less than 100 percent power, so slow speed pumps are not used in commercial utility applications). Recalling the pump affinity laws for two-speed pumps, slow speed operation consumes one-eighth of the power needed to drive the pumps at fast speed. This also means that the pumps will move one-half of the volumetric flow and have one-quarter of the pressure rise across the pumps. Tactically and important for submarines, slow speed pump operation reduces sound signature and improves stealth. Using less energy also extends the life of the reactor core. Acoustic stealth is not a concept typically associated with an aircraft carrier, but core life is important to a carrier as well, so reduced speed MCP operation is used when ship speed and flight operations do not require the additional reactor power.

Motor speed can either be controlled by the number of poles in the circuit in the motor or by varying the electrical frequency to the motor. Two-speed motors run at full or half speed. Some vessels have the capability of powering the driving motor with a variable frequency drive (VFD). A VFD receives nominal 60 Hz input and modifies it electrically to some lower frequency. This allows reduced pump rotational speed. Recalling that a motor's rotational speed is governed by the equation $Np = 120 f$, if the electrical frequency is reduced from 60 to 50 Hz, the speed of rotation will become 50/60 of the normal speed. Knowing the power and pressure rise and the ratio of electrical frequencies, the pump affinity laws will predict the pressure rise across the pump and the power required when operating with reduced electrical frequency.

8.6.1 Primary System Integrated Operations

Assuming balanced operation between the two primary system loops allows us to simplify the power plant schematic from Figure 8.2 to that shown in Figure 8.1. This simplification treats parallel S/Gs as one S/G and parallel MCPs as one pump.

Figure 8.3 illustrates the concepts of a simple nuclear propulsion plant (this schematic ignores the steam used for electrical generation, for illustration purposes). State points 11, 12, and 13 are in the primary loop, and SPs 21, 22, 23, and 24 are in the secondary loop. The seawater cooling loop is not illustrated on the T-s graph. Let's look more closely at some of the assumptions included in the T-s graph. Note that the primary loop's MCP is shown as ideal and pressure drops are shown through both the reactor and the steam generator; the secondary loop is shown with non-ideal pump and turbine, condensate depression

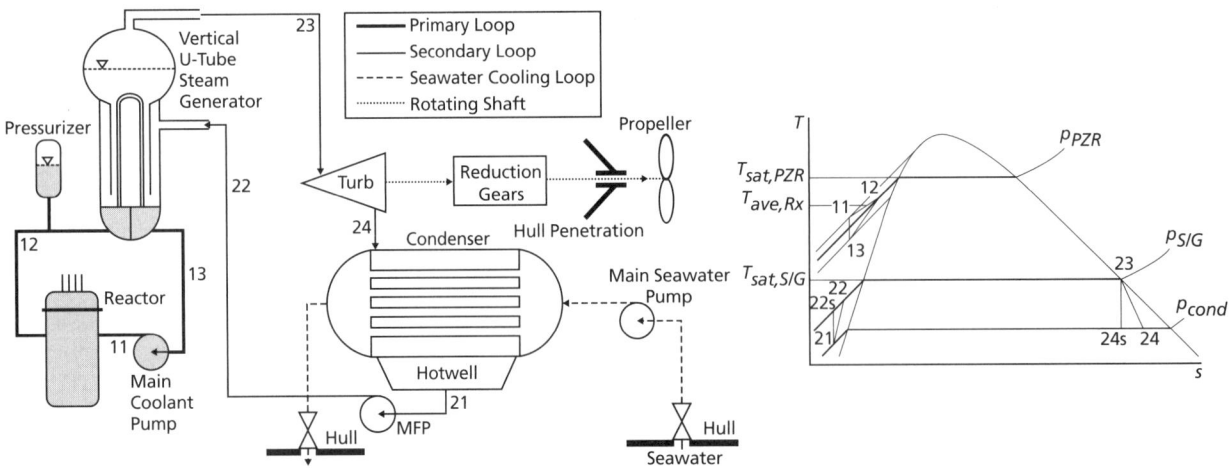

FIGURE 8.3: *Simple Nuclear Steam Plant Schematic with Associated T-s Concept Graph (Showing Primary and Secondary Loop Conditions)*

entering the pump, and both heat exchangers are isobaric. Note that $p_{12} = p_{PZR}$ in both the schematic and on the T-s graph. Recall that the isobars are exaggerated in the compression liquid region in order to illustrate the process more clearly. In reality the isobars are very close to the SL line of the vapor dome.

The reactor is operated so that the average temperature of the coolant stays about the same, regardless of power level as illustrated in Figure 8.4. With the average temperature fixed and the coolant flow rate determined by the speed of the MCP, increases in reactor power show up as increases in the change in enthalpy across the reactor—the hot leg gets hotter and the cold leg gets colder. Note that since the primary system coolant is at CL conditions, unless compression liquid tables are available the enthalpy of the fluid may be determined from temperature using the "CL approximation" technique discussed in the previous unit. T_{hot} must be below the saturation conditions for pressure in the reactor, therefore must be somewhat lower than the conditions in the pressurizer. Thinking about the temperature gradient conditions necessary to transfer the heat from the primary to the secondary, T_{cold} must also be above T_{sat} for the secondary side of the S/G.

Another simplification in this analysis comes from considering the enthalpy change in the coolant across the MCP. As was shown when analyzing the feedwater pump in the previous unit, when pumping an incompressible fluid, the ideal pump work and power are:

$$w_{p,s} = \Delta h_s = v_f \Delta p \qquad (8.2)$$

$$WHP = \dot{m}\Delta h_s = \dot{m}v_f \Delta p = \dot{V}\Delta p \qquad (8.3)$$

EXAMPLE 8.1: MCP WORK VS REACTOR HEAT

GIVEN: MCPs operating in fast speed. The MCPs are in the cold leg with Δp of 120 psid and $T_{cold} = 560\ °F$ out of the S/G.

FIND: Δh for these conditions. How does the pump work compare to the heat input of the reactor if $T_{hot} = 600\ °F$?

SOLUTION: The primary system water is CL except in the pressurizer. Even though the more accurate schematic is Figure 8.2, all MCPs will be lumped together and the S/Gs will be lumped so that the equivalent schematic is Figure 8.1. The steam tables used in this course do not have CL data, so using the CL approximation technique and the 2000 edition Steam Tables, v_f (@ 560 °F) = 0.02208 ft³/lb$_m$ and h_f (@ 560 °F) = 562.3 Btu/lb$_m$. The reactor down-

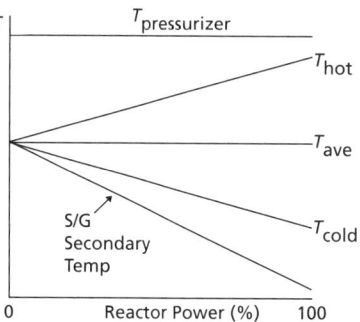

FIGURE 8.4: *Temperatures vs Reactor Power (Single Speed MCP Operation)*

stream conditions are h_f (@ 600 °F) = 616.9 Btu/lb$_m$. Another important point: the single pump as we're modeling the system and the reactor are seeing the same flow, so it is OK to solve this on the specific basis.

Substituting into the pump work equation for an incompressible fluid, and evaluating for required conversion factors:

$$w_{p,s} = \Delta h_s = v_f \Delta p$$

$$= \left(0.02208 \frac{ft^3}{lb_m}\right)\left(120 \frac{lb_f}{in^2}\right)\left(\frac{144\ in^2}{ft^2}\right)\left(\frac{1\ Btu}{778\ ft-lb_f}\right)$$

$$= 0.49 \frac{Btu}{lb_m}$$

$$\boxed{w_{p,s} = \Delta h_s = 0.49 \frac{Btu}{lb_m}}$$

No information was given about the pump efficiency. Neither can it be calculated without knowing downstream conditions, but note the value of ideal pump work at these conditions. Even if the pump were only 25% efficient (a conservative low value), the Δh_{real} would be about 2 Btu/lb$_m$.

Considering the reactor heat addition in specific terms:

$$q_{reactor} = \Delta h_{reactor} = h_{out} - h_{in}$$

$$= \left(616.9 \frac{Btu}{lb_m}\right) - (562.3 + 0.49)\frac{Btu}{lb_m} = 54.1 \frac{Btu}{lb_m}$$

$$\boxed{q_{reactor} = 54.1 \frac{Btu}{lb_m}}$$

Notice that using ideal pump work input (0.5 Btu/lb_m) vs reactor heat input (54.1 Btu/lb_m), we're talking about two orders of magnitude difference. Even if we were talking about the difference of ~50 Btu/lb_m vs ~2 Btu/lb_m (for real pump work), the reactor heat is still a factor of ~25 higher than the pump's work input to the primary system fluid. So, it is not unreasonable to <u>ignore the pump's contribution to the primary system's energy</u> and then, using this simplification, it follows that the heat input by the reactor is the heat transfer of the S/G. And since the reactor and single S/G in the simplified schematic see the same primary system mass flow rate, the total rate basis expressions also hold true.

$$q_{reactor} \cong q_{S/G,primary\,side} \cong h_{hot\,leg} - h_{cold\,leg}$$

$$\dot{Q}_{reactor} = \dot{m}_{primary}\Delta h_{reactor} \cong \dot{m}_{primary}\Delta h_{S/G,primary} = \dot{Q}_{S/G}$$

and it follows that

$$\dot{Q}_{S/G} = \dot{m}_{primary}\Delta h_{S/G,primary} = \dot{m}_{secondary}\Delta h_{S/G,secondary}$$

Note that the Δh on the primary side of the S/G is <u>not</u> the same as the Δh on the secondary side due to different primary and secondary mass flow rates, but the total heat rate is equal.

EXAMPLE 8.2: MAIN COOLANT PUMP SPEED SHIFT

GIVEN: The reactor is rated at 80 MW and T_{ave} is controlled at 475 °F. At 50% power and with MCPs in slow speed, T_{hot} is 500 °F, T_{cold} is 450 °F, and Δp across the MCPs is 30 psid. The MCPs are shifted to fast speed.

FIND: Assuming the reactor power is constant, what is:
a) the mass flow rate of primary coolant?
b) the pressure difference across the MCPs in fast speed? and
c) the volumetric flow rate of the MCPs (GPM) in fast speed?

SOLUTION: a) The primary coolant is a CL, so use the CL approximation technique to determine h. The reactor inlet and outlet enthalpies are:

h_f (500 °F) = 487.9 Btu/lb_m; h_f (450 °F) = 430.2 Btu/lb_m (interpolated)
(from 2000 edition *Steam Tables*)

Since the reactor power is 50% and thermal power is on a rate basis (time in the denominator), use the rate basis form of the thermal power equation to solve for mass flow rate.

$$\dot{Q}_{reactor} = \dot{m}_{primary}\Delta h_{reactor} \Rightarrow \dot{m}_{primary} = \frac{\dot{Q}_{reactor}}{\Delta h_{reactor}}$$

$$\dot{m}_{primary} = \frac{\dot{Q}_{reactor}}{\Delta h_{reactor}} = (40 MW)\left(\frac{1}{(487.9 - 430.2)Btu/lb_m}\right)$$
$$\left(3.412E6\frac{Btu}{MW-hr}\right) = 2.37E6\,lb_m/hr$$

$$\boxed{\dot{m}_{primary} = 2.37E6\,lb_m/hr = 3.95E4\,lb_m/min = 658\,lb_m/sec}$$

b) The MCPs are centrifugal pumps, so their performance follows the pump affinity laws. The MCP motor speed ratio between fast speed and slow speed is ~2.

$$\frac{\Delta p_{fast}}{\Delta p_{slow}} = \left(\frac{N_{fast}}{N_{slow}}\right)^2 \Rightarrow \Delta p_{fast} = \Delta p_{slow}\left(\frac{N_{fast}}{N_{slow}}\right)^2$$

$$= \Delta p_{slow}(2)^2 = 30\,psid\,(4) = 120\,psid$$

$$\boxed{\Delta p_{fast} = 120\,psid}$$

c) The volumetric flow rate ($\dot{V}$) of the MCP may be calculated from

$$\dot{m} = \rho\dot{V} \Rightarrow \dot{V} = \frac{\dot{m}}{\rho} = \dot{m}v$$

Recall that the primary coolant is at CL conditions, then $v = v_f(T)$. Now, the best temperature to use is probably the cold leg temperature, which is what is going into the reactor. Otherwise, recognizing that the temperature will change slightly going through the pump, the temperature coming out of the S/G and going to the MCP will be slightly lower than the reactor's cold leg.

v_f (450 °F) = 0.019438 ft³/lb_m (Interpolated)

$$\dot{V} = \left(3.95E4\frac{lb_m}{min}\right)\left(0.019438\frac{ft^3}{lb_m}\right)\left(7.48\frac{gal}{ft^3}\right)$$

$$= 5,743\,GPM$$

$$\boxed{\dot{V} = 5,743\,GPM}$$

That's a lot of water being forced through the reactor to transfer its heat to the S/G (and indirectly to the secondary side of the system). If we corrected for the temperature of the primary coolant coming out of the S/G and then took the average of that, the calculated value would be close to this because the density hasn't changed much.

EXAMPLE 8.3: NUCLEAR PLANT PRIMARY WITH PRESSURIZER AND SECONDARY OPERATION

In a submarine, two S/Gs each receive 20 lb$_m$/sec of secondary plant feedwater. The reactor is operating at a pressurizer temperature of 625 °F. Primary hot leg temperature T_H = 576 °F and primary cold leg temperature T_C = 528 °F. Steam generator secondary temperature is 480 °F. Condenser pressure is 2 psia. There is 4 °F of condensate depression. Neglect work done by the main feed pump.

a) What is the pressure inside the pressurizer? Pressurizer operates under saturation conditions. From Steam Table 1, 625 °F corresponds to a saturation pressure of approximately 1850 psia (1851.95 psia interpolated).

$\boxed{p_{pzr} = 1850 \text{ psia}}$

b) Will pressure on the suction of MCPs be greater or less than the pressurizer pressure? Explain.

> The pressurizer surge line connects to the reactor coolant hot leg prior to that loop's steam generator. Since there is a pressure drop across the steam generator, which is after the surge line but before the MCPs, the pressure at the suction of the MCPs will be less than the pressurizer pressure. Refer back to Figure 8.3 to see this on T-s coordinates.

c) Neglecting work done by the main feed pump (MFP) and assuming ideal turbines, what is the heat produced by the reactor (in MW)?

We'll model the two-loop primary system as a single loop and assume that all of the reactor's thermal power is transferred to the secondary system via the S/G. This part of the problem, then, becomes a simple steam plant analysis with schematic and T-s as shown. I expanded that state point numbering to double digits with the first digit (the "tens" place) indicating whether the fluid being analyzed is in the primary or secondary loop.

T_{SIG} = 480 °F (temperature of steam exiting the S/G)

From Steam Table 1 for 480 °F, saturated vapor (SV) enthalpy h_g = 1204.4 Btu/lb$_m$.

P_{cond} = 2 psia → T_{sat} = 126 °F → T_{cond}
  = 126 °F - 4 °F = 122 °F

From Steam Table 1 for 122 °F, saturated liquid enthalpy h_f = 90.0 Btu/lb$_m$.

Using a partial state point table for the secondary system to summarize the data:

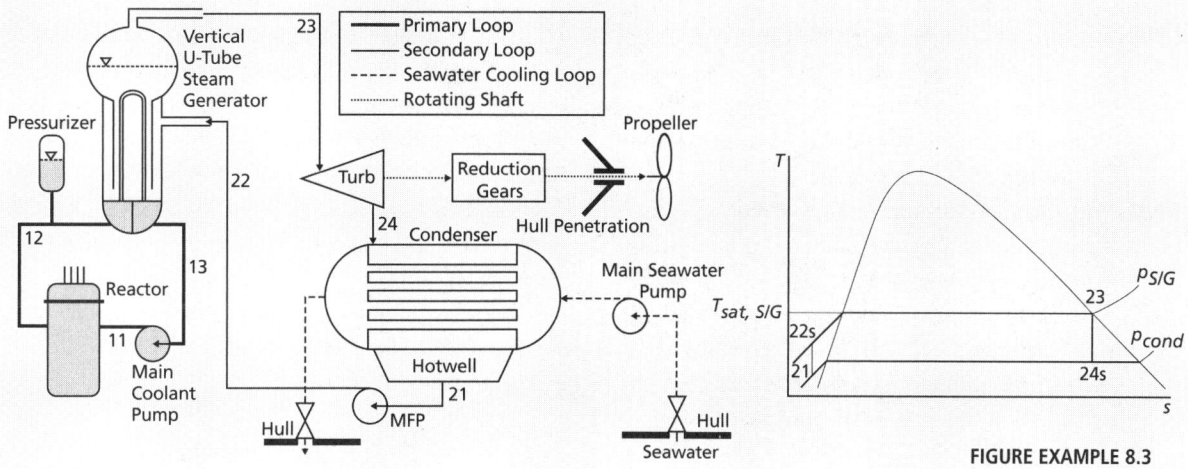

FIGURE EXAMPLE 8.3

	21	22s	23
p (psia)	2		
T (°F)	122		552
h (Btu/lb$_m$)	90.0	90.0	1204.4
x (%), State	CL	CL	SV*

*Assumed; Givens via preliminary analysis **BOLD**.

$$\dot{Q}_{S/G} = \dot{m}_{steam}(h_{23} - h_{22s})$$

$$= \left(2 \times \frac{20 lb_m}{sec}\right)\left(1204.4 - 90 \frac{BTU}{lb_m}\right) = 44576 \frac{BTU}{sec}$$

$$\dot{Q}_{S/G} = 44,576 \frac{Btu}{sec}$$

$$\dot{Q}_{Rx} = \dot{Q}_{S/G} = \left(\frac{44,576 Btu}{sec}\right)\left(\frac{hp}{0.707 Btu/sec}\right)$$

$$\left(\frac{0.746 kW}{hp}\right)\left(\frac{MW}{1000 kW}\right) = 47.0 MW$$

$$\boxed{\dot{Q}_{RX} = 47.0 MW}$$

Now let's take on a more complicated example.

EXAMPLE 8.4: COMBINED PROPULSION AND ELECTRICAL GENERATION NUKE PLANT

GIVEN: A nuclear submarine's power plant, running at full power, delivers 15,000 SHP while the electrical loads total 2500 kW. The *mechanical efficiencies* of the SSTG turbine, the propulsion turbine (HPT and LPT combination), and the reduction gears are 89%, 92% and 85%, respectively. The SSTG electrical generator efficiency is 95%. The main feed pump (MFP) may be assumed to operate ideally. The steam generator pressure gage reads 485 psi and the condenser pressure and hotwell temperature gages read 26.5 in-Hg and 121 °F, respectively. The reactor's thermal power is 78 MW at full power and the primary system's nominal pressure is 2500 psi. With the MCP running in F/S, the primary system's hot leg temperature is 630 °F and the cold leg temperature is 570 °F. The barometer inside the boat reads 30.53 in-Hg. You may assume that the steam throttle, governor, hotwell level control, and feed regulating valves are wide open in this full-power scenario and have negligible pressure drops.

FIND:
a) The condensate pump discharge pressure gage reads 52 psi and the pump's discharge thermometer reads 121.5 °F. What is the isentropic efficiency of the condensate pump?
b) If the pumps are driven by motors with 94% efficiency, what electrical power is required to run the condensate pump? The MFP? The MCP? What is the rated power of each pump's motor in hp? What portion of the SSTG electrical output power goes to run these pumps? Is the power output of the SSTG enough?
c) What is the condensate depression in this operating condition?
d) Find the thermal efficiency of the submarine's power plant in this operating condition.
e) What is the overall efficiency of the submarine's power plant in this operating condition?
f) Assuming all of the energy losses that are due to <u>mechanical</u> inefficiencies are absorbed as heat by the lube oil system {and assuming that the SSTG generator is cooled by the auxiliary seawater (ASW) system, so ignore the oil-cooled generator bearings}, what is the heat load of the lube oil system?
g) What is the temperature in the pressurizer?
h) Determine the flow rate in the primary loop in GPM.

SOLUTION: OK, so this is a "gut cruncher." Break it down and work it sequentially. It may be necessary to skip certain parts and come back to them after working other areas.

I'll start by identifying state points on the schematic and *T-s* graph of the secondary system, then do the preliminary pressure analyses.

Nuclear Power | **189**

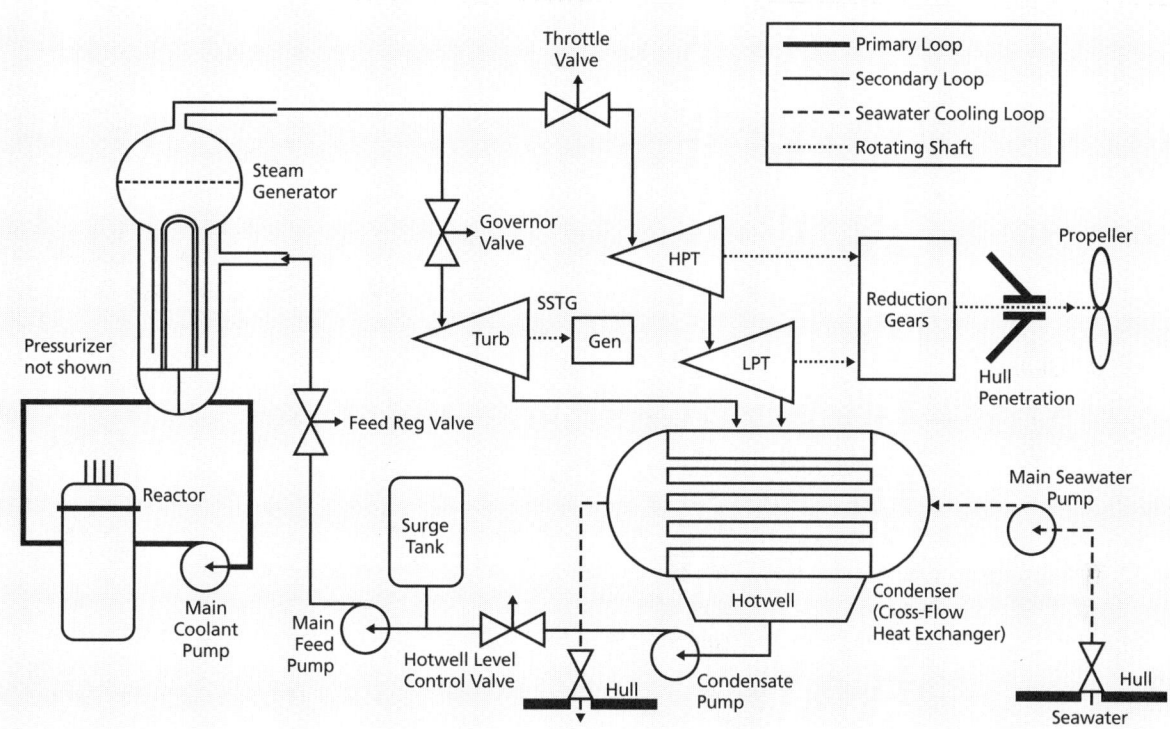

FIGURE EXAMPLE 8.4A

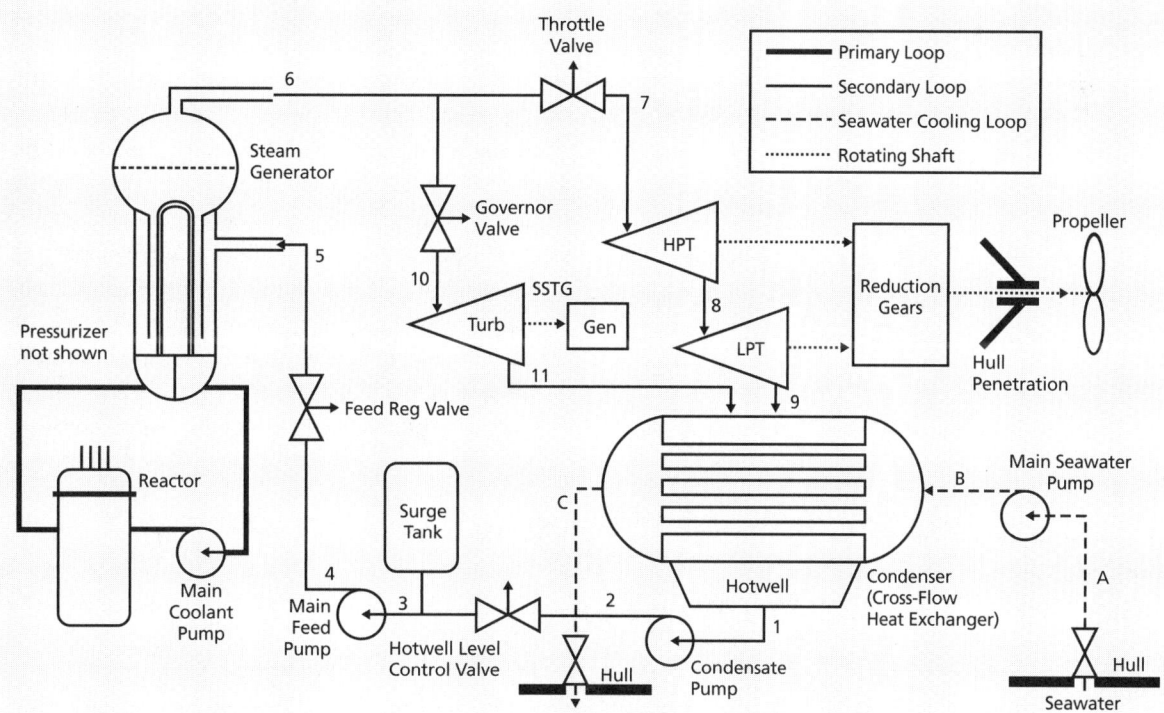

FIGURE EXAMPLE 8.4B

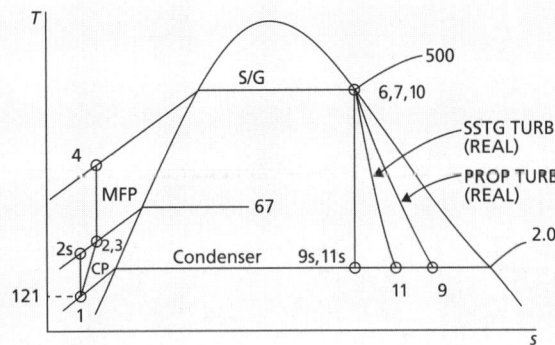

FIGURE EXAMPLE 8.4C

The secondary system was numbered sequentially around from the condenser outlet, not really pausing to analyze the individual components. Then the components were analyzed for processes. The *T-s* graph and SP Table ignore all components that do not change the properties of the secondary system's working fluid (pure water). Some components can be combined (HPT and LPT). I'm using my 1967 edition of *Steam Tables* to work this example out. If you use the 2000 edition, the values will vary slightly, but the answers won't change significantly.

Preliminary Pressure Analyses:

$p_{atm} = p_{bar}$ = 30.53 in-Hg (14.7 psi / 29.92 in-Hg)
= 15.0 psia

p_1 = 15.0 psia − 26.5 in-Hg (14.7 psi / 29.92 in-Hg)
= 2 psia

p_2 = 15.0 psia + 52 psig = 67 psia

$p_4 = p_6$ = 15.0 psia + 485 psig = 500 psia

Note : It is OK to round to first digit at high pressures, but not at low pressures. Round low pressures to the tenth or first decimal. (See Secondary System table below.)

The "?" entries in the SP Table are <u>not able to be determined</u> due to multiple unknown parameters (turbine isentropic efficiency and mass flow rate). We know SPs 9 and 11 will be WV, but do not know the quality and can not determine the real enthalpies (they may not be the same for both turbines).

a) The condensate pump discharge pressure gage reads 52 psi and the pump's discharge thermometer reads 121.5 °F. What is the isentropic efficiency of the condensate pump?

$h_1 = h_f$ (121 °F) = 88.965 Btu/lb$_m$ {interpolated}

$w_{p,s} = v_f \Delta p = h_{2s} − h_1$ = (0.016209 ft^3/lb$_m$) (65 lb$_f$/in^2)
(144 in^2/ft^2) (1 Btu/778 ft-lb$_f$)

$w_{p,s}$ = <u>0.195</u> Btu/lb$_m$ I'm showing this value by itself specifically to make a point.

$h_{2s} = h_1 + v_f \Delta p$ = 89.16 Btu/lb$_m$

$h_2 = h_f$ (121.5 °F) = 89.463 Btu/lb$_m$ {interpolated}

$w_{p,real} = h_{2s} − h_1$ = 89.463 − 88.965 = <u>0.498</u> Btu/lb$_m$

$$\boxed{h_{CP} = w_{p,s} / w_{p,real} = 39.1\%}$$

b) If the pumps are driven by motors with 94% efficiency, what electrical power is required to run the condensate pump? The MFP? The MCP? What is the rated power of each pump's motor in hp? What portion of the SSTG electrical output power goes to run these pumps? Is the power output of the SSTG enough?

The concept sketch showing the energy conversions for motor-driven pumps is shown below. We need to analyze three pumps. Note that there is a fourth pump in the system, the main seawater pump (MSW Pump), but no data points were given and no calculations required for this component. Like the other pumps

Secondary System	1	2s	2	4	6	9s	9	11s	11
p (psia)	**2.0**	**67**	**67**	**500**	**500**	2.0	2.0	2.0	2.0
T (°F)	**121**	NR	**121.5**	NR	NR	126	126	126	126
h (Btu/lb$_m$)	88.97	89.16	89.46	90.77	1204.7	848.9	?	848.9	?
s (Btu/lb$_m$-°R)	NR	NR	NR	NR	1.4639	1.4639	NR	1.4639	NR
v (ft^3/lb$_m$)	0.016209	0.016209	0.016211	0.016211	NR	NR	NR	NR	NR
State or Quality	CL	CL	CL	CL	SV	73.9%	?%	73.9%	?%

Givens and preliminary pressure analyses are shown in **BOLD**.

shown in the schematic, this pump represents a parasitic electrical load that must be supplied or the whole power plant will not operate. To be fully correct in solving this problem's electrical loads, the MSW Pump power requirements should be included.

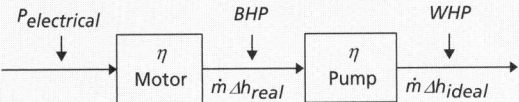

FIGURE EXAMPLE 8.4D

We need $\dot{m}$ to determine power. We have three powers given: $\dot{Q}_{Rx}$; $P_{propulsion}$ (at the SHP level); and $P_{electrical}$ (at the generator output). The best place to calculate the secondary flow rate is at the single component that also shares the full flow rate of the two secondary pumps, the S/G. Considering the primary loop energy flow, a valid simplification is that $\dot{Q}_{Rx} \approx \dot{Q}_{S/G}$. The heat given up by the primary coolant flowing through the S/G is transferred to the secondary side fluid. Therefore, $\dot{Q}_{S/G} = \dot{m}_{S/G,secondary} \Delta h_{S/G,secondary} \approx 78 MW$. Now, focusing on the secondary side, we can easily determine the $h_{S/G,out}$ because we know (1) $p_{S/G}$ and (2) the steam produced is a SV. This is h_6 on the SP Table. We need to calculate the S/G inlet, which is the same as the MFP discharge. Calculate the discharge enthalpy of the MFP. It is given as an ideal pump, so we need h_{4s}.

$v_f (121.5 °F) = 0.016211$ ft³/lb$_m$ {interpolated}

$w_{p,s} = v_f \Delta p = h_{4s} - h_2 = (0.016211$ ft³/lb$_m)$
$(500 - 65$ lb$_f$/in²$) (144$ in²/ft²$) (1$ Btu/778 ft-lb$_f)$

$w_{p,s} = \underline{1.31}$ Btu/lb$_m$

$h_{4s} = h_2 + v_f \Delta p = 89.46 + 1.31 = \underline{90.77}$ Btu/lb$_m$

Then, solving for the secondary flow rate,

$\dot{Q}_{S/G} = \dot{m}_{secondary} \Delta h_{S/G,secondary} \Rightarrow \dot{m}_{secondary} = \dot{Q}_{S/G} / \Delta h_{S/G,secondary}$

$\dot{m}_{secondary} = \dot{Q}_{S/G} / \Delta h_{S/G,secondary}$
$= \left(\dfrac{78 MW}{(1204.7 - 90.77) Btu/lb_m} \right) \left(\dfrac{3412 Btu}{kW-hr} \right) \left(\dfrac{1000 kW}{MW} \right)$

$\dot{m}_{secondary} = 2.39 E5 lb_m / hr$

Note that this is 100% of the secondary flow rate. It will be less through each turbine, due to splitting the 100% steam flow between the propulsion and the electrical generation turbines.

The MCP is a special case in this problem. We know that the temperature into the pump is slightly below the reactor's cold leg temperature, but the change in pressure through the pump is not known. There is no other way to calculate it without knowing either the Δp or the real T_{in} and T_{out} for the pump. So, this leaves me to assume a reasonable value. All of the homework and text examples have F/S MCP pressure differences on the order of 100 to 140 psi. I'll use 120 for this calculation. We'll also need the $\dot{m}$ for the primary flow through the MCP, which is available by evaluating the conditions across the reactor.

Primary Loop	Rx HL	Rx CL	MCP in
p (psia)	2500	<2500	>2500
T (°F)	630	570	<570
h (Btu/lb$_m$)	662.7*	575.6*	<575.6
v (ft³/lb$_m$)	NR	0.02242*	<0.02242
State	CL	CL	CL

* interpolated using CL Approx technique.

$\dot{Q}_{Rx} = \dot{m}_{Rx} \Delta h_{Rx} = \dot{m}_{Rx} (h_{hot\,leg} - h_{cold\,leg})_{Rx} \Rightarrow \dot{m}_{Rx} = \dot{Q}_{Rx} / \Delta h_{Rx}$

$\dot{m}_{Rx} = \dot{Q}_{Rx} / \Delta h_{Rx} = \left(\dfrac{78 MW}{(662.7 - 575.6) Btu/lb_m} \right)$

$= \left(\dfrac{78 MW}{(662.7 - 575.6) Btu/lb_m} \right)$
$\left(\dfrac{3412 Btu}{kW-hr} \right) \left(\dfrac{1000 kW}{MW} \right) = 3.06 E6 lb_m / hr$

Summarizing the approach and data in a table format, for the three pumps (calculating from <u>right-to-left</u>): (see table on next page.)

Note that the $w_{p,s} = \Delta h$ across the MCP is fairly small. Factoring this into the Primary Loop SP Table produces the following entries.

Primary Loop	MCP in	MCP out
T (°F)	569.6*	570
h (Btu/lb$_m$)	**575.1**	575.6
v (ft³/lb$_m$)	0.02241*	0.02242

* interpolated from the entering value shown in **BOLD**.

Note that comparing the T, h, and v values for the MCP in and out locations shows that specific work input to the primary coolant by the MCP is fairly small.

Pump	$P_{electrical}$ (kW)	h_{motor}	BHP (hp)	h_{pump}	WHP (hp)	$\dot{m}$ (lb$_m$/hr)	$w_{p,s}$ (Btu/lb$_m$)	v_f (ft^3/lb$_m$)	Dp (psid)
CP	37.2	**94%**	46.9	39.1%	18.3	2.39E5	0.195	0.016209	65
MFP	97.7	**94%**	123.1	100%*	123.1	2.39E5	1.31	0.016211	433
MCP	496.3	**94%**	625.5	100%*	625.5	3.06E6	0.52	0.02242	125*
Total	631.3								

* Assumed. CFs used: 144 in^2/ft^2 778 ft-lb$_f$/Btu   2544 Btu/HP-hr   746 W/HP

However, the power is fairly large due to the large mass flow rate.

Now, ATQ...

The individual pump electrical power is double-boxed in the table above. Motors are rated by the BHP produced, so that answer is double-boxed.

The percentage of the electrical power produced by the SSTG that is consumed by these pumps is:

$$P_{e,pumps} / P_{e,total} = 631.3 \text{ kW} / 2500 \text{ kW} = 25.3\%$$

Obviously, the power required to run these pumps is less than the output of the SSTG. That's the way things are supposed to be when you're steaming and not connected to shore power.

c) What is the condensate depression in this operating condition?

Condensate depression applies at the discharge of the condenser. Its purpose is to help meet the NPSHR of the CP and thereby prevent cavitation in the CP impeller. Refer to the T-s graph and note where SP$_1$ is graphed.

T_{sat} (2 psia) = 126 °F

$T_{condensate} = T_1 = 121$ °F

$$CD = T_{sat} - T_{cond} = 5 \text{ °F}$$

d) Find the thermal efficiency of the submarine's power plant in this operating condition.

Thermal efficiency is measured by the heat supplied to and the work net out predicted by thermo-physical analysis of the working fluid.

$$\eta_{th} = \underbrace{\frac{w_{net}}{q_s}}_{\text{specific basis}} = \underbrace{\frac{P_{net}}{\dot{Q}_s}}_{\text{total rate basis}}$$

The other name for P_{net} is *indicated net power* (power indicated by the thermodynamics of the fluid). We have called the power out of a turbine IHP during this course. But we have a split flow situation and parasitic loads to consider, so we need to adapt the approach somewhat.

Refer back to the secondary SP Table and note the "?" locations. If we knew the isentropic efficiencies of the PT and SSTG turbines, we could calculate the real discharge h and then the $\dot{m}$ through each turbine and use that to calculate the IHP of each turbine. We don't have this information, so we look for another way. We have the mechanical and electrical conversion efficiencies for the other power trains and their outputs, so we can work backwards to IHP. We'll still have to account for the parasitic loads to get to IHP$_{net}$.

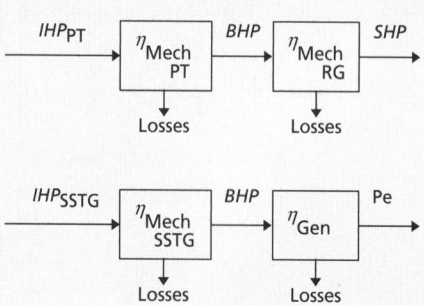

FIGURE EXAMPLE 8.4E

A table will be useful for this purpose. Since the SSTG carries the parasitic loads of the pumps, the net electrical power is used: $P_{electrical, net} = 2.5 - 0.631 = 1.869$ MW. (See table on next page.)

The losses are the differences between "ins" and "outs." The generator losses do not go to lube oil, so they were not calculated.

$$\eta_{th} = IHP_{net} / \dot{Q}_s = \frac{(14.31 + 2.21) \text{ MW}}{78 \text{ MW}} = 21.2\%$$

Propulsion	IHP	$h_{PT, Mech}$	BHP	$h_{red\,gears}$	SHP
P (hp)	19,182	92%	17,647	85%	15,000
P (MW)	14.31	92%	13.16	85%	11.19
P (Btu/hr)	4.88E7	92%	4.49E7	85%	3.82E7
Electrical	**IHP**	$h_{SSTG, Mech}$	**BHP**	$h_{Generator}$	P_{elect}
P_{net} (MW)	2.210	89%	1.967	95%	1.869
P_{total} (MW)	2.957	89%	2.632	95%	2.500
P_{total} (Btu/hr) based on total	1.01E7	89%	8.97E6	95%	8.53E6
Losses to Lube Oil					
Propulsion Train (Btu/hr)			3.9E6		6.7E6
Electrical Train (Btu/hr)			1.1E6		NR

CFs: 0.746 kW/hp 1000 kW/MW 3412 Btu/kW-hr

e) What is the overall efficiency of the submarine's power plant in this operating condition?

Overall efficiency is measured as far downstream for which we have information. In this case, at the *SHP* and $P_{elec,net}$ levels. Both must be in the same units, as must the thermal input power. That's why the propulsion train values were converted to MW units. Due to the different mechanical and electrical efficiencies given, the split flow, and the lack of turbine isentropic efficiencies, overall efficiency cannot be determined by multiplying the thermal and mechanical efficiencies together.

$$\eta_{OA} = P_{net}/\dot{Q}_s = \frac{(11.19 + 1.87)MW}{78 MW} = 16.7\%$$

Some thoughts about overall efficiency. Typically, we only count the propulsion train and calculate the "*overall propulsion efficiency*" and measure the output at the *SHP* level. For this plant, that would be only 14.3%! However, this problem was evaluating the whole plant, so the net electrical generation counted in the net power output.

One more step in thinking about the electrical power: All of the electrical power used inside the vessel ultimately is converted to heat. All of that heat, plus the real heat losses through the insulation surrounding the hot steam pipes, valves, pumps, and turbines, the hot primary system components, and the heat given off by the crew, must be removed from the interior of the vessel. If this isn't accomplished the interior temperature will climb to uninhabitable levels. The paths for removing this heat are heat transfer through the hull (which is typically insulated with foam) and to the auxiliary seawater system via the boat's air conditioning system. The air conditioning system is covered in Unit 9.

f) Assuming all of the energy losses that are due to mechanical inefficiencies are absorbed as heat by the lube oil system {and assuming that the SSTG generator is cooled by the auxiliary seawater (ASW) system, therefore ignore the oil-cooled generator bearings}, what is the heat load of the lube oil system?

The losses to the lube oil system are based upon the total electrical power output and not the net electrical, which was used in calculating IHP_{net}.

$$\dot{Q}_{lube\,oil} = (3.9E6 + 6.7E6 + 1.1E6) Btu/hr$$
$$= 11.7E6\ Btu/hr = 1.95E5 Btu/min$$
$$= 3.25E3 Btu/s$$

g) What is the temperature in the pressurizer?

We know the "nominal" primary system pressure. That is typically taken as the pressure in the pressurizer since the pressurizer establishes the primary system pressure. We don't know if the given info is gage or absolute pressure, so, we'll assume that it is absolute. Since the pressurizer is a two-phase system, it is a saturated system. Therefore, the temperature is the saturation temperature for the given pressure.

$$T_{sat}(2500\ psia) = 668.11\ °F$$

h) Determine the flow rate in the primary loop in GPM.

The $\dot{m}$ will be the same throughout the primary loop. This was calculated above. The $\dot{V}$ will depend on the density of the primary coolant, which changes with temperature. So, the answer will depend on where it is calculated and could be at the Rx inlet, outlet, average; or the MCP.

$$\dot{m} = \rho \dot{V} = \dot{V}/v \Rightarrow \dot{V} = \dot{m}v$$

Recall that $\dot{m}_{primary} = 3.06E6 lb_m/hr$ was calculated above.

	Rx, average	Rx, HL	Rx, CL	MCP in
T (°F)	600	630	570	575.1
v (ft³/lb$_m$)	0.02364	0.02527	0.02242	0.02241
$\dot{V}$ (GPM)	9018	9640	8553	8549

CFs: 7.48 gal/ft³ 60 min/hr

Notice that the density ($\rho = 1/v$) for these various conditions is <u>not</u> 62.4 lb$_m$/ft³. Many students forget that density is temperature-dependent and assume the room temperature value from the equation sheet. The density and percent errors are shown for the various values of the primary loop as compared to the room temperature value of 62.4 lb$_m$/ft³.

	Rx, average	Rx, HL	Rx, CL	MCP in
T (°F)	600	630	570	575.1
ρ (ft³/lb$_m$)	42.30	39.57	44.60	44.62
% error	47.5%	57.7%	39.9%	39.8%

REFERENCES

DOE Fundamentals Handbook: Thermodynamics, Heat Transfer, and Fluid Flow. Vol 1. (DOE-HDBK-1012/1-92).
Jane's Fighting Ships, 2006–2007, Jane's Information Group, Surrey, UK.

Practice Problems

"Every accomplishment, great or small, starts with the same decision—I'll try."

—Anonymous

8.1 Write the nuclear fission reaction for U-235. What is the definition of enrichment as it applies to this topic? What is depleted uranium (DU) and where does it come from?

8.2 Refer to the objectives for this unit. Then sketch and label the major components of a pressurized water reactor (PWR) propulsion system. If you use a cut-and-paste graphic that is not hand-drawn, then be sure to annotate the sketch with the names of all significant components and list their primary functions. Cite all sources.

8.3 List the active classes of ships and submarines in the U.S. Fleet that are nuclear powered and: (1) their nuclear power plant designation (e.g., S5W); (2) the rated reactor power (MW); (3) the rated shaft horsepower (*SHP*) of their propulsion plants; and (4) their overall propulsion efficiency. Use only unclassified sources and cite sources. Report this in table format.

8.4 A main coolant pump (MCP) in a pressurized water reactor's (PWR) primary system pumps 600 °F liquid water. The differential pressure across this centrifugal pump in fast speed is 125 psid. If the pump is 75% efficient and pumps 2000 GPM in fast speed, what is the rated power of the pump (hp)? What is the pump's brake horsepower in slow speed (fast speed is twice slow speed)? If the slow speed pump is capable of operating at a reduced electrical frequency of 45 Hz (for additional acoustic security), what is the expected mass flow rate through the MCP at reduced frequency?

8.5 Given a nuclear power plant with a primary loop operating at a T_{hot} and T_{cold} of 535 °F and 495 °F, respectively. The temperature difference between the average primary temperature and the secondary temperature in the steam generator is 35 °F. The mass flow rate of steam through the secondary is 55 lb_m/s, and the main condenser is operating at 1 psia. Assume a turbine isentropic efficiency of 90%, a condensate depression of 3.7 °F, and a drivetrain mechanical efficiency of 85%.

  a) What is the real specific work of the turbine?

  b) What is the power output at the screw?

  c) What is the real pump work?

  d) What is the cycle thermal efficiency?

	1	2	3	4s	4
p (psia)					
T (°F)					
h (Btu/lb_m)					
s (Btu/lb_m-°R)					
v (ft³/lb_m)					
State or Quality					

8.6 Given a nuclear power plant with a four primary loops. The reactor core is rated at 500 MW, and the pressurizer temperature is maintained at 650 °F. While answering an ahead full bell, each steam generator (of the four) draws 60 lb_m/s of saturated steam. Temperature of the hot and cold leg are 532 °F and 472 °F, respectively. The main condensers are operating at a pressure of 1 psia with a condensate depression of 5.7 degrees. The temperature difference between average primary temperature and secondary temperature across the steam generator is 62 °F. If the main propulsion turbines have an isentropic efficiency of 90% and the main feed pumps have an isentropic efficiency of 75%, answer the following:

a) What is the pressure of the primary loop? (psia)

b) What is the pressure in the main steam header? (psia)

c) What is the real work of the main feed pump? (Btu/lb$_m$)

d) What is the horse power being supplied by the turbines? (hp)

e) What is the mass flow rate in the primary? (lb$_m$/s)

f) What is the cycle efficiency?

8.7 The reactor core is rated at 120 MW, and the pressurizer temperature is maintained at 648 °F. While answering an ahead full bell, the total steam flow is 193,700 lb$_m$/hr of saturated steam at 460 °F. Temperatures of the hot and cold legs are 540 °F and 488 °F, respectively, with a primary mass flow rate of 3,500,000 lb$_m$/hr. The main condensers are operating at a pressure of 1 psia with a condensate depression of 3.7 degrees. If the main propulsion turbines have an isentropic efficiency of 90% and the main feed pumps have an isentropic efficiency of 75%, answer the following:

a) What is the pressure of the primary loop? (psia)

b) What is the reactor power? (%)

c) What is the pressure in the main steam header? (psia)

d) What is the real work of the main feed pump? (Btu/lb$_m$)

e) What is the horse power being supplied by the turbines? (hp)

f) What is the cycle efficiency?

g) What is the log mean temperature difference in the steam generator? (°F)

8.8 In a submarine, two steam generators (S/G) each receive 30 lb$_m$/sec of secondary plant feedwater. Enthalpy out of all turbines in the secondary plant averages 867 Btu/lb$_m$. The primary hot leg temperature T_H = 548 °F and the primary cold leg temperature T_C = 484 °F. 90% of the steam goes to the main propulsion turbines (MPT) with η_{mech} = 92%. What is the power produced (*BHP*) by the MPT? The S/G secondary side temperature is 428 °F.

8.9 All of the operating reactor coolant pumps (RCPs) in a pressurized water reactor's (PWR) primary system may be modeled as a single pump in a single loop consisting of the reactor (Rx), the steam generator (S/G), and the RCP, all connected by pipes. The pumps handle liquid water that is nominally at 500 °F. The differential pressure across this centrifugal pump in fast speed is 100 psid.

a) If the pump is 90% efficient, what is the real change in enthalpy of the water through the pump?

b) If the pump has a volumetric flow rate of 2000 GPM in fast speed, what is the rated power of the pump (hp)?

c) Using your *Steam Tables*, estimate *Cp* for water at this temperature. What is the change in temperature of the water as it moves through the pump?

d) The RCP is shifted to slow speed ($N_{S/S}$ = ½ $N_{F/S}$). What is the RCP's brake horsepower in slow speed?

e) To minimize radiated noise, the RCP is shifted to reduced frequency ($N_{R/F}$ = ¾ $N_{S/S}$). What is the mass flow rate through the primary system in reduced frequency?

8.10 Refer to the simplified schematics, Figures HW 8.11A and HW 8.11B below. Note that typical redundancy and flow control devices are not shown. The pressure of the primary loop is nominally 1900 psi. The RCPs are two-speed pumps and both are run at the same speed. In fast speed, the differential pressure across each RCP is 125 psid. Under fast speed operation, the pressure drop in the primary loop across the reactor is 100 psid. The reactor is rated at 100 MW of thermal power. The hot leg water temperature is 600 °F and the cold leg water temperature is 500 °F. The DT between the average temperature on the primary side of the steam generator (S/G) and the saturation temperature on the secondary side of the S/G is 50 °F. Both S/Gs provide steam to a cross-connected steam distribution system that powers both SSTG and propulsion turbines, as shown in Figure HW 8.11B.

a) What is the temperature in the pressurizer?

b) What causes the water to spray into the pressurizer when the normally closed spray valve is opened?

c) Ignoring the head losses in the piping and fittings, what is the expected head loss across the reactor and steam generator if the RCPs are run in slow speed ($N_{S/S} = \frac{1}{2} N_{F/S}$) (psid)?

d) Considering the secondary side of the plant, what is $\dot{Q}_s$ at full power (Btu/hr)?

e) What is the pressure in the S/G secondary side (psia)?

f) What is the enthalpy of the steam out of the S/G (Btu/lb_m)?

g) If the condenser pressure gage reads 27.88 in-Hg and assuming the condensate and feed pumps are ideal, what is the total steam flow rate of the secondary system (lb_m/hr)?

8.11 Given a submarine nuclear power plant with a primary loop shown in Figure HW 8.11A. The reactor core is rated at 100 MW, and the pressurizer temperature is maintained at 624 °F. While answering an ahead full bell, each steam generator draws 35 lb_m/s of saturated steam. Temperatures of hot and cold legs are 504°F and 477°F, respectively. Pressure in the secondary side of the steam generator is 500 psia. The main condensers are operating at a pressure of 1 psia. 20% of the total mass flow rate of the steam is consumed by the submarines 2 ships service turbine generators (SSTGs). If the main propulsion turbines have an isentropic efficiency of 90% and the reduction gears and drive train have a total mechanical efficiency of 92%, answer the following:

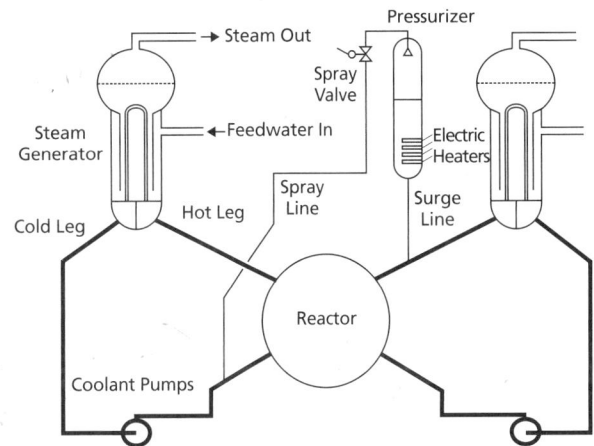

FIGURE HW 8.11A: PWR Simplified Primary System

a) What is the pressure of the primary loop? (psia)

b) What is the shaft horsepower being supplied to the ocean by the screw? (hp)

c) What is the mass flow rate in the primary? (lb_m/s)

d) What is the operating reactor power, neglecting inputs and losses of the main coolant pumps and primary piping? (%)

e) What is the log mean temperature difference across the steam generator? (°F)

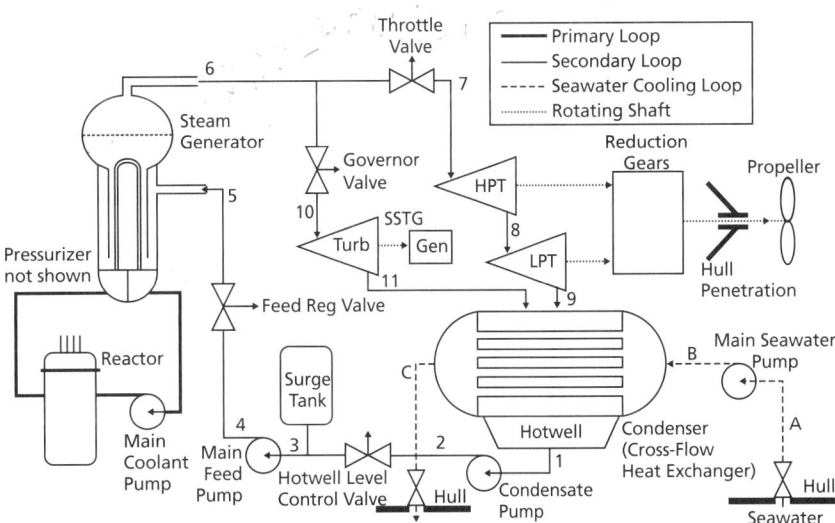

FIGURE HW 8.11B: Nuclear Submarine Power Plant

9 Heating, Ventilation, Air Conditioning & Refrigeration (HVAC&R)

The nuclear-powered aircraft carrier USS *George H. W. Bush* (CVN-77) is the tenth and final ship of the *Nimitz*-class of aircraft carriers. The cooling capacity of the installed air conditioning equipment is sufficient to cool several thousand homes.

9.1 UNIT LEARNING OBJECTIVES

9.1.1 Terminology: definitions, variables, and typical units

9.1.1.1 relative humidity, wet bulb temperature, dry bulb temperature, dew point, sensible heat, latent heat, tons, refrigeration effect, coefficient of performance, compressor, evaporator, condenser, thermostatic expansion valve (TXV), orifice

9.1.2 Concepts: ideas and engineering expressions of them

9.1.2.1 List and describe the major categories of HVAC&R technology.

9.1.2.2 Describe the major air conditioning processes. Show the effect of each process on a psychrometric chart.

9.1.2.3 Describe the operation of a basic vapor compression refrigeration system. Sketch a component schematic and plot the cycle on both the *T-s* graph and the *p-h* graph.

9.1.3 Skills: procedures, practices, or methods that enable reasoning

9.1.3.1 Calculate the refrigerating effect (RE) and capacity (CAP) of a refrigeration cycle.

9.1.3.2 Calculate the coefficient of performance (COP) for a refrigeration cycle.

9.1.3.3 Calculate power requirements of a refrigeration cycle.

9.1.3.4 Determine properties of air using a psychrometric chart.

9.1.3.5 Determine properties of a refrigerant using a *p-h* diagram.

9.2 INTRODUCTION

The technology of heating, ventilation, air conditioning and refrigeration (HVAC&R) is fundamental to ship, submarine, aircraft, and land vehicle habitability and comfort. Refrigeration systems are used to keep foodstuffs from spoiling by chilling or freezing, but also have industrial uses within the vessel, such as the maintenance technique of "freeze sealing" water pipes where valves are not available to isolate the maintenance area from the rest of the system. An extreme example of refrigeration technology is cryogenics, which produces liquid oxygen, via fractional distillation from air, for use at high altitudes by aircraft pilots and liquid nitrogen for freeze seals and inerting pipes containing flammable vapors. Heating and air conditioning are intended to modify the environment inside of the "conditioned space" for human occupants and equipment. Ventilation maintains the composition of air within the bounds of human health needs and may, under some outside conditions, reduce the need for air conditioning. Removing dust particles or harmful vapors is another element of air conditioning and ventilation. In naval applications, HVAC technology also contributes to a vessel's ability to function in a nuclear, biological, and chemical (NBC) warfare environment. The closed atmosphere of a submerged submarine represents an environment with unique requirements. Specialized atmosphere control equipment enables submarines to stay submerged for extended periods of time while maintaining acceptable air quality. While heating and ventilating techniques will be covered, we will concentrate on the use of vapor compression technology for air conditioning (cooling and dehumidification) and refrigeration for food preservation.

9.3 HISTORICAL OVERVIEW

Historically, ventilation was the only method of providing the air quality necessary for human occupancy inside a ship hull. In the age of sail, ships were rarely heated, except when the galley fires were lit for cooking. Ventilation was limited to what the wind delivered when the sea state and weather allowed hatches to be open. With the advent of steam engines in marine applications, the heat given off to the operating spaces was significant, but there was also an available power source to run ventilation fans. Ventilation fans were then added to the steam-powered ship's auxiliary systems.

On land, architectural features were adapted to support ventilation. For example, homes in the southern portion of the United States often incorporated high ceilings, shady porches, large windows, and cupolas that could be opened. These features aided the conditions for natural-draft air circulation and minimized solar heat gain. Conversely, homes in the north tended to have lower ceilings and smaller windows to hold the heat in and closer to the floor level. Heating was largely by fireplaces. Much trial and error went into improving both the open fireplace and iron stoves used for heating purposes and how to distribute the heat to various rooms. Eventually, steam heat was adapted to buildings. A steam distribution system allowed heat to be carried greater distances from a central boiler and thus eliminated the need for multiple fireplaces in a structure. There were some innovative attempts at cooling air, but none of these were particularly practical. The first true air conditioning system was designed and built by Willis Carrier in 1902 in a printing business in Buffalo, New York. For the first time, a system was built that could effectively adjust both temperature and humidity in a "conditioned space."

On the refrigeration side, by the late 1800s two technologies existed for ice production—vapor compression and absorption chillers—and these replaced the harvesting and storage of ice from ponds and rivers, a technique that had been used for centuries. Harvested ice was even shipped to southern Amerian cities via ships from the north. Ice-making machinery allowed ice to be made in more areas of the country and in areas where ice traditionally wasn't readily available. However, early vapor compression and absorption technology both used chemicals that were poisonous or flammable when the system leaked. In response to the hazards of the early refrigerants, chemists at the DuPont Corporation introduced "Freon" for use in vapor compression refrigerators in 1930. Mechanical refrigerators replaced iceboxes in homes for food preservation. A prototype air conditioning system for motor vehicles was built by Cadillac in 1939, but wasn't commonly available in motor vehicles until several decades later.

Refrigeration and, later, air conditioning technology was adapted to shipboard use. Vapor compression systems gradually proved preferable to steam-powered absorption chillers, especially as the Fleet shifted away from steam propulsion plants in the 1970s. Vapor compression refrigeration is the most prevalent refrigeration and air conditioning technology in shipboard, commercial, and residential use today.

Concern over possible depletion of stratospheric ozone by chlorinated refrigerants, which was first observed in the laboratory in the 1970s, led to the "Montreal Protocol," a United Nations–sponsored treaty that went into effect in 1989. This treaty phased out or is in the process of phasing out refrigerants that have a significant ozone depletion potential (ODP). Freon (R-12) stopped production at the end of 1995 as a result of this treaty. Other refrigerants are scheduled to be phased out of production in the coming years. Development of substitute and replacement refrigerants, lubricants, and associated equipment modifications has been an active area of research. Overall, this issue has impacted the Navy and Marine Corps. AC&R systems are being converted to new refrigerants. Hardware compatibility issues with new refrigerants have also forced equipment change out rather than just changing the refrigerant in legacy AC&R systems.

Similarly, the Kyoto Protocol, even though it has not as yet been ratified by the United States, is also influencing refrigerant technology. The Kyoto Protocol addresses global climate change with particular emphasis on "greenhouse gases." Greenhouse gases are gases or vapors that help to retain heat that is given off to space from the earth in the form of radiation heat transfer, similar to the glass or plastic used in greenhouses for plants. Refrigerants that are considered greenhouse gases are coming under increased scrutiny, especially by the countries that have signed on to the Kyoto Protocol.

So, today's naval officers will be impacted by the environmental issues associated with refrigerants, and they'll see more equipment conversions and additional controls placed on handling refrigerants as compared to their predecessors.

9.3.1 Charcteristics of Refrigerants

There literally are hundreds of compounds and mixtures of compounds that may be used in vapor compression refrigeration. Even water (R-718) can be used as a refrigerant in a vapor compression system. The R-# numbering system was developed by DuPont and systematically identifies the molecular structure of refrigerants made with a single halogenated (chlorine or fluorine) hydrocarbon. The alternate abbreviations: HFC means the refrigerant is a hydrofluorocarbon; CFC means chlorofluorocarbon; and HCFC stands for hydrochlorofluorocarbon. Chlorine is the component implicated in stratospheric ozone depletion.

$$R - \underbrace{X}_{\text{\# double bonds}} \underbrace{X}_{C \text{ atoms} - 1} \underbrace{X}_{H \text{ atoms} + 1} \underbrace{X}_{F \text{ atoms}} \quad (9.1)$$

A suffix with a capital B and a number indicates the number of bromine atoms, when present. This is uncommon. Remaining bonds not accounted for are occupied by chlorine atoms. *A suffix of a lowercase letter a, b, or c* indicates increasingly unbalanced isomers.

As a special case, the R-400 series is made up of zeotropic blends (those where the boiling point of constituent compounds differs enough to lead to changes in relative concentration because of fractional distillation) and the R-500 series is made up of so-called azeotropic blends. The rightmost digit is assigned arbitrarily by ASHRAE, an industry organization.

$$R - 134a = R - \underbrace{1}_{C \text{ atoms} - 1} \underbrace{3}_{H \text{ atoms} + 1} \underbrace{4}_{F \text{ atoms}} a$$

So, R-134a has four F atoms, two H atoms, and two C atoms. There are no double bonds. The "a" suffix indicates that the isomer is unbalanced by one atom, giving an equivalent 1,1,1,2-Tetrafluoroethane. R-134 without the "a" suffix would have a molecular structure of 1,1,2,2-Tetrafluoroethane—a compound not especially effective as a refrigerant.

R-12, dichlorodifluoromethane, is CCl_2F_2. R-12 and R-134a are shown with the most basic hydrocarbon, methane, in Figure 9.1.

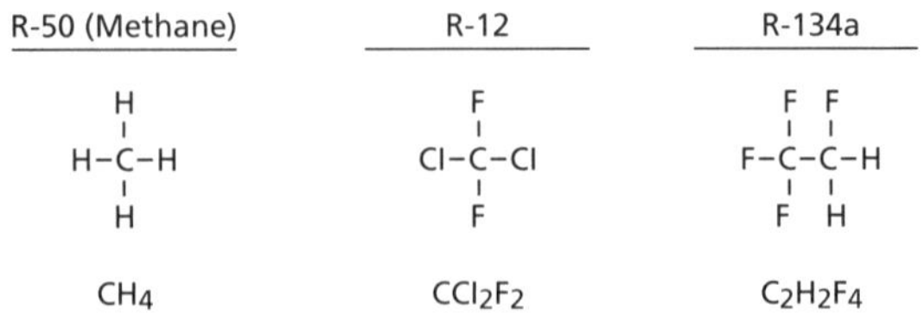

R-50 (Methane)	R-12	R-134a
H | H–C–H | H	F | Cl–C–Cl | F	F F | | F–C–C–H | | F H
CH_4	CCl_2F_2	$C_2H_2F_4$

FIGURE 9.1: *Basic Refrigerant Chemical Configurations*

Key attributes in which refrigerants are selected for any specific application include:

- the **saturation pressure and temperature** characteristics: The ideal refrigerant will achieve the temperatures necessary to absorb or reject heat at pressures that will not be excessively high (and require thick walls on the components) and will be near to atmospheric pressure at the seals to the compressor. (Seals are a potential leak path. We don't want air to leak into the refrigerant and we don't want refrigerant to leak out.)

- **thermal conductivity** (k): A high k, in both liquid and gaseous states, will ensure that the refrigerant carries the heat and can absorb or reject heat effectively.
- **viscosity** (v): The ideal refrigerant will compress as a gas easily and flow well in all states. The cost of running the compressor is not insignificant.
- **chemical stability:** The ideal refrigerant will not break down chemically and will be compatible with the lubricants used to protect the compressor.
- **corrosivity:** The ideal refrigerant will not corrode the components through which it flows.
- **flammability:** The ideal refrigerant will not burn or explode when it leaks. (Some refrigerants are ideal fire suppression chemicals—halon will work as a refrigerant.)
- **human health effects:** The ideal refrigerant will not be hazardous to the health of humans (some are classified as hazardous material due to being poisonous or carcinogenic; some change into poisons, e.g., phosgene gas, when exposed to high temperatures and flames; many are asphyxiants and displace air if they leak out).
- **environmental impacts:** The ideal refrigerant will not harm the environment when it leaks (the two main issues here are stratospheric ozone depletion and greenhouse gas potential; a secondary issue is whether leaked refrigerant contributes to acid rain).
- **cost:** The ideal refrigerant will not be expensive. (In reality, these materials can be cost prohibitive, so systems are typically designed to minimize the refrigerant inventory necessary to run the machine.)

The problem with this list of attributes is that, it could be argued, there are no ideal refrigerants. We can only optimize the choice, given all of these things to consider. While there are numerous legacy refrigerants in use, we will focus on one, R-134a. As discussed above, R-134a is a hydrofluorocarbon (HFC) compound and is the common replacement for R-12 (a chorofluorocarbon, CFC); and the thermo-physical properties for R-134a are included as an appendix. These tables are laid out just like steam tables.

9.4 PSYCHROMETRICS

Before covering mechanical refrigeration technology and its analysis, it will be useful to discuss the nature and characteristics of the humid air in which we live.

Dry atmospheric air is assumed to be only oxygen and nitrogen. This is done by grouping all of the small fractional constituents under the nitrogen category. This assumption doesn't change the thermal properties of dry air very much, because the properties of the trace and inert gases are fairly close to nitrogen. Dry atmospheric air by weight or mass is 23.15 percent O_2 and 76.85 percent N_2 and by volume is 20.9 percent O_2 and 79.1 percent N_2 with a molecular weight of 28.9.

Real air is a solution of water in a mixture of gases and often also is a suspension of solid particles and with additional gases in solution that we refer to as pollutants. The pollutants may affect the chemistry and health impacts of the air, but their effect on the thermodynamic properties of air is typically negligible.

The thermodynamics of humid air is known as "psychrometrics." Most students will have some experiential basis for humidity and the role of perspiration in cooling the body. Perspiration evaporates and removes latent heat in the process and thereby cools the skin and the blood flowing through shallow blood vessels. If the humidity of the air is high, then the air can not evaporate the liquid perspiration as effectively and the cooling effect of perspiration is then greatly reduced.

Operational military units are required to manage heat stress on their crews caused by a combination of high temperature and high humidity (see OPNAVINST 5100.19). Physiological heat exposure limits (PHEL) are determined from a combination of three different temperature measurements—"dry bulb," "wet bulb," and "globe" temperatures. The tool used to determine dry bulb and wet bulb temperature inputs for the heat stress determination is a "psychrometer." While most Navy ships have a motorized psychrometer, the manual version is simply two thermometers attached to a handle such that the thermometers may be swung around in the air by the operator. One of the thermometers has a cloth that covers the bulb and is thoroughly wet with room temperature water before being swung around. In the process of swinging the thermometers around in the air, the water on the wet thermometer evaporates and removes latent heat from the water. This heat transfer removes heat from the thermometer fluid and lowers the temperature reading. The resulting reading is called the "wet bulb temperature" (T_{wb}). The reading from the other thermometer is the "dry bulb temperature" (T_{db}). The third component of heat stress is related to the amount of radiation heat transfer, typically from sunlight, but also possible from other sources such as hot steam pipes. For example, working in the shade is cooler than working in the sunlight, even if the air temperature at both locations is the same.

Those who watch the weather report on television news programs may have noticed that the meteorologist reports the dew point temperature (T_{dp}) or relative humidity (ϕ or RH). T_{dp} is the temperature at which liquid water starts to condense out of the humid air—because the air is saturated with water and can't hold any more of it. Many students who have driven a car with the defroster turned on may have created conditions where the windshield was below T_{dp} and they had to run the windshield wipers to clear the accumulated moisture on the <u>outside</u> of the glass. Similarly, students should realize that when T_{wb} and T_{db} converge, conditions favor fog. Recall from the discussion of steam that humans do not see water vapor, but can see water in the solid or liquid phases. Clouds are suspensions of liquid water droplets or ice crystals. Humidity is vapor that may change the light transmission characteristics of air, but humans don't see the water vapor in humidity.

Humidity ratio (w) is the ratio of water vapor (subscript $_{wv}$) mass to dry air mass.

$$\omega = \frac{m_{wv}}{m_{air}} \quad (9.2)$$

Since $m = \rho V$ and $V_{air} = V_{wv}$ and from the ideal gas law ($pV = mRT$) with $T_{air} = T_{wv}$, and $R_{air} = 53.3$ ft-lb$_f$/lb$_m$-°R and

$$R_{wv} = \frac{R_u}{mw_{water}} = \frac{1545.3 \; \frac{ft-lb_f}{pmole-°R}}{18 \; \frac{lb_m}{pmole}} = 85.9 \; \frac{ft-lb_f}{lb_m - °R}$$

the following additional ratios are developed.

$$\omega = \frac{\rho_{wv}}{\rho_{air}} = \frac{R_{air} p_{wv}}{R_{wv} p_{air}} = \frac{53.3 p_{wv}}{85.9 p_{air}} = 0.62 \frac{p_{wv}}{p_{air}}$$

Caution: *Do not attempt to apply the ideal <u>gas</u> law to water in the liquid (CL, SL) or the wet vapor (WV) form. The ideal <u>gas</u> constant for water only applies to <u>water vapor</u> (aka "superheated" vapor). And if the conditions are even close to saturation, the ideal gas law used on superheated vapor will include unacceptable errors.*

Recall from your chemistry course that in a mixture of gases the total pressure of the mixture is comprised of the sum of the partial pressures of the constituents. For humid air, the components are water vapor and dry air. The total pressure becomes

$$p_{atmosphere} = p_{air} + p_{wv} \quad (9.3)$$

Relative humidity represents how close the water vapor in the humid air is to saturation conditions, which is when it will condense. This can be related to pressure and through the ideal gas law ($pv = RT$; $p/\rho = RT$; $\rho = p/RT$) with the

$$\phi = \frac{p_{wv}}{p_{sat}} \quad (9.4)$$

where, p_{sat} may be obtained from the steam tables using T_{db} as the saturation temperature.

9.5 PSYCHROMETRIC CHART

Fortunately, it is not always necessary to resort to higher mathematical analysis of humid air. The properties of humid air are shown graphically on a "psychrometric chart." Despite the apparent complexity, the chart is relatively easy to use. The thermodynamic state is defined by the intersection of any two lines. A representative psychrometric chart is reproduced in Figure 9.2 below to aid in orientation to a full psychrometric chart shown in Figure 9.3.

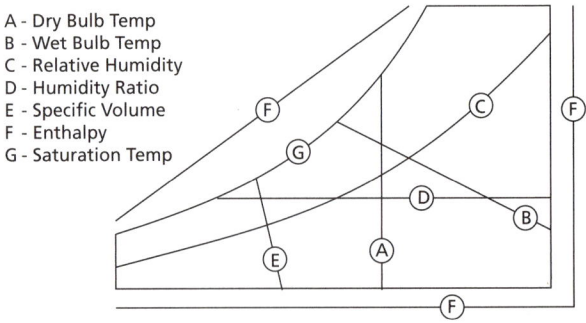

FIGURE 9.2: *Psychrometric Chart with Representative Data Lines*

There are many versions of the psychrometric chart, some including various additional pieces of information. The charts are prepared for nominal atmospheric pressure, and the charts will change with altitude. In order to familiarize yourself with the chart, highlight the words on the chart and notice the relationship between the words and their associated lines. Pay attention to the units. Depending

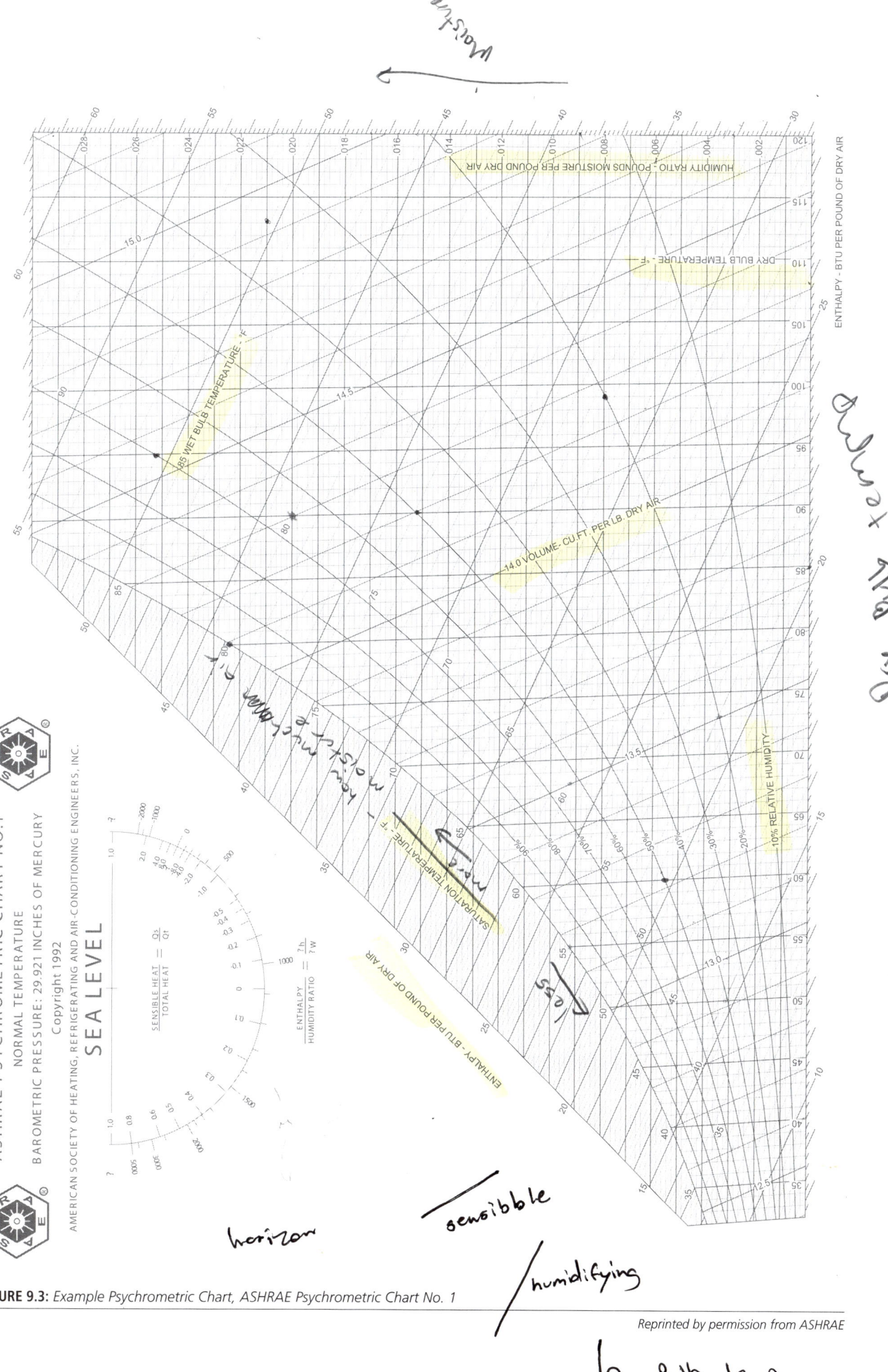

FIGURE 9.3: Example Psychrometric Chart, ASHRAE Psychrometric Chart No. 1

Reprinted by permission from ASHRAE

upon the source of the chart, it may also include the partial pressure of the water vapor and sensible heat ratio or additional graphical design aids. Vapor pressure (p_{wv}) may be given in psia or in-Hg; humidity ratio (w) may be given in $lb_{m_{moisture}}/lb_{m_{da}}$ or in grains/$lb_{m_{da}}$, where 7000 gr = 1 lb_m.

There are a number of websites with psychrometric charts that may be downloaded free of charge. Noteworthy is Hands Down Software (www.handsdownsoftware.com), which also has interactive Psychrometric charts that are useful for checking your answers.

EXAMPLE 9.1: PSYCHROMETRIC CHART PRACTICE

GIVEN: A dry bulb temperature of 90 °F and a wet bulb temperature of 75 °F.

FIND:
a) the relative humidity (ϕ)
b) the humidity ratio (w)
c) the dew point temperature (T_{dp})
d) the enthalpy (h)
e) the specific volume (v)
f) and, the vapor pressure of the water vapor (p_{wv})

SOLUTION: It is a recommended practice to circle the given parameters on their scales and use a see-through straight edge to draw along that line. This reduces the probability of tracking off and obtaining wrong answers. Graphical interpolation may be necessary. Find the intersection of the two pieces of given information ($T_{db} = 90$; $T_{wb} = 75$); place a dot there and circle it so that it is easier to see. This is the "state point."

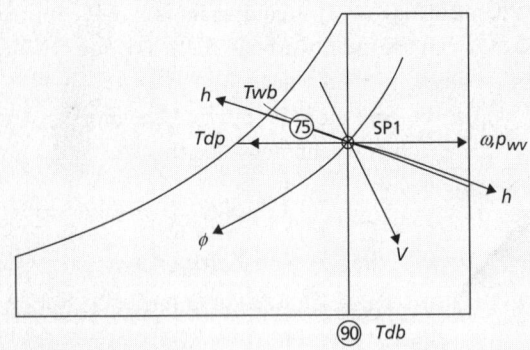

FIGURE EXAMPLE 9.1

Then read the following:
a) Follow the relative humidity curve down and to the left and read $\phi = 50\%$. Notice that no interpolation is necessary on this part of the problem. However, students are expected to be able to perform graphical interpolations to be able to arrive at a more precise answer.
b) Follow the line of constant humidity ratio horizontally to the right and read the humidity ratio, $w = 0.0153\ lb_{mwv}/lb_{mdac}$. Some charts record this in pounds of moisture per pound of dry air. Others record this in grains (gr) of moisture per pound of dry air. Conversion factor: 7000 gr = 1 lb_m. So, $w = 107\ gr_{wv}/lb_{mda}$ by conversion and this compares to the reading on the psychrometric chart with this scale.
c) Follow the line of constant humidity ratio (w) horizontally to the left and read the saturation temperature, $T_{dp} = 69$ °F. Some charts will also have the dew point shown on its own scale along the right side of the chart. Notice that constant w means that the amount of water in the air does not change.
d) Notice that the enthalpy scale goes around three sides of the graph. This is to allow better precision in reading the value. Drawing a straight line through the given point and making sure that both enthalpy scales have the same reading (to ensure the straight edge is parallel to the enthalpy scale) $h = 38.4\ Btu/lb_{mda}$.
e) By direct reading or graphical interpolation between specific volume lines $v = 14.2\ ft^3/lb_{mda}$.
f) Vapor pressure is related to w and for those charts with a vapor pressure scale it is a reading off a horizontal line through the state point to obtain $p_{wv} = 0.71$ in-Hg and using the conversion factor of 29.92 in-Hg = 14.696 psi $p_{wv} = 0.35$ psia. Using the steam tables and entering it using $T_{dp} = 69$ °F as the saturation pressure yields $p_{wv} = 0.351$ psia. That's good agreement!

Here are some additional practice problems in a state point table format. Again, find the intersec-

	T_{db} (°F)	T_{wb} (°F)	f (%)	T_{dp} (°F)	h (Btu/lb_{mda})	v (ft³/lb_{mda})	w ($lb_{m_{moisture}}/lb_{mda}$)	p_{wv} (psia)
1	75	62.5	50	55.1	28.2	13.68	0.0093	0.22
2	83	60	24.2	42.8	26.3	13.80	0.0058	0.14
3	101.8	80.3	40	72.9	43.8	14.55	0.0175	0.40

tion of the given information (**bolded** and boxed in the table) on the psychrometric chart (previous page) and then read the other values.

9.6 HVAC PROCESSES

HVAC processes may also be graphed, and the psychrometric chart is useful in determining data and solving problems. The processes are illustrated on Figure 9.4.

A - Humidifying Only
B - Heating and Humidifying
C - Sensible Heating Only
D - Chemical Dehumidifying
E - Dehumidifying Only
F - Cooling & Dehumidifying
G - Sensible Cooling Only
H - Evaporative Cooling Only

FIGURE 9.4: *Basic Psychrometric Processes*

9.7 HUMAN COMFORT ZONE

Human comfort includes a personal metabolism component that also depends upon a number of factors affecting heat loss from the body, such as the amount of insulating clothing and air movement in the conditioned space. A person who has been engaging in strenuous physical activity will have raised his or her metabolism and will feel comfortable with a colder environment. Someone who is sedentary will not feel as comfortable in the same environment. Otherwise, most sedentary people feel comfortable in the ranges set out in Table 9.1.

TABLE 9.1: *Human Comfort Zone*

Summer		Winter	
T_{db} = 72 °F	T_{dp} = 36 °F	T_{db} = 69 °F	T_{dp} = 36 °F
T_{db} = 81 °F	T_{dp} = 36 °F	T_{db} = 76 °F	T_{dp} = 36 °F
T_{db} = 79 °F	T_{wb} = 68 °F	T_{db} = 74 °F	T_{wb} = 64 °F
T_{db} = 73 °F	T_{wb} = 68 °F	T_{db} = 68 °F	T_{wb} = 64 °F

These comfort zones graph as trapezoidal shapes on the psychrometric chart. Graph these limits on the psychrometric chart in your notes; label and shade in the areas.

On a hardcopy psychrometric chart, circle the SPs and highlight the various lines that show how these answers were obtained. Remember that copying isn't the same as doing. Develop your skills.

9.8 HVAC TECHNOLOGY

There are several basic types of air conditioning and refrigeration technology of which the student should be aware. A few Navy ships still have absorption chillers. These devices require a heat input to and a solution of chemicals, typically lithium bromide (LiBr). Other than knowing that this class of technology exists, the specific operation of these units will not be described in this text.

Many aircraft use a reverse Brayton air conditioning scheme. This uses bleed air from the compressor; then cools it in an air-air heat exchanger (the cooling air is usually rammed through this heat exchanger during flight); then expands it through a turbine, cooling the air even further. The turbine discharge is then either run through another heat exchanger to cool it off further still or returned to the cabin where it is blown into the cabin to cool the passengers.

Evaporative cooling is used in areas where the relative humidity is low. This equipment is suitable for desert climates, but not usually around the sea due to the higher relative humidity conditions.

Vapor compression is by far the most common mechanical cooling technology and this will be discussed thoroughly below.

9.8.1 Vapor Compression Refrigeration

Refrigeration technology is at the heart of both air conditioning and refrigeration systems. Vapor compression technology is the most common type of refrigeration technology in general use today.

Recall that the goal is to remove heat from the cool conditioned space and transfer it to a hotter environment. This is shown in the thermodynamic concept sketch, Figure 9.5, below.

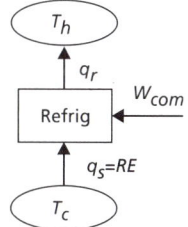

FIGURE 9.5: *Thermodynamic Concept Sketch for Refrigeration*

In this case and in specific terms, q_s is the heat supplied to the working fluid in the system, the "refrigerant," q_r is the heat rejected to the environment, and w_{comp} is the work put into the working fluid by the compressor. Since the first law is still satisfied in that the sum of the energy into the system is equal to the sum of the energy out, or $q_s + w_{comp} = q_r$. The second law is also satisfied in that it is impossible to move the heat without putting work into the system. The concept of thermodynamic efficiency also applies, but because the numerical value may exceed one (1), it is now referred to as "coefficient of performance" to reduce the potential for confusion with efficiency. As with efficiency, COP relates the desired energy effect to the required energy cause. And as was done with η_{th} for heat engines, this may be calculated in specific, total, or total rate basis terms.

$$COP = \frac{E_{desired}}{E_{required}} = \underbrace{\frac{q_s}{w_{comp}} = \frac{RE}{w_{comp}}}_{\text{specific basis}} = \underbrace{\frac{\dot{m}RE}{\dot{m}w_{comp}} = \frac{\dot{Q}_s}{\dot{W}_{comp}}}_{\text{total rate basis}} \quad (9.5)$$

The box representing the refrigerator in Figure 9.5 is comprised of a compressor, a condenser, a throttling device, and an evaporator. In the schematic shown in Figure 9.6 the throttling device is shown as a *thermostatic expansion valve* (TXV). Other throttling devices may also be used, such as an *orifice* or a *capillary tube*. We're familiar with an orifice from our unit on fluid flow. Automotive air conditioners typically use orifice tubes for throttling. A capillary tube is simply a tube with a small inner diameter that essentially has a very high friction factor. Capillary tubes tend to be very noticeable for their length and are often coiled. Household refrigerators typically use capillary tubes for the throttling process. The thermodynamic process for all of these throttling devices is modeled as isenthalpic expansion. Since the process is isenthalpic ($\Delta h = 0$), there is no work or heat involved, but entropy (s) will increase as the refrigerant molecules travel through the throttling device. Note that this is yet another special case on the *T-s* diagram shown below. There is some area under the *T-s* curve, but it does not represent heat as it

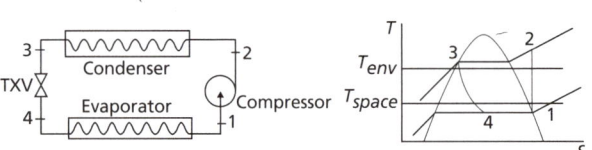

FIGURE 9.6: *Basic, Ideal Vapor Compression Refrigeration*

would in other uses of the *T-s* diagram. Table 9.2 summarizes the components and their associated processes.

Notice that in contrast to all of the heat engines covered to this point, which proceed around the cycle graph in a clockwise direction, the refrigeration cycle processes proceed <u>counterclockwise</u>.

Starting with the steady flow energy equation (SFEE), the derivation is the same as with the Rankine cycle—the ke and pe terms cancel, resulting in either work (*w*) or heat (*q*) being the change in enthalpy (*h*) across each component. And we can generally assume that each component is either a *w* or a *q* component—both processes are not going on at the same time in an individual component. The primary data source for enthalpy involving refrigerants is a pressure-enthalpy (*p-h*) graph. Sometimes a table will be available. The thermophysical property data for R-134a are included as an appendix.

Figure 9.7 shows *T-s* and *p-h* concept graphs illustrating the same state points for an <u>ideal</u> vapor compression refrigeration system. "Ideal" means that the evaporator sends an SV to the compressor in order to ensure that the compressor does not see any liquid; the compression process is modeled as isentropic; and the condenser produces an SL.

The "dome" on the *p-h* graph represents the saturation conditions with saturated liquid (SL) to the left of the

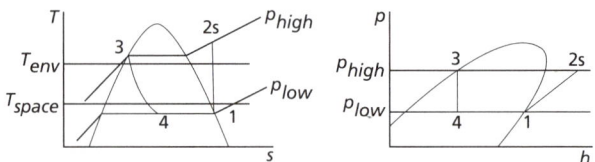

FIGURE 9.7: *T-s and p-h Graphs for an Ideal Vapor Compression Refrigeration System*

TABLE 9.2: *Refrigeration Components and Processes*

	Component	Process Model	Specific Term	Total Rate Term
1-2	Compressor	Isentropic ($Ds = 0$)	w_{comp}	$\dot{P}_{comp}$
2-3	Condenser	Isobaric ($Dp = 0$)	q_r	$\dot{Q}_r$
3-4	Expansion Device	Isenthalpic ($Dh = 0$)		
4-1	Evaporator	Isobaric ($Dp = 0$)	$q_s = RE$	$\dot{Q}_s = Capacity$

peak and saturated vapor (SV) to the right, just as with the *T-s* graph. Notice that an isentropic compression process curves out to the right on *p-h* coordinates. Also take note that the isenthalpic throttling process is vertical on these coordinates. Figure 9.6 also shows the relationship of the environmental temperature to the refrigerant in the condenser and the conditioned space temperature to the refrigerant in the evaporator.

As shown in Figure 9.8, <u>real</u> vapor compression equipment will have the refrigerant enter the compressor slightly superheated. As with the ideal cycle, this still ensures that the compressor does not have liquid refrigerant going through it. An SHV set point also allows for a feedback control range for modulating the TXV either open or closed as necessary to maintain the compressor inlet with the desired amount of superheated vapor. As with previous cycles, the real compressor will not be isentropic. And the discharge of the condenser will typically be slightly sub-cooled to ensure the inlet to the throttling process (TXV) is a liquid.

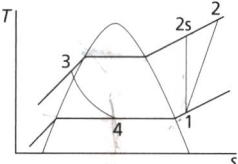 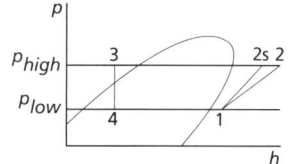

FIGURE 9.8: T-s *and* p-h *Graphs for a Real Vapor Compression Refrigeration System*

Figure 9.9 shows a *p-h* graph for the refrigerant R-134a. This is a post–Montreal Protocol refrigerant that is replacing R-12 and several other refrigerants that are being phased out. R-134a is used in modern motor vehicle air conditioning, residential refrigerators, and large chiller units. As with the Mollier chart and the psychrometric chart, students may find that highlighting the words on the chart will help in orientation.

Figure 9.9 is from the DuPont publication "Thermodynamic Properties of HFC-134a." This document is available at http://www2.dupont.com/Refrigerants/en_US/assets/downloads/h47751_hfc134a_thermo_prop_eng.pdf or use the key words "DuPont SUVA 134a" in your web search engine. Print only page 35 of 36 to get this graph without printing the entire document. Note the rest of the document is organized similar to the familiar steam tables. More precise data are available in the tables than can be determined from the graphical source. However, proficiency with the graph is the expected skill for this course.

Now, let's practice using this graphical data source.

EXAMPLE 9.2: IDEAL VAPOR COMPRESSION REFRIGERATION SYSTEM ANALYSIS

GIVEN: An ideal vapor compression air conditioner unit using R-134a and operating between 40 and 200 psia.

FIND:
a) the *RE*
b) the w_{comp}
c) *COP*
d) the mass flow rate of refrigerant necessary to achieve a capacity of 2 tons

SOLUTION: Draw a schematic that represents this system, and label the parts and state points. Sketch the representative *T-s* and *p-h* graphs, noting the "control points" (SV @ SP1; SL @ SP3). Complete the state point table for the refrigerant (SP1 is at the compressor inlet) using the *p-h* graph for R-134a as the source for *h*. To use the graph, find the high and low pressures and draw horizontally across the graph (we're employing the isobaric heat exchanger simplification). Then circle the SPs and read up or down to obtain the enthalpies. Note the line of constant entropy (*s*) in the SHV region of the *p-h* graph. You should be able to read the enthalpies summarized in the SP Table and be within +/- 1.

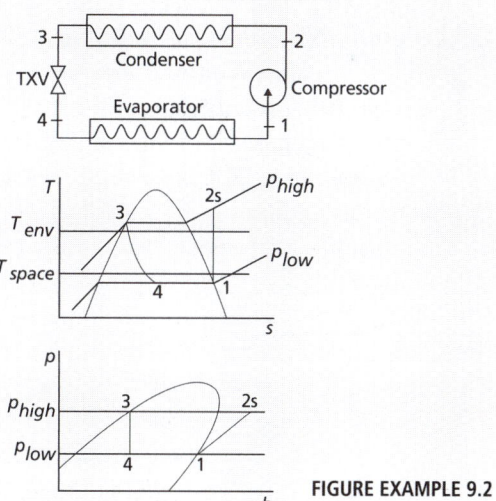

FIGURE EXAMPLE 9.2

R-134a Data	1	2s	3	4
p (psia)	40	200	200	40
T (°F)	29	~~142~~	105	29
h (Btu/lb$_m$)	107	123	54	54
s (Btu/lb$_m$-°F)	0.2225	0.2225	NR	NR
State	SV*	SHV	SL*	WV

* "Ideal" system control points.

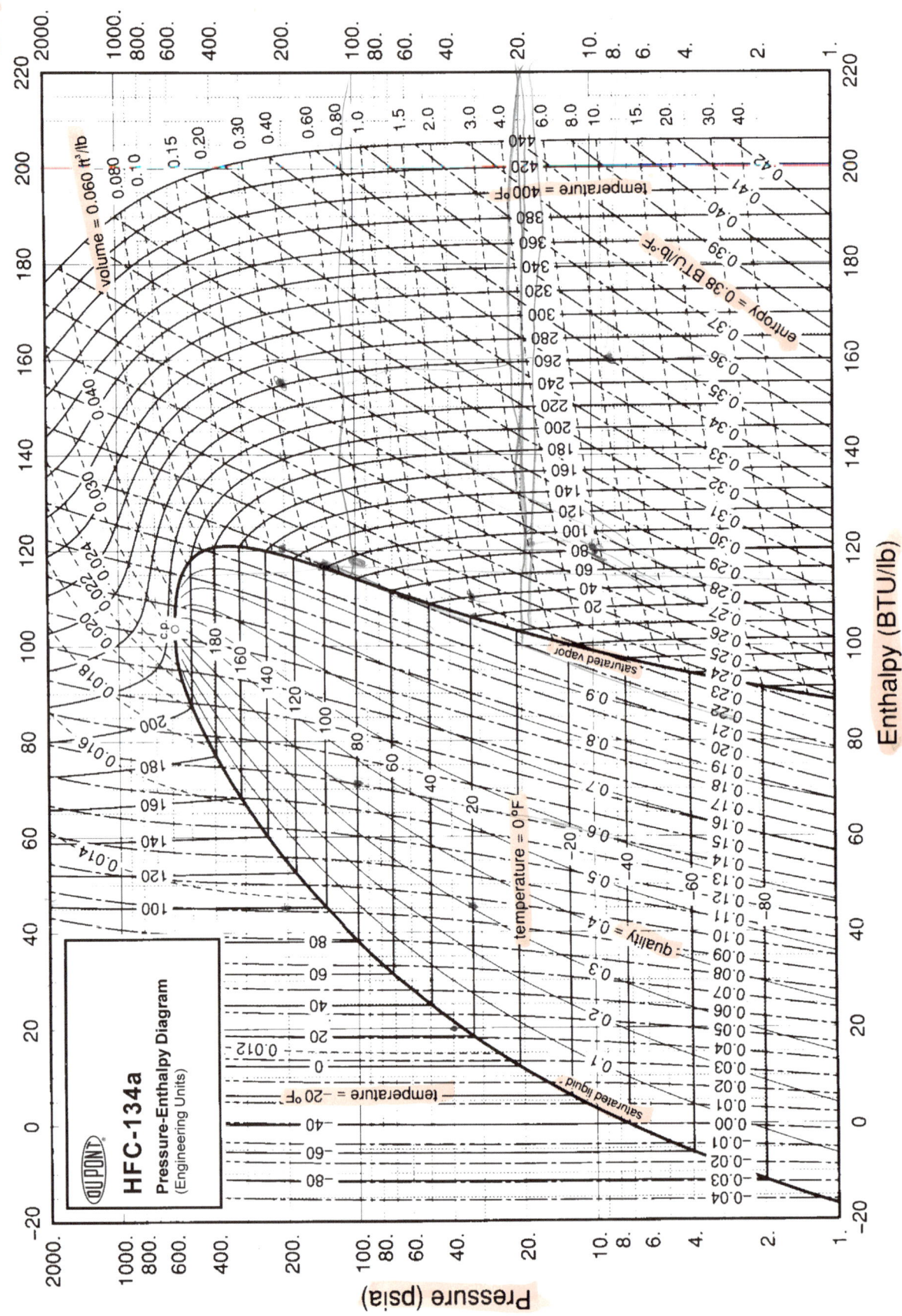

FIGURE 9.9: p-h *Graph for R-134a*

Reprinted by permission from DuPont

a) $RE = \Delta h_{evaporator} = h_1 - h_4 = 107 - 54 = 53$ Btu/lb$_m$

$\boxed{RE = 53 \text{ Btu/lb}_m}$ (positive, heat into the refrigerant)

b) $w_{comp} = h_1 - h_{2s} = 107 - 123 = -16$ Btu/lb$_m$

$\boxed{w_{comp} = -16 \text{ Btu/lb}_m}$ (negative due to work in)

c) $COP = \dfrac{RE}{w_{comp}} = \dfrac{53 \text{ Btu}/lb_m}{16 \text{ Btu}/lb_m} = 3.31$

$\boxed{COP = 3.31}$ No units. Answer must be positive.

d) $CAP = \dot{Q}_s = \dot{m}_{refrigerant} RE \Rightarrow$

$\dot{m} = \dfrac{CAP}{RE} = \dfrac{(2 \text{ tons})}{\left(53 \frac{Btu}{lb_m}\right)} \left(12000 \dfrac{Btu}{hr - ton}\right)$

$\boxed{\dot{m}_{ref} = 453 \dfrac{lb_m}{hr} = 7.55 \dfrac{lb_m}{min} = 0.13 \dfrac{lb_m}{sec}}$

Time units weren't specified by the problem.

We briefly touched on the throttling function of the TXV, above, and stated that the design of this valve allows it to throttle open or closed (it typically cannot totally shut) based upon the temperature signal at the exit of the evaporator. Physically, this is either accomplished with electronics or, more simply, with a bulb actuator employing a volatile fluid. The bulb contains a small amount of a saturated fluid (a wet vapor, WV). As the temperature goes up, it raises the pressure in the fluid. A small tube connects the bulb to the TXV actuator and causes it to throttle in the open direction to increase the flow of refrigerant as the evaporator outlet temperature rises. This feedback mechanism requires the evaporator discharge to be slightly superheated in order for the sensing bulb to modulate the TXV throttle position. If the evaporator discharge were saturated, the sensing bulb temperature and pressure would not change.

Now let's ratchet it up a notch and include some reality elements.

EXAMPLE 9.3: REAL VAPOR COMPRESSION REFRIGERATION SYSTEM ANALYSIS

GIVEN: A vapor compression air conditioner unit using R-134a and operating between 40 and 200 psia. The evaporator outlet temperature is 40 °F and the compressor's isentropic efficiency is 66%. The condenser discharge is sub-cooled by 20 °F. Draw a schematic that represents this system and label the parts and state points. Complete the state point table for the refrigerant.

FIND:
a) RE
b) COP
c) the temperature drop as a result of the expansion process

R-134a Data	1	2s	2	3	4
p (psia)	40	200	200	200	40
T (°F)	40	145	NR	125	29
h (Btu/lb$_m$)	110	125	132.7	47	47
s (Btu/lb$_m$-°F)	0.2275	0.2275	NR	NR	NR
State	SHV	SHV	SHV	CL	WV

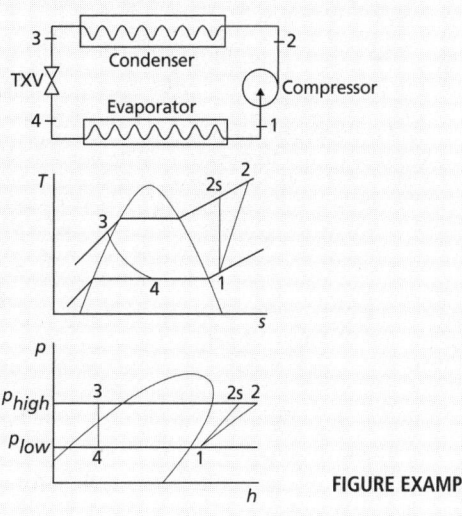

FIGURE EXAMPLE 9.3

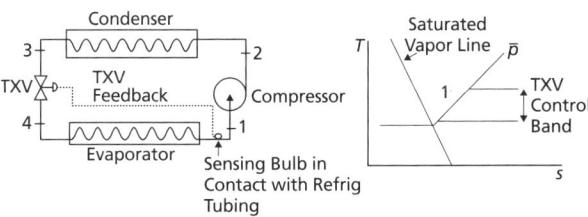

FIGURE 9.10: *Real Refrigeration TXV Feedback and Control Band Shown on Partial T-s Graph*

SOLUTION:
$h_1 = h$ (40 °F; 40 psia) Remember that it is an SHV.
h_{2s} comes from following s up to 200 psia.
We will need to calculate h_2 from h_{comp}.

$$\eta_{compressor} = \frac{w_{comp,ideal}}{w_{comp,real}} = \frac{h_{2s} - h_1}{h_2 - h_1} \Rightarrow h_2 = h_1 + \frac{h_{2s} - h_1}{\eta_{comp}}$$

$$h_2 = h_1 + \frac{h_{2s} - h_1}{\eta_{comp}} = 110 + \frac{125 - 110}{0.66} = 132.7 \frac{Btu}{lb_m}$$

We need T_{sat} (200 psia) to determine T_3

T_{sat} (200 psia) = 125 °F

$T_3 = T_{sat}$ (200 psia) − 20 = 105 °F

Remember that the refrigerant is sub-cooled at this point.

Use the *p-h* graph to determine h_3; $h_4 = h_3$
Now we can get on with answering the questions.

a) What is the *RE* for this air conditioner?

$RE = h_1 - h_4 = 110 - 47$

$\boxed{RE = 63 \text{ Btu/lb}_m}$

b) What is the *COP* for this air conditioner?

$COP = RE / w_{comp} = 63 / (132.7 - 110)$

$\boxed{COP = 2.77}$

Notice that the compressor work is the real work and not the ideal work. I've also solved it in absolute value terms since *COP* is positive.

c) What is the temperature drop as a result of the expansion process?

$\Delta T_{TXV} = T_3 - T_4 = 105 - 29$

$\boxed{\Delta T_{TXV} = 76 \text{ °F}}$

A note about evaporator temperature in Example 9.3: The evaporator typically either cools air directly or water in chill water systems. If the refrigerant is operated below 32 °F, then we could expect humidity in the air to frost up on the coils and impede or eventually block further air flow (or potentially the chill water could freeze if an antifreeze solution is not employed and potentially burst the pipes in chiller-type refrigeration systems). I chose 40 psia for the evaporator pressure in Example 9.3 to make use of the *p-h* graph a bit easier for the student. At least 29 °F is close to OK. In reality, air conditioner systems would be designed to operate above freezing or frosting conditions. However, if an air conditioner system leaks, it often shows up as a lower operating pressure in the evaporator (and consequently a lower refrigeration *T* there, since the evaporator is mostly a saturated system) and frost forms on the air-side of the evaporator coils.

EXAMPLE 9.4: REAL VAPOR COMPRESSION AIR CONDITIONER ANALYSIS

GIVEN: The refrigeration system of Example 9.3 is being used in a "direct expansion" (DX) air conditioner. (DX is typical of most residential and automotive air conditioners. The air being cooled is giving up its heat directly to the refrigerant in the evaporator.) The air conditioner is cooling the air in a space that has a dry bulb temp of 85 °F and a wet bulb temp of 80 °F. The A/C air discharge $T_{db} = 68$ °F. The air flow rate going into the AC unit is 500 CFM.

FIND:
a) What is the mass flow rate of the air?
b) How much condensate is produced?
c) What is the heat transfer rate to the refrigerant?
d) What is the capacity of this AC unit?

SOLUTION: We're going to perform an energy balance on the evaporator, so let's draw it. We'll also need a psychrometric chart. Check to see if humidity is condensed out of the air.

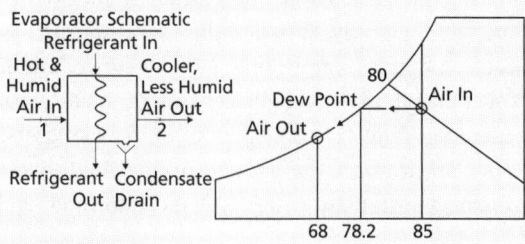

FIGURE EXAMPLE 9.4A

As the air moves past the coils of the evaporator, the air will cool off at constant *w* until it reaches the dew point. Notice the two-step process for the air on the concept psychrometric chart. Draw this on a copy of the full chart you are using to analyze this example. The accepted practice is to assume that the condensate drips off the evaporator tubes at the same temperature as the exiting air temperature unless other data are available. However, refer back to the note after Example 9.3 about frosting. The condensate enthalpy value is from the *Steam Tables* at 68 °F.

Air-side Data	1	2	Condensate
T_{db} (°F)	85	68	NR
T_{wb} (°F)	80	68	NR
T_{dp} (°F)	78.2	68	68
ϕ (%)	80	100	NR
w (lb$_{mwv}$/lb$_{mda}$)	0.0210	0.0148	NR
h (Btu/lb$_{mda}$)	43.5	32.4	36.1
v (ft³/lb$_{mda}$)	14.20	NR (13.62)	NR
$\dot{V}$ (CFM)	500	NR	NR

a) What is the mass flow rate of the air?

$$\dot{m}_{dry\,air} = \rho_{air}\dot{V}_{air} = \frac{\dot{V}_{air}}{v_{air}} = \frac{500\,ft^3/min}{14.20\,ft^3/lb_{mda}}$$

$$= 35.21\,lb_{mda}/min$$

$$\boxed{\dot{m}_{air} = 35.21\,lb_{mda}/min}$$

b) How much condensate is produced?

$$\dot{m}_w = \dot{m}_{air}(\omega_1 - \omega_2)$$

$$= 35.21\,lb_{mda}/min\,(0.0210 - 0.0148)\,lb_{mwv}/lb_{mda}$$

$$= 0.2\,lb_{mwv}/min$$

$$\boxed{\dot{m}_w = 0.2\,lb_{mwv}/min}$$

c) What is the heat transfer rate to the refrigerant? Perform an energy balance on the <u>evaporator</u>.

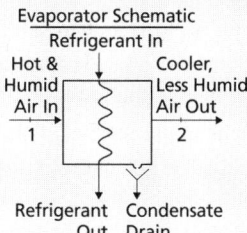

FIGURE EXAMPLE 9.4B

$$\sum \dot{E}_{in} = \sum \dot{E}_{out}$$

$$\dot{m}_{air,in}h_{air,in} + \dot{m}_{ref,in}h_{ref,in}$$
$$= \dot{m}_{air,out}h_{air,out} + \dot{m}_{ref,out}h_{ref,out} + \dot{m}_{w,out}h_{w,out}$$

where $\dot{m}_{ref,in} = \dot{m}_{ref,out}$ and $\dot{m}_{dry\,air,in} = \dot{m}_{dry\,air,out}$

Whenever we have analyzed heat exchangers thus far in this course, we have assumed that the heat given off by one fluid is absorbed by the other fluid. That basically means that there is no heat absorbed from or given off to the environment in which the equipment sits. We'll use that same approach here as well.

$$\dot{Q}_{evaporator} = \dot{Q}_{refrigerant\,side} = \dot{Q}_{air\,side} = \dot{m}_{ref}(h_{ref,out} - h_{ref,in})$$
$$= \dot{m}_{air}(h_{air,in} - h_{air,out}) - \dot{m}_{w,out}h_{w,out}$$

We don't know the refrigerant flow rate, even though we know the refrigerant enthalpies. We do know the mass flow rate of the air and the air-side enthalpies, so use the air side of the evaporator to calculate the answer to this part.

$$\dot{Q}_{evaporator} = \dot{m}_{air}(h_{air,in} - h_{air,out}) - [\dot{m}_{w,out}h_{w,out}]$$

$$\dot{Q}_{evaporator} = 35.21\frac{lb_{mda}}{min}(43.5 - 32.5)\frac{Btu}{lb_{mda}}$$
$$- \left[0.22\frac{lb_{mwv}}{min} 36.1\frac{Btu}{lb_m}\right]$$

$$\boxed{\dot{Q}_{evaporator} = 387.31\frac{Btu}{min} - 7.94\frac{Btu}{min} = 379.37\frac{Btu}{min}}$$

d) What is the Capacity of this AC unit?
CAP is $\dot{Q}$ of the evaporator, but with units of refrigeration-tons (tons for short).

$$\dot{Q}_{evaporator} = \left(387.3\frac{Btu}{min}\right)\left(\frac{1\,min - ton}{200\,Btu}\right)$$

$$\boxed{= 1.94\,tons|_{ignoring\,condensate} = 1.90\,tons|_{including\,condensate}}$$

Notice that ignoring the condensate in the energy balance doesn't change the answer very much (~2.1%). So, we can just use the air enthalpy values and get a good enough answer from the equation below.

$$\dot{Q}_{evaporator} = \dot{m}_{air}(h_{air,in} - h_{air,out})$$

9.8.2 Evaporative Coolers

"Swamp coolers" operate on the principle of blowing air through a water mist. The evaporation of the liquid water transfers the latent heat of vaporization of the liquid water from the air, reducing its temperature. The amount of temperature drop depends on the entering air relative humidity, with lower humidity producing a larger temperature drop. The suitable atmospheric conditions tend to be in desert areas. This is Process "H" shown in Figure 9.3.

The effectiveness of a swamp cooler will range from 50 percent to 95 percent. Operation and effectiveness is best illustrated by the following example.

EXAMPLE 9.5: SWAMP COOLER ANALYSIS

GIVEN: A 50% effective swamp cooler is installed supplying air to a tent in the desert with dry bulb temperature (T_{db}) of 110 °F and relative humidity (ϕ) of 20%.

FIND: The conditions of the air supplied to the tent.

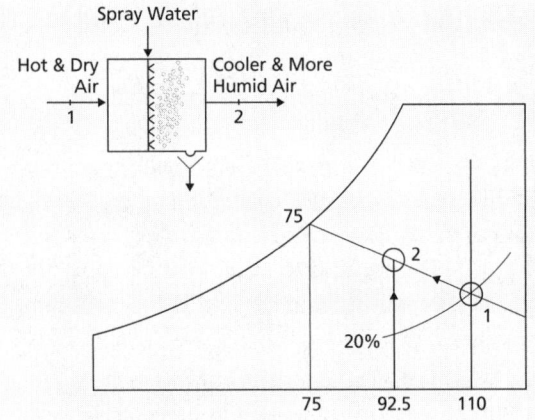

FIGURE EXAMPLE 9.5

SOLUTION: The spray humidification process follows a constant T_{wb} model.

Following the line of constant T_{wb}, the 100% effective cooler could reach 75 °F. However, we only get 50% there, so the supply air conditions are $T_{db,2}$ = 92.5 °F. This is from $\{T_{db,1} - 0.5(110 - 75)\}$. Then looking up the other properties from the psychrometric chart:

$$T_{db,2} = 92.5 \text{ °F}; T_{wb} = 75 \text{ °F}; \phi = 45\%; w = 0.015$$

Still, that's a lot better than 110 °F!

9.8.3 Chill Water Systems

The expense of purchasing and maintaining a piped system full of refrigerant has resulted in the development of chill water systems. Vapor compression chillers use cold water to absorb heat from various spaces and then to conduct that heat to the chiller unit. This cascaded system is used to create the temperature conditions by which heat is moved from the air, transferred to chill water, transferred to the refrigerant in the evaporator section, and then transferred to the environment via the condenser water system. The enthalpy difference of the air and chill water as these fluids move through the AHU fan and chill water pump is much smaller than the heat transferred in the heat exchangers. This means the work (or power) of the fans may be neglected, as may the work (or power) of the liquid water pumps. However, the compressor contributes in a more significant way to the energy content of the refrigerant, so the energy addition by the compressor should be accounted for.

A submarine recirculates the air inside the hull continuously as shown in Figure 9.11.

Other components, not shown in the schematic, that a submarine uses to maintain a suitable environment are:

(1) CO_2 scrubbers, which use a chemical that absorbs CO_2 from the air at room temperature and then releases the CO_2 when heated to a higher temperature; the CO_2 is pumped overboard;
(2) CO-H_2 burners, which at high temperature catalytically cause CO to react with O_2 to turn to CO_2 and H_2 to turn to water;
(3) Electrolytic oxygen generators, which split water into O_2 and H_2 at high pressure, with the H_2 then being diffused into the seawater and the O_2 being stored in high pressure flasks; and
(4) Filters and electrostatic precipitators, which remove dust and other particulates.

The chemical balance of the air is constantly monitored and balanced within the parameters required for human occupancy.

The submarine can bring in fresh outside air via open hatches when on the surface, or via the snorkel mast, which can be extended above the surface while the submarine is submerged, but near the surface. The snorkel mast has a valve in the top that senses water and will snap shut if the valve is crested by any waves. The valve will re-open when the top of the snorkel is above the water. Exhaust air is diffused out into the water while snorkeling. The snorkel mast is also used to allow the sub's crew to operate the emergency diesel generator, which provides

Heating, Ventilation, Air Conditioning & Refrigeration (HVAC&R)

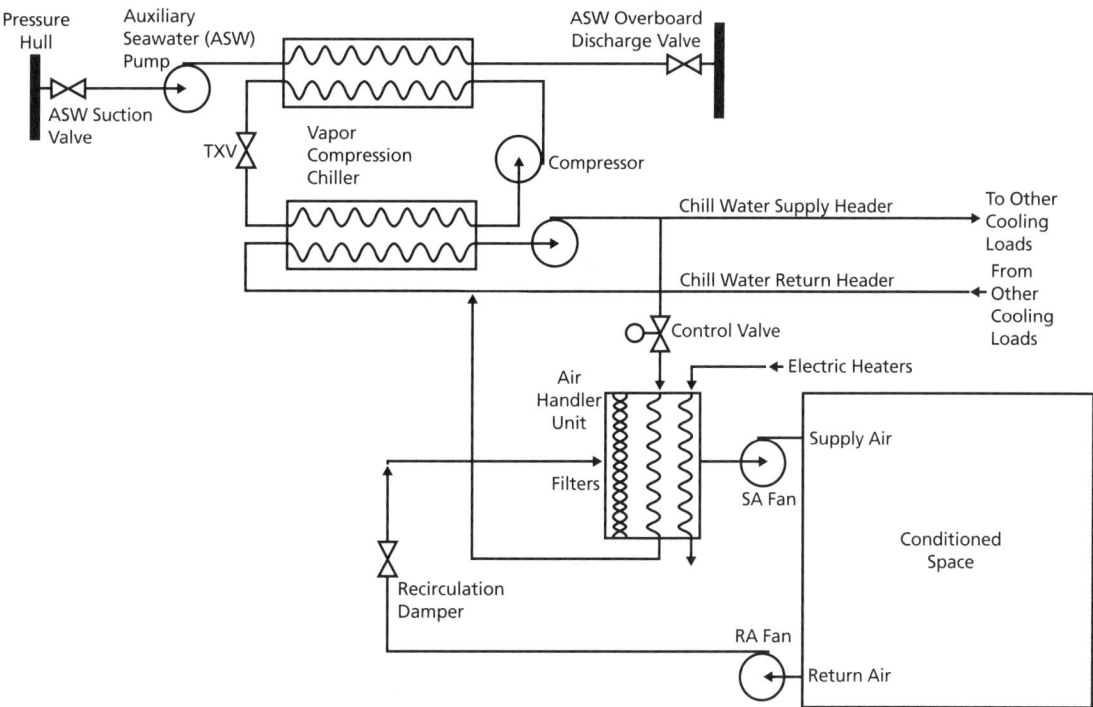

FIGURE 9.11: *Submarine HVAC System Simplified Schematic*

electrical power should the nuclear power system be offline due to casualty or normal shutdown.

The similarity of a commercial HVAC system to a shipboard system can clearly be seen by comparing the two figures. The main difference is that the condenser water that is used to carry the rejected heat away is seawater on the ship, but freshwater for the land-based unit. Further, evaporative cooling towers are used to reject the condenser water heat to the environment. These units function by spraying the condenser water into a large volume of moving air. The air is not saturated coming into the tower, so some of the condenser water evaporates. The latent heat of vaporization absorbed in the evaporation process cools off the condenser water, which then returns to the chiller via a pump. Make-up water is added to the pan under the cooling tower to make up for the lost inventory that was evaporated and blown into the local atmosphere.

EXAMPLE 9.6: AC&R EVAPORATOR HEAT EXCHANGER ANALYSIS

GIVEN: The air conditioner operating with the parameters described in Example 9.3.

FIND: a) Draw the temperature profile graph and
b) determine the *LMTD* for the evaporator.

SOLUTION: The refrigerant is colder than the air, in order for the heat transfer to proceed from the air to the refrigerant in this heat exchanger. The evaporator is a latent/sensible indirect heat exchanger and uses a cross-flow configuration. Recall that we draw a cross-flow HX like a counter-flow HX for the purpose of developing the temperature profile graph. I recommend picking the direction of refrigerant flow and the shape from the *T-s* graph and then drawing the air temperature profile. The evaporator is between SP4 and SP1.

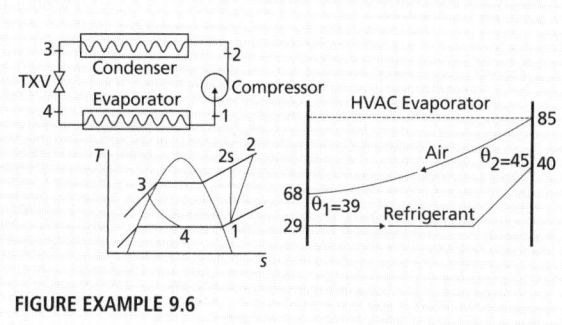

FIGURE EXAMPLE 9.6

$$\theta_m = \frac{\theta_1 - \theta_2}{\ln\left(\frac{\theta_1}{\theta_2}\right)} = \frac{39-45}{\ln\left(\frac{39}{45}\right)} = 41.9 F°$$

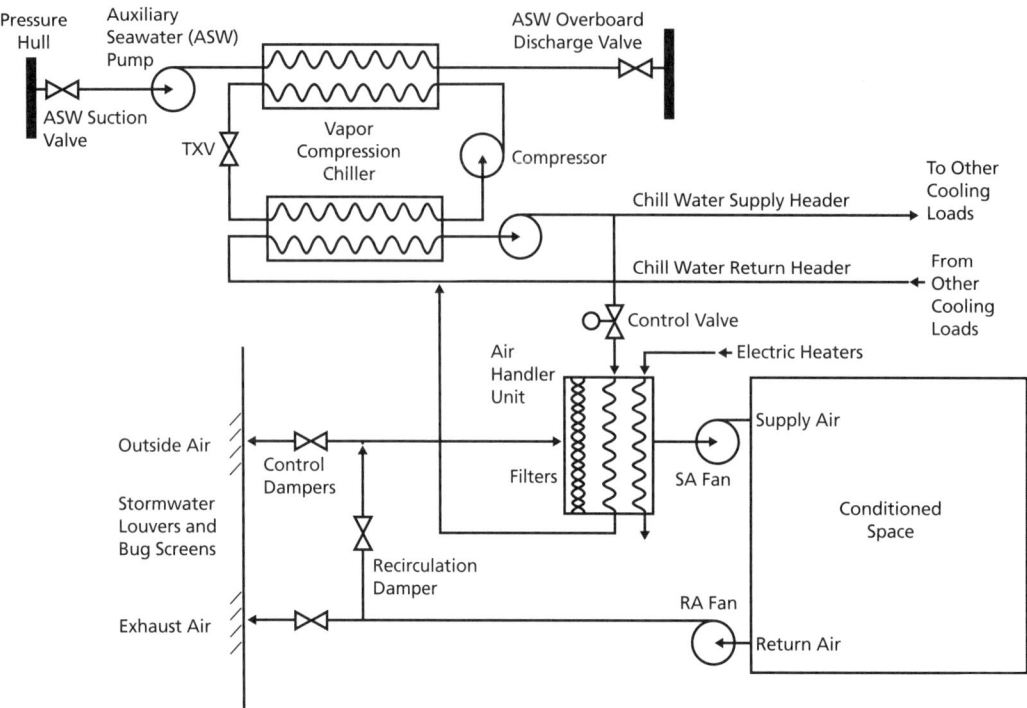

FIGURE 9.12: *Ship HVAC System Simplified Schematic*

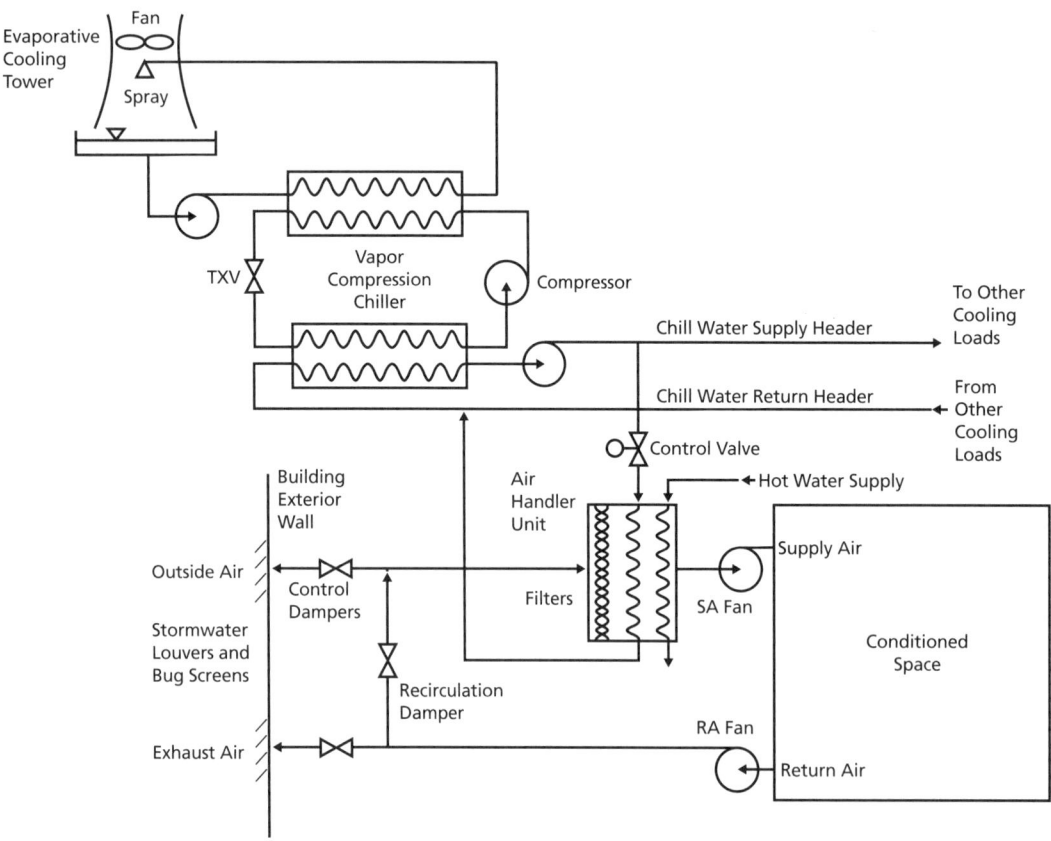

FIGURE 9.13: *Commercial Building HVAC System Simplified Concept Schematic*

EXAMPLE 9.7: MIXING AIR STREAMS

GIVEN: A shipboard HVAC system, as shown in the schematic below. Note: In order to reduce the cost of conditioning air, it is a common practice to recirculate some of the return air with the outside air that is being drawn into the ship. Yet because the ship does not have the type of equipment allowing 100% recirculation, like a submarine, some outside air must be brought in to maintain indoor air quality within healthy limits. The mixture of outside air and recirculation air is sent through the cooling coils to meet the supply air temperature and humidity requirements.

The outside air has the following conditions: 90 °F DB; 70% RH. Return air has the same characteristics as recirculation air: 75 °F DB; 60% RH. The recirculation air rate is 1000 CFM. The outside air make-up rate is 500 CFM.

FIND: The conditions of the mixture $\{T_{DB}, \phi, \dot{V}(CFM)\}$.

SOLUTION: The first step is to understand the schematic. The applicable portion is redrawn and state points are added to aid in organizing the solution.

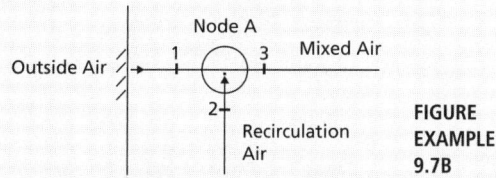

FIGURE EXAMPLE 9.7B

The two applicable equations are <u>conservation of mass</u> ($\Sigma \dot{m} = 0$) and <u>conservation of energy</u> ($\Sigma \dot{E} = 0$) with both analyzed on the "rate basis" due to this being a flowing system. They are written for this "system" as follows:

$$\dot{m}_1 + \dot{m}_2 = \dot{m}_3$$

and

$$\dot{m}_1 h_1 + \dot{m}_2 h_2 = \dot{m}_3 h_3$$

In order to determine mass flow rate from volumetric flow rate, use the psychrometric chart to determine the specific volume (v). Also use the psychrometric chart to determine the needed enthalpies (h).

$$v_1 = 14.33 \, ft^3/lb_m \qquad v_2 = 13.72 \, ft^3/lb_m$$

Use the volumetric flow rates and specific volumes to calculate the mass flow rates.

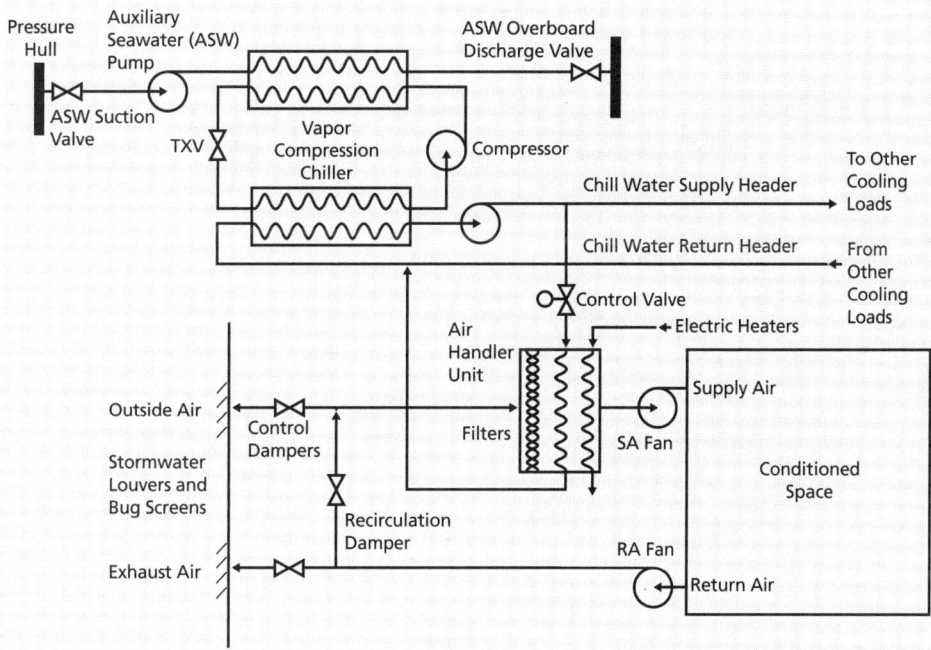

FIGURE EXAMPLE 9.7A

$$\dot{m}_1 = \frac{\dot{V}_1}{v_1} = \frac{500\, ft^3/min}{14.3\, ft^3/lb_m} = 34.89\, lb_m/min$$

$$\dot{m}_2 = \frac{\dot{V}_2}{v_2} = \frac{1000\, ft^3/min}{13.72\, ft^3/lb_m} = 72.89\, lb_m/min$$

$$\dot{m}_3 = \dot{m}_1 + \dot{m}_2 = 34.89 + 72.89 = 107.78\, lb_m/min$$

Combine the enthalpies from the psyhrometric chart with the conservation of energy and solve for the unknown enthalpy, h_3.

$$h_1 = h_{OA} = 45.3\, Btu/lb_m \quad h_2 = h_{RA} = 30.3\, Btu/lb_m$$

$$\dot{m}_1 h_1 + \dot{m}_2 h_2 = \dot{m}_3 h_3$$

$$(34.89\, lb_m/min)(45.3\, Btu/lb_m) + (72.89)(30.3)$$
$$= 107.78 h_3$$

$$h_3 = 35.2\, Btu/lb_m$$

Now, the quickest way to determine the final answers is to use the psychrometric chart. Draw a straight line between the state points 1 and 2 on the psychrometric chart. This is known as the "process line." Determine where h_3 crosses the process line and read T_{DB}, f, and v.

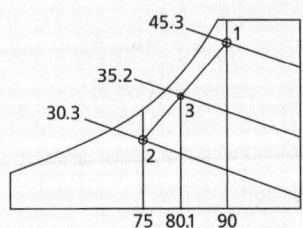

FIGURE EXAMPLE 9.7C

$$\boxed{T_{DB3} = 80.1\,°F} \quad \boxed{\phi_3 = 65\%} \quad v_3 = 13.93\, ft^3/lb_m$$

$$\dot{V}_3 = \dot{m}_3 v_3 = (107.78\, lb_m/min)(13.93\, ft^3/lb_m)$$
$$= 1501.4\, CFM$$

$$\boxed{\dot{V}_3 = 1501.4\, CFM}$$

Note that the volumetric flow rate is **not** the sum of the two input flows, because of the density difference. However, the mass flows do add up.

Practice Problems

"The problem is not that there are problems. The problem is expecting otherwise and thinking that having problems is a problem."

—Theodore Rubin

9.1 Using a psychrometric chart, complete the following table. Submit this table and a copy of the chart with the sub-problem located and numbered on the chart. (Highlight and number the state points on the chart.)

	ϕ (%)	T_{db} (°F)	T_{wb} (°F)	h (Btu/lb$_m$)	v (ft^3/lb$_m$)
1	50	95.0			
2	10		40.0		
3	40			40.0	
4	10				13.00
5		70.0	60.0		
6		81.4		29.6	
7		113.3			15.00
8			71.7		13.95
9				26.2	14.10

9.2 Using a psychrometric chart, complete the following table. Submit this table and a copy of the chart with the sub-problem located and numbered on the chart. (Highlight and number the state points on the chart.)

	T_{db} (°F)	T_{wb} (°F)	h (Btu/lb$_m$)	v (ft^3/lb$_m$)	ω (lb$_{m_m}$/lb$_{m_a}$)
1	55	55			
2	50		15.4		
3	95			14.55	
4	108				0.021
5		60		13.75	
6			21.2	13.5	
7		68.5			0.008
8			38.4		0.0152

9.3 Using a *p-h* diagram for R-134a, complete the following table. Submit this table and a copy of the chart with the sub-problem located and numbered on the chart. (Highlight and number each state point on the *p-h* diagram.)

	p (psia)	T (°F)	h (Btu/lb$_m$)	s (Btu/lb$_m$-°F)	State
1	40		20		
2	200	260			
3			160	0.34	
4	100			0.14	
5		100			SV
6	10		120		

9.4 Using a *p-h* diagram for R-134a, complete the following table. Submit this table and a copy of the chart with the sub-problem located and numbered on the chart. (Highlight and number each state point on the *p-h* diagram.)

	p (psia)	T (°F)	h (Btu/lb$_m$)	s (Btu/lb$_m$-°F)	State
1	50		25		
2	100	260			
3			140	0.31	
4	8			0.06	
5		80			SV
6	20		120		

9.5 Compare the density of air using the ideal gas laws at 70 °F and standard pressure with the same temperature at 50% relative humidity using the psychrometric chart. Then compare the Δh of ideal gas air using the Air Tables vs humid air using the psychrometric chart for ΔT of 10 degrees at these conditions. (Use constant humidity ratio on the psych chart.)

9.6 A vapor compression air conditioner has a capacity of 20 tons and a coefficient of performance of 3.5 when operating with a refrigerating effect of 61.4 Btu per pound of refrigerant. Draw the First Law Concept sketch and calculate the

   a) refrigerant flow rate (lb$_m$/min)

   b) work done on the refrigerant by the compressor (Btu/lb$_m$)

   c) compressor internal power (hp)

   d) rate of heat rejection from the system (Btu/min)

9.7 A refrigeration system has a capacity of 25 tons and rejects heat at a rate of 8560 Btu/min. Draw the First Law Concept sketch, the system schematic, the process graph on *T-s* coordinates, and calculate the

   a) rate of heat absorption by the refrigerant (Btu/min)

   b) power required as input to the system (Btu/min)

   c) coefficient of performance of the system

9.8 A refrigeration system has a capacity of 15 tons and a compressor rated at 25 hp. Draw the First Law Concept sketch, the system schematic, the process graph on *T-s* coordinates, and calculate the

   a) rate of heat absorption by the refrigerant (Btu/min)

   b) rate of heat rejection from the system (Btu/min)

   c) coefficient of performance of the system

9.9 A vapor compression air conditioner has a coefficient of performance of 3.0 when operating with a refrigerating effect of 75 Btu per pound of refrigerant and a refrigerant flow rate of 45 lb$_m$/min. Draw the First Law Concept sketch and calculate the

   a) capacity (tons)

   b) work done on the refrigerant by the compressor (Btu/lb$_m$)

   c) compressor internal power (hp)

   d) rate of heat rejection from the system (Btu/min)

9.10 You purchase a 12-oz. can of R-134a refrigerant from the local automotive supply store. The steel can has a "dished" bottom and top, but may be modeled as a cylinder 2-7/8 inches in diameter and 3-3/4 inches in length. The booklet *Thermodynamic Properties of HFC-134a Refrigerant*, published by DuPont (available free on the Internet), lists the v_f (70 °F) = 0.0131 ft^3/lb$_m$ and v_g (70 °F) = 0.5570. What is the density of the refrigerant mixture? What is the quality (x) of the can's contents at 70 °F? What should a pressure gage read if connected to the can and the can's seal is punctured? A warning is printed on the side of the can that states that the can may burst if left inside a hot car at temperatures above 120 °F. What is the bursting pressure of the can?

9.11 Draw *p-h* and *T-s* concept graphs side-by-side (8 total graphs) for the following scenarios:

a) ideal vapor compression refrigeration cycle

b) ideal vapor compression refrigeration cycle with a real (non-isentropic) compressor

c) real vapor compression refrigeration cycle with superheated vapor into the compressor, sub-cooled liquid out of the condenser, and an ideal compressor.

d) real vapor compression refrigeration cycle with superheated vapor into the compressor, sub-cooled liquid out of the condenser, and a real compressor.

9.12 Air with a dry bulb temperature 65 °F and 50% relative humidity is supplied to the operating spaces of a submarine as shown in the accompanying figure. Return air (RA) measurements have a dry bulb temperature of 85 °F and a wet bulb temperature of 77 °F. The ventilation supply fan takes suction on the cooling coils and discharges the conditioned air to the operating spaces. The fan is moving 5000 CFM. (a) What is the mass flow rate of air (lb$_m$/min)? (b) What are the enthalpies of the air upstream and downstream of the coils (Btu/lb$_m$)? Submit a marked-up psychrometric chart with your solution showing how you obtained all values.

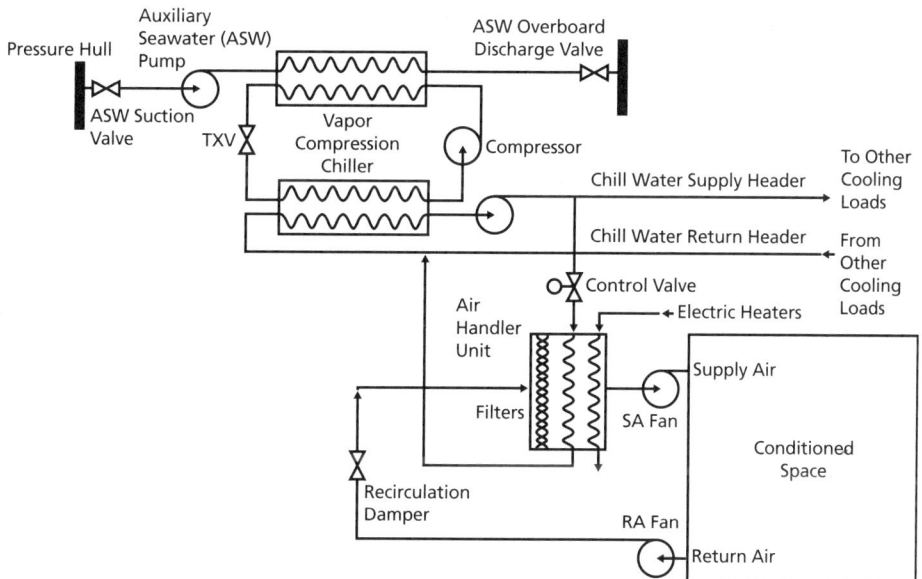

FIGURE HW 9.12, 9.13, 9.14, 9.15, AND 9.16: *Submarine Combined HVAC System Concept Schematic*

9.13 The supply air (SA) fan shown in the accompanying figure is delivering 1,500 CFM to the conditioned space. The air conditions upstream of the cooling coil are 90 °F dry-bulb and 80 °F wet-bulb. The air conditions downstream of the cooling coil are 60 °F dry-bulb and 50% relative humidity. (a) Find the mass flow rate of air through the cooling coil. (b) Find the heat transfer rate of the cooling coil in the air handler unit. You may assume that the thermostat is calling for cooling, and as a result, the electric heaters are secured and the chill water control valve is open. The R134a chiller operates between 20 and 180 psia. You may assume an ideal vapor compression refrigeration cycle. (c) What is the capacity of the chiller (tons) due to the conditioned space described above?

(d) How much heat is rejected to the condenser water system (BtuH)? (e) Submit a marked-up psychrometric chart and *p-h* graph with your solution showing how you obtained all values.

9.14 The chiller shown in the accompanying figure uses R-134a. The compressor discharge pressure is 400 psia and the compressor's isentropic efficiency is 87%. The pressure at the inlet of the evaporator is 40 psia. The evaporator discharge includes 30 degrees of superheat in the refrigerant. The condenser refrigerant discharge temperature is 120 °F. The flow rate of refrigerant is 30 lb_m/min. (a) List the enthalpies determined for the four state points. (b) What is the COP of the chiller? (c) What is the maximum seawater injection temperature in order to accept the rejected heat? Explain your logic. (d) If this chiller provides supply air at 55 °F and 55% relative humidity, and receives return air at 85 °F dry-bulb and 75 °F wet-bulb, determine mass flow rate of air that can be conditioned by this system. (e) Submit a marked-up *p-h* graph with your solution clearly showing how you obtained all values.

9.15 The chiller shown in the accompanying figure uses R-134a. The compressor discharge pressure is 200 psia and the compressor's isentropic efficiency is 85%. The temperature at the inlet of the evaporator is 20 °F. The evaporator discharge includes 20 degrees of superheat in the refrigerant. The condenser refrigerant discharge temperature is 100 °F. (a) List the enthalpies determined for the four state points in a state point table. (b) What is the *COP* of the chiller? (c) What is the maximum seawater injection temperature in order to accept the rejected heat? Explain your logic. (d) Submit a marked-up *p-h* graph with your solution clearly showing how you obtained all values.

9.16 The R-134a chiller shown in the accompanying figure is operating at a capacity of 150 tons and with a refrigerant low pressure of 20 psia and a high pressure of 200 psia. The evaporator discharge temperature is 20 °F. The compressor discharge temperature is 160 °F. Seawater injection temperature is 80 °F.

   a) What are the enthalpies determined from the R-134a *p-h* graph? What is the condenser refrigerant discharge temperature?

   b) What is the mass flow rate of the refrigerant?

   c) What is the heat transfer rate of the condenser?

   d) What is the required seawater flow rate (GPM) if the overboard discharge is limited to 100 °F?

   e) Sketch the temperature profile graph for the condenser.

   f) The condenser has 500 SF of heat transfer surface area. What is the *LMTD* for the chiller's condenser in this operating condition?

   g) What is the overall heat transfer coefficient for the condenser?

   h) What is the COP of the chiller?

To obtain psychrometric and *p-h* charts:
Recommend bookmarking the source websites once you found them.
(1) To obtain a high-quality psychrometric chart, conduct a web search "psychrometric chart" and find one that prints clearly on your printer. There are a number of websites with psychrometric charts that may be downloaded free of charge. Noteworthy is Hands Down Software (www.handsdownsoftware.com), which also has interactive psychrometric charts that are useful for checking your answers.

   Adding the key word "Heatcraft" will also bring up a site with a good psychrometric chart. Be sure to print a psych chart for the appropriate elevation.

   The answers given for this unit were looked up manually using the Heatcraft Psychrometric Chart and then checked using HDPsyChart by Trane, both available online.

(2) To obtain a p-h chart for R-134a refrigerant, print page 35 of 36 from the following website: http://www.refrigerants.dupont.com/Suva/en_US/pdf/h47751.pdf or use the key words "DuPont SUVA 134a" in your web search engine.

10 Desalination

The Seabees in this photograph are setting up a reverse osmosis water purification unit. Ships, submarines, and land-based units use this technology to produce water that can be used for drinking, bathing, or propulsion plant purposes from seawater or other water sources containing a variety of impurities.

10.1 UNIT LEARNING OBJECTIVES

10.1.1 Terminology: definitions, variables, and typical units

10.1.1.1 desalination, deionization, demineralization, distillation, reverse osmosis, vapor compression distillation, flash distillation, ion exchanger, potable water

10.1.2 Concepts: ideas and engineering expressions of them

10.1.2.1 Describe the major categories and operating principles of water purification technology.

10.1.2.2 List the major classifications (grades) of water used on U.S. Navy ships.

10.1.2.3 Elaborate the major shipboard uses of water.

10.1.3 Skills: procedures, practices, or methods that enable reasoning

10.1.3.1 No unit-specific skills (general course skills will enable learning and reasoning about any particular water production technology).

10.2 INTRODUCTION

Oceans cover nearly three-quarters of the earth's surface and are the basic source of all water on our planet. Seawater contains about 3.5 percent of various impurities, the concentration of which is customarily expressed in terms of the *parts (by weight) of the component per million parts of water*. Ocean water contains, on average, 35,000 parts per million (ppm) of dissolved solid impurities. These materials consist primarily of various salts (chlorides and sulfates) of alkaline elements (sodium, calcium, magnesium, and potassium). Brackish water, typically found at the nexus of fresh and salt bodies of water, also contains substantial amounts of bicarbonates, irons, and silicates. The composition of brackish water varies so widely that an average analysis is not practical.

Freshwater is required for numerous applications aboard naval vessels but generally falls into two categories: freshwater suitable for personnel consumption, known as *potable water*, and water for the engineering plant. Potable water has chemical, biological, and radiologic limits for maintenance of crew health. The *feedwater* required for injection to a steam plant and *demineralized water* (aka deionized water) suitable for injection into a nuclear primary water system both have higher purity requirements than potable water. The level of impurities in the propulsion system water must be reduced to avoid corrosion and the degradation of performance that follows. Excessive feedwater *salinity*, or salt concentration, can lead to corrosion of boiler components and must be minimized. In addition, engineering plant water must be "softened" or demineralized to remove remaining impurities. Water softening, via chemical treatment or *ion exchange*, is critical to prevent scale, an incrustation of alkaline sulfates or silicates deposited on the interior surfaces of metal components in which water is heated (e.g., boilers). On the other hand, potable water often has oxidizers such as chlorine or bromine added to it to kill off biological growth in the water. Sometimes ultra-violet (UV) light is used to sanitize the water.

Since storage facilities are space-limited, every deployable ship must be self-sufficient in producing freshwater. With a readily available and abundant supply of seawater, ships incorporate various methods of purification to meet, or exceed, their freshwater supply demands.

Purification of seawater normally requires a reduction of the dissolved salts by a factor of 70, from around 35,000 ppm to less than 500 ppm. Brackish water purification requirements vary widely with the composition of the water and can range from a factor of 3 up to a reduction in dissolved salts by a factor of 30.

Likewise, ground operations require a source of freshwater to support personnel and equipment operation. Many of the places where U.S. forces operate have polluted or brackish water that must be treated before it should be consumed by people or added to vehicle systems. From a community relations perspective, provision of facilities for clean freshwater for the local populace is generally appreciated. (Likewise, operational sewage collection and treatment infrastructure is also appreciated by the crew and local populace.)

The most general techniques for purification of impure water involve (1) boiling it and condensing the vapor; (2) filtering it with ultra-fine filters (e.g., reverse osmosis); or (3) electrically forcing movement of impurities to where they can be flushed away, thereby leaving cleaner water.

The oldest method of purification of seawater is *distillation* (boiling). Three general types of distilling plants are installed in naval ships:

1. Vapor compression
2. Low-pressure steam
3. Heat recovery

The major differences among the types of plants are the form of energy used for operation and the pressure under which distillation occurs. Vapor compression units employ electrical energy for heaters and compressors and boil seawater at or slightly above atmospheric pressure. Low-pressure steam plants use steam from the auxiliary exhaust steam system or the auxiliary steam system as a heat source. Heat recovery units use waste heat from other systems, such as diesel engine cooling jacket water or gas turbine exhaust. Both heat recovery and vapor compression plants depend on a relatively high vacuum (and therefore lower saturation temperature to boil water) for operation.

Within the past few decades, additional methods of water purification have been developed, commercialized, and used by naval vessels and land units. *Reverse osmosis* (RO) purifies seawater by forcing it through a membrane that permits water to pass through but holds back dissolved solids larger than specified dimensions.

Stationary land facilities typically use sand filtration and chemical treatment to remove solids and kill bacteria. This type of treatment is beyond the scope of this course. Likewise, sewage treatment will not be covered.

10.3 DISTILLATION BASICS

Although several different types of distilling plants exist, all generate freshwater from seawater through employment of the same basic principles. The distillation process can be represented by the schematic on the following page.

A heat source is applied to seawater. When boiled, the seawater releases vapor that is relatively free of any dissolved solids. The vapor is then cooled and condenses into freshwater or distillate. The concentrated seawater, now termed brine due to its higher salinity, is typically discharged.

The rated output capacity of a distilling plant is the principal manner in which a unit is selected for a particular application. Standard ratings are provided in gallons per hour (GPH) or gallons per day (GPD) and indicate the number of gallons of purified water generated during the specified time period.

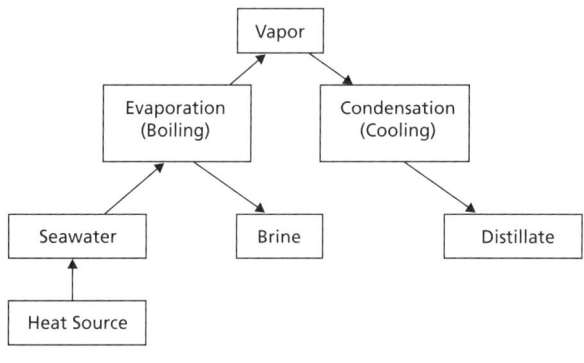

FIGURE 10.1: *Basic Steam Distillation*

10.4 METHODS OF DISTILLATION

As stated in the introduction, there are three general types of distilling plants employed aboard naval vessels: low-pressure steam, vapor compression, and heat recovery. A brief discussion of each type of plant, to include background information, general principles of operation, and naval application, is found below:

10.4.1 Low-Pressure Steam

There are three basic types of low-pressure steam distilling plants:

1. Submerged-tube
2. Vertical-basket
3. Flash

The first two types differ only in the type of heat transfer surface used to boil the incoming seawater. In submerged-tube plants steam passes through tube bundles, boiling the surrounding water. In vertical-basket units the heating steam is contained within deeply corrugated baskets, and the seawater is boiled outside the basket between the corrugations. In the flash-type plant, the seawater feed is not boiled by an external heat source, but is flashed to vapor through a pressure drop.

To achieve greater distillate output capacities at a reasonable economy, typical low-pressure steam distillers will incorporate several successive stages of feed vaporization and condensation. These *multiple-effect plants*, as they are commonly termed, operate on identical principles as single stage plants, with additional stages added between the feed inlet and brine outlet.

SUBMERGED-TUBE AND VERTICAL-BASKET

Since these two types of distilling plants differ only in the geometry of the device within which the seawater feed is boiled, a schematic of a submerged-tube unit is shown below and will be discussed.

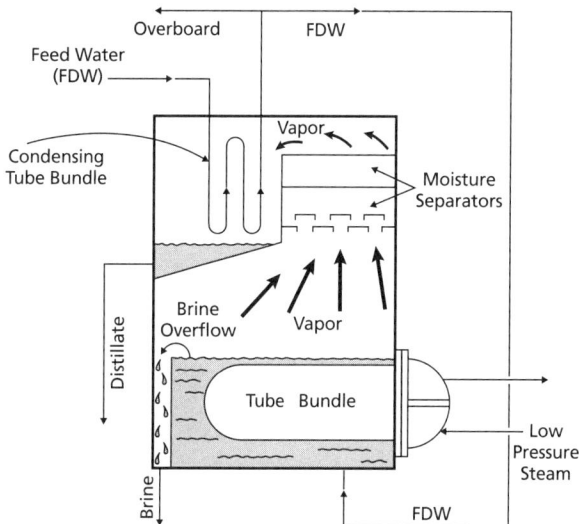

FIGURE 10.2: *Submerged-Tube Type Distillation Plant Schematic*

A feed pump inducts seawater and directs it through a strainer and condensing tube bundle, where the transfer of latent heat from condensing vapor heats it. A portion of the heated feed is directed through a flow meter and into the bottom of the unit. Unneeded feedwater, as determined by operator requirements, is discharged overboard. The seawater surrounds tubes that contain circulating low-pressure steam. The steam in the tubes causes the surrounding feed to boil and produce steam. The low-pressure steam surrenders its latent heat as it condenses and is drawn out of the evaporator section as hot water. The rising vapor is drawn through moisture separators, or demisters, within which entrained particulates are removed. The clean vapor then passes into the condenser section and is cooled by incoming seawater feed, condensing into distillate. The distillate is collected and pumped to storage tanks. Brine remains at the bottom of the unit, is forced toward a collection area by the incoming seawater feed, and pumped overboard. In a vertical-basket plant, the process is virtually identical, with the exception that the preheated seawater feed collects in a corrugated basket surrounded by steam tubes.

FLASH-TYPE

A schematic of a two-effect flash-type distilling plant is shown in Figure 10.3.

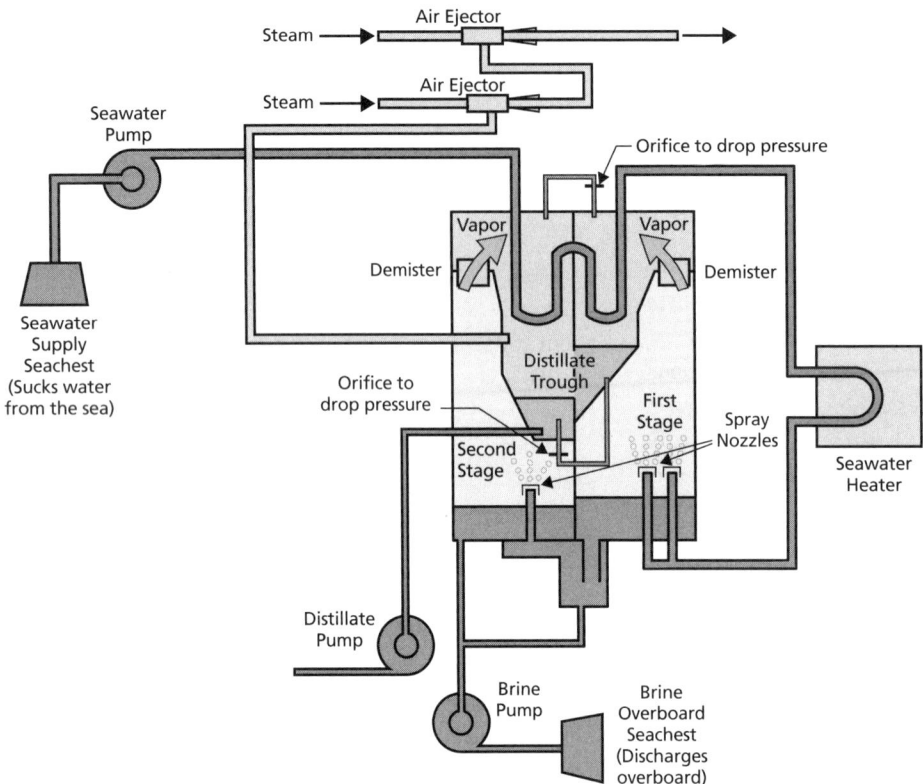

FIGURE 10.3: *Flash-Type Distillation Plant Schematic*

The seawater pump inducts feed through a strainer and discharges it at approximately 30 psig to the distillate cooler. The seawater is heated by the latent heat released by the condensation of steam vapor. From the condenser, the seawater is directed into the seawater heater where it is heated to approximately 170 °F, a temperature high enough to allow seawater to flash when exposed to the vacuum inside the evaporation chamber but low enough to prevent scale formation. The preheated seawater is then directed into the first stage evaporation shell where it flashes into steam as it encounters the vacuum inside. Steam rises to the top of the evaporation chamber through the moisture separators, or demisters. The demisters remove any particulates that may be entrained within the steam. Steam condenses as it comes in contact with the condenser tubes that are filled with incoming seawater feed. The distillate collects in a trough and is pumped into a storage tank. Seawater that did not flash in the first stage is drawn into the second stage, due to its higher vacuum, where the distillation process is repeated. The brine pump removes any accumulated brine from the bottom of the flash chambers and discharges it overboard.

The condensation of the steam into a much smaller volume of water creates most of the vacuum within the evaporator shells. In addition, air ejectors maintain the vacuum and are designed (and oriented) to ensure that the vacuum within the second stage evaporator shell is higher than that in the first stage.

NAVAL APPLICATIONS

The distillation process is similar for all submerged-tube and vertical-basket plants, regardless of the number of stages. The normal rated capacity for these units installed on U.S. Navy ships varies between 4,000 and 50,000 gallons per day. The single stage vertical-basket distilling plants found on some older ships have a rated capacity of 8,000 GPD and are sometimes referred to as the "8K evaporators." These distillers are typically larger than other types of plants of similar capacity and require significant maintenance and treatment for scale removal and prevention. For these reasons, coupled with the advent of distilling plants capable of delivering greater capacity at a fraction of the size and cost, these units are being phased out of use in naval vessels.

The flash-type distilling plant, on the other hand, is widely used throughout the Navy. Two- through six-stage flash-type distilling plants are used in various Navy surface craft and submarines, ranging in capacity from 6,000 to 100,000 GPD. Flash-type distilling plants offer numerous advantages over boiling distillers. Since no boiling occurs directly on heat transfer surfaces and brine is only concentrated to 5 to 10 percent above normal seawater, scale and corrosion rates are significantly lower than in submerged-tube and vertical-basket plants. Flash-type distillers are capable of operating at full capacity for extended periods. Additionally, there is no brine level or density control required. Flash-type distilling plants are also easy to automate.

Flash-type distillers, however, have certain drawbacks. Since optimum operation effectiveness is achieved with a low-pressure steam heat source, they do not readily adapt to diesel engine heat recovery applications. Furthermore, the large quantities of noncondensable gases released in the first stage require larger air ejectors. Finally, increased seawater feed rates produce larger quantities of distillate, which results in higher consumption of treatment chemicals.

10.4.2 Vapor Compression

Vapor compression distilling plants are implemented when low-pressure steam or waste heat is unavailable or insufficient to operate the other types of distilling plants. The only energy required to maintain this process is electrical power for the compressor motor and electric immersion heaters, which replace heat lost within the physical plant. Since they only require electrical power and a seawater source to operate, vapor compression distillers are well suited as supplemental units to main distillers.

PRINCIPLES OF OPERATION

The basic operation of a vapor compression plant is shown in Figure 10.4.

Seawater passes through a strainer, flow meter, and preheater prior to entering the system. The preheater is thermostatically controlled and supplies heat to the feed only during cold water (below 55 °F) operation. The incoming feed is then heated in a heat exchanger and mixed with brine already in the evaporator section of the plant. The brine mixture enters the heat transfer tubes and a portion of it is boiled at about 1 psig. The wet steam, now at approximately 215 °F, rises through a moisture separator and enters a compressor, which pressurizes and heats the dry steam. Moisture removed from the steam returns to the feed-brine mixture and noncondensable gases are vented from the evaporator. Across the compressor, the steam pressure increases about 5 psig and the temperature increases approximately 50 °F. The compressed, heated steam is directed to the outside of the heat transfer tubes, where it condenses, transferring its heat energy to the brine inside the tubes. Any excess brine is pumped out of the evaporator section and also passes through the heat exchanger, transferring heat to the incoming feed and promoting the boiling process.

NAVAL APPLICATIONS

Naval vapor compression distilling plants were first developed for submarines and later applied to diesel-driven surface craft. Currently, most submarine vapor compression distillers serve as backup units for use when the main desalination equipment is unavailable. The Navy currently uses vapor compression distilling plants of four different capacities: 1,600 GPD, 2,000 GPD, 3,000 GPD, and 9,600 GPD. The small capacity plants are used mainly as backup distillers on submarines and are termed "vertical-tube" due to their design. The 9,600 GPD plants, termed "rising film" and used as main distillers on some surface ships, are similar to the other three plants in basic operation but differ in design detail.

10.4.3 Heat Recovery

Heat recovery distilling plants are used in vessels with diesel main propulsion or auxiliary engines. Two variations of heat recovery units are employed; both use the heat from engine cooling systems for vaporization of seawater feed. In one model of a heat recovery plant, the heat of the diesel engine cooling jacket water is transferred to the seawater in a heat exchanger. The heated seawater is then flashed to vapor through exposure to vacuum, as in the flash-type distilling unit. Alternatively, the hot diesel

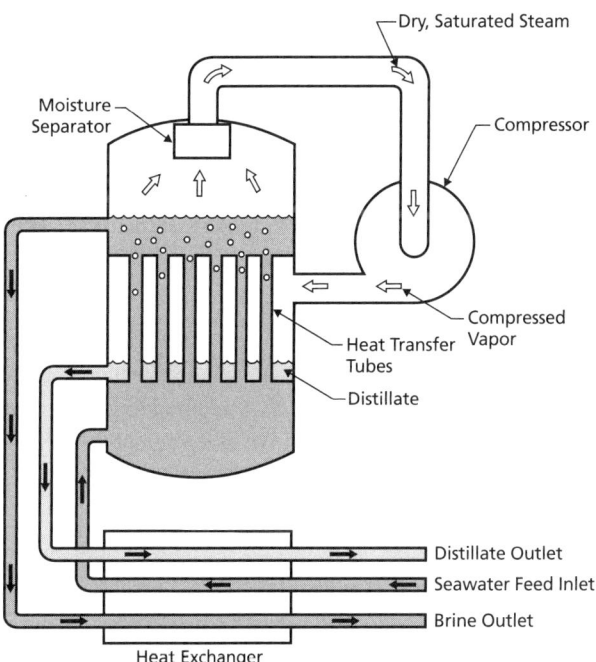

FIGURE 10.4: *Vapor Compression Type Distillation Plant Schematic*

engine cooling jacket water is circulated through a tube bundle that is submerged in seawater. The seawater is then boiled in a chamber held under vacuum, much like the submerged-tube distilling unit.

PRINCIPLES OF OPERATION

The schematic of the submerged-tube distilling plant (Figure 10.2) should be referenced in the discussion of the operation of the heat recovery plant. Rather than utilize low-pressure steam, the heat source is that contained within the water circulated through the cooling jackets of the diesel main propulsion engines or ship's service diesel generators. To supplement the heat in the cooling jacket water when engines are operating at low speeds, the plant also implements electric heating modules and/or steam heaters. This ensures that the cooling jacket water will be at an appropriate temperature when it enters the heat exchanger.

A circulating pump directs the jacket water into the inlet of the submerged-tube bundle, in which heat is transferred to the seawater feed in the evaporator section. The water then exits the tube bundle and circulates back to the engine cooling jacket. The seawater surrounding the tube bundle evaporates, and the resulting vapor is drawn through moisture separators to the distillate condensing tube bundle. The seawater feed in the condensing tubes cools the rising vapor, and the distillate is collected, chemically treated, and pumped to storage.

Most heat recovery distillers aboard Navy ships have a secondary heat exchanger between the engine cooling jacket water system and the distiller unit. This heat exchanger isolates the engine coolant, with all its chemical additives, from the distiller. Systems that do not employ this secondary heat exchanger utilize heat from the engine coolant to support the distiller. This is called a single-loop system and must be monitored continuously to ensure that no engine coolant leaks through the heat exchanger and contaminates the distillate.

NAVAL APPLICATIONS

The nominal rating of 12,000 GPD provides sufficient capacity to serve as a main distiller for smaller, diesel-powered surface craft or as a supplemental distiller on board larger vessels with auxiliary diesel engines.

10.4.4 Reverse Osmosis

A natural starting point for a discussion on reverse osmosis is normal osmosis. By definition, osmosis is the movement of a solvent through a semi-permeable membrane into a solution of higher solute concentration that tends to equalize the concentrations of solute on both sides of the membrane. A semi-permeable membrane is a membrane that permits passage of some atoms or molecules (solvent) while preventing passage of others (solute). Common examples of semi-permeable membranes are a cell wall, the lining of your intestines, or Gore-Tex fabric. A pictorial representation of osmosis is provided below:

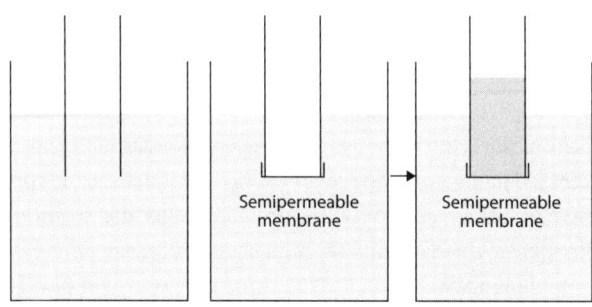

FIGURE 10.5: *Osmosis Basics*

On the left is a water-filled container with a tube that is partially submerged. As expected, the water level in the tube is identical to that in the container. In the middle figure, the end of the tube has been sealed with a semi-permeable membrane and the tube has been partially filled with a salt-water solution and submerged. Initially, the level of the salt solution and the water are equal, but over time, the water level in the tube will rise. The increased water level is due to osmotic pressure since the membrane allows passage of water molecules but prevents the passage of salt molecules. To understand osmotic pressure, think of the water molecules on both sides of the membrane. On the salt solution side, some of the pores get "plugged" with salt molecules but on the other, pure-water, side this does not occur. Therefore, more water passes from the pure-water side to the salt solution side, as there are more pores on the pure-water side through which the water molecules may pass. The water level on the salt solution side rises until one of two things occurs: the salt concentration equalizes across the membrane or the water pressure of the column of salt-water solution equals the osmotic pressure.

In reverse osmosis, the membrane acts as an extremely fine filter to purify seawater. Pressure is applied to the seawater on one side of the membrane to overcome osmotic pressure and reverse the osmotic process by separating the freshwater from the dissolved salts in the seawater. A larger difference between the applied pressure and the osmotic pressure will result in larger quantities of water permeating the membrane and higher purity of the permeated water. In practice, a pressure of 700 to 1000 psi is required to overcome the osmotic pressure of 350 psi and obtain an acceptable flow of purified water.

Principles of Operation

The basic operation of a reverse osmosis plant is shown in Figure 10.6.

Seawater is inducted into the system via a through-hull fitting and is filtered through a strainer to remove debris and sediment. The pressure of the water is then boosted to 30 psi and filtered through 20- and 3-micron filters in series to remove any remaining suspended particles. The oil/water separator then removes any oil that may be present. The cleaned water next enters the high-pressure pump, which operates at high pressure (typically in the range of 800 psi) and supplies the required force to drive the water through the semi-permeable RO membrane modules. As the high-pressure seawater passes over the membranes, it becomes more concentrated as a portion of it permeates through the membranes as freshwater. The remaining concentrated salt water (brine) is then discharged overboard through a back pressure regulator, which controls the pressure level in the membrane modules, and brine flow meter. An automatic high-pressure safety shutdown switch protects the system if pressure exceeds recommended levels.

The freshwater permeate leaves the RO membrane modules up to 99.2 percent free of salts, minerals, and other ions. A salinity probe registers the salt content of the product water. If the permeate is out of specifications, it is automatically dumped to bilge. If the product water salinity is acceptable, it flows through the product flow meter where the amount of potable water produced per hour is registered. The water then passes through a charcoal filter to remove any unpleasant odor or taste. An optional ultra-violet sterilizer can complete the filtration process where 99.8 percent of all microorganisms, including viruses and bacteria, are destroyed. Alternatively, the water is treated with bromine and sent to potable water storage tanks. Additional chemical treatment, or softening, is required prior to utilization as steam generator feedwater.

Naval Applications

Reverse osmosis desalination can be used wherever sufficient electrical power is available. A pre-production 9,000 GPD unit was first installed on board USS *Fletcher* (DD-992) in 1988. Due to the positive outcome of this testing,

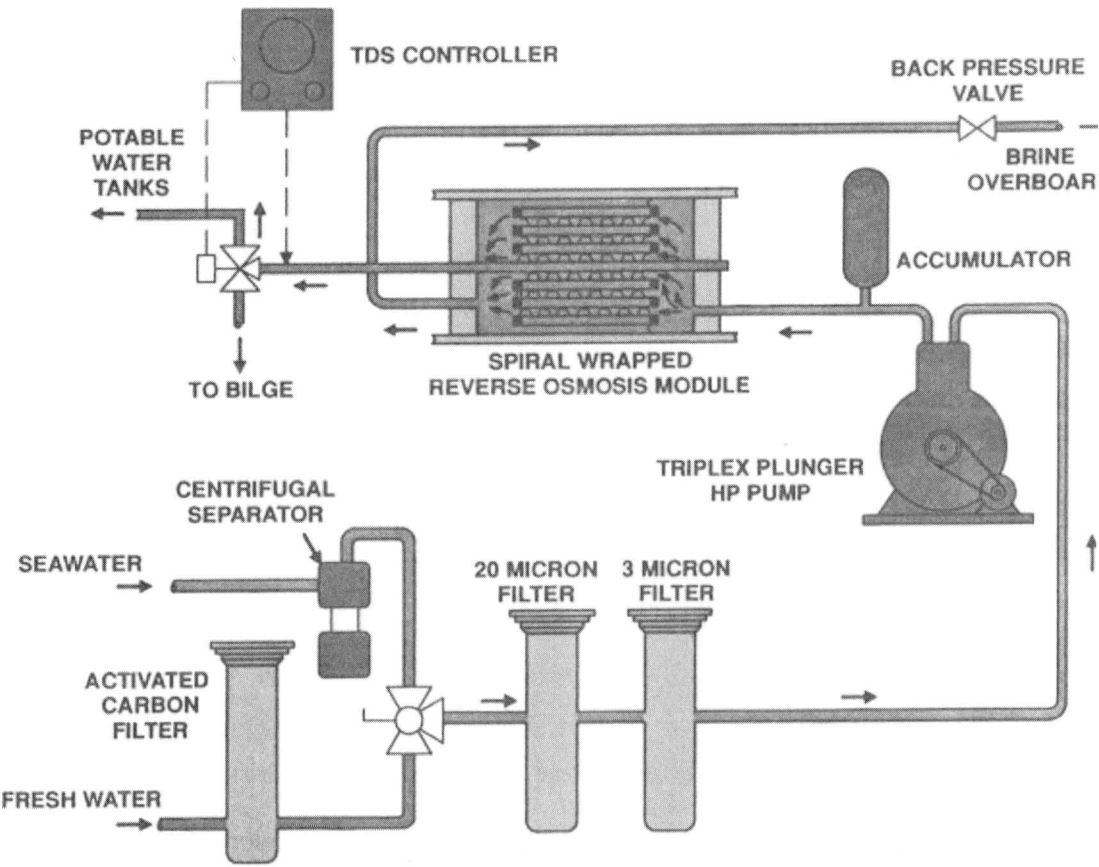

FIGURE 10.6: *Reverse Osmosis Desalination Plant Schematic*

gas turbine powered ships (FFG-7 class, DD-963 class, and DDG-51 class) were retrofit or built with reverse osmosis plants. Ballistic missile submarines currently utilize RO desalination plants of varying capacities. Naval Sea Systems Command (NAVSEA) is evaluating applications with rated capacities up to 100,000 GPD on board several (larger) operational vessels.

10.5 CONCLUSIONS

The treatment best suited, or economically justified, for any given vessel depends upon the characteristics of the water supply, the amount of makeup water needed by equipment and crew, and the availability of electricity, steam, or waste heat from other processes. While rated capacity has an obvious impact on desalination equipment selection, the most important requirements are effectiveness and reliability.

Distilling plants installed in naval ships are of three general types: (1) vapor compression, (2) low-pressure steam, and (3) heat recovery. The major differences between the three types are the kinds of energy used to operate the units and the pressure under which distillation takes place. Vapor compression units use electrical energy for heaters and a compressor and boil the seawater feed slightly above atmospheric pressure. Although used on submarines, the vapor compression type of distilling plant has all but been replaced on surface craft by the heat recovery distilling units. Low-pressure steam units use auxiliary or auxiliary exhaust steam as the heat source and vaporize the feedwater at less than atmospheric pressure. Heat recovery units use water from main propulsion or auxiliary diesel cooling jackets as a heat source and also operate with a relatively high vacuum.

The reverse osmosis units, rather than heating the incoming seawater, force the seawater through membranes at high pressure. The membranes permit passage of purified water while entraining and discharging dissolved salts with the brine. The RO systems are being used more extensively through the Fleet and Marine Corps.

REFERENCES

Avallone, Eugene A. and Theodore Baumeister III. *Marks' Standard Handbook for Mechanical Engineers* (9th Edition). New York: McGraw-Hill, 1987. (Ch. 6 192-196, Ch. 9 28-36)

NAVEDTRA 14151: Machinist Mate 3 & 2 (Surface). Non-resident Training Course Manual, Chapter 9 "Steam Operated Distilling Plants."

Naval Ships Technical Manuals, Chapter 531 (Vol. 1-3).

Practice Problems

10.1 Explain the operation of and list the major types of desalination systems.

10.2 Describe the reverse osmosis process.

Equation Summary

USEFUL CONSTANTS

$g_c = 32.2 \dfrac{\text{ft-lb}_m}{\text{lb}_f\text{-s}^2}$	$\dfrac{g}{g_c} = \dfrac{\text{lb}_f}{\text{lb}_m}$	$g = 32.2 \dfrac{\text{ft}}{\text{s}^2}$	$R_u = 1545 \dfrac{\text{ft-lb}_f}{\text{lbmole-R}^\circ}$

AIR PROPERTIES
(at Standard Temperature Conditions)

$R_{air} = 53.3 \dfrac{\text{ft-lb}_f}{\text{lb}_m\text{-R}^\circ}$	$c_{p,air} = 0.24 \dfrac{\text{BTU}}{\text{lb}_m\text{-R}^\circ}$	$c_{v,air} = 0.171 \dfrac{\text{BTU}}{\text{lb}_m\text{-R}^\circ}$	$k_{air} = \dfrac{c_{p,air}}{c_{v,air}} = 1.4$

WATER PROPERTIES
(at Standard Temperature Conditions)

$\rho_{fw} = 62.4 \dfrac{\text{lb}_m}{\text{ft}^3}$	$c_{p,fw} = 1.0 \dfrac{\text{BTU}}{\text{lb}_m\text{-R}^\circ}$	$\rho_{sw} = 64.0 \dfrac{\text{lb}_m}{\text{ft}^3}$	$c_{p,sw} = 0.94 \dfrac{\text{BTU}}{\text{lb}_m\text{-R}^\circ}$

UNIT CONVERSION FACTORS

$^\circ R = ^\circ F + 460^\circ$		$1\,\text{atm} = 14.7\,\text{psia} = 29.92\,"\text{Hg absolute}$	
$1\,\text{mil} = 0.001\,\text{in}$	$1\,\text{ft}^2 = 144\,\text{in}^2$	$1\,\text{ft}^3 = 7.48\,\text{gal}$	$1\,\text{nm} = 6080\,\text{ft}$
$1\,\text{hp} = 0.707 \dfrac{\text{BTU}}{\text{s}}$	$1\,\text{hp} = 42.42 \dfrac{\text{BTU}}{\text{min}}$	$1\,\text{hp} = 2545 \dfrac{\text{BTU}}{\text{hr}}$	$1\,\text{hp} = 550 \dfrac{\text{ft} \cdot \text{lb}_f}{\text{s}}$
$1\,\text{BTU} = 778\,\text{ft} \cdot \text{lb}_f$	$1\,\text{hp} = 0.746\,\text{kW}$	$1\,\text{MW} = 3.41 \times 10^6 \dfrac{\text{BTU}}{\text{hr}}$	$1\,\text{hp} = 33{,}000 \dfrac{\text{ft} \cdot \text{lb}_f}{\text{min}}$
$1\,\text{ref ton} = 200 \dfrac{\text{BTU}}{\text{min}} = 12000 \dfrac{\text{BTU}}{\text{hr}}$		$1\,\text{long ton} = 2200\,\text{lb}_f$	$1\,\text{short ton} = 2000\,\text{lb}_f$

See other conversion factors in the *Steam Tables*.

GENERAL THERMODYNAMIC EQUATIONS

Steady Flow Energy Equation (SFEE) i=initial f=final	$PE_i + KE_i + U_i + p_i V_i + Q_{if} = PE_f + KE_f + U_f + p_f V_f + W_{if}$ $pe_i + ke_i + u_i + p_i v_i + q_{if} = pe_f + ke_f + u_f + p_f v_f + w_{if}$ $\left(\dfrac{g}{g_c}\right) z_i + \left(\dfrac{1}{2g_c}\right)\vec{v}_i^2 + u_i + p_i v_i + q_{if} = \left(\dfrac{g}{g_c}\right) z_f + \left(\dfrac{1}{2g_c}\right)\vec{v}_f^2 + u_f + p_f v_f + w_{if}$		
Non-Flow Energy Equation	$u_i + q_{if} = u_f + w_{if}$ $U_i + Q_{if} = U_f + W_{if}$	Enthalpy	$h = u + pv$ $H = U + pV$
Internal Energy (no phase change)	$\Delta u = c_v \Delta T$ $\Delta U = m c_v \Delta T$	Enthalpy (no phase change)	$\Delta h = c_p \Delta T$ $\Delta H = m c_p \Delta T$
Reversible Work	$w_{if} = \int_i^f p \cdot dv$	Reversible Heat	$q_{if} = \int_i^f T \cdot ds$
Mechanical Power	$IHP = \dot{W} = \dot{m} \cdot w$	Thermal Power	$\dot{Q} = \dot{m} \cdot q$
Specific Gravity (ref fluid is std water for liquids; std air for gases)	$s.g._{fluid} = \dfrac{\rho_{fluid}}{\rho_{reference\,fluid}}$	Density and Specific Volume	$\rho = \dfrac{1}{v}$
Specific Weight	$\gamma = \rho \dfrac{g}{g_c}$	Pressure	$p = \dfrac{\bar{F}}{A}$
Hydrostatic Pressure	$\Delta p = \gamma \cdot \Delta z$	Gage Pressure	$p_{abs} = p_{atm} \pm p_{gage}$
Continuity	$\dot{V} = \vec{v} A$ $\dot{m} = \rho \dot{V} = \rho \vec{v} A$ $\sum \dot{m}_{in} = \sum \dot{m}_{out}$	Vacuum Pressure	$p_{abs} = p_{atm} - p_{vac}$
Moles and Molecular Weight	$n = \dfrac{m}{mw}$	Ideal Gas Constant	$R = \dfrac{R_u}{mw}$

FLUID FLOW EQUATIONS

Pump Head Eqn [ft units]	$w_{p,s} \dfrac{g_c}{g} = (z_f - z_i) + \left(\dfrac{\vec{v}_f^2 - \vec{v}_i^2}{2g}\right) + \dfrac{p_f - p_i}{\gamma} + H_{L,if}$ Note: In this equation, pump work is *positive* into the fluid.	Pump Affinity Laws	$\dot{V} \propto N$ $\Delta p \propto N^2$ $\dot{W} \propto N^3$
Total Head Loss [ft units]	$H_{L,if} = \left(\dfrac{fl}{d} + \sum k\right) \dfrac{\vec{v}_{pipe}^2}{2g}$	Pipe Cross-Sectional Area	$A = \dfrac{\pi d^2}{4}$
Bernoulli's Eqn w/ Head Loss [ft units]	$z_i + \dfrac{\vec{v}_i^2}{2g} + \dfrac{p_i}{\gamma} = z_f + \dfrac{\vec{v}_f^2}{2g} + \dfrac{p_f}{\gamma} + \left(\dfrac{g_c}{g}\right) w_{if,s} + H_{L,if}$	Pump Efficiency	$\eta_{pump} = \dfrac{WHP}{BHP}$ $\eta_{pump} = \dfrac{w_{p,s}}{w_{p,real}}$

FLUID FLOW EQUATIONS—CONTINUED

Isentropic Incompressible Pump Work	$w_{p,s} = v\Delta p_{pump}$ $w_{p,s} = \left|w_{in,out\,s}\right| = \left|h_{in} - h_{out,s}\right|$	Water Horsepower	$WHP = \dot{m}\, w_{p,s}$ $WHP = \dot{V}\, \Delta p_{pump}$

VARIOUS ENGINE EQUATIONS

Efficiency Concept	$\eta = \dfrac{\text{desired output}}{\text{required input}}$	Thermal Efficiency	$\eta_{th} = \dfrac{w_{net}}{q_s} = \dfrac{\Sigma w}{q_s} = \dfrac{w_{by} - \left|w_{on}\right|}{q_s}$ $\eta_{th} = \dfrac{q_{net}}{q_s} = \dfrac{\Sigma q}{q_s} = \dfrac{q_s - \left|q_r\right|}{q_s}$
Combustion Engine Overall Efficiency	$\eta_{OA} = \dfrac{BHP}{\dot{m}_{fuel} \cdot HV}$ $\eta_{OA} = \eta_{comb}\,\eta_{th}\,\eta_{mech}$	Boiler Efficiency	$\eta_{boiler} = \dfrac{\dot{m}_{stm} \cdot q_s}{\dot{m}_{fuel} \cdot HV}$
Combustion Efficiency	$\eta_{comb} = \eta_{cc} = \dfrac{\dot{m}_{air} \cdot q_s}{\dot{m}_{fuel} \cdot HV}$	Compressor Efficiency	$\eta_{comp} = \dfrac{w_{comp,isen}}{w_{comp,real}} = \dfrac{h_{in} - h_{out,s}}{h_{in} - h_{out}}$
Turbine Efficiency	$\eta_{turbine} = \dfrac{w_{turbine,real}}{w_{turbine,isen}} = \dfrac{h_{in} - h_{out}}{h_{in} - h_{out,s}}$	Back Work Ratio	$BWR = \left|\dfrac{w_{in}}{w_{out}}\right|$
AC Motor Synchronous Speed [units]	$N[RPM]\,P[poles] = 120\,f[Hz]$	Engine Mechanical Efficiency	$\eta_{mech} = \dfrac{BHP}{IHP} = \dfrac{BHP}{\dot{m} \cdot w_{engine}}$
Jet Thrust	$\vec{F}_{thrust} = \dfrac{\dot{m}\Delta \vec{v}}{g_c}$	Jet Power	$\dot{W} = \vec{F} \cdot \vec{v}_{aircraft}$
Air-to-Fuel Ratio	$AFR = \dfrac{\dot{m}_{air}}{\dot{m}_{fuel}}$	Air Rate	$AR = \dfrac{\dot{m}_{air}}{BHP}$
Brake Specific Fuel Consumption	$bsfc = \dfrac{\dot{m}_{fuel}}{BHP}$	Brake Horsepower	$BHP = 2\pi\tau N$
Recip Engine Compression Ratio	$r_v = \dfrac{v_{bdc}}{v_{tdc}} = \dfrac{v_{start\,of\,compression}}{v_{end\,of\,compression}}$	Diesel Fuel Cutoff Ratio	$\beta = \dfrac{v_{fuel\,cutout}}{v_{start\,of\,fuel\,injection}}$
Compression Ratio	$CR = r_v = \dfrac{v_i}{v_f} = \dfrac{V_i}{V_f}$	Pressure Ratio	$PR = r_p = \dfrac{p_{compressor\,discharge}}{p_{compressor\,inlet}}$
Saturated Steam Properties	$v_x = v_f + x \cdot v_{fg}$ $h_x = h_f + x \cdot h_{fg}$ $s_x = s_f + x \cdot s_{fg}$	Relative Pressure ($\Delta s = 0$)	$\left.\dfrac{\Pr_f}{\Pr_i}\right|_s = \left.\dfrac{p_f}{p_i}\right|_s$

HEAT TRANSFER EQUATIONS

Refrigeration Capacity	$CAP = \dot{Q}_S = \dot{m} \cdot q_s = \dot{m} \cdot RE$	Coefficient of Performance (Refrigeration)	$COP_r = \dfrac{RE}{w_{comp}} = \dfrac{\dot{Q}_S}{\dot{W}_{comp}}$
Heat Transfer for Single Fluid in HX	$\dot{Q} = \dot{m}\lvert h_{out} - h_{in}\rvert = \dot{m}\cdot c_p \lvert T_{out} - T_{in}\rvert$	Heat Transfer Rate Between HX Fluids	$\dot{Q} = UA\theta_m$
Logarithmic Mean Temp Difference (*LMTD*)	$\theta_m = \dfrac{\theta_1 - \theta_2}{\ln\left(\dfrac{\theta_1}{\theta_2}\right)}$	Mixed Flow Heat Balance	$\sum \dot{E}_{in} = \sum \dot{E}_{out}$ $\sum (\dot{m}h)_{in} = \sum (\dot{m}h)_{out}$
Conductive Heat Transfer	$\dot{Q} = \dfrac{kA \cdot \Delta T}{L}$	Convective Heat Transfer	$\dot{Q} = hA(T_{surf} - T_{bulk})$
Overall Heat Transfer Coefficient for Combined Conduction and Convection	$U = \dfrac{1}{\sum \dfrac{1}{h} + \sum \dfrac{L}{k}}$	Radiative Heat Transfer	$\dot{Q} = \varepsilon \sigma A(T_H^4 - T_C^4)$

IDEAL GAS LAW AND RELATED EQUATIONS

Ideal Gas Law Variations	$pv = RT$ $pV = mRT$ $pV = nR_u T$	Constant Specific Heat Relations	$c_p = c_v + R$ $k = \dfrac{c_p}{c_v}$

THERMODYNAMIC RELATIONS FOR AN IDEAL GAS
(Assuming Constant Specific Heat Capacity)

	Polytropic	Isometric	Isobaric	Isentropic	Isothermal
Polytropic exponent	n	∞	0	k	1
p-v Relations	$p_i v_i^n = p_f v_f^n$	$p_i v_i^\infty = p_f v_f^\infty$	$p_i v_i^0 = p_f v_f^0$	$p_i v_i^k = p_f v_f^k$	$p_i v_i^1 = p_f v_f^1$
p-v-T Relations	$\dfrac{T_f}{T_i} = \left(\dfrac{p_f}{p_i}\right)^{\frac{n-1}{n}}$ $\dfrac{T_f}{T_i} = \left(\dfrac{v_i}{v_f}\right)^{n-1}$	$\dfrac{T_f}{T_i} = \dfrac{p_f}{p_i}$	$\dfrac{T_f}{T_i} = \dfrac{v_f}{v_i}$	$\dfrac{T_f}{T_i} = \left(\dfrac{p_f}{p_i}\right)^{\frac{k-1}{k}}$ $\dfrac{T_f}{T_i} = \left(\dfrac{v_i}{v_f}\right)^{k-1}$	$\dfrac{p_f}{p_i} = \dfrac{v_i}{v_f}$
Δu	$c_v(T_f - T_i)$	$c_v(T_f - T_i)$	$c_v(T_f - T_i)$	$c_v(T_f - T_i)$	0
Δh	$c_p(T_f - T_i)$	$c_p(T_f - T_i)$	$c_p(T_f - T_i)$	$c_p(T_f - T_i)$	0
Δs	$\left(c_v + \dfrac{R}{1-n}\right)\ln\left(\dfrac{T_f}{T_i}\right)$	$c_v \ln\left(\dfrac{T_f}{T_i}\right)$	$c_p \ln\left(\dfrac{T_f}{T_i}\right)$	0	$\dfrac{q_{if}}{T} = R\ln\left(\dfrac{p_i}{p_f}\right)$

IDEAL GAS – HEAT AND WORK RELATIONS
(Assuming Closed System and Constant Specific Heat Capacity)

w_{if}	$\dfrac{p_f v_f - p_i v_i}{1-n}$ $\dfrac{R(T_f - T_i)}{1-n}$	0	$p_f v_f - p_i v_i$ $R(T_f - T_i)$	$c_v(T_i - T_f)$ $\dfrac{p_f v_f - p_i v_i}{1-k}$ $\dfrac{R(T_f - T_i)}{1-k}$	$p_i v_i \ln\left(\dfrac{v_f}{v_i}\right)$ $RT \ln\left(\dfrac{p_i}{p_f}\right)$
q_{if}	$\left(c_v + \dfrac{R}{1-n}\right)(T_f - T_i)$	$c_v(T_f - T_i)$	$c_p(T_f - T_i)$	0	$p_i v_i \ln\left(\dfrac{v_f}{v_i}\right)$ $RT \ln\left(\dfrac{p_i}{p_f}\right)$

Nomenclature

VARIABLE NOMENCLATURE (with Typical Units)

Quantity	Symbol	Units	Quantity	Symbol	Units
Specific Volume	v	$\frac{ft^3}{lb_m}$	Total Volume	V	gal, ft^3
Volumetric Flow Rate	$\dot{V}$	$GPM, \frac{ft^3}{min}$	Velocity	$\bar{v}$	$\frac{ft}{s}$
Mass	m	lb_m	Mass Flow Rate	$\dot{m}$	$\frac{lb_m}{min}$
Specific Enthalpy	h	$\frac{BTU}{lb_m}$	Total Enthalpy	H	BTU
Specific Internal Energy	u	$\frac{BTU}{lb_m}$	Total Internal Energy	U	BTU
Specific Entropy	s	$\frac{BTU}{lb_m \cdot R°}$	Total Entropy	S	$\frac{BTU}{R°}$
Work (in Specific Terms)	w	$\frac{ft \cdot lb_f}{lb_m}, \frac{BTU}{lb_m}$	Total Work	W	$ft \cdot lb_f, BTU$
Heat (in Specific Terms)	q	$\frac{BTU}{lb_m}$	Total Heat	Q	BTU
Mechanical Power	$\dot{W}$ P	$hp, \frac{BTU}{hr}$	Thermal Power	$\dot{Q}$	$\frac{BTU}{hr}$
Length	l L	ft	Diameter	d D	ft
Area	A	ft^2	Temperature	T	$°R, °F$
Gravitational Acceleration	g	$\frac{ft}{s^2}$	Gravitational Constant	g_c	$\frac{ft \cdot lb_m}{lb_f \cdot s^2}$
Density	ρ	$\frac{lb_m}{ft^3}$	Specific Weight	γ	$\frac{lb_f}{ft^3}$
Pressure	p	$\frac{lb_f}{in^2}, psi$	Specific Gravity	$s.g.$	dimension-less

VARIABLE NOMENCLATURE (with Typical Units)—CONTINUED

Weight	w	lb_f	Head Loss	H_L	ft
Elevation	z, h	ft	Cutoff Ratio	β	dimension-less
Rotational speed	N	RPM	Pressure Ratio	R, r_p	dimension-less
Compression Ratio	CR, r_v	dimension-less	Brake Horsepower	BHP	hp
Internal or Indicated Horsepower	IHP	hp	Shaft Horsepower	SHP	hp
Water Horsepower	WHP	hp	Torque	τ	$ft \cdot lb_f$
Specific Heat (Isobaric)	c_p	$\dfrac{BTU}{lb_m \cdot R°}$	Specific Heat (Isometric)	c_v	$\dfrac{BTU}{lb_m \cdot R°}$
Gas Constant (unique to each gas or gas mixture)	R	$\dfrac{ft \cdot lb_f}{lb_m \cdot R°}$	Universal Gas Constant	R_u	$\dfrac{ft \cdot lb_f}{lbmole \cdot R°}$
Thermal Conductivity	k	$\dfrac{BTU}{hr \cdot ft \cdot R°}$	Overall Heat Transfer Coefficient	U	$\dfrac{BTU}{hr \cdot ft^2 \cdot R°}$
Minor Head Loss Coefficient; Fitting Friction Factor	k	dimension-less	Convective Heat Transfer Coefficient	h	$\dfrac{BTU}{hr \cdot ft^2 \cdot F°}$
Pipe Flow Friction Factor	f	dimension-less	Steam Quality	x	%
Kinematic Viscosity	υ	$\dfrac{ft^2}{s}$	Dynamic Viscosity	μ	$\dfrac{lb_f \cdot s}{ft^3}$
Pipe Surface Roughness	ε	mil, in	Reynolds Number	N_{RE}	dimension-less
emissivity (radiation)	ε	unit-less	Boltzmann Constant	σ	$\dfrac{BTU}{hr \cdot ft^2 \cdot °R^4}$
Refrigerant Effect	RE	$\dfrac{BTU}{lb_m}$	Coefficient of Performance	COP	dimension-less
Refrigeration Capacity	$\dot{Q}_s$, CAP	$\dfrac{BTU}{hr}$, ref ton	Molecular "Weight"; Molecular Mass	mw	$\dfrac{lb_m}{lbmole}$
Efficiency	η	%	Log Mean Temp Difference (LMTD)	θ_m	F°, R°
Force	$\vec{F}$	lb_f	Polytropic Constant	n	dimension-less
Lower Heating Value	LHV	$\dfrac{BTU}{lb_m}, \dfrac{BTU}{gal}$	Higher Heating Value	HHV	$\dfrac{BTU}{lb_m}, \dfrac{BTU}{gal}$

Appendix
Answers to Practice Problems

1 FUNDAMENTALS OF THERMODYNAMICS

1.1   {a) 2; b) 0.5; c) 0.5}
1.2   13.4
1.3   {annotated graph}
1.4   {a) graph; b) 4.2; c) 8.5}
1.5   {a) graphs; b) 489.2; c) 1.75}
1.6   {17.44}
1.7   {28.0; 13.3}
1.8   {210}
1.9   {3.73}
1.10 {54 added}
1.11 {1125.3}
1.12 {7144}
1.13 {5.10}
1.14 {30.3}
1.15 {a) Hint: The definition is analogous to the "calorie"; b) Hint: This comes from the ice-making industry; c) 3413 Btu/kW-hr; d) 778 ft-lb_f/Btu}
1.16 {a) 342.86 b) 1142.9}
1.17 {sketch; 14.3%; 2.28E8}
1.18 {sketches; 54280 hp; 225 rpm; 1.267E6 ft-lb_f}
1.19 {315,120; 4250; 16,304; 10,074; 3.32E6 Btu/hr}
1.20 { a) 775; b) 525; c) 250; d) 20; e) 32.3}
1.21 {3010.3 psig; 2555.3 psig; assumes std atmospheric pressure and 70 °F at sea level.}
1.22 {6.0 +/- 0.02 lb_m; 676 +/- 2}
1.23 {148 psia; 47 lb_m; 41 lb_m} Answers assume 70 °F seawater and ignore the weight of air in calculating buoyancy.
1.24 {4719; +4.28 Btu/lb_m; isometric heat addition}
1.25 {23.16; graph}
1.26 {6.9%}

2 INCOMPRESSIBLE FLUID FLOW

2.1   {980.2 lb_m/s; 5 ft/s}
2.2   {28.4}
2.3   {0.55 psid; 5020 lb_f; 489.5 lb_m/ft^3; sketches}
2.4   {0.0197}
2.5   {250.3 lb_m/sec; 11.5 ft/sec; 25.3}
2.6   {68%; sketches; derivations}
2.7   {5.82}
2.8   {written answer}
2.9   {125; 1.44}
2.10 {1 graph with 4 curves and 8 terminal values}
2.11 {2 graphs}
2.12 {graph with three curves and 6 values shown}
2.13 {1.114; 0.349; 3.19; 55.7; 0.158; 16.55; 295.35; 29.9; 49.8; 102.6}
2.14 {IS adequate: a) 6 bhp, b) 8.3 bhp, c) 8.6 bhp}
2.15 {10.0, 150., 34.8}
2.16 {2503; 5.54; 7.5; 74%; ; ~75.2%}
2.17 {1715 kW; 1355 kW, yes}
2.18 { a) 6.1 hp; b) 8 hp; c) 750 gpm; 112 psid}
2.19 {0.55 psid; 132.6 psia; $7.44; 59.8 HP; 250 gpm, 25.6 psid}
2.20 {derivation}
2.21 {152.4}
2.22 {2.93}
2.23 {15 ft-lb_f}

3 HYDRAULIC AND PNEUMATIC SYSTEMS

3.1 {182.1; written answer}
3.2 {1571 lb_f toward the right}
3.3 {14,400 lb_f; 69 mils}
3.4 {497 psi}
3.5 {sketch; 2749 lb_f; discussion}
3.6 {Sketch and logic}
3.7 {92,900 lb_f; 40,800 lb_f}
3.8 {Logic}
3.9 {739 psig; 8011 in^3}

4 INTERNAL COMBUSTION ENGINES

4.1 {a) graphs; b) $q_{12} = 0$, $w_{12} = -48,700$; c) $q_{23} = -38.5$, $w_{23} = 0$}
4.2 {graphs}
4.3 {7.5:1; -112.3; 199.9; 222.9; -89.3; 110.6; 55.5}
4.4 {8.5; 329; -139.8; 57.5; 189.3; 310.8}
4.5 {table; 467.4; 258.6}

	1	2	3	4
p (psia)	12	201.5	665.7	39.6
T (°R)	530	1187	3920	1751
v (ft³/lb_m)	16.35	2.18	2.18	16.35
r_v	7.5	1	0.133	1
process	$\bar{s}$	$\bar{v}$	$\bar{s}$	$\bar{v}$
q (Btu/lb_m)	X	467.4	X	-208.8
w (Btu/lb_m)	-112.3	X	370.9	X

4.6 {table; 8.6; 57.7}

	1	2	3	4
p (psia)	14.7	300.0	679.0	33.4
T (°R)	535	1265	2873	1215
v (ft³/lb_m)	13.47	1.57	1.57	13.47
r_v	8.6	1	0.116	1
Process	Isentropic	Isometric	Isentropic	Isometric
q (Btu/lb_m)	0	275.0	0	-116.3
w (Btu/lb_m)	-125.0	0	283.6	0

4.7 {table; 9.0; 337 Btu/lb_m; 58.8%; 418 hp}
4.8 {5.2; 520, 1012.2, 2960, 1520.6; 147.8; 48.6}
4.9 {graphs}
4.10 {16; -177.2; 273.4; 348.2; -101.7; 171.7; 62.6; 1.74}
4.11 {16; 249; -91.3; 63.2; 157.2; 342.7; 1.64}

	1	2	3	4
p (psia)	14.7	713.0	713.0	29.4
T (°R)	535	1622	2663	1071
v (ft³/lb_m)	13.47	0.84	1.38	13.47
r_v	16	β =1.64	0.103	1
process	Isentropic	Isobaric	Isentropic	Isometric
q (Btu/lb_m)	0	+250	0	-91.7
w (Btu/lb_m)	-186.0	+71.4	+272.2	0

4.12 {table; 346.2; 98.8; 215; 62; 1.85}

	1	2	3	4
p (psia)	17	824.5	824.5	40.2
T (°R)	**560**	1698	**3140**	1325
v (ft³/lb$_m$)	12.19	0.76	1.41	12.19
r_v	16	$\beta = 1.85$	0.116	1
process	Isentropic	Isobaric	Isentropic	Isometric
q (Btu/lb$_m$)	0	+346.2	0	-130.8
w (Btu/lb$_m$)	-194.5	+98.8	+310.4	0

4.13 {2547; 665; 101.7; 198; 66; 140; 89}
4.14 {15; -180.4; 142.8; 331.6; 500.2; 205.2}
4.15 {2805; 242.8 Btu/lb$_m$; 62.5%}
4.16 {321; 360; 5.625; 50.625}
4.17 {8.11E-4; 1.16E-2; 14.3; 14.75; 1272}

5 GAS TURBINES

5.1 {Table Practice}
5.2 {a) 150, b) 15, c) 151, d) 28,833, e) 46}
5.3 {a) 255, b) 15, c) 153.2, d) 29,253, e) 53.3}
5.4 {a) -102.9, b) 248.3, c) 145.4, d) 42.8, e) 41.4}
5.5 {a) -128.6, b) 233.0, c) 104.4, d) 30.7, e) 55.2}
5.6 {a) Sketch, b) Table, c) 7.5, d) 89.9, e) 28.2}
5.7 {a) -175.3, b) 175.3, c) 513.0, d) 148.4}
5.8 {a) Table, b) 96.5, c) 3.11, d) 0.93}
5.9 {a) Sketches/Table, b) 454.4, c) 160.1, d) 64.4, e) 34.9, f) 1.097 lb$_m$/hp-hr, g) 14.2}
5.10 {a) 17.99, b) 207.4, c) 4.53, d) 0.129, e) 0.137, f) 986.4, g) -}
5.11 {a) Diagram, b) 7, c) -96.1, d) - , e) 2510}
5.12 {a) 3, b) 270.5, c) 14.9, d) - , e) 15.34, f) -}
5.13 {a 2,327.3 b) 17,710.6 c) 20,930.7}
5.14 {676; 14.9:1; lean; 200 lb$_m$/sec; 12,700 lb$_f$; 15,600 hp}

6 STEAM POWER

6.1 Givens are listed in **BOLD**. Undefined cells have a dash in them. Item 7 is listed directly in the © 2000 *Steam Tables* and can be interpolated from the ©1967 tables

	p (psia)	T (°F)	v (ft³/lb$_m$)	h (Btu/lb$_m$)	s (Btu/lb$_m$-°R)	x (%)	°Sh	State
1	0.95	**100**	0.01613	68.0	0.1296	0	-	SL
2	80	312.03	4.381	1003.0	1.387	**80**	-	WV
3	150	358.4	1.4927	**755.6**	1.0336	49.2	-	WV
4	**600**	**1000**	1.41	1518	1.716	-	513.75	SHV
5	**150**	**340**	0.01787	311.3	0.4902	-	-	SCL
6	30	250.3	**13.722**	1162.4	1.697	99.8	-	WV
7	**50**	**320**	9.0379	1194.9	1.6860	-	39	SHV
8	**40**	**110**	0.016165	78	0.147	-	-	SCL
9	67.0	300	1.3072	451.7	0.6767	**20**	-	WV
10	80	312.0	5.4711	1183.1	1.621	100	-	**SV**
11	**1000**	**1000**	0.83	1506	1.653	-	455.4	SHV
12	**80**	**700**	8.56	**1380.5**	1.8289	-	388	SHV
13	11.53	**200**	33.3	1136.5	**1.762**	99.0	-	WV
14	**50**	**500**	11.306	1284.1	1.789	-	219	SHV
15	**20**	227.9	20.09	**1156.3**	1.732	100	-	SV

6.2 Givens listed in **BOLD**. Undefined cells have a dash in them.

	p (psia)	T (°F)	h (Btu/lb$_m$)	s (Btu/lb$_m$°R)	m (%)	x (%)	°Sh	State
1	4	155	1052.5	**1.74**	7.4	92.6	-	WV
2	80	312	1020	1.41	**18**	82	-	WV
3	**100**	**600**	1328	1.76	-	-	273	SH
4	70	**400**	1233	1.695	-	-	100	SH
5	35	660	1362	**1.90**	-	-	400	SH
6	**600**	485	**1100**	1.37	14	86	-	WV
7	**300**	**550**	1287	**1.60**	-	-	135	SH
8	245	**400**	1118	1.43	10	90	-	WV
9	14.6	212	**1150**	1.757	0	100	-	SV
9 (alt)	1800	620	1150	1.305	**0**	100	-	SV
10	2	126	952	**1.64**	16	84	-	WV
11	**300**	422	1203	1.51	0	100	-	**SV**
12	133	547	**1300**	1.70	-	-	200	SH

6.3 {152 °F}
6.4 {0.01674; 26.804; 1603.7 }
6.5 {graphs}
6.6 {h @ SP (Btu/lb$_m$) 1 = 93.992, 2 = 95.788, 3 = 1407.9, 4 = 949.0}

	1	2s	3	4s
p (psia)	= 2	600 =	**600**	2
T (°F)	126	127.8	**800**	126
h (Btu/lb$_m$)	93.992	95.788	1407.9	949.0
s (Btu/lb$_m$-°R)	NR	NR	1.6348	= 1.6348
v (ft³/lb$_m$)	0.01623	= 0.01623	NR	NR
x (%)	NA	NA	NA	83.7
State	SL*	CL	SHV	WV

* Assumed based upon lack of other information.

6.7 {a) Yes, 4.2°F, b) No, c) h@ SP(Btu/lb$_m$) 1 = 125.96, 2 = 127.8, 3 = 1203.7, 4 = 883.9}

	1	2s	3	4s
p (psia)	= 5	600 =	**600**	5
T (°F)	**158**		486.2	162.24
h (Btu/lb$_m$)	124.96			883.9
s (Btu/lb$_m$-°R)	NR	NR	1.4461	=1.4461
v (ft³/lb$_m$)	0.016384			

6.8 {a) $w_{p,s}$ = 3.0 Btu/lb$_m$, $w_{p,R}$ = 3.55 Btu/lb$_m$ b) $w_{t,s}$ = 565 Btu/lb$_m$}

	1	2s	2	3	4s
p (psia)	= 1	1000	<=1000	<=1000	1
T (°F)	101.74			**950**	101.74
h (Btu/lb$_m$)	69.73		73.28	1476.9	
s (Btu/lb$_m$-°R)				1.6328	
v (ft³/lb$_m$)	0.016136				
x (%)					81.29

6.9 {a} $\eta_{turb} = 80\%$}

	3	4s	4
p (psia)	**600**	**2**	= 2
T (°F)	486.25	126	126
h (Btu/lb$_m$)	1203.9		
s (Btu/lb$_m$-°R)	1.4464	1.4464	
x (%)	100		80

Per the 2000 *Steam Tables*.

6.10 {a) 2.4; b) 1440.0; c) 869.3; d) 39.6; e) 16.1}

	1	2s	3	4s
p (psia)	= 1	800 =	**800**	1
T (°F)	101.74	104.13	**1000**	101.74
h (Btu/lb$_m$)	69.73	72.1	1511.4	938.9
s (Btu/lb$_m$-°R)	NR	NR	1.6807	= 1.6807
v (ft³/lb$_m$)	0.016136	= 0.016136	NR	NR
x (%)	NA	NA	NA	83.9
State	SL*	CL	SHV	WV

* Assumed based upon lack of other information.

6.11 {a) 3.0; b) 1375.8; c) 476.7; d) 902.1; e) 34.4; f) 53,388 lb$_m$/hr}
6.12 {a) 71.3; b) 365 Btu/lb$_m$; c) 260.6 Btu/lb$_m$; d) statement}
6.13 {a) 3.8 b) 7° c) 486.2° d) real e) isentropic f) 111.95, 113.75, 114.2, 1203.7, 869.8, 903.2 g) 93.8}
6.14 {a) 1; b) 61; c) 335; d) 1.79; e) 2.39; f) 808 ; g) 1899}
6.15 {a) CD = 5°F; b) 61.69 lb$_m$/ft3, 4124 lb$_m$/min; c) 39.1%; d) 38 kW; e) 48.5 hp}

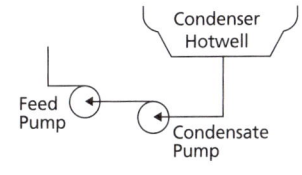

	1	2s	2
p (psia)	2	67	67
T (°F)	121		121.5
h (Btu/lb$_m$)	89.0		89.499
v (ft³/lb$_m$)	0.016209		

enthalpy and specific volume interpolated from Table 1

6.16 {a) 2; b) 437; c) 29,157; d) 4.4; e) 346; f) 233; g) 33.7; h) 1.31 E7 }

	1	2s	2	3	4s	4
p (psia)	= 2	1250	=1250	**1200**	2	26.08"Hg
T (°F)	T$_{SAT}$ – 4°F (122)			**800**		T$_{sat}$=126
h (Btu/lb$_m$)	90	93.7	94.4	1379	894	942.5
s (Btu/lb$_m$-°R)				1.5408		
v (ft³/lb$_m$)	0.016213					

6.17 {a) 16,129 hp; 3,753 hp; b) 25.85; c) 91,462; d) 22,823; e) 32.94%; f) 9,553; g) 63,619}

7 HEAT EXCHANGERS

7.1 {36.4; yes; 37}
7.2 {281,497}
7.3 {2084.6}
7.4 {1800}
7.5 {112}
7.6 {a) 3700; b) 152.7}
7.7 {a) schematic & graph; b) 1915; c) 17.34}
7.8 {a) 1.57×10^8; b) 8.34×10^6; c) 30.66; d) 8521; e) 5425}
7.9 {a) 58.4 Btu/ hr·ft^2·°F; b) 272 ft^2}
7.10 {a) 1.24E7 Btu/hr; b) 952.5 gpm; c) 40 °F; d) 141.1 Btu/hr·ft^2·°F}
7.11 {a) 33,000 lb$_m$/hr; b) 86.32 °F; c) 95.83 ft^2; d) 86.91 ft^2 increase in surface area}
7.12 {19.4 Btu/ft^2-hr-°F}
7.13 {a.) 1.274E5 lb$_m$/hr; 917 °F; b.) 1.419E5 lb$_m$/hr; c.) 486.2 °F; d.) 199,000 GPD}

8 NUCLEAR POWER

8.4 {194.4 HP; 24.3 HP; 70.7 lb$_m$/sec}
8.5 {354.5 Btu/lb$_m$; 23,400 hp; 1.69 Btu/lb$_m$; 31.0%}
8.6 {2210; 381; 1.51; 114,000; 3825; 29.4%}
8.7 {2177.3; 53.8%; 466.7; 1.85; 26,300; 30.2%; 195.1}
8.8 {23,700}
8.9 { a) 0.420, b) 129.5, c) 1.19, d) 16.2, e) 81.8}
8.10 { a) 628.6, b) verbal, c) 25, 6.25, d) 3.41E8, e) 680.53, f) 1202.2, g) 3.01E5}
8.11 { a.) 1838.3; b.) 28,300; c.) 2420; d.) 83.8%; e.) 145.4}

9 HEATING, VENTILATION, AIR CONDITIONING & REFRIGERATION

9.1 Psychrometric Chart Practice Answers

	φ (%)	T_{db} (°F)	T_{wb} (°F)	h (Btu/lb$_m$)	v (ft^3/lb$_m$)
1	**50**	**95.0**	79.1	42.4	14.38
2	**10**	58.6	**40.0**	15.2	13.09
3	**40**	97.2	76.8	**40.0**	14.38
4	**10**	55.5	38.1	14.3	**13.00**
5	56	**70.0**	**60.0**	26.3	13.54
6	40	**81.4**	64.6	**29.6**	13.84
7	38	**113.3**	88.4	53.3	**15.00**
8	64	81.0	**71.7**	35.4	**13.95**
9	7	97.3	60.1	**26.2**	**14.10**

"Givens" are shown in **BOLD**. These answers are provided for student feedback. Copying them for submitted work is cheating.

9.2 Psychrometric Chart Practice Answers

	T_{db} (°F)	T_{wb} (°F)	h (Btu/lb$_m$)	v (ft^3/lb$_m$)	ω (lb$_{m_m}$/lb$_{m_a}$)
1	**55**	**55**	23.2	13.175	0.0092
2	**50**	40	**15.4**	12.91	0.003
3	**95**	86	50.6	**14.55**	0.0252
4	**108**	85	49.2	14.79	**0.021**
5	80	**60**	26.2	**13.75**	0.0066
6	73	52	**21.2**	13.5	0.0034
7	99	**68.5**	32.6	14.27	**0.008**
8	90	75	**38.4**	14.2	**0.0152**

"Givens" are shown in **BOLD**. These answers are provided for student feedback. Copying them for submitted work is cheating.

9.3   R-134a *p-h* Graph Practice Answers

	p (psia)	*T* (°F)	*h* (Btu/lb$_m$)	*s* (Btu/lb$_m$-°F)	State
1	**40**	30 (29)	20	0.043 (0.0467)	CL
2	**200**	**260**	154 (154.7)	0.273 (0.2731)	SHV
3	**8**	**260**	160	0.34	SHV
4	**100**	80 (79)	71	**0.14**	WV
5	140 (139)	**100**	115 (116.3)	0.22 (0.219)	**SV**
6	**10**	**80**	120	0.272 (0.273)	SHV

"Givens" are shown **BOLD**. Copying them for submitted work is cheating. Values in parentheses were from tables for comparison with *p-h* graph results.

9.4   R-134a *p-h* Graph Practice Answers

	p (psia)	*T* (°F)	*h* (Btu/lb$_m$)	*s* (Btu/lb$_m$-°F)	State
1	**50**	**40**	25	0.055	SL
2	**100**	**260**	157	0.29	SHV
3	**8.5**	175	**140**	**0.31**	SHV
4	**8**	-37	25	**0.06**	SM
5	**100**	**80**	115	0.22	**SV**
6	**20**	82	**120**	0.26	SHV

"Givens" are shown **BOLD**. Copying them for submitted work is cheating.

9.5   {0.0749 vs 0.0740 - 1.2% error; 2.4 vs 2.5 – 4% error}
9.6   {a) 65.2 b) 17.5 c) 26.9 d) 5143}
9.7   {a) 5000 b) 3560 c) 1.4}
9.8   {a) 3000; b) 4060.4; c) 2.8}
9.9   {a) 16.9; b) 25; c) 26.5; d) 4500}
9.10 {53.20 +/- 0.04 lb$_m$/ft^3; 1%; 71.3 +/- 2 psig at 70 °F; 186 +/- 5 psia}
9.11 {Graphs}
9.12 {a) 374.25; b) 40.5, 22.8}
9.13 {a) 113.5 lb$_m$/min; b) 2645 Btu/min; c) 13.2 tons; d) 2.21x10^5; e) charts}
9.14 {a) 115, 139, 52; b) 2.63; c) 120; d) 94.5; e) graph}
9.15 {a); b) 3.25; c) 100; d) graph}
9.16 {a) 55, 107, 128; 125; b) 3.46E4 lb$_m$/hr; c) 2.53E6 BtuH; d) 262; e) graph; f) 52.14; g) 97; h) 2.48}

Appendix: Refrigerant 134a Data

Table 1
HFC-134a Saturation Properties—Temperature Table

TEMP. °F	PRESSURE psia	VOLUME ft³/lb		DENSITY lb/ft³		ENTHALPY Btu/lb			ENTROPY Btu/(lb)(°R)		TEMP. °F
		LIQUID v_f	VAPOR v_g	LIQUID $1/v_f$	VAPOR $1/v_g$	LIQUID h_f	LATENT h_{fg}	VAPOR h_g	LIQUID s_f	VAPOR s_g	
-30	9.851	0.0115	4.4366	87.31	0.2254	3.0	95.7	98.7	0.0070	0.2297	-30
-29	10.126	0.0115	4.3234	87.21	0.2313	3.3	95.5	98.8	0.0077	0.2296	-29
-28	10.407	0.0115	4.2141	87.11	0.2373	3.6	95.4	99.0	0.0084	0.2294	-28
-27	10.694	0.0115	4.1068	87.01	0.2435	3.9	95.2	99.1	0.0091	0.2292	-27
-26	10.987	0.0115	4.0032	86.91	0.2498	4.2	95.1	99.3	0.0098	0.2291	-26
-25	11.287	0.0115	3.9032	86.81	0.2562	4.5	94.9	99.4	0.0105	0.2289	-25
-24	11.594	0.0115	3.8066	86.71	0.2627	4.8	94.8	99.6	0.0112	0.2288	-24
-23	11.906	0.0115	3.7120	86.61	0.2694	5.1	94.6	99.7	0.0119	0.2286	-23
-22	12.226	0.0116	3.6206	86.51	0.2762	5.4	94.5	99.9	0.0126	0.2284	-22
-21	12.553	0.0116	3.5323	86.41	0.2831	5.7	94.3	100.0	0.0132	0.2283	-21
-20	12.885	0.0116	3.4471	86.30	0.2901	6.0	94.2	100.2	0.0139	0.2281	-20
-19	13.225	0.0116	3.3625	86.20	0.2974	6.3	94.0	100.3	0.0146	0.2280	-19
-18	13.572	0.0116	3.2819	86.10	0.3047	6.6	93.9	100.5	0.0153	0.2278	-18
-17	13.927	0.0116	3.2031	86.00	0.3122	6.9	93.7	100.6	0.0160	0.2277	-17
-16	14.289	0.0116	3.1270	85.89	0.3198	7.2	93.6	100.8	0.0167	0.2276	-16
-15	14.659	0.0117	3.0525	85.79	0.3276	7.5	93.4	100.9	0.0174	0.2274	-15
-14	15.035	0.0117	2.9806	85.69	0.3355	7.8	93.3	101.1	0.0180	0.2273	-14
-13	15.420	0.0117	2.9104	85.59	0.3436	8.1	93.1	101.2	0.0187	0.2271	-13
-12	15.812	0.0117	2.8417	85.48	0.3519	8.4	92.9	101.4	0.0194	0.2270	-12
-11	16.212	0.0117	2.7762	85.38	0.3602	8.7	92.8	101.5	0.0201	0.2269	-11
-10	16.620	0.0117	2.7115	85.28	0.3688	9.0	92.6	101.7	0.0208	0.2267	-10
-9	17.037	0.0117	2.6490	85.17	0.3775	9.3	92.5	101.8	0.0214	0.2266	-9
-8	17.461	0.0118	2.5880	85.07	0.3864	9.7	92.3	102.0	0.0221	0.2265	-8
-7	17.893	0.0118	2.5291	84.97	0.3954	10.0	92.1	102.1	0.0228	0.2264	-7
-6	18.334	0.0118	2.4716	84.86	0.4046	10.3	92.0	102.3	0.0235	0.2262	-6
-5	18.784	0.0118	2.4155	84.76	0.4140	10.6	91.8	102.4	0.0241	0.2261	-5
-4	19.242	0.0118	2.3613	84.65	0.4235	10.9	91.7	102.5	0.0248	0.2260	-4
-3	19.709	0.0118	2.3084	84.55	0.4332	11.2	91.5	102.7	0.0255	0.2259	-3
-2	20.184	0.0118	2.2568	84.44	0.4431	11.5	91.3	102.8	0.0262	0.2258	-2
-1	20.669	0.0119	2.2065	84.34	0.4532	11.8	91.2	103.0	0.0268	0.2256	-1
0	21.163	0.0119	2.1580	84.23	0.4634	12.1	91.0	103.1	0.0275	0.2255	0
1	21.666	0.0119	2.1106	84.13	0.4738	12.4	90.9	103.3	0.0282	0.2254	1
2	22.178	0.0119	2.0644	84.02	0.4844	12.7	90.7	103.4	0.0288	0.2253	2
3	22.700	0.0119	2.0194	83.91	0.4952	13.0	90.5	103.6	0.0295	0.2252	3
4	23.231	0.0119	1.9755	83.81	0.5062	13.4	90.4	103.7	0.0302	0.2251	4
5	23.772	0.0119	1.9327	83.70	0.5174	13.7	90.2	103.9	0.0308	0.2250	5
6	24.322	0.0120	1.8911	83.60	0.5288	14.0	90.0	104.0	0.0315	0.2249	6
7	24.883	0.0120	1.8508	83.49	0.5403	14.3	89.9	104.2	0.0322	0.2248	7
8	25.454	0.0120	1.8113	83.38	0.5521	14.6	89.7	104.3	0.0328	0.2247	8
9	26.034	0.0120	1.7727	83.27	0.5641	14.9	89.5	104.4	0.0335	0.2245	9
10	26.625	0.0120	1.7355	83.17	0.5762	15.2	89.4	104.6	0.0342	0.2244	10
11	27.227	0.0120	1.6989	83.06	0.5886	15.5	89.2	104.7	0.0348	0.2244	11
12	27.839	0.0121	1.6633	82.95	0.6012	15.9	89.0	104.9	0.0355	0.2243	12
13	28.462	0.0121	1.6287	82.84	0.6140	16.2	88.9	105.0	0.0362	0.2242	13
14	29.095	0.0121	1.5949	82.73	0.6270	16.5	88.7	105.2	0.0368	0.2241	14
15	29.739	0.0121	1.5620	82.63	0.6402	16.8	88.5	105.3	0.0375	0.2240	15
16	30.395	0.0121	1.5298	82.52	0.6537	17.1	88.3	105.5	0.0381	0.2239	16
17	31.061	0.0121	1.4984	82.41	0.6674	17.4	88.2	105.6	0.0388	0.2238	17
18	31.739	0.0122	1.4678	82.30	0.6813	17.7	88.0	105.7	0.0395	0.2237	18
19	32.428	0.0122	1.4380	82.19	0.6954	18.1	87.8	105.9	0.0401	0.2236	19
20	33.129	0.0122	1.4090	82.08	0.7097	18.4	87.7	106.0	0.0408	0.2235	20
21	33.841	0.0122	1.3806	81.97	0.7243	18.7	87.5	106.2	0.0414	0.2234	21
22	34.566	0.0122	1.3528	81.86	0.7392	19.0	87.3	106.3	0.0421	0.2233	22
23	35.302	0.0122	1.3259	81.75	0.7542	19.3	87.1	106.5	0.0427	0.2233	23
24	36.050	0.0122	1.2995	81.64	0.7695	19.6	87.0	106.6	0.0434	0.2232	24
25	36.810	0.0123	1.2737	81.52	0.7851	20.0	86.8	106.7	0.0440	0.2231	25
26	37.583	0.0123	1.2486	81.41	0.8009	20.3	86.6	106.9	0.0447	0.2230	26
27	38.368	0.0123	1.2240	81.30	0.8170	20.6	86.4	107.0	0.0453	0.2229	27
28	39.166	0.0123	1.2000	81.19	0.8333	20.9	86.2	107.2	0.0460	0.2229	28
29	39.977	0.0123	1.1766	81.08	0.8499	21.2	86.1	107.3	0.0467	0.2228	29

Reprinted by permission from DuPont

Table 1 (continued)
HFC-134a Saturation Properties—Temperature Table

TEMP. °F	PRESSURE psia	VOLUME ft³/lb		DENSITY lb/ft³		ENTHALPY Btu/lb			ENTROPY Btu/(lb)(°R)		TEMP. °F
		LIQUID v_f	VAPOR v_g	LIQUID $1/v_f$	VAPOR $1/v_g$	LIQUID h_f	LATENT h_{fg}	VAPOR h_g	LIQUID s_f	VAPOR s_g	
30	40.800	0.0124	1.1538	80.96	0.8667	21.6	85.9	107.4	0.0473	0.2227	30
31	41.636	0.0124	1.1315	80.85	0.8838	21.9	85.7	107.6	0.0480	0.2226	31
32	42.486	0.0124	1.1098	80.74	0.9011	22.2	85.5	107.7	0.0486	0.2226	32
33	43.349	0.0124	1.0884	80.62	0.9188	22.5	85.3	107.9	0.0492	0.2225	33
34	44.225	0.0124	1.0676	80.51	0.9367	22.8	85.2	108.0	0.0499	0.2224	34
35	45.115	0.0124	1.0472	80.40	0.9549	23.2	85.0	108.1	0.0505	0.2223	35
36	46.018	0.0125	1.0274	80.28	0.9733	23.5	84.8	108.3	0.0512	0.2223	36
37	46.935	0.0125	1.0080	80.17	0.9921	23.8	84.6	108.4	0.0518	0.2222	37
38	47.866	0.0125	0.9890	80.05	1.0111	24.1	84.4	108.6	0.0525	0.2221	38
39	48.812	0.0125	0.9705	79.94	1.0304	24.5	84.2	108.7	0.0531	0.2221	39
40	49.771	0.0125	0.9523	79.82	1.0501	24.8	84.1	108.8	0.0538	0.2220	40
41	50.745	0.0125	0.9346	79.70	1.0700	25.1	83.9	109.0	0.0544	0.2219	41
42	51.733	0.0126	0.9173	79.59	1.0902	25.4	83.7	109.1	0.0551	0.2219	42
43	52.736	0.0126	0.9003	79.47	1.1107	25.8	83.5	109.2	0.0557	0.2218	43
44	53.754	0.0126	0.8837	79.35	1.1316	26.1	83.3	109.4	0.0564	0.2217	44
45	54.787	0.0126	0.8675	79.24	1.1527	26.4	83.1	109.5	0.0570	0.2217	45
46	55.835	0.0126	0.8516	79.12	1.1742	26.7	82.9	109.7	0.0576	0.2216	46
47	56.898	0.0127	0.8361	79.00	1.1960	27.1	82.7	109.8	0.0583	0.2216	47
48	57.976	0.0127	0.8210	78.88	1.2181	27.4	82.5	109.9	0.0589	0.2215	48
49	59.070	0.0127	0.8061	78.76	1.2405	27.7	82.3	110.1	0.0596	0.2214	49
50	60.180	0.0127	0.7916	78.64	1.2633	28.0	82.1	110.2	0.0602	0.2214	50
51	61.305	0.0127	0.7774	78.53	1.2864	28.4	81.9	110.3	0.0608	0.2213	51
52	62.447	0.0128	0.7634	78.41	1.3099	28.7	81.8	110.5	0.0615	0.2213	52
53	63.604	0.0128	0.7498	78.29	1.3337	29.0	81.6	110.6	0.0621	0.2212	53
54	64.778	0.0128	0.7365	78.16	1.3578	29.4	81.4	110.7	0.0628	0.2211	54
55	65.963	0.0128	0.7234	78.04	1.3823	29.7	81.2	110.9	0.0634	0.2211	55
56	67.170	0.0128	0.7106	77.92	1.4072	30.0	81.0	111.0	0.0640	0.2210	56
57	68.394	0.0129	0.6981	77.80	1.4324	30.4	80.8	111.1	0.0647	0.2210	57
58	69.635	0.0129	0.6859	77.68	1.4579	30.7	80.6	111.3	0.0653	0.2209	58
59	70.892	0.0129	0.6739	77.56	1.4839	31.0	80.4	111.4	0.0659	0.2209	59
60	72.167	0.0129	0.6622	77.43	1.5102	31.4	80.2	111.5	0.0666	0.2208	60
61	73.459	0.0129	0.6507	77.31	1.5369	31.7	80.0	111.6	0.0672	0.2208	61
62	74.769	0.0130	0.6394	77.19	1.5640	32.0	79.7	111.8	0.0678	0.2207	62
63	76.096	0.0130	0.6283	77.06	1.5915	32.4	79.5	111.9	0.0685	0.2207	63
64	77.440	0.0130	0.6175	76.94	1.6194	32.7	79.3	112.0	0.0691	0.2206	64
65	78.803	0.0130	0.6069	76.81	1.6477	33.0	79.1	112.2	0.0698	0.2206	65
66	80.184	0.0130	0.5965	76.69	1.6764	33.4	78.9	112.3	0.0704	0.2205	66
67	81.582	0.0131	0.5863	76.56	1.7055	33.7	78.7	112.4	0.0710	0.2205	67
68	83.000	0.0131	0.5764	76.44	1.7350	34.0	78.5	112.5	0.0717	0.2204	68
69	84.435	0.0131	0.5666	76.31	1.7649	34.4	78.3	112.7	0.0723	0.2204	69
70	85.890	0.0131	0.5570	76.18	1.7952	34.7	78.1	112.8	0.0729	0.2203	70
71	87.363	0.0131	0.5476	76.05	1.8260	35.1	77.9	112.9	0.0735	0.2203	71
72	88.855	0.0132	0.5384	75.93	1.8573	35.4	77.6	113.0	0.0742	0.2202	72
73	90.366	0.0132	0.5294	75.80	1.8889	35.7	77.4	113.2	0.0748	0.2202	73
74	91.897	0.0132	0.5206	75.67	1.9210	36.1	77.2	113.3	0.0754	0.2201	74
75	93.447	0.0132	0.5119	75.54	1.9536	36.4	77.0	113.4	0.0761	0.2201	75
76	95.016	0.0133	0.5034	75.41	1.9866	36.8	76.8	113.5	0.0767	0.2200	76
77	96.606	0.0133	0.4950	75.28	2.0201	37.1	76.6	113.7	0.0773	0.2200	77
78	98.215	0.0133	0.4868	75.15	2.0541	37.4	76.3	113.8	0.0780	0.2200	78
79	99.844	0.0133	0.4788	75.02	2.0885	37.8	76.1	113.9	0.0786	0.2199	79
80	101.494	0.0134	0.4709	74.89	2.1234	38.1	75.9	114.0	0.0792	0.2199	80
81	103.164	0.0134	0.4632	74.75	2.1589	38.5	75.7	114.1	0.0799	0.2198	81
82	104.855	0.0134	0.4556	74.62	2.1948	38.8	75.4	114.3	0.0805	0.2198	82
83	106.566	0.0134	0.4482	74.49	2.2312	39.2	75.2	114.4	0.0811	0.2197	83
84	108.290	0.0134	0.4409	74.35	2.2681	39.5	75.0	114.5	0.0817	0.2197	84
85	110.050	0.0135	0.4337	74.22	2.3056	39.9	74.8	114.6	0.0824	0.2196	85
86	111.828	0.0135	0.4267	74.08	2.3436	40.2	74.5	114.7	0.0830	0.2196	86
87	113.626	0.0135	0.4198	73.95	2.3821	40.5	74.3	114.9	0.0836	0.2196	87
88	115.444	0.0135	0.4130	73.81	2.4211	40.9	74.1	115.0	0.0843	0.2195	88
89	117.281	0.0136	0.4064	73.67	2.4607	41.2	73.8	115.1	0.0849	0.2195	89

Reprinted by permission from DuPont

Table 1 (continued)
HFC–134a Saturation Properties—Temperature Table

TEMP. °F	PRESSURE psia	VOLUME ft³/lb		DENSITY lb/ft³		ENTHALPY Btu/lb			ENTROPY Btu/(lb)(°R)		TEMP. °F
		LIQUID v_f	VAPOR v_g	LIQUID $1/v_f$	VAPOR $1/v_g$	LIQUID h_f	LATENT h_{fg}	VAPOR h_g	LIQUID s_f	VAPOR s_g	
90	119.138	0.0136	0.3999	73.54	2.5009	41.6	73.6	115.2	0.0855	0.2194	90
91	121.024	0.0136	0.3935	73.40	2.5416	41.9	73.4	115.3	0.0861	0.2194	91
92	122.930	0.0137	0.3872	73.26	2.5829	42.3	73.1	115.4	0.0868	0.2193	92
93	124.858	0.0137	0.3810	73.12	2.6247	42.6	72.9	115.5	0.0874	0.2193	93
94	126.809	0.0137	0.3749	72.98	2.6672	43.0	72.7	115.7	0.0880	0.2193	94
95	128.782	0.0137	0.3690	72.84	2.7102	43.4	72.4	115.8	0.0886	0.2192	95
96	130.778	0.0138	0.3631	72.70	2.7539	43.7	72.2	115.9	0.0893	0.2192	96
97	132.798	0.0138	0.3574	72.56	2.7981	44.1	71.9	116.0	0.0899	0.2191	97
98	134.840	0.0138	0.3517	72.42	2.8430	44.4	71.7	116.1	0.0905	0.2191	98
99	136.906	0.0138	0.3462	72.27	2.8885	44.8	71.4	116.2	0.0912	0.2190	99
100	138.996	0.0139	0.3408	72.13	2.9347	45.1	71.2	116.3	0.0918	0.2190	100
101	141.109	0.0139	0.3354	71.99	2.9815	45.5	70.9	116.4	0.0924	0.2190	101
102	143.247	0.0139	0.3302	71.84	3.0289	45.8	70.7	116.5	0.0930	0.2189	102
103	145.408	0.0139	0.3250	71.70	3.0771	46.2	70.4	116.6	0.0937	0.2189	103
104	147.594	0.0140	0.3199	71.55	3.1259	46.6	70.2	116.7	0.0943	0.2188	104
105	149.804	0.0140	0.3149	71.40	3.1754	46.9	69.9	116.9	0.0949	0.2188	105
106	152.039	0.0140	0.3100	71.25	3.2256	47.3	69.7	117.0	0.0955	0.2187	106
107	154.298	0.0141	0.3052	71.11	3.2765	47.6	69.4	117.1	0.0962	0.2187	107
108	156.583	0.0141	0.3005	70.96	3.3282	48.0	69.2	117.2	0.0968	0.2186	108
109	158.893	0.0141	0.2958	70.81	3.3806	48.4	68.9	117.3	0.0974	0.2186	109
110	161.227	0.0142	0.2912	70.66	3.4337	48.7	68.6	117.4	0.0981	0.2185	110
111	163.588	0.0142	0.2867	70.51	3.4876	49.1	68.4	117.5	0.0987	0.2185	111
112	165.974	0.0142	0.2823	70.35	3.5423	49.5	68.1	117.6	0.0993	0.2185	112
113	168.393	0.0142	0.2780	70.20	3.5977	49.8	67.8	117.7	0.0999	0.2184	113
114	170.833	0.0143	0.2737	70.05	3.6539	50.2	67.6	117.8	0.1006	0.2184	114
115	173.298	0.0143	0.2695	69.89	3.7110	50.5	67.3	117.9	0.1012	0.2183	115
116	175.790	0.0143	0.2653	69.74	3.7689	50.9	67.0	117.9	0.1018	0.2183	116
117	178.297	0.0144	0.2613	69.58	3.8276	51.3	66.8	118.0	0.1024	0.2182	117
118	180.846	0.0144	0.2573	69.42	3.8872	51.7	66.5	118.1	0.1031	0.2182	118
119	183.421	0.0144	0.2533	69.26	3.9476	52.0	66.2	118.2	0.1037	0.2181	119
120	186.023	0.0145	0.2494	69.10	4.0089	52.4	65.9	118.3	0.1043	0.2181	120
121	188.652	0.0145	0.2456	68.94	4.0712	52.8	65.6	118.4	0.1050	0.2180	121
122	191.308	0.0145	0.2419	68.78	4.1343	53.1	65.4	118.5	0.1056	0.2180	122
123	193.992	0.0146	0.2382	68.62	4.1984	53.5	65.1	118.6	0.1062	0.2179	123
124	196.703	0.0146	0.2346	68.46	4.2634	53.9	64.8	118.7	0.1068	0.2178	124
125	199.443	0.0146	0.2310	68.29	4.3294	54.3	64.5	118.8	0.1075	0.2178	125
126	202.211	0.0147	0.2275	68.13	4.3964	54.6	64.2	118.8	0.1081	0.2177	126
127	205.008	0.0147	0.2240	67.96	4.4644	55.0	63.9	118.9	0.1087	0.2177	127
128	207.834	0.0147	0.2206	67.80	4.5334	55.4	63.6	119.0	0.1094	0.2176	128
129	210.688	0.0148	0.2172	67.63	4.6034	55.8	63.3	119.1	0.1100	0.2176	129
130	213.572	0.0148	0.2139	67.46	4.6745	56.2	63.0	119.2	0.1106	0.2175	130
131	216.485	0.0149	0.2107	67.29	4.7467	56.5	62.7	119.2	0.1113	0.2174	131
132	219.429	0.0149	0.2075	67.12	4.8200	56.9	62.4	119.3	0.1119	0.2174	132
133	222.402	0.0149	0.2043	66.95	4.8945	57.3	62.1	119.4	0.1125	0.2173	133
134	225.405	0.0150	0.2012	66.77	4.9700	57.7	61.8	119.5	0.1132	0.2173	134
135	228.438	0.0150	0.1981	66.60	5.0468	58.1	61.5	119.6	0.1138	0.2172	135
136	231.502	0.0151	0.1951	66.42	5.1248	58.5	61.2	119.6	0.1144	0.2171	136
137	234.597	0.0151	0.1922	66.24	5.2040	58.8	60.8	119.7	0.1151	0.2171	137
138	237.723	0.0151	0.1892	66.06	5.2844	59.2	60.5	119.8	0.1157	0.2170	138
139	240.880	0.0152	0.1864	65.88	5.3661	59.6	60.2	119.8	0.1163	0.2169	139
140	244.068	0.0152	0.1835	65.70	5.4491	60.0	59.9	119.9	0.1170	0.2168	140
141	247.288	0.0153	0.1807	65.52	5.5335	60.4	59.6	120.0	0.1176	0.2168	141
142	250.540	0.0153	0.1780	65.34	5.6192	60.8	59.2	120.0	0.1183	0.2167	142
143	253.824	0.0153	0.1752	65.15	5.7064	61.2	58.9	120.1	0.1189	0.2166	143
144	257.140	0.0154	0.1726	64.96	5.7949	61.6	58.6	120.1	0.1195	0.2165	144
145	260.489	0.0154	0.1699	64.78	5.8849	62.0	58.2	120.2	0.1202	0.2165	145
146	263.871	0.0155	0.1673	64.59	5.9765	62.4	57.9	120.3	0.1208	0.2164	146
147	267.270	0.0155	0.1648	64.39	6.0695	62.8	57.5	120.3	0.1215	0.2163	147
148	270.721	0.0156	0.1622	64.20	6.1642	63.2	57.2	120.4	0.1221	0.2162	148
149	274.204	0.0156	0.1597	64.01	6.2604	63.6	56.8	120.4	0.1228	0.2161	149

Reprinted by permission from DuPont

Table 1 (continued)
HFC-134a Saturation Properties—Temperature Table

TEMP. °F	PRESSURE psia	VOLUME ft³/lb		DENSITY lb/ft³		ENTHALPY Btu/lb			ENTROPY Btu/(lb)(°R)		TEMP. °F
		LIQUID v_f	VAPOR v_g	LIQUID $1/v_f$	VAPOR $1/v_g$	LIQUID h_f	LATENT h_{fg}	VAPOR h_g	LIQUID s_f	VAPOR s_g	
150	277.721	0.0157	0.1573	63.81	6.3584	64.0	56.5	120.5	0.1234	0.2160	150
151	281.272	0.0157	0.1548	63.61	6.4580	64.4	56.1	120.5	0.1240	0.2159	151
152	284.857	0.0158	0.1525	63.41	6.5593	64.8	55.7	120.6	0.1247	0.2158	152
153	288.477	0.0158	0.1501	63.21	6.6625	65.2	55.4	120.6	0.1253	0.2157	153
154	292.131	0.0159	0.1478	63.01	6.7675	65.6	55.0	120.6	0.1260	0.2156	154
155	295.820	0.0159	0.1455	62.80	6.8743	66.0	54.6	120.7	0.1266	0.2155	155
156	299.544	0.0160	0.1432	62.59	6.9831	66.4	54.3	120.7	0.1273	0.2154	156
157	303.304	0.0160	0.1410	62.38	7.0940	66.9	53.9	120.7	0.1279	0.2153	157
158	307.100	0.0161	0.1388	62.17	7.2068	67.3	53.5	120.8	0.1286	0.2152	158
159	310.931	0.0161	0.1366	61.96	7.3218	67.7	53.1	120.8	0.1293	0.2151	159
160	314.800	0.0162	0.1344	61.74	7.4390	68.1	52.7	120.8	0.1299	0.2150	160
161	318.704	0.0163	0.1323	61.52	7.5584	68.5	52.3	120.9	0.1306	0.2149	161
162	322.646	0.0163	0.1302	61.30	7.6801	69.0	51.9	120.9	0.1312	0.2148	162
163	326.625	0.0164	0.1281	61.08	7.8042	69.4	51.5	120.9	0.1319	0.2146	163
164	330.641	0.0164	0.1261	60.86	7.9308	69.8	51.1	120.9	0.1326	0.2145	164
165	334.696	0.0165	0.1241	60.63	8.0600	70.2	50.7	120.9	0.1332	0.2144	165
166	338.788	0.0166	0.1221	60.40	8.1917	70.7	50.3	120.9	0.1339	0.2142	166
167	342.919	0.0166	0.1201	60.16	8.3262	71.1	49.8	120.9	0.1346	0.2141	167
168	347.089	0.0167	0.1182	59.93	8.4635	71.5	49.4	120.9	0.1352	0.2140	168
169	351.298	0.0168	0.1162	59.69	8.6037	72.0	49.0	120.9	0.1359	0.2138	169
170	355.547	0.0168	0.1143	59.45	8.7470	72.4	48.5	120.9	0.1366	0.2137	170
171	359.835	0.0169	0.1124	59.20	8.8934	72.8	48.1	120.9	0.1373	0.2135	171
172	364.164	0.0170	0.1106	58.95	9.0431	73.3	47.6	120.9	0.1380	0.2133	172
173	368.533	0.0170	0.1087	58.70	9.1961	73.7	47.2	120.9	0.1386	0.2132	173
174	372.942	0.0171	0.1069	58.45	9.3527	74.2	46.7	120.9	0.1393	0.2130	174
175	377.393	0.0172	0.1051	58.19	9.5129	74.6	46.2	120.8	0.1400	0.2128	175
176	381.886	0.0173	0.1033	57.92	9.6770	75.1	45.7	120.8	0.1407	0.2126	176
177	386.421	0.0173	0.1016	57.66	9.8451	75.5	45.2	120.8	0.1414	0.2125	177
178	390.998	0.0174	0.0998	57.39	10.0173	76.0	44.7	120.7	0.1421	0.2123	178
179	395.617	0.0175	0.0981	57.11	10.1939	76.5	44.2	120.7	0.1428	0.2121	179
180	400.280	0.0176	0.0964	56.83	10.3750	76.9	43.7	120.7	0.1435	0.2119	180
181	404.987	0.0177	0.0947	56.55	10.5609	77.4	43.2	120.6	0.1442	0.2116	181
182	409.738	0.0178	0.0930	56.26	10.7518	77.9	42.7	120.5	0.1449	0.2114	182
183	414.533	0.0179	0.0913	55.96	10.9481	78.4	42.1	120.5	0.1456	0.2112	183
184	419.373	0.0180	0.0897	55.66	11.1498	78.8	41.6	120.4	0.1464	0.2109	184
185	424.258	0.0181	0.0880	55.35	11.3575	79.3	41.0	120.3	0.1471	0.2107	185
186	429.189	0.0182	0.0864	55.04	11.5713	79.8	40.4	120.2	0.1478	0.2104	186
187	434.167	0.0183	0.0848	54.72	11.7916	80.3	39.8	120.1	0.1486	0.2102	187
188	439.192	0.0184	0.0832	54.40	12.0189	80.8	39.2	120.0	0.1493	0.2099	188
189	444.264	0.0185	0.0816	54.07	12.2536	81.3	38.6	119.9	0.1501	0.2096	189
190	449.384	0.0186	0.0800	53.73	12.4962	81.8	38.0	119.8	0.1508	0.2093	190
191	454.552	0.0187	0.0784	53.38	12.7472	82.3	37.4	119.7	0.1516	0.2090	191
192	459.757	0.0189	0.0769	53.02	13.0072	82.8	36.7	119.5	0.1523	0.2087	192
193	465.026	0.0190	0.0753	52.65	13.2769	83.4	36.0	119.4	0.1531	0.2083	193
194	470.346	0.0191	0.0738	52.27	13.5570	83.9	35.3	119.2	0.1539	0.2080	194
195	475.717	0.0193	0.0722	51.88	13.8484	84.4	34.6	119.1	0.1547	0.2076	195
196	481.139	0.0194	0.0707	51.48	14.1522	85.0	33.9	118.9	0.1555	0.2072	196
197	486.614	0.0196	0.0691	51.07	14.4693	85.5	33.1	118.7	0.1563	0.2068	197
198	492.142	0.0197	0.0676	50.64	14.8012	86.1	32.3	118.4	0.1572	0.2063	198
199	497.724	0.0199	0.0660	50.20	15.1493	86.7	31.5	118.2	0.1580	0.2059	199
200	503.361	0.0201	0.0645	49.73	15.5155	87.3	30.7	118.0	0.1589	0.2054	200
201	509.054	0.0203	0.0629	49.25	15.9020	87.9	29.8	117.7	0.1597	0.2049	201
202	514.805	0.0205	0.0613	48.75	16.3113	88.5	28.9	117.4	0.1606	0.2043	202
203	520.613	0.0207	0.0597	48.22	16.7466	89.1	27.9	117.0	0.1616	0.2037	203
204	526.481	0.0210	0.0581	47.66	17.2121	89.8	26.9	116.7	0.1625	0.2031	204
205	532.410	0.0212	0.0565	47.06	17.7129	90.5	25.8	116.3	0.1635	0.2024	205
206	538.402	0.0215	0.0548	46.42	18.2558	91.2	24.7	115.8	0.1645	0.2016	206
207	544.458	0.0219	0.0531	45.73	18.8499	91.9	23.4	115.3	0.1656	0.2008	207
208	550.581	0.0222	0.0513	44.98	19.5084	92.7	22.1	114.8	0.1667	0.1998	208
209	556.773	0.0227	0.0494	44.14	20.2504	93.5	20.6	114.1	0.1680	0.1988	209

Reprinted by permission from DuPont

Table 1 (continued)
HFC–134a Saturation Properties—Temperature Table

TEMP. °F	PRESSURE psia	VOLUME ft³/lb		DENSITY lb/ft³		ENTHALPY Btu/lb			ENTROPY Btu/(lb)(°R)		TEMP. °F
		LIQUID v_f	VAPOR v_g	LIQUID $1/v_f$	VAPOR $1/v_g$	LIQUID h_f	LATENT h_{fg}	VAPOR h_g	LIQUID s_f	VAPOR s_g	
210	563.037	0.0232	0.0474	43.19	21.1071	94.4	18.9	113.4	0.1693	0.1976	210
211	569.378	0.0238	0.0452	42.07	22.1329	95.5	17.0	112.4	0.1708	0.1961	211
212	575.801	0.0246	0.0427	40.67	23.4420	96.7	14.5	111.2	0.1726	0.1942	212
213	582.316	0.0259	0.0394	38.65	25.3583	98.4	11.1	109.5	0.1750	0.1915	213

Reprinted by permission from DuPont

Table 2
HFC-134a Superheated Vapor—Constant Pressure Tables

V = Volume in ft³/lb   H = Enthalpy in Btu/lb   S = Entropy in Btu/(lb) (°R) v_s = Velocity of Sound in ft/sec
Cp = Heat Capacity at Constant Pressure in Btu/(lb) (°F)   Cp/Cv = Heat Capacity Ratio (Dimensionless)

TEMP °F	PRESSURE = 1.00 PSIA							PRESSURE = 2.00 PSIA						TEMP °F
	V	H	S	Cp	Cp/Cv	v_s		V	H	S	Cp	Cp/Cv	v_s	
−97.6	0.01065	−16.6	−0.0426	0.2813	1.5295	3163.1	SAT LIQ	0.01085	−11.6	−0.0290	0.2857	1.5183	3004.1	−79.9
−97.6	37.87879	88.5	0.2478	0.1570	1.1488	447.2	SAT VAP	19.76285	91.2	0.2416	0.1629	1.1470	455.7	−79.9
−90	38.61004	89.7	0.2511	0.1588	1.1462	451.6		—	—	—	—	—	—	−90
−80	39.68254	91.3	0.2554	0.1612	1.1429	457.3		—	—	—	—	—	—	−80
−70	40.81633	92.9	0.2596	0.1637	1.1398	463.0		20.28398	92.8	0.2458	0.1651	1.1435	461.4	−70
−60	41.84100	94.6	0.2638	0.1662	1.1370	468.5		20.83333	94.4	0.2500	0.1674	1.1401	467.1	−60
−50	42.91845	96.3	0.2679	0.1686	1.1342	474.0		21.36752	96.1	0.2542	0.1698	1.1370	472.6	−50
−40	44.05286	98.0	0.2720	0.1711	1.1317	479.3		21.92982	97.8	0.2583	0.1721	1.1341	478.1	−40
−30	45.04505	99.7	0.2761	0.1736	1.1292	484.7		22.47191	99.6	0.2624	0.1745	1.1314	483.5	−30
−20	46.08295	101.4	0.2801	0.1760	1.1269	489.9		22.98851	101.3	0.2664	0.1768	1.1288	488.8	−20
−10	47.16981	103.2	0.2841	0.1785	1.1247	495.1		23.52941	103.1	0.2704	0.1792	1.1264	494.1	−10
0	48.30918	105.0	0.2880	0.1809	1.1226	500.2		24.03846	104.9	0.2744	0.1816	1.1242	499.3	0
10	49.26108	106.8	0.2919	0.1834	1.1206	505.3		24.57002	106.7	0.2783	0.1839	1.1220	504.4	10
20	50.25126	108.7	0.2958	0.1858	1.1187	510.3		25.12563	108.6	0.2822	0.1863	1.1200	509.5	20
30	51.28205	110.5	0.2997	0.1882	1.1169	515.2		25.64103	110.5	0.2861	0.1887	1.1180	514.5	30
40	52.35602	112.4	0.3035	0.1906	1.1151	520.1		26.17801	112.4	0.2899	0.1910	1.1161	519.4	40
50	53.47594	114.4	0.3073	0.1930	1.1134	525.0		26.73797	114.3	0.2937	0.1933	1.1144	524.3	50
60	54.64481	116.3	0.3111	0.1953	1.1118	529.8		27.24796	116.2	0.2975	0.1957	1.1127	529.2	60
70	55.55556	118.3	0.3148	0.1977	1.1103	534.5		27.77778	118.2	0.3013	0.1980	1.1110	533.9	70
80	56.81818	120.2	0.3186	0.2000	1.1088	539.3		28.32861	120.2	0.3050	0.2003	1.1095	538.7	80
90	57.80347	122.3	0.3223	0.2023	1.1073	543.9		28.81844	122.2	0.3087	0.2026	1.1080	543.4	90
100	58.82353	124.3	0.3259	0.2046	1.1059	548.5		29.32551	124.2	0.3124	0.2049	1.1065	548.0	100
110	59.88024	126.4	0.3296	0.2069	1.1046	553.1		29.85075	126.3	0.3160	0.2071	1.1051	552.7	110
120	60.97561	128.4	0.3332	0.2092	1.1033	557.7		30.39514	128.4	0.3196	0.2094	1.1038	557.2	120
130	62.11180	130.5	0.3368	0.2114	1.1020	562.2		30.95975	130.5	0.3232	0.2116	1.1025	561.7	130
140	62.89308	132.7	0.3404	0.2137	1.1008	566.6		31.44654	132.6	0.3268	0.2139	1.1013	566.2	140
150	64.10256	134.8	0.3439	0.2159	1.0997	571.1		32.05128	134.8	0.3304	0.2161	1.1001	570.7	150
160	64.93506	137.0	0.3474	0.2181	1.0985	575.5		32.57329	136.9	0.3339	0.2182	1.0989	575.1	160
170	66.22517	139.2	0.3509	0.2203	1.0974	579.8		33.11258	139.1	0.3374	0.2204	1.0978	579.5	170
180	67.11409	141.4	0.3544	0.2224	1.0964	584.2		33.55705	141.4	0.3409	0.2226	1.0967	583.8	180
190	68.49315	143.6	0.3579	0.2246	1.0953	588.5		34.12969	143.6	0.3444	0.2247	1.0956	588.1	190
200	69.44444	145.9	0.3613	0.2267	1.0943	592.7		34.60208	145.8	0.3478	0.2268	1.0946	592.4	200
210	70.42254	148.2	0.3648	0.2288	1.0934	597.0		35.21127	148.1	0.3512	0.2289	1.0936	596.7	210
200	—	—	—	—	—	—		35.71429	150.4	0.3547	0.2310	1.0927	600.9	220

TEMP °F	PRESSURE = 3.00 PSIA							PRESSURE = 4.00 PSIA						TEMP °F
	V	H	S	Cp	Cp/Cv	v_s		V	H	S	Cp	Cp/Cv	v_s	
−68.5	0.01098	−8.4	−0.0206	0.2887	1.5126	2903.8	SAT LIQ	0.01108	−5.9	−0.0143	0.2910	1.5092	2829.0	−60
−68.5	13.51351	92.9	0.2382	0.1669	1.1466	460.7	SAT VAP	10.31992	94.2	0.2360	0.1700	1.1467	464.1	−60
−60	13.83126	94.3	0.2418	0.1687	1.1434	465.6		—	—	—	—	—	—	−60
−50	14.18440	96.0	0.2460	0.1709	1.1399	471.3		10.60445	95.9	0.2402	0.1720	1.1428	469.9	−50
−40	14.55604	97.7	0.2502	0.1731	1.1366	476.9		10.86957	97.6	0.2444	0.1741	1.1392	475.6	−40
−30	14.92537	99.5	0.2543	0.1754	1.1336	482.4		11.14827	99.4	0.2485	0.1763	1.1359	481.2	−30
−20	15.26718	101.2	0.2583	0.1776	1.1308	487.8		11.42857	101.1	0.2526	0.1784	1.1328	486.7	−20
−10	15.62500	103.0	0.2624	0.1799	1.1282	493.1		11.69591	102.9	0.2566	0.1806	1.1300	492.1	−10
0	16.00000	104.8	0.2663	0.1822	1.1257	498.4		11.96172	104.7	0.2606	0.1829	1.1273	497.4	0
10	16.33987	106.7	0.2703	0.1845	1.1234	503.6		12.23990	106.6	0.2646	0.1851	1.1248	502.7	10
20	16.72241	108.5	0.2742	0.1868	1.1212	508.7		12.50000	108.4	0.2685	0.1873	1.1225	507.9	20
30	17.06485	110.4	0.2781	0.1891	1.1191	513.7		12.77139	110.3	0.2724	0.1896	1.1203	513.0	30
40	17.42160	112.3	0.2819	0.1914	1.1172	518.7		13.03781	112.2	0.2762	0.1919	1.1182	518.0	40
50	17.79359	114.2	0.2857	0.1937	1.1153	523.6		13.31558	114.2	0.2800	0.1941	1.1162	523.0	50
60	18.14882	116.2	0.2895	0.1960	1.1135	528.5		13.58696	116.1	0.2838	0.1964	1.1144	527.9	60
70	18.48429	118.1	0.2933	0.1983	1.1118	533.4		13.85042	118.1	0.2876	0.1987	1.1126	532.8	70
80	18.83239	120.1	0.2970	0.2006	1.1102	538.1		14.10437	120.1	0.2913	0.2009	1.1109	537.6	80
90	19.19386	122.2	0.3007	0.2029	1.1086	542.9		14.38849	122.1	0.2951	0.2031	1.1093	542.3	90
100	19.56947	124.2	0.3044	0.2051	1.1071	547.5		14.64129	124.2	0.2987	0.2054	1.1077	547.0	100
110	19.92032	126.3	0.3081	0.2074	1.1057	552.2		14.90313	126.2	0.3024	0.2076	1.1063	551.7	110
120	20.24291	128.3	0.3117	0.2096	1.1043	556.8		15.17451	128.3	0.3060	0.2098	1.1048	556.3	120
130	20.61856	130.5	0.3153	0.2118	1.1030	561.3		15.45595	130.4	0.3096	0.2120	1.1035	560.9	130
140	20.96436	132.6	0.3189	0.2140	1.1017	565.8		15.72327	132.5	0.3132	0.2142	1.1022	565.4	140
150	21.32196	134.7	0.3224	0.2162	1.1005	570.3		15.97444	134.7	0.3168	0.2164	1.1009	569.9	150
160	21.69197	136.9	0.3260	0.2184	1.0993	574.7		16.23377	136.9	0.3203	0.2186	1.0997	574.4	160
170	22.02643	139.1	0.3295	0.2206	1.0981	579.1		16.50165	139.1	0.3238	0.2207	1.0985	578.8	170
180	22.37136	141.3	0.3330	0.2227	1.0970	583.5		16.77852	141.3	0.3273	0.2228	1.0974	583.2	180
190	22.72727	143.6	0.3364	0.2248	1.0960	587.8		17.03578	143.5	0.3308	0.2250	1.0963	587.5	190
200	23.09469	145.8	0.3399	0.2270	1.0949	592.1		17.30104	145.8	0.3343	0.2271	1.0952	591.8	200
210	23.41920	148.1	0.3433	0.2290	1.0939	596.4		17.57469	148.1	0.3377	0.2292	1.0942	596.1	210
220	23.80952	150.4	0.3467	0.2311	1.0929	600.6		17.82531	150.4	0.3411	0.2312	1.0932	600.3	220
230	24.15459	152.7	0.3501	0.2332	1.0920	604.8		18.08318	152.7	0.3445	0.2333	1.0923	604.5	230
240	24.50980	155.1	0.3535	0.2352	1.0911	609.0		18.34862	155.0	0.3479	0.2353	1.0913	608.7	240
250	—	—	—	—	—	—		18.62197	157.4	0.3512	0.2373	1.0904	612.8	250

Reprinted by permission from DuPont

Table 2 (continued)
HFC-134a Superheated Vapor—Constant Pressure Tables

V = Volume in ft³/lb   H = Enthalpy in Btu/lb   S = Entropy in Btu/(lb) (°R) v_s = Velocity of Sound in ft/sec
Cp = Heat Capacity at Constant Pressure in Btu/(lb) (°F)   Cp/Cv = Heat Capacity Ratio (Dimensionless)

TEMP °F	PRESSURE = 5.00 PSIA							PRESSURE = 6.00 PSIA						TEMP °F
	V	H	S	Cp	Cp/Cv	v_s		V	H	S	Cp	Cp/Cv	v_s	
−53	0.01116	−3.8	−0.0093	0.2929	1.5071	2768.7	SAT LIQ	0.01123	−2.1	−0.0051	0.2946	1.5057	2717.9	−47.1
−53	8.37521	95.2	0.2343	0.1726	1.1471	466.8	SAT VAP	7.06215	96.1	0.2330	0.1749	1.1476	468.9	−47.1
−50	8.44595	95.7	0.2356	0.1732	1.1458	468.6		—	—	—	—	—	—	−50
−40	8.66551	97.5	0.2398	0.1752	1.1418	474.4		7.19424	97.4	0.2360	0.1762	1.1445	473.1	−40
−30	8.88889	99.2	0.2439	0.1772	1.1382	480.0		7.38007	99.1	0.2402	0.1781	1.1406	478.9	−30
−20	9.10747	101.0	0.2480	0.1793	1.1349	485.6		7.57002	100.9	0.2443	0.1801	1.1369	484.5	−20
−10	9.32836	102.8	0.2521	0.1814	1.1318	491.1		7.75194	102.7	0.2484	0.1821	1.1336	490.1	−10
0	9.55110	104.7	0.2561	0.1835	1.1289	496.5		7.93651	104.6	0.2524	0.1842	1.1306	495.6	0
10	9.76563	106.5	0.2601	0.1857	1.1263	501.8		8.11688	106.4	0.2564	0.1863	1.1277	500.9	10
20	9.98004	108.4	0.2640	0.1879	1.1238	507.0		8.29876	108.3	0.2603	0.1884	1.1251	506.2	20
30	10.20408	110.3	0.2679	0.1901	1.1215	512.2		8.48176	110.2	0.2643	0.1906	1.1226	511.4	30
40	10.41667	112.2	0.2718	0.1923	1.1193	517.3		8.66551	112.1	0.2681	0.1927	1.1203	516.6	40
50	10.62699	114.1	0.2756	0.1945	1.1172	522.3		8.84173	114.0	0.2720	0.1949	1.1182	521.6	50
60	10.84599	116.1	0.2794	0.1968	1.1152	527.3		9.02527	116.0	0.2758	0.1971	1.1161	526.6	60
70	11.06195	118.0	0.2832	0.1990	1.1134	532.2		9.20810	118.0	0.2796	0.1993	1.1142	531.6	70
80	11.27396	120.0	0.2869	0.2012	1.1116	537.0		9.38086	120.0	0.2833	0.2015	1.1124	536.4	80
90	11.49425	122.1	0.2906	0.2034	1.1100	541.8		9.56023	122.0	0.2870	0.2037	1.1106	541.3	90
100	11.70960	124.1	0.2943	0.2056	1.1084	546.5		9.73710	124.1	0.2907	0.2059	1.1090	546.0	100
110	11.91895	126.2	0.2980	0.2078	1.1068	551.2		9.92063	126.1	0.2944	0.2081	1.1074	550.7	110
120	12.13592	128.3	0.3016	0.2100	1.1054	555.9		10.10101	128.2	0.2980	0.2103	1.1059	555.4	120
130	12.34568	130.4	0.3052	0.2122	1.1040	560.5		10.27749	130.3	0.3016	0.2124	1.1044	560.0	130
140	12.56281	132.5	0.3088	0.2144	1.1026	565.0		10.44932	132.5	0.3052	0.2146	1.1031	564.6	140
150	12.77139	134.7	0.3124	0.2166	1.1013	569.5		10.62699	134.6	0.3088	0.2167	1.1017	569.1	150
160	12.98701	136.8	0.3159	0.2187	1.1001	574.0		10.81081	136.8	0.3123	0.2189	1.1005	573.6	160
170	13.19261	139.0	0.3194	0.2209	1.0989	578.4		10.98901	139.0	0.3159	0.2210	1.0992	578.1	170
180	13.40483	141.3	0.3229	0.2230	1.0977	582.8		11.16071	141.2	0.3194	0.2231	1.0981	582.5	180
190	13.62398	143.5	0.3264	0.2251	1.0966	587.2		11.33787	143.5	0.3228	0.2252	1.0969	586.9	190
200	13.83126	145.8	0.3299	0.2272	1.0955	591.5		11.52074	145.7	0.3263	0.2273	1.0958	591.2	200
210	14.04494	148.0	0.3333	0.2293	1.0945	595.8		11.69591	148.0	0.3297	0.2294	1.0948	595.5	210
220	14.24501	150.3	0.3367	0.2313	1.0935	600.0		11.87648	150.3	0.3331	0.2314	1.0937	599.8	220
230	14.47178	152.7	0.3401	0.2334	1.0925	604.3		12.04819	152.6	0.3365	0.2335	1.0928	604.0	230
240	14.68429	155.0	0.3435	0.2354	1.0916	608.4		12.22494	155.0	0.3399	0.2355	1.0918	608.2	240
250	14.90313	157.4	0.3468	0.2374	1.0907	612.6		12.40695	157.3	0.3433	0.2375	1.0909	612.4	250
260	—	—	—	—	—	—		12.57862	159.7	0.3466	0.2395	1.0900	616.5	260

TEMP °F	PRESSURE = 7.00 PSIA							PRESSURE = 8.00 PSIA						TEMP °F
	V	H	S	Cp	Cp/Cv	v_s		V	H	S	Cp	Cp/Cv	v_s	
−42	0.01130	−0.6	−0.0014	0.2960	1.5048	2673.7	SAT LIQ	0.01136	0.8	0.0018	0.2973	1.5042	2634.5	−37.4
−42	6.11247	96.9	0.2320	0.1770	1.1482	470.6	SAT VAP	5.39374	97.6	0.2311	0.1788	1.1489	472.1	−37.4
−40	6.14251	97.2	0.2328	0.1773	1.1473	471.8		—	—	—	—	—	—	−40
−30	6.30517	99.0	0.2370	0.1791	1.1430	477.7		5.49753	98.9	0.2342	0.1800	1.1455	476.5	−30
−20	6.46831	100.8	0.2411	0.1809	1.1391	483.5		5.64016	100.7	0.2383	0.1818	1.1412	482.3	−20
−10	6.62691	102.6	0.2452	0.1829	1.1355	489.1		5.78035	102.5	0.2424	0.1836	1.1374	488.1	−10
0	6.78426	104.5	0.2493	0.1849	1.1322	494.6		5.92066	104.4	0.2465	0.1855	1.1339	493.7	0
10	6.94444	106.3	0.2533	0.1869	1.1292	500.1		6.06061	106.2	0.2505	0.1875	1.1307	499.2	10
20	7.09723	108.2	0.2572	0.1890	1.1264	505.4		6.19965	108.1	0.2545	0.1895	1.1278	504.6	20
30	7.25689	110.1	0.2611	0.1911	1.1238	510.7		6.33714	110.0	0.2584	0.1916	1.1251	509.9	30
40	7.41290	112.0	0.2650	0.1932	1.1214	515.9		6.47249	112.0	0.2623	0.1936	1.1225	515.1	40
50	7.57002	114.0	0.2689	0.1953	1.1191	521.0		6.60939	113.9	0.2662	0.1957	1.1201	520.3	50
60	7.72201	115.9	0.2727	0.1975	1.1170	526.0		6.74764	115.9	0.2700	0.1979	1.1179	525.4	60
70	7.88022	117.9	0.2765	0.1996	1.1150	531.0		6.88231	117.9	0.2738	0.2000	1.1158	530.4	70
80	8.03213	119.9	0.2802	0.2018	1.1131	535.9		7.01754	119.9	0.2776	0.2021	1.1138	535.3	80
90	8.18331	122.0	0.2840	0.2040	1.1113	540.7		7.15308	121.9	0.2813	0.2043	1.1120	540.2	90
100	8.34028	124.0	0.2876	0.2061	1.1096	545.5		7.28863	124.0	0.2850	0.2064	1.1102	545.0	100
110	8.49618	126.1	0.2913	0.2083	1.1080	550.3		7.42390	126.0	0.2887	0.2085	1.1085	549.8	110
120	8.65052	128.2	0.2950	0.2105	1.1064	555.0		7.55858	128.1	0.2923	0.2107	1.1069	554.5	120
130	8.80282	130.3	0.2986	0.2126	1.1049	559.6		7.69231	130.2	0.2959	0.2128	1.1054	559.2	130
140	8.95255	132.4	0.3022	0.2148	1.1035	564.2		7.82473	132.4	0.2995	0.2150	1.1040	563.8	140
150	9.10747	134.6	0.3057	0.2169	1.1022	568.7		7.96178	134.5	0.3031	0.2171	1.1026	568.4	150
160	9.25926	136.8	0.3093	0.2190	1.1009	573.3		8.09061	136.7	0.3066	0.2192	1.1012	572.9	160
170	9.40734	139.0	0.3128	0.2211	1.0996	577.7		8.23045	138.9	0.3102	0.2213	1.1000	577.4	170
180	9.56023	141.2	0.3163	0.2233	1.0984	582.1		8.36120	141.2	0.3137	0.2234	1.0987	581.8	180
190	9.71817	143.4	0.3198	0.2253	1.0972	586.5		8.49618	143.4	0.3172	0.2255	1.0976	586.2	190
200	9.86193	145.7	0.3232	0.2274	1.0961	590.9		8.62813	145.7	0.3206	0.2275	1.0964	590.6	200
210	10.02004	148.0	0.3267	0.2295	1.0951	595.2		8.76424	147.9	0.3241	0.2296	1.0953	594.9	210
220	10.17294	150.3	0.3301	0.2315	1.0940	599.5		8.89680	150.3	0.3275	0.2316	1.0943	599.2	220
230	10.31992	152.6	0.3335	0.2336	1.0930	603.7		9.02527	152.6	0.3309	0.2337	1.0933	603.5	230
240	10.47120	155.0	0.3369	0.2356	1.0920	607.9		9.15751	154.9	0.3343	0.2357	1.0923	607.7	240
250	10.62699	157.3	0.3402	0.2376	1.0911	612.1		9.29368	157.3	0.3376	0.2377	1.0913	611.9	250
260	10.77586	159.7	0.3436	0.2396	1.0902	616.3		9.42507	159.7	0.3409	0.2397	1.0904	616.0	260
270	—	—	—	—	—	—		9.56023	162.1	0.3443	0.2416	1.0895	620.2	270

Reprinted by permission from DuPont

Table 2 (continued)
HFC-134a Superheated Vapor—Constant Pressure Tables

V = Volume in ft³/lb   H = Enthalpy in Btu/lb   S = Entropy in Btu/(lb) (°R) v_s = Velocity of Sound in ft/sec
Cp = Heat Capacity at Constant Pressure in Btu/(lb) (°F)   Cp/Cv = Heat Capacity Ratio (Dimensionless)

PRESSURE = 9.00 PSIA

TEMP °F	V	H	S	Cp	Cp/Cv	v_s
-33.2 SAT LIQ	0.01141	2.0	0.0047	0.2985	1.5039	2599.2
-33.2 SAT VAP	4.82859	98.2	0.2303	0.1805	1.1496	473.3
-30	4.87092	98.8	0.2317	0.1810	1.1480	475.3
-20	4.99750	100.6	0.2359	0.1827	1.1435	481.2
-10	5.12295	102.4	0.2400	0.1844	1.1394	487.0
0	5.24934	104.3	0.2441	0.1862	1.1357	492.7
10	5.37346	106.2	0.2481	0.1881	1.1323	498.3
20	5.49753	108.1	0.2521	0.1901	1.1292	503.8
30	5.62114	110.0	0.2560	0.1921	1.1263	509.1
40	5.74383	111.9	0.2599	0.1941	1.1236	514.4
50	5.86510	113.8	0.2638	0.1961	1.1211	519.6
60	5.98802	115.8	0.2676	0.1982	1.1188	524.7
70	6.10874	117.8	0.2714	0.2003	1.1166	529.8
80	6.23053	119.8	0.2752	0.2024	1.1146	534.7
90	6.34921	121.9	0.2789	0.2045	1.1127	539.7
100	6.47249	123.9	0.2826	0.2067	1.1108	544.5
110	6.59196	126.0	0.2863	0.2088	1.1091	549.3
120	6.71141	128.1	0.2900	0.2109	1.1075	554.0
130	6.83060	130.2	0.2936	0.2130	1.1059	558.7
140	6.94927	132.3	0.2972	0.2151	1.1044	563.4
150	7.06714	134.5	0.3008	0.2172	1.1030	568.0
160	7.18907	136.7	0.3043	0.2194	1.1016	572.5
170	7.30994	138.9	0.3078	0.2214	1.1003	577.0
180	7.42942	141.1	0.3113	0.2235	1.0991	581.5
190	7.54717	143.4	0.3148	0.2256	1.0979	585.9
200	7.66284	145.6	0.3183	0.2277	1.0967	590.3
210	7.78210	147.9	0.3217	0.2297	1.0956	594.6
220	7.89889	150.2	0.3252	0.2318	1.0945	598.9
230	8.01925	152.6	0.3285	0.2338	1.0935	603.2
240	8.13670	154.9	0.3319	0.2358	1.0925	607.4
250	8.25764	157.3	0.3353	0.2378	1.0916	611.6
260	8.37521	159.7	0.3386	0.2398	1.0906	615.8
270	8.48896	162.1	0.3420	0.2417	1.0897	619.9
280	—	—	—	—	—	—

PRESSURE = 10.00 PSIA

TEMP °F	V	H	S	Cp	Cp/Cv	v_s
-29.5 SAT LIQ	0.01146	3.1	0.0074	0.2996	1.5038	2567.0
-29.5 SAT VAP	4.37445	98.8	0.2297	0.1821	1.1503	474.4
-20	4.48430	100.5	0.2336	0.1836	1.1457	480.1
-10	4.59770	102.3	0.2378	0.1852	1.1414	486.0
0	4.71254	104.2	0.2419	0.1869	1.1374	491.8
10	4.82393	106.1	0.2459	0.1888	1.1338	497.4
20	4.93583	108.0	0.2499	0.1906	1.1306	502.9
30	5.04796	109.9	0.2539	0.1926	1.1275	508.3
40	5.15996	111.8	0.2578	0.1945	1.1247	513.7
50	5.26870	113.8	0.2616	0.1966	1.1222	518.9
60	5.37924	115.8	0.2655	0.1986	1.1197	524.1
70	5.48847	117.8	0.2693	0.2007	1.1175	529.2
80	5.59910	119.8	0.2731	0.2027	1.1154	534.2
90	5.70776	121.8	0.2768	0.2048	1.1134	539.1
100	5.81734	123.9	0.2805	0.2069	1.1115	544.0
110	5.92417	125.9	0.2842	0.2090	1.1097	548.8
120	6.03500	128.0	0.2879	0.2111	1.1080	553.6
130	6.14251	130.2	0.2915	0.2132	1.1064	558.3
140	6.25000	132.3	0.2951	0.2153	1.1049	563.0
150	6.35728	134.5	0.2987	0.2174	1.1034	567.6
160	6.46412	136.7	0.3022	0.2195	1.1020	572.1
170	6.57030	138.9	0.3057	0.2216	1.1007	576.7
180	6.68003	141.1	0.3093	0.2237	1.0994	581.1
190	6.78887	143.3	0.3127	0.2257	1.0982	585.6
200	6.89180	145.6	0.3162	0.2278	1.0970	590.0
210	6.99790	147.9	0.3196	0.2298	1.0959	594.3
220	7.10732	150.2	0.3231	0.2319	1.0948	598.6
230	7.21501	152.5	0.3265	0.2339	1.0938	602.9
240	7.32064	154.9	0.3298	0.2359	1.0928	607.2
250	7.42942	157.2	0.3332	0.2379	1.0918	611.4
260	7.53580	159.6	0.3366	0.2398	1.0908	615.6
270	7.63942	162.0	0.3399	0.2418	1.0899	619.7
280	7.74593	164.5	0.3432	0.2437	1.0890	623.8

PRESSURE = 11.00 PSIA

TEMP °F	V	H	S	Cp	Cp/Cv	v_s
-26 SAT LIQ	0.01151	4.2	0.0098	0.3007	1.5038	2537.3
-26 SAT VAP	4.00000	99.3	0.2291	0.1836	1.1511	475.3
-20	4.06174	100.4	0.2316	0.1845	1.1481	479.0
-10	4.16840	102.2	0.2357	0.1860	1.1434	485.0
0	4.27168	104.1	0.2399	0.1876	1.1392	490.8
10	4.37445	106.0	0.2439	0.1894	1.1354	496.5
20	4.47828	107.9	0.2479	0.1912	1.1320	502.1
30	4.58085	109.8	0.2519	0.1931	1.1288	507.6
40	4.68165	111.8	0.2558	0.1950	1.1259	512.9
50	4.78240	113.7	0.2597	0.1970	1.1232	518.2
60	4.88281	115.7	0.2635	0.1990	1.1207	523.4
70	4.98256	117.7	0.2674	0.2010	1.1183	528.5
80	5.08388	119.7	0.2711	0.2030	1.1161	533.6
90	5.18135	121.8	0.2749	0.2051	1.1141	538.6
100	5.28262	123.8	0.2786	0.2072	1.1121	543.5
110	5.37924	125.9	0.2823	0.2093	1.1103	548.3
120	5.47945	128.0	0.2859	0.2113	1.1086	553.1
130	5.57724	130.1	0.2896	0.2134	1.1069	557.9
140	5.67537	132.3	0.2932	0.2155	1.1054	562.5
150	5.77367	134.4	0.2968	0.2176	1.1039	567.2
160	5.87199	136.6	0.3003	0.2197	1.1024	571.8
170	5.97015	138.8	0.3039	0.2217	1.1011	576.3
180	6.06796	141.1	0.3074	0.2238	1.0998	580.8
190	6.16523	143.3	0.3108	0.2259	1.0985	585.2
200	6.26174	145.6	0.3143	0.2279	1.0973	589.7
210	6.36132	147.9	0.3178	0.2299	1.0962	594.0
220	6.45578	150.2	0.3212	0.2320	1.0951	598.4
230	6.55308	152.5	0.3246	0.2340	1.0940	602.7
240	6.64894	154.9	0.3280	0.2360	1.0930	606.9
250	6.74764	157.2	0.3313	0.2380	1.0920	611.1
260	6.84463	159.6	0.3347	0.2399	1.0911	615.3
270	6.94444	162.0	0.3380	0.2419	1.0901	619.5
280	7.03730	164.4	0.3413	0.2438	1.0892	623.6

PRESSURE = 12.00 PSIA

TEMP °F	V	H	S	Cp	Cp/Cv	v_s
-22.7 SAT LIQ	0.01155	5.2	0.0121	0.3017	1.5039	2509.8
-22.7 SAT VAP	3.68460	99.8	0.2286	0.1850	1.1519	476.2
-20	3.71195	100.3	0.2297	0.1854	1.1504	477.8
-10	3.80952	102.1	0.2339	0.1868	1.1455	483.9
0	3.90472	104.0	0.2380	0.1884	1.1410	489.8
10	4.00160	105.9	0.2421	0.1900	1.1370	495.6
20	4.09500	107.8	0.2461	0.1918	1.1334	501.2
30	4.18936	109.7	0.2501	0.1936	1.1301	506.8
40	4.28266	111.7	0.2540	0.1955	1.1270	512.2
50	4.37637	113.6	0.2579	0.1974	1.1242	517.5
60	4.46828	115.6	0.2618	0.1994	1.1216	522.8
70	4.55996	117.6	0.2656	0.2013	1.1192	527.9
80	4.65333	119.7	0.2694	0.2034	1.1169	533.0
90	4.74383	121.7	0.2731	0.2054	1.1148	538.0
100	4.83559	123.8	0.2768	0.2074	1.1128	543.0
110	4.92611	125.9	0.2805	0.2095	1.1109	547.9
120	5.01756	128.0	0.2842	0.2116	1.1091	552.7
130	5.10725	130.1	0.2878	0.2136	1.1074	557.4
140	5.19751	132.2	0.2914	0.2157	1.1058	562.1
150	5.28821	134.4	0.2950	0.2178	1.1043	566.8
160	5.37924	136.6	0.2986	0.2198	1.1028	571.4
170	5.46747	138.8	0.3021	0.2219	1.1015	575.9
180	5.55864	141.0	0.3056	0.2239	1.1001	580.5
190	5.64653	143.3	0.3091	0.2260	1.0989	584.9
200	5.73723	145.5	0.3126	0.2280	1.0977	589.3
210	5.82751	147.8	0.3160	0.2301	1.0965	593.7
220	5.91716	150.1	0.3195	0.2321	1.0954	598.1
230	6.00601	152.5	0.3229	0.2341	1.0943	602.4
240	6.09385	154.8	0.3262	0.2361	1.0932	606.7
250	6.18429	157.2	0.3296	0.2380	1.0922	610.9
260	6.27353	159.6	0.3330	0.2400	1.0913	615.1
270	6.36132	162.0	0.3363	0.2420	1.0903	619.3
280	6.44745	164.4	0.3396	0.2439	1.0894	623.4

Reprinted by permission from DuPont

Refrigerant 134a Data

Table 2 (continued)
HFC-134a Superheated Vapor—Constant Pressure Tables

V = Volume in ft³/lb H = Enthalpy in Btu/lb S = Entropy in Btu/(lb) (°R) v_s = Velocity of Sound in ft/sec
Cp = Heat Capacity at Constant Pressure in Btu/(lb) (°F) Cp/Cv = Heat Capacity Ratio (Dimensionless)

TEMP °F	PRESSURE = 13.00 PSIA							PRESSURE = 14.00 PSIA						TEMP °F
	V	H	S	Cp	Cp/Cv	v_s		V	H	S	Cp	Cp/Cv	v_s	
-19.7	0.01159	6.1	0.0142	0.3026	1.5041	2484.1	SAT LIQ	0.01163	7.0	0.0161	0.3035	1.5044	2459.9	-16.8
-19.7	3.41763	100.2	0.2281	0.1864	1.1527	476.9	SAT VAP	3.18776	100.7	0.2277	0.1876	1.1535	477.6	-16.8
-10	3.50508	102.0	0.2322	0.1876	1.1476	482.9		3.24570	101.9	0.2305	0.1885	1.1498	481.8	-10
0	3.59454	103.9	0.2363	0.1891	1.1429	488.8		3.32889	103.8	0.2347	0.1898	1.1448	487.9	0
10	3.68460	105.8	0.2404	0.1907	1.1387	494.7		3.41180	105.7	0.2388	0.1913	1.1404	493.8	10
20	3.77216	107.7	0.2444	0.1924	1.1349	500.4		3.49406	107.7	0.2429	0.1929	1.1363	499.5	20
30	3.85951	109.7	0.2484	0.1941	1.1314	506.0		3.57654	109.6	0.2468	0.1946	1.1327	505.2	30
40	3.94633	111.6	0.2523	0.1959	1.1282	511.5		3.65764	111.5	0.2508	0.1964	1.1294	510.7	40
50	4.03226	113.6	0.2562	0.1978	1.1253	516.8		3.73832	113.5	0.2547	0.1982	1.1263	516.1	50
60	4.11862	115.6	0.2601	0.1997	1.1225	522.1		3.81825	115.5	0.2586	0.2001	1.1235	521.5	60
70	4.20345	117.6	0.2639	0.2017	1.1200	527.3		3.89712	117.5	0.2624	0.2020	1.1209	526.7	70
80	4.28816	119.6	0.2677	0.2037	1.1177	532.4		3.97614	119.6	0.2662	0.2040	1.1185	531.9	80
90	4.37254	121.7	0.2715	0.2057	1.1155	537.5		4.05515	121.6	0.2700	0.2060	1.1162	536.9	90
100	4.45831	123.7	0.2752	0.2077	1.1134	542.5		4.13394	123.7	0.2737	0.2080	1.1141	541.9	100
110	4.54133	125.8	0.2789	0.2097	1.1115	547.4		4.21408	125.8	0.2774	0.2100	1.1121	546.9	110
120	4.62535	127.9	0.2826	0.2118	1.1097	552.2		4.29185	127.9	0.2811	0.2120	1.1102	551.7	120
130	4.71032	130.0	0.2862	0.2138	1.1079	557.0		4.36872	130.0	0.2847	0.2140	1.1084	556.6	130
140	4.79386	132.2	0.2898	0.2159	1.1063	561.7		4.44642	132.2	0.2883	0.2161	1.1068	561.3	140
150	4.87567	134.4	0.2934	0.2179	1.1047	566.4		4.52489	134.3	0.2919	0.2181	1.1052	566.0	150
160	4.96032	136.5	0.2970	0.2200	1.1033	571.0		4.60193	136.5	0.2955	0.2202	1.1037	570.6	160
170	5.04286	138.8	0.3005	0.2220	1.1018	575.6		4.67946	138.7	0.2990	0.2222	1.1022	575.2	170
180	5.12558	141.0	0.3040	0.2241	1.1005	580.1		4.75737	141.0	0.3026	0.2242	1.1008	579.8	180
190	5.20833	143.2	0.3075	0.2261	1.0992	584.6		4.83325	143.2	0.3060	0.2262	1.0995	584.3	190
200	5.29101	145.5	0.3110	0.2281	1.0980	589.0		4.91159	145.5	0.3095	0.2283	1.0983	588.7	200
210	5.37346	147.8	0.3144	0.2302	1.0968	593.4		4.98753	147.8	0.3130	0.2303	1.0971	593.1	210
220	5.45852	150.1	0.3179	0.2322	1.0956	597.8		5.06329	150.1	0.3164	0.2323	1.0959	597.5	220
230	5.54017	152.4	0.3213	0.2342	1.0945	602.1		5.14139	152.4	0.3198	0.2343	1.0948	601.9	230
240	5.62114	154.8	0.3247	0.2362	1.0935	606.4		5.21921	154.8	0.3232	0.2362	1.0937	606.1	240
250	5.70451	157.2	0.3280	0.2381	1.0925	610.7		5.29381	157.1	0.3266	0.2382	1.0927	610.4	250
260	5.78704	159.6	0.3314	0.2401	1.0915	614.9		5.37057	159.5	0.3299	0.2402	1.0917	614.6	260
270	5.86854	162.0	0.3347	0.2420	1.0905	619.0		5.44662	161.9	0.3332	0.2421	1.0908	618.8	270
280	5.95238	164.4	0.3380	0.2440	1.0896	623.2		5.52486	164.4	0.3365	0.2440	1.0898	623.0	280
290	6.03500	166.9	0.3413	0.2459	1.0887	627.3		5.59910	166.8	0.3398	0.2460	1.0889	627.1	290

TEMP °F	PRESSURE = 14.696 PSIA							PRESSURE = 15.00 PSIA						TEMP °F
	V	H	S	Cp	Cp/Cv	v_s		V	H	S	Cp	Cp/Cv	v_s	
-14.9	0.01166	7.5	0.0174	0.3041	1.5046	2443.9	SAT LIQ	0.01167	7.8	0.0180	0.3043	1.5047	2437.1	-14.1
-14.9	3.04507	100.9	0.2274	0.1885	1.1540	478.0	SAT VAP	2.98686	101.1	0.2273	0.1888	1.1543	478.1	-14.1
-10	3.08547	101.9	0.2295	0.1891	1.1513	481.0		3.01932	101.8	0.2290	0.1893	1.1520	480.7	-10
0	3.16556	103.8	0.2336	0.1904	1.1461	487.2		3.09885	103.7	0.2332	0.1906	1.1467	486.9	0
10	3.24465	105.7	0.2378	0.1918	1.1415	493.1		3.17662	105.6	0.2373	0.1920	1.1420	492.8	10
20	3.32336	107.6	0.2418	0.1934	1.1374	498.9		3.25415	107.6	0.2414	0.1935	1.1378	498.7	20
30	3.40136	109.5	0.2458	0.1950	1.1336	504.6		3.33111	109.5	0.2454	0.1952	1.1340	504.4	30
40	3.47947	111.5	0.2498	0.1967	1.1302	510.2		3.40716	111.5	0.2494	0.1969	1.1306	510.0	40
50	3.55619	113.5	0.2537	0.1985	1.1271	515.7		3.48189	113.5	0.2533	0.1987	1.1274	515.5	50
60	3.63240	115.5	0.2576	0.2004	1.1242	521.0		3.55745	115.4	0.2571	0.2005	1.1245	520.8	60
70	3.70920	117.5	0.2614	0.2023	1.1215	526.3		3.63240	117.5	0.2610	0.2024	1.1218	526.1	70
80	3.78501	119.5	0.2652	0.2042	1.1190	531.5		3.70645	119.5	0.2648	0.2043	1.1193	531.3	80
90	3.85951	121.6	0.2690	0.2062	1.1167	536.6		3.78072	121.5	0.2686	0.2063	1.1169	536.4	90
100	3.93546	123.6	0.2727	0.2082	1.1145	541.6		3.85356	123.6	0.2723	0.2082	1.1147	541.4	100
110	4.01123	125.7	0.2764	0.2102	1.1125	546.5		3.92773	125.7	0.2760	0.2102	1.1127	546.4	110
120	4.08497	127.8	0.2801	0.2122	1.1106	551.4		4.00160	127.8	0.2797	0.2122	1.1108	551.3	120
130	4.15973	130.0	0.2837	0.2142	1.1088	556.2		4.07332	130.0	0.2833	0.2142	1.1090	556.1	130
140	4.23370	132.1	0.2874	0.2162	1.1071	561.0		4.14594	132.1	0.2869	0.2163	1.1072	560.9	140
150	4.30849	134.3	0.2910	0.2182	1.1055	565.7		4.21941	134.3	0.2905	0.2183	1.1056	565.6	150
160	4.38212	136.5	0.2945	0.2203	1.1039	570.4		4.29185	136.5	0.2941	0.2203	1.1041	570.3	160
170	4.45633	138.7	0.2981	0.2223	1.1025	575.0		4.36300	138.7	0.2977	0.2223	1.1026	574.9	170
180	4.52899	140.9	0.3016	0.2243	1.1011	579.5		4.43656	140.9	0.3012	0.2244	1.1012	579.4	180
190	4.60193	143.2	0.3051	0.2263	1.0998	584.1		4.50857	143.2	0.3047	0.2264	1.0999	584.0	190
200	4.67727	145.5	0.3086	0.2284	1.0985	588.5		4.58085	145.5	0.3081	0.2284	1.0986	588.4	200
210	4.74834	147.8	0.3120	0.2304	1.0973	592.9		4.65333	147.7	0.3116	0.2304	1.0974	592.8	210
220	4.82393	150.1	0.3154	0.2324	1.0961	597.3		4.72387	150.1	0.3150	0.2324	1.0962	597.2	220
230	4.89476	152.4	0.3188	0.2343	1.0950	601.7		4.79616	152.4	0.3184	0.2344	1.0951	601.6	230
240	4.97018	154.8	0.3222	0.2363	1.0939	606.0		4.86618	154.7	0.3218	0.2363	1.0940	605.9	240
250	5.04286	157.1	0.3256	0.2383	1.0929	610.2		4.93827	157.1	0.3252	0.2383	1.0929	610.2	250
260	5.11509	159.5	0.3289	0.2402	1.0919	614.5		5.01002	159.5	0.3285	0.2403	1.0919	614.4	260
270	5.18672	161.9	0.3323	0.2422	1.0909	618.7		5.08130	161.9	0.3319	0.2422	1.0910	618.6	270
280	5.26039	164.4	0.3356	0.2441	1.0900	622.8		5.15198	164.4	0.3352	0.2441	1.0900	622.8	280
290	5.33333	166.8	0.3389	0.2460	1.0891	627.0		5.22466	166.8	0.3385	0.2460	1.0891	626.9	290

Reprinted by permission from DuPont

Table 2 (continued)
HFC-134a Superheated Vapor—Constant Pressure Tables

V = Volume in ft³/lb H = Enthalpy in Btu/lb S = Entropy in Btu/(lb) (°R) v_s = Velocity of Sound in ft/sec
Cp = Heat Capacity at Constant Pressure in Btu/(lb) (°F) Cp/Cv = Heat Capacity Ratio (Dimensionless)

TEMP °F	PRESSURE = 16.00 PSIA							PRESSURE = 17.00 PSIA						TEMP °F
	V	H	S	Cp	Cp/Cv	v_s		V	H	S	Cp	Cp/Cv	v_s	
-11.5	0.01170	8.6	0.0197	0.3051	1.5051	2415.5	SAT LIQ	0.01174	9.3	0.0214	0.3059	1.5056	2395.0	-9.1
-11.5	2.81057	101.4	0.2269	0.1900	1.1551	478.7	SAT VAP	2.65463	101.8	0.2266	0.1911	1.1559	479.1	-9.1
-10	2.82247	101.7	0.2276	0.1902	1.1542	479.6		—	—	—	—	—	—	-10
0	2.89687	103.6	0.2318	0.1913	1.1487	485.9		2.71887	103.5	0.2305	0.1921	1.1507	484.9	0
10	2.97089	105.6	0.2359	0.1927	1.1438	491.9		2.78862	105.5	0.2346	0.1934	1.1455	491.0	10
20	3.04321	107.5	0.2400	0.1941	1.1394	497.8		2.85796	107.4	0.2387	0.1947	1.1409	497.0	20
30	3.11624	109.4	0.2440	0.1957	1.1354	503.6		2.92654	109.4	0.2427	0.1963	1.1368	502.8	30
40	3.18776	111.4	0.2480	0.1974	1.1318	509.2		2.99401	111.3	0.2467	0.1979	1.1330	508.5	40
50	3.25839	113.4	0.2519	0.1991	1.1285	514.7		3.06185	113.3	0.2506	0.1995	1.1296	514.0	50
60	3.33000	115.4	0.2558	0.2009	1.1254	520.2		3.12891	115.3	0.2545	0.2013	1.1264	519.5	60
70	3.40020	117.4	0.2597	0.2028	1.1226	525.5		3.19489	117.3	0.2584	0.2031	1.1235	524.9	70
80	3.46981	119.4	0.2635	0.2046	1.1201	530.7		3.26158	119.4	0.2622	0.2050	1.1209	530.1	80
90	3.53982	121.5	0.2672	0.2066	1.1177	535.8		3.32668	121.4	0.2660	0.2069	1.1184	535.3	90
100	3.60881	123.6	0.2710	0.2085	1.1154	540.9		3.39213	123.5	0.2697	0.2088	1.1161	540.4	100
110	3.67782	125.7	0.2747	0.2105	1.1133	545.9		3.45781	125.6	0.2734	0.2107	1.1139	545.4	110
120	3.74672	127.8	0.2784	0.2125	1.1113	550.8		3.52237	127.7	0.2771	0.2127	1.1119	550.4	120
130	3.81534	129.9	0.2820	0.2145	1.1095	555.7		3.58809	129.9	0.2808	0.2147	1.1100	555.2	130
140	3.88350	132.1	0.2856	0.2165	1.1077	560.5		3.65230	132.0	0.2844	0.2167	1.1082	560.1	140
150	3.95257	134.2	0.2892	0.2185	1.1061	565.2		3.71609	134.2	0.2880	0.2186	1.1065	564.8	150
160	4.01929	136.4	0.2928	0.2205	1.1045	569.9		3.78072	136.4	0.2916	0.2206	1.1049	569.5	160
170	4.08831	138.7	0.2964	0.2225	1.1030	574.5		3.84468	138.6	0.2951	0.2226	1.1034	574.2	170
180	4.15628	140.9	0.2999	0.2245	1.1016	579.1		3.90930	140.9	0.2987	0.2246	1.1019	578.8	180
190	4.22297	143.1	0.3034	0.2265	1.1002	583.6		3.97298	143.1	0.3022	0.2266	1.1005	583.3	190
200	4.29185	145.4	0.3069	0.2285	1.0989	588.1		4.03551	145.4	0.3056	0.2286	1.0992	587.8	200
210	4.35920	147.7	0.3103	0.2305	1.0977	592.6		4.10004	147.7	0.3091	0.2306	1.0979	592.3	210
220	4.42674	150.0	0.3137	0.2325	1.0965	597.0		4.16320	150.0	0.3125	0.2326	1.0967	596.7	220
230	4.49438	152.4	0.3171	0.2345	1.0953	601.3		4.22654	152.3	0.3159	0.2346	1.0956	601.0	230
240	4.55996	154.7	0.3205	0.2364	1.0942	605.6		4.29000	154.7	0.3193	0.2365	1.0945	605.4	240
250	4.62749	157.1	0.3239	0.2384	1.0932	609.9		4.35350	157.1	0.3227	0.2385	1.0934	609.7	250
260	4.69484	159.5	0.3273	0.2403	1.0921	614.2		4.41696	159.5	0.3260	0.2404	1.0924	613.9	260
270	4.76190	161.9	0.3306	0.2423	1.0912	618.4		4.48029	161.9	0.3294	0.2423	1.0914	618.1	270
280	4.82859	164.3	0.3339	0.2442	1.0902	622.5		4.54339	164.3	0.3327	0.2443	1.0904	622.3	280
290	4.89716	166.8	0.3372	0.2461	1.0893	626.7		4.60617	166.8	0.3360	0.2462	1.0895	626.5	290
300	—	—	—	—	—	—		4.66853	169.2	0.3393	0.2480	1.0886	630.6	300

TEMP °F	PRESSURE = 18.00 PSIA							PRESSURE = 19.00 PSIA						TEMP °F
	V	H	S	Cp	Cp/Cv	v_s		V	H	S	Cp	Cp/Cv	v_s	
-6.8	0.01177	10.0	0.0230	0.3067	1.5060	2375.4	SAT LIQ	0.01181	10.7	0.0245	0.3074	1.5065	2356.7	-4.5
-6.8	2.51509	102.1	0.2263	0.1922	1.1568	479.5	SAT VAP	2.38949	102.5	0.2261	0.1933	1.1576	479.9	-4.5
0	2.56016	103.4	0.2292	0.1929	1.1527	483.8		2.41896	103.3	0.2280	0.1937	1.1548	482.8	0
10	2.62674	105.4	0.2333	0.1940	1.1473	490.1		2.48201	105.3	0.2321	0.1947	1.1491	489.1	10
20	2.69251	107.3	0.2374	0.1954	1.1425	496.1		2.54518	107.2	0.2363	0.1960	1.1441	495.2	20
30	2.75786	109.3	0.2415	0.1968	1.1382	502.0		2.60688	109.2	0.2403	0.1974	1.1396	501.2	30
40	2.82247	111.3	0.2455	0.1984	1.1342	507.7		2.66880	111.2	0.2443	0.1988	1.1355	507.0	40
50	2.88600	113.3	0.2494	0.2000	1.1307	513.3		2.72926	113.2	0.2483	0.2004	1.1318	512.6	50
60	2.94985	115.3	0.2533	0.2017	1.1274	518.8		2.78940	115.2	0.2522	0.2021	1.1284	518.2	60
70	3.01296	117.3	0.2572	0.2035	1.1244	524.2		2.84981	117.2	0.2561	0.2038	1.1253	523.6	70
80	3.07503	119.3	0.2610	0.2053	1.1217	529.5		2.90951	119.3	0.2599	0.2056	1.1225	529.0	80
90	3.13775	121.4	0.2648	0.2072	1.1191	534.7		2.96824	121.3	0.2637	0.2075	1.1199	534.2	90
100	3.20000	123.5	0.2686	0.2091	1.1168	539.9		3.02755	123.4	0.2674	0.2093	1.1174	539.4	100
110	3.26158	125.6	0.2723	0.2110	1.1145	544.9		3.08642	125.5	0.2712	0.2112	1.1152	544.4	110
120	3.32336	127.7	0.2760	0.2129	1.1125	549.9		3.14465	127.7	0.2749	0.2131	1.1130	549.4	120
130	3.38524	129.8	0.2796	0.2149	1.1105	554.8		3.20307	129.8	0.2785	0.2151	1.1110	554.4	130
140	3.44590	132.0	0.2833	0.2168	1.1087	559.6		3.26158	132.0	0.2821	0.2170	1.1092	559.2	140
150	3.50631	134.2	0.2869	0.2188	1.1069	564.4		3.31895	134.1	0.2858	0.2190	1.1074	564.0	150
160	3.56761	136.4	0.2904	0.2208	1.1053	569.1		3.37724	136.3	0.2893	0.2210	1.1057	568.8	160
170	3.62845	138.6	0.2940	0.2228	1.1038	573.8		3.43407	138.6	0.2929	0.2230	1.1041	573.4	170
180	3.68868	140.8	0.2975	0.2248	1.1023	578.4		3.49162	140.8	0.2964	0.2249	1.1026	578.1	180
190	3.74813	143.1	0.3010	0.2268	1.1009	583.0		3.54862	143.1	0.2999	0.2269	1.1012	582.7	190
200	3.80952	145.4	0.3045	0.2288	1.0995	587.5		3.60620	145.3	0.3034	0.2289	1.0998	587.2	200
210	3.86997	147.7	0.3079	0.2307	1.0982	592.0		3.66300	147.6	0.3069	0.2308	1.0985	591.7	210
220	3.92927	150.0	0.3114	0.2327	1.0970	596.4		3.72024	149.9	0.3103	0.2328	1.0973	596.1	220
230	3.99042	152.3	0.3148	0.2347	1.0958	600.8		3.77786	152.3	0.3137	0.2348	1.0961	600.5	230
240	4.04858	154.7	0.3182	0.2366	1.0947	605.1		3.83436	154.6	0.3171	0.2367	1.0950	604.9	240
250	4.10846	157.0	0.3216	0.2386	1.0936	609.4		3.89105	157.0	0.3205	0.2387	1.0939	609.2	250
260	4.17014	159.4	0.3249	0.2405	1.0926	613.7		3.94789	159.4	0.3238	0.2406	1.0928	613.5	260
270	4.22833	161.9	0.3282	0.2424	1.0916	617.9		4.00481	161.8	0.3272	0.2425	1.0918	617.7	270
280	4.28816	164.3	0.3316	0.2443	1.0906	622.1		4.06174	164.3	0.3305	0.2444	1.0908	621.9	280
290	4.34783	166.7	0.3348	0.2462	1.0897	626.3		4.11692	166.7	0.3338	0.2463	1.0899	626.1	290
300	4.40917	169.2	0.3381	0.2481	1.0888	630.4		4.17362	169.2	0.3370	0.2482	1.0890	630.2	300

Reprinted by permission from DuPont

Table 2 (continued)
HFC-134a Superheated Vapor—Constant Pressure Tables

V = Volume in ft³/lb H = Enthalpy in Btu/lb S = Entropy in Btu/(lb) (°R) v_s = Velocity of Sound in ft/sec
Cp = Heat Capacity at Constant Pressure in Btu/(lb) (°F) Cp/Cv = Heat Capacity Ratio (Dimensionless)

TEMP °F	PRESSURE = 20.00 PSIA							PRESSURE = 21.00 PSIA						TEMP °F
	V	H	S	Cp	Cp/Cv	v_s		V	H	S	Cp	Cp/Cv	v_s	
-2.4	0.01184	11.4	0.0259	0.3081	1.5071	2338.7	SAT LIQ	0.01187	12.0	0.0273	0.3088	1.5076	2321.4	-0.3
-2.4	2.27635	102.8	0.2258	0.1943	1.1585	480.2	SAT VAP	2.17391	103.1	0.2256	0.1953	1.1593	480.5	-0.3
0	2.29095	103.3	0.2268	0.1945	1.1569	481.8		2.17581	103.2	0.2257	0.1953	1.1591	480.8	0
10	2.35183	105.2	0.2310	0.1955	1.1510	488.2		2.23414	105.1	0.2299	0.1962	1.1529	487.2	10
20	2.41196	107.2	0.2351	0.1966	1.1457	494.3		2.29148	107.1	0.2341	0.1972	1.1474	493.5	20
30	2.47158	109.1	0.2392	0.1979	1.1410	500.3		2.34852	109.1	0.2381	0.1985	1.1425	499.5	30
40	2.52972	111.1	0.2432	0.1994	1.1368	506.2		2.40442	111.0	0.2422	0.1999	1.1381	505.4	40
50	2.58799	113.1	0.2472	0.2009	1.1329	511.9		2.46063	113.1	0.2461	0.2013	1.1341	511.2	50
60	2.64550	115.1	0.2511	0.2025	1.1295	517.5		2.51572	115.1	0.2501	0.2029	1.1305	516.8	60
70	2.70270	117.2	0.2550	0.2042	1.1263	523.0		2.57003	117.1	0.2539	0.2046	1.1272	522.4	70
80	2.76014	119.2	0.2588	0.2060	1.1233	528.4		2.62467	119.2	0.2578	0.2063	1.1242	527.8	80
90	2.81690	121.3	0.2626	0.2078	1.1206	533.6		2.67881	121.2	0.2616	0.2081	1.1214	533.1	90
100	2.87274	123.4	0.2664	0.2096	1.1181	538.8		2.73224	123.3	0.2654	0.2099	1.1188	538.3	100
110	2.92826	125.5	0.2701	0.2115	1.1158	543.9		2.78552	125.4	0.2691	0.2117	1.1164	543.4	110
120	2.98418	127.6	0.2738	0.2134	1.1136	549.0		2.83930	127.6	0.2728	0.2136	1.1142	548.5	120
130	3.03951	129.7	0.2775	0.2153	1.1116	553.9		2.89184	129.7	0.2765	0.2155	1.1121	553.5	130
140	3.09502	131.9	0.2811	0.2172	1.1097	558.8		2.94464	131.9	0.2801	0.2174	1.1101	558.4	140
150	3.14961	134.1	0.2847	0.2192	1.1078	563.6		2.99760	134.1	0.2837	0.2194	1.1083	563.2	150
160	3.20513	136.3	0.2883	0.2211	1.1061	568.4		3.04971	136.3	0.2873	0.2213	1.1066	568.0	160
170	3.26052	138.5	0.2918	0.2231	1.1045	573.1		3.10270	138.5	0.2909	0.2233	1.1049	572.7	170
180	3.31455	140.8	0.2954	0.2251	1.1030	577.7		3.15457	140.7	0.2944	0.2252	1.1034	577.4	180
190	3.36927	143.0	0.2989	0.2270	1.1015	582.3		3.20616	143.0	0.2979	0.2272	1.1019	582.0	190
200	3.42349	145.3	0.3024	0.2290	1.1002	586.9		3.25839	145.3	0.3014	0.2291	1.1005	586.6	200
210	3.47826	147.6	0.3058	0.2310	1.0988	591.4		3.31016	147.6	0.3048	0.2311	1.0991	591.1	210
220	3.53232	149.9	0.3093	0.2329	1.0976	595.8		3.36247	149.9	0.3083	0.2330	1.0979	595.5	220
230	3.58680	152.3	0.3127	0.2349	1.0964	600.2		3.41413	152.2	0.3117	0.2350	1.0966	600.0	230
240	3.64033	154.6	0.3161	0.2368	1.0952	604.6		3.46500	154.6	0.3151	0.2369	1.0955	604.3	240
250	3.69413	157.0	0.3195	0.2387	1.0941	608.9		3.51741	157.0	0.3185	0.2388	1.0943	608.7	250
260	3.74813	159.4	0.3228	0.2407	1.0930	613.2		3.56888	159.4	0.3218	0.2408	1.0932	613.0	260
270	3.80228	161.8	0.3261	0.2426	1.0920	617.5		3.61925	161.8	0.3252	0.2427	1.0922	617.2	270
280	3.85654	164.2	0.3295	0.2445	1.0910	621.7		3.67107	164.2	0.3285	0.2446	1.0912	621.5	280
290	3.91083	166.7	0.3328	0.2464	1.0901	625.9		3.72162	166.7	0.3318	0.2464	1.0902	625.7	290
300	3.96354	169.2	0.3360	0.2482	1.0891	630.0		3.77358	169.1	0.3351	0.2483	1.0893	629.8	300

TEMP °F	PRESSURE = 22.00 PSIA							PRESSURE = 23.00 PSIA						TEMP °F
	V	H	S	Cp	Cp/Cv	v_s		V	H	S	Cp	Cp/Cv	v_s	
1.7	0.01190	12.6	0.0286	0.3094	1.5082	2304.8	SAT LIQ	0.01193	13.2	0.0299	0.3101	1.5088	2288.8	3.6
1.7	2.07987	103.4	0.2253	0.1962	1.1602	480.8	SAT VAP	1.99402	103.7	0.2251	0.1972	1.1610	481.1	3.6
10	2.12675	105.0	0.2289	0.1969	1.1548	486.2		2.02881	104.9	0.2278	0.1976	1.1567	485.3	10
20	2.18198	107.0	0.2330	0.1979	1.1490	492.6		2.08247	106.9	0.2320	0.1985	1.1507	491.7	20
30	2.23664	109.0	0.2371	0.1991	1.1439	498.7		2.13493	108.9	0.2361	0.1996	1.1454	497.9	30
40	2.29095	111.0	0.2411	0.2004	1.1394	504.7		2.18675	110.9	0.2402	0.2009	1.1407	503.9	40
50	2.34412	113.0	0.2451	0.2018	1.1352	510.5		2.23814	112.9	0.2442	0.2023	1.1364	509.8	50
60	2.39693	115.0	0.2491	0.2033	1.1315	516.2		2.28885	114.9	0.2481	0.2037	1.1326	515.5	60
70	2.44978	117.1	0.2530	0.2049	1.1281	521.7		2.33918	117.0	0.2520	0.2053	1.1291	521.1	70
80	2.50188	119.1	0.2568	0.2066	1.1250	527.2		2.38949	119.1	0.2559	0.2070	1.1259	526.6	80
90	2.55363	121.2	0.2606	0.2084	1.1222	532.5		2.43902	121.1	0.2597	0.2087	1.1229	532.0	90
100	2.60485	123.3	0.2644	0.2102	1.1195	537.8		2.48818	123.2	0.2635	0.2104	1.1202	537.3	100
110	2.65604	125.4	0.2681	0.2120	1.1171	542.9		2.53743	125.3	0.2672	0.2122	1.1177	542.4	110
120	2.70709	127.5	0.2718	0.2138	1.1148	548.0		2.58665	127.5	0.2709	0.2141	1.1154	547.6	120
130	2.75786	129.7	0.2755	0.2157	1.1126	553.0		2.63505	129.6	0.2746	0.2159	1.1132	552.6	130
140	2.80820	131.8	0.2791	0.2176	1.1106	558.0		2.68384	131.8	0.2782	0.2178	1.1111	557.5	140
150	2.85878	134.0	0.2828	0.2195	1.1088	562.8		2.73149	134.0	0.2819	0.2197	1.1092	562.4	150
160	2.90867	136.2	0.2864	0.2215	1.1070	567.6		2.78009	136.2	0.2854	0.2216	1.1074	567.2	160
170	2.95946	138.4	0.2899	0.2234	1.1053	572.4		2.82805	138.4	0.2890	0.2236	1.1057	572.0	170
180	3.00842	140.7	0.2934	0.2254	1.1037	577.0		2.87604	140.7	0.2925	0.2255	1.1041	576.7	180
190	3.05904	143.0	0.2970	0.2273	1.1022	581.7		2.92312	142.9	0.2961	0.2274	1.1026	581.3	190
200	3.10849	145.2	0.3004	0.2293	1.1008	586.2		2.97177	145.2	0.2995	0.2294	1.1011	585.9	200
210	3.15756	147.5	0.3039	0.2312	1.0994	590.8		3.01932	147.5	0.3030	0.2313	1.0997	590.5	210
220	3.20821	149.9	0.3074	0.2331	1.0981	595.3		3.06654	149.8	0.3065	0.2332	1.0984	595.0	220
230	3.25627	152.2	0.3108	0.2351	1.0969	599.7		3.11333	152.2	0.3099	0.2352	1.0972	599.4	230
240	3.30579	154.6	0.3142	0.2370	1.0957	604.1		3.16056	154.5	0.3133	0.2371	1.0960	603.8	240
250	3.35570	156.9	0.3175	0.2389	1.0946	608.4		3.20821	156.9	0.3167	0.2390	1.0948	608.2	250
260	3.40483	159.3	0.3209	0.2408	1.0935	612.7		3.25521	159.3	0.3200	0.2409	1.0937	612.5	260
270	3.45304	161.8	0.3242	0.2427	1.0924	617.0		3.30251	161.7	0.3233	0.2428	1.0926	616.8	270
280	3.50263	164.2	0.3276	0.2446	1.0914	621.3		3.34896	164.2	0.3267	0.2447	1.0916	621.0	280
290	3.55240	166.7	0.3309	0.2465	1.0904	625.5		3.39674	166.6	0.3300	0.2466	1.0906	625.2	290
300	3.60101	169.1	0.3341	0.2484	1.0895	629.6		3.44234	169.1	0.3332	0.2484	1.0897	629.4	300
310	3.64964	171.6	0.3374	0.2502	1.0886	633.8		3.48918	171.6	0.3365	0.2503	1.0888	633.6	310

Table 2 (continued)
HFC-134a Superheated Vapor—Constant Pressure Tables

V = Volume in ft³/lb   H = Enthalpy in Btu/lb   S = Entropy in Btu/(lb) (°R) v_s = Velocity of Sound in ft/sec
Cp = Heat Capacity at Constant Pressure in Btu/(lb) (°F)   Cp/Cv = Heat Capacity Ratio (Dimensionless)

TEMP °F	PRESSURE = 24.00 PSIA							PRESSURE = 25.00 PSIA						TEMP °F
	V	H	S	Cp	Cp/Cv	v_s		V	H	S	Cp	Cp/Cv	v_s	
5.4	0.01195	13.8	0.0311	0.3107	1.5094	2273.3	SAT LIQ	0.01198	14.4	0.0323	0.3113	1.5100	2258.3	7.2
5.4	1.91534	103.9	0.2249	0.1981	1.1619	481.3	SAT VAP	1.84264	104.2	0.2247	0.1990	1.1627	481.5	7.2
10	1.93911	104.8	0.2269	0.1984	1.1587	484.3		1.85667	104.7	0.2259	0.1992	1.1607	483.3	10
20	1.99045	106.8	0.2311	0.1992	1.1525	490.8		1.90621	106.7	0.2301	0.1999	1.1542	489.9	20
30	2.04123	108.8	0.2352	0.2002	1.1469	497.0		1.95542	108.7	0.2343	0.2008	1.1485	496.2	30
40	2.09118	110.8	0.2392	0.2014	1.1420	503.1		2.00361	110.8	0.2383	0.2019	1.1434	502.3	40
50	2.14087	112.9	0.2432	0.2027	1.1376	509.0		2.05128	112.8	0.2423	0.2032	1.1388	508.3	50
60	2.18962	114.9	0.2472	0.2042	1.1336	514.8		2.09864	114.8	0.2463	0.2046	1.1347	514.1	60
70	2.23814	116.9	0.2511	0.2057	1.1300	520.5		2.14546	116.9	0.2502	0.2061	1.1310	519.8	70
80	2.28624	119.0	0.2550	0.2073	1.1267	526.0		2.19202	118.9	0.2541	0.2077	1.1276	525.4	80
90	2.33427	121.1	0.2588	0.2090	1.1237	531.4		2.23764	121.0	0.2579	0.2093	1.1245	530.8	90
100	2.38152	123.2	0.2626	0.2107	1.1209	536.7		2.28363	123.1	0.2617	0.2110	1.1216	536.2	100
110	2.42895	125.3	0.2663	0.2125	1.1183	541.9		2.32937	125.2	0.2655	0.2128	1.1190	541.4	110
120	2.47586	127.4	0.2700	0.2143	1.1160	547.1		2.37417	127.4	0.2692	0.2145	1.1166	546.6	120
130	2.52270	129.6	0.2737	0.2162	1.1137	552.1		2.41955	129.5	0.2729	0.2164	1.1143	551.7	130
140	2.56937	131.8	0.2774	0.2180	1.1116	557.1		2.46427	131.7	0.2765	0.2182	1.1121	556.7	140
150	2.61575	133.9	0.2810	0.2199	1.1097	562.0		2.50878	133.9	0.2801	0.2201	1.1101	561.6	150
160	2.66170	136.1	0.2846	0.2218	1.1078	566.9		2.55363	136.1	0.2837	0.2220	1.1083	566.5	160
170	2.70856	138.4	0.2881	0.2237	1.1061	571.6		2.59740	138.3	0.2873	0.2239	1.1065	571.3	170
180	2.75406	140.6	0.2917	0.2256	1.1045	576.4		2.64201	140.6	0.2908	0.2258	1.1048	576.0	180
190	2.80034	142.9	0.2952	0.2276	1.1029	581.0		2.68601	142.9	0.2944	0.2277	1.1033	580.7	190
200	2.84495	145.2	0.2987	0.2295	1.1014	585.6		2.73000	145.1	0.2979	0.2296	1.1018	585.3	200
210	2.89101	147.5	0.3021	0.2314	1.1000	590.2		2.77316	147.5	0.3013	0.2315	1.1003	589.9	210
220	2.93686	149.8	0.3056	0.2334	1.0987	594.7		2.81690	149.8	0.3048	0.2335	1.0990	594.4	220
230	2.98240	152.1	0.3090	0.2353	1.0974	599.2		2.86123	152.1	0.3082	0.2354	1.0977	598.9	230
240	3.02755	154.5	0.3124	0.2372	1.0962	603.6		2.90529	154.5	0.3116	0.2373	1.0965	603.3	240
250	3.07220	156.9	0.3158	0.2391	1.0950	607.9		2.94811	156.9	0.3150	0.2392	1.0953	607.7	250
260	3.11818	159.3	0.3192	0.2410	1.0939	612.3		2.99222	159.3	0.3183	0.2411	1.0941	612.0	260
270	3.16356	161.7	0.3225	0.2429	1.0928	616.6		3.03490	161.7	0.3217	0.2430	1.0931	616.3	270
280	3.20821	164.1	0.3258	0.2448	1.0918	620.8		3.07882	164.1	0.3250	0.2449	1.0920	620.6	280
290	3.25309	166.6	0.3291	0.2467	1.0908	625.0		3.12207	166.6	0.3283	0.2467	1.0910	624.8	290
300	3.29815	169.1	0.3324	0.2485	1.0899	629.2		3.16456	169.1	0.3316	0.2486	1.0900	629.0	300
310	3.34336	171.6	0.3357	0.2504	1.0889	633.4		3.20821	171.6	0.3348	0.2504	1.0891	633.2	310

TEMP °F	PRESSURE = 26.00 PSIA							PRESSURE = 27.00 PSIA						TEMP °F
	V	H	S	Cp	Cp/Cv	v_s		V	H	S	Cp	Cp/Cv	v_s	
8.9	0.01201	14.9	0.0335	0.3119	1.5106	2243.7	SAT LIQ	0.01203	15.4	0.0346	0.3125	1.5112	2229.6	10.6
8.9	1.77494	104.4	0.2246	0.1999	1.1636	481.6	SAT VAP	1.71233	104.7	0.2244	0.2007	1.1644	481.8	10.6
10	1.78031	104.7	0.2250	0.1999	1.1628	482.3		—	—	—	—	—	—	10
20	1.82849	106.7	0.2292	0.2005	1.1560	489.0		1.75623	106.6	0.2284	0.2012	1.1578	488.0	20
30	1.87582	108.7	0.2334	0.2014	1.1500	495.4		1.80213	108.6	0.2325	0.2020	1.1516	494.5	30
40	1.92271	110.7	0.2375	0.2025	1.1448	501.6		1.84740	110.6	0.2366	0.2030	1.1462	500.8	40
50	1.96850	112.7	0.2415	0.2037	1.1400	507.6		1.89215	112.6	0.2406	0.2042	1.1413	506.9	50
60	2.01450	114.8	0.2454	0.2050	1.1358	513.5		1.93648	114.7	0.2446	0.2054	1.1369	512.8	60
70	2.05973	116.8	0.2494	0.2065	1.1320	519.2		1.98020	116.8	0.2485	0.2068	1.1330	518.6	70
80	2.10438	118.9	0.2532	0.2080	1.1285	524.8		2.02347	118.8	0.2524	0.2083	1.1294	524.2	80
90	2.14869	121.0	0.2571	0.2096	1.1253	530.3		2.06654	120.9	0.2563	0.2099	1.1261	529.7	90
100	2.19298	123.1	0.2609	0.2113	1.1224	535.7		2.10926	123.0	0.2601	0.2116	1.1231	535.1	100
110	2.23714	125.2	0.2646	0.2130	1.1197	540.9		2.15146	125.2	0.2638	0.2133	1.1203	540.4	110
120	2.28050	127.3	0.2684	0.2148	1.1171	546.1		2.19346	127.3	0.2676	0.2150	1.1178	545.7	120
130	2.32396	129.5	0.2720	0.2166	1.1148	551.2		2.23514	129.5	0.2713	0.2168	1.1154	550.8	130
140	2.36686	131.7	0.2757	0.2184	1.1126	556.3		2.27739	131.6	0.2749	0.2186	1.1132	555.8	140
150	2.41022	133.9	0.2793	0.2203	1.1106	561.2		2.31857	133.8	0.2785	0.2205	1.1111	560.8	150
160	2.45278	136.1	0.2829	0.2222	1.1087	566.1		2.36016	136.0	0.2821	0.2223	1.1091	565.7	160
170	2.49563	138.3	0.2865	0.2240	1.1069	570.9		2.40154	138.3	0.2857	0.2242	1.1073	570.6	170
180	2.53807	140.6	0.2900	0.2259	1.1052	575.7		2.44260	140.5	0.2893	0.2261	1.1056	575.3	180
190	2.58065	142.8	0.2936	0.2278	1.1036	580.4		2.48324	142.8	0.2928	0.2280	1.1040	580.0	190
200	2.62329	145.1	0.2971	0.2298	1.1021	585.0		2.52398	145.1	0.2963	0.2299	1.1024	584.7	200
210	2.66525	147.4	0.3005	0.2317	1.1006	589.6		2.56476	147.4	0.2998	0.2318	1.1009	589.3	210
220	2.70783	149.7	0.3040	0.2336	1.0993	594.1		2.60552	149.7	0.3032	0.2337	1.0996	593.8	220
230	2.74952	152.1	0.3074	0.2355	1.0980	598.6		2.64620	152.1	0.3066	0.2356	1.0982	598.3	230
240	2.79174	154.5	0.3108	0.2374	1.0967	603.1		2.68673	154.4	0.3100	0.2375	1.0970	602.8	240
250	2.83366	156.8	0.3142	0.2393	1.0955	607.4		2.72702	156.8	0.3134	0.2394	1.0957	607.2	250
260	2.87522	159.2	0.3175	0.2412	1.0944	611.8		2.76702	159.2	0.3168	0.2413	1.0946	611.6	260
270	2.91715	161.7	0.3209	0.2431	1.0933	616.1		2.80741	161.6	0.3201	0.2431	1.0935	615.9	270
280	2.95858	164.1	0.3242	0.2449	1.0922	620.4		2.84819	164.1	0.3235	0.2450	1.0924	620.2	280
290	3.00030	166.6	0.3275	0.2468	1.0912	624.6		2.88850	166.5	0.3268	0.2469	1.0914	624.4	290
300	3.04136	169.0	0.3308	0.2487	1.0902	628.8		2.92826	169.0	0.3300	0.2487	1.0904	628.6	300
310	3.08356	171.5	0.3341	0.2505	1.0893	633.0		2.96824	171.5	0.3333	0.2506	1.0895	632.8	310
320	—	—	—	—	—	—		3.00842	174.0	0.3365	0.2524	1.0885	636.9	320

Reprinted by permission from DuPont

Refrigerant 134a Data

Table 2 (continued)
HFC-134a Superheated Vapor—Constant Pressure Tables

V = Volume in ft³/lb   H = Enthalpy in Btu/lb   S = Entropy in Btu/(lb) (°R) v_s = Velocity of Sound in ft/sec
Cp = Heat Capacity at Constant Pressure in Btu/(lb) (°F)   Cp/Cv = Heat Capacity Ratio (Dimensionless)

TEMP °F	PRESSURE = 28.00 PSIA							PRESSURE = 29.00 PSIA						TEMP °F
	V	H	S	Cp	Cp/Cv	v_s		V	H	S	Cp	Cp/Cv	v_s	
12.3	0.01206	15.9	0.0357	0.3131	1.5119	2216.0	SAT LIQ	0.01208	16.4	0.0367	0.3137	1.5125	2202.6	13.9
12.3	1.65426	104.9	0.2242	0.2016	1.1653	481.9	SAT VAP	1.59974	105.1	0.2241	0.2024	1.1661	482.0	13.9
20	1.68919	106.5	0.2275	0.2019	1.1597	487.1		1.62681	106.4	0.2267	0.2026	1.1616	486.2	20
30	1.73400	108.5	0.2317	0.2026	1.1533	493.7		1.67029	108.4	0.2309	0.2032	1.1549	492.8	30
40	1.77778	110.5	0.2358	0.2035	1.1476	500.0		1.71292	110.5	0.2350	0.2041	1.1490	499.2	40
50	1.82116	112.6	0.2398	0.2046	1.1425	506.1		1.75500	112.5	0.2390	0.2051	1.1438	505.4	50
60	1.86393	114.6	0.2438	0.2059	1.1380	512.1		1.79662	114.6	0.2430	0.2063	1.1392	511.4	60
70	1.90621	116.7	0.2478	0.2072	1.1340	517.9		1.83756	116.6	0.2470	0.2076	1.1350	517.3	70
80	1.94818	118.8	0.2516	0.2087	1.1303	523.6		1.87829	118.7	0.2509	0.2091	1.1312	523.0	80
90	1.98965	120.9	0.2555	0.2102	1.1269	529.2		1.91865	120.8	0.2547	0.2106	1.1277	528.6	90
100	2.03128	123.0	0.2593	0.2119	1.1238	534.6		1.95848	122.9	0.2585	0.2122	1.1246	534.1	100
110	2.07254	125.1	0.2631	0.2135	1.1210	539.9		1.99800	125.1	0.2623	0.2138	1.1217	539.4	110
120	2.11282	127.2	0.2668	0.2153	1.1184	545.2		2.03791	127.2	0.2661	0.2155	1.1190	544.7	120
130	2.15378	129.4	0.2705	0.2170	1.1159	550.3		2.07684	129.4	0.2698	0.2173	1.1165	549.9	130
140	2.19394	131.6	0.2742	0.2188	1.1137	555.4		2.11595	131.5	0.2734	0.2190	1.1142	555.0	140
150	2.23364	133.8	0.2778	0.2207	1.1116	560.4		2.15471	133.7	0.2771	0.2208	1.1120	560.0	150
160	2.27428	136.0	0.2814	0.2225	1.1096	565.3		2.19346	136.0	0.2807	0.2227	1.1100	565.0	160
170	2.31374	138.2	0.2850	0.2244	1.1077	570.2		2.23214	138.2	0.2842	0.2245	1.1081	569.8	170
180	2.35349	140.5	0.2885	0.2262	1.1060	575.0		2.27066	140.5	0.2878	0.2264	1.1063	574.6	180
190	2.39292	142.8	0.2920	0.2281	1.1043	579.7		2.30894	142.7	0.2913	0.2283	1.1047	579.4	190
200	2.43250	145.1	0.2955	0.2300	1.1027	584.4		2.34687	145.0	0.2948	0.2301	1.1031	584.1	200
210	2.47158	147.4	0.2990	0.2319	1.1013	589.0		2.38493	147.3	0.2983	0.2320	1.1016	588.7	210
220	2.51130	149.7	0.3025	0.2338	1.0998	593.6		2.42307	149.7	0.3018	0.2339	1.1001	593.3	220
230	2.55037	152.0	0.3059	0.2357	1.0985	598.1		2.46063	152.0	0.3052	0.2358	1.0988	597.8	230
240	2.58933	154.4	0.3093	0.2376	1.0972	602.5		2.49875	154.4	0.3086	0.2377	1.0975	602.3	240
250	2.62881	156.8	0.3127	0.2395	1.0960	607.0		2.53614	156.8	0.3120	0.2396	1.0962	606.7	250
260	2.66738	159.2	0.3161	0.2413	1.0948	611.3		2.57400	159.2	0.3153	0.2414	1.0950	611.1	260
270	2.70636	161.6	0.3194	0.2432	1.0937	615.7		2.61165	161.6	0.3187	0.2433	1.0939	615.4	270
280	2.74499	164.1	0.3227	0.2451	1.0926	620.0		2.64901	164.0	0.3220	0.2452	1.0928	619.7	280
290	2.78396	166.5	0.3260	0.2469	1.0916	624.2		2.68673	166.5	0.3253	0.2470	1.0918	624.0	290
300	2.82247	169.0	0.3293	0.2488	1.0906	628.4		2.72405	169.0	0.3286	0.2489	1.0908	628.2	300
310	2.86123	171.5	0.3326	0.2506	1.0896	632.6		2.76167	171.5	0.3319	0.2507	1.0898	632.4	310
320	2.90023	174.0	0.3358	0.2524	1.0887	636.8		2.79877	174.0	0.3351	0.2525	1.0889	636.6	320

TEMP °F	PRESSURE = 30.00 PSIA							PRESSURE = 31.00 PSIA						TEMP °F
	V	H	S	Cp	Cp/Cv	v_s		V	H	S	Cp	Cp/Cv	v_s	
15.4	0.01211	16.9	0.0377	0.3142	1.5132	2189.7	SAT LIQ	0.01213	17.4	0.0387	0.3148	1.5138	2177.1	16.9
15.4	1.54895	105.4	0.2239	0.2032	1.1670	482.1	SAT VAP	1.50128	105.6	0.2238	0.2040	1.1679	482.2	16.9
20	1.56863	106.3	0.2259	0.2033	1.1635	485.3		1.51400	106.2	0.2251	0.2041	1.1654	484.3	20
30	1.61082	108.3	0.2301	0.2039	1.1566	491.9		1.55521	108.3	0.2293	0.2045	1.1582	491.1	30
40	1.65235	110.4	0.2342	0.2046	1.1505	498.4		1.59566	110.3	0.2335	0.2052	1.1520	497.6	40
50	1.69319	112.4	0.2383	0.2056	1.1451	504.6		1.63532	112.4	0.2375	0.2061	1.1464	503.9	50
60	1.73340	114.5	0.2423	0.2068	1.1403	510.7		1.67448	114.4	0.2416	0.2072	1.1414	510.0	60
70	1.77336	116.6	0.2462	0.2080	1.1360	516.6		1.71350	116.5	0.2455	0.2084	1.1370	516.0	70
80	1.81291	118.7	0.2501	0.2094	1.1321	522.4		1.75162	118.6	0.2494	0.2098	1.1330	521.8	80
90	1.85185	120.8	0.2540	0.2109	1.1285	528.0		1.78987	120.7	0.2533	0.2112	1.1294	527.4	90
100	1.89107	122.9	0.2578	0.2124	1.1253	533.5		1.82749	122.8	0.2571	0.2127	1.1261	533.0	100
110	1.92938	125.0	0.2616	0.2141	1.1223	538.9		1.86498	125.0	0.2609	0.2143	1.1230	538.4	110
120	1.96773	127.2	0.2653	0.2157	1.1196	544.2		1.90223	127.1	0.2646	0.2160	1.1202	543.8	120
130	2.00562	129.3	0.2690	0.2175	1.1171	549.4		1.93911	129.3	0.2684	0.2177	1.1176	549.0	130
140	2.04332	131.5	0.2727	0.2192	1.1147	554.6		1.97550	131.5	0.2720	0.2194	1.1152	554.1	140
150	2.08117	133.7	0.2764	0.2210	1.1125	559.6		2.01248	133.7	0.2757	0.2212	1.1130	559.2	150
160	2.11864	135.9	0.2800	0.2228	1.1105	564.6		2.04834	135.9	0.2793	0.2230	1.1109	564.2	160
170	2.15610	138.2	0.2835	0.2247	1.1085	569.5		2.08464	138.1	0.2829	0.2248	1.1089	569.1	170
180	2.19346	140.4	0.2871	0.2265	1.1067	574.3		2.12089	140.4	0.2864	0.2267	1.1071	573.9	180
190	2.23015	142.7	0.2906	0.2284	1.1050	579.0		2.15703	142.7	0.2900	0.2285	1.1054	578.7	190
200	2.26706	145.0	0.2941	0.2303	1.1034	583.7		2.19250	145.0	0.2935	0.2304	1.1037	583.4	200
210	2.30415	147.3	0.2976	0.2321	1.1019	588.4		2.22816	147.3	0.2969	0.2323	1.1022	588.1	210
220	2.34082	149.6	0.3011	0.2340	1.1004	593.0		2.26398	149.6	0.3004	0.2341	1.1007	592.7	220
230	2.37756	152.0	0.3045	0.2359	1.0990	597.5		2.29938	152.0	0.3038	0.2360	1.0993	597.2	230
240	2.41429	154.4	0.3079	0.2378	1.0977	602.0		2.33481	154.3	0.3072	0.2379	1.0980	601.8	240
250	2.45038	156.7	0.3113	0.2396	1.0965	606.5		2.37023	156.7	0.3106	0.2397	1.0967	606.2	250
260	2.48694	159.1	0.3147	0.2415	1.0953	610.9		2.40553	159.1	0.3140	0.2416	1.0955	610.6	260
270	2.52334	161.6	0.3180	0.2434	1.0941	615.2		2.44081	161.5	0.3173	0.2435	1.0943	615.0	270
280	2.55951	164.0	0.3213	0.2452	1.0930	619.5		2.47647	164.0	0.3207	0.2453	1.0932	619.3	280
290	2.59605	166.5	0.3246	0.2471	1.0920	623.8		2.51130	166.4	0.3240	0.2472	1.0922	623.5	290
300	2.63227	169.0	0.3279	0.2489	1.0910	628.0		2.54647	168.9	0.3273	0.2490	1.0911	627.8	300
310	2.66809	171.5	0.3312	0.2507	1.0900	632.2		2.58131	171.4	0.3305	0.2508	1.0901	632.0	310
320	2.70416	174.0	0.3344	0.2526	1.0890	636.4		2.61643	173.9	0.3338	0.2526	1.0892	636.2	320

Reprinted by permission from DuPont

Table 2 (continued)
HFC-134a Superheated Vapor—Constant Pressure Tables

V = Volume in ft³/lb   H = Enthalpy in Btu/lb   S = Entropy in Btu/(lb) (°R) v_s = Velocity of Sound in ft/sec
Cp = Heat Capacity at Constant Pressure in Btu/(lb) (°F)   Cp/Cv = Heat Capacity Ratio (Dimensionless)

TEMP °F	PRESSURE = 32.00 PSIA							PRESSURE = 33.00 PSIA						TEMP °F
	V	H	S	Cp	Cp/Cv	v_s		V	H	S	Cp	Cp/Cv	v_s	
18.4	0.01216	17.9	0.0397	0.3153	1.5145	2164.7	SAT LIQ	0.01218	18.3	0.0406	0.3158	1.5152	2152.7	19.8
18.4	1.45645	105.8	0.2237	0.2048	1.1687	482.2	SAT VAP	1.41423	106.0	0.2235	0.2055	1.1696	482.3	19.8
20	1.46306	106.1	0.2244	0.2048	1.1674	483.4		1.41483	106.0	0.2236	0.2055	1.1694	482.4	20
30	1.50308	108.2	0.2286	0.2052	1.1600	490.2		1.45412	108.1	0.2279	0.2058	1.1617	489.3	30
40	1.54250	110.2	0.2327	0.2058	1.1535	496.8		1.49254	110.2	0.2320	0.2064	1.1550	496.0	40
50	1.58128	112.3	0.2368	0.2066	1.1477	503.1		1.53022	112.2	0.2361	0.2071	1.1491	502.4	50
60	1.61943	114.4	0.2408	0.2077	1.1426	509.3		1.56740	114.3	0.2402	0.2081	1.1438	508.6	60
70	1.65700	116.5	0.2448	0.2088	1.1380	515.3		1.60411	116.4	0.2441	0.2092	1.1391	514.7	70
80	1.69434	118.5	0.2487	0.2101	1.1339	521.2		1.64069	118.5	0.2481	0.2105	1.1349	520.6	80
90	1.73130	120.7	0.2526	0.2115	1.1302	526.9		1.67645	120.6	0.2519	0.2119	1.1310	526.3	90
100	1.76772	122.8	0.2564	0.2130	1.1268	532.5		1.71233	122.7	0.2558	0.2133	1.1276	531.9	100
110	1.80440	124.9	0.2602	0.2146	1.1237	537.9		1.74764	124.9	0.2596	0.2149	1.1244	537.4	110
120	1.84060	127.1	0.2640	0.2162	1.1208	543.3		1.78285	127.0	0.2633	0.2165	1.1215	542.8	120
130	1.87617	129.2	0.2677	0.2179	1.1182	548.5		1.81752	129.2	0.2670	0.2181	1.1188	548.1	130
140	1.91205	131.4	0.2714	0.2196	1.1157	553.7		1.85219	131.4	0.2707	0.2199	1.1163	553.3	140
150	1.94780	133.6	0.2750	0.2214	1.1135	558.8		1.88679	133.6	0.2744	0.2216	1.1140	558.4	150
160	1.98295	135.9	0.2786	0.2232	1.1113	563.8		1.92086	135.8	0.2780	0.2234	1.1118	563.4	160
170	2.01776	138.1	0.2822	0.2250	1.1094	568.7		1.95503	138.1	0.2816	0.2252	1.1098	568.4	170
180	2.05297	140.4	0.2858	0.2268	1.1075	573.6		1.98926	140.3	0.2851	0.2270	1.1079	573.2	180
190	2.08768	142.6	0.2893	0.2287	1.1057	578.4		2.02306	142.6	0.2887	0.2288	1.1061	578.1	190
200	2.12269	144.9	0.2928	0.2305	1.1041	583.1		2.05677	144.9	0.2922	0.2306	1.1044	582.8	200
210	2.15750	147.2	0.2963	0.2324	1.1025	587.8		2.09030	147.2	0.2957	0.2325	1.1028	587.5	210
220	2.19154	149.6	0.2997	0.2342	1.1010	592.4		2.12404	149.5	0.2991	0.2343	1.1013	592.1	220
230	2.22618	151.9	0.3032	0.2361	1.0996	597.0		2.15796	151.9	0.3026	0.2362	1.0999	596.7	230
240	2.26040	154.3	0.3066	0.2380	1.0982	601.5		2.19106	154.3	0.3060	0.2381	1.0985	601.2	240
250	2.29516	156.7	0.3100	0.2398	1.0969	606.0		2.22469	156.7	0.3094	0.2399	1.0972	605.7	250
260	2.32937	159.1	0.3134	0.2417	1.0957	610.4		2.25734	159.1	0.3127	0.2418	1.0960	610.1	260
270	2.36351	161.5	0.3167	0.2435	1.0946	614.8		2.29095	161.5	0.3161	0.2436	1.0948	614.5	270
280	2.39751	164.0	0.3200	0.2454	1.0934	619.1		2.32396	163.9	0.3194	0.2455	1.0936	618.9	280
290	2.43191	166.4	0.3233	0.2472	1.0924	623.4		2.35682	166.4	0.3227	0.2473	1.0925	623.2	290
300	2.46609	168.9	0.3266	0.2491	1.0913	627.6		2.39006	168.9	0.3260	0.2491	1.0915	627.4	300
310	2.49938	171.4	0.3299	0.2509	1.0903	631.9		2.42307	171.4	0.3293	0.2509	1.0905	631.7	310
320	2.53357	173.9	0.3331	0.2527	1.0894	636.0		2.45640	173.9	0.3325	0.2527	1.0895	635.9	320

TEMP °F	PRESSURE = 34.00 PSIA							PRESSURE = 35.00 PSIA						TEMP °F
	V	H	S	Cp	Cp/Cv	v_s		V	H	S	Cp	Cp/Cv	v_s	
21.2	0.01220	18.8	0.0416	0.3164	1.5159	2141.0	SAT LIQ	0.01223	19.2	0.0425	0.3169	1.5165	2129.5	22.6
21.2	1.37438	106.2	0.2234	0.2063	1.1704	482.3	SAT VAP	1.33572	106.4	0.2233	0.2070	1.1713	482.4	22.6
30	1.40786	108.0	0.2271	0.2065	1.1635	488.4		1.36444	107.9	0.2265	0.2071	1.1653	487.6	30
40	1.44550	110.1	0.2313	0.2069	1.1565	495.2		1.40095	110.0	0.2306	0.2075	1.1581	494.3	40
50	1.48236	112.2	0.2354	0.2076	1.1504	501.6		1.43699	112.1	0.2348	0.2082	1.1518	500.9	50
60	1.51860	114.2	0.2395	0.2086	1.1450	507.9		1.47254	114.2	0.2388	0.2090	1.1462	507.2	60
70	1.55448	116.3	0.2435	0.2096	1.1402	514.0		1.50761	116.3	0.2428	0.2101	1.1412	513.3	70
80	1.58983	118.4	0.2474	0.2109	1.1358	519.9		1.54202	118.4	0.2467	0.2112	1.1368	519.3	80
90	1.62496	120.5	0.2513	0.2122	1.1319	525.7		1.57629	120.5	0.2506	0.2125	1.1328	525.1	90
100	1.65975	122.7	0.2551	0.2136	1.1283	531.4		1.61005	122.6	0.2545	0.2139	1.1291	530.8	100
110	1.69405	124.8	0.2589	0.2152	1.1251	536.9		1.64366	124.8	0.2583	0.2154	1.1258	536.4	110
120	1.72831	127.0	0.2627	0.2167	1.1221	542.3		1.67701	126.9	0.2620	0.2170	1.1227	541.8	120
130	1.76211	129.2	0.2664	0.2184	1.1194	547.6		1.71028	129.1	0.2658	0.2186	1.1199	547.2	130
140	1.79598	131.3	0.2701	0.2201	1.1168	552.9		1.74307	131.3	0.2695	0.2203	1.1173	552.4	140
150	1.82949	133.6	0.2737	0.2218	1.1144	558.0		1.77557	133.5	0.2731	0.2220	1.1149	557.6	150
160	1.86289	135.8	0.2774	0.2235	1.1122	563.0		1.80832	135.7	0.2767	0.2237	1.1127	562.6	160
170	1.89645	138.0	0.2809	0.2253	1.1102	568.0		1.84026	138.0	0.2803	0.2255	1.1106	567.6	170
180	1.92938	140.3	0.2845	0.2271	1.1083	572.9		1.87266	140.3	0.2839	0.2273	1.1086	572.5	180
190	1.96194	142.6	0.2880	0.2289	1.1064	577.7		1.90476	142.5	0.2874	0.2291	1.1068	577.4	190
200	1.99521	144.9	0.2916	0.2308	1.1047	582.5		1.93686	144.8	0.2910	0.2309	1.1051	582.2	200
210	2.02758	147.2	0.2950	0.2326	1.1031	587.2		1.96850	147.2	0.2944	0.2327	1.1034	586.9	210
220	2.06016	149.5	0.2985	0.2345	1.1016	591.8		2.00040	149.5	0.2979	0.2346	1.1019	591.6	220
230	2.09293	151.9	0.3019	0.2363	1.1001	596.4		2.03211	151.8	0.3013	0.2364	1.1004	596.2	230
240	2.12540	154.2	0.3054	0.2382	1.0987	601.0		2.06356	154.2	0.3048	0.2383	1.0990	600.7	240
250	2.15796	156.6	0.3087	0.2400	1.0974	605.5		2.09512	156.6	0.3082	0.2401	1.0977	605.2	250
260	2.19010	159.0	0.3121	0.2419	1.0962	609.9		2.12630	159.0	0.3115	0.2419	1.0964	609.7	260
270	2.22272	161.5	0.3155	0.2437	1.0950	614.3		2.15796	161.4	0.3149	0.2438	1.0952	614.1	270
280	2.25428	163.9	0.3188	0.2455	1.0938	618.7		2.18962	163.9	0.3182	0.2456	1.0940	618.4	280
290	2.28676	166.4	0.3221	0.2474	1.0927	623.0		2.22074	166.4	0.3215	0.2474	1.0929	622.8	290
300	2.31857	168.9	0.3254	0.2492	1.0917	627.2		2.25175	168.8	0.3248	0.2493	1.0919	627.0	300
310	2.35073	171.4	0.3287	0.2510	1.0907	631.5		2.28311	171.3	0.3281	0.2511	1.0908	631.3	310
320	2.38265	173.9	0.3319	0.2528	1.0897	635.7		2.31428	173.9	0.3313	0.2529	1.0899	635.5	320
330	2.41488	176.4	0.3352	0.2546	1.0888	639.8		2.34522	176.4	0.3346	0.2547	1.0889	639.7	330

Reprinted by permission from DuPont

Table 2 (continued)
HFC-134a Superheated Vapor—Constant Pressure Tables

V = Volume in ft³/lb   H = Enthalpy in Btu/lb   S = Entropy in Btu/(lb) (°R) v_s = Velocity of Sound in ft/sec
Cp = Heat Capacity at Constant Pressure in Btu/(lb) (°F)   Cp/Cv = Heat Capacity Ratio (Dimensionless)

TEMP °F	PRESSURE = 36.00 PSIA							PRESSURE = 37.00 PSIA						TEMP °F
	V	H	S	Cp	Cp/Cv	v_s		V	H	S	Cp	Cp/Cv	v_s	
23.9	0.01225	19.6	0.0433	0.3174	1.5172	2118.3	SAT LIQ	0.01227	20.0	0.0442	0.3179	1.5179	2107.3	25.2
23.9	1.30124	106.6	0.2232	0.2078	1.1722	482.4	SAT VAP	1.26743	106.8	0.2231	0.2085	1.1730	482.4	25.2
30	1.32328	107.9	0.2258	0.2078	1.1671	486.7		1.28436	107.8	0.2251	0.2085	1.1690	485.8	30
40	1.35906	109.9	0.2300	0.2081	1.1597	493.5		1.31944	109.9	0.2293	0.2087	1.1613	492.7	40
50	1.39431	112.0	0.2341	0.2087	1.1532	500.1		1.35391	111.9	0.2335	0.2092	1.1546	499.3	50
60	1.42898	114.1	0.2382	0.2095	1.1474	506.5		1.38793	114.0	0.2375	0.2100	1.1487	505.8	60
70	1.46327	116.2	0.2422	0.2105	1.1423	512.7		1.42126	116.1	0.2415	0.2109	1.1434	512.0	70
80	1.49701	118.3	0.2461	0.2116	1.1377	518.7		1.45412	118.3	0.2455	0.2120	1.1387	518.1	80
90	1.53022	120.4	0.2500	0.2129	1.1336	524.6		1.48677	120.4	0.2494	0.2132	1.1345	524.0	90
100	1.56323	122.6	0.2539	0.2142	1.1299	530.3		1.51906	122.5	0.2533	0.2146	1.1307	529.7	100
110	1.59617	124.7	0.2577	0.2157	1.1265	535.9		1.55087	124.7	0.2571	0.2160	1.1272	535.4	110
120	1.62866	126.9	0.2614	0.2172	1.1234	541.4		1.58278	126.8	0.2608	0.2175	1.1240	540.9	120
130	1.66085	129.1	0.2652	0.2188	1.1205	546.7		1.61394	129.0	0.2646	0.2191	1.1211	546.3	130
140	1.69291	131.3	0.2689	0.2205	1.1179	552.0		1.64555	131.2	0.2683	0.2207	1.1184	551.6	140
150	1.72473	133.5	0.2725	0.2222	1.1154	557.2		1.67645	133.4	0.2719	0.2224	1.1159	556.8	150
160	1.75623	135.7	0.2762	0.2239	1.1132	562.3		1.70736	135.7	0.2756	0.2241	1.1136	561.9	160
170	1.78763	138.0	0.2797	0.2256	1.1110	567.3		1.73792	137.9	0.2792	0.2258	1.1114	566.9	170
180	1.81917	140.2	0.2833	0.2274	1.1090	572.2		1.76866	140.2	0.2827	0.2276	1.1094	571.8	180
190	1.85048	142.5	0.2869	0.2292	1.1072	577.1		1.79921	142.5	0.2863	0.2294	1.1075	576.7	190
200	1.88147	144.8	0.2904	0.2310	1.1054	581.9		1.82949	144.8	0.2898	0.2312	1.1057	581.5	200
210	1.91241	147.1	0.2939	0.2329	1.1037	586.6		1.85977	147.1	0.2933	0.2330	1.1041	586.3	210
220	1.94363	149.5	0.2973	0.2347	1.1022	591.3		1.89000	149.4	0.2968	0.2348	1.1025	591.0	220
230	1.97433	151.8	0.3008	0.2365	1.1007	595.9		1.91975	151.8	0.3002	0.2366	1.1010	595.6	230
240	2.00521	154.2	0.3042	0.2384	1.0993	600.5		1.95008	154.2	0.3036	0.2385	1.0995	600.2	240
250	2.03583	156.6	0.3076	0.2402	1.0979	605.0		1.97981	156.6	0.3070	0.2403	1.0982	604.7	250
260	2.06654	159.0	0.3110	0.2420	1.0966	609.4		2.00965	159.0	0.3104	0.2421	1.0969	609.2	260
270	2.09688	161.4	0.3143	0.2439	1.0954	613.9		2.03915	161.4	0.3138	0.2439	1.0956	613.6	270
280	2.12766	163.9	0.3176	0.2457	1.0942	618.2		2.06911	163.8	0.3171	0.2458	1.0945	618.0	280
290	2.15796	166.3	0.3210	0.2475	1.0931	622.6		2.09864	166.3	0.3204	0.2476	1.0933	622.3	290
300	2.18818	168.8	0.3242	0.2493	1.0921	626.8		2.12811	168.8	0.3237	0.2494	1.0922	626.6	300
310	2.21828	171.3	0.3275	0.2511	1.0910	631.1		2.15796	171.3	0.3270	0.2512	1.0912	630.9	310
320	2.24871	173.8	0.3308	0.2529	1.0900	635.3		2.18723	173.8	0.3302	0.2530	1.0902	635.1	320
330	2.27894	176.4	0.3340	0.2547	1.0891	639.5		2.21680	176.4	0.3335	0.2548	1.0892	639.3	330

TEMP °F	PRESSURE = 38.00 PSIA							PRESSURE = 39.00 PSIA						TEMP °F
	V	H	S	Cp	Cp/Cv	v_s		V	H	S	Cp	Cp/Cv	v_s	
26.5	0.01229	20.4	0.0450	0.3184	1.5186	2096.5	SAT LIQ	0.01231	20.9	0.0459	0.3189	1.5193	2086.0	27.8
26.5	1.23533	107.0	0.2230	0.2092	1.1739	482.4	SAT VAP	1.20496	107.1	0.2229	0.2099	1.1748	482.4	27.8
30	1.24750	107.7	0.2245	0.2092	1.1709	484.9		1.21256	107.6	0.2238	0.2099	1.1728	484.0	30
40	1.28189	109.8	0.2287	0.2093	1.1629	491.8		1.24626	109.7	0.2281	0.2099	1.1646	491.0	40
50	1.31562	111.9	0.2328	0.2098	1.1560	498.6		1.27926	111.8	0.2322	0.2103	1.1574	497.8	50
60	1.34880	114.0	0.2369	0.2104	1.1499	505.1		1.31182	113.9	0.2363	0.2109	1.1512	504.3	60
70	1.38141	116.1	0.2409	0.2113	1.1445	511.4		1.34372	116.0	0.2403	0.2117	1.1456	510.7	70
80	1.41383	118.2	0.2449	0.2124	1.1397	517.5		1.37533	118.1	0.2443	0.2127	1.1407	516.8	80
90	1.44550	120.3	0.2488	0.2136	1.1354	523.4		1.40647	120.3	0.2482	0.2139	1.1363	522.8	90
100	1.47710	122.5	0.2527	0.2149	1.1315	529.2		1.43740	122.4	0.2521	0.2152	1.1323	528.6	100
110	1.50830	124.6	0.2565	0.2163	1.1279	534.8		1.46778	124.6	0.2559	0.2165	1.1287	534.3	110
120	1.53941	126.8	0.2603	0.2177	1.1247	540.4		1.49813	126.7	0.2597	0.2180	1.1253	539.9	120
130	1.56986	129.0	0.2640	0.2193	1.1217	545.8		1.52812	128.9	0.2634	0.2195	1.1223	545.3	130
140	1.60051	131.2	0.2677	0.2209	1.1190	551.1		1.55788	131.1	0.2672	0.2211	1.1195	550.7	140
150	1.63079	133.4	0.2714	0.2226	1.1164	556.3		1.58755	133.4	0.2708	0.2227	1.1169	555.9	150
160	1.66085	135.6	0.2750	0.2242	1.1141	561.5		1.61708	135.6	0.2745	0.2244	1.1145	561.1	160
170	1.69090	137.9	0.2786	0.2260	1.1119	566.5		1.64636	137.8	0.2781	0.2261	1.1123	566.2	170
180	1.72058	140.2	0.2822	0.2277	1.1098	571.5		1.67560	140.1	0.2816	0.2279	1.1102	571.1	180
190	1.75039	142.4	0.2857	0.2295	1.1079	576.4		1.70445	142.4	0.2852	0.2296	1.1083	576.1	190
200	1.77999	144.7	0.2893	0.2313	1.1061	581.2		1.73310	144.7	0.2887	0.2314	1.1064	580.9	200
210	1.80963	147.1	0.2927	0.2331	1.1044	586.0		1.76211	147.0	0.2922	0.2332	1.1047	585.7	210
220	1.83891	149.4	0.2962	0.2349	1.1028	590.7		1.79083	149.4	0.2957	0.2350	1.1031	590.4	220
230	1.86646	151.8	0.2997	0.2367	1.1012	595.3		1.81951	151.7	0.2991	0.2368	1.1015	595.1	230
240	1.89753	154.1	0.3031	0.2386	1.0998	599.9		1.84775	154.1	0.3025	0.2387	1.1001	599.7	240
250	1.92678	156.5	0.3065	0.2404	1.0984	604.5		1.87652	156.5	0.3059	0.2405	1.0987	604.2	250
260	1.95580	158.9	0.3099	0.2422	1.0971	609.0		1.90476	158.9	0.3093	0.2423	1.0973	608.7	260
270	1.98491	161.4	0.3132	0.2440	1.0959	613.4		1.93311	161.4	0.3127	0.2441	1.0961	613.2	270
280	2.01369	163.8	0.3165	0.2458	1.0947	617.8		1.96117	163.8	0.3160	0.2459	1.0949	617.6	280
290	2.04290	166.3	0.3199	0.2477	1.0935	622.1		1.98926	166.3	0.3193	0.2477	1.0937	621.9	290
300	2.07168	168.8	0.3232	0.2495	1.0924	626.5		2.01776	168.8	0.3226	0.2495	1.0926	626.3	300
310	2.10040	171.3	0.3264	0.2513	1.0914	630.7		2.04583	171.3	0.3259	0.2513	1.0916	630.5	310
320	2.12902	173.8	0.3297	0.2531	1.0904	634.9		2.07383	173.8	0.3292	0.2531	1.0905	634.8	320
330	2.15796	176.3	0.3329	0.2548	1.0894	639.1		2.10172	176.3	0.3324	0.2549	1.0896	639.0	330

Reprinted by permission from DuPont

Table 2 (continued)
HFC-134a Superheated Vapor—Constant Pressure Tables

V = Volume in ft³/lb   H = Enthalpy in Btu/lb   S = Entropy in Btu/(lb) (°R) v_s = Velocity of Sound in ft/sec
Cp = Heat Capacity at Constant Pressure in Btu/(lb) (°F)   Cp/Cv = Heat Capacity Ratio (Dimensionless)

TEMP °F	PRESSURE = 40.00 PSIA							PRESSURE = 41.00 PSIA						TEMP °F
	V	H	S	Cp	Cp/Cv	v_s		V	H	S	Cp	Cp/Cv	v_s	
29	0.01233	21.2	0.0467	0.3194	1.5200	2075.7	SAT LIQ	0.01236	21.6	0.0475	0.3198	1.5207	2065.5	30.2
29	1.17606	107.3	0.2228	0.2106	1.1756	482.3	SAT VAP	1.14837	107.5	0.2227	0.2113	1.1765	482.3	30.2
30	1.17925	107.5	0.2232	0.2106	1.1748	483.0		—	—	—	—	—	—	30
40	1.21242	109.6	0.2275	0.2105	1.1663	490.2		1.18008	109.5	0.2269	0.2112	1.1680	489.3	40
50	1.24471	111.7	0.2316	0.2108	1.1589	497.0		1.21197	111.7	0.2310	0.2114	1.1604	496.2	50
60	1.27665	113.8	0.2357	0.2114	1.1525	503.6		1.24316	113.8	0.2352	0.2119	1.1538	502.9	60
70	1.30787	116.0	0.2398	0.2122	1.1468	510.0		1.27389	115.9	0.2392	0.2126	1.1479	509.3	70
80	1.33887	118.1	0.2437	0.2131	1.1417	516.2		1.30412	118.0	0.2432	0.2135	1.1427	515.6	80
90	1.36930	120.2	0.2477	0.2142	1.1372	522.2		1.33404	120.2	0.2471	0.2146	1.1381	521.6	90
100	1.39958	122.4	0.2515	0.2155	1.1331	528.1		1.36351	122.3	0.2510	0.2158	1.1339	527.5	100
110	1.42939	124.5	0.2554	0.2168	1.1294	533.8		1.39276	124.5	0.2548	0.2171	1.1301	533.3	110
120	1.45900	126.7	0.2592	0.2183	1.1260	539.4		1.42167	126.7	0.2586	0.2185	1.1267	538.9	120
130	1.48832	128.9	0.2629	0.2198	1.1229	544.9		1.45033	128.8	0.2624	0.2200	1.1235	544.4	130
140	1.51745	131.1	0.2666	0.2213	1.1201	550.3		1.47885	131.1	0.2661	0.2215	1.1206	549.8	140
150	1.54655	133.3	0.2703	0.2229	1.1174	555.5		1.50739	133.3	0.2698	0.2231	1.1179	555.1	150
160	1.57505	135.6	0.2739	0.2246	1.1150	560.7		1.53539	135.5	0.2734	0.2248	1.1155	560.3	160
170	1.60359	137.8	0.2775	0.2263	1.1127	565.8		1.56348	137.8	0.2770	0.2265	1.1132	565.4	170
180	1.63212	140.1	0.2811	0.2280	1.1106	570.8		1.59109	140.0	0.2806	0.2282	1.1110	570.4	180
190	1.66058	142.4	0.2847	0.2298	1.1086	575.7		1.61891	142.3	0.2841	0.2299	1.1090	575.4	190
200	1.68890	144.7	0.2882	0.2316	1.1068	580.6		1.64636	144.6	0.2877	0.2317	1.1071	580.3	200
210	1.71674	147.0	0.2917	0.2333	1.1050	585.4		1.67392	147.0	0.2912	0.2335	1.1053	585.1	210
220	1.74520	149.3	0.2952	0.2351	1.1034	590.1		1.70126	149.3	0.2946	0.2353	1.1037	589.8	220
230	1.77305	151.7	0.2986	0.2369	1.1018	594.8		1.72861	151.7	0.2981	0.2370	1.1021	594.5	230
240	1.80050	154.1	0.3020	0.2388	1.1003	599.4		1.75593	154.1	0.3015	0.2389	1.1006	599.1	240
250	1.82849	156.5	0.3054	0.2406	1.0989	604.0		1.78317	156.5	0.3049	0.2407	1.0992	603.7	250
260	1.85598	158.9	0.3088	0.2424	1.0976	608.5		1.80995	158.9	0.3083	0.2425	1.0978	608.2	260
270	1.88359	161.3	0.3122	0.2442	1.0963	612.9		1.83688	161.3	0.3117	0.2443	1.0965	612.7	270
280	1.91131	163.8	0.3155	0.2460	1.0951	617.4		1.86393	163.8	0.3150	0.2461	1.0953	617.1	280
290	1.93911	166.2	0.3188	0.2478	1.0939	621.7		1.89072	166.2	0.3183	0.2479	1.0941	621.5	290
300	1.96657	168.7	0.3221	0.2496	1.0928	626.1		1.91755	168.7	0.3216	0.2497	1.0930	625.9	300
310	1.99362	171.2	0.3254	0.2514	1.0917	630.3		1.94439	171.2	0.3249	0.2515	1.0919	630.1	310
320	2.02102	173.8	0.3286	0.2532	1.0907	634.6		1.97122	173.7	0.3281	0.2532	1.0909	634.4	320
330	2.04834	176.3	0.3319	0.2550	1.0897	638.8		1.99760	176.3	0.3314	0.2550	1.0899	638.6	330
340	—	—	—	—	—	—		2.02429	178.8	0.3346	0.2568	1.0889	642.8	340

TEMP °F	PRESSURE = 42.00 PSIA							PRESSURE = 43.00 PSIA						TEMP °F
	V	H	S	Cp	Cp/Cv	v_s		V	H	S	Cp	Cp/Cv	v_s	
31.4	0.01238	22.0	0.0482	0.3203	1.5214	2055.6	SAT LIQ	0.01240	22.4	0.0490	0.3208	1.5221	2045.8	32.6
31.4	1.12208	107.6	0.2226	0.2120	1.1774	482.3	SAT VAP	1.09685	107.8	0.2225	0.2127	1.1783	482.2	32.6
40	1.14943	109.5	0.2263	0.2118	1.1697	488.5		1.12007	109.4	0.2257	0.2125	1.1715	487.6	40
50	1.18064	111.6	0.2305	0.2119	1.1619	495.5		1.15075	111.5	0.2299	0.2125	1.1634	494.7	50
60	1.21124	113.7	0.2346	0.2124	1.1551	502.2		1.18078	113.6	0.2340	0.2129	1.1564	501.4	60
70	1.24131	115.8	0.2386	0.2130	1.1491	508.7		1.21036	115.8	0.2381	0.2135	1.1503	508.0	70
80	1.27113	118.0	0.2426	0.2139	1.1438	514.9		1.23946	117.9	0.2421	0.2143	1.1448	514.3	80
90	1.30039	120.1	0.2466	0.2149	1.1390	521.0		1.26823	120.1	0.2460	0.2153	1.1399	520.5	90
100	1.32926	122.3	0.2505	0.2161	1.1347	527.0		1.29668	122.2	0.2499	0.2164	1.1356	526.4	100
110	1.35777	124.4	0.2543	0.2174	1.1309	532.8		1.32468	124.4	0.2538	0.2177	1.1316	532.3	110
120	1.38639	126.6	0.2581	0.2188	1.1273	538.4		1.35245	126.6	0.2576	0.2190	1.1280	537.9	120
130	1.41443	128.8	0.2618	0.2202	1.1241	544.0		1.37988	128.8	0.2613	0.2205	1.1247	543.5	130
140	1.44238	131.1	0.2656	0.2218	1.1212	549.4		1.40726	131.0	0.2650	0.2220	1.1217	548.9	140
150	1.46994	133.2	0.2692	0.2233	1.1185	554.7		1.43431	133.2	0.2687	0.2235	1.1190	554.3	150
160	1.49745	135.5	0.2729	0.2250	1.1159	559.9		1.46156	135.4	0.2724	0.2252	1.1164	559.5	160
170	1.52509	137.7	0.2765	0.2266	1.1136	565.0		1.48810	137.7	0.2760	0.2268	1.1140	564.7	170
180	1.55207	140.0	0.2801	0.2283	1.1114	570.1		1.51469	140.0	0.2796	0.2285	1.1118	569.7	180
190	1.57928	142.3	0.2836	0.2301	1.1094	575.1		1.54131	142.3	0.2831	0.2302	1.1097	574.7	190
200	1.60617	144.6	0.2872	0.2318	1.1075	580.0		1.56789	144.6	0.2867	0.2320	1.1078	579.6	200
210	1.63292	146.9	0.2907	0.2336	1.1057	584.8		1.59388	146.9	0.2902	0.2337	1.1060	584.5	210
220	1.65975	149.3	0.2941	0.2354	1.1040	589.5		1.62022	149.3	0.2937	0.2355	1.1043	589.3	220
230	1.68663	151.7	0.2976	0.2372	1.1024	594.2		1.64636	151.6	0.2971	0.2373	1.1026	594.0	230
240	1.71292	154.0	0.3010	0.2390	1.1008	598.9		1.67224	154.0	0.3005	0.2391	1.1011	598.6	240
250	1.73943	156.4	0.3044	0.2408	1.0994	603.5		1.69808	156.4	0.3039	0.2408	1.0997	603.3	250
260	1.76616	158.8	0.3078	0.2426	1.0980	608.0		1.72414	158.8	0.3073	0.2426	1.0983	607.8	260
270	1.79244	161.3	0.3112	0.2444	1.0967	612.5		1.75009	161.3	0.3107	0.2444	1.0970	612.3	270
280	1.81851	163.7	0.3145	0.2462	1.0955	616.9		1.77557	163.7	0.3140	0.2462	1.0957	616.7	280
290	1.84502	166.2	0.3178	0.2480	1.0943	621.3		1.80148	166.2	0.3173	0.2480	1.0945	621.1	290
300	1.87126	168.7	0.3211	0.2497	1.0932	625.7		1.82715	168.7	0.3206	0.2498	1.0934	625.5	300
310	1.89717	171.2	0.3244	0.2515	1.0921	630.0		1.85254	171.2	0.3239	0.2516	1.0923	629.8	310
320	1.92345	173.7	0.3277	0.2533	1.0910	634.2		1.87829	173.7	0.3272	0.2534	1.0912	634.0	320
330	1.94970	176.3	0.3309	0.2551	1.0900	638.4		1.90331	176.2	0.3304	0.2551	1.0902	638.3	330
340	1.97550	178.8	0.3341	0.2568	1.0891	642.6		1.92901	178.8	0.3336	0.2569	1.0892	642.5	340

Reprinted by permission from DuPont

Table 2 (continued)
HFC-134a Superheated Vapor—Constant Pressure Tables

V = Volume in ft³/lb H = Enthalpy in Btu/lb S = Entropy in Btu/(lb) (°R) v_s = Velocity of Sound in ft/sec
Cp = Heat Capacity at Constant Pressure in Btu/(lb) (°F) Cp/Cv = Heat Capacity Ratio (Dimensionless)

TEMP °F	PRESSURE = 44.00 PSIA							PRESSURE = 45.00 PSIA						TEMP °F
	V	H	S	Cp	Cp/Cv	v_s		V	H	S	Cp	Cp/Cv	v_s	
33.7	0.01242	22.8	0.0497	0.3212	1.5228	2036.2	SAT LIQ	0.01244	23.1	0.0505	0.3217	1.5235	2026.8	34.9
33.7	1.07285	108.0	0.2224	0.2133	1.1791	482.2	SAT VAP	1.04987	108.1	0.2223	0.2140	1.1800	482.1	34.9
40	1.09206	109.3	0.2251	0.2131	1.1733	486.8		1.06530	109.2	0.2246	0.2138	1.1751	485.9	40
50	1.12221	111.4	0.2293	0.2131	1.1650	493.9		1.09493	111.4	0.2288	0.2137	1.1666	493.1	50
60	1.15181	113.6	0.2335	0.2134	1.1578	500.7		1.12397	113.5	0.2329	0.2139	1.1591	500.0	60
70	1.18078	115.7	0.2376	0.2139	1.1515	507.3		1.15260	115.6	0.2370	0.2144	1.1527	506.6	70
80	1.20934	117.8	0.2416	0.2147	1.1459	513.7		1.18064	117.8	0.2410	0.2151	1.1469	513.0	80
90	1.23762	120.0	0.2455	0.2157	1.1409	519.9		1.20831	119.9	0.2450	0.2160	1.1418	519.3	90
100	1.26534	122.2	0.2494	0.2168	1.1364	525.9		1.23564	122.1	0.2489	0.2171	1.1372	525.3	100
110	1.29299	124.3	0.2533	0.2180	1.1324	531.7		1.26263	124.3	0.2528	0.2183	1.1331	531.2	110
120	1.32013	126.5	0.2571	0.2193	1.1287	537.4		1.28916	126.5	0.2566	0.2196	1.1294	537.0	120
130	1.34698	128.7	0.2608	0.2207	1.1254	543.0		1.31562	128.7	0.2603	0.2209	1.1260	542.6	130
140	1.37382	130.9	0.2645	0.2222	1.1223	548.5		1.34192	130.9	0.2641	0.2224	1.1229	548.1	140
150	1.40036	133.2	0.2682	0.2237	1.1195	553.9		1.36799	133.1	0.2677	0.2239	1.1200	553.5	150
160	1.42694	135.4	0.2719	0.2253	1.1169	559.1		1.39392	135.4	0.2714	0.2255	1.1174	558.7	160
170	1.45307	137.7	0.2755	0.2270	1.1145	564.3		1.41965	137.6	0.2750	0.2271	1.1149	563.9	170
180	1.47929	139.9	0.2791	0.2287	1.1122	569.4		1.44509	139.9	0.2786	0.2288	1.1126	569.0	180
190	1.50534	142.2	0.2827	0.2304	1.1101	574.4		1.47059	142.2	0.2822	0.2305	1.1105	574.1	190
200	1.53092	144.6	0.2862	0.2321	1.1082	579.3		1.49589	144.5	0.2857	0.2322	1.1085	579.0	200
210	1.55666	146.9	0.2897	0.2338	1.1063	584.2		1.52138	146.9	0.2892	0.2340	1.1066	583.9	210
220	1.58253	149.2	0.2932	0.2356	1.1046	589.0		1.54607	149.2	0.2927	0.2357	1.1049	588.7	220
230	1.60798	151.6	0.2966	0.2374	1.1029	593.7		1.57134	151.6	0.2962	0.2375	1.1032	593.4	230
240	1.63319	154.0	0.3001	0.2392	1.1014	598.4		1.59617	153.9	0.2996	0.2393	1.1016	598.1	240
250	1.65865	156.4	0.3035	0.2409	1.0999	603.0		1.62101	156.4	0.3030	0.2410	1.1002	602.7	250
260	1.68407	158.8	0.3068	0.2427	1.0985	607.5		1.64582	158.8	0.3064	0.2428	1.0987	607.3	260
270	1.70940	161.2	0.3102	0.2445	1.0972	612.0		1.67056	161.2	0.3097	0.2446	1.0974	611.8	270
280	1.73430	163.7	0.3135	0.2463	1.0959	616.5		1.69549	163.7	0.3131	0.2464	1.0961	616.3	280
290	1.75963	166.2	0.3169	0.2481	1.0947	620.9		1.71999	166.1	0.3164	0.2482	1.0949	620.7	290
300	1.78444	168.6	0.3202	0.2499	1.0935	625.3		1.74429	168.6	0.3197	0.2499	1.0937	625.1	300
310	1.80995	171.2	0.3234	0.2517	1.0924	629.6		1.76897	171.1	0.3230	0.2517	1.0926	629.4	310
320	1.83453	173.7	0.3267	0.2534	1.0914	633.9		1.79308	173.7	0.3262	0.2535	1.0916	633.7	320
330	1.85977	176.2	0.3299	0.2552	1.0904	638.1		1.81785	176.2	0.3295	0.2552	1.0905	637.9	330
340	1.88466	178.8	0.3332	0.2569	1.0894	642.3		1.84196	178.8	0.3327	0.2570	1.0895	642.1	340

TEMP °F	PRESSURE = 46.00 PSIA							PRESSURE = 47.00 PSIA						TEMP °F
	V	H	S	Cp	Cp/Cv	v_s		V	H	S	Cp	Cp/Cv	v_s	
36	0.01246	23.5	0.0512	0.3222	1.5242	2017.5	SAT LIQ	0.01248	23.8	0.0519	0.3226	1.5249	2008.4	37.1
36	1.02775	108.3	0.2223	0.2146	1.1809	482.0	SAT VAP	1.00664	108.4	0.2222	0.2153	1.1818	481.9	37.1
40	1.03972	109.1	0.2240	0.2144	1.1770	485.0		1.01513	109.1	0.2235	0.2151	1.1788	484.2	40
50	1.06895	111.3	0.2282	0.2142	1.1682	492.3		1.04395	111.2	0.2277	0.2148	1.1698	491.5	50
60	1.09745	113.4	0.2324	0.2144	1.1605	499.2		1.07204	113.4	0.2319	0.2149	1.1619	498.5	60
70	1.12549	115.6	0.2365	0.2148	1.1539	505.9		1.09963	115.5	0.2360	0.2153	1.1551	505.2	70
80	1.15314	117.7	0.2405	0.2155	1.1480	512.4		1.12676	117.7	0.2400	0.2159	1.1491	511.8	80
90	1.18022	119.9	0.2445	0.2164	1.1428	518.7		1.15340	119.8	0.2440	0.2167	1.1437	518.1	90
100	1.20715	122.1	0.2484	0.2174	1.1381	524.8		1.17980	122.0	0.2479	0.2177	1.1390	524.2	100
110	1.23350	124.2	0.2523	0.2186	1.1339	530.7		1.20584	124.2	0.2518	0.2189	1.1347	530.2	110
120	1.25976	126.4	0.2561	0.2198	1.1301	536.5		1.23153	126.4	0.2556	0.2201	1.1308	536.0	120
130	1.28568	128.6	0.2598	0.2212	1.1266	542.1		1.25723	128.6	0.2594	0.2214	1.1272	541.6	130
140	1.31148	130.8	0.2636	0.2226	1.1234	547.6		1.28222	130.8	0.2631	0.2229	1.1240	547.2	140
150	1.33726	133.1	0.2673	0.2241	1.1205	553.0		1.30736	133.0	0.2668	0.2243	1.1211	552.6	150
160	1.36240	135.3	0.2709	0.2257	1.1178	558.3		1.33227	135.3	0.2705	0.2259	1.1183	557.9	160
170	1.38773	137.6	0.2746	0.2273	1.1153	563.6		1.35704	137.6	0.2741	0.2275	1.1158	563.2	170
180	1.41283	139.9	0.2782	0.2290	1.1130	568.7		1.38160	139.8	0.2777	0.2291	1.1134	568.3	180
190	1.43761	142.2	0.2817	0.2306	1.1109	573.7		1.40607	142.1	0.2813	0.2308	1.1112	573.4	190
200	1.46263	144.5	0.2853	0.2324	1.1088	578.7		1.43041	144.5	0.2848	0.2325	1.1092	578.4	200
210	1.48699	146.8	0.2888	0.2341	1.1070	583.6		1.45455	146.8	0.2883	0.2342	1.1073	583.3	210
220	1.51172	149.2	0.2922	0.2358	1.1052	588.4		1.47863	149.1	0.2918	0.2359	1.1055	588.1	220
230	1.53633	151.5	0.2957	0.2376	1.1035	593.1		1.50263	151.5	0.2953	0.2377	1.1038	592.9	230
240	1.56055	153.9	0.2991	0.2394	1.1019	597.8		1.52648	153.9	0.2987	0.2395	1.1022	597.6	240
250	1.58504	156.3	0.3025	0.2411	1.1004	602.5		1.55063	156.3	0.3021	0.2412	1.1007	602.2	250
260	1.60927	158.7	0.3059	0.2429	1.0990	607.0		1.57431	158.7	0.3055	0.2430	1.0992	606.8	260
270	1.63345	161.2	0.3093	0.2447	1.0976	611.6		1.59795	161.2	0.3089	0.2448	1.0978	611.3	270
280	1.65755	163.6	0.3126	0.2465	1.0963	616.0		1.62153	163.6	0.3122	0.2465	1.0965	615.8	280
290	1.68180	166.1	0.3160	0.2482	1.0951	620.5		1.64528	166.1	0.3155	0.2483	1.0953	620.3	290
300	1.70561	168.6	0.3193	0.2500	1.0939	624.9		1.66889	168.6	0.3188	0.2501	1.0941	624.7	300
310	1.72980	171.1	0.3225	0.2518	1.0928	629.2		1.69233	171.1	0.3221	0.2519	1.0930	629.0	310
320	1.75377	173.6	0.3258	0.2536	1.0917	633.5		1.71556	173.6	0.3254	0.2536	1.0919	633.3	320
330	1.77778	176.2	0.3290	0.2553	1.0907	637.7		1.73913	176.2	0.3286	0.2554	1.0909	637.6	330
340	1.80148	178.7	0.3323	0.2571	1.0897	642.0		1.76243	178.7	0.3318	0.2571	1.0899	641.8	340

Reprinted by permission from DuPont

Table 2 (continued)
HFC-134a Superheated Vapor—Constant Pressure Tables

V = Volume in ft³/lb   H = Enthalpy in Btu/lb   S = Entropy in Btu/(lb) (°R) v_s = Velocity of Sound in ft/sec
Cp = Heat Capacity at Constant Pressure in Btu/(lb) (°F)   Cp/Cv = Heat Capacity Ratio (Dimensionless)

TEMP °F	PRESSURE = 48.00 PSIA							PRESSURE = 49.00 PSIA						TEMP °F
	V	H	S	Cp	Cp/Cv	v_s		V	H	S	Cp	Cp/Cv	v_s	
38.1	0.01249	24.2	0.0526	0.3231	1.5257	1999.4	SAT LIQ	0.01251	24.5	0.0533	0.3235	1.5264	1990.5	39.2
38.1	0.98629	108.6	0.2221	0.2159	1.1827	481.9	SAT VAP	0.96684	108.7	0.2220	0.2166	1.1835	481.8	39.2
40	0.99167	109.0	0.2229	0.2158	1.1808	483.3		0.96909	108.9	0.2224	0.2165	1.1827	482.4	40
50	1.01999	111.1	0.2272	0.2154	1.1714	490.7		0.99701	111.1	0.2267	0.2160	1.1731	489.8	50
60	1.04767	113.3	0.2314	0.2154	1.1634	497.7		1.02428	113.2	0.2309	0.2160	1.1648	497.0	60
70	1.07481	115.4	0.2355	0.2158	1.1563	504.5		1.05097	115.4	0.2350	0.2162	1.1576	503.8	70
80	1.10144	117.6	0.2395	0.2163	1.1502	511.1		1.07724	117.5	0.2390	0.2167	1.1513	510.5	80
90	1.12765	119.8	0.2435	0.2171	1.1447	517.5		1.10302	119.7	0.2430	0.2175	1.1457	516.9	90
100	1.15354	121.9	0.2474	0.2181	1.1398	523.6		1.12854	121.9	0.2470	0.2184	1.1407	523.1	100
110	1.17911	124.1	0.2513	0.2191	1.1354	529.6		1.15367	124.1	0.2508	0.2194	1.1362	529.1	110
120	1.20453	126.3	0.2551	0.2204	1.1315	535.5		1.17841	126.3	0.2547	0.2206	1.1322	535.0	120
130	1.22941	128.5	0.2589	0.2217	1.1279	541.2		1.20294	128.5	0.2584	0.2219	1.1285	540.7	130
140	1.25439	130.8	0.2626	0.2231	1.1246	546.7		1.22745	130.7	0.2622	0.2233	1.1252	546.3	140
150	1.27877	133.0	0.2663	0.2245	1.1216	552.2		1.25141	133.0	0.2659	0.2248	1.1221	551.8	150
160	1.30327	135.3	0.2700	0.2261	1.1188	557.6		1.27567	135.2	0.2696	0.2263	1.1193	557.2	160
170	1.32749	137.5	0.2736	0.2277	1.1162	562.8		1.29938	137.5	0.2732	0.2278	1.1167	562.4	170
180	1.35172	139.8	0.2772	0.2293	1.1138	568.0		1.32310	139.8	0.2768	0.2294	1.1143	567.6	180
190	1.37571	142.1	0.2808	0.2309	1.1116	573.0		1.34662	142.1	0.2804	0.2311	1.1120	572.7	190
200	1.39958	144.4	0.2844	0.2326	1.1096	578.0		1.37005	144.4	0.2839	0.2328	1.1099	577.7	200
210	1.42328	146.8	0.2879	0.2343	1.1076	583.0		1.39334	146.7	0.2874	0.2345	1.1079	582.6	210
220	1.44697	149.1	0.2914	0.2361	1.1058	587.8		1.41643	149.1	0.2909	0.2362	1.1061	587.5	220
230	1.47037	151.5	0.2948	0.2378	1.1041	592.6		1.43968	151.5	0.2944	0.2379	1.1044	592.3	230
240	1.49410	153.9	0.2983	0.2396	1.1024	597.3		1.46263	153.8	0.2978	0.2397	1.1027	597.0	240
250	1.51722	156.3	0.3017	0.2413	1.1009	602.0		1.48566	156.2	0.3012	0.2414	1.1012	601.7	250
260	1.54059	158.7	0.3051	0.2431	1.0995	606.6		1.50852	158.7	0.3046	0.2432	1.0997	606.3	260
270	1.56372	161.1	0.3084	0.2449	1.0981	611.1		1.53139	161.1	0.3080	0.2449	1.0983	610.9	270
280	1.58705	163.6	0.3118	0.2466	1.0968	615.6		1.55400	163.6	0.3113	0.2467	1.0970	615.4	280
290	1.61005	166.1	0.3151	0.2484	1.0955	620.1		1.57679	166.0	0.3147	0.2485	1.0957	619.8	290
300	1.63319	168.6	0.3184	0.2502	1.0943	624.5		1.59949	168.5	0.3180	0.2502	1.0945	624.3	300
310	1.65645	171.1	0.3217	0.2519	1.0932	628.8		1.62206	171.0	0.3213	0.2520	1.0933	628.6	310
320	1.67926	173.6	0.3249	0.2537	1.0921	633.1		1.64447	173.6	0.3245	0.2537	1.0922	632.9	320
330	1.70242	176.1	0.3282	0.2554	1.0910	637.4		1.66722	176.1	0.3278	0.2555	1.0912	637.2	330
340	1.72533	178.7	0.3314	0.2572	1.0900	641.6		1.68947	178.7	0.3310	0.2572	1.0902	641.5	340

TEMP °F	PRESSURE = 50.00 PSIA							PRESSURE = 55.00 PSIA						TEMP °F
	V	H	S	Cp	Cp/Cv	v_s		V	H	S	Cp	Cp/Cv	v_s	
40.2	0.01253	24.9	0.0539	0.3239	1.5271	1981.8	SAT LIQ	0.01262	26.5	0.0571	0.3261	1.5307	1940.2	45.2
40.2	0.94805	108.9	0.2220	0.2172	1.1844	481.7	SAT VAP	0.86423	109.5	0.2217	0.2202	1.1889	481.1	45.2
50	0.97494	111.0	0.2262	0.2166	1.1748	489.0		0.87650	110.6	0.2237	0.2198	1.1836	484.9	50
60	1.00180	113.1	0.2304	0.2165	1.1663	496.2		0.90163	112.8	0.2280	0.2193	1.1739	492.4	60
70	1.02807	115.3	0.2345	0.2167	1.1589	503.1		0.92610	115.0	0.2322	0.2191	1.1655	499.6	70
80	1.05396	117.5	0.2386	0.2172	1.1524	509.8		0.95012	117.2	0.2363	0.2193	1.1582	506.5	80
90	1.07933	119.7	0.2426	0.2178	1.1467	516.3		0.97371	119.4	0.2403	0.2198	1.1518	513.2	90
100	1.10436	121.8	0.2465	0.2187	1.1416	522.5		0.99691	121.6	0.2443	0.2204	1.1461	519.6	100
110	1.12905	124.0	0.2504	0.2197	1.1370	528.6		1.01978	123.8	0.2482	0.2213	1.1411	525.9	110
120	1.15354	126.2	0.2542	0.2209	1.1329	534.5		1.04232	126.0	0.2520	0.2223	1.1365	532.0	120
130	1.17772	128.5	0.2580	0.2222	1.1292	540.2		1.06474	128.2	0.2559	0.2234	1.1325	537.9	130
140	1.20178	130.7	0.2617	0.2235	1.1258	545.9		1.08660	130.5	0.2596	0.2247	1.1288	543.6	140
150	1.22534	132.9	0.2655	0.2250	1.1226	551.4		1.10852	132.7	0.2634	0.2260	1.1254	549.3	150
160	1.24891	135.2	0.2691	0.2265	1.1198	556.8		1.13033	135.0	0.2670	0.2274	1.1223	554.8	160
170	1.27226	137.5	0.2728	0.2280	1.1171	562.1		1.15181	137.3	0.2707	0.2289	1.1194	560.2	170
180	1.29550	139.7	0.2764	0.2296	1.1147	567.3		1.17330	139.6	0.2743	0.2304	1.1168	565.5	180
190	1.31874	142.0	0.2799	0.2312	1.1124	572.4		1.19432	141.9	0.2779	0.2320	1.1143	570.7	190
200	1.34174	144.4	0.2835	0.2329	1.1103	577.4		1.21551	144.2	0.2815	0.2336	1.1121	575.8	200
210	1.36444	146.7	0.2870	0.2346	1.1083	582.3		1.23655	146.5	0.2850	0.2352	1.1100	580.8	210
220	1.38735	149.1	0.2905	0.2363	1.1064	587.2		1.25723	148.9	0.2885	0.2369	1.1080	585.8	220
230	1.41004	151.4	0.2940	0.2380	1.1046	592.0		1.27812	151.3	0.2920	0.2386	1.1061	590.6	230
240	1.43266	153.8	0.2974	0.2398	1.1030	596.8		1.29887	153.7	0.2954	0.2403	1.1044	595.5	240
250	1.45518	156.2	0.3008	0.2415	1.1014	601.5		1.31944	156.1	0.2988	0.2420	1.1027	600.2	250
260	1.47754	158.6	0.3042	0.2433	1.0999	606.1		1.33976	158.5	0.3022	0.2437	1.1011	604.9	260
270	1.50015	161.1	0.3076	0.2450	1.0985	610.7		1.36054	161.0	0.3056	0.2454	1.0997	609.5	270
280	1.52230	163.5	0.3109	0.2468	1.0972	615.2		1.38083	163.4	0.3089	0.2472	1.0983	614.1	280
290	1.54440	166.0	0.3142	0.2485	1.0959	619.6		1.40115	165.9	0.3123	0.2489	1.0969	618.6	290
300	1.56691	168.5	0.3176	0.2503	1.0947	624.1		1.42126	168.4	0.3156	0.2507	1.0956	623.1	300
310	1.58907	171.0	0.3208	0.2521	1.0935	628.4		1.44175	170.9	0.3189	0.2524	1.0944	627.5	310
320	1.61108	173.6	0.3241	0.2538	1.0924	632.8		1.46177	173.5	0.3221	0.2541	1.0933	631.8	320
330	1.63292	176.1	0.3273	0.2555	1.0913	637.0		1.48192	176.0	0.3254	0.2558	1.0922	636.2	330
340	1.65509	178.7	0.3306	0.2573	1.0903	641.3		1.50218	178.6	0.3286	0.2576	1.0911	640.5	340
350	1.67701	181.2	0.3338	0.2590	1.0893	645.5		1.52230	181.2	0.3318	0.2593	1.0901	644.7	350

Reprinted by permission from DuPont

Table 2 (continued)
HFC-134a Superheated Vapor—Constant Pressure Tables

V = Volume in ft³/lb   H = Enthalpy in Btu/lb   S = Entropy in Btu/(lb)(°R) v_s = Velocity of Sound in ft/sec
Cp = Heat Capacity at Constant Pressure in Btu/(lb)(°F)   Cp/Cv = Heat Capacity Ratio (Dimensionless)

TEMP °F	PRESSURE = 60.00 PSIA							PRESSURE = 65.00 PSIA						TEMP °F
	V	H	S	Cp	Cp/Cv	v_s		V	H	S	Cp	Cp/Cv	v_s	
49.8	0.01271	28.0	0.0601	0.3282	1.5343	1901.3	SAT LIQ	0.01280	29.4	0.0629	0.3302	1.5380	1864.8	54.2
49.8	0.79390	110.2	0.2214	0.2232	1.1934	480.5	SAT VAP	0.73400	110.7	0.2211	0.2261	1.1979	479.7	54.2
50	0.79428	110.2	0.2215	0.2232	1.1932	480.6		—	—	—	—	—	—	50
60	0.81793	112.4	0.2258	0.2222	1.1820	488.5		0.74699	112.1	0.2237	0.2253	1.1908	484.5	60
70	0.84104	114.6	0.2300	0.2217	1.1725	496.0		0.76888	114.3	0.2280	0.2243	1.1800	492.3	70
80	0.86356	116.9	0.2342	0.2215	1.1643	503.2		0.79020	116.5	0.2321	0.2239	1.1708	499.8	80
90	0.88566	119.1	0.2382	0.2217	1.1572	510.1		0.81103	118.8	0.2363	0.2238	1.1628	506.9	90
100	0.90728	121.3	0.2422	0.2222	1.1509	516.7		0.83139	121.0	0.2403	0.2240	1.1558	513.8	100
110	0.92868	123.5	0.2462	0.2229	1.1453	523.2		0.85143	123.3	0.2443	0.2245	1.1497	520.4	110
120	0.94967	125.8	0.2500	0.2237	1.1403	529.4		0.87116	125.5	0.2482	0.2252	1.1443	526.8	120
130	0.97040	128.0	0.2539	0.2247	1.1359	535.5		0.89071	127.8	0.2520	0.2260	1.1394	533.1	130
140	0.99098	130.3	0.2577	0.2258	1.1318	541.4		0.90975	130.0	0.2558	0.2270	1.1350	539.1	140
150	1.01133	132.5	0.2614	0.2271	1.1282	547.1		0.92894	132.3	0.2596	0.2282	1.1311	545.0	150
160	1.03135	134.8	0.2651	0.2284	1.1248	552.8		0.94751	134.6	0.2633	0.2294	1.1275	550.7	160
170	1.05130	137.1	0.2688	0.2298	1.1218	558.3		0.96628	136.9	0.2670	0.2307	1.1242	556.4	170
180	1.07101	139.4	0.2724	0.2312	1.1189	563.7		0.98464	139.2	0.2707	0.2321	1.1212	561.9	180
190	1.09075	141.7	0.2760	0.2327	1.1163	569.0		1.00291	141.5	0.2743	0.2335	1.1184	567.3	190
200	1.11012	144.0	0.2796	0.2343	1.1139	574.2		1.02114	143.9	0.2778	0.2350	1.1158	572.5	200
210	1.12943	146.4	0.2831	0.2359	1.1117	579.3		1.03918	146.2	0.2814	0.2365	1.1134	577.7	210
220	1.14903	148.8	0.2866	0.2375	1.1096	584.3		1.05719	148.6	0.2849	0.2381	1.1112	582.8	220
230	1.16795	151.1	0.2901	0.2391	1.1076	589.2		1.07492	151.0	0.2884	0.2397	1.1091	587.8	230
240	1.18723	153.5	0.2936	0.2408	1.1057	594.1		1.09266	153.4	0.2919	0.2413	1.1072	592.8	240
250	1.20627	156.0	0.2970	0.2425	1.1040	598.9		1.11037	155.8	0.2953	0.2430	1.1053	597.7	250
260	1.22519	158.4	0.3004	0.2442	1.1024	603.7		1.12816	158.3	0.2987	0.2446	1.1036	602.5	260
270	1.24409	160.8	0.3038	0.2459	1.1008	608.4		1.14561	160.7	0.3021	0.2463	1.1020	607.2	270
280	1.26295	163.3	0.3071	0.2476	1.0993	613.0		1.16306	163.2	0.3055	0.2480	1.1004	611.9	280
290	1.28172	165.8	0.3105	0.2493	1.0979	617.5		1.18036	165.7	0.3088	0.2497	1.0990	616.5	290
300	1.30039	168.3	0.3138	0.2510	1.0966	622.0		1.19804	168.2	0.3121	0.2514	1.0976	621.0	300
310	1.31891	170.8	0.3171	0.2527	1.0953	626.5		1.21521	170.7	0.3154	0.2531	1.0963	625.5	310
320	1.33761	173.4	0.3204	0.2544	1.0941	630.9		1.23259	173.2	0.3187	0.2548	1.0950	630.0	320
330	1.35630	175.9	0.3236	0.2561	1.0930	635.3		1.24969	175.8	0.3220	0.2564	1.0938	634.4	330
340	1.37457	178.5	0.3268	0.2578	1.0919	639.6		1.26678	178.4	0.3252	0.2581	1.0927	638.8	340
350	1.39334	181.1	0.3301	0.2595	1.0908	643.9		1.28403	181.0	0.3284	0.2598	1.0916	643.1	350
360	—	—	—	—	—	—		1.30124	183.6	0.3316	0.2615	1.0905	647.4	360

TEMP °F	PRESSURE = 70.00 PSIA							PRESSURE = 75.00 PSIA						TEMP °F
	V	H	S	Cp	Cp/Cv	v_s		V	H	S	Cp	Cp/Cv	v_s	
58.3	0.01288	30.8	0.0655	0.3321	1.5417	1830.2	SAT LIQ	0.01296	32.1	0.0680	0.3341	1.5454	1797.5	62.2
58.3	0.68236	111.3	0.2209	0.2288	1.2026	478.9	SAT VAP	0.63739	111.8	0.2207	0.2316	1.2072	478.0	62.2
60	0.68601	111.7	0.2217	0.2285	1.2002	480.3		—	—	—	—	—	—	60
70	0.70691	114.0	0.2260	0.2272	1.1880	488.5		0.65304	113.6	0.2241	0.2302	1.1966	484.6	70
80	0.72717	116.2	0.2302	0.2264	1.1777	496.3		0.67245	115.9	0.2284	0.2289	1.1850	492.7	80
90	0.74694	118.5	0.2344	0.2260	1.1688	503.7		0.69132	118.2	0.2326	0.2282	1.1751	500.4	90
100	0.76628	120.7	0.2385	0.2259	1.1611	510.8		0.70972	120.5	0.2367	0.2279	1.1666	507.7	100
110	0.78524	123.0	0.2425	0.2262	1.1543	517.6		0.72775	122.7	0.2408	0.2280	1.1592	514.8	110
120	0.80386	125.3	0.2464	0.2267	1.1484	524.2		0.74543	125.0	0.2448	0.2283	1.1526	521.6	120
130	0.82217	127.5	0.2503	0.2274	1.1431	530.6		0.76283	127.3	0.2487	0.2288	1.1469	528.1	130
140	0.84034	129.8	0.2541	0.2283	1.1383	536.8		0.77997	129.6	0.2525	0.2295	1.1417	534.5	140
150	0.85807	132.1	0.2579	0.2293	1.1341	542.8		0.79675	131.9	0.2563	0.2304	1.1371	540.6	150
160	0.87573	134.4	0.2617	0.2304	1.1302	548.7		0.81347	134.2	0.2601	0.2314	1.1330	546.6	160
170	0.89326	136.7	0.2654	0.2316	1.1266	554.4		0.83008	136.5	0.2638	0.2326	1.1292	552.5	170
180	0.91058	139.0	0.2690	0.2329	1.1234	560.0		0.84631	138.9	0.2675	0.2338	1.1257	558.2	180
190	0.92773	141.4	0.2726	0.2343	1.1205	565.5		0.86259	141.2	0.2711	0.2351	1.1226	563.8	190
200	0.94473	143.7	0.2762	0.2357	1.1177	570.9		0.87850	143.6	0.2747	0.2364	1.1197	569.2	200
210	0.96172	146.1	0.2798	0.2372	1.1152	576.2		0.89453	145.9	0.2783	0.2379	1.1170	574.6	210
220	0.97847	148.5	0.2833	0.2387	1.1128	581.4		0.91033	148.3	0.2818	0.2393	1.1145	579.9	220
230	0.99512	150.9	0.2868	0.2403	1.1106	586.4		0.92618	150.7	0.2853	0.2408	1.1122	585.0	230
240	1.01184	153.3	0.2903	0.2419	1.1086	591.5		0.94171	153.1	0.2888	0.2424	1.1100	590.1	240
250	1.02828	155.7	0.2937	0.2435	1.1067	596.4		0.95730	155.6	0.2922	0.2440	1.1080	595.1	250
260	1.04482	158.1	0.2971	0.2451	1.1049	601.2		0.97267	158.0	0.2957	0.2456	1.1061	600.0	260
270	1.06135	160.6	0.3005	0.2467	1.1032	606.0		0.98814	160.5	0.2991	0.2472	1.1043	604.9	270
280	1.07747	163.1	0.3039	0.2484	1.1015	610.8		1.00341	163.0	0.3024	0.2488	1.1027	609.6	280
290	1.09385	165.6	0.3072	0.2501	1.1000	615.4		1.01864	165.5	0.3058	0.2504	1.1011	614.4	290
300	1.11012	168.1	0.3106	0.2517	1.0986	620.0		1.03402	168.0	0.3091	0.2521	1.0996	619.0	300
310	1.12625	170.6	0.3139	0.2534	1.0972	624.6		1.04910	170.5	0.3124	0.2537	1.0982	623.6	310
320	1.14233	173.1	0.3172	0.2551	1.0959	629.1		1.06417	173.0	0.3157	0.2554	1.0968	628.2	320
330	1.15835	175.7	0.3204	0.2567	1.0947	633.5		1.07921	175.6	0.3190	0.2571	1.0955	632.6	330
340	1.17426	178.3	0.3237	0.2584	1.0935	637.9		1.09433	178.2	0.3222	0.2587	1.0943	637.1	340
350	1.19033	180.9	0.3269	0.2601	1.0923	642.3		1.10926	180.8	0.3254	0.2604	1.0931	641.5	350
360	1.20642	183.5	0.3301	0.2618	1.0913	646.6		1.12410	183.4	0.3287	0.2620	1.0920	645.8	360
370	—	—	—	—	—	—		1.13908	186.0	0.3318	0.2637	1.0909	650.1	370

Reprinted by permission from DuPont

Table 2 (continued)
HFC-134a Superheated Vapor—Constant Pressure Tables

V = Volume in ft³/lb H = Enthalpy in Btu/lb S = Entropy in Btu/(lb) (°R) v_s = Velocity of Sound in ft/sec
Cp = Heat Capacity at Constant Pressure in Btu/(lb) (°F) Cp/Cv = Heat Capacity Ratio (Dimensionless)

TEMP °F	PRESSURE = 80.00 PSIA							PRESSURE = 85.00 PSIA						TEMP °F
	V	H	S	Cp	Cp/Cv	v_s		V	H	S	Cp	Cp/Cv	v_s	
65.9	0.01304	33.3	0.0703	0.3359	1.5492	1766.4	SAT LIQ	0.01311	34.5	0.0725	0.3378	1.5530	1736.6	69.4
65.9	0.59787	112.3	0.2205	0.2342	1.2120	477.0	SAT VAP	0.56284	112.7	0.2204	0.2369	1.2168	476.0	69.4
70	0.60580	113.2	0.2224	0.2333	1.2058	480.7		0.56395	112.9	0.2206	0.2367	1.2158	476.6	70
80	0.62445	115.6	0.2267	0.2317	1.1928	489.1		0.58200	115.2	0.2250	0.2345	1.2011	485.4	80
90	0.64255	117.9	0.2309	0.2306	1.1818	497.1		0.59945	117.6	0.2293	0.2331	1.1888	493.6	90
100	0.66020	120.2	0.2351	0.2300	1.1723	504.7		0.61641	119.9	0.2335	0.2322	1.1784	501.5	100
110	0.67741	122.5	0.2392	0.2298	1.1642	511.9		0.63295	122.2	0.2376	0.2317	1.1695	509.0	110
120	0.69430	124.8	0.2432	0.2299	1.1571	518.9		0.64910	124.5	0.2417	0.2316	1.1617	516.2	120
130	0.71083	127.1	0.2471	0.2303	1.1508	525.6		0.66494	126.8	0.2456	0.2318	1.1549	523.1	130
140	0.72711	129.4	0.2510	0.2308	1.1453	532.1		0.68055	129.1	0.2495	0.2322	1.1489	529.8	140
150	0.74322	131.7	0.2548	0.2316	1.1403	538.4		0.69585	131.5	0.2534	0.2328	1.1436	536.2	150
160	0.75896	134.0	0.2586	0.2325	1.1358	544.6		0.71093	133.8	0.2572	0.2336	1.1388	542.5	160
170	0.77465	136.3	0.2623	0.2335	1.1318	550.5		0.72574	136.1	0.2609	0.2345	1.1345	548.6	170
180	0.79020	138.7	0.2660	0.2347	1.1281	556.4		0.74052	138.5	0.2646	0.2356	1.1306	554.5	180
190	0.80548	141.0	0.2696	0.2359	1.1248	562.0		0.75506	140.9	0.2683	0.2367	1.1270	560.3	190
200	0.82068	143.4	0.2733	0.2372	1.1217	567.6		0.76953	143.2	0.2719	0.2380	1.1237	565.9	200
210	0.83577	145.8	0.2768	0.2386	1.1188	573.0		0.78388	145.6	0.2755	0.2393	1.1207	571.5	210
220	0.85063	148.2	0.2804	0.2400	1.1162	578.4		0.79802	148.0	0.2790	0.2406	1.1180	576.9	220
230	0.86558	150.6	0.2839	0.2414	1.1138	583.6		0.81208	150.4	0.2826	0.2420	1.1154	582.2	230
240	0.88020	153.0	0.2874	0.2429	1.1115	588.8		0.82617	152.9	0.2861	0.2435	1.1130	587.4	240
250	0.89501	155.4	0.2908	0.2445	1.1094	593.8		0.84005	155.3	0.2895	0.2450	1.1108	592.5	250
260	0.90950	157.9	0.2943	0.2460	1.1074	598.8		0.85383	157.8	0.2930	0.2465	1.1087	597.6	260
270	0.92404	160.4	0.2977	0.2476	1.1056	603.7		0.86760	160.2	0.2964	0.2481	1.1068	602.5	270
280	0.93853	162.8	0.3011	0.2492	1.1038	608.5		0.88137	162.7	0.2998	0.2496	1.1050	607.4	280
290	0.95302	165.3	0.3044	0.2508	1.1021	613.3		0.89493	165.2	0.3031	0.2512	1.1032	612.2	290
300	0.96721	167.9	0.3078	0.2525	1.1006	618.0		0.90851	167.7	0.3065	0.2528	1.1016	617.0	300
310	0.98155	170.4	0.3111	0.2541	1.0991	622.6		0.92200	170.3	0.3098	0.2544	1.1001	621.7	310
320	0.99582	172.9	0.3144	0.2557	1.0977	627.2		0.93545	172.8	0.3131	0.2561	1.0986	626.3	320
330	1.01010	175.5	0.3176	0.2574	1.0964	631.8		0.94886	175.4	0.3164	0.2577	1.0972	630.9	330
340	1.02417	178.1	0.3209	0.2590	1.0951	636.2		0.96219	178.0	0.3196	0.2593	1.0959	635.4	340
350	1.03821	180.7	0.3241	0.2606	1.0939	640.7		0.97570	180.6	0.3228	0.2609	1.0947	639.9	350
360	1.05230	183.3	0.3273	0.2623	1.0927	645.0		0.98892	183.2	0.3260	0.2625	1.0935	644.3	360
370	1.06644	185.9	0.3305	0.2639	1.0916	649.4		1.00220	185.8	0.3292	0.2642	1.0923	648.6	370

TEMP °F	PRESSURE = 90.00 PSIA							PRESSURE = 95.00 PSIA						TEMP °F
	V	H	S	Cp	Cp/Cv	v_s		V	H	S	Cp	Cp/Cv	v_s	
72.8	0.01319	35.7	0.0747	0.3396	1.5568	1708.1	SAT LIQ	0.01326	36.8	0.0767	0.3415	1.5607	1680.6	76
72.8	0.53155	113.1	0.2202	0.2395	1.2217	475.0	SAT VAP	0.50345	113.5	0.2200	0.2420	1.2266	473.9	76
80	0.54416	114.9	0.2234	0.2376	1.2100	481.6		0.51020	114.5	0.2218	0.2408	1.2196	477.6	80
90	0.56107	117.2	0.2278	0.2357	1.1964	490.2		0.52662	116.9	0.2262	0.2384	1.2044	486.6	90
100	0.57743	119.6	0.2320	0.2344	1.1849	498.3		0.54248	119.3	0.2305	0.2368	1.1917	495.0	100
110	0.59333	121.9	0.2361	0.2337	1.1751	506.0		0.55785	121.6	0.2347	0.2357	1.1809	503.0	110
120	0.60887	124.3	0.2402	0.2333	1.1666	513.4		0.57284	124.0	0.2388	0.2351	1.1717	510.6	120
130	0.62406	126.6	0.2442	0.2333	1.1592	520.5		0.58751	126.3	0.2428	0.2349	1.1637	517.9	130
140	0.63906	128.9	0.2481	0.2336	1.1527	527.4		0.60183	128.7	0.2468	0.2350	1.1567	524.9	140
150	0.65368	131.3	0.2520	0.2341	1.1470	534.0		0.61588	131.0	0.2507	0.2353	1.1505	531.7	150
160	0.66814	133.6	0.2558	0.2347	1.1418	540.4		0.62980	133.4	0.2545	0.2359	1.1450	538.2	160
170	0.68236	136.0	0.2596	0.2355	1.1372	546.6		0.64346	135.8	0.2583	0.2366	1.1401	544.6	170
180	0.69643	138.3	0.2633	0.2365	1.1331	552.6		0.65690	138.1	0.2620	0.2374	1.1356	550.7	180
190	0.71023	140.7	0.2670	0.2376	1.1293	558.5		0.67024	140.5	0.2657	0.2384	1.1316	556.7	190
200	0.72401	143.1	0.2706	0.2387	1.1258	564.2		0.68339	142.9	0.2694	0.2395	1.1280	562.6	200
210	0.73768	145.5	0.2742	0.2400	1.1227	569.9		0.69643	145.3	0.2730	0.2407	1.1246	568.3	210
220	0.75126	147.9	0.2778	0.2413	1.1197	575.4		0.70932	147.7	0.2765	0.2419	1.1215	573.9	220
230	0.76470	150.3	0.2813	0.2426	1.1170	580.8		0.72218	150.1	0.2801	0.2433	1.1187	579.3	230
240	0.77791	152.7	0.2848	0.2441	1.1145	586.0		0.73486	152.6	0.2836	0.2446	1.1161	584.7	240
250	0.79108	155.2	0.2883	0.2455	1.1122	591.2		0.74744	155.0	0.2871	0.2460	1.1137	589.9	250
260	0.80431	157.6	0.2917	0.2470	1.1101	596.3		0.76005	157.5	0.2905	0.2475	1.1114	595.1	260
270	0.81746	160.1	0.2951	0.2485	1.1080	601.4		0.77250	160.0	0.2940	0.2490	1.1093	600.2	270
280	0.83036	162.6	0.2985	0.2501	1.1061	606.3		0.78493	162.5	0.2974	0.2505	1.1073	605.2	280
290	0.84331	165.1	0.3019	0.2516	1.1043	611.2		0.79726	165.0	0.3007	0.2520	1.1054	610.1	290
300	0.85616	167.6	0.3052	0.2532	1.1026	616.0		0.80952	167.5	0.3041	0.2536	1.1037	614.9	300
310	0.86904	170.2	0.3086	0.2548	1.1010	620.7		0.82169	170.1	0.3074	0.2551	1.1020	619.7	310
320	0.88191	172.7	0.3119	0.2564	1.0995	625.4		0.83389	172.6	0.3107	0.2567	1.1005	624.4	320
330	0.89461	175.3	0.3151	0.2580	1.0981	630.0		0.84595	175.2	0.3140	0.2583	1.0990	629.1	330
340	0.90728	177.9	0.3184	0.2596	1.0967	634.5		0.85815	177.8	0.3173	0.2599	1.0976	633.7	340
350	0.91988	180.5	0.3216	0.2612	1.0955	639.0		0.87017	180.4	0.3205	0.2615	1.0962	638.2	350
360	0.93257	183.1	0.3249	0.2628	1.0942	643.5		0.88215	183.0	0.3237	0.2631	1.0950	642.7	360
370	0.94518	185.7	0.3280	0.2644	1.0930	647.9		0.89405	185.7	0.3269	0.2647	1.0938	647.2	370
380	0.95767	188.4	0.3312	0.2660	1.0919	652.3		0.90613	188.3	0.3301	0.2663	1.0926	651.5	380

Reprinted by permission from DuPont

Table 2 (continued)
HFC-134a Superheated Vapor—Constant Pressure Tables

V = Volume in ft³/lb   H = Enthalpy in Btu/lb   S = Entropy in Btu/(lb) (°R) v_s = Velocity of Sound in ft/sec
Cp = Heat Capacity at Constant Pressure in Btu/(lb) (°F)   Cp/Cv = Heat Capacity Ratio (Dimensionless)

TEMP °F	PRESSURE = 100.00 PSIA							PRESSURE = 110.00 PSIA						TEMP °F
	V	H	S	Cp	Cp/Cv	v_s		V	H	S	Cp	Cp/Cv	v_s	
79.1	0.01333	37.8	0.0787	0.3433	1.5646	1654.2	SAT LIQ	0.01347	39.8	0.0824	0.3469	1.5726	1604.1	85
79.1	0.47803	113.9	0.2199	0.2446	1.2317	472.8	SAT VAP	0.43391	114.6	0.2196	0.2496	1.2420	470.4	85
80	0.47952	114.1	0.2203	0.2442	1.2300	473.6		—	—	—	—	—	—	80
90	0.49552	116.6	0.2248	0.2413	1.2129	482.9		0.44156	115.9	0.2219	0.2477	1.2319	475.4	90
100	0.51093	119.0	0.2291	0.2393	1.1989	491.7		0.45627	118.3	0.2264	0.2446	1.2146	484.8	100
110	0.52587	121.4	0.2333	0.2379	1.1871	499.9		0.47043	120.8	0.2307	0.2425	1.2004	493.6	110
120	0.54037	123.7	0.2375	0.2370	1.1770	507.8		0.48414	123.2	0.2349	0.2410	1.1884	502.0	120
130	0.55451	126.1	0.2415	0.2366	1.1683	515.3		0.49746	125.6	0.2390	0.2401	1.1782	509.9	130
140	0.56838	128.5	0.2455	0.2365	1.1608	522.5		0.51044	128.0	0.2430	0.2395	1.1694	517.5	140
150	0.58194	130.8	0.2494	0.2366	1.1541	529.4		0.52309	130.4	0.2470	0.2394	1.1618	524.7	150
160	0.59527	133.2	0.2533	0.2370	1.1482	536.1		0.53562	132.8	0.2509	0.2395	1.1550	531.7	160
170	0.60835	135.6	0.2570	0.2376	1.1430	542.5		0.54786	135.2	0.2547	0.2399	1.1491	538.5	170
180	0.62135	137.9	0.2608	0.2384	1.1383	548.8		0.55979	137.6	0.2585	0.2404	1.1438	545.0	180
190	0.63416	140.3	0.2645	0.2393	1.1340	554.9		0.57166	140.0	0.2622	0.2411	1.1390	551.3	190
200	0.64675	142.7	0.2682	0.2403	1.1301	560.9		0.58340	142.4	0.2659	0.2420	1.1347	557.4	200
210	0.65924	145.1	0.2718	0.2414	1.1266	566.7		0.59503	144.8	0.2696	0.2429	1.1308	563.4	210
220	0.67155	147.6	0.2754	0.2426	1.1234	572.3		0.60643	147.3	0.2732	0.2440	1.1272	569.3	220
230	0.68385	150.0	0.2789	0.2439	1.1204	577.9		0.61774	149.7	0.2768	0.2452	1.1239	575.0	230
240	0.69604	152.4	0.2825	0.2452	1.1177	583.3		0.62893	152.2	0.2803	0.2464	1.1209	580.5	240
250	0.70806	154.9	0.2859	0.2466	1.1151	588.6		0.64012	154.6	0.2838	0.2477	1.1181	586.0	250
260	0.72015	157.4	0.2894	0.2480	1.1128	593.9		0.65121	157.1	0.2873	0.2490	1.1156	591.4	260
270	0.73196	159.9	0.2928	0.2495	1.1106	599.0		0.66212	159.6	0.2907	0.2504	1.1132	596.6	270
280	0.74388	162.4	0.2962	0.2509	1.1085	604.1		0.67308	162.1	0.2941	0.2518	1.1109	601.8	280
290	0.75569	164.9	0.2996	0.2524	1.1066	609.0		0.68385	164.6	0.2975	0.2533	1.1088	606.9	290
300	0.76740	167.4	0.3030	0.2540	1.1047	613.9		0.69464	167.2	0.3009	0.2547	1.1069	611.9	300
310	0.77906	170.0	0.3063	0.2555	1.1030	618.7		0.70542	169.7	0.3042	0.2562	1.1050	616.8	310
320	0.79064	172.5	0.3096	0.2571	1.1014	623.5		0.71613	172.3	0.3076	0.2577	1.1033	621.6	320
330	0.80225	175.1	0.3129	0.2586	1.0999	628.2		0.72680	174.9	0.3109	0.2593	1.1017	626.4	330
340	0.81387	177.7	0.3162	0.2602	1.0984	632.8		0.73725	177.5	0.3141	0.2608	1.1001	631.1	340
350	0.82535	180.3	0.3194	0.2618	1.0970	637.4		0.74783	180.1	0.3174	0.2624	1.0987	635.8	350
360	0.83675	182.9	0.3226	0.2634	1.0957	641.9		0.75832	182.7	0.3206	0.2639	1.0973	640.4	360
370	0.84818	185.6	0.3258	0.2649	1.0945	646.4		0.76870	185.4	0.3238	0.2655	1.0959	644.9	370
380	0.85955	188.2	0.3290	0.2665	1.0933	650.8		0.77924	188.0	0.3270	0.2670	1.0947	649.4	380
390	—	—	—	—	—	—		0.78958	190.7	0.3302	0.2686	1.0935	653.8	390

TEMP °F	PRESSURE = 120.00 PSIA							PRESSURE = 130.00 PSIA						TEMP °F
	V	H	S	Cp	Cp/Cv	v_s		V	H	S	Cp	Cp/Cv	v_s	
90.5	0.01361	41.8	0.0858	0.3504	1.5808	1557.1	SAT LIQ	0.01374	43.6	0.0890	0.3540	1.5893	1512.8	95.6
90.5	0.39689	115.3	0.2194	0.2546	1.2528	467.9	SAT VAP	0.36538	115.8	0.2192	0.2596	1.2640	465.4	95.6
100	0.41044	117.7	0.2238	0.2506	1.2326	477.6		0.37136	117.0	0.2212	0.2573	1.2531	470.1	100
110	0.42402	120.2	0.2282	0.2475	1.2153	487.1		0.38453	119.5	0.2257	0.2531	1.2321	480.3	110
120	0.43710	122.6	0.2324	0.2453	1.2010	496.0		0.39712	122.0	0.2301	0.2501	1.2150	489.8	120
130	0.44976	125.1	0.2366	0.2438	1.1890	504.4		0.40925	124.5	0.2344	0.2479	1.2009	498.7	130
140	0.46206	127.5	0.2407	0.2428	1.1788	512.4		0.42100	127.0	0.2385	0.2464	1.1890	507.1	140
150	0.47405	129.9	0.2447	0.2423	1.1700	520.0		0.43241	129.5	0.2426	0.2454	1.1786	515.1	150
160	0.48577	132.3	0.2487	0.2421	1.1623	527.3		0.44350	131.9	0.2466	0.2448	1.1701	522.7	160
170	0.49724	134.8	0.2525	0.2422	1.1555	534.3		0.45442	134.4	0.2505	0.2446	1.1624	530.0	170
180	0.50860	137.2	0.2564	0.2425	1.1495	541.1		0.46507	136.8	0.2544	0.2447	1.1557	537.1	180
190	0.51967	139.6	0.2601	0.2430	1.1442	547.6		0.47556	139.2	0.2582	0.2450	1.1497	543.9	190
200	0.53064	142.0	0.2639	0.2437	1.1394	554.0		0.48584	141.7	0.2619	0.2455	1.1444	550.5	200
210	0.54139	144.5	0.2675	0.2445	1.1351	560.1		0.49601	144.2	0.2656	0.2461	1.1396	556.8	210
220	0.55206	146.9	0.2712	0.2454	1.1311	566.2		0.50602	146.6	0.2692	0.2469	1.1352	563.0	220
230	0.56268	149.4	0.2747	0.2465	1.1275	572.0		0.51597	149.1	0.2729	0.2478	1.1313	569.1	230
240	0.57307	151.9	0.2783	0.2476	1.1243	577.8		0.52576	151.6	0.2764	0.2489	1.1277	575.0	240
250	0.58350	154.4	0.2818	0.2488	1.1212	583.4		0.53550	154.1	0.2800	0.2500	1.1244	580.7	250
260	0.59379	156.8	0.2853	0.2501	1.1184	588.9		0.54511	156.6	0.2835	0.2511	1.1214	586.3	260
270	0.60394	159.4	0.2888	0.2514	1.1158	594.2		0.55460	159.1	0.2870	0.2524	1.1186	591.8	270
280	0.61406	161.9	0.2922	0.2527	1.1134	599.5		0.56408	161.6	0.2904	0.2536	1.1160	597.2	280
290	0.62414	164.4	0.2956	0.2541	1.1112	604.7		0.57353	164.2	0.2938	0.2550	1.1136	602.5	290
300	0.63408	167.0	0.2990	0.2555	1.1091	609.8		0.58282	166.7	0.2972	0.2563	1.1113	607.7	300
310	0.64404	169.5	0.3023	0.2570	1.1071	614.8		0.59207	169.3	0.3006	0.2577	1.1092	612.8	310
320	0.65389	172.1	0.3057	0.2584	1.1052	619.8		0.60129	171.9	0.3039	0.2592	1.1072	617.9	320
330	0.66375	174.7	0.3090	0.2599	1.1035	624.6		0.61039	174.5	0.3072	0.2606	1.1054	622.8	330
340	0.67354	177.3	0.3122	0.2614	1.1019	629.4		0.61962	177.1	0.3105	0.2621	1.1036	627.7	340
350	0.68329	179.9	0.3155	0.2629	1.1003	634.1		0.62865	179.7	0.3138	0.2635	1.1020	632.5	350
360	0.69300	182.6	0.3187	0.2645	1.0988	638.8		0.63771	182.4	0.3170	0.2650	1.1004	637.2	360
370	0.70264	185.2	0.3220	0.2660	1.0974	643.4		0.64666	185.0	0.3202	0.2665	1.0989	641.9	370
380	0.71225	187.9	0.3252	0.2675	1.0961	648.0		0.65569	187.7	0.3234	0.2680	1.0975	646.5	380
390	0.72192	190.6	0.3283	0.2690	1.0948	652.5		0.66458	190.4	0.3266	0.2695	1.0962	651.1	390
400	0.73142	193.3	0.3315	0.2706	1.0936	656.9		0.67345	193.1	0.3298	0.2710	1.0949	655.6	400

Table 2 (continued)
HFC-134a Superheated Vapor—Constant Pressure Tables

V = Volume in ft³/lb   H = Enthalpy in Btu/lb   S = Entropy in Btu/(lb) (°R) v_s = Velocity of Sound in ft/sec
Cp = Heat Capacity at Constant Pressure in Btu/(lb) (°F)   Cp/Cv = Heat Capacity Ratio (Dimensionless)

TEMP °F	PRESSURE = 140.00 PSIA							PRESSURE = 150.00 PSIA						TEMP °F
	V	H	S	Cp	Cp/Cv	v_s		V	H	S	Cp	Cp/Cv	v_s	
100.5	0.01388	45.3	0.0921	0.3576	1.5980	1470.8	SAT LIQ	0.01401	46.9	0.0950	0.3612	1.6070	1430.7	105.1
100.5	0.33818	116.4	0.2190	0.2646	1.2757	462.7	SAT VAP	0.31448	116.9	0.2188	0.2697	1.2878	460.0	105.1
110	0.35042	118.9	0.2234	0.2593	1.2512	473.2		0.32062	118.2	0.2211	0.2664	1.2732	465.7	110
120	0.36266	121.4	0.2279	0.2553	1.2307	483.3		0.33259	120.8	0.2257	0.2610	1.2484	476.6	120
130	0.37438	124.0	0.2322	0.2523	1.2140	492.8		0.34400	123.4	0.2301	0.2571	1.2285	486.7	130
140	0.38568	126.5	0.2364	0.2502	1.2001	501.7		0.35495	126.0	0.2344	0.2543	1.2123	496.1	140
150	0.39662	129.0	0.2406	0.2487	1.1884	510.1		0.36550	128.5	0.2386	0.2523	1.1988	505.0	150
160	0.40727	131.5	0.2446	0.2477	1.1784	518.1		0.37573	131.0	0.2427	0.2508	1.1874	513.3	160
170	0.41764	133.9	0.2486	0.2472	1.1697	525.7		0.38565	133.5	0.2467	0.2499	1.1776	521.3	170
180	0.42779	136.4	0.2524	0.2470	1.1622	533.0		0.39538	136.0	0.2506	0.2494	1.1691	528.9	180
190	0.43777	138.9	0.2563	0.2470	1.1555	540.1		0.40491	138.5	0.2545	0.2492	1.1616	536.2	190
200	0.44755	141.3	0.2601	0.2473	1.1496	546.9		0.41418	141.0	0.2583	0.2493	1.1550	543.3	200
210	0.45714	143.8	0.2638	0.2478	1.1443	553.5		0.42337	143.5	0.2621	0.2496	1.1492	550.1	210
220	0.46659	146.3	0.2675	0.2485	1.1395	559.9		0.43232	146.0	0.2658	0.2501	1.1440	556.7	220
230	0.47596	148.8	0.2711	0.2492	1.1352	566.1		0.44125	148.5	0.2694	0.2507	1.1393	563.1	230
240	0.48520	151.3	0.2747	0.2501	1.1313	572.1		0.44996	151.0	0.2730	0.2515	1.1350	569.3	240
250	0.49432	153.8	0.2782	0.2511	1.1277	578.0		0.45861	153.5	0.2766	0.2524	1.1311	575.3	250
260	0.50342	156.3	0.2818	0.2522	1.1244	583.8		0.46718	156.0	0.2801	0.2533	1.1276	581.2	260
270	0.51235	158.8	0.2853	0.2534	1.1214	589.4		0.47574	158.6	0.2836	0.2544	1.1243	587.0	270
280	0.52121	161.4	0.2887	0.2546	1.1186	594.9		0.48412	161.1	0.2871	0.2555	1.1213	592.6	280
290	0.53011	163.9	0.2921	0.2558	1.1160	600.3		0.49242	163.7	0.2906	0.2567	1.1185	598.1	290
300	0.53888	166.5	0.2955	0.2572	1.1136	605.6		0.50073	166.3	0.2940	0.2580	1.1159	603.6	300
310	0.54750	169.1	0.2989	0.2585	1.1113	610.9		0.50893	168.9	0.2973	0.2593	1.1135	608.9	310
320	0.55611	171.7	0.3022	0.2599	1.1092	616.0		0.51706	171.5	0.3007	0.2606	1.1113	614.1	320
330	0.56478	174.3	0.3056	0.2613	1.1073	621.0		0.52521	174.1	0.3040	0.2620	1.1092	619.2	330
340	0.57330	176.9	0.3089	0.2627	1.1054	626.0		0.53319	176.7	0.3073	0.2633	1.1072	624.2	340
350	0.58184	179.5	0.3121	0.2641	1.1036	630.9		0.54121	179.3	0.3106	0.2647	1.1054	629.2	350
360	0.59028	182.2	0.3154	0.2656	1.1020	635.7		0.54918	182.0	0.3139	0.2662	1.1036	634.1	360
370	0.59866	184.8	0.3186	0.2670	1.1004	640.4		0.55710	184.7	0.3171	0.2676	1.1020	638.9	370
380	0.60716	187.5	0.3218	0.2685	1.0989	645.1		0.56500	187.3	0.3203	0.2690	1.1004	643.7	380
390	0.61542	190.2	0.3250	0.2700	1.0975	649.7		0.57290	190.0	0.3235	0.2705	1.0989	648.4	390
400	0.62375	192.9	0.3282	0.2715	1.0962	654.3		0.58069	192.7	0.3267	0.2719	1.0975	653.0	400
410	0.63203	195.6	0.3313	0.2730	1.0949	658.8		0.58851	195.5	0.3298	0.2734	1.0962	657.6	410

TEMP °F	PRESSURE = 160.00 PSIA							PRESSURE = 170.00 PSIA						TEMP °F
	V	H	S	Cp	Cp/Cv	v_s		V	H	S	Cp	Cp/Cv	v_s	
109.5	0.01414	48.5	0.0977	0.3648	1.6163	1392.4	SAT LIQ	0.01427	50.1	0.1004	0.3685	1.6259	1355.6	113.7
109.5	0.29362	117.3	0.2186	0.2748	1.3006	457.2	SAT VAP	0.27512	117.7	0.2184	0.2800	1.3139	454.4	113.7
110	0.29426	117.5	0.2188	0.2744	1.2988	457.9		—	—	—	—	—	—	110
120	0.30608	120.2	0.2235	0.2675	1.2685	469.6		0.28247	119.5	0.2214	0.2748	1.2915	462.3	120
130	0.31726	122.8	0.2281	0.2624	1.2448	480.4		0.29350	122.2	0.2261	0.2683	1.2631	473.8	130
140	0.32792	125.4	0.2324	0.2588	1.2258	490.4		0.30395	124.9	0.2305	0.2636	1.2406	484.4	140
150	0.33817	128.0	0.2367	0.2561	1.2101	499.7		0.31395	127.5	0.2349	0.2602	1.2225	494.3	150
160	0.34806	130.5	0.2409	0.2541	1.1970	508.5		0.32356	130.1	0.2391	0.2577	1.2075	503.5	160
170	0.35767	133.1	0.2449	0.2528	1.1859	516.8		0.33282	132.6	0.2432	0.2559	1.1949	512.2	170
180	0.36700	135.6	0.2489	0.2519	1.1764	524.7		0.34189	135.2	0.2472	0.2546	1.1842	520.5	180
190	0.37611	138.1	0.2528	0.2515	1.1681	532.3		0.35066	137.7	0.2512	0.2538	1.1749	528.4	190
200	0.38504	140.6	0.2566	0.2513	1.1608	539.6		0.35923	140.3	0.2550	0.2534	1.1669	535.9	200
210	0.39378	143.1	0.2604	0.2514	1.1544	546.7		0.36762	142.8	0.2588	0.2533	1.1598	543.2	210
220	0.40238	145.7	0.2641	0.2517	1.1486	553.4		0.37591	145.3	0.2626	0.2534	1.1535	550.2	220
230	0.41085	148.2	0.2678	0.2522	1.1435	560.0		0.38400	147.9	0.2663	0.2538	1.1479	556.9	230
240	0.41916	150.7	0.2715	0.2528	1.1388	566.4		0.39200	150.4	0.2700	0.2543	1.1428	563.5	240
250	0.42746	153.2	0.2751	0.2536	1.1346	572.6		0.39987	152.9	0.2736	0.2549	1.1383	569.9	250
260	0.43554	155.8	0.2786	0.2545	1.1308	578.6		0.40758	155.5	0.2771	0.2557	1.1341	576.0	260
270	0.44364	158.3	0.2821	0.2555	1.1273	584.5		0.41533	158.1	0.2807	0.2566	1.1304	582.1	270
280	0.45165	160.9	0.2856	0.2565	1.1241	590.3		0.42287	160.6	0.2842	0.2575	1.1269	588.0	280
290	0.45954	163.5	0.2891	0.2577	1.1211	595.9		0.43042	163.2	0.2876	0.2586	1.1237	593.7	290
300	0.46729	166.0	0.2925	0.2588	1.1183	601.4		0.43787	165.8	0.2911	0.2597	1.1208	599.3	300
310	0.47508	168.6	0.2959	0.2601	1.1158	606.9		0.44528	168.4	0.2945	0.2609	1.1181	604.8	310
320	0.48281	171.2	0.2992	0.2613	1.1134	612.2		0.45265	171.0	0.2979	0.2621	1.1155	610.3	320
330	0.49053	173.9	0.3026	0.2627	1.1112	617.4		0.45996	173.6	0.3012	0.2634	1.1132	615.6	330
340	0.49816	176.5	0.3059	0.2640	1.1091	622.5		0.46718	176.3	0.3045	0.2647	1.1110	620.8	340
350	0.50566	179.1	0.3092	0.2654	1.1071	627.6		0.47432	178.9	0.3078	0.2660	1.1089	625.9	350
360	0.51324	181.8	0.3124	0.2667	1.1053	632.5		0.48144	181.6	0.3111	0.2673	1.1069	630.9	360
370	0.52067	184.5	0.3157	0.2681	1.1035	637.4		0.48854	184.3	0.3143	0.2687	1.1051	635.9	370
380	0.52818	187.2	0.3189	0.2696	1.1019	642.2		0.49564	187.0	0.3176	0.2701	1.1034	640.8	380
390	0.53565	189.9	0.3221	0.2710	1.1003	647.0		0.50269	189.7	0.3208	0.2715	1.1018	645.6	390
400	0.54301	192.6	0.3253	0.2724	1.0988	651.7		0.50968	192.4	0.3240	0.2729	1.1002	650.4	400
410	0.55030	195.3	0.3284	0.2738	1.0974	656.3		0.51674	195.1	0.3271	0.2743	1.0987	655.0	410
420	—	—	—	—	—	—		0.52362	197.9	0.3303	0.2757	1.0973	659.7	420

Reprinted by permission from DuPont

Table 2 (continued)
HFC-134a Superheated Vapor—Constant Pressure Tables

V = Volume in ft³/lb H = Enthalpy in Btu/lb S = Entropy in Btu/(lb)(°R) v_s = Velocity of Sound in ft/sec
Cp = Heat Capacity at Constant Pressure in Btu/(lb)(°F) Cp/Cv = Heat Capacity Ratio (Dimensionless)

TEMP °F	PRESSURE = 180.00 PSIA							PRESSURE = 190.00 PSIA						TEMP °F
	V	H	S	Cp	Cp/Cv	v_s		V	H	S	Cp	Cp/Cv	v_s	
117.7	0.01439	51.5	0.1029	0.3723	1.6360	1320.2	SAT LIQ	0.01452	53.0	0.1053	0.3761	1.6465	1286.0	121.5
117.7	0.25856	118.1	0.2182	0.2854	1.3280	451.5	SAT VAP	0.24371	118.5	0.2180	0.2909	1.3427	448.5	121.5
120	0.26125	118.8	0.2193	0.2831	1.3184	454.6		—	—	—	—	—	—	120
130	0.27221	121.6	0.2241	0.2749	1.2838	467.0		0.25296	120.9	0.2221	0.2823	1.3076	459.8	130
140	0.28251	124.3	0.2287	0.2690	1.2572	478.3		0.26319	123.7	0.2268	0.2749	1.2759	471.9	140
150	0.29231	126.9	0.2331	0.2647	1.2361	488.7		0.27284	126.4	0.2313	0.2695	1.2511	483.0	150
160	0.30170	129.6	0.2374	0.2615	1.2189	498.4		0.28205	129.1	0.2357	0.2656	1.2313	493.2	160
170	0.31073	132.2	0.2415	0.2591	1.2046	507.5		0.29085	131.7	0.2399	0.2626	1.2150	502.7	170
180	0.31948	134.8	0.2456	0.2575	1.1925	516.1		0.29943	134.3	0.2440	0.2605	1.2014	511.7	180
190	0.32803	137.3	0.2496	0.2563	1.1822	524.3		0.30763	136.9	0.2480	0.2590	1.1899	520.2	190
200	0.33629	139.9	0.2535	0.2556	1.1732	532.2		0.31569	139.5	0.2520	0.2580	1.1800	528.3	200
210	0.34438	142.4	0.2573	0.2553	1.1654	539.7		0.32356	142.1	0.2559	0.2573	1.1714	536.1	210
220	0.35227	145.0	0.2611	0.2552	1.1586	546.9		0.33122	144.7	0.2597	0.2570	1.1639	543.5	220
230	0.36006	147.5	0.2648	0.2554	1.1524	553.8		0.33872	147.2	0.2634	0.2570	1.1572	550.7	230
240	0.36776	150.1	0.2685	0.2557	1.1470	560.6		0.34609	149.8	0.2671	0.2572	1.1513	557.6	240
250	0.37536	152.7	0.2722	0.2562	1.1420	567.1		0.35337	152.4	0.2708	0.2576	1.1459	564.3	250
260	0.38278	155.2	0.2757	0.2569	1.1376	573.4		0.36052	154.9	0.2744	0.2582	1.1411	570.8	260
270	0.39014	157.8	0.2793	0.2577	1.1335	579.6		0.36755	157.5	0.2780	0.2588	1.1368	577.1	270
280	0.39739	160.4	0.2828	0.2586	1.1298	585.6		0.37453	160.1	0.2815	0.2596	1.1328	583.2	280
290	0.40461	163.0	0.2863	0.2596	1.1264	591.5		0.38143	162.7	0.2850	0.2605	1.1292	589.2	290
300	0.41168	165.6	0.2897	0.2606	1.1233	597.2		0.38829	165.3	0.2884	0.2615	1.1259	595.1	300
310	0.41873	168.2	0.2931	0.2617	1.1204	602.8		0.39502	168.0	0.2919	0.2626	1.1228	600.8	310
320	0.42577	170.8	0.2965	0.2629	1.1177	608.3		0.40175	170.6	0.2953	0.2637	1.1199	606.4	320
330	0.43269	173.4	0.2999	0.2641	1.1152	613.7		0.40840	173.2	0.2986	0.2648	1.1173	611.9	330
340	0.43964	176.1	0.3032	0.2653	1.1129	619.0		0.41501	175.9	0.3020	0.2660	1.1148	617.3	340
350	0.44645	178.7	0.3065	0.2666	1.1107	624.2		0.42157	178.5	0.3053	0.2673	1.1125	622.6	350
360	0.45325	181.4	0.3098	0.2679	1.1086	629.3		0.42799	181.2	0.3086	0.2685	1.1104	627.8	360
370	0.46007	184.1	0.3131	0.2693	1.1067	634.4		0.43452	183.9	0.3118	0.2698	1.1083	632.9	370
380	0.46677	186.8	0.3163	0.2706	1.1049	639.3		0.44098	186.6	0.3151	0.2712	1.1064	637.9	380
390	0.47344	189.5	0.3195	0.2720	1.1032	644.2		0.44731	189.3	0.3183	0.2725	1.1047	642.8	390
400	0.48015	192.2	0.3227	0.2734	1.1016	649.0		0.45366	192.1	0.3215	0.2739	1.1030	647.7	400
410	0.48681	195.0	0.3259	0.2748	1.1000	653.8		0.46007	194.8	0.3247	0.2752	1.1013	652.5	410
420	0.49336	197.7	0.3290	0.2762	1.0986	658.5		0.46633	197.6	0.3278	0.2766	1.0998	657.3	420
430	—	—	—	—	—	—		0.47257	200.3	0.3310	0.2780	1.0984	661.9	430

TEMP °F	PRESSURE = 200.00 PSIA							PRESSURE = 220.00 PSIA						TEMP °F
	V	H	S	Cp	Cp/Cv	v_s		V	H	S	Cp	Cp/Cv	v_s	
125.2	0.01465	54.3	0.1076	0.3801	1.6574	1253.0	SAT LIQ	0.01491	57.0	0.1120	0.3884	1.6807	1189.8	132.2
125.2	0.23026	118.8	0.2178	0.2966	1.3582	445.6	SAT VAP	0.20685	119.3	0.2174	0.3085	1.3919	439.5	132.2
130	0.23544	120.2	0.2202	0.2908	1.3352	452.3		—	—	—	—	—	—	130
140	0.24565	123.0	0.2250	0.2816	1.2969	465.3		0.21486	121.7	0.2213	0.2975	1.3488	451.1	140
150	0.25521	125.8	0.2296	0.2749	1.2678	477.0		0.22440	124.6	0.2262	0.2874	1.3075	464.5	150
160	0.26428	128.5	0.2340	0.2700	1.2449	487.8		0.23331	127.5	0.2308	0.2801	1.2764	476.6	160
170	0.27294	131.2	0.2383	0.2664	1.2263	497.8		0.24174	130.2	0.2352	0.2748	1.2520	487.6	170
180	0.28129	133.9	0.2425	0.2637	1.2110	507.2		0.24979	133.0	0.2395	0.2708	1.2324	497.9	180
190	0.28930	136.5	0.2466	0.2618	1.1981	516.1		0.25749	135.7	0.2437	0.2678	1.2163	507.5	190
200	0.29712	139.1	0.2505	0.2604	1.1872	524.5		0.26498	138.3	0.2478	0.2657	1.2027	516.5	200
210	0.30477	141.7	0.2545	0.2595	1.1777	532.5		0.27217	141.0	0.2517	0.2641	1.1913	525.1	210
220	0.31217	144.3	0.2583	0.2590	1.1694	540.2		0.27919	143.6	0.2557	0.2631	1.1814	533.3	220
230	0.31946	146.9	0.2621	0.2587	1.1622	547.5		0.28605	146.2	0.2595	0.2624	1.1728	541.1	230
240	0.32658	149.5	0.2658	0.2588	1.1557	554.6		0.29278	148.9	0.2633	0.2621	1.1652	548.6	240
250	0.33360	152.1	0.2695	0.2590	1.1500	561.5		0.29932	151.5	0.2670	0.2620	1.1585	555.8	250
260	0.34046	154.7	0.2731	0.2594	1.1448	568.1		0.30584	154.1	0.2707	0.2621	1.1525	562.8	260
270	0.34727	157.3	0.2767	0.2600	1.1401	574.6		0.31217	156.7	0.2743	0.2625	1.1472	569.5	270
280	0.35398	159.9	0.2802	0.2607	1.1359	580.9		0.31842	159.3	0.2778	0.2630	1.1423	576.1	280
290	0.36063	162.5	0.2837	0.2615	1.1320	587.0		0.32462	162.0	0.2814	0.2636	1.1379	582.5	290
300	0.36716	165.1	0.2872	0.2624	1.1285	593.0		0.33069	164.6	0.2849	0.2643	1.1339	588.7	300
310	0.37365	167.7	0.2906	0.2634	1.1252	598.8		0.33677	167.3	0.2883	0.2652	1.1303	594.7	310
320	0.38008	170.4	0.2941	0.2645	1.1222	604.5		0.34274	169.9	0.2918	0.2661	1.1269	600.6	320
330	0.38646	173.0	0.2974	0.2656	1.1194	610.1		0.34863	172.6	0.2952	0.2671	1.1238	606.4	330
340	0.39282	175.7	0.3008	0.2667	1.1168	615.5		0.35447	175.3	0.2985	0.2682	1.1209	612.0	340
350	0.39909	178.3	0.3041	0.2679	1.1144	620.9		0.36032	177.9	0.3019	0.2693	1.1182	617.6	350
360	0.40532	181.0	0.3074	0.2692	1.1121	626.2		0.36605	180.6	0.3052	0.2704	1.1157	623.0	360
370	0.41151	183.7	0.3107	0.2704	1.1100	631.3		0.37182	183.4	0.3085	0.2716	1.1134	628.3	370
380	0.41762	186.4	0.3139	0.2717	1.1080	636.4		0.37743	186.1	0.3117	0.2728	1.1112	633.5	380
390	0.42375	189.2	0.3171	0.2730	1.1061	641.5		0.38314	188.8	0.3150	0.2741	1.1091	638.7	390
400	0.42989	191.9	0.3203	0.2743	1.1044	646.4		0.38873	191.6	0.3182	0.2753	1.1072	643.7	400
410	0.43590	194.7	0.3235	0.2757	1.1027	651.3		0.39434	194.3	0.3214	0.2766	1.1054	648.7	410
420	0.44191	197.4	0.3267	0.2770	1.1011	656.1		0.39987	197.1	0.3246	0.2779	1.1037	653.6	420
430	0.44791	200.2	0.3298	0.2784	1.0996	660.8		0.40538	199.9	0.3277	0.2792	1.1020	658.5	430
440	—	—	—	—	—	—		0.41088	202.7	0.3308	0.2806	1.1005	663.2	440

Table 2 (continued)
HFC-134a Superheated Vapor—Constant Pressure Tables

V = Volume in ft³/lb   H = Enthalpy in Btu/lb   S = Entropy in Btu/(lb) (°R)   v_s = Velocity of Sound in ft/sec
Cp = Heat Capacity at Constant Pressure in Btu/(lb) (°F)   Cp/Cv = Heat Capacity Ratio (Dimensionless)

TEMP °F	PRESSURE = 240.00 PSIA							PRESSURE = 260.00 PSIA						TEMP °F
	V	H	S	Cp	Cp/Cv	v_s		V	H	S	Cp	Cp/Cv	v_s	
138.7	0.01517	59.5	0.1162	0.3972	1.7064	1130.0	SAT LIQ	0.01543	61.9	0.1201	0.4067	1.7349	1073.0	144.9
138.7	0.18715	119.8	0.2169	0.3214	1.4297	433.3	SAT VAP	0.17030	120.2	0.2165	0.3355	1.4725	426.9	144.9
140	0.18846	120.2	0.2176	0.3189	1.4201	435.4		—	—	—	—	—	—	140
150	0.19819	123.3	0.2227	0.3032	1.3589	450.9		0.17540	121.9	0.2193	0.3240	1.4284	435.9	150
160	0.20712	126.3	0.2276	0.2924	1.3154	464.6		0.18452	125.0	0.2244	0.3077	1.3653	451.6	160
170	0.21545	129.2	0.2322	0.2846	1.2829	476.9		0.19288	128.1	0.2292	0.2965	1.3207	465.5	170
180	0.22330	132.0	0.2366	0.2789	1.2575	488.1		0.20068	131.0	0.2338	0.2884	1.2874	477.9	180
190	0.23083	134.8	0.2409	0.2747	1.2371	498.6		0.20805	133.8	0.2383	0.2826	1.2614	489.3	190
200	0.23799	137.5	0.2451	0.2716	1.2204	508.3		0.21507	136.6	0.2425	0.2782	1.2405	499.9	200
210	0.24495	140.2	0.2492	0.2692	1.2064	517.5		0.22178	139.4	0.2467	0.2749	1.2234	509.8	210
220	0.25165	142.9	0.2531	0.2675	1.1945	526.2		0.22824	142.1	0.2507	0.2724	1.2092	519.0	220
230	0.25818	145.6	0.2570	0.2664	1.1843	534.5		0.23450	144.9	0.2547	0.2707	1.1970	527.8	230
240	0.26457	148.2	0.2609	0.2656	1.1754	542.5		0.24061	147.6	0.2586	0.2694	1.1866	536.2	240
250	0.27079	150.9	0.2646	0.2652	1.1677	550.1		0.24657	150.2	0.2624	0.2685	1.1775	544.2	250
260	0.27685	153.5	0.2683	0.2650	1.1608	557.4		0.25233	152.9	0.2662	0.2680	1.1696	551.9	260
270	0.28289	156.2	0.2720	0.2651	1.1546	564.4		0.25804	155.6	0.2699	0.2678	1.1626	559.3	270
280	0.28878	158.8	0.2756	0.2653	1.1491	571.3		0.26362	158.3	0.2735	0.2678	1.1563	566.4	280
290	0.29459	161.5	0.2792	0.2658	1.1441	577.9		0.26915	161.0	0.2771	0.2680	1.1507	573.3	290
300	0.30028	164.1	0.2827	0.2663	1.1396	584.3		0.27456	163.6	0.2807	0.2684	1.1456	580.0	300
310	0.30593	166.8	0.2862	0.2670	1.1355	590.6		0.27988	166.3	0.2842	0.2689	1.1410	586.5	310
320	0.31156	169.5	0.2896	0.2678	1.1318	596.7		0.28517	169.0	0.2877	0.2696	1.1368	592.8	320
330	0.31709	172.2	0.2931	0.2687	1.1283	602.7		0.29033	171.7	0.2911	0.2703	1.1330	598.9	330
340	0.32253	174.8	0.2965	0.2696	1.1251	608.5		0.29553	174.4	0.2945	0.2711	1.1295	605.0	340
350	0.32794	177.5	0.2998	0.2706	1.1222	614.2		0.30059	177.1	0.2979	0.2721	1.1262	610.8	350
360	0.33332	180.3	0.3031	0.2717	1.1194	619.8		0.30565	179.9	0.3012	0.2730	1.1232	616.6	360
370	0.33866	183.0	0.3064	0.2728	1.1169	625.3		0.31066	182.6	0.3045	0.2741	1.1204	622.2	370
380	0.34395	185.7	0.3097	0.2740	1.1145	630.6		0.31560	185.3	0.3078	0.2751	1.1178	627.7	380
390	0.34921	188.5	0.3130	0.2751	1.1122	635.9		0.32052	188.1	0.3111	0.2762	1.1154	633.1	390
400	0.35443	191.2	0.3162	0.2764	1.1101	641.1		0.32540	190.9	0.3143	0.2774	1.1131	638.5	400
410	0.35962	194.0	0.3194	0.2776	1.1081	646.2		0.33026	193.7	0.3175	0.2786	1.1110	643.7	410
420	0.36478	196.8	0.3226	0.2788	1.1063	651.2		0.33509	196.4	0.3207	0.2798	1.1090	648.8	420
430	0.36990	199.6	0.3257	0.2801	1.1045	656.2		0.33988	199.2	0.3239	0.2810	1.1071	653.9	430
440	0.37501	202.4	0.3289	0.2814	1.1028	661.0		0.34465	202.1	0.3271	0.2822	1.1053	658.9	440
450	—	—	—	—	—	—		0.34939	204.9	0.3302	0.2835	1.1036	663.8	450

TEMP °F	PRESSURE = 280.00 PSIA							PRESSURE = 300.00 PSIA						TEMP °F
	V	H	S	Cp	Cp/Cv	v_s		V	H	S	Cp	Cp/Cv	v_s	
150.6	0.01570	64.3	0.1238	0.4172	1.7668	1018.2	SAT LIQ	0.01598	66.5	0.1274	0.4287	1.8027	965.6	156.1
150.6	0.15571	120.5	0.2160	0.3512	1.5212	420.5	SAT VAP	0.14292	120.7	0.2154	0.3688	1.5771	413.9	156.1
160	0.16463	123.7	0.2211	0.3275	1.4317	437.5		0.14674	122.1	0.2177	0.3549	1.5251	421.9	160
170	0.17318	126.8	0.2262	0.3111	1.3684	453.2		0.15567	125.5	0.2232	0.3298	1.4305	440.1	170
180	0.18101	129.9	0.2310	0.2998	1.3235	467.1		0.16367	128.7	0.2282	0.3135	1.3683	455.7	180
190	0.18832	132.9	0.2356	0.2916	1.2899	479.6		0.17099	131.8	0.2330	0.3024	1.3241	469.5	190
200	0.19519	135.7	0.2400	0.2857	1.2637	491.1		0.17783	134.8	0.2375	0.2943	1.2907	482.0	200
210	0.20178	138.6	0.2443	0.2812	1.2427	501.7		0.18434	137.7	0.2419	0.2883	1.2647	493.4	210
220	0.20809	141.4	0.2484	0.2778	1.2254	511.7		0.19050	140.6	0.2462	0.2838	1.2437	504.1	220
230	0.21413	144.1	0.2525	0.2753	1.2110	521.0		0.19641	143.4	0.2503	0.2805	1.2265	514.0	230
240	0.22001	146.9	0.2564	0.2735	1.1988	529.8		0.20210	146.2	0.2543	0.2779	1.2121	523.4	240
250	0.22573	149.6	0.2603	0.2722	1.1882	538.3		0.20763	148.9	0.2582	0.2761	1.1999	532.2	250
260	0.23127	152.3	0.2641	0.2713	1.1791	546.3		0.21302	151.7	0.2621	0.2747	1.1893	540.7	260
270	0.23672	155.0	0.2678	0.2707	1.1711	554.0		0.21823	154.4	0.2659	0.2738	1.1801	548.8	270
280	0.24208	157.7	0.2715	0.2704	1.1639	561.5		0.22332	157.2	0.2696	0.2732	1.1721	556.5	280
290	0.24728	160.4	0.2751	0.2704	1.1576	568.6		0.22834	159.9	0.2733	0.2729	1.1649	564.0	290
300	0.25246	163.1	0.2787	0.2706	1.1519	575.6		0.23327	162.6	0.2769	0.2729	1.1586	571.2	300
310	0.25753	165.9	0.2823	0.2709	1.1468	582.3		0.23810	165.4	0.2804	0.2730	1.1529	578.2	310
320	0.26250	168.6	0.2858	0.2714	1.1422	588.9		0.24289	168.1	0.2840	0.2733	1.1477	584.9	320
330	0.26744	171.3	0.2892	0.2720	1.1379	595.2		0.24756	170.8	0.2875	0.2738	1.1430	591.5	330
340	0.27230	174.0	0.2927	0.2727	1.1340	601.4		0.25219	173.6	0.2909	0.2743	1.1387	597.9	340
350	0.27712	176.7	0.2960	0.2735	1.1305	607.5		0.25677	176.3	0.2943	0.2750	1.1348	604.1	350
360	0.28188	179.5	0.2994	0.2744	1.1272	613.4		0.26129	179.1	0.2977	0.2758	1.1312	610.2	360
370	0.28661	182.2	0.3027	0.2753	1.1241	619.2		0.26577	181.8	0.3010	0.2766	1.1279	616.1	370
380	0.29128	185.0	0.3060	0.2763	1.1213	624.8		0.27020	184.6	0.3044	0.2776	1.1248	621.9	380
390	0.29592	187.7	0.3093	0.2774	1.1186	630.4		0.27460	187.4	0.3077	0.2785	1.1220	627.6	390
400	0.30053	190.5	0.3126	0.2785	1.1162	635.8		0.27896	190.2	0.3109	0.2795	1.1193	633.2	400
410	0.30509	193.3	0.3158	0.2796	1.1138	641.2		0.28328	193.0	0.3142	0.2806	1.1168	638.6	410
420	0.30964	196.1	0.3190	0.2807	1.1117	646.4		0.28757	195.8	0.3174	0.2817	1.1145	644.0	420
430	0.31414	198.9	0.3222	0.2819	1.1096	651.6		0.29183	198.6	0.3206	0.2828	1.1123	649.3	430
440	0.31862	201.8	0.3253	0.2831	1.1077	656.7		0.29607	201.4	0.3237	0.2840	1.1102	654.5	440
450	0.32308	204.6	0.3285	0.2843	1.1059	661.7		0.30027	204.3	0.3269	0.2851	1.1083	659.6	450
460	0.32751	207.4	0.3316	0.2855	1.1042	666.6		0.30446	207.1	0.3300	0.2863	1.1064	664.6	460

Reprinted by permission from DuPont

Appendix: Air Tables

	Air Tables developed by K. W. Lindler - U. S. Naval Academy								
T	t	h	Pr	u	vr	φ	Cp	Cv	k
°R	°F	Btu/lbm		Btu/lbm		Btu/lbm°R	Btu/lbm°R	Btu/lbm°R	
400	-60	95.5	0.485	68.1	305.239	0.5289	0.2398	0.1713	1.400
405	-55	96.7	0.507	69.0	295.912	0.5318	0.2398	0.1712	1.400
410	-50	97.9	0.529	69.8	286.981	0.5348	0.2398	0.1712	1.400
415	-45	99.1	0.552	70.7	278.424	0.5377	0.2397	0.1712	1.400
420	-40	100.3	0.576	71.5	270.221	0.5406	0.2397	0.1712	1.400
425	-35	101.5	0.600	72.4	262.353	0.5434	0.2397	0.1711	1.401
430	-30	102.7	0.625	73.3	254.803	0.5462	0.2397	0.1711	1.401
435	-25	103.9	0.651	74.1	247.555	0.5490	0.2397	0.1711	1.401
440	-20	105.1	0.677	75.0	240.592	0.5517	0.2397	0.1711	1.401
445	-15	106.3	0.705	75.8	233.902	0.5544	0.2396	0.1711	1.401
450	-10	107.5	0.733	76.7	227.469	0.5571	0.2396	0.1711	1.401
455	-5	108.7	0.762	77.5	221.281	0.5597	0.2396	0.1711	1.401
460	0	109.9	0.791	78.4	215.327	0.5624	0.2396	0.1711	1.401
465	5	111.1	0.822	79.2	209.595	0.5649	0.2396	0.1711	1.401
470	10	112.3	0.853	80.1	204.075	0.5675	0.2396	0.1711	1.401
475	15	113.5	0.885	81.0	198.755	0.5700	0.2396	0.1711	1.401
480	20	114.7	0.918	81.8	193.628	0.5726	0.2396	0.1711	1.401
485	25	115.9	0.952	82.7	188.684	0.5750	0.2396	0.1711	1.401
490	30	117.1	0.987	83.5	183.915	0.5775	0.2397	0.1711	1.401
495	35	118.3	1.023	84.4	179.313	0.5799	0.2397	0.1711	1.401
500	40	119.5	1.059	85.2	174.870	0.5823	0.2397	0.1711	1.401
505	45	120.7	1.097	86.1	170.580	0.5847	0.2397	0.1711	1.401
510	50	121.9	1.135	86.9	166.435	0.5871	0.2397	0.1712	1.400
515	55	123.1	1.174	87.8	162.429	0.5894	0.2397	0.1712	1.400
520	60	124.3	1.215	88.7	158.557	0.5917	0.2398	0.1712	1.400
525	65	125.5	1.256	89.5	154.812	0.5940	0.2398	0.1712	1.400
530	70	126.7	1.299	90.4	151.189	0.5963	0.2398	0.1712	1.400
535	75	127.9	1.342	91.2	147.684	0.5986	0.2398	0.1713	1.400
540	80	129.1	1.386	92.1	144.291	0.6008	0.2399	0.1713	1.400
545	85	130.3	1.432	92.9	141.005	0.6030	0.2399	0.1713	1.400
550	90	131.5	1.478	93.8	137.823	0.6052	0.2399	0.1714	1.400
555	95	132.7	1.526	94.6	134.739	0.6074	0.2399	0.1714	1.400
560	100	133.9	1.574	95.5	131.751	0.6095	0.2400	0.1714	1.400
565	105	135.1	1.624	96.4	128.854	0.6116	0.2400	0.1715	1.400
570	110	136.3	1.675	97.2	126.045	0.6138	0.2401	0.1715	1.400
575	115	137.5	1.727	98.1	123.320	0.6159	0.2401	0.1716	1.400
580	120	138.7	1.780	98.9	120.677	0.6179	0.2401	0.1716	1.399
585	125	139.9	1.835	99.8	118.111	0.6200	0.2402	0.1716	1.399
590	130	141.1	1.890	100.7	115.621	0.6220	0.2402	0.1717	1.399
595	135	142.3	1.947	101.5	113.202	0.6241	0.2403	0.1717	1.399
600	140	143.5	2.005	102.4	110.853	0.6261	0.2403	0.1718	1.399
605	145	144.7	2.064	103.2	108.571	0.6281	0.2404	0.1718	1.399
610	150	145.9	2.125	104.1	106.354	0.6301	0.2404	0.1719	1.399
615	155	147.1	2.186	104.9	104.199	0.6320	0.2405	0.1719	1.399
620	160	148.3	2.249	105.8	102.104	0.6340	0.2405	0.1720	1.399
625	165	149.5	2.314	106.7	100.066	0.6359	0.2406	0.1720	1.398
630	170	150.7	2.379	107.5	98.085	0.6378	0.2407	0.1721	1.398
635	175	151.9	2.446	108.4	96.157	0.6397	0.2407	0.1722	1.398
640	180	153.1	2.515	109.2	94.281	0.6416	0.2408	0.1722	1.398
645	185	154.3	2.584	110.1	92.455	0.6435	0.2408	0.1723	1.398

Air Tables developed by K. W. Lindler - U. S. Naval Academy									
T	t	h	Pr	u	vr	φ	Cp	Cv	k
°R	°F	Btu/lbm		Btu/lbm		Btu/lbm°R	Btu/lbm°R	Btu/lbm°R	
650	190	155.5	2.655	111.0	90.677	0.6453	0.2409	0.1724	1.398
655	195	156.7	2.728	111.8	88.947	0.6472	0.2410	0.1724	1.398
660	200	157.9	2.802	112.7	87.261	0.6490	0.2410	0.1725	1.397
665	205	159.1	2.877	113.6	85.619	0.6508	0.2411	0.1726	1.397
670	210	160.3	2.954	114.4	84.019	0.6526	0.2412	0.1726	1.397
675	215	161.6	3.032	115.3	82.460	0.6544	0.2413	0.1727	1.397
680	220	162.8	3.112	116.1	80.941	0.6562	0.2413	0.1728	1.397
685	225	164.0	3.193	117.0	79.460	0.6580	0.2414	0.1729	1.397
690	230	165.2	3.276	117.9	78.015	0.6597	0.2415	0.1729	1.396
695	235	166.4	3.361	118.7	76.607	0.6615	0.2416	0.1730	1.396
700	240	167.6	3.447	119.6	75.233	0.6632	0.2416	0.1731	1.396
705	245	168.8	3.534	120.5	73.893	0.6649	0.2417	0.1732	1.396
710	250	170.0	3.623	121.3	72.585	0.6667	0.2418	0.1733	1.396
715	255	171.2	3.714	122.2	71.308	0.6683	0.2419	0.1733	1.395
720	260	172.4	3.807	123.1	70.063	0.6700	0.2420	0.1734	1.395
725	265	173.6	3.901	123.9	68.846	0.6717	0.2421	0.1735	1.395
730	270	174.8	3.997	124.8	67.659	0.6734	0.2421	0.1736	1.395
735	275	176.1	4.094	125.7	66.499	0.6750	0.2422	0.1737	1.395
740	280	177.3	4.194	126.5	65.366	0.6767	0.2423	0.1738	1.394
745	285	178.5	4.295	127.4	64.259	0.6783	0.2424	0.1739	1.394
750	290	179.7	4.397	128.3	63.178	0.6799	0.2425	0.1740	1.394
755	295	180.9	4.502	129.2	62.121	0.6815	0.2426	0.1741	1.394
760	300	182.1	4.609	130.0	61.088	0.6831	0.2427	0.1742	1.394
765	305	183.3	4.717	130.9	60.079	0.6847	0.2428	0.1743	1.393
770	310	184.5	4.827	131.8	59.092	0.6863	0.2429	0.1743	1.393
775	315	185.8	4.939	132.6	58.127	0.6879	0.2430	0.1744	1.393
780	320	187.0	5.053	133.5	57.183	0.6894	0.2431	0.1745	1.393
785	325	188.2	5.169	134.4	56.260	0.6910	0.2432	0.1747	1.392
790	330	189.4	5.286	135.3	55.357	0.6925	0.2433	0.1748	1.392
795	335	190.6	5.406	136.1	54.473	0.6941	0.2434	0.1749	1.392
800	340	191.8	5.528	137.0	53.609	0.6956	0.2435	0.1750	1.392
805	345	193.1	5.652	137.9	52.763	0.6971	0.2436	0.1751	1.392
810	350	194.3	5.777	138.8	51.935	0.6986	0.2437	0.1752	1.391
815	355	195.5	5.905	139.6	51.124	0.7001	0.2438	0.1753	1.391
820	360	196.7	6.035	140.5	50.331	0.7016	0.2439	0.1754	1.391
825	365	197.9	6.167	141.4	49.553	0.7031	0.2441	0.1755	1.391
830	370	199.2	6.301	142.3	48.793	0.7046	0.2442	0.1756	1.390
835	375	200.4	6.438	143.1	48.047	0.7060	0.2443	0.1757	1.390
840	380	201.6	6.576	144.0	47.317	0.7075	0.2444	0.1758	1.390
845	385	202.8	6.717	144.9	46.602	0.7090	0.2445	0.1760	1.390
850	390	204.0	6.860	145.8	45.902	0.7104	0.2446	0.1761	1.389
855	395	205.3	7.005	146.7	45.215	0.7118	0.2447	0.1762	1.389
860	400	206.5	7.152	147.5	44.542	0.7133	0.2449	0.1763	1.389
865	405	207.7	7.302	148.4	43.883	0.7147	0.2450	0.1764	1.389
870	410	208.9	7.454	149.3	43.236	0.7161	0.2451	0.1765	1.388
875	415	210.2	7.608	150.2	42.603	0.7175	0.2452	0.1767	1.388
880	420	211.4	7.765	151.1	41.981	0.7189	0.2453	0.1768	1.388
885	425	212.6	7.924	152.0	41.372	0.7203	0.2455	0.1769	1.387
890	430	213.9	8.085	152.8	40.775	0.7217	0.2456	0.1770	1.387
895	435	215.1	8.249	153.7	40.189	0.7230	0.2457	0.1772	1.387

Air Tables developed by K. W. Lindler - U. S. Naval Academy									
T	t	h	Pr	u	vr	φ	Cp	Cv	k
°R	°F	Btu/lbm		Btu/lbm		Btu/lbm°R	Btu/lbm°R	Btu/lbm°R	
900	440	216.3	8.416	154.6	39.614	0.7244	0.2458	0.1773	1.387
905	445	217.5	8.585	155.5	39.051	0.7258	0.2460	0.1774	1.386
910	450	218.8	8.756	156.4	38.497	0.7271	0.2461	0.1775	1.386
915	455	220.0	8.930	157.3	37.955	0.7285	0.2462	0.1777	1.386
920	460	221.2	9.107	158.2	37.422	0.7298	0.2463	0.1778	1.386
925	465	222.5	9.286	159.1	36.900	0.7312	0.2465	0.1779	1.385
930	470	223.7	9.468	159.9	36.387	0.7325	0.2466	0.1781	1.385
935	475	224.9	9.652	160.8	35.883	0.7338	0.2467	0.1782	1.385
940	480	226.2	9.839	161.7	35.389	0.7351	0.2469	0.1783	1.384
945	485	227.4	10.029	162.6	34.904	0.7364	0.2470	0.1785	1.384
950	490	228.6	10.222	163.5	34.427	0.7377	0.2471	0.1786	1.384
955	495	229.9	10.417	164.4	33.960	0.7390	0.2473	0.1787	1.384
960	500	231.1	10.615	165.3	33.500	0.7403	0.2474	0.1789	1.383
965	505	232.3	10.816	166.2	33.049	0.7416	0.2475	0.1790	1.383
970	510	233.6	11.020	167.1	32.606	0.7429	0.2477	0.1791	1.383
975	515	234.8	11.227	168.0	32.171	0.7442	0.2478	0.1793	1.382
980	520	236.1	11.436	168.9	31.743	0.7454	0.2480	0.1794	1.382
985	525	237.3	11.649	169.8	31.323	0.7467	0.2481	0.1795	1.382
990	530	238.5	11.864	170.7	30.910	0.7480	0.2482	0.1797	1.382
995	535	239.8	12.083	171.6	30.504	0.7492	0.2484	0.1798	1.381
1000	540	241.0	12.304	172.5	30.106	0.7505	0.2485	0.1800	1.381
1005	545	242.3	12.529	173.4	29.714	0.7517	0.2487	0.1801	1.381
1010	550	243.5	12.756	174.3	29.329	0.7529	0.2488	0.1802	1.380
1015	555	244.8	12.987	175.2	28.950	0.7542	0.2489	0.1804	1.380
1020	560	246.0	13.221	176.1	28.578	0.7554	0.2491	0.1805	1.380
1025	565	247.2	13.458	177.0	28.212	0.7566	0.2492	0.1807	1.379
1030	570	248.5	13.699	177.9	27.853	0.7578	0.2494	0.1808	1.379
1035	575	249.7	13.942	178.8	27.499	0.7590	0.2495	0.1810	1.379
1040	580	251.0	14.189	179.7	27.151	0.7602	0.2497	0.1811	1.378
1045	585	252.2	14.439	180.6	26.809	0.7614	0.2498	0.1813	1.378
1050	590	253.5	14.692	181.5	26.473	0.7626	0.2500	0.1814	1.378
1055	595	254.7	14.949	182.4	26.142	0.7638	0.2501	0.1816	1.378
1060	600	256.0	15.209	183.3	25.817	0.7650	0.2503	0.1817	1.377
1065	605	257.2	15.473	184.2	25.496	0.7662	0.2504	0.1819	1.377
1070	610	258.5	15.740	185.1	25.181	0.7673	0.2505	0.1820	1.377
1075	615	259.7	16.011	186.1	24.872	0.7685	0.2507	0.1821	1.376
1080	620	261.0	16.285	187.0	24.567	0.7697	0.2508	0.1823	1.376
1085	625	262.3	16.563	187.9	24.267	0.7708	0.2510	0.1824	1.376
1090	630	263.5	16.844	188.8	23.971	0.7720	0.2511	0.1826	1.375
1095	635	264.8	17.129	189.7	23.681	0.7731	0.2513	0.1828	1.375
1100	640	266.0	17.417	190.6	23.395	0.7743	0.2515	0.1829	1.375
1105	645	267.3	17.710	191.5	23.113	0.7754	0.2516	0.1831	1.374
1110	650	268.5	18.006	192.4	22.836	0.7766	0.2518	0.1832	1.374
1115	655	269.8	18.305	193.4	22.563	0.7777	0.2519	0.1834	1.374
1120	660	271.1	18.609	194.3	22.295	0.7788	0.2521	0.1835	1.374
1125	665	272.3	18.916	195.2	22.030	0.7799	0.2522	0.1837	1.373
1130	670	273.6	19.228	196.1	21.770	0.7811	0.2524	0.1838	1.373
1135	675	274.8	19.543	197.0	21.514	0.7822	0.2525	0.1840	1.373
1140	680	276.1	19.862	198.0	21.261	0.7833	0.2527	0.1841	1.372
1145	685	277.4	20.185	198.9	21.013	0.7844	0.2528	0.1843	1.372

| \multicolumn{10}{c}{Air Tables developed by K. W. Lindler - U. S. Naval Academy} |
T °R	t °F	h Btu/lbm	Pr	u Btu/lbm	vr	φ Btu/lbm°R	Cp Btu/lbm°R	Cv Btu/lbm°R	k
1150	690	278.6	20.512	199.8	20.768	0.7855	0.2530	0.1844	1.372
1155	695	279.9	20.843	200.7	20.527	0.7866	0.2531	0.1846	1.371
1160	700	281.2	21.179	201.6	20.289	0.7877	0.2533	0.1848	1.371
1165	705	282.4	21.518	202.6	20.055	0.7888	0.2535	0.1849	1.371
1170	710	283.7	21.862	203.5	19.825	0.7899	0.2536	0.1851	1.370
1175	715	285.0	22.209	204.4	19.598	0.7909	0.2538	0.1852	1.370
1180	720	286.2	22.561	205.3	19.374	0.7920	0.2539	0.1854	1.370
1185	725	287.5	22.918	206.3	19.154	0.7931	0.2541	0.1855	1.369
1190	730	288.8	23.278	207.2	18.937	0.7942	0.2542	0.1857	1.369
1195	735	290.0	23.643	208.1	18.723	0.7952	0.2544	0.1859	1.369
1200	740	291.3	24.012	209.1	18.512	0.7963	0.2546	0.1860	1.369
1205	745	292.6	24.386	210.0	18.304	0.7973	0.2547	0.1862	1.368
1210	750	293.9	24.765	210.9	18.099	0.7984	0.2549	0.1863	1.368
1215	755	295.1	25.147	211.9	17.897	0.7995	0.2550	0.1865	1.368
1220	760	296.4	25.535	212.8	17.699	0.8005	0.2552	0.1867	1.367
1225	765	297.7	25.926	213.7	17.502	0.8015	0.2554	0.1868	1.367
1230	770	299.0	26.323	214.7	17.309	0.8026	0.2555	0.1870	1.367
1235	775	300.2	26.724	215.6	17.119	0.8036	0.2557	0.1871	1.366
1240	780	301.5	27.130	216.5	16.931	0.8047	0.2558	0.1873	1.366
1245	785	302.8	27.541	217.5	16.746	0.8057	0.2560	0.1875	1.366
1250	790	304.1	27.956	218.4	16.563	0.8067	0.2562	0.1876	1.365
1255	795	305.4	28.377	219.3	16.383	0.8077	0.2563	0.1878	1.365
1260	800	306.7	28.802	220.3	16.205	0.8088	0.2565	0.1879	1.365
1265	805	307.9	29.232	221.2	16.030	0.8098	0.2566	0.1881	1.364
1270	810	309.2	29.667	222.2	15.858	0.8108	0.2568	0.1883	1.364
1275	815	310.5	30.107	223.1	15.687	0.8118	0.2570	0.1884	1.364
1280	820	311.8	30.552	224.0	15.519	0.8128	0.2571	0.1886	1.364
1285	825	313.1	31.002	225.0	15.354	0.8138	0.2573	0.1887	1.363
1290	830	314.4	31.458	225.9	15.190	0.8148	0.2574	0.1889	1.363
1295	835	315.6	31.918	226.9	15.029	0.8158	0.2576	0.1891	1.363
1300	840	316.9	32.384	227.8	14.870	0.8168	0.2578	0.1892	1.362
1305	845	318.2	32.855	228.8	14.714	0.8178	0.2579	0.1894	1.362
1310	850	319.5	33.331	229.7	14.559	0.8188	0.2581	0.1895	1.362
1315	855	320.8	33.813	230.7	14.406	0.8197	0.2582	0.1897	1.361
1320	860	322.1	34.300	231.6	14.256	0.8207	0.2584	0.1899	1.361
1325	865	323.4	34.792	232.6	14.107	0.8217	0.2586	0.1900	1.361
1330	870	324.7	35.290	233.5	13.961	0.8227	0.2587	0.1902	1.360
1335	875	326.0	35.794	234.5	13.816	0.8237	0.2589	0.1903	1.360
1340	880	327.3	36.303	235.4	13.673	0.8246	0.2591	0.1905	1.360
1345	885	328.6	36.818	236.4	13.532	0.8256	0.2592	0.1907	1.360
1350	890	329.9	37.338	237.3	13.393	0.8265	0.2594	0.1908	1.359
1355	895	331.2	37.864	238.3	13.256	0.8275	0.2595	0.1910	1.359
1360	900	332.5	38.396	239.2	13.121	0.8285	0.2597	0.1911	1.359
1365	905	333.8	38.934	240.2	12.987	0.8294	0.2599	0.1913	1.358
1370	910	335.1	39.477	241.1	12.855	0.8304	0.2600	0.1915	1.358
1375	915	336.4	40.027	242.1	12.725	0.8313	0.2602	0.1916	1.358
1380	920	337.7	40.582	243.1	12.596	0.8323	0.2603	0.1918	1.357
1385	925	339.0	41.144	244.0	12.470	0.8332	0.2605	0.1919	1.357
1390	930	340.3	41.711	245.0	12.344	0.8341	0.2607	0.1921	1.357
1395	935	341.6	42.285	245.9	12.221	0.8351	0.2608	0.1923	1.357

Air Tables developed by K. W. Lindler - U. S. Naval Academy									
T	t	h	Pr	u	vr	φ	Cp	Cv	k
°R	°F	Btu/lbm		Btu/lbm		Btu/lbm°R	Btu/lbm°R	Btu/lbm°R	
1400	940	342.9	42.865	246.9	12.099	0.8360	0.2610	0.1924	1.356
1405	945	344.2	43.450	247.9	11.978	0.8369	0.2611	0.1926	1.356
1410	950	345.5	44.043	248.8	11.859	0.8379	0.2613	0.1927	1.356
1415	955	346.8	44.641	249.8	11.742	0.8388	0.2614	0.1929	1.355
1420	960	348.1	45.246	250.8	11.626	0.8397	0.2616	0.1931	1.355
1425	965	349.4	45.857	251.7	11.511	0.8406	0.2618	0.1932	1.355
1430	970	350.7	46.475	252.7	11.398	0.8416	0.2619	0.1934	1.354
1435	975	352.0	47.099	253.7	11.286	0.8425	0.2621	0.1935	1.354
1440	980	353.3	47.730	254.6	11.176	0.8434	0.2622	0.1937	1.354
1445	985	354.6	48.367	255.6	11.067	0.8443	0.2624	0.1938	1.354
1450	990	356.0	49.011	256.6	10.959	0.8452	0.2626	0.1940	1.353
1455	995	357.3	49.662	257.5	10.853	0.8461	0.2627	0.1942	1.353
1460	1000	358.6	50.319	258.5	10.748	0.8470	0.2629	0.1943	1.353
1465	1005	359.9	50.983	259.5	10.644	0.8479	0.2630	0.1945	1.352
1470	1010	361.2	51.654	260.5	10.542	0.8488	0.2632	0.1946	1.352
1475	1015	362.5	52.332	261.4	10.441	0.8497	0.2633	0.1948	1.352
1480	1020	363.9	53.017	262.4	10.341	0.8506	0.2635	0.1949	1.352
1485	1025	365.2	53.709	263.4	10.242	0.8515	0.2636	0.1951	1.351
1490	1030	366.5	54.408	264.4	10.144	0.8524	0.2638	0.1952	1.351
1495	1035	367.8	55.115	265.3	10.048	0.8532	0.2640	0.1954	1.351
1500	1040	369.1	55.828	266.3	9.953	0.8541	0.2641	0.1956	1.351
1505	1045	370.4	56.549	267.3	9.859	0.8550	0.2643	0.1957	1.350
1510	1050	371.8	57.276	268.3	9.766	0.8559	0.2644	0.1959	1.350
1515	1055	373.1	58.012	269.2	9.674	0.8568	0.2646	0.1960	1.350
1520	1060	374.4	58.754	270.2	9.583	0.8576	0.2647	0.1962	1.349
1525	1065	375.7	59.504	271.2	9.494	0.8585	0.2649	0.1963	1.349
1530	1070	377.1	60.262	272.2	9.405	0.8594	0.2650	0.1965	1.349
1535	1075	378.4	61.027	273.2	9.317	0.8602	0.2652	0.1966	1.349
1540	1080	379.7	61.800	274.2	9.231	0.8611	0.2653	0.1968	1.348
1545	1085	381.0	62.580	275.1	9.145	0.8619	0.2655	0.1969	1.348
1550	1090	382.4	63.369	276.1	9.061	0.8628	0.2656	0.1971	1.348
1555	1095	383.7	64.165	277.1	8.977	0.8637	0.2658	0.1972	1.348
1560	1100	385.0	64.969	278.1	8.895	0.8645	0.2659	0.1974	1.347
1565	1105	386.4	65.780	279.1	8.813	0.8654	0.2661	0.1975	1.347
1570	1110	387.7	66.600	280.1	8.732	0.8662	0.2662	0.1977	1.347
1575	1115	389.0	67.428	281.1	8.653	0.8671	0.2664	0.1978	1.347
1580	1120	390.4	68.264	282.0	8.574	0.8679	0.2665	0.1980	1.346
1585	1125	391.7	69.108	283.0	8.496	0.8687	0.2667	0.1981	1.346
1590	1130	393.0	69.960	284.0	8.419	0.8696	0.2668	0.1983	1.346
1595	1135	394.4	70.820	285.0	8.343	0.8704	0.2670	0.1984	1.345
1600	1140	395.7	71.689	286.0	8.267	0.8713	0.2671	0.1986	1.345
1605	1145	397.0	72.566	287.0	8.193	0.8721	0.2672	0.1987	1.345
1610	1150	398.4	73.452	288.0	8.120	0.8729	0.2674	0.1988	1.345
1615	1155	399.7	74.346	289.0	8.047	0.8738	0.2675	0.1990	1.344
1620	1160	401.0	75.248	290.0	7.975	0.8746	0.2677	0.1991	1.344
1625	1165	402.4	76.160	291.0	7.904	0.8754	0.2678	0.1993	1.344
1630	1170	403.7	77.079	292.0	7.833	0.8762	0.2680	0.1994	1.344
1635	1175	405.1	78.008	293.0	7.764	0.8771	0.2681	0.1996	1.344
1640	1180	406.4	78.946	294.0	7.695	0.8779	0.2682	0.1997	1.343
1645	1185	407.7	79.892	295.0	7.627	0.8787	0.2684	0.1998	1.343

Air Tables developed by K. W. Lindler - U. S. Naval Academy

T	t	h	Pr	u	vr	φ	Cp	Cv	k
°R	°F	Btu/lbm		Btu/lbm		Btu/lbm°R	Btu/lbm°R	Btu/lbm°R	
1650	1190	409.1	80.847	296.0	7.560	0.8795	0.2685	0.2000	1.343
1655	1195	410.4	81.811	297.0	7.494	0.8803	0.2687	0.2001	1.343
1660	1200	411.8	82.784	298.0	7.428	0.8811	0.2688	0.2003	1.342
1665	1205	413.1	83.767	299.0	7.363	0.8819	0.2689	0.2004	1.342
1670	1210	414.5	84.758	300.0	7.299	0.8827	0.2691	0.2005	1.342
1675	1215	415.8	85.759	301.0	7.235	0.8835	0.2692	0.2007	1.342
1680	1220	417.1	86.769	302.0	7.172	0.8844	0.2693	0.2008	1.341
1685	1225	418.5	87.789	303.0	7.110	0.8852	0.2695	0.2009	1.341
1690	1230	419.8	88.817	304.0	7.048	0.8859	0.2696	0.2011	1.341
1695	1235	421.2	89.856	305.0	6.988	0.8867	0.2698	0.2012	1.341
1700	1240	422.5	90.903	306.0	6.927	0.8875	0.2699	0.2013	1.340
1705	1245	423.9	91.961	307.0	6.868	0.8883	0.2700	0.2015	1.340
1710	1250	425.2	93.028	308.0	6.809	0.8891	0.2701	0.2016	1.340
1715	1255	426.6	94.105	309.0	6.751	0.8899	0.2703	0.2017	1.340
1720	1260	427.9	95.192	310.0	6.693	0.8907	0.2704	0.2019	1.340
1725	1265	429.3	96.288	311.0	6.636	0.8915	0.2705	0.2020	1.339
1730	1270	430.6	97.395	312.1	6.580	0.8923	0.2707	0.2021	1.339
1735	1275	432.0	98.511	313.1	6.524	0.8931	0.2708	0.2022	1.339
1740	1280	433.4	99.638	314.1	6.469	0.8938	0.2709	0.2024	1.339
1745	1285	434.7	100.774	315.1	6.414	0.8946	0.2711	0.2025	1.339
1750	1290	436.1	101.921	316.1	6.360	0.8954	0.2712	0.2026	1.338
1755	1295	437.4	103.078	317.1	6.307	0.8962	0.2713	0.2028	1.338
1760	1300	438.8	104.246	318.1	6.254	0.8969	0.2714	0.2029	1.338
1765	1305	440.1	105.424	319.1	6.202	0.8977	0.2716	0.2030	1.338
1770	1310	441.5	106.612	320.2	6.150	0.8985	0.2717	0.2031	1.337
1775	1315	442.9	107.811	321.2	6.099	0.8992	0.2718	0.2032	1.337
1780	1320	444.2	109.020	322.2	6.048	0.9000	0.2719	0.2034	1.337
1785	1325	445.6	110.240	323.2	5.998	0.9008	0.2720	0.2035	1.337
1790	1330	446.9	111.471	324.2	5.948	0.9015	0.2722	0.2036	1.337
1795	1335	448.3	112.713	325.3	5.899	0.9023	0.2723	0.2037	1.336
1800	1340	449.7	113.965	326.3	5.851	0.9030	0.2724	0.2038	1.336
1805	1345	451.0	115.229	327.3	5.803	0.9038	0.2725	0.2040	1.336
1810	1350	452.4	116.503	328.3	5.755	0.9045	0.2726	0.2041	1.336
1815	1355	453.7	117.789	329.3	5.708	0.9053	0.2727	0.2042	1.336
1820	1360	455.1	119.086	330.4	5.661	0.9061	0.2729	0.2043	1.336
1825	1365	456.5	120.393	331.4	5.615	0.9068	0.2730	0.2044	1.335
1830	1370	457.8	121.713	332.4	5.570	0.9075	0.2731	0.2045	1.335
1835	1375	459.2	123.043	333.4	5.524	0.9083	0.2732	0.2046	1.335
1840	1380	460.6	124.385	334.4	5.480	0.9090	0.2733	0.2048	1.335
1845	1385	461.9	125.738	335.5	5.435	0.9098	0.2734	0.2049	1.335
1850	1390	463.3	127.103	336.5	5.392	0.9105	0.2735	0.2050	1.334
1855	1395	464.7	128.480	337.5	5.348	0.9113	0.2736	0.2051	1.334
1860	1400	466.0	129.868	338.5	5.305	0.9120	0.2737	0.2052	1.334
1865	1405	467.4	131.268	339.6	5.263	0.9127	0.2739	0.2053	1.334
1870	1410	468.8	132.680	340.6	5.221	0.9135	0.2740	0.2054	1.334
1875	1415	470.2	134.104	341.6	5.179	0.9142	0.2741	0.2055	1.334
1880	1420	471.5	135.540	342.7	5.138	0.9149	0.2742	0.2056	1.333
1885	1425	472.9	136.987	343.7	5.097	0.9157	0.2743	0.2057	1.333
1890	1430	474.3	138.447	344.7	5.057	0.9164	0.2744	0.2058	1.333
1895	1435	475.6	139.919	345.7	5.017	0.9171	0.2745	0.2059	1.333

Air Tables developed by K. W. Lindler - U. S. Naval Academy									
T	t	h	Pr	u	vr	φ	Cp	Cv	k
°R	°F	Btu/lbm		Btu/lbm		Btu/lbm°R	Btu/lbm°R	Btu/lbm°R	
1900	1440	477.0	141.403	346.8	4.977	0.9178	0.2746	0.2061	1.333
1905	1445	478.4	142.900	347.8	4.938	0.9185	0.2747	0.2062	1.332
1910	1450	479.8	144.409	348.8	4.899	0.9193	0.2748	0.2063	1.332
1915	1455	481.1	145.931	349.9	4.861	0.9200	0.2749	0.2064	1.332
1920	1460	482.5	147.466	350.9	4.823	0.9207	0.2750	0.2065	1.332
1925	1465	483.9	149.013	351.9	4.785	0.9214	0.2751	0.2066	1.332
1930	1470	485.3	150.573	353.0	4.748	0.9221	0.2752	0.2067	1.332
1935	1475	486.6	152.145	354.0	4.711	0.9228	0.2754	0.2068	1.331
1940	1480	488.0	153.731	355.0	4.675	0.9236	0.2755	0.2069	1.331
1945	1485	489.4	155.330	356.1	4.638	0.9243	0.2756	0.2070	1.331
1950	1490	490.8	156.941	357.1	4.603	0.9250	0.2757	0.2071	1.331
1955	1495	492.1	158.566	358.1	4.567	0.9257	0.2758	0.2072	1.331
1960	1500	493.5	160.205	359.2	4.532	0.9264	0.2759	0.2073	1.331
1965	1505	494.9	161.856	360.2	4.497	0.9271	0.2760	0.2074	1.330
1970	1510	496.3	163.521	361.2	4.463	0.9278	0.2761	0.2075	1.330
1975	1515	497.7	165.199	362.3	4.429	0.9285	0.2762	0.2076	1.330
1980	1520	499.0	166.891	363.3	4.395	0.9292	0.2763	0.2077	1.330
1985	1525	500.4	168.597	364.4	4.361	0.9299	0.2764	0.2079	1.330
1990	1530	501.8	170.316	365.4	4.328	0.9306	0.2765	0.2080	1.330
1995	1535	503.2	172.049	366.4	4.295	0.9313	0.2766	0.2081	1.329
2000	1540	504.6	173.796	367.5	4.263	0.9320	0.2767	0.2082	1.329
2005	1545	506.0	175.557	368.5	4.231	0.9327	0.2768	0.2083	1.329
2010	1550	507.3	177.332	369.6	4.199	0.9333	0.2769	0.2084	1.329
2015	1555	508.7	179.121	370.6	4.167	0.9340	0.2770	0.2085	1.329
2020	1560	510.1	180.924	371.6	4.136	0.9347	0.2771	0.2086	1.329
2025	1565	511.5	182.741	372.7	4.105	0.9354	0.2772	0.2087	1.329
2030	1570	512.9	184.573	373.7	4.074	0.9361	0.2773	0.2088	1.328
2035	1575	514.3	186.420	374.8	4.044	0.9368	0.2774	0.2089	1.328
2040	1580	515.7	188.281	375.8	4.014	0.9375	0.2775	0.2090	1.328
2045	1585	517.0	190.156	376.9	3.984	0.9381	0.2776	0.2091	1.328
2050	1590	518.4	192.047	377.9	3.954	0.9388	0.2777	0.2092	1.328
2055	1595	519.8	193.952	379.0	3.925	0.9395	0.2778	0.2093	1.328
2060	1600	521.2	195.872	380.0	3.896	0.9402	0.2779	0.2094	1.327
2065	1605	522.6	197.807	381.1	3.867	0.9408	0.2780	0.2095	1.327
2070	1610	524.0	199.757	382.1	3.839	0.9415	0.2781	0.2096	1.327
2075	1615	525.4	201.722	383.1	3.810	0.9422	0.2782	0.2097	1.327
2080	1620	526.8	203.703	384.2	3.782	0.9429	0.2783	0.2098	1.327
2085	1625	528.2	205.698	385.2	3.755	0.9435	0.2784	0.2099	1.327
2090	1630	529.6	207.710	386.3	3.727	0.9442	0.2785	0.2100	1.326
2095	1635	531.0	209.736	387.3	3.700	0.9449	0.2786	0.2101	1.326
2100	1640	532.3	211.779	388.4	3.673	0.9455	0.2787	0.2102	1.326
2105	1645	533.7	213.837	389.4	3.647	0.9462	0.2788	0.2103	1.326
2110	1650	535.1	215.910	390.5	3.620	0.9468	0.2789	0.2104	1.326
2115	1655	536.5	218.000	391.5	3.594	0.9475	0.2790	0.2104	1.326
2120	1660	537.9	220.105	392.6	3.568	0.9482	0.2791	0.2105	1.326
2125	1665	539.3	222.227	393.7	3.542	0.9488	0.2792	0.2106	1.325
2130	1670	540.7	224.365	394.7	3.517	0.9495	0.2793	0.2107	1.325
2135	1675	542.1	226.519	395.8	3.491	0.9501	0.2794	0.2108	1.325
2140	1680	543.5	228.689	396.8	3.466	0.9508	0.2795	0.2109	1.325
2145	1685	544.9	230.876	397.9	3.442	0.9514	0.2796	0.2110	1.325

Air Tables developed by K. W. Lindler - U. S. Naval Academy									
T	t	h	Pr	u	vr	φ	Cp	Cv	k
°R	°F	Btu/lbm		Btu/lbm		Btu/lbm°R	Btu/lbm°R	Btu/lbm°R	
2150	1690	546.3	233.079	398.9	3.417	0.9521	0.2797	0.2111	1.325
2155	1695	547.7	235.299	400.0	3.393	0.9527	0.2798	0.2112	1.325
2160	1700	549.1	237.535	401.0	3.368	0.9534	0.2799	0.2113	1.324
2165	1705	550.5	239.788	402.1	3.345	0.9540	0.2800	0.2114	1.324
2170	1710	551.9	242.058	403.2	3.321	0.9547	0.2800	0.2115	1.324
2175	1715	553.3	244.345	404.2	3.297	0.9553	0.2801	0.2116	1.324
2180	1720	554.7	246.650	405.3	3.274	0.9560	0.2802	0.2117	1.324
2185	1725	556.1	248.971	406.3	3.251	0.9566	0.2803	0.2118	1.324
2190	1730	557.5	251.309	407.4	3.228	0.9572	0.2804	0.2119	1.324
2195	1735	558.9	253.665	408.4	3.205	0.9579	0.2805	0.2120	1.323
2200	1740	560.3	256.039	409.5	3.183	0.9585	0.2806	0.2121	1.323
2205	1745	561.7	258.429	410.6	3.161	0.9592	0.2807	0.2122	1.323
2210	1750	563.1	260.838	411.6	3.139	0.9598	0.2808	0.2122	1.323
2215	1755	564.5	263.264	412.7	3.117	0.9604	0.2809	0.2123	1.323
2220	1760	565.9	265.708	413.8	3.095	0.9611	0.2810	0.2124	1.323
2225	1765	567.3	268.170	414.8	3.073	0.9617	0.2811	0.2125	1.323
2230	1770	568.7	270.650	415.9	3.052	0.9623	0.2812	0.2126	1.322
2235	1775	570.1	273.148	416.9	3.031	0.9630	0.2812	0.2127	1.322
2240	1780	571.6	275.664	418.0	3.010	0.9636	0.2813	0.2128	1.322
2245	1785	573.0	278.199	419.1	2.989	0.9642	0.2814	0.2129	1.322
2250	1790	574.4	280.752	420.1	2.969	0.9648	0.2815	0.2130	1.322
2255	1795	575.8	283.323	421.2	2.948	0.9655	0.2816	0.2131	1.322
2260	1800	577.2	285.913	422.3	2.928	0.9661	0.2817	0.2132	1.322
2265	1805	578.6	288.522	423.3	2.908	0.9667	0.2818	0.2132	1.321
2270	1810	580.0	291.150	424.4	2.888	0.9673	0.2819	0.2133	1.321
2275	1815	581.4	293.796	425.5	2.868	0.9680	0.2820	0.2134	1.321
2280	1820	582.8	296.462	426.5	2.849	0.9686	0.2821	0.2135	1.321
2285	1825	584.2	299.147	427.6	2.829	0.9692	0.2821	0.2136	1.321
2290	1830	585.6	301.850	428.7	2.810	0.9698	0.2822	0.2137	1.321
2295	1835	587.1	304.574	429.7	2.791	0.9704	0.2823	0.2138	1.321
2300	1840	588.5	307.316	430.8	2.772	0.9710	0.2824	0.2139	1.321
2305	1845	589.9	310.078	431.9	2.754	0.9717	0.2825	0.2139	1.320
2310	1850	591.3	312.860	432.9	2.735	0.9723	0.2826	0.2140	1.320
2315	1855	592.7	315.662	434.0	2.717	0.9729	0.2827	0.2141	1.320
2320	1860	594.1	318.483	435.1	2.698	0.9735	0.2828	0.2142	1.320
2325	1865	595.5	321.324	436.2	2.680	0.9741	0.2828	0.2143	1.320
2330	1870	596.9	324.186	437.2	2.662	0.9747	0.2829	0.2144	1.320
2335	1875	598.4	327.067	438.3	2.645	0.9753	0.2830	0.2145	1.320
2340	1880	599.8	329.969	439.4	2.627	0.9759	0.2831	0.2146	1.319
2345	1885	601.2	332.891	440.4	2.609	0.9765	0.2832	0.2146	1.319
2350	1890	602.6	335.834	441.5	2.592	0.9771	0.2833	0.2147	1.319
2355	1895	604.0	338.797	442.6	2.575	0.9777	0.2834	0.2148	1.319
2360	1900	605.4	341.780	443.7	2.558	0.9783	0.2834	0.2149	1.319
2365	1905	606.9	344.785	444.7	2.541	0.9789	0.2835	0.2150	1.319
2370	1910	608.3	347.811	445.8	2.524	0.9795	0.2836	0.2151	1.319
2375	1915	609.7	350.857	446.9	2.507	0.9801	0.2837	0.2152	1.319
2380	1920	611.1	353.925	448.0	2.491	0.9807	0.2838	0.2152	1.318
2385	1925	612.5	357.013	449.0	2.475	0.9813	0.2839	0.2153	1.318
2390	1930	614.0	360.124	450.1	2.458	0.9819	0.2840	0.2154	1.318
2395	1935	615.4	363.255	451.2	2.442	0.9825	0.2840	0.2155	1.318

Air Tables developed by K. W. Lindler - U. S. Naval Academy

T	t	h	Pr	u	vr	φ	Cp	Cv	k
°R	°F	Btu/lbm		Btu/lbm		Btu/lbm°R	Btu/lbm°R	Btu/lbm°R	
2400	1940	616.8	366.408	452.3	2.426	0.9831	0.2841	0.2156	1.318
2405	1945	618.2	369.583	453.4	2.411	0.9837	0.2842	0.2157	1.318
2410	1950	619.6	372.780	454.4	2.395	0.9843	0.2843	0.2157	1.318
2415	1955	621.1	375.998	455.5	2.379	0.9849	0.2844	0.2158	1.318
2420	1960	622.5	379.239	456.6	2.364	0.9855	0.2845	0.2159	1.317
2425	1965	623.9	382.501	457.7	2.348	0.9860	0.2845	0.2160	1.317
2430	1970	625.3	385.786	458.8	2.333	0.9866	0.2846	0.2161	1.317
2435	1975	626.7	389.093	459.8	2.318	0.9872	0.2847	0.2162	1.317
2440	1980	628.2	392.422	460.9	2.303	0.9878	0.2848	0.2162	1.317
2445	1985	629.6	395.775	462.0	2.288	0.9884	0.2849	0.2163	1.317
2450	1990	631.0	399.149	463.1	2.274	0.9890	0.2849	0.2164	1.317
2455	1995	632.4	402.547	464.2	2.259	0.9895	0.2850	0.2165	1.317
2460	2000	633.9	405.967	465.2	2.245	0.9901	0.2851	0.2166	1.317
2465	2005	635.3	409.411	466.3	2.230	0.9907	0.2852	0.2166	1.316
2470	2010	636.7	412.877	467.4	2.216	0.9913	0.2853	0.2167	1.316
2475	2015	638.1	416.367	468.5	2.202	0.9919	0.2853	0.2168	1.316
2480	2020	639.6	419.880	469.6	2.188	0.9924	0.2854	0.2169	1.316
2485	2025	641.0	423.417	470.7	2.174	0.9930	0.2855	0.2170	1.316
2490	2030	642.4	426.977	471.7	2.160	0.9936	0.2856	0.2170	1.316
2495	2035	643.9	430.561	472.8	2.147	0.9942	0.2857	0.2171	1.316
2500	2040	645.3	434.168	473.9	2.133	0.9947	0.2857	0.2172	1.316
2505	2045	646.7	437.800	475.0	2.120	0.9953	0.2858	0.2173	1.315
2510	2050	648.1	441.456	476.1	2.106	0.9959	0.2859	0.2174	1.315
2515	2055	649.6	445.136	477.2	2.093	0.9964	0.2860	0.2174	1.315
2520	2060	651.0	448.840	478.3	2.080	0.9970	0.2861	0.2175	1.315
2525	2065	652.4	452.569	479.3	2.067	0.9976	0.2861	0.2176	1.315
2530	2070	653.9	456.322	480.4	2.054	0.9981	0.2862	0.2177	1.315
2535	2075	655.3	460.100	481.5	2.041	0.9987	0.2863	0.2177	1.315
2540	2080	656.7	463.902	482.6	2.028	0.9993	0.2864	0.2178	1.315
2545	2085	658.2	467.730	483.7	2.016	0.9998	0.2865	0.2179	1.315
2550	2090	659.6	471.582	484.8	2.003	1.0004	0.2865	0.2180	1.314
2555	2095	661.0	475.460	485.9	1.991	1.0010	0.2866	0.2181	1.314
2560	2100	662.5	479.363	487.0	1.978	1.0015	0.2867	0.2181	1.314
2565	2105	663.9	483.291	488.1	1.966	1.0021	0.2868	0.2182	1.314
2570	2110	665.3	487.245	489.2	1.954	1.0026	0.2868	0.2183	1.314
2575	2115	666.8	491.224	490.2	1.942	1.0032	0.2869	0.2184	1.314
2580	2120	668.2	495.229	491.3	1.930	1.0037	0.2870	0.2184	1.314
2585	2125	669.6	499.260	492.4	1.918	1.0043	0.2871	0.2185	1.314
2590	2130	671.1	503.317	493.5	1.906	1.0049	0.2871	0.2186	1.314
2595	2135	672.5	507.400	494.6	1.894	1.0054	0.2872	0.2187	1.313
2600	2140	673.9	511.510	495.7	1.883	1.0060	0.2873	0.2187	1.313
2605	2145	675.4	515.646	496.8	1.871	1.0065	0.2874	0.2188	1.313
2610	2150	676.8	519.808	497.9	1.860	1.0071	0.2874	0.2189	1.313
2615	2155	678.3	523.997	499.0	1.849	1.0076	0.2875	0.2190	1.313
2620	2160	679.7	528.212	500.1	1.837	1.0082	0.2876	0.2190	1.313
2625	2165	681.1	532.455	501.2	1.826	1.0087	0.2877	0.2191	1.313
2630	2170	682.6	536.725	502.3	1.815	1.0093	0.2877	0.2192	1.313
2635	2175	684.0	541.021	503.4	1.804	1.0098	0.2878	0.2193	1.313
2640	2180	685.4	545.345	504.5	1.793	1.0104	0.2879	0.2193	1.313
2645	2185	686.9	549.696	505.6	1.782	1.0109	0.2880	0.2194	1.312

Air Tables developed by K. W. Lindler - U. S. Naval Academy									
T	t	h	Pr	u	vr	φ	Cp	Cv	k
°R	°F	Btu/lbm		Btu/lbm		Btu/lbm°R	Btu/lbm°R	Btu/lbm°R	
2650	2190	688.3	554.075	506.7	1.772	1.0114	0.2880	0.2195	1.312
2655	2195	689.8	558.482	507.8	1.761	1.0120	0.2881	0.2196	1.312
2660	2200	691.2	562.916	508.9	1.750	1.0125	0.2882	0.2196	1.312
2665	2205	692.6	567.378	510.0	1.740	1.0131	0.2882	0.2197	1.312
2670	2210	694.1	571.869	511.1	1.730	1.0136	0.2883	0.2198	1.312
2675	2215	695.5	576.387	512.2	1.719	1.0141	0.2884	0.2198	1.312
2680	2220	697.0	580.934	513.3	1.709	1.0147	0.2885	0.2199	1.312
2685	2225	698.4	585.509	514.4	1.699	1.0152	0.2885	0.2200	1.312
2690	2230	699.9	590.113	515.5	1.689	1.0158	0.2886	0.2201	1.312
2695	2235	701.3	594.745	516.6	1.679	1.0163	0.2887	0.2201	1.311
2700	2240	702.7	599.406	517.7	1.669	1.0168	0.2888	0.2202	1.311
2705	2245	704.2	604.097	518.8	1.659	1.0174	0.2888	0.2203	1.311
2710	2250	705.6	608.816	519.9	1.649	1.0179	0.2889	0.2203	1.311
2715	2255	707.1	613.564	521.0	1.639	1.0184	0.2890	0.2204	1.311
2720	2260	708.5	618.342	522.1	1.629	1.0190	0.2890	0.2205	1.311
2725	2265	710.0	623.150	523.2	1.620	1.0195	0.2891	0.2206	1.311
2730	2270	711.4	627.987	524.3	1.610	1.0200	0.2892	0.2206	1.311
2735	2275	712.9	632.854	525.4	1.601	1.0206	0.2892	0.2207	1.311
2740	2280	714.3	637.751	526.5	1.591	1.0211	0.2893	0.2208	1.311
2745	2285	715.8	642.677	527.6	1.582	1.0216	0.2894	0.2208	1.310
2750	2290	717.2	647.634	528.7	1.573	1.0221	0.2895	0.2209	1.310
2755	2295	718.6	652.622	529.8	1.564	1.0227	0.2895	0.2210	1.310
2760	2300	720.1	657.639	530.9	1.555	1.0232	0.2896	0.2210	1.310
2765	2305	721.5	662.688	532.0	1.546	1.0237	0.2897	0.2211	1.310
2770	2310	723.0	667.767	533.1	1.537	1.0242	0.2897	0.2212	1.310
2775	2315	724.4	672.877	534.2	1.528	1.0248	0.2898	0.2212	1.310
2780	2320	725.9	678.018	535.3	1.519	1.0253	0.2899	0.2213	1.310
2785	2325	727.3	683.190	536.4	1.510	1.0258	0.2899	0.2214	1.310
2790	2330	728.8	688.393	537.5	1.501	1.0263	0.2900	0.2215	1.310
2795	2335	730.2	693.628	538.6	1.493	1.0268	0.2901	0.2215	1.309
2800	2340	731.7	698.895	539.7	1.484	1.0274	0.2901	0.2216	1.309
2805	2345	733.1	704.193	540.9	1.476	1.0279	0.2902	0.2217	1.309
2810	2350	734.6	709.523	542.0	1.467	1.0284	0.2903	0.2217	1.309
2815	2355	736.0	714.885	543.1	1.459	1.0289	0.2903	0.2218	1.309
2820	2360	737.5	720.279	544.2	1.450	1.0294	0.2904	0.2219	1.309
2825	2365	738.9	725.706	545.3	1.442	1.0299	0.2905	0.2219	1.309
2830	2370	740.4	731.165	546.4	1.434	1.0305	0.2905	0.2220	1.309
2835	2375	741.9	736.656	547.5	1.426	1.0310	0.2906	0.2221	1.309
2840	2380	743.3	742.180	548.6	1.417	1.0315	0.2907	0.2221	1.309
2845	2385	744.8	747.737	549.7	1.409	1.0320	0.2907	0.2222	1.309
2850	2390	746.2	753.328	550.8	1.401	1.0325	0.2908	0.2223	1.308
2855	2395	747.7	758.951	552.0	1.393	1.0330	0.2909	0.2223	1.308
2860	2400	749.1	764.607	553.1	1.386	1.0335	0.2909	0.2224	1.308
2865	2405	750.6	770.297	554.2	1.378	1.0340	0.2910	0.2224	1.308
2870	2410	752.0	776.021	555.3	1.370	1.0345	0.2911	0.2225	1.308
2875	2415	753.5	781.778	556.4	1.362	1.0350	0.2911	0.2226	1.308
2880	2420	754.9	787.569	557.5	1.355	1.0355	0.2912	0.2226	1.308
2885	2425	756.4	793.395	558.6	1.347	1.0361	0.2913	0.2227	1.308
2890	2430	757.9	799.254	559.7	1.339	1.0366	0.2913	0.2228	1.308
2895	2435	759.3	805.148	560.9	1.332	1.0371	0.2914	0.2228	1.308

| \multicolumn{10}{c}{Air Tables developed by K. W. Lindler - U. S. Naval Academy} |
| T | t | h | Pr | u | vr | φ | Cp | Cv | k |
°R	°F	Btu/lbm		Btu/lbm		Btu/lbm°R	Btu/lbm°R	Btu/lbm°R	
2900	2440	760.8	811.076	562.0	1.324	1.0376	0.2914	0.2229	1.308
2905	2445	762.2	817.039	563.1	1.317	1.0381	0.2915	0.2230	1.307
2910	2450	763.7	823.037	564.2	1.310	1.0386	0.2916	0.2230	1.307
2915	2455	765.1	829.069	565.3	1.302	1.0391	0.2916	0.2231	1.307
2920	2460	766.6	835.137	566.4	1.295	1.0396	0.2917	0.2232	1.307
2925	2465	768.1	841.240	567.6	1.288	1.0401	0.2918	0.2232	1.307
2930	2470	769.5	847.379	568.7	1.281	1.0406	0.2918	0.2233	1.307
2935	2475	771.0	853.552	569.8	1.274	1.0411	0.2919	0.2233	1.307
2940	2480	772.4	859.762	570.9	1.267	1.0416	0.2920	0.2234	1.307
2945	2485	773.9	866.007	572.0	1.260	1.0421	0.2920	0.2235	1.307
2950	2490	775.4	872.289	573.1	1.253	1.0426	0.2921	0.2235	1.307
2955	2495	776.8	878.606	574.3	1.246	1.0430	0.2921	0.2236	1.307
2960	2500	778.3	884.960	575.4	1.239	1.0435	0.2922	0.2237	1.306
2965	2505	779.7	891.351	576.5	1.232	1.0440	0.2923	0.2237	1.306
2970	2510	781.2	897.778	577.6	1.225	1.0445	0.2923	0.2238	1.306
2975	2515	782.7	904.242	578.7	1.219	1.0450	0.2924	0.2238	1.306
2980	2520	784.1	910.742	579.8	1.212	1.0455	0.2924	0.2239	1.306
2985	2525	785.6	917.280	581.0	1.205	1.0460	0.2925	0.2240	1.306
2990	2530	787.0	923.855	582.1	1.199	1.0465	0.2926	0.2240	1.306
2995	2535	788.5	930.468	583.2	1.192	1.0470	0.2926	0.2241	1.306
3000	2540	790.0	937.118	584.3	1.186	1.0475	0.2927	0.2241	1.306
3005	2545	791.4	943.805	585.4	1.179	1.0480	0.2928	0.2242	1.306
3010	2550	792.9	950.531	586.6	1.173	1.0484	0.2928	0.2243	1.306
3015	2555	794.4	957.295	587.7	1.167	1.0489	0.2929	0.2243	1.306
3020	2560	795.8	964.096	588.8	1.160	1.0494	0.2929	0.2244	1.306
3025	2565	797.3	970.937	589.9	1.154	1.0499	0.2930	0.2244	1.305
3030	2570	798.8	977.815	591.1	1.148	1.0504	0.2931	0.2245	1.305
3035	2575	800.2	984.733	592.2	1.142	1.0509	0.2931	0.2246	1.305
3040	2580	801.7	991.689	593.3	1.136	1.0513	0.2932	0.2246	1.305
3045	2585	803.2	998.684	594.4	1.129	1.0518	0.2932	0.2247	1.305
3050	2590	804.6	1005.719	595.5	1.123	1.0523	0.2933	0.2247	1.305
3055	2595	806.1	1012.793	596.7	1.117	1.0528	0.2933	0.2248	1.305
3060	2600	807.6	1019.906	597.8	1.111	1.0533	0.2934	0.2249	1.305
3065	2605	809.0	1027.059	598.9	1.105	1.0537	0.2935	0.2249	1.305
3070	2610	810.5	1034.251	600.0	1.100	1.0542	0.2935	0.2250	1.305
3075	2615	812.0	1041.484	601.2	1.094	1.0547	0.2936	0.2250	1.305
3080	2620	813.4	1048.757	602.3	1.088	1.0552	0.2936	0.2251	1.305
3085	2625	814.9	1056.070	603.4	1.082	1.0557	0.2937	0.2251	1.304
3090	2630	816.4	1063.424	604.5	1.076	1.0561	0.2938	0.2252	1.304
3095	2635	817.8	1070.818	605.7	1.071	1.0566	0.2938	0.2253	1.304
3100	2640	819.3	1078.253	606.8	1.065	1.0571	0.2939	0.2253	1.304
3105	2645	820.8	1085.729	607.9	1.059	1.0576	0.2939	0.2254	1.304
3110	2650	822.2	1093.247	609.1	1.054	1.0580	0.2940	0.2254	1.304
3115	2655	823.7	1100.805	610.2	1.048	1.0585	0.2940	0.2255	1.304
3120	2660	825.2	1108.405	611.3	1.043	1.0590	0.2941	0.2255	1.304
3125	2665	826.7	1116.047	612.4	1.037	1.0594	0.2942	0.2256	1.304
3130	2670	828.1	1123.730	613.6	1.032	1.0599	0.2942	0.2257	1.304
3135	2675	829.6	1131.456	614.7	1.026	1.0604	0.2943	0.2257	1.304
3140	2680	831.1	1139.223	615.8	1.021	1.0609	0.2943	0.2258	1.304
3145	2685	832.5	1147.033	617.0	1.016	1.0613	0.2944	0.2258	1.304

\multicolumn{10}{c}{Air Tables developed by K. W. Lindler - U. S. Naval Academy}									
T	t	h	Pr	u	vr	φ	Cp	Cv	k
°R	°F	Btu/lbm		Btu/lbm		Btu/lbm°R	Btu/lbm°R	Btu/lbm°R	
3150	2690	834.0	1154.886	618.1	1.010	1.0618	0.2944	0.2259	1.303
3155	2695	835.5	1162.781	619.2	1.005	1.0623	0.2945	0.2259	1.303
3160	2700	837.0	1170.719	620.3	1.000	1.0627	0.2945	0.2260	1.303
3165	2705	838.4	1178.700	621.5	0.995	1.0632	0.2946	0.2261	1.303
3170	2710	839.9	1186.724	622.6	0.989	1.0637	0.2947	0.2261	1.303
3175	2715	841.4	1194.792	623.7	0.984	1.0641	0.2947	0.2262	1.303
3180	2720	842.8	1202.903	624.9	0.979	1.0646	0.2948	0.2262	1.303
3185	2725	844.3	1211.058	626.0	0.974	1.0650	0.2948	0.2263	1.303
3190	2730	845.8	1219.257	627.1	0.969	1.0655	0.2949	0.2263	1.303
3195	2735	847.3	1227.500	628.3	0.964	1.0660	0.2949	0.2264	1.303
3200	2740	848.7	1235.787	629.4	0.959	1.0664	0.2950	0.2264	1.303
3205	2745	850.2	1244.118	630.5	0.954	1.0669	0.2950	0.2265	1.303
3210	2750	851.7	1252.494	631.7	0.949	1.0674	0.2951	0.2265	1.303
3215	2755	853.2	1260.915	632.8	0.944	1.0678	0.2951	0.2266	1.303
3220	2760	854.6	1269.381	633.9	0.940	1.0683	0.2952	0.2267	1.302
3225	2765	856.1	1277.892	635.1	0.935	1.0687	0.2953	0.2267	1.302
3230	2770	857.6	1286.448	636.2	0.930	1.0692	0.2953	0.2268	1.302
3235	2775	859.1	1295.049	637.3	0.925	1.0696	0.2954	0.2268	1.302
3240	2780	860.6	1303.697	638.5	0.921	1.0701	0.2954	0.2269	1.302
3245	2785	862.0	1312.390	639.6	0.916	1.0706	0.2955	0.2269	1.302
3250	2790	863.5	1321.129	640.7	0.911	1.0710	0.2955	0.2270	1.302
3255	2795	865.0	1329.914	641.9	0.907	1.0715	0.2956	0.2270	1.302
3260	2800	866.5	1338.746	643.0	0.902	1.0719	0.2956	0.2271	1.302
3265	2805	867.9	1347.625	644.1	0.897	1.0724	0.2957	0.2271	1.302
3270	2810	869.4	1356.550	645.3	0.893	1.0728	0.2957	0.2272	1.302
3275	2815	870.9	1365.522	646.4	0.888	1.0733	0.2958	0.2272	1.302
3280	2820	872.4	1374.541	647.5	0.884	1.0737	0.2958	0.2273	1.302
3285	2825	873.9	1383.607	648.7	0.879	1.0742	0.2959	0.2273	1.302
3290	2830	875.3	1392.721	649.8	0.875	1.0746	0.2959	0.2274	1.301
3295	2835	876.8	1401.883	650.9	0.871	1.0751	0.2960	0.2274	1.301
3300	2840	878.3	1411.092	652.1	0.866	1.0755	0.2960	0.2275	1.301
3305	2845	879.8	1420.349	653.2	0.862	1.0760	0.2961	0.2275	1.301
3310	2850	881.3	1429.655	654.4	0.858	1.0764	0.2961	0.2276	1.301
3315	2855	882.7	1439.009	655.5	0.853	1.0769	0.2962	0.2276	1.301
3320	2860	884.2	1448.412	656.6	0.849	1.0773	0.2962	0.2277	1.301
3325	2865	885.7	1457.863	657.8	0.845	1.0778	0.2963	0.2277	1.301
3330	2870	887.2	1467.363	658.9	0.841	1.0782	0.2963	0.2278	1.301
3335	2875	888.7	1476.913	660.1	0.836	1.0786	0.2964	0.2278	1.301
3340	2880	890.1	1486.512	661.2	0.832	1.0791	0.2964	0.2279	1.301
3345	2885	891.6	1496.160	662.3	0.828	1.0795	0.2965	0.2279	1.301
3350	2890	893.1	1505.858	663.5	0.824	1.0800	0.2965	0.2280	1.301
3355	2895	894.6	1515.606	664.6	0.820	1.0804	0.2966	0.2280	1.301
3360	2900	896.1	1525.404	665.8	0.816	1.0809	0.2966	0.2281	1.301
3365	2905	897.6	1535.253	666.9	0.812	1.0813	0.2967	0.2281	1.300
3370	2910	899.0	1545.151	668.0	0.808	1.0817	0.2967	0.2282	1.300
3375	2915	900.5	1555.101	669.2	0.804	1.0822	0.2968	0.2282	1.300
3380	2920	902.0	1565.101	670.3	0.800	1.0826	0.2968	0.2283	1.300
3385	2925	903.5	1575.153	671.5	0.796	1.0831	0.2969	0.2283	1.300
3390	2930	905.0	1585.255	672.6	0.792	1.0835	0.2969	0.2284	1.300
3395	2935	906.5	1595.409	673.7	0.788	1.0839	0.2970	0.2284	1.300

Index

absolute pressure, 13, 14, 15, 16, 27
absorption chillers, 199, 200, 205
accumulators, 72
actuators: bulb, 209; hydraulic and pneumatic systems, 67, 68–69, 72–73; valve, 36, 39
adiabatic process, 22
air: composition of, 98–99, 201; air properties, 231; heat exchangers, 165, 166, 174; humid air and humidity, 201–2; ideal gas behavior and ideal gas laws, 27–30; submarines and chemical balance of air, 212
air compressors: applications and uses, 81; dieseling, 74; oil-free, 74; performance of, 81; pneumatic systems, 71, 72, 73; receivers, 72, 73; safety issues, 73–74
air conditioning: absorption chillers, 199, 200, 205; chill water systems, 210, 212–16; comfort and human comfort zone, 199, 205; cooling capacity of equipment, 198; evaporative cooling (swamp coolers), 205, 212; freezing and frosting conditions, 210; historical overview, 199–200; reverse Brayton air conditioning scheme, 205. *See also* refrigerants; vapor compression refrigeration
air standard analysis, 87, 88, 93, 107, 109
air tables, 109, 169, 271–82
aircraft: catapult-launched, 183; gas turbine engines, 106, 120–25; hydraulic system controls, 65, 72–73; piston-engined aircraft, 79, 120
aircraft carriers: air conditioning and cooling capacity, 198; catapult steam, 183; desalination capacity, 183; nuclear power, 133, 178, 179, 180, 183
air-to-fuel ratio, 99, 100, 233
algebra review, 6–8
alternating current (AC) motors: pumps driven by, 55; synchronous speed equation, 55, 233
altitude and atmospheric pressure, 13, 14
Arleigh Burke–class destroyers, 1, 106
asphaltic bitumens, 98

atmospheric pressure: altitude and, 13, 14; definition and concept, 13–14; gas turbine engines, 112; internal combustion engines and, 91, 95–96, 97, 98; measurement of, 13–14; pressure relationships, 14; steam power and, 132; weather and, 13, 14
atomic bomb, 180
Austin-class vessels, 133
auto-ignition temperature, 74, 80, 93
auxiliary steam, 154
auxiliary systems, 8
axial flow pump, 46, 47

back work ratio, 107, 108, 233
backpressure, 68, 70–71
Bainbridge, 179
barometers and barometric pressure, 13–14
Beau de Rochas, Alphonse, 80
Bernoulli's equation: fluid rate calculation, 55; frictionless form, 7, 41–43; with head loss, 43, 45, 51, 232; net positive suction head (NPSH), 48–50; pump power, 52
bio-fouling/biological growth, 163, 164, 165, 167–68
biofuels, 100
blowdown, 37
boilers: corrosion, 222; efficiency, 149–50, 233; salinity of water, 222; steam plant operation, 133, 144, 147, 149, 154–56, 174
boiling point (saturation temperature), 11, 71, 133–36, 142, 144, 200–201
boiling water reactors, 182
Bosch, Robert, 80
boundary layer, 45
boundary work, 22, 81, 83, 109
Bourdon tube, 16
brackish water, 222
brake horsepower (BHP), 25, 52, 53, 92, 116, 233, 237
brake specific fuel consumption (bsfc), 26, 92, 233
Brayton, George, 106
Brayton cycle, 106–8, 110–15, 205
breeder reactors, 181

brine, 222–25, 227, 228
British Thermal Units (Btu): conversion factor, 17; definition and concept, 17; mechanical equivalent to heat, 6–7, 17
bulb actuator, 209
butane, 97, 98, 99
bypass air, 123–24

CAER engine processes, 23
California, 179
cams, 87
capillary tube, 206
carbon dioxide (CO_2), 98, 99, 100, 212
carbon dioxide (CO_2) scrubbers, 212
carbon monoxide (CO), 99, 100, 212
carbon monoxide (CO)-hydrogen burners, 212
Carrier, Willis, 199
catapults, 183
cavitation, 48–50, 147
Celsius scale, 13
centrifugal pumps, 46–48, 53, 70
centrifuge enrichment process, 180
chain reaction, 180–81
charging, 95
check valves, 36, 37, 58
chill water systems, 210
chlorine, 200
chlorofluorocarbon (CFC), 200
cladding, 180
clearance volume, 90
Clerk, Dugald, 80
Clermont, 132
closed systems, 10, 21, 22, 80–81, 87–88
coefficient of performance, 206, 234, 237
cold legs (PWR), 183, 184, 185
combustion chamber/combustor (gas turbine engine), 106–8, 115
combustion efficiency and combustion efficiency equation, 91–92, 100, 233
combustion process: gas turbine engines, 112, 163; hydrocarbon combustion, 97–100; SI engines, 87
combustion product, 100
comfort zone and human comfort zone, 199, 205

compressed liquid (sub-cooled liquid), 134–35, 137, 138, 139, 144, 147, 148
Compressed Liquid Approximation Technique, 139
compression ignition (CI) engines: definition and concept, 79; diesel cycle, 79, 92–95; 4-stroke engines, 79; fuels, 79, 95; historical overview, 79–80; 2-stroke engines, 79. *See also* diesel engines
compression ratio, 88, 233, 237
compressor, air. *See* air compressors
compressor (gas generator), 106–8, 111–12, 115
compressor (vapor compression refrigeration), 206, 207
compressor efficiency, 233
compressor pressure ratio, 107
condensate depression (sub-cooling), 141–42, 147–48, 183
condensation (hydrocarbon combustion process), 100
condenser (steam plant components), 132, 144, 147, 167, 174
condenser (vapor compression refrigeration), 174, 206, 207
conduction/conductive heat transfer, 163, 164–66, 234
connecting rod, 85, 86
conservation of energy, 16, 51, 215
conservation of mass, 51, 215
constant C methodology, 88, 108–9
constant pressure, 19
constant specific heat relations, 234
constant volume, 19
continuity equation, 39–41, 42, 46, 50, 55, 232
control rods and control rod drive motors, 181
control valves (hydraulic), 70, 72–73
convection/convective heat transfer, 163–66, 234
corrosion: boilers, 222; flash-type distillation and, 225; heat exchangers, 163, 164, 165, 167; refrigerants, 201; seawater, salinity of water, and, 167, 222
counter-flow heat exchanger, 167, 169
crank, 85
crankshaft, 85, 86
critical nuclear reactor, 181
cross-flow heat exchanger, 167, 169
cryogenics, 199
cutoff ratio, 93, 233, 237
cycles: definition and concept, 11, 22–23; first law of thermodynamics, 23
cylinders: displacement, 90; number in engines, 85–86

Daimler, Gottlieb, 80
dampers, 36, 73
Darcy's equation, 43
Day, Joseph, 80
decay heat, 181
degrees of superheat, 142
degrees Rankine, conversion of, 13
delivered horsepower (DHP), 25

demineralized (deionized) water, 222
Demologos (*Fulton*), 132
density: definition and concept, 12; intensive system properties, 11; specific gravity and, 12–13; specific volume and, 12–13, 232; specific weight and, 12–13; units and nomenclature, 236
derived units, 11
desalination: aircraft carriers, 183; applications, 221–22, 224–25, 226, 227–28; distillation, 222–26, 228; practice problems, 229; rated output capacity, 222–23, 224, 225, 226, 227–28; reverse osmosis water purification, 221, 222, 226–28
dew point temperature, 202
Diesel, Rudolph, 80
diesel engines: advantages, 95; applications, 8, 80, 132–33; CI engine, 79, 92–95; efficiency, 95; electrical generators, 8, 80, 212–13; 4-stroke engines, 94–95, 97, 98; fuel injectors, 80; historical overview, 8, 80; ideal, air standard diesel analysis constant C method, 93–94; operation of, 92; opposed piston engines, 95, 97; Otto cycle compared to, 95–97, 98; propulsion methods, 8; torque and power, 95, 98; 2-stroke engines, 94–95, 96, 97
diesel fuel, 25, 79, 80, 95, 98, 99
diesel fuel cutoff ratio, 93, 233, 237
dieseling, 74
diffusers and nozzles, 118–20, 121
direct contact (mixing) heat exchangers, 162, 166, 168
directional control valves, 70
displacement, 90
distillation: applications, 224–25, 226; concept and principles, 222–23; flash-type distilling plants, 223–25; heat recovery distillation, 222, 225–26, 228; low-pressure steam distillation, 222, 223–25, 228; multiple-effects distilling plants, 223; rated output capacity, 222–23, 224, 225, 226; submerged-tube distilling plants, 223, 224, 226; vapor compression distillation, 222, 225, 228; vertical-basket distilling plants, 223, 224
double-acting actuators, 68–69
downcomers, 154, 155
dry atmospheric air, 201
dry bulb thermometer and temperature, 201
dry saturated steam. *See* saturated vapor (dry saturated steam)
dynamic viscosity, 237

economizer, 100, 154, 155
effective horsepower, 25
efficiency: coefficient of performance, 206, 234, 237; combustion efficiency and combustion efficiency equation, 91–92, 100, 233; conversion factors, 23; definition and concept, 23–25; engine mechanical efficiency, 25, 233; engine

overall efficiency, 24–25, 92; equations, 233; first law of thermodynamics, 23–24; multiple energy conversion concepts, 24–25; propulsive efficiency, 25; pump efficiency, 50–53, 54, 232; second law of thermodynamics, 22, 23; thermal units and nomenclature, 237; unity, 23, 25
electric generating plants, 133, 182
electric hot water heater, 163, 174
electric motors: AC motors, 55, 233; equations, 233; motor slip, 55; pump motors, 50, 53, 55, 184; ratings, 53; variable frequency drive (VFD), 55, 184
electrical generators: diesel engines, 8, 80, 212–13; gas turbines, 8, 106; nuclear power, 183, 188–94; steam power, 154–55, 156
electrical horsepower (EHP), 53
electrical systems and power, 65
electric-drive vessels, 8
elevation: atmospheric pressure and, 13, 14; Bernoulli's equation, 7; net positive suction head (NPSH), 49–50; units and nomenclature, 237
elevation head, 42
emissivity (radiation), 237
energy: conservation of, 16, 51, 215; multiple energy conversion concepts, 24–25; stored energies, 17; thermodynamic system, 10–11, 18–19; thermodynamics, 10, 16; transitional energies, 17; units, 17
energy forms: flow work, 16–17, 18, 21; internal, 16–17; kinetic, 16–17, 21; potential, 16–17, 21
energy rate, 18. *See also* horsepower
engines: CAER engine processes, 23; cycles, 22–23; equations, 233; first law of thermodynamics, 23–24; irreversibilities and mechanical losses, 24, 25; mechanical efficiency, 25, 233; overall efficiency, 24–25, 92, 233; pumps driven by, 50; reduction gears, 25; thermodynamic processes, 23. *See also* compression ignition (CI) engines; gas turbine engines; internal combustion (IC) engines; spark ignition (SI) engines
Enterprise, 133, 179, 180
enthalpy: air tables, 109; definition and concept, 18; equations, 18, 232; Mollier diagrams/chart, 142–44; psychrometric charts, 202–4; pump, thermodynamic work of, 51; specific enthalpy, 12, 18, 236; *Steam Tables*, 136–37; total enthalpy, 12, 18, 236
entropy: change in, 21–22; definition and concept, 21; Mollier diagrams/chart, 142–44; saturated liquid, 136; specific entropy, 236; *Steam Tables*, 136; total entropy, 236
equations: dimensionally homogeneous, 6; problem solving and, 4; railroad

track analysis, 6, 7, 11; units and unit conversion, 6–8, 11, 231
equilibrium conditions (steady state), 10
Ericsson, John, 106
Ericsson cycle, 106
evaporative cooling (swamp coolers), 205, 212
evaporative cooling towers, 213, 214
evaporator (vapor compression refrigeration), 167, 174, 206, 209, 210
exhaust gases: composition of, 98–99, 100; scavenging, 95, 96; split-shaft gas turbine engines, 115–16; turbocharging, 96
exhaust valves, 86, 87, 91
expansion, thermodynamic process, 22
expeditionary fighting vehicle (EFV), 80
extensive (total) system properties, 11, 12

Fahrenheit scale, 13
feedwater, 100, 144, 148, 154, 222
finned-surface heat exchangers, 167
fissionable elements, 180
Fitch, John, 132
flash-type distilling plants, 223–25
Fletcher, 227
flow work energy, 16–17, 18, 21
fluid (substance): heat convection, 163–64; properties, 11, 12–16; relatively incompressible fluids, 35; working fluid (working substance), 11
fluid flow/incompressible fluid flow: applications of principles, 35; cavitation, 48–50, 147; equations, 232–33; flow rate calculation, 55; laminar flow, 45, 164; practice problems, 60–63, 239; theory development and problem solving methodology, 39–46; turbulent flow, 45, 164, 167; venturi meters, 55; viscosity, 43, 45, 47, 237
fluid power laws, 65
fluid systems, 8
force: derived unit example, 11; fundamental units, 11; Newton's second law of motion, 11–12; units and nomenclature, 11–12, 237
forced induction, 95–97
fouling, 163, 164, 165, 167–68
4-stroke engines, 79, 80, 85, 86–87, 94–95, 97, 98
Freon (R-12), 199, 200, 201, 207
freshwater, 222
friction factors, 43–45, 237
fuel economy, 25
fuel injectors, 80, 92
fuel oil, 25, 79, 132, 144. *See also* diesel fuel; kerosene
fuel pumps, 92
fuels: additives, 97, 99; alternative fuels, 79; biofuels, 100; cetane rating, 99; CI engines, 79, 95; costs of, 25; diesel fuel, 25, 79, 80, 95, 98, 99; gasoline, 25, 80, 95, 98, 99; heating values, 24, 91, 100, 149–50, 180; hydrocarbon combustion, 97–100; octane ratings, 99; refining process, 99–100; SI engines, 79, 95; steam engines, 132
Fulton (*Demologos*), 132
Fulton, Robert, 132
fundamental units, 11

gage pressure: absolute pressure and, 13, 14, 15, 16, 27; equations, 14, 15, 232; pressure relationships, 14; units and nomenclature, 14
gages: barometers, 13–14; environmental influences on, 13, 16; manometers, 14–16, 55; measurement with, 13; pressure gages, 16; symbols, 59; units for readings from, 10, 11, 13, 14
gas binding, 48
gas constant: characteristic gas constant, 27; definition and concept, 27; equations, 231; units and nomenclature, 27, 237; universal gas constant, 27
gas generator, 106–8, 115–16, 118
gas turbine engines: advantages, 106; applications, 8, 80, 106, 120–25, 132–33; combustion process, 112, 163; components of, 107, 115, 118–20; compression process, 111–12; constant C methodology, 108–9; diffusers, 118–20, 121; electrical generators, 8, 106; equations, 233; expansion process, 108, 112; historical background, 8, 106; ideal Brayton cycle, 106–8, 114–15; ideal Brayton cycle state point table, 110–11; nozzles, 118–20, 121; practice problems, 126–30, 241; propulsion methods, 8; real Brayton cycle, 111–15; reduction gears, 25; single-shaft engines, 115; split-shaft engines, 115–18, 121; variable C approach, 109
gaseous diffusion enrichment process, 180
gases: pressure related to depth of submergence, 29–30; relatively incompressible, 35. *See also* air; exhaust gases; ideal gas
gasoline, 25, 80, 95, 98, 99
gasoline engines, 79, 80. *See also* Otto cycle
gear pumps, 71
George H. W. Bush, 198
globe temperature, 201
governor (throttle) valve, 144, 147, 153–54, 156
graphite-moderated reactor, 180–81
gravitational acceleration: equations, 231; Newton's second law of motion, 12; units and nomenclature, 11–12, 236
gravitational constant: equations, 231; Newton's second law of motion, 11–12; units and nomenclature, 11–12, 236

head: definition and concept, 42; pump curves, 46–47; pump shutoff head, 46
head (waterbox), 167
head loss: combined head loss, 45; definition and concept, 43; major losses, pipes, 43–45; minor losses and fitting loss coefficient, 43, 45, 237; net positive suction head (NPSH), 49–50; units and nomenclature, 43, 237
health hazards, 74, 99, 201
heat: definition and concept, 10, 18; energy and thermodynamics, 10, 18; energy sign convention and mnemonics, 19, 163; entropy change, 21–22; ideal gas heat capacities, 27; ideal gas processes, 81–85; ideal gas relations equations, 234; mechanical equivalent to heat, 6–7, 17; rejected, 23–24, 144, 163; second law of thermodynamics, 21–22; steady flow energy equation (SFEE), 16–21; supplied (addition), 19, 21–22, 23–25; units and nomenclature, 236
heat engines: concept of, 19; efficiency, 22; energy concept sketches, 19, 22; first law of thermodynamics, 22, 23–24; processes, 23. *See also* internal combustion (IC) engines
heat exchangers: analysis and problem solving methodology, 168–73; applications, 144, 154, 155, 163, 166–67, 168, 174; baffles, 167, 168; bio-fouling/biological growth, 163, 164, 165, 167–68; classifications, 166–67; corrosion, 163, 164, 165, 167; counter-flow, 167, 169; cross-flow, 167, 169; direct contact (mixing), 162, 166, 168; flow arrangements, 167, 169; indirect contact (surface), 163, 164, 166–69; logarithmic mean temp difference (LMTD), 166, 234, 237; maintenance of, 163, 165, 166; nuclear power, 181, 182–85; parallel-flow, 167, 169; performance of, 163, 166; practice problems, 175–77, 244; seawater-cooled, 162, 167–68; shell and tube heat exchanger, 165–66, 167–68; sludge, 163, 164, 165; solids, heat conduction through, 163, 164; steam plant components, 144, 154, 155, 171–73, 174; surface geometry, 167; tube material, 164
heat flux, 164, 169
heat recovery distillation, 222, 225–26, 228
heat sink, 19, 23, 108, 163
heat stress, 201–2
heat transfer: boundary layer film, 164; coefficient for naval applications, 164–65; equations, 234; HX fluids, rate between, 234; indirect, 163; mixed flow heat balance equation, 234; modes, 163–66; overall heat transfer coefficient for combined conduction and convection equation, 164–65, 234; processes, 163; rate, 19, 164; single fluid in HX equation, 234; solids, heat conduction through, 163, 164; temperature difference and, 164–66; units and nomenclature, 237
heating, ventilation, air conditioning & refrigeration: comfort and human comfort zone, 199, 205; heat exchangers, 174; heat transfer process,

163; historical overview, 199–200; NBC warfare environment and, 199; pneumatic system controls, 66, 73; practice problems, 217–20, 244–45; processes, 205; steam heat, 199; submarines and atmosphere control equipment, 199, 212–13; symbols, 59; technology, 205–16. *See also* air conditioning; refrigerants; refrigeration
heating values, 24, 91, 100, 149–50, 180
heavy water reactor, 180–81
high mobility multi-purpose wheeled vehicle (HMMWV), 65–66, 80
high temperature gas-cooled reactors, 182
higher heating value, 100, 147–50, 237
horsepower: brake horsepower, 25, 52, 53, 92, 116, 233, 237; definition and concept, 18; delivered horsepower, 25; effective horsepower, 25; electrical horsepower, 53; indicated horsepower, 25, 91–92, 116, 237; internal horsepower, 237; shaft horsepower, 25, 78, 131, 133, 180, 237; torque, speed, and, 26; units, 18; water horsepower, 52, 233, 237
hot water heaters, 163, 174
HowStuffWorks, 79, 92
human comfort zone, 199, 205
humid air and humidity, 201–2; dry bulb thermometer and temperature, 201; humidity ratio, 202; psychrometric charts, 169, 174, 202–5; relative humidity, 202; wet bulb thermometer and temperature, 201
Huygens, Christian, 79–80
hydraulic jack, 67
hydraulic press, 66–68
hydraulic systems: accumulators, 72; actuators, 67, 68–69, 72–73; advantages and disadvantages, 65, 74; applications and examples, 65–66, 72–73; definition and concept, 64–65; practice problems, 75–77, 240; pumps, 53, 70–71, 72–73; theory development and problem solving methodology, 66–69; valves, 69–70, 72–73
hydrocarbon combustion: heating values, 100; isomers, 98; processes, 97–100; single bond hydrocarbon chains, 97
hydrochlorofluorocarbon (HCFC), 200
hydrofluorocarbon (HFC), 200
hydrogen, 212
hydrostatic pressure: equation, 232; ideal gas laws and, 29–30; seawater cooling loop intake, 183; submarines, 167
hydrostatics: definition and concept, 15; manometers, 14–16, 55

ideal Brayton cycle, 106–8, 110–11, 114–15
ideal gas: heat capacities of, 27; heat relations equations, 235; specific heat capacity, 19; thermodynamic relations equations, 234; work relations equations, 235
ideal gas constant equation, 232

ideal gas law: absolute pressure and, 27; absolute temperature and, 27; applicability, 202; constant specific heat relations, 234; discussion of and equations, 27–30, 80, 234; hydrostatic pressure and, 29–30
ideal gas processes, 80–81; direction of the process, 81, 85; isentropic processes, 80, 81, 82, 83–84, 107, 109, 234; isobaric processes, 80, 81, 83, 107–8, 234; isometric/isochoric processes, 80, 81, 82–83, 234; isothermal processes, 80, 81, 82, 83, 85, 234; polytropic processes, 81–85, 234, 237
ideal pump work, 42, 43, 51–52
inches of mercury, 14, 15
inches of water, 15
incompressible fluid flow. *See* fluid flow/incompressible fluid flow
indicated horsepower (IHP), 25, 91–92, 116, 237
indicator card, 88
indirect contact (surface) heat exchangers, 163, 164, 166–69
instruments and equipment, 10
intake valves, 86, 87, 88, 91
intensive (specific) system properties, 11, 12–16
internal combustion (IC) engines: applications, 79, 80; auxiliary loads, 25, 92; categories of, 79; efficiency, 24–25, 92, 233; energy conversion and transmission processes, 24; equations, 233; forced induction, 95–97; 4-stroke engines, 79, 80, 85, 86–87, 94–95, 97, 98; gasoline engines, 79, 80; historical overview, 79–80; lean fuel mixture, 99; number of cylinders, 85–86; orientation of reciprocating action, 79; parasitic loads, 24, 25, 92; practice problems, 101–4, 240–41; pumps driven by, 50; rich fuel mixture, 99; 2-stroke engines, 79, 80, 86–87, 92, 94–95, 96, 97. *See also* compression ignition (CI) engines; diesel engines; gasoline engines; spark ignition (SI) engines
internal energy, 16–17, 232; specific internal energy, 11, 236; total internal energy, 236
internal horsepower, 237
ion exchange, 222
irreversibilities, 21–22, 23, 24, 25
isenthalpic process, 153, 206, 207
isentropic expansion, 108
isentropic (incompressible) pump work, 43, 233
isentropic processes: definition and concept, 22; ideal gas processes, 80, 81, 82, 83–84, 107, 109, 234
isentropic pump head, 42
isentropic pump work, 43
isobaric processes, 22, 80, 81, 83, 107–8
isobars, 14
isobutane, 98
isolated system, 10

isometric/isochoric processes, 22, 80, 81, 82–83
isothermal processes, 22, 80, 81, 82, 83, 85
isotopes, 180
Iwo Jima, 131

jet power, 233
jet thrust, 108, 121–25, 233

Kelvin-Planck Statement, 22
Kelvins (degrees), 13
kerosene, 98
kinematic viscosity, 43, 237
kinetic energy, 16–17, 21
Kyoto Protocol, 200

laminar flow, 45, 164
latent heat and latent heat transfer, 100, 135, 167, 169
lean fuel mixture, 99
length, 11, 236
Lenoir, J. J. Étienne, 80
lever rule/principle, 66
light armored vehicle (LAV-25), 80
light water reactor, 180–81
liquid sodium cooled reactor/liquid metal cooled reactors, 179, 181–82
LM2500 split-shaft gas turbine engine, 106, 115–18, 121
logarithmic mean temp difference (LMTD), 166, 234, 237
Long Beach, 179
long tons, 30
Los Angeles–class vessels, 133, 151, 180
lower heating value, 100, 147–50, 237
low-pressure steam distillation, 222, 223–25, 228
lube oil cooler, 166, 167, 170–71, 174

M1 Abrams tank, 67, 80, 106
major head losses, 43–45
manometers, 14–16, 55
mass: conservation of, 51, 215; extensive system properties, 11; fundamental unit, 11; Newton's second law of motion, 11–12; units and nomenclature, 11–12, 236
mass flow rate: of air, 90–91; brake specific fuel consumption (bsfc), 26, 92; fluid flow, 18, 19, 39–41, 48, 236; heat exchangers and, 166
mechanical efficiency, 25, 233
mechanical energy, 7. *See also* work (mechanical energy)
mechanical equivalent to heat, 6–7
mechanical losses, 25
mechanical power: definition and concept, 18, 19; equation, 18, 232; units and nomenclature, 236
mechanical systems, 65
medium tactical replacement vehicle (MTRV), 66
methane, 97, 98, 100
minor head losses and fitting loss coefficient, 43, 45, 237

mixed flow pump, 46
mixing (direct contact) heat exchangers, 162, 166, 168
moisture content, 134; air, 202
molecular mass, 237
moles and molecular weight, 232, 237
Mollier diagrams/chart, 142–44
Monitor, 106
Montreal Protocol, 200, 207
Moody Friction Factor Chart, 43, 44, 45
motion, Newton's second law of, 11–12
multiple-effects distilling plants, 223

napthas, 98
Nautilus, 179
Navy Knowledge Online (NKO), 8
net positive suction head (NPSH), 48–50, 53, 147
net work, 107, 116
neutrons and neutron moderation, 180–81
New York, 78
Newcomen, Thomas, 132
Newton's laws of motion: second law, 11–12; third law, 66
Nimitz and *Nimitz*-class vessels, 133, 179, 180, 198
nitrogen (N_2), 98–99, 100
nitrogen oxides (NO_x), 99, 100
non-flow energy equation (NFEE), 21, 232
nozzle continuity, 40–41
nozzles and diffusers, 118–20, 121
nuclear, biological, and chemical (NBC) warfare environment, 199
nuclear fission: chain reaction, 180–81; decay heat, 181; mass/matter conversion to energy, 18; neutron moderation, 180–81
nuclear fusion, 18, 181
nuclear power: advantages and disadvantages, 179; applications, 8, 133, 178, 179, 180; cold legs, 183, 184, 185; electrical generators, 183, 188–94; heat exchangers, 181, 182–85; heat transfer process, 163, 182–94; heating values, 180; historical background, 8; hot legs, 182–83, 184, 185; mass/matter conversion to energy, 16; overall efficiency, 25, 179; practice problems, 195–97, 244; pressurizer, 183, 187–88; primary loop/primary coolant, 182–85; primary water, 181; propulsion power plant, 182–94; pumps/main coolant pumps, 182, 183, 184–94; seawater cooling loop, 182, 183; secondary loop, 182, 183, 184–85; secondary water, 181; steam plant (steam generator), 133, 144, 154–56, 167, 174, 181, 182–94; surge line, 183; water, demineralized, 222
nuclear reactor: chain reaction, 180–81; control rods and control rod drive motors, 181; critical, 181; decay heat, 181; designation number system, 179; fuel element assemblies, 180; heat transfer process, 163; reactor core, 180, 182; thermal power, 182
nuclear reactor types: boiling water reactors, 182; breeder reactors, 181; graphite-moderated reactor, 180–81; heavy water reactor, 180–81; high temperature gas-cooled reactors, 182; light water reactor, 180–81; liquid sodium cooled reactor/liquid metal cooled reactors, 179, 181–82; pebble bed reactors, 182; power reactors, 181; pressurized water reactors, 179, 181

oceans and ocean water. *See* seawater
Ohio-class vessels, 133, 180
open system, 10, 21
opposed piston engines, 95, 97
orifice, 206
osmosis and osmotic pressure, 226–27
Otto, Nikolaus A., 80
Otto cycle: development of, 80; diesel engine compared to, 95–97, 98; 4-stroke cycle, 86, 87–88, 95, 98; ideal, air standard Otto analysis constant C method, 88–90; real Otto cycle engines, 90–92; SI engines, 79, 85–92; state point table, 88–89; 2-stroke cycle, 88, 92
oxygen: generation of, 212; hydrocarbon combustion, 98–99, 100

paraffin waxes, 98
parallel pump configuration, 53
parallel-flow heat exchanger, 167, 169
Pascal's Law, 65, 66–68
passenger fuel economy, 25
pebble bed reactors, 182
performance measures, 25–26. *See also* efficiency
Perry-class frigates, 106
pipe fittings: friction factor, 237; minor losses and fitting loss coefficient, 43, 45, 237; symbols, 58
pipes and piping systems: cross-sectional area, 232; friction factor, 43–45, 237; head losses, major, 43–45; identification colors and markings, 56–57; pipe stock, 40; relative roughness, 44, 45; surface roughness, 43, 237; symbols, 58
piping & instrument drawing (P&ID): interpretation of, 39; symbols, 39, 58–59
piston, 85, 86, 90
piston-engined aircraft, 79, 120
pneumatic systems: actuators, 68–69, 73; advantages and disadvantages, 65, 74; applications and examples, 66, 73; compressors, 71, 72, 73; definition and concept, 64–65; practice problems, 240; receivers, 72, 73
pneumatic tools, 66, 71
polytropic processes/polytropism, 81–85, 234, 237
positive displacement pumps, 46, 53, 70

potable water, 35, 222, 227
potential energy, 16–17, 21
power: definition and concept, 18; torque and, 26; total rate basis, 52
pressure: absolute pressure, 13, 14, 15, 16, 27; Bernoulli's equation, 7; definition and concept, 13; equation, 232; intensive system properties, 11; net positive suction head (NPSH), 49–50; pressure relationships, 14; units and nomenclature, 13–14, 15, 236. *See also* atmospheric pressure
pressure head, 42
pressure ratio, 109, 233, 237
pressure relief valves, 69–70, 72–73
pressurized water reactors, 179, 181
pressurizer, 183, 187–88
problem solving, 3–5
process graphs, character of, 4, 6
propeller (axial) pump, 46, 47
propellers: delivered horsepower, 25; gas turbine engines and direction of rotation, 64; shaft seals, 25; shaft support bearings, 25; torque and power, 26
properties: conversion between extensive and intensive, 12; extensive (total) system properties, 11, 12; intensive (specific) system properties, 11, 12–16
propulsion systems: historical background, 8; overall efficiency, 25. *See also* gas turbine engines; internal combustion (IC) engines; nuclear power; steam power/steam engines
propulsion turbine throttle, 156
propulsive efficiency, 25
psychrometer, 201
psychrometric charts, 169, 174, 202–5
psychrometrics, 201–2
pump affinity laws, 47–48, 184, 232
pump curves, 46–47, 53, 54, 70
pump head equation, 43, 232
pump head equation without friction, 42–43
pump motors, 50, 53, 55, 184
pump runout, 46–47
pump shutoff head, 46
pumps and pumping systems: cavitation, 48–50, 147; energy, power, and efficiency, 50–53, 54, 232; hydraulic systems, 53, 70–71, 72–73; installation of, 35; multiple variable displacement pumps, 53; net positive suction head (NPSH), 48–50, 53, 147; nuclear power, 182, 183, 184–94; parallel configuration, 53; portable pumps, 35, 50; positive displacement pumps, 46, 53, 70; pump curves, 46–47, 53, 54, 70; purpose and uses, 35, 46; ratings, 53; series configuration, 53; shutoff head/shutoff conditions, 46, 70–71; steam plant component, 144, 147, 148; variable displacement pumps, 46–48, 53, 70–71
push rods, 87

PWR, 179; quality, 181; steam, 136–37, 140–41, 237

radial engines, 79
radial flow pump, 46, 47
radiation/radiative heat transfer, 163, 164, 202, 234
radiator, motor vehicle, 166, 167
railroad track analysis, 6, 7, 11
ram air induction, 91
ramjet engines, 120, 121
Rankine cycle, 144–45, 150–51, 206
rate, 18
receivers, 72, 73
reciprocating engines: compression ratio equation, 233; definition and concept, 79; mechanical efficiency, 25, 233; orientation of reciprocating action, 79
reduction gears, 25, 26, 132
refrigerant R-12 (Freon), 199, 200, 201, 207
refrigerant R-134a: molecular structure, 200; R-12 replacement, 201; thermodynamic properties graph, 207, 208; thermophysical property data, 201, 206, 247–70
refrigerants: CFC designation, 200; characteristics, 200–201; conversion of equipment for new, 200; development of, 200; environmental concerns and, 8, 200, 201; HFC designation, 200; Kyoto Protocol, 200; molecular structure, 200; Montreal Protocol, 200, 207; R-# numbering system, 200; R-400 series, 200; R-500 series, 200; refrigerant effect, 237
refrigeration: coefficient of performance equation, 234; heat engine and, 19; historical overview, 199–200; uses, 199. *See also* vapor compression refrigeration
refrigeration capacity, 234, 237
refrigeration equipment symbols, 59
rejected heat, 23–24, 144, 163
relative humidity, 202
relative pressure, 109, 233
relief valves, 36, 37, 38, 58, 69–70, 72–73
reverse Brayton air conditioning scheme, 205
reverse osmosis water purification, 221, 222, 226–28
reversibilities, 21–22, 232
Reynolds number, 43, 44, 45, 237
rich fuel mixture, 99
rotational power, 26
rotational speed, 237
rotative engines, 25
rudders, 65, 72–73

safety valves, 36, 37, 38, 58
salinity, 222. *See also* desalination
San Antonio–class vessels, 133
saturated liquid: condensate depression (sub-cooling), 141–42, 147–48, 183; definition and properties, 134, 135; determining state of water, 137–39; enthalpy, 136; entropy, 136
saturated steam properties equation, 233
saturated steam tables, 136–37
saturated vapor (dry saturated steam): definition and properties, 134, 135, 136, 141, 149; determining state of water, 137–39; nuclear power, 154, 181, 183; production of, 144, 149, 154–56
saturation, 133–36
saturation pressure, 48, 133–36, 142, 144, 200–201
saturation temperature (boiling point), 11, 71, 133–36, 142, 144, 200–201
Savannah, 133, 179
Savery, Thomas, 132
scavenging, 95, 96
scavenging pumps, 95–96
schematics, 4, 6
scram, 181
scramjet, 121
seawater, 167, 221–22. *See also* desalination; distillation
seawater-cooled heat exchangers, 162, 167–68
Seawolf and *Seawolf*-class vessels, 133, 179, 180
semi-permeable membrane, 226, 227
sensible heat and sensible heat transfer, 134–35, 136, 166, 167, 169
series pump configuration, 53
shaft horsepower (SHP), 25, 78, 131, 133, 180, 237
shell and tube heat exchanger, 165–66, 167–68
ship propulsion. *See* vessel propulsion
single-acting actuators, 68–69
siphons, 55–56
sludge, 163, 164, 165
slug, 11, 48
sodium cooled reactor, liquid, 179, 181–82
solids, heat conduction through, 163, 164
spark ignition (SI) engines: boost, 96–97; combustion, 87; cycle of, 86, 87; definition and concept, 79; diesel engine compared to, 95–97, 98; displacement, 90; efficiency, 91–92; exhaust valves, 86, 87; forced induction, 95–97; 4-stroke engines, 79, 80, 85, 86–87; fuels, 79, 95; historical overview, 79–80; intake valves, 86, 87; Otto cycle, 79, 85–92; parts of, 85, 86; torque and power, 95, 98; 2-stroke engines, 79, 80, 86–87, 92. *See also* Otto cycle
specific (intensive) system properties, 11, 12–16
specific enthalpy, 12, 18, 236
specific entropy, 236
specific gravity, 12–13, 232, 236
specific heat (isobaric), 237
specific heat (isometric), 237
specific heat capacity, 18–19, 234
specific heat capacity ratio, 27
specific internal energy, 11, 236
specific volume: definition and concept, 11; density and specific volume equation, 12–13, 232; units and nomenclature, 236
specific weight, 12–13, 232, 236
specific work, 18
split-shaft gas turbine engine, 106, 115–18, 121
state point table, 88–89, 110–11
steady flow energy equation (SFEE), 16–21, 41–42, 51, 107, 119, 232
steady state (equilibrium conditions), 10
steam: intensive properties, 137; Mollier diagrams/chart, 142–44; quality, 136–37; state and properties of, 137–42; water states, 133–36
steam dome, 134–36, 142, 144
steam generator (nuclear), 133, 144
steam heat, 199
steam plant components: boiler, 133, 144, 147, 149, 154–56, 174, 222; condenser, 132, 144, 147, 167, 174; downcomers, 154, 155; economizer, 100, 154, 155; heat exchangers, 144, 154, 155, 171–73, 174; nuclear (steam generator), 133, 144, 154–56, 174, 181, 182–94; propulsion turbine throttle, 156; pump, 147, 148; pump (feed pump), 144; reduction gears, 25, 132; steam drum, 154, 155; throttle (governor) valve, 144, 147, 153–54, 156; turbine, 132, 144, 147, 148–49, 155, 156
steam power/steam engines: analysis of, 149–54; applications, 8, 131, 132–33, 154–56; auxiliary steam, 154; boiler efficiency, 149–50, 233; electrical generators, 154–55, 156; feedwater, 100, 144, 148, 154, 222; fuels, 132, 133, 144; heat transfer process, 163; historical overview, 8, 79–80, 132–33; mechanical efficiency, 25, 233; non-ideal steam plant, 147–49; nuclear steam plant, 133, 144, 154–56, 174, 181, 182–94; operation of, 100; practice problems, 157–61, 241–43; pumps driven by steam turbines, 50; Rankine cycle, 144–45, 150–51; real steam plant analysis, 151–53; reciprocating engines, 132; steam cycle analysis, 145–47; thermal efficiency, 147–48, 149
steam quality, 136–37, 140–41, 237
Steam Tables, 136–42, 144, 169, 174
Stefan-Boltzmann constant, 164, 237
stoichiometric combustion, 98–99
stored energies, 17
stroke (movement), 85
sub-cooled liquid (compressed liquid), 134–35, 137, 138, 139, 144, 147, 148
sub-cooling (condensate depression), 141–42, 147–48, 183
submarines: air and ideal gas laws, 28–30; atmosphere control equipment, 199, 212–13; chemical balance of air, 212; HVAC systems, 212–13;

hydrostatic pressure, 167; main ballast tanks, 29–30; nuclear power, 8, 133, 178, 179, 180, 183; reverse osmosis water purification, 228; steam condenser, 167; steam power, 8; vapor compression distillation, 225, 228
submerged-tube distilling plants, 223, 224, 226
suction, 48–50, 53, 147
sulfur oxides (SO_x), 99, 100
supercharging, 91, 95–97
superheat, degrees of, 142
superheated vapor: definition and properties, 134, 135, 136, 141, 149; determining state of water, 138; production of, 144, 149
supplied (addition) heat, 19, 21–22, 23–25
surface (indirect contact) heat exchangers, 163, 164, 166–69
surge line, 183
swamp coolers (evaporative cooling), 205, 212
swept volume, 90
system curve (head), 47
Système International (SI) units, 10, 11

Tarawa-class vessels, 133
tars, 98
temperature: absolute temperature, 27; absolute temperature scales, 13; definition and concept, 13; dry bulb thermometer and temperature, 201; entropy change, 21–22; heat transfer and temperature difference, 164–66; intensive system properties, 11; net positive suction head (NPSH), 49–50; relative temperature scales, 13; second law of thermodynamics, 21–22; units and nomenclature, 13, 236; wet bulb thermometer and temperature, 201
text: learning outcomes, 2–3; prerequisite skills, 3, 6
thermal conductivity: heat exchangers, 163, 164, 167, 168; refrigerants, 201; units and nomenclature, 237; uranium dioxide cladding, 180
thermal efficiency: Brayton cycle, 108; equation, 233; steam plant operation, 147–48, 149
thermal energy, 10, 18, 45. *See also* heat
thermal power: definition and concept, 19; equation, 19, 232; heat flux calculation, 164; nuclear reactors, 182; units and nomenclature, 236
thermodynamic laws: zeroth law, 13; first law, 16, 18, 21, 22, 23–24; second law, 21–22, 23
thermodynamic processes: adiabatic, 22; cycles, 11, 22–23; definition and concept, 11; engine processes, 23; expansion process, 22; irreversibilities, 21–22, 23, 24, 25; isentropic, 22; isobaric, 22; isochoric, 22; isothermal, 22; reversibilities, 21–22, 232

thermodynamic system: closed systems, 10, 21, 22, 80–81, 87–88; cycles, 11, 22–23; energy, 10–11, 18–19; entropy change, 21–22; extensive system properties, 11, 12; inlets, 17, 19; intensive system properties, 11, 12–16; isolated system, 10; open system, 10, 21; outlets, 17, 19; process (*see* thermodynamic processes); working substance (working fluid), 11
thermodynamics: definition and concept, 10; equations, 232; practice problems, 31–33, 239
thermostatic expansion valve (TXV), 206, 207, 209
throttle (governor) valve, 144, 147, 153–54, 156
throttling devices and processes (vapor compression refrigeration), 206, 207, 209
thrust, 108, 121–25, 233
Ticonderoga-class guided missile cruisers, 106
timing gears, 87
torque: definition and concept, 26; diesel engines, 95, 98; power and, 26; reduction gears and, 25; SI engines, 95, 98; units and nomenclature, 26, 237
total energy, 11
total enthalpy, 12, 18, 236
total entropy, 236
total head loss, 232
total heat, 236
total internal energy, 236
total volume, 236
total work, 236
transitional energies, 17
transmission, 25, 26
traps, 59
Truxtun, 179
turbine (gas turbine engine), 106–8, 115–16
turbine (nuclear propulsion plant), 181, 182, 183
turbine (steam plant component), 132, 144, 147, 148–49, 155, 156
turbine efficiency equation, 233
turbocharging, 91, 95–97
turbocharging cycle, 25
turbofan engines, 120, 121, 123–25
turbo-generators, 25, 233
turbojet engines, 118–20, 121–23
turboprop engines, 120, 121
turboshaft engines, 121
turbulent flow, 45, 164, 167
2-stroke engines, 79, 80, 86–87, 92, 94–95, 96, 97

ultra-violet (UV) light sanitization, 222, 227
United States Customary System (USCS) units, 10, 11
units: conversion exercise, 41; conversion of units and conversion factors, 6–8, 11, 231; derived units, 11;

dimensionally homogeneous equations, 6; fundamental units, 11; railroad track analysis, 6, 7, 11; Système International (SI), 10, 11; United States Customary System (USCS), 10, 11
unity (efficiency), 23, 25
universal gas constant, 27, 237
uranium: depleted, 180; enriched, 180; enrichment processes, 180; as fissionable element, 180; U-235 isotope, 180; U-238 isotope, 180
uranium dioxide, 180
uranium hexafluoride gas, 180

vacuum pressure: definition and concept, 14; equation, 14, 232; manometers, 15; units and nomenclature, 14
valves: applications and types, 36–39, 58–59; blowdown, 37; hydraulic systems, 69–70, 72–73; methods of controlling flow, 38; symbols, 38, 58–59
valves (engines): exhaust valves, 86, 87, 91; intake valves, 86, 87, 88, 91
vapor binding, 48, 147
vapor compression distillation, 222, 225, 228
vapor compression refrigeration: coefficient of performance, 206, 234, 237; compressor, 206, 207; condenser, 174, 206, 207; evaporator, 167, 174, 206, 209, 210; expansion device, 206; heat exchangers, 167, 174; historical overview, 199–200; prevalence of use of, 200, 205; refrigeration cycle processes, 206–11; thermodynamic concept, 205–6; thermostatic expansion valve (TXV), 206, 207, 209; throttling devices and processes, 206, 207, 209
vapor dome, 142, 185
vapor pressure curve, 133
variable: definition, 6; units and nomenclature, 236–37; units and unit conversion, 6–8, 11, 231
variable C approach, 109
variable displacement pumps, 70–71. *See also* centrifugal pumps
variable frequency drive (VFD) motors, 55, 184
velocity: Bernoulli's equation, 7, 41–45; net positive suction head (NPSH), 49–50; units and nomenclature, 236
velocity head, 42
ventilation: comfort and human comfort zone, 199, 205; historical overview, 199
venturi meters, 55
vertical-basket distilling plants, 223, 224
vessel propulsion: diesel engines, 8, 132–33; electric drive, 8; gas turbine engines, 8, 106, 115, 132–33; multiple energy conversion concepts, 24–25; nuclear power, 8, 133, 178, 179, 180; steam power, 8, 131, 132–33, 154–56
Virginia and *Virginia*-class vessels, 133, 178, 179, 180

viscosity: dynamic viscosity, 237; fluid flow, 45, 47; kinematic viscosity, 43, 237; refrigerants, 201
volumetric efficiency, 91
volumetric flow rate, 40, 46–47, 53, 236

Wasp-class vessels, 133
water: density, 41; heat exchangers, 165, 166, 174; heat transfer properties, 167; hydrocarbon combustion, 98–99, 100; incompressible nature, 41; moisture content, 134; purification requirements and methods, 222, 227; quality, 134; refrigerant value, 200; salinity, 222; sanitization of, 222, 227; softening, 222; states and properties of, 133–36, 137–39

water horsepower, 52, 233, 237
water systems, 35
water types: brackish water, 222; brine 222–25, 227, 228; feedwater, 100, 144, 148, 154, 222; freshwater, 222; potable water, 35, 222, 227; seawater, 167, 221–22
waterbox (head), 167
Watt, James, 132
weather and atmospheric pressure, 13, 14
weight: moles and molecular weight, 232, 237; Newton's second law of motion, 11–12; specific weight, 12–13, 232, 236; units and nomenclature, 11–12, 237
wet bulb thermometer and temperature, 201
wet vapor, 134, 135, 137–39, 140

Winston Churchill, 1
work (mechanical energy): boundary work, 22, 81, 83, 109; definition and concept, 10, 18; energy and thermodynamics, 10; energy sign convention and mnemonics, 19; equation, 18; ideal gas relations equations, 234; mechanical equivalent to heat, 6–7, 17; net work, 107, 116; specific work, 18; total work, 236; units and nomenclature, 236
working fluid (working substance): definition and concept, 11; extensive system properties, 11, 12; intensive system properties, 11, 12–16
Wright Double Cyclone engine, 79

About the Editor

CDR Matthew A. Carr, USN, entered the submarine service upon graduation from the U.S. Naval Academy in 1981 with a BS in ocean engineering. Following nuclear power training, he served on the submarines USS *Tecumseh* (SSBN-628) and USS *Sunfish* (SSN-649) and on the submarine tender USS *Simon Lake* (AS-33) in a variety of engineering assignments. Shore assignments included service at the Submarine Force–Atlantic, Eastern Atlantic, and Mediterranean Operations Control Centers as a submarine operating authority and battle staff watch officer. A licensed professional engineer with engineering design experience at a large commercial nuclear power plant, with master's degrees in engineering administration and environmental engineering and a PhD in mechanical engineering, Commander Carr is currently a permanent military professor of mechanical engineering at the Naval Academy.

Pierre Haury & Jean-Pierre Lacroux

A Passion For Pens

Translated by Fred Gorstein, M.D.

GREENTREE
PUBLICATIONS

of the text

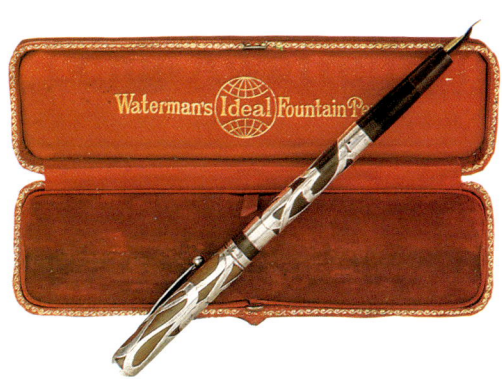

Beautiful example

of a standard Waterman

fountain pen with

an 0.999 fine silver filigree body,

very rare (1905).

Family Portrait

In order to appreciate the contribution of the fountain pen, we must consider its place in the larger family of writing instruments. The following is not a detailed study of each member, but rather a brief overview. Since the dawn of time, when man drew on cave walls with pieces of rock or charred wood, only two basic types of writing tool have been available - that is, until the present-day invention of tape recorders, computers, and the like made possible a new type of communication. It is a long story...one in which the descendants of the piece of charred wood have not yet written their final word.

copy the gospel of St. John. Nibs or reed pens of gold or silver were designed for this purpose. They did not write as well as their organic counterparts, but that did not matter. Utility was not their attraction.

Other inventors worked with copper, brass, steel, silver, and gold in the hope of creating a durable, permanently sharpened nib that would withstand changes in humidity. By the end of the 17th century, considerable progress had been made, but despite triumphant announcements, the results were less than convincing. Yes, the nibs did last longer; however, they had a distressing tendency to tear the paper. When it was humid, unlike the quill that became only temporarily weakened, the metal

The covers of metal nib boxes were offered in a variety of imaginative colors. Unfortunately, the dip pens to which these nibs were affixed did not always exhibit a similar use of color and design.

nib became rusty. This was in addition to the damage caused by the acidity of the ink used. Only gold and silver could be used to create satisfactory instruments, yet even these nibs were inferior to a well sharpened quill. They were well-made, but little used. Their price did not make them a bargain either. One mediocre metal nib was far more expensive than a fistful of high-quality feather quills.

It was not craftsmanship but rather the Industrial Revolution that would replace geese. With the advent of the latter in the 19th century, steam engines proliferated and manufacturing facilities were developed, creating a need for factories and office workers. Society was changing and becoming more urban. Governments required more civil servants. While it was advantageous to keep the peasants and even the proletariat class illiterate, civil servants and office workers needed to write. Schools were created. In addition, post-1815 Europe had been shaken by revolutions. Education was now a right as well as a necessity. An industrialized Europe needed pens in many hands and a democratic Europe wanted them in all hands.

By the end of the 18th century, the first steam engines had appeared in England. Into this cradle of industry the steel nib was born. It

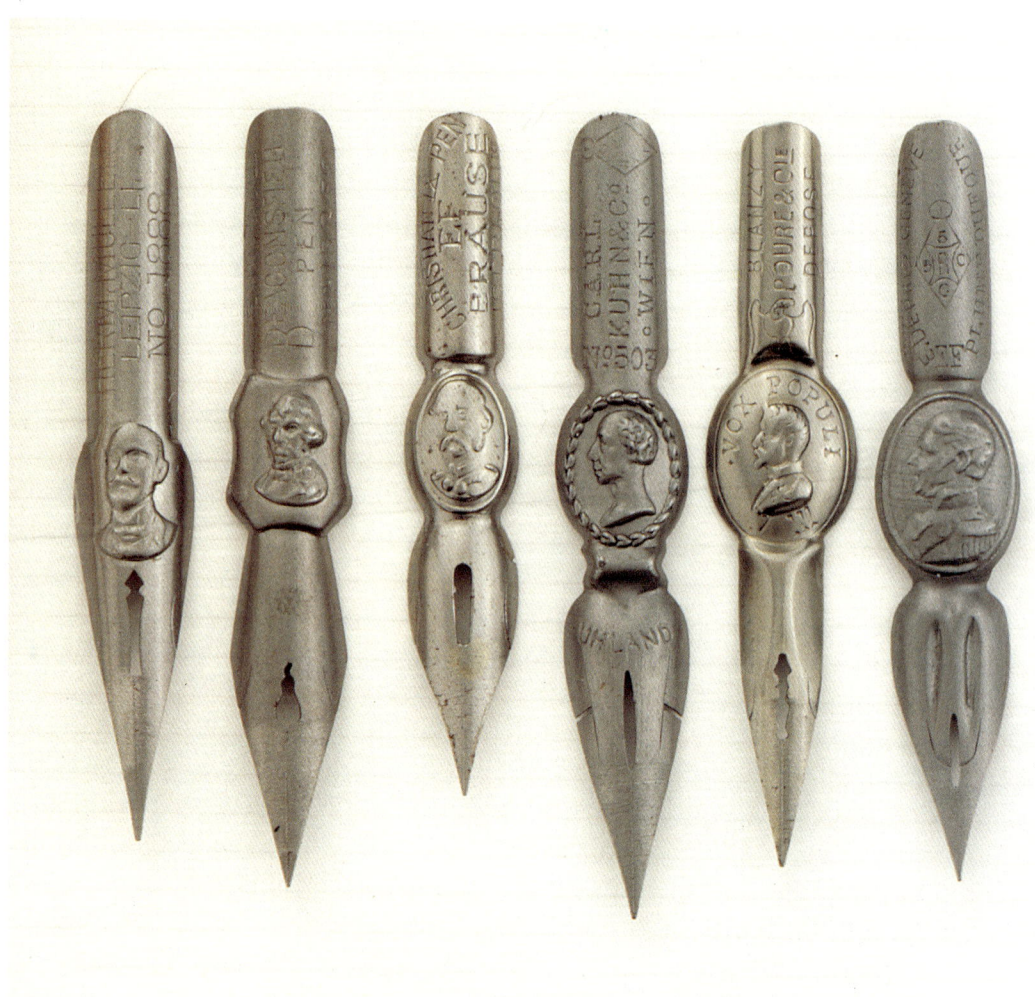

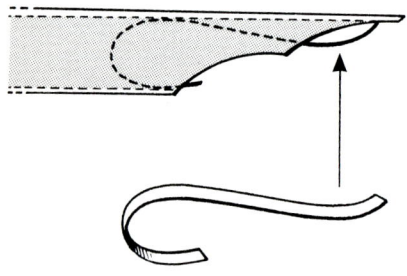

Fine examples of steel nibs with figurine bust engravings.

One patent among thousands, demonstrating a reservoir chamber for a metal nib. This one, designed by C.F. Clarke, dates from 1919 and represents a futile attempt to compete with the increasing popularity of the fountain pen.

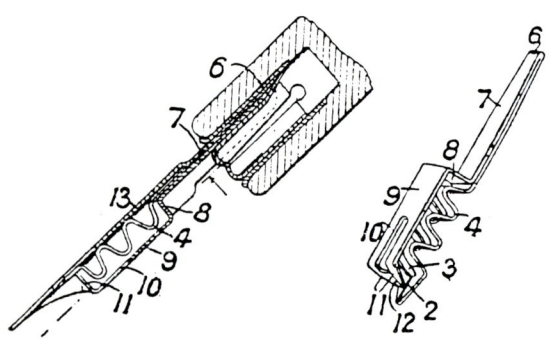

first appeared in Birmingham, England around 1820. Its creators were Joseph Gillott, John and William Mitchell, and Josiah Mason. These men were apprentices to craftsmen who produced, among other things, metal nibs. Gillott worked in the Sheffield workshop of the cutler John Skinner, the Mitchells worked for Spittle, and Mason worked for Samuel Harrison in Birmingham.

James Perry was another type of pioneer - a financier. With Josiah Mason's help, his company became prosperous. Between 1820 and 1840, patent after patent was filed as improvements continued to be made to the nib. Flexibility of the nib was increased by adding slits, a hole, and incisions. Tubular nibs using a large amount of steel were gradually replaced by models with shanks. The metal nib became a quality instrument whose price continued to drop. The feather quill was doomed to extinction. In 1846, the French, with the aid of some specialists from Birmingham, created a flourishing industry at Boulogne-sur-Mer. The Germans followed suit. Thousands of different nibs were produced, some of them marvelous. The steel nib conquered Europe and would become the weapon with which the battle for democracy in writing would be won. For more than a century, it was used to make the civilized art of writing a reality for the majority of people.

Throughout the 19th century, new companies were created, many of which remained in business until the middle of the 20th century before being acquired, more or less successfully. For those who learned to write with a steel nib, the major names will evoke memories of the gentle rustling of pen on graph paper, of the mastering of downstrokes and upstrokes, of the odor of ink, and of blotches. The important names in order of their appearance are: Gillott (1820), John Mitchell (1822), Perry (1824), William Mitchell (1825), Hinks & Wells (1839), Myers (1842), Blanzy-Poure (1846), Heintze & Blankertz (1849), Baignol & Farjon (Sau-

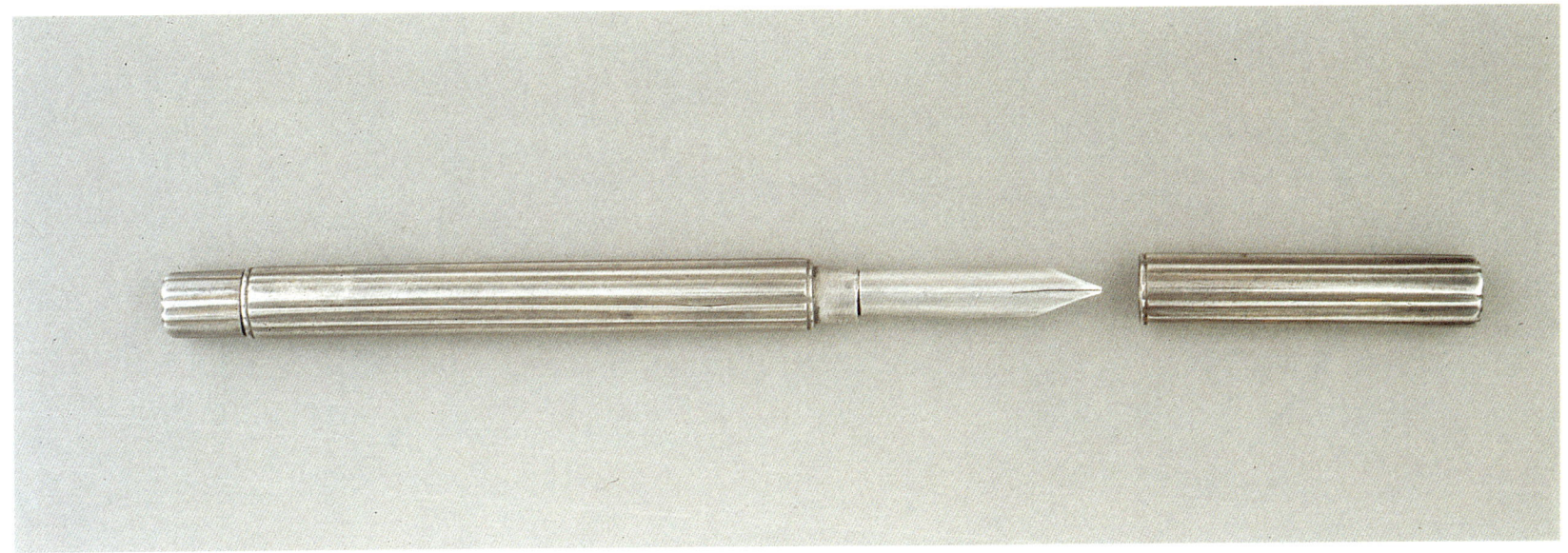

"Eternal pen," French (circa 1780). Silver nib and body. Model is similar to that of Bion.

vage 1850, Lebeau 1856), Leonardt (1856), Brandauer (1862), G.-W. Hughes (1865), Soennecken (1875), and the Compagnie Française (Delpierre 1880).

It is difficult to obtain a clear picture of an industry that has virtually disappeared today, but a few figures might help. At the end of the 19th century, 5,000 people were employed in Birmingham and 1,600 in Boulogne-sur-Mer. Annual production reached 2,800 million nibs: 1,800 million destined for England, 500 million for France, and 500 million for the rest of the world. Between 1830 and 1900, estimated total production was 125,000 million nibs. Business was good. In Birmingham and Boulogne-sur-Mer, everyone was convinced it would go on forever. The future was bright but had crossed the Atlantic with no warning. The steel nib would die, but not until a descendant had been born.

The Pen and The Ink Well

The following does not really belong to the history of the fountain pen, but rather to its prehistory. No fountain pen worthy of the name was manufactured before the 1880s. Nevertheless, the first true fountain pens did not appear out of the blue. A long series of more or less successful attempts paved the way.

While it may be an exaggeration to claim that the Egyptian reed pen filled with dried ink was a real fountain pen, it is evident that the idea of creating a reserve of ink inside the pen is as old as the pen itself. From the reed stem to the feather quill, the pen was always envisioned as an independent tool, not tied to the ink well (that tyrannical watering trough manacling the very instrument that in its greatest days would write of liberty). Its freedom was slow in coming. Nevertheless, numerous well-intentioned inventors had tried to free the pen. Two different goals were pursued throughout the centuries and are described here.

The first goal was to supply the pen with a sufficient reserve of ink so as to permit it to write without interruption and frequent refills. In ancient times, reed pens were filled with ink-soaked fibers and, later, the same method was tried with the shafts of goose feathers. A simpler and more efficient method was to insert a small metal nib under the tip of the quill. The nib retained enough ink to write long sentences. In the 17th century, many artisans of varying ability tackled the problem and customized metal nibs with ink reservoirs. The products of their imagination have, for the most part, disappeared; the only traces left are the descriptions of these "eternal pens" found in specialized journals and newsletters. According to such articles, these "eternal pens" were amazing instruments. Nevertheless, they all

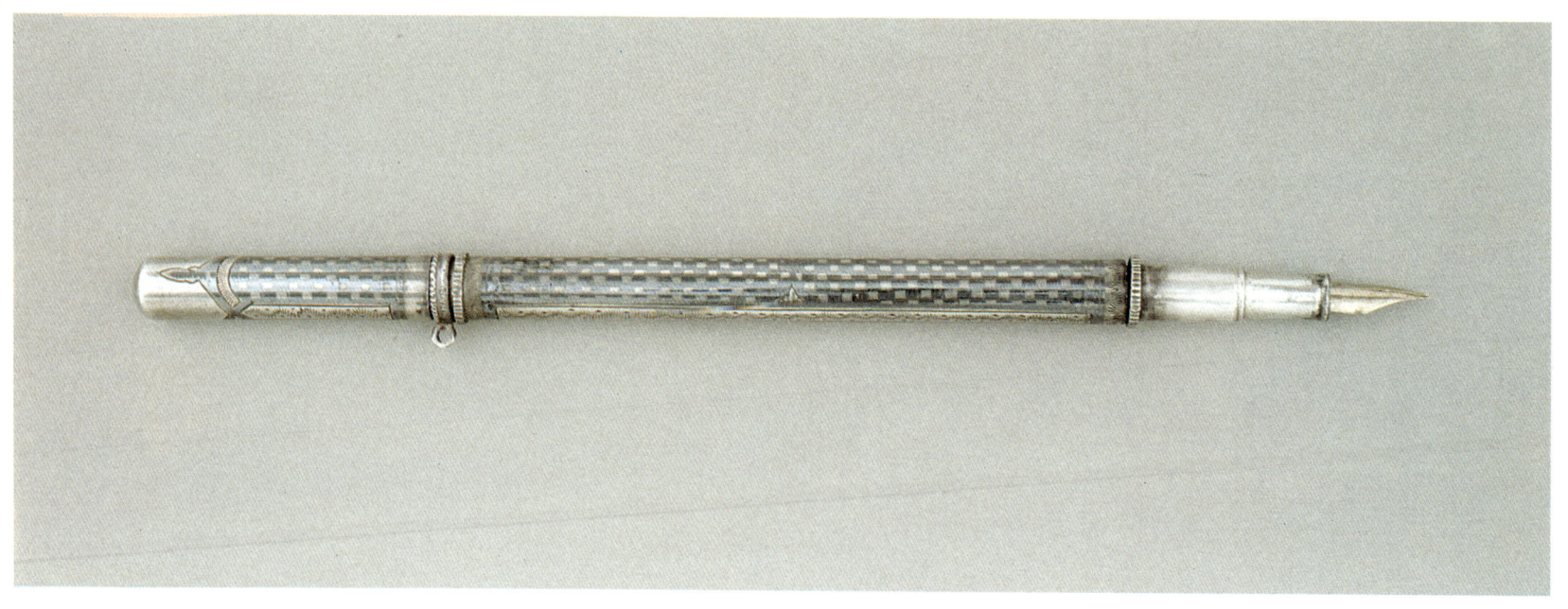

Barrel with a section equipped with a "hand-made" gold nib. Engraved silver body. The mechanism feeds through a small silver duct. The ink is sucked up by the rotation of a ring, which opens or closes an air intake valve (circa 1880).

The Syphoïde
The Syphoïde *was designed by Jean Benoît Mallat (1864). Not quite a real fountain pen, but one of its rare mass-produced ancestors. The pen worked in a conventional manner...on the condition that the instructions were followed very carefully (C.P.).*

Patent for the "eternal pen" of Nicolas Bion (1707).

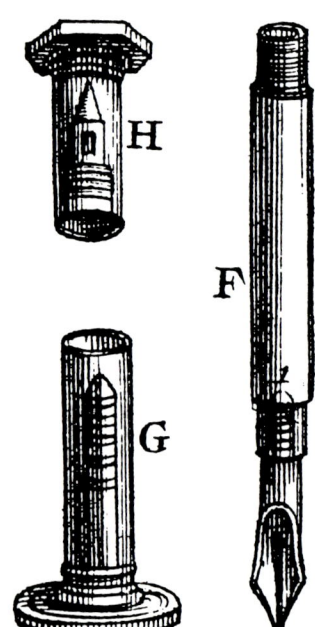

must have had major flaws since they appear to have been completely disdained by quill users. In 1657, in Paris, there were pens with reservoirs that could be used to write more than half a quire of paper, or 12 to 13 pages. Few were sold. In 1707, Nicolas Bion, the "King's Engineer for Mathematical Instruments," developed a perpetual needle for compasses. In England, Germany, and Italy, there was similar activity but no success. Diderot and d'Alembert's *Encyclopédie* mentions "a sort of pen made so that it contains a quantity of ink that flows little by little, allowing one to continue writing without a refill." An annotated note written in the margin of the page reads "poor instrument." Those people really knew what writing meant.

In the 19th century, the quest continued and intensified. Hundreds of patents were filed. Watchmakers, cutlers, mechanics, pharmacists, metalworkers, jewelers, gunsmiths, physicians, and teachers all tried. The instruments that were created began to resemble what we would today call a fountain pen, but only in appearance. Aside from their questionable designs, most had a metal body, often brass, which the ink corroded. This caused the composition of the ink to be altered. The feeds were poorly conceived. Some were as simple as shaking the pen to cause the ink to flow downwards, but nothing prevented the user from being splattered.

Wiser from such an experience, inventors thought to fill the body with a spongy or fibrous material to slow down the flow, but nothing provided a steady flow. Finally, there were patents inspired by plumbing or hydraulic systems with valves and faucets, one faucet for ink and another for air. Among these largely useless ideas, two techniques were actually noteworthy. The first involved applying pressure with one's finger to a diaphragm or rubber sac. The other method involved a piston. While it is true that both of these techniques worked on the same principle as a faucet, they did contain the seeds of what would become excellent systems of filling the pen reservoir.

In 1905, James P. Maginnis made a list of all the patents for pens filed in the civilized world, that is to say, the Anglo-Saxon world. More than 50 pages of the *Journal of the Society of Arts* were filled with this incomplete and one-sided list, which ever since has been considered by many to be a definitive and exhaustive source document on early fountain pens. In 1911, Georges Sénéchal produced the same type of list, citing only those patents filed in France. He did, however, inform the reader of the limitations of his book. Patents were filed in countries by inventors who were not nationals of the country in which the patent was filed. Both Maginnis' and Sénéchal's lists contain some of the most important patents, including

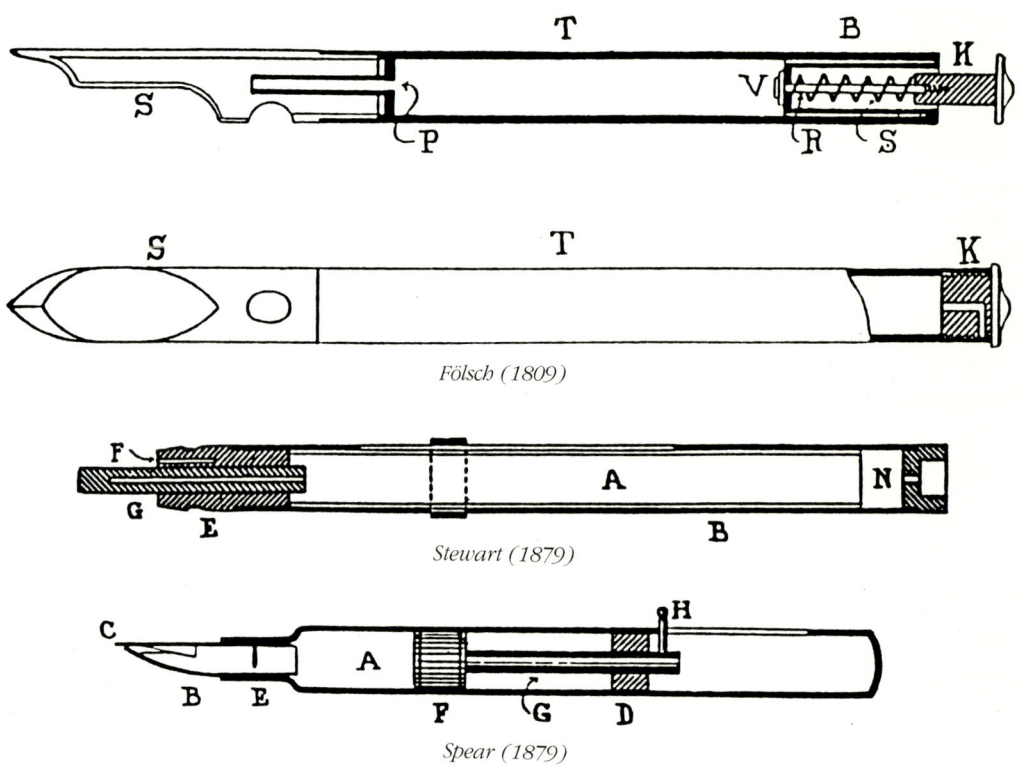

Fölsch (1809)

Stewart (1879)

Spear (1879)

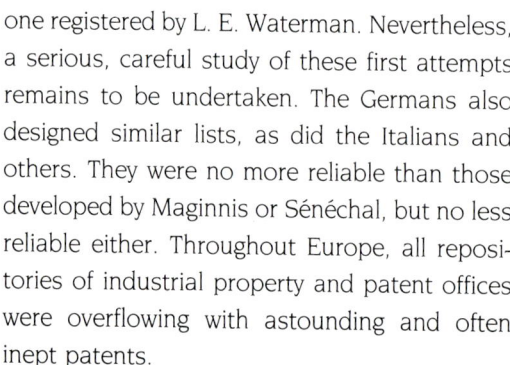

*Patent for the **Syphoïde** dip pen by Mallat (1864).*

one registered by L. E. Waterman. Nevertheless, a serious, careful study of these first attempts remains to be undertaken. The Germans also designed similar lists, as did the Italians and others. They were no more reliable than those developed by Maginnis or Sénéchal, but no less reliable either. Throughout Europe, all repositories of industrial property and patent offices were overflowing with astounding and often inept patents.

Most of these early instruments were merely hollow tubes with a nib at the end. From this basic design, a variety of imaginative methods of filling pens and controlling ink flow were possible, using faucets, pistons, valves, springs, siphons, bulbs, and sponges. It is amazing that even after 1890, some would-be inventors desperately continued to propose the same types of systems that clearly had been proven unworkable decades earlier. For over two centuries, all attempts to add a reasonably sized reservoir to the nib failed. The initial goal was too ambitious. However, ultimately a nib containing a small reservoir did become a reliable adjunct. Many inventors filed hundreds of useless patents, whereas the major manufacturers offered remarkable models with hollows, ridges, tabs, covers, small receptacles, and fixed or removable reservoirs. The best of these enabled an individual to write several pages with a safe, even flow of ink.

The desire to escape incessant trips back and forth between the page and the ink well seems to be the principal reason for the success of the fountain pen. Its importance should not be overstated. "Incessant" is not a completely accurate term. A well-sharpened quill or a good steel nib allowed an individual to write a sentence longer than the grammatical capacity of most people. Only someone who has never used a quill would think it necessary to return to the ink well after writing only five or six words. Besides, as we have seen, there were nibs with a reservoir that held enough ink for even the most verbose writers. Of course, one might consider the need for even infrequent refills to be a hindrance, an annoyance, or an impediment to the written expression of the ideas of a

Historical meeting on a construction site of an insurance broker, a businessman, and a capricious fountain pen.

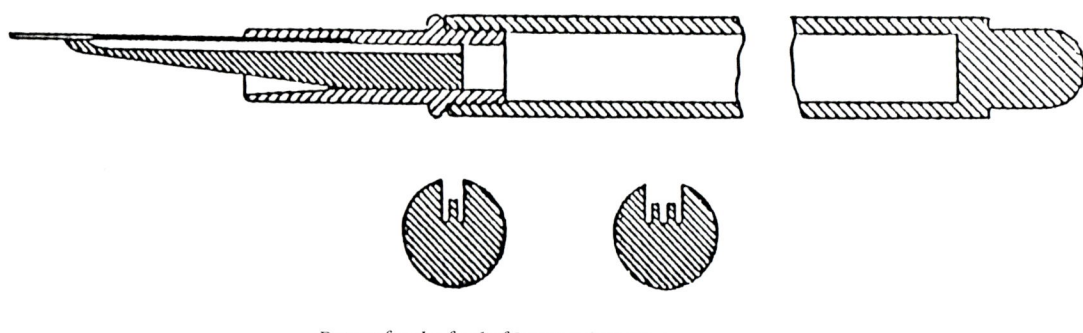

Patent for the feed of Lewis Edson Waterman - an excellent example of efficient simplicity.

Ferris Pump
The Pump Filler *was the first Waterman fountain pen with an automatic filling system. This W.I. Ferris patent, registered in 1897, was granted in 1900. The original pump filler had an overly long pump mechanism and required a lengthy taper at the barrel end to house it. This was clearly not successful for Waterman and apparently was laid aside by the company until 1914 when it was re-introduced in a substantially modified form. However, Waterman was soon to introduce a lever filling system and this second generation pump filler went out of production in approximately 1 year. Starting as an errand boy at L.E. Waterman in 1885, William Irving Ferris became General Secretary of the company 20 years later. In the interim, he invented the Pump Filler, the* Spoon Feed *(1899), and the* Clip-Cap *(1905), as well as improving the production machinery; in brief, ensuring the pre-eminence of L.E. Waterman in the registration of new patents.*

understanding of the complexity of the principle of capillary attraction combined with the effects of atmospheric pressure. If the body of the fountain pen is compared to an ink bottle, then the feed must play the role of a flow regulator. Waterman directed his efforts to that end, and he quickly realized that a conduit allowing simultaneous air and ink passage would be the solution. There still remained the challenges of manufacturing and testing. At this point, he went to see his brother Elijah, a wheelmaker in Kankakee, a small town to the south of Chicago. Legend has it that Elijah Waterman manufactured a fountain pen out of a spoke salvaged from an old wheel. With a saw, a rasp, and a pocket knife as his only tools, Lewis Edson Waterman began carving and shaping small wooden and black hard rubber rods. The first attempts were disappointing, but knowing that he was headed in the right direction, he continued designing and testing his ink conduits. Finally, he succeeded in building a working model - a cylinder with a canal having a square cross section and two thin grooves located at the bottom. It worked, but later Waterman wanted to perfect his invention and added a third groove in the middle of the canal. It was with this system that he registered his patent on February 12, 1884, a historical date because it is without question this event that marks the birth of the fountain pen as we know it today. All feed systems with which pens were subsequently equipped were based on this same principle. Nothing in Lewis Edson Waterman's life could have predicted this success. Born in 1837 in Decatur, Otsego County, New York State, he had only a minimal education. At the age of 16, he moved to Illinois with his parents. During the summer months, he worked as a carpenter. He became, in turn, a publishing agent, a stenography teacher, and finally found his calling in insurance sales. Nothing he had done up to this point demonstrated any scientific gift nor was he a skilled mechanic, but

INSTRUCTIONS POUR L'USAGE
de Watermans "IDEAL" Pomp-filling Pen

Ce porte-plume à **Remplissage automatique** est établi sur un principe nouveau qui n'a jamais été appliqué aux porte-plume à reservoir. Il est le plus simple et le plus ancien de tous les systèmes à remplissage automatique, rien ne peut s'y déranger et il permet de s'assurer que le remplissage est complet.

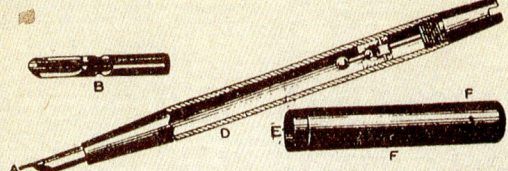

Les instructions suivantes sont à observer pour le maniement de ce porte-plume :

1. — Le porter dans la poche pointe en haut ; le déposer la plume légèrement relevée.

2. — Ouvrir les trous d'air du capuchon, si le porte-plume est destiné à être porté dans la poche ; les boucher à la cire au contraire, si celui-ci doit rester sur un bureau ou pupitre, autrement l'encre sècherait et la plume n'écrirait pas de suite.

3. — Pour enlever le capuchon, lui donner un léger coup de gauche à droite ; pour le remettre un tour de droite à gauche. Si le capuchon était sale à l'intérieur, *l'essuyer soigneusement*, autrement on salirait l'autre bout du porte-plume et ensuite les doigts.

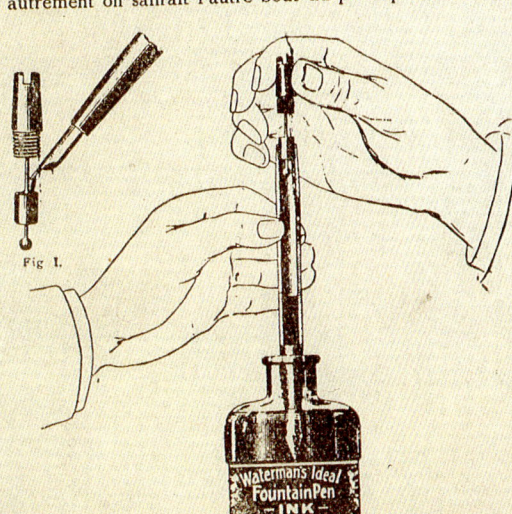

Fig. 1.

Le remplissage se fait automatiquement dans ce porte-plume par un petit système de pompe très ingénieux. Pour y procéder, enlever le capuchon, tremper la plume entière et une petite partie du réservoir (1 à 2 m/m) dans l'encre, dévissez le bout opposé de la plume avec les doigts ou avec une pièce de monnaie et humecter préalablement la petite pompe pour établir l'adhésion si le porte-plume n'avait pas encore servi.

Faire actionner ensuite la petite pompe par un mouvement rapide de va-et-vient dans le réservoir sans toutefois le retirer entièrement de celui-ci. Quand on voit arriver l'encre au-dessus de la petite pompe on arrête le mouvement et on revisse rapidement le tampon sans retirer la plume de l'encre. Seulement lorsque cette partie est vissée à bloc, le porte-plume tiendra l'encre et pourra être retiré. Après un petit essuyage de la plume et de la jonction, la plume est prête à écrire. Le petit tampon ne doit jamais être dévissé sans que la plume se trouve trempée dans la bouteille ou tenue au-dessus d'un récipient quelconque autrement le réservoir se viderait, comme il est facile à comprendre, et un accident pourrait s'en suivre.

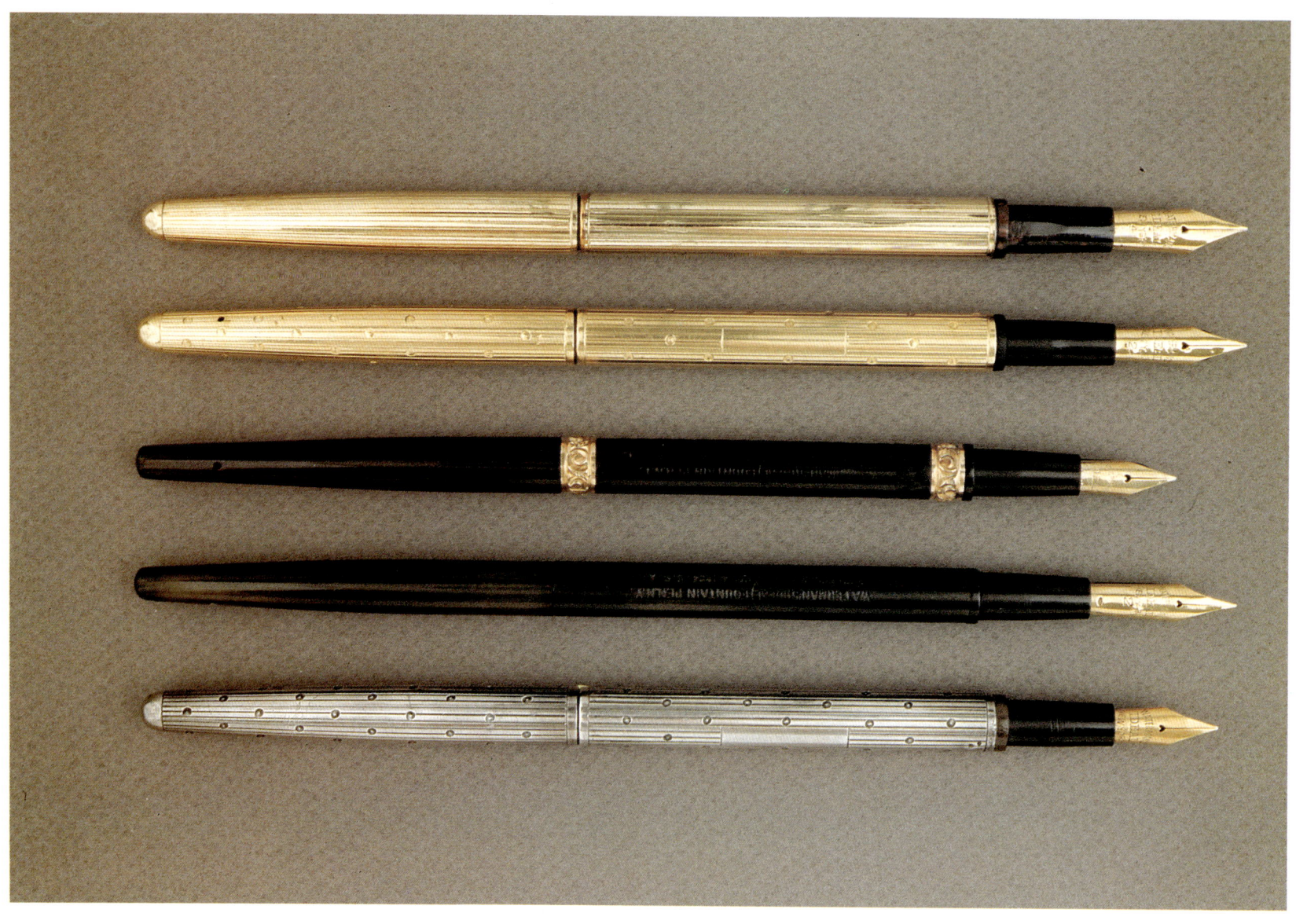

Standard Waterman Taper Cap, French and American bodies (circa 1904).

Banded Pens
Retractable pen by Waterman in red and black hard rubber (1907-1920), with plain bands and French clips, ornate American bands. Very simple yet elegant bodies. The bands were used to personalize the pen and could be engraved with the name or the initials of its owner.

rather a self-taught jack-of-all-trades. It might be said that to have created a few grooves and a canal in a segment of black rubber was not a terribly complicated process. This may be true, but the fact remains that no one had conceived of it before. Hundreds of intellects, some sharper than Waterman's, had for centuries pondered the question, but with no success. Mechanisms both more and less complex had been invented by brilliant scientific minds, but millions of ink blots had also been produced. The most striking thing about Waterman's early pens was their incredible simplicity - the beauty of just five components. There was no complex mechanism, merely a closed tube as a cap, another to hold the ink, a feed, a nib, and a shank to keep the whole thing together.

Lewis Edson Waterman built his first fountain pen for himself in order to have an instrument that would allow him to pursue his career as an insurance broker without mishaps, leaks, ill-timed ink blots, and ruined contracts. The first "Waterman" worked so well, at least so much better than any of its predecessors, that it would have been a shame, even unethical, to deprive the rest of humanity of its benefits. Close friends, colleagues, and clients all wanted to have a Waterman fountain pen. It was difficult to deny them; hence, Waterman's career as a door-to-door insurance agent came to an end and, as a consequence, a new industry was born.

Waterman opened a shop on Fulton Street, located behind a cigar store, with a sign

Top: "Knock on Wood". Hard rubber and wood talismans were given away as advertising gimmicks.

Right: Standard Waterman fountain pens with screw caps, called "securities" (circa 1913). This pen became the P.S.F. (Pocket Self-Filler), Waterman's first lever filler, with the addition of a rubber sac and a lever.

Filigree
From 1900 to 1925, this type of trim was very popular. Most manufacturers offered several models in their catalogs, in gold, gold-plate, or silver. They are beautiful but erroneously named. Filigree is an English term that means assembled with metal wires, which has nothing to do with the cut-out metal designs shown here.

that read: **"Waterman's Ideal Fountain Pen - Guaranteed for 5 Years"**. In his first year of business, he made 200 black hard rubber fountain pens on his kitchen table; in the second year, 500. Although he sold all that he produced, profits were lean. The pen, at least initially, proved to be less profitable than insurance policies. Fortunately for him, his landlady advanced him credit.

The third year was decisive. Advertising came into play through the efforts of a very shrewd salesman named E.T. Howard, who offered Waterman a quarter page advertisement in a well known magazine at a rate of $62.00 per issue...a mere trifle, but too much for the impoverished Lewis Edson. So Howard did something rather unusual, even for those days. He financed the advertisement himself, asking to be paid only if it was successful. Orders flocked in beyond expectations. The first phase was ending but another one was beginning. Waterman had to move to larger premises, he had to find factory facilites that would enable him to meet the demand. As a result, the Waterman Pen Company was founded, and Lewis Edson Waterman managed it until his death in 1901.

Waterman had given the company its foundation and impetus so that it could survive without him and dominate the world of fountain pens for three decades, while at the same time remaining a family business. In 1901, his nephew Frank D. Waterman succeeded him. In the early 1930s, the chief management officers were: President, Frank D. Waterman; Vice-President, L.E. Waterman, Jr.; General Secretary, Fred S. Waterman; Treasurer, Frank D. Waterman, Jr.; Sales Manager, Clyde H. Waterman. Unfortunately, genius is not hereditary, which partially explains the decline of the Waterman Company in the United States. Today, the Waterman name is still inscribed on fountain pens, but on the other side of the Atlantic, in Paris. But that is another story.

Standard Waterman, hard rubber overlay, (from top to bottom) gold-plated, gold-filled, solid gold (circa 1900).

Thin line Waterman Secretary pens in marbled and black hard rubber, American made bodies, silver and gold-filled (1908-1916).

Cone Cap
This style first appeared in 1893, but was called the Cone Cap only from 1898 on. The ends of the body and of the section are truncated, which allows for better handling and an improved fit for the cap. This illustration shows several sizes, from 12 1/2 to 18, and one large size 20.

Retractable nib (safety) Waterman, 42 1/2 v. Among many American and European models (gold, silver, and gold-filled), three silver lacquered French models are shown. The "coquille d'oeuf" (circa 1930) and two models designed for export to French Indochina are also shown.

Standard Waterman pens, black hard rubber, silver (English and French), gold-filled (French). The tip of the cap is embellished with semi-precious stones, which could be engraved and used as seals (1906-1908).

Ladies Pens

Retractable nib Waterman pens (42 1/2 v.) in black hard rubber and red marbled. They were manufactured in the United States from 1908 until 1929 and in France until World War II. French models in gold, silver, and gold-filled. These pens cover two periods. Until 1930, threading for the cap was short and protuberant; in the 1930s, the threads were longer and deeply grooved.

(Preceding page, left)
Large Sizes
Waterman 58 (1915-1929). This is the largest lever model that Waterman made. Black hard rubber, red or red marbled. The clip model (circa 1926) is the most recent of this group. Its cap band is original. The large reinforcing bands on the other two pens were mounted in France.

(Preceding page, right)
P.S.F.
Waterman pens with lever, from 51 to 55 v., black hard rubber and red marbled (1915-1932). Waterman wisely bought the old Barnes' lever filling patent in 1903. With improvements, he produced excellent P.S.F. (Pocket Self-Filler) pens without having to contend with Sheaffer's patent.

George Parker's first patent (December 1889), registering a feed.

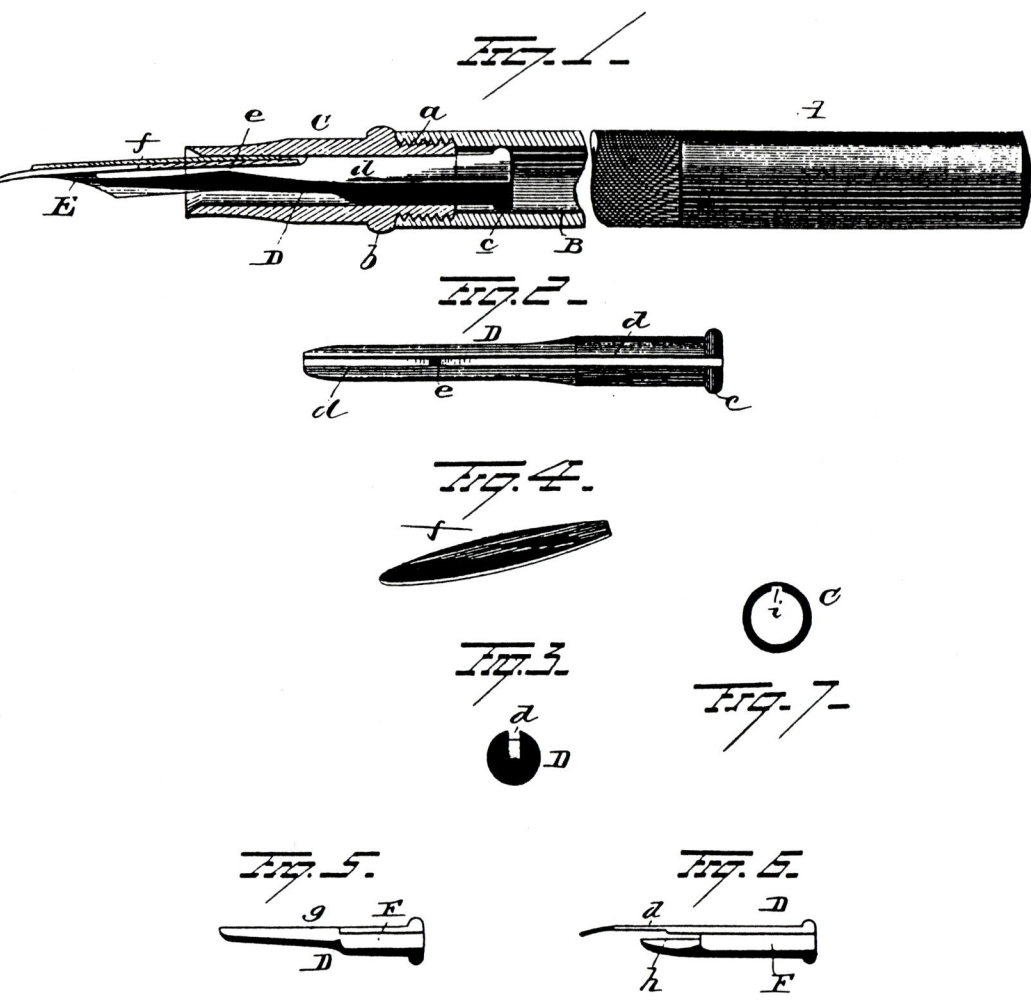

Parker
With the exception of the Duofold *and* Vacumatic, *shown elsewhere, this illustration traces the history of Parker from 1899 to 1942.*
Top, left to right: Nine Lucky Curve *pens in marbled hard rubber (1900), in black hard rubber (1899), in marbled hard rubber (1905), No. 46 mother-of-pearl body with gold-filled cap (1905-1914), No. 45 mother-of-pearl alternating with abalone body (1905-1914), No. 16 gold-filled filigree (1905-1914), black hard rubber (1905-1914) with push-button* Jack Knife Safety *in silver (1918), standard model in transparent bakelite (1917).*
Bottom, left to right: Three Lucky Curve-Jack Knife Safety *push-button pens, black hard rubber (1916-1920),* Parkette *with lever (1933), English* Victory *(1935),* Victory *(1942), Canadian* Televisor *(1935),* Royal Challenger *(1934-1938) toothbrush motif (1936), (P.H.-K.T.).*

George Parker (1863-1937)

In the latter part of the 19th century at the Valentine School in Janesville, Wisconsin, there was a young teacher of telegraphy named George Parker, who, in order to supplement his meager salary, sold John Holland fountain pens to his students. These young men were demanding consumers and they brought him their pens whenever they were in need of repair. The conscientious Parker undertook the servicing of the problematic instruments. With the exception of the contemporary Waterman models, a fountain pen's ink flow was interrupted whenever air failed to enter the barrel chamber. Initially, Parker's competence with repairs was almost nil and his knowledge even sketchier. He learned on the job, by dismantling and then reassembling his students' pens. He soon became discouraged because it seemed that the more pens he sold, the more he had to repair. Nevertheless, this vicious cycle afforded him the opportunity to acquire a thorough understanding of the mechanisms of pens used at that time. Like Waterman, with whose invention he apparently was not familiar, he sought to resolve the problem of air-ink exchange. He bought a few tools, a lathe, a chain saw, a small drill, and tinkered endlessly in his rented room.

Parker created a small feed with an air channel onto which he fitted a nib. The miracle was that it worked! He proceeded to fit John Holland pens with his system. His students were delighted, to the point where he seriously thought about making his own pens. In that way, he could see to it that they would be so well made that they would not be returned immediately to his shop for repair.

In 1889, he obtained his basic patent. He purchased black hard rubber rods and 14 karat nibs, and began manufacturing. To minimize the risk, he retained his teaching position. While

Lucky Curve
George Parker patented this feed on December 4, 1894. Its principle is simple. When the fountain pen is in a vertical position with the nib upward, gravity causes the ink to flow in a downward direction towards the reservoir but capillary attraction causes a little ink to remain in the feed. This small amount of ink is pushed upward if heat creates an increase in air pressure; when the pen is taken out of the pocket and the cap removed, a leak will almost certainly occur. In this system, the curvature of the feed is in close contact with the reservoir, thereby creating additional capillary attraction, and the ink contained in the feed flows back into the reservoir. Parker manufactured the Lucky Curve *for more than 30 years.*

Waterman had started his business in a back alley, Parker started his in a rented room. Without financial support or a method for distribution, his future was, at best, uncertain. Competition was fierce and small manufacturers like Parker were a dime a dozen. He knew he had to find a gimmick. In those days, there were very few hotels in Janesville, Wisconsin, and the one in which Parker was lodged also housed most of the traveling salesmen of the area. Parker distributed some of his pens to a few of them. From that time on, sales accelerated at a rapid pace, but not without significant financial problems. An insurance salesman named W.F. Palmer called on Parker to sell him a policy, but poor George was too broke. So Palmer, who had a keen interest in the Parker's fountain pen, seized the opportunity to offer Parker a partnership. Parker sold Palmer a 50% share in the business (including the patents) for $1,000. The combination of the two men's talents worked well, with Parker running the manufacturing, sales, and advertising and Palmer managing the money and the bookkeeping.

Parker registered a succession of patents during the first few years, including the **Lucky Curve** feed (1894), the **Jointless Pen** (1899), a filling system with a compression bar (1904), and the **Arrowhead** feed (1905). Subsequent new models appeared every 10 years or so, with names that reflect fountain pen history: the **Jack Knife** (1909), the **Duofold** (1921), the **Vacumatic** (1933), and the **51** (1939).

In time, George Parker involved his two sons in the management of the company: Russell, who was the elder, in 1914 and Kenneth in 1919. In this instance, nepotism proved to be a wise move because the two sons played a very active role in the development of the company. Kenneth Parker championed the **Duofold** despite strong objections and was instrumental in fostering research that led to the development of the **Vacumatic**. George Parker travelled the world promoting his pens and opened new branches of the business.

Russell Parker died in 1932 and George Parker, who never recovered from the death of his son, died 5 years later in 1937. Kenneth Parker continued the innovative work and contributed tremendously to the success of the company, but that is another story. In 1986, the Parker Pen Company was sold to English investors, and in 1993, to the Gillette Company.

▸ **Duofold I**
A significant name in the history of the fountain pen, Parker's Duofold *was created in 1921. Until 1925, it was manufactured in black or red hard rubber. Thereafter, Parker produced colored models using permanite (derived from nitrocellulose). The career of the* Duofold *spans two periods. From 1925 to 1929, the ends (cap and push-button blind cap) were corrugated and the section had a ridge. From 1929 to 1932, its ends were smooth and truncated, and the section curved. The* Duofolds *shown on the opposite page are from the first period. Two examples (red) of the second period illustrate the differences.*

Top: *French Parker pens (in display case) and* Plexor *(made by Fernand Laureau).*

Bottom: *Ladies* Duofold *by Parker, green permanite, made in Canada (1930-1933).*

▸ **Duofold**
Duofold *by Parker made in black, red, yellow, orange, burgundy, lapis blue, jade green, and moderne green permanite (1925-1932).*

Fountain and ball-point pen set, Parker 51, with "aerometric" filling mechanism (1949-1978). Moholy-Nagy was a prolific inventor and the Bauhaus influence is evident. The special character of the fountain pen was no longer apparent, due to its lack of an obvious nib.

Saved By the Vacuum
At the beginning of the 1930s, the Great Depression and Sheaffer's Balance *placed Parker in a difficult economic situation. Kenneth Parker decided to attempt a recovery by launching the* Vacumatic *in 1933. The conventional rubber sac was eliminated and filling was accomplished using an air column controlled by a plunger with a diaphragm.*

Conklin pen in black hard rubber with a Crescent Filler *system (circa 1920).*

Sheaffer, The Story of a Little White Dot
Top: *Four* Lifetime *pens with lever, jade green and black radite (1924-1929). Center right: Two pens with gold-filled trim (circa 1920). Center left: Three* Lifetime Balance *pens with lever, black radite (1930-1934), one of which is a fountain pen and mechanical pencil combination. Bottom:* Military Clip *(permitting pocket flap to be closed), brown and gold (1941-1946),* Triumph 1250 *(1942-1945),* Triumph Snorkel *(1952-1954). (P.H.-K.T.).*

Walter Sheaffer (1867-1946)

In the early part of the 20th century, in Fort Madison, Iowa, there lived a jeweler named Walter A. Sheaffer. Life in the midwest at that time was a far cry from the elegant streets of New York or Paris. Sheaffer primarily sold watches and, occasionally, a fountain pen or two to neighboring farmers. He was secure, but not wealthy, and his business flourished. Sheaffer could have spent his life selling his wares, but one fateful evening in 1907, Providence intervened in the form of a Conklin pen advertisement in a local newspaper.

At that time, most fountain pens were filled using an eye-dropper, a delicate, clumsy, and sometimes risky operation. Roy Conklin had partially solved that problem with the invention of his **Crescent Filler**.

Conklin's pens were fitted with a rubber sac that was compressed by a pressure bar fitted to a semi-circular crescent-shaped metal extension that protruded through the barrel of the pen, and broke the graceful lines of the pen. Around the pen was a circular "doughnut", which slipped through the crescent, to avoid any unintentional compression of the sac.

Walter Sheaffer felt that there must be a way to improve Conklin's system; something perhaps a little less simple, but more elegant and practical. He was able to devise a mechanism that improved upon Conklin's system. By the next day, he had come up with a brilliant idea. He planned to replace the metal crescent with a lever, which, on closing, would disappear into the body of the pen. He set to work in the small workshop of his jewelry store and, after a few unsucccessful attempts, designed a workable system. On August 25, 1908, he registered his first patent. Knowing that he was on the right track, he returned to selling watches, while still working to improve his idea. By 1912, he had registered another patent in which the lever was not directly linked to the rubber sac. He had a few samples manufactured and distributed them to his friends for testing. Everyone agreed the automatic filling system was a fine idea.

This was all well and good, but Sheaffer, who was a devoted family man already in his forties, owned a stable business and, at the time, there were already many fountain pen manufacturers in the industry. His small lever was clever, but most of his friends advised him

Eversharp Doric *fountain pen and mechanical pencil set (1935-1941). Nothing doric about it, nevertheless, it is beautiful.*

L.E.C. Half Size

L.E.C. (lower end covered) casings on Waterman half size, with lever (52 v.). All are French, except for the top one, which is Italian.

sold to Remington. Between 1925 and 1940, Wahl closely paralleled the big three pen manufacturers in production. While the company's pens were generally as good and as handsome as the others' (notably the **Personal Point** in 1929 and the **Doric** in 1931), sometimes of even better quality, they were not as inventive.

In 1940, Wahl merged with its subsidiary Eversharp and new management took over. Finally, an extraordinary man, Martin Straus, was hired, who, in 1941, launched a line of fountain pens that is considered by some to be among the ugliest and the best sellers of all time. Called the **Skyline**, it had a squat cap with a line that looked like the nose of a flying fortress. After the introduction of Sheaffer's **Balance** in 1929, aerodynamic shapes were the rage, and, of course, everyone recognized that wind resistance slowed down your handwriting. Thanks to the **Skyline**, Wahl-Eversharp briefly took the lead in the market, but could not hold it for long.

The period immediately following World War II saw the beginning of the ball-point pen hysteria. Straus joined the trend and the company lost millions of dollars. The losses continued until 1957, when Parker acquired the writing instrument division of Eversharp and converted it to the manufacture of low-end products. About a year later, the Wahl-Eversharp name was dropped and, in 1962, the company was sold.

Epilogue

Once upon a time there were three great American inventors and four great fountain pen

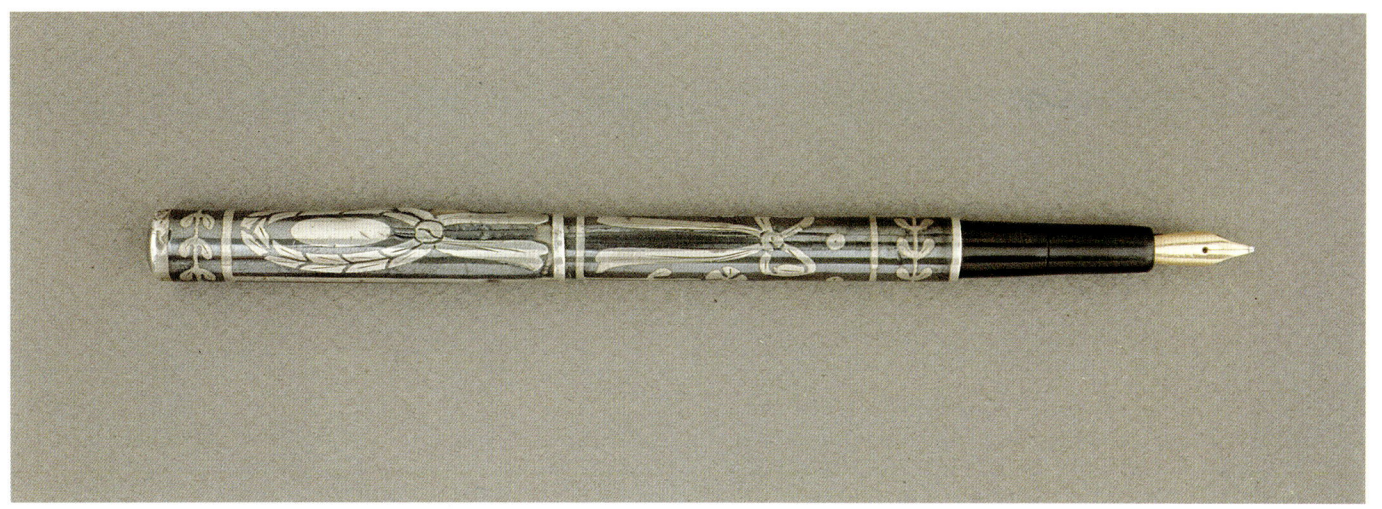

Standard model in enameled silver (circa 1905).

Gemini
Two nibs, two inks within the same pen; is that useful? Well, at any rate, it is possible. The Duocolor, *made by Unic (1932, Zerollo patent), is one of the most complicated fountain pens ever created. Its mechanism retracts one nib when the other is extended. To fill it, the lever corresponding to the extended nib is pressed through a small pin placed in the upper part of the cap. As is the case with most Unic models, the finish is extraordinary. Other manufacturers, such as Omas (circa 1925), were equally stymied by the challenge.*

Mallat and Unic
Top: Mallat: Plexigraf *(1938)*, 150 *(1943)*, 120 *(1943)*, 250 *(1947)*, 300 *(1947)*, Plexigraf 225 *(1947)*, Leda *(1950)*. Bottom: Unic: *(1917)*, *(1928)*, *(1930)*, *(1932)*, *(1932)*, *(1938)*, *(1940)*, Mondial *(1939)*, Luxe *(1946)*, *(P.H-B.J.)*.

manufacturers. The fourth has disappeared, the third is still American, the second is now British, and the pens of the first are now entirely manufactured in France, but owned by an American corporation.

French Production

It is rather difficult to retrace the birth and first steps of the fountain pen industry in France. Whether they were made in France or abroad, many fountain pens were sold under the trade name of the retailer. Consequently, many small manufacturers, such as Sabon, Charcellay, and Laureau, have been virtually forgotten today. Their pens were of a quality comparable to those of the "big" names. Some manufacturers subcontracted for parts that they could not produce themselves. Companies specializing in the cutting and molding of metal parts supplied the levers, compression bars, push buttons, clips, etc... Others were merely pen assemblers, since they could procure the bodies, caps, and feeds from small manufacturers located in or around the environs of Paris or in the Jura area (Eastern France). They could buy the cork for the safeties and rubber sacs from those specializing in the production of such materials. However, only the major manufacturers produced nibs. As a result, it is difficult today to determine whether any given company was a manufacturer of fountain pens or merely assembled the component parts.

Bayard and Edacoto
Top: Bayard: (1920), (1930), Special 8 (1932), Super Luxe (1934), (1935), Excelsior (1938), Special Luxe (1938), (1947), (1949). Bottom: Edacoto: Edac (1920), Edac Cadet (1925), Edacoto 104 (1932), Jewelers series, gold-filled (1934), (1934), 37 Luxe (1937), 200 (1938), Super Edacoto Visible (1939), 33 (1939), 87 (1943). (P.H.-B.J.).

Abbreviated Chronology of the Fountain Pen in France

1864. Jean Benoît Mallat obtained a patent for his **Syphoïde**. The feed and filling (by piston) systems were not in the least revolutionary, but it was the first "instrument" with a reservoir to be mass produced and marketed. It was not quite a fountain pen, but it was the only conventional operative ancestor of the fountain pen to be sold.

From 1886 until the end of World War I, the fountain pen market was dominated by the Americans and the British. There were a few French companies but their production was limited. From 1910 on, the Germans manufactured moderately priced pens, generally in celluloid. Quality fountain pens in black hard rubber were still imported from the United States and Great Britain.

In 1887, in the Didot Bottin Almanac, under the category, **Writing Nibs (natural quills, metal nibs)**, there appeared an advertisement for a fountain pen nib, placed by the Anglo-American Co, the first exclusive licensee of the L.E. Waterman Company.

1889. During the Universal Exposition in Paris, bronze medals were awarded to L.E. Waterman & Co. and to Caw's Nib and Pen Co. At that time, Waterman's world production had reached 30,000.

1898. De La Rue (Great Britain) obtained a distributor in Paris. In the same year, E.L. Moreau started business (in 1904, E.L. Moreau changed its trade name to J.M. Paillard).

1900. L.E. Waterman, Blair, and Caw's

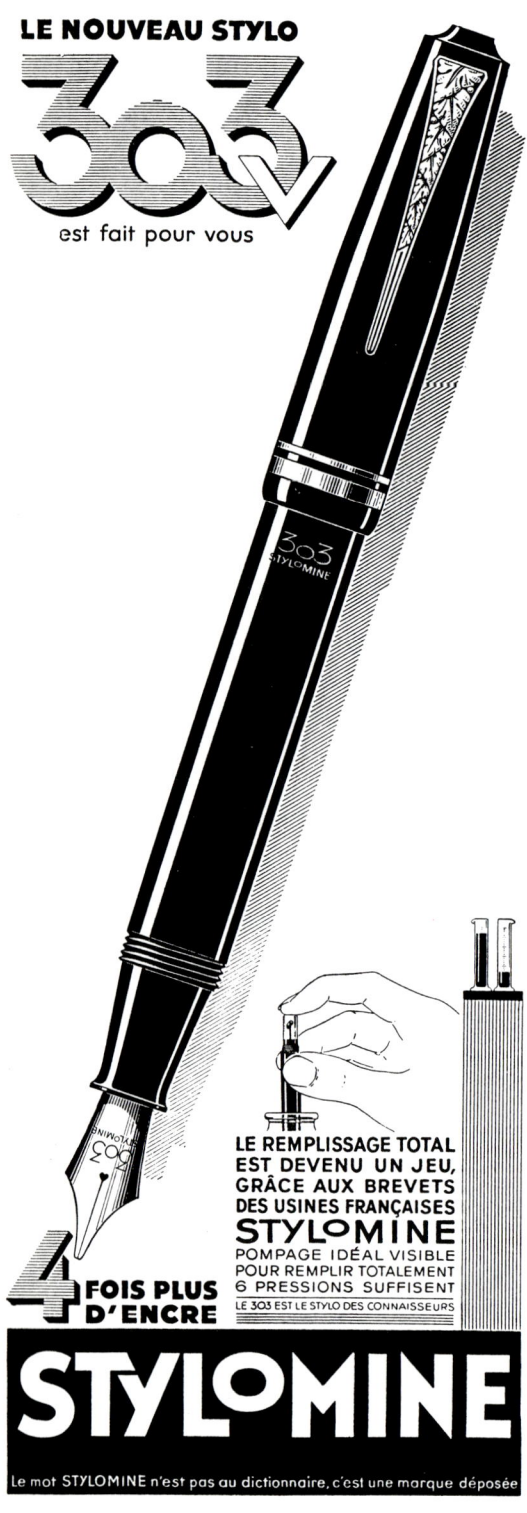

Stylomine
Top: *One of the first 303s, 303B small, 303B medium, 303B large, 303B medium, 303V medium.* Bottom: *Three 303V large and three 303D pens, (P.H.-B.J.).*

were represented in the American pavilion of the Paris Universal Exposition. Waterman was the big winner, with a gold medal for the company, a silver medal for Lewis Edson himself, and a bronze medal for William J. Ferris, the other inventive genius of the company. Caw's received a silver medal.

1904. L. and C. Hardtmuth, agents for Waterman in Europe, opened an office in Paris.

1909. Maurice Jandelle began distributing Conway Stewart (Great Britain) pens under the brand name, Gold Star. Unfortunately, this name was already registered, and the brand name was changed to Gold Starry in 1912.

1910. Charcellay founded Franco-British Pen Manufacturing.

1912. The Forbin company, already a distributor of American brands, started manu-

Triomphe, *completely metal, except for the section and the feed, which are in hard rubber, push-button filling (circa 1920).*

Right: Rool's. *Unusual model with a "flattened" body and cap, transparent section. Clip and compression bar are mounted on a center rod surrounded by a spring. Made by De Soultrait in the late 1940s.*

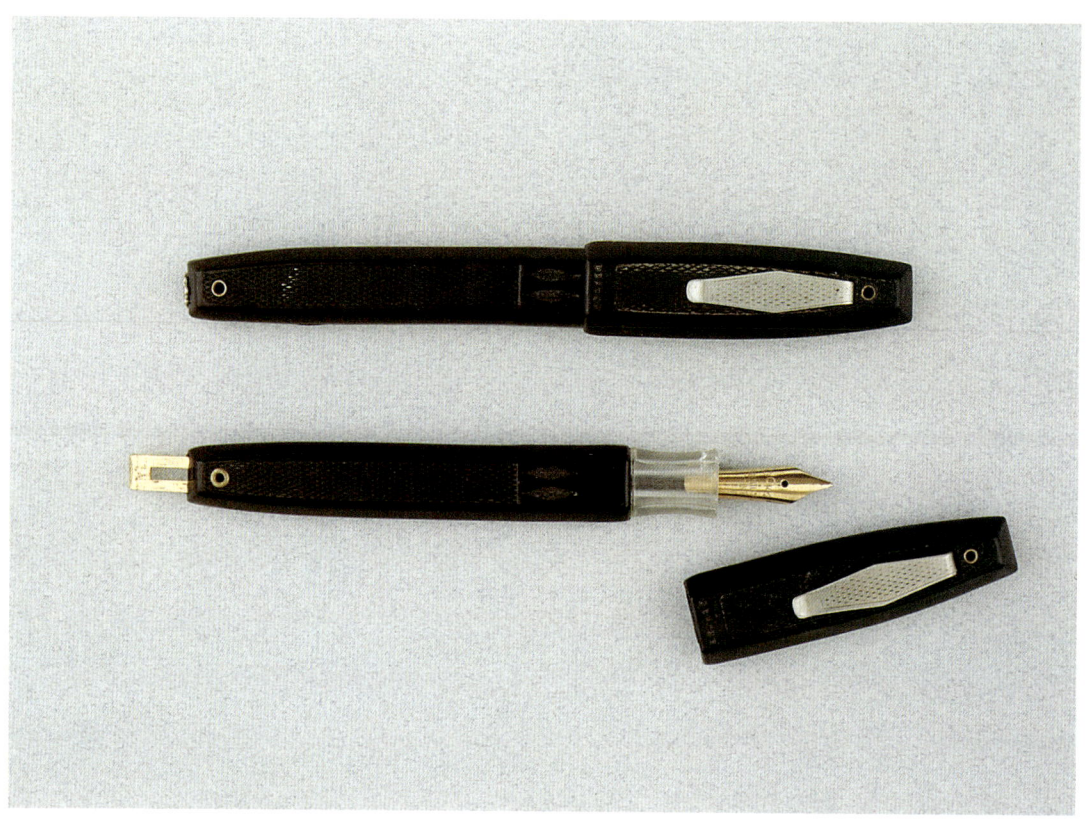

facturing its own pens under the trademark, **Bayard**.

1914. Jules Fagard became Waterman's distributor for Belgium, France, and their African and Asian colonies.

From 1915 to 1922, fountain pen development rapidly evolved. It is probable that American and British troops contributed to the introduction of an instrument that most French soldiers knew little about. After the war, a few veterans, former jewelers, went into the fountain pen industry. In Toulon, Joseph Mercier founded La Palice Company and sold his fountain pens under the trademark, **Grand Aigle**. Joseph Beaufils founded the Stellor company. Demilly and Degen created La Manufacture Parisienne de P.P.R. company, which would become La Plume d'Or and would sell pens under the brand name, Le Météore. The Mallat company reinstituted the manufacture of fountain pens. Founded in 1842, it was the oldest French company making writing instruments (the **Syphoïde**, invented in 1864). Even so, this still was not enough to justify the company's advertisements stating that Mallat was "the oldest P.P.R. in the world" (porte-plume réservoir). Paul Janvrin and André Petit began manufacturing quality fountain pens in a Paris suburb. In 1921, they went into partnership with Maurice Jandelle, who distributed Gold Starry pens that were no longer imported, but manufactured in France. Kothe and Vannier founded the Unic company. Jacques Bonhomme opened a factory in Issy-les-Moulineaux, creating the Edac company. The Panici brothers, nephews of Étienne Joseph Forbin, acquired the Bayard name. Y.E. Zuber took the brand name, Stylomine, for its mechanical pencils, and, in 1925, started making fountain pens.

1926. Jules Fagard founded JiF, a name derived from his initials.

The 1920s and 1930s mark the golden age of the French fountain pen. For the most part, they were made of black hard rubber until 1935;

Pneumatic Pens

Top right (horizontally):
Styloplum (1925) and Plumoto (1929),
silver-plated, with filling system
based on the same principle
as that of the Stylo-Pneu (1932)
and the Babel (1948).

Slogans

If the Gold Starry *is the pen that works*
The Unic *fountain pen works better.*

The Valco *is long-lasting*
The Semper *lasts a lifetime.*

The Stephens *promotes itself as a stylish fountain pen*
But the Matador *is the nib of the elite.*

Matcher Pen is the interpreter of your mind
The Rallye *rallies all the votes.*

Mallat offers quality without compromise
While Météore *puts quality first.*

The Bayard *is without reproach*
And the Stellor, *the perfect pen.*

The Stylomine *holds four times more ink*
The Globe d'Or *six times more ink.*

Brand Names

The trademarks of large or small manufacturers and retailers, and the names of the models are often the fruits of rhetoric, certainly a little old-fashioned, but charming.
The superiority of King, Emperor, *and other* Ruler(s) *is clear, as is the size or inaccessibility of* Everest, Canigou, Himalaya, Mont Blanc, Mont d'Or, Simplo...*on the other hand,* Bambin *and* Lilliput *rival the* Mignon *and the* Miniature *in their diminuitive size.*
Beginning in 1910, the Age of Aviation saw an abundance of Aviator, Aero, Aeroplane, Aeroman's, Aeronautique, *etc. models. During World War I, there was a veritable mobilization of* Glorieux, Revanche, Triomphal, Vainqueur, Victorieux *and, of course, the* Stylo des Alliés *and the* Piou-Piou. *Pens named after animals would fill a good-sized zoo:* Condor, Cygne, Eagle, Furet, Gazelle, Grand Aigle, Griffon, Hirondelle, Merle Blanc, Oiseau Bleu, Paon, Pelican *(English)*, Pelican *(German)*, Roitelet, Scarabée, Seal, Swallow, Swan, Zèbre, *etc.*
In addition, there are the facetious names, such as the KIHE-CRI *and its contemporary* KI-E-CRI, *as well as the* IVA BIEN *and* ELVA BIEN *set. Other models made use of all possible variations of the word "stylo":* Stylobloc, Stylochap, Stylomine, Stylophore, Styloplum, Stylopompe, Stylotube...

From Head To Toe
Waterman lever pens for men with E.C. (end covered) casing covering the part of the body opposite the nib. These models were produced from 1924 until the late 1930s, while their predecessors, without such trim, were still being sold. From top to bottom: gold, silver, gold, silver, gold-filled, and silver.

Pullman, *made by La Plume d'Or (1932). The nib can be extended or retracted using only one hand. When the rear tip of the pen is pushed, a trap opens and the nib appears. Filling is accomplished through a compression bar and push-button.*

they were subsequently made of cellulose acetate and, by 1941, of plexiglas.

World War II considerably changed the picture. An increase in demand, combined with an interruption in supply from the Allied countries, brought about the creation of many small companies. Unfortunately, the poor quality of raw materials, often made with recycled products, resulted in the production of mediocre fountain pens. There even appeared wooden fountain pens (only the feed and the section were made of plastic) and others made from aluminum that was recycled from armament factories.

Even after 1945, it was illegal to use gold in the manufacture of nibs and fountain pens, a restriction that was not lifted until 1949. In the 1950s, pre-war companies had to face dual competition, namely, newcomers in a declining market and the new ball-point pens.

In an effort to combat these unfavorable conditions, some of the oldest French companies joined forces and offered co-production models. In 1956, Stylomine, Météore, Paillard, and Unic launched the **Pulsa Pen**. In 1959, Gold Starry, Edac, Mallat, and Evergood produced the **Visor Pen**, a name that will be remembered by scores of students from the 1960s.

In 1958, Waterman went out of business in the United States. Fortunately, the JiF/Waterman partnership continued operations during rough times and survived the 1960s reasonably well. The year 1971 marks an important date in French fountain pen history. JiF/Waterman became Waterman S.A. and acquired the American name, Waterman from BiC. The following year, it acquired the English Waterman name.

From that time on, all pens bearing the name of Waterman have been and continue to be manufactured in France. Waterman is currently the second largest worldwide producer of quality fountain pens.

NÉ A PARIS
FABRIQUÉ PAR LA GRANDE
MARQUE FRANÇAISE DES
"CRAYONS CONTÉ"
LE STYLOGRAPHE
CONTÉ
EST ADOPTÉ PAR TOUS CEUX
QUI EXIGENT UN PORTEPLUME
SOLIDE ÉLÉGANT ET D'UN
FONCTIONNEMENT PARFAIT

Une arme
pour la vie
Semper
Le Porte-plume
"Semper"
J.M.PAILLARD
PUB. J. GUILHOT
Régulier
Safety
Automatique
de 27 à 125 francs
est toujours prêt
à écrire.
Dans
toutes les
bonnes Maisons
de Papeterie,
Grands Magasins, etc.

Le Météore
90 fr. 80
DÉTAIL : Papeteries, Spécialités, Gds Magasins –
GROS : Sté La Plume d'Or, 63, r. des Archives · PARIS ·
UNIS FRANCE

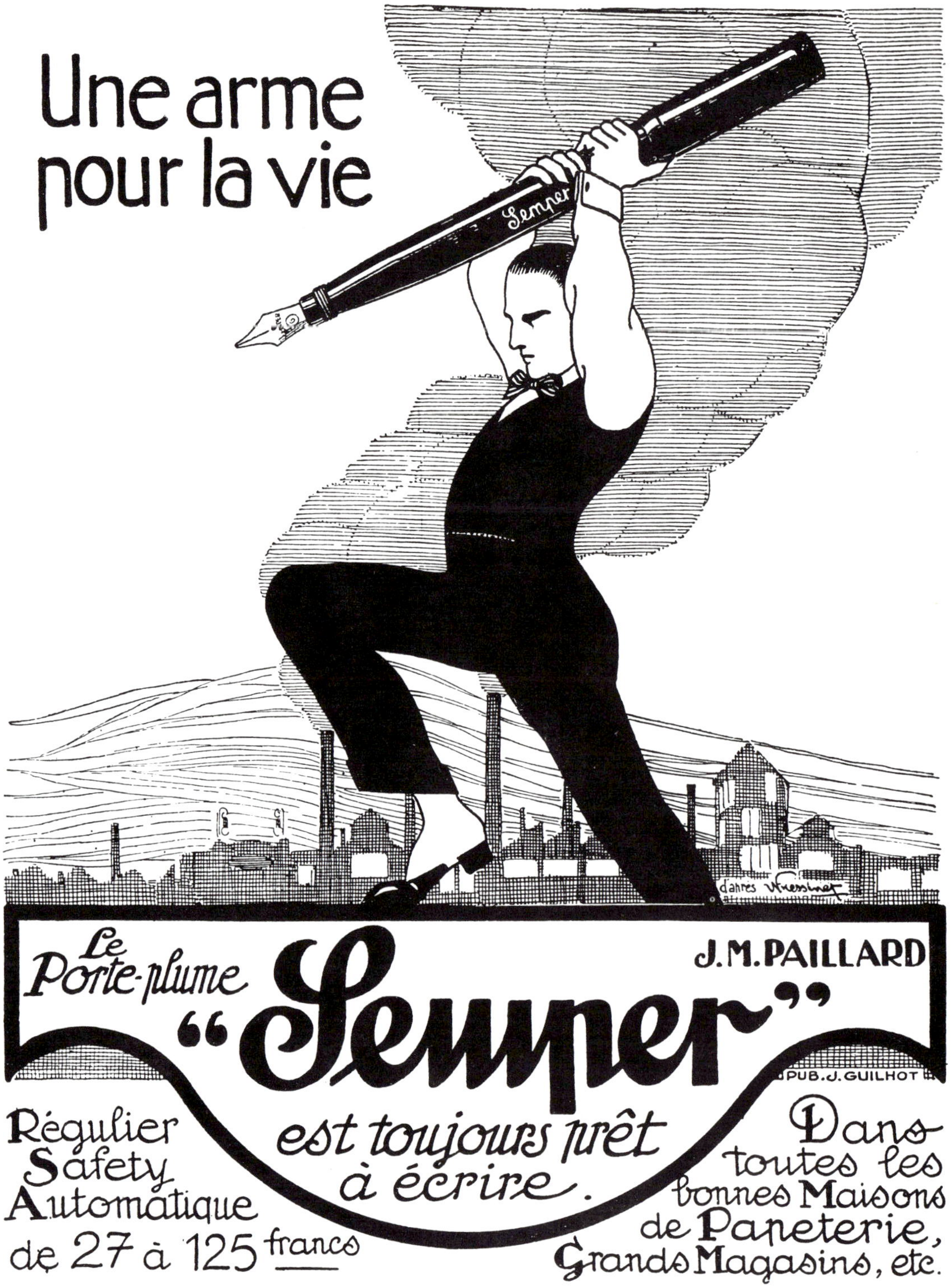

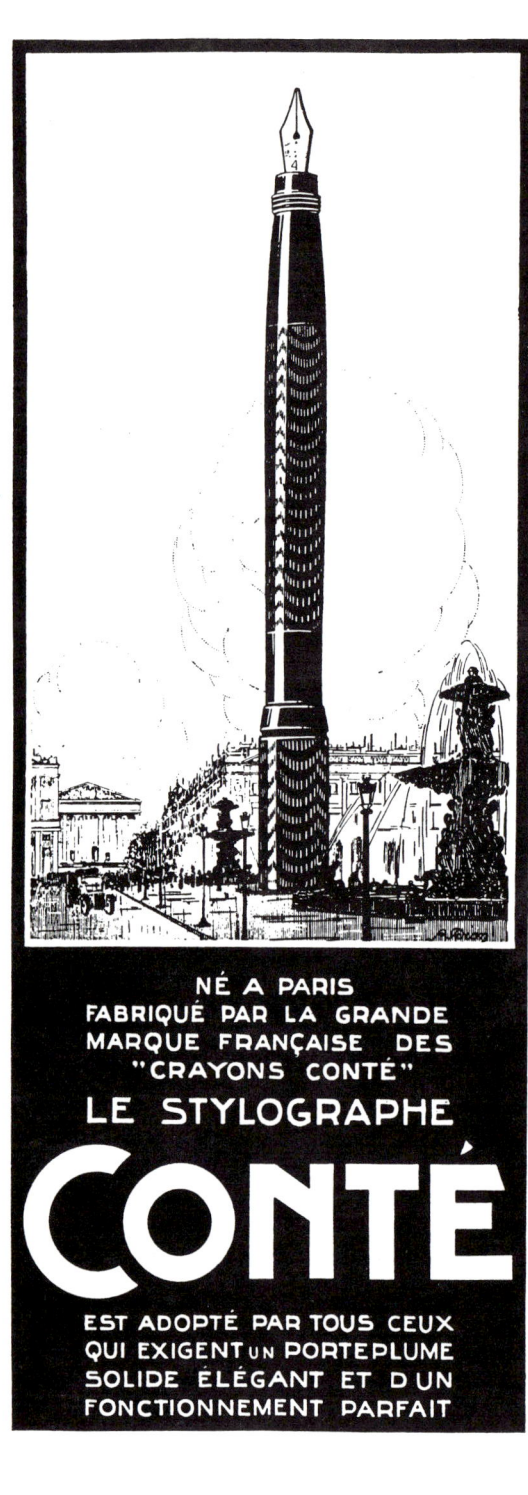

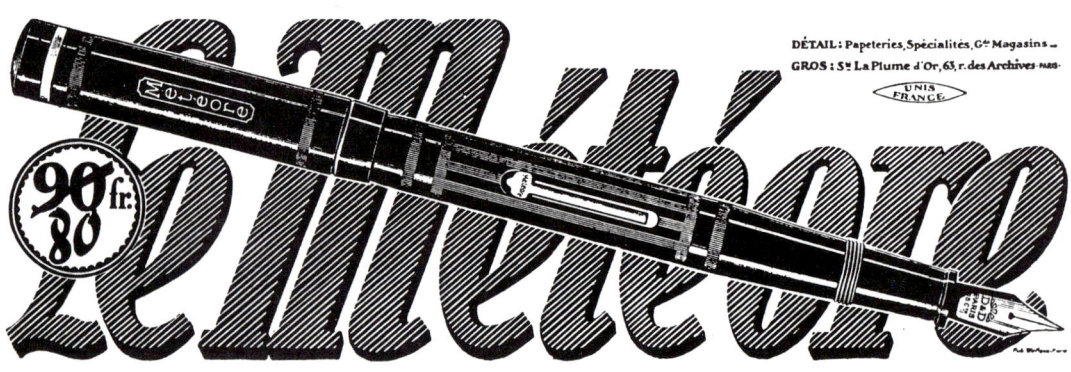

Veterans
For many soldiers in World War I, the fountain pen was a very precious companion, one of the rare civilian objects that they carried into the trenches, a small piece of hard rubber that permitted them to be heard beyond the roar of battle, by a wife, a child, a friend. In those years, the only pleasant words to be heard were those conveyed by the fountain pen. The pen did not win or lose the war, it was only used to carry a little love from a soldier under fire. Once peace was restored, some veterans threw away their fountain pens, others kept them, and today some of these pens enjoy a peaceful retirement with collectors.

L'ONOTO est l'arme moderne dans la lutte des affaires

Aujourd'hui on ne peut plus se passer d'un Porte-Plume Réservoir.

Achetez un ONOTO, c'est le plus parfait.

L'ONOTO ne fuit jamais, se porte dans toutes les positions et se remplit automatiquement, sans compte-gouttes, en 3 secondes.

Demandez l'ONOTO dans sa boîte verte avec la notice.

INDISPENSABILI MI SONO IL FUCILE E LA Waterman's Ideal Fountain Pen

Porte-Plume Ideal Waterman

LE PLUS APPRÉCIÉ SUR LE FRONT

En Vente dans toutes les Bonnes Maisons et chez
KIRBY, BEARD & C° L^d
Catalogue Spécial 201 franco.   5, Rue Auber, Paris.

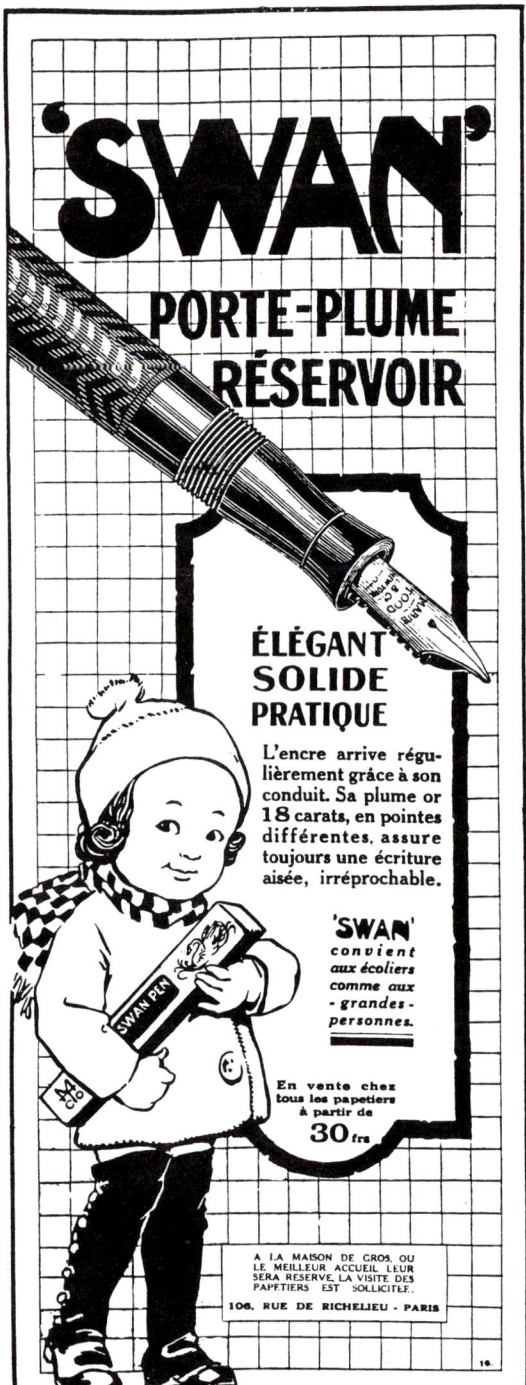
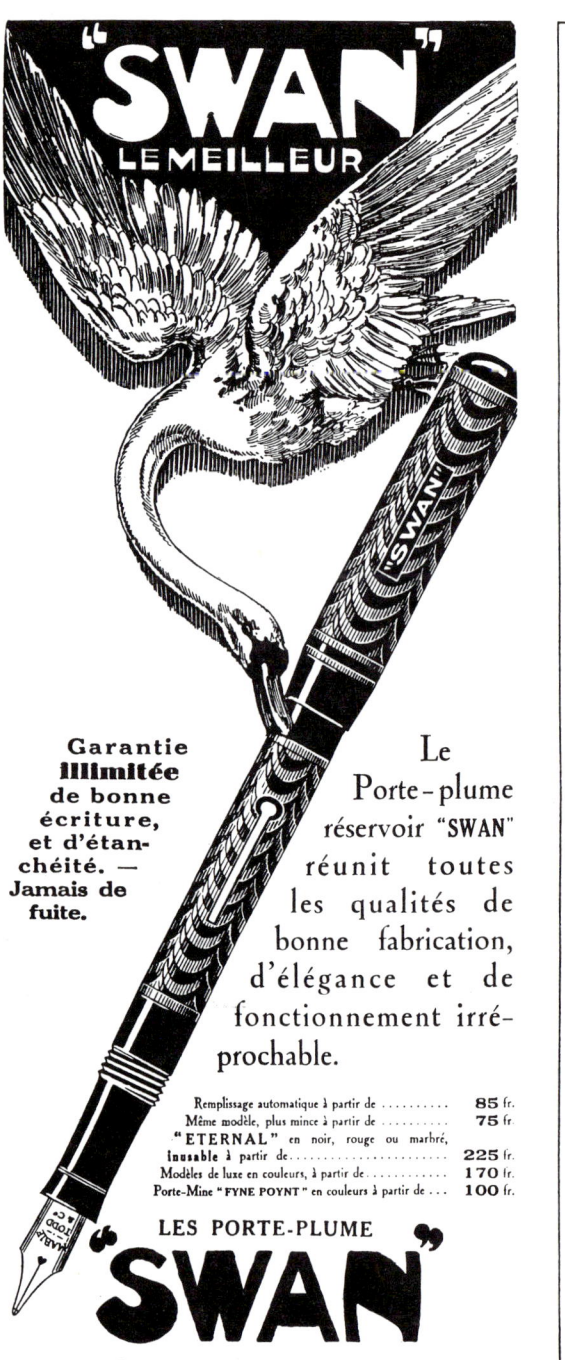

Half a Century of Swans
Swan *fountain pens manufactured by Mabie Todd. The collection on the opposite page groups together American and British made Mabie Todds (1905-1947).*

ONOTO

L'ONOTO a conquis l'univers ! Tous : hommes du monde, officiers, magistrats, savants, hommes d'affaires, etc., etc., se serviront de l'ONOTO, ce merveilleux Porte-Plume-Réservoir qui ne fuit jamais dans la poche, quelle que soit la position où il se trouve, et se remplit automatiquement sans compte-gouttes en trois secondes.

Chez tous les Papetiers.   Gros : DE LA RUE, PARIS.

Onoto
de quoi écrire pour la vie !

De La Rue
With the exception of the lever filling model (1920), all of these fountain pens were made by Onoto. Filling by plunger, dual feed, black hard rubber with precious metal trim (1910-1915). Lower left: Silver (1908). Lower right: Solid gold (1910).

Montblanc
Top, left to right: Pen with lever, black hard rubber *(circa 1920)*; pen with lever, marbled *(circa 1924)*; pen with push-button *(circa 1930)*; Meisterstuck *in black hard rubber, push-button filler (1931)*; Meisterstuck *in celluloid, marbled and black, push-button filler (1931)*; Spider *standard (circa 1920)*, with push-button *(1936)*, with piston *(1936)*; Meisterstuck 149 *(1952)*.
Bottom, left to right: Three Red and Black safeties *(1910)*, Simplo-Montblanc *(1911-1914)*, Montblanc Red and Black *(circa 1920)*, black hard rubber *(1925)*, gold *(1925)*, hexagonal No. 1 in black hard rubber *(1927)*, No. 0 gold-filled *(1925)*, *(P.H.-K.T.)*.

The Telescopic Fountain Pen
For a pen to fit in a ladies purse or in a vest pocket, it had to be very short. However, for it to write well, it had to be long. A clever mechanism took care of the problem. A telescopic device allowed the pen to extend to a sufficient length. It also permitted a mechanical pencil to be hidden at the other end of a fountain pen.

In a fountain pen capillary attraction is influenced by the nib in two ways, as is the case with both the reed pen and the feather quill. The first occurs via the contact between the ink and the medium on which writing takes place, and the second occurs during the feeding process. (Ink viscosity naturally has a considerable influence on both aspects.)

The mastery of ink flow should result in the end of leaks, stains, and worries. However, this is not quite so simple, since as we write the reservoir empties, creating a relative vacuum. The feed must, in addition, permit an air/ink exchange. Not only must the ink output be compensated by equal air intake, but the effects of gravity and atmospheric pressure must also be considered. Atmospheric pressure is not uniform; it varies with altitude, and the higher we climb, the lower the air pressure. However, at any given time and place, it has a specific value. Thus, it is likely that the air contained in the pen's reservoir will be exposed to an air pressure fairly close to that of the prevailing atmospheric pressure. Assuming that the fountain pen is placed horizontally and out of the direct heat of the sun, when we write our body heat will slightly warm the air contained in the reservoir, causing some expansion. If a writer, inspired by the fresh mountain air, decides to create a poem and grabs a half empty fountain pen in his warm hand, heating the air inside the pen, the air expands and creates a pressure much higher than that of the prevailing atmospheric pressure. As a result, a virtual flood of ink will be released all over his budding masterpiece. To prevent this sort of thing from happening, most feeds are designed to contain some overflow; the ink in the slit of the nib should be exhausted before the ink in the reservoir is allowed to flow.

To this end, materials that are poor heat conductors are preferable. Black rubber has this quality, whereas most of the metals used in the fountain pen industry do not. When metal is

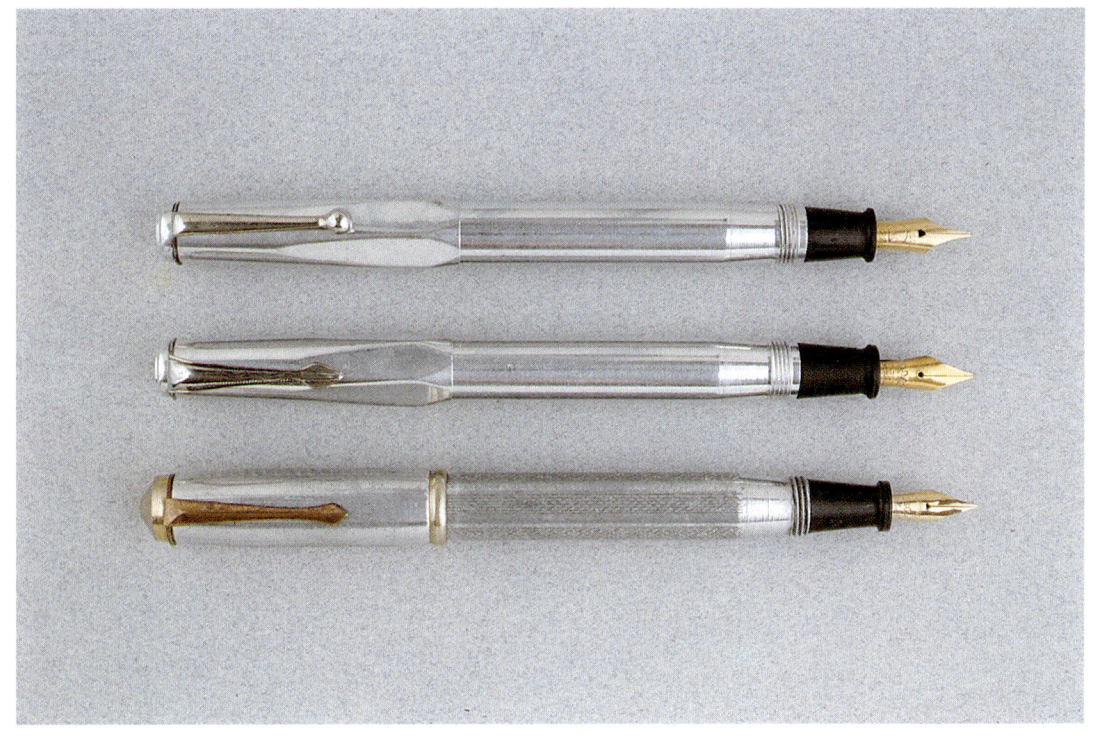

Metal pens made in France during World War II from aluminum scrap metal. The section and feed were made of hard rubber. Push-button filling system.

Bibax
Instrument made by La Plume d'Or that looks like a fountain pen, but is not really one. Launched in 1924, it can be considered the "missing link" or a very late transition between the dip pen and the fountain pen. It is filled using the principle of capillary attraction without any absorbent material. At the end of a hard rubber body, a silver nib with an iridium point is mounted on a feed. That feed, which is an essential part of the Bibax, is a hard rubber cylinder with four parallel grooves through which the ink flows when the nib is placed in an ink well. There is no reservoir, but a reserve that provides a flow of ink adequate to write six to seven pages, which is more than enough for most people.

used as an external ornament, it has little effect. However, a fountain pen with a body made of solid gold or silver would be not only very expensive, but very messy as well. The same can be said of pens with large reservoirs. Even half empty, such a reservoir can hold a very impressive amount of air. One should not confuse well-proportioned pens equipped with well-designed reservoirs and big pens merely having large reservoirs. There is a difference. Cartridges, as we will see later, soon replaced the massive, heavy pens on the market.

Feeds

All feeds operate on the same basic principle as the one invented by Waterman. In most cases, the feed is located under the nib, but some pens were equipped with feeds placed above the nib (first Parkers), and others, which are quite rare (such as a few early **Swans**), were equipped with dual feeds, one on either side of the nib (over-under feeds).

For a long time, feeds were cut and bored from cylindrical black rubber rods. When synthetic resins became available, feeds were made by injection molding. Their convex shape is adapted precisely to the concavity of the nib (for upper feeds, the reverse is true). The external side is beveled under the tip of the nib, in order to avoid any friction with the paper. The other end of the feed is inserted into the section at the base of the nib. Thus wedged, the nib is maintained in close contact with the convex surface of the feed.

In the usual feed, the side beneath the nib has a grooved channel shaped like a trench, one end of which is inserted into the ink reservoir with the other extending to within 1 mm of the tip of the feed. When the feed is in place, its end lies under the nib, half-way between the eye and the tip.

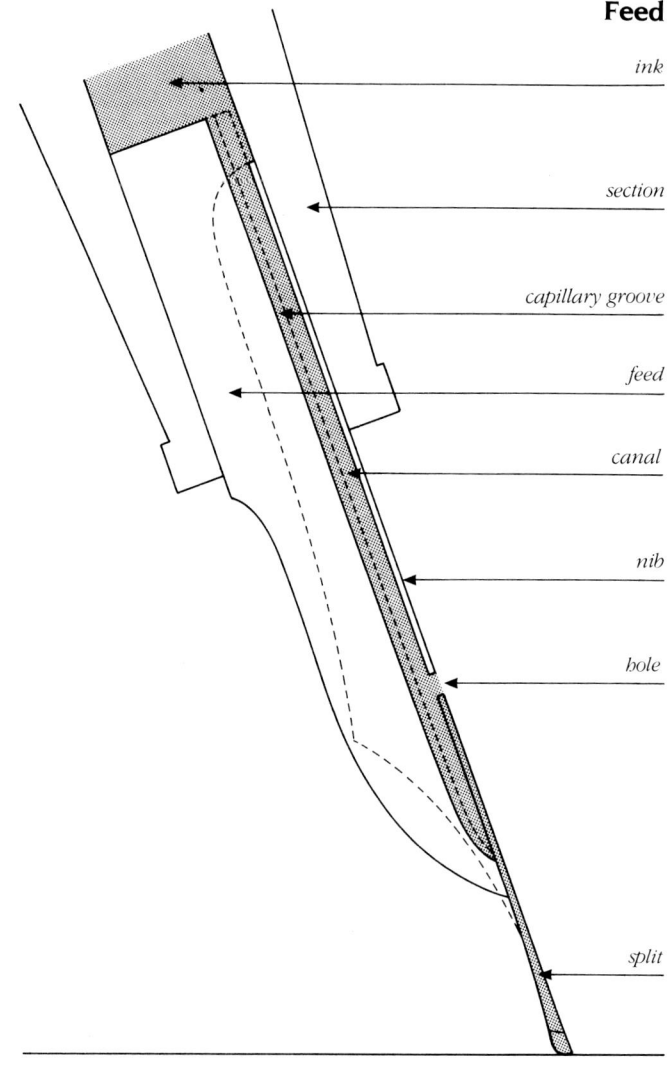

Feed

- ink
- section
- capillary groove
- feed
- canal
- nib
- hole
- split

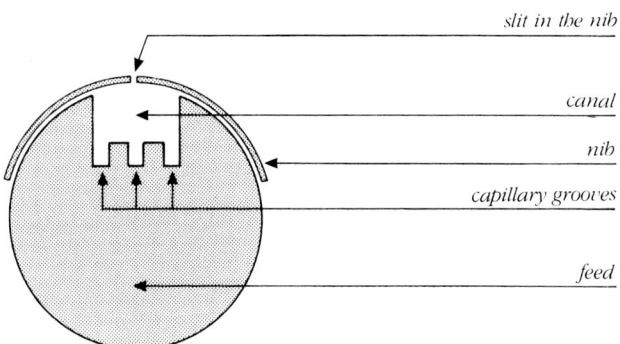

Cross-section of the Feed

- slit in the nib
- canal
- nib
- capillary grooves
- feed

During the process of writing, the flow of ink results in a decrease in the air pressure within the reservoir. To permit pressure equalization, there must be an intake of air, which acts as an air valve. The upper level canal is usually filled with ink, but it is through this channel that air bubbles travel up through the reservoir to equalize the pressure. Meanwhile, the lower grooves remain filled with ink, ensuring flow between the reservoir and the tip of the nib, even while air is passing in the opposite direction. The capillary attraction thus can remain constant.

Designed to collect the surplus ink resulting from variation in air pressure (altitude, body heat) are small pockets on either side of the central grooves, often represented by depressions or a series of fissures. There were none in Waterman's original patent. They first appeared in 1899 (Spoon Feed). The ink thus passes from the barrel reservoir to the tip of the nib, thanks to a series of capillary spaces, which include grooves, contact areas between the nib and the feed where the ink accumulates, lateral pockets, and a slit in the nib. When writing, the pressure applied by the hand causes these areas to fill and the tips of the nib to spread while the nib moves away from the feed and the ink flow increases. To obtain maximum capillary attraction at the tip, the width of the slit decreases from the eye to the point of contact with the paper.

Self-filling Swan *(1913). Slightly more practical than the eyedropper. After the blue cap is removed, the pen is placed inside the opening. The bottle is turned upside down and the rubber half-sphere is squeezed to fill the pen.*

Bottom: Travel ink wells. Some are equipped with an eyedropper (circa 1915).

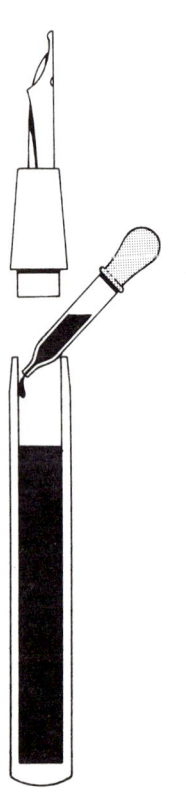
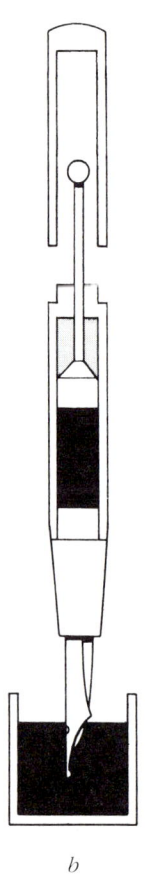
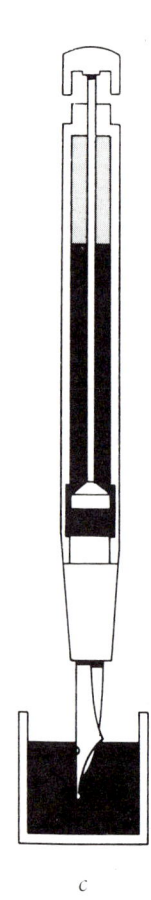
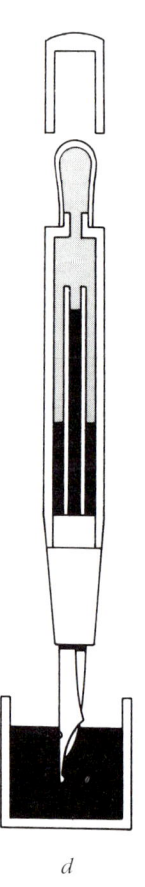
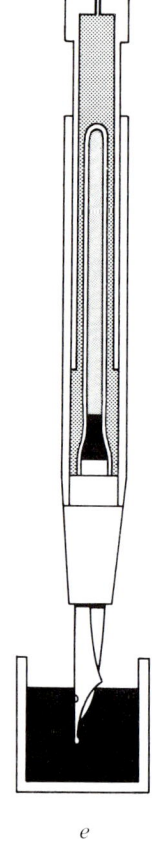

a   *b*   *c*   *d*   *e*

Filling Systems

To ensure the flow of ink to the nib, virtually all manufacturers developed feeds that operated on the principle of capillary attraction. Technical differences focused largely on the filling systems. From the end of the 19th century to the present, there have been two basic types of filling systems: the ink is either contained directly in the body of the fountain pen or in a separate reservoir.

The even flow of the ink is a critical element in the operation of a pen. Once that problem is satisfactorily solved, the filling system affects only the ease of operation. Manufacturers attempted to offer systems that ensured ease of operation, while at the same time they strove to be different from their competitors. Hence, there existed a multiplicity of systems, the major purpose of which was to be different and distinctive.

The filling systems in which the body is the receptacle can be divided into four major categories:

– Direct filling with an eyedropper or similar instrument (Fig. a). Used with the first standard fountain pens and later with the **"safeties"** with retractable nibs, it is the oldest method, the crudest, and the least convenient. It requires considerable dexterity on the part of the user.

– Filling using a simple piston (Fig. b). The piston may be mounted on a smooth shaft, as in a syringe, or on a threaded rod activated by the rotation of a knob. In both cases, air is ejected during the downward movement of the piston; the ink is then sucked up and always remains on the same side of the piston. A smooth shaft has the advantage of simplicity, but it reduces the capacity of the reservoir.

– Filling by plunger (Fig. c). Although there is a piston in this system, this process is very different from the preceding one. In its downward course, the piston expels the air in the cylinder and a decrease in pressure is created behind it. Near the end of its course, it reaches a portion of the body with a wider inner diameter, resulting in an upward ink flow that compensates for the pressure drop, and the reservoir fills without any additional movement of the piston. The reserve ink is behind the piston.

– Filling utilizing a vacuum tube (Fig. d). The process may be initiated using a button or compressible bulb acting on the membrane of a simple rubber tube (Parker **Vacumatic**) or by any other system resulting in the expulsion of a small amount of air to be replaced with a similar volume of ink. The tube is long enough so that its opening remains clear until fully filled. Such tubes are also used in fountain pens fitted with flexible reservoirs (Stylomine).

Systems with reservoirs independent from the body of the pen have an obvious advantage. They permit the use of materials that are not necessarily compatible with ink. These systems can be divided into two major categories:

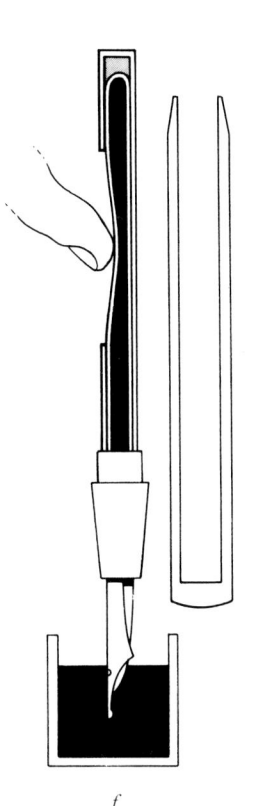 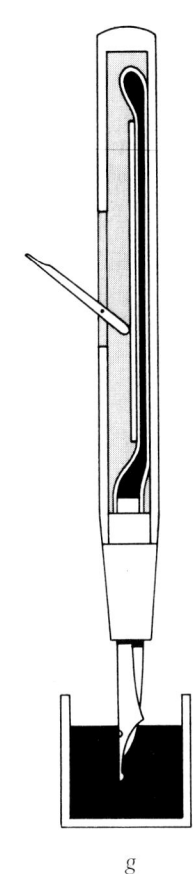 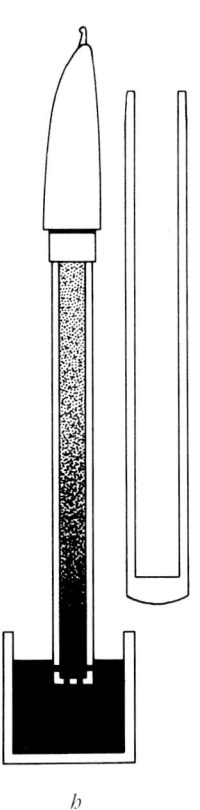 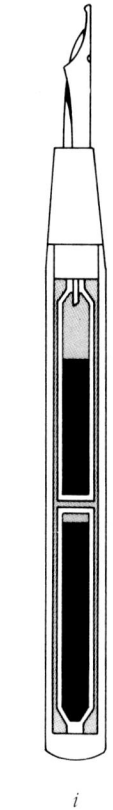

f   g   h   i

built-in reservoirs or removable reservoirs. When it is built-in, the reservoir is filled as a result of the expulsion of the air it holds or by capillary attraction. The built-in reservoirs are generally made of flexible material. Replaced by plastic materials today, the rubber sac was formerly the unchallenged master in that category. Since filling occurs following air expulsion, the difference between the systems rests only in the way the rubber sac is compressed. Four examples illustrate the systems that operate through the compression of a flexible reservoir:

— By torsion of the reservoir. A crude, little used method. (1903, A.A. Waterman's adaptation of Moseley's 1859 patent and Kollisch's 1841 patent).

— By the pressure of an air column compressing the reservoir (Fig. e). At the end of the downward course, when the air is released, the reservoir expands and fills (Chilton, Sheaffer, Pneu).

Safety First:
The Waterman safety pen, "the pen that can be carried in any position," held an important place on the French market for 30 years, from 1908 to 1939. It was produced in the United States until the 1940s, although it was largely replaced in the 1920s by lever filling models with which it had been competing since 1915. It is reputed that the safety pen patent recorded by Waterman was inspired by the mechanism of the mechanical pencil, registered in 1903. The safety appeared on the market in 1907, and interestingly enough, the mechanism of the first models was not the same as that described in the patent. This situation prevailed for 5 years, and it was not until 1912 that its manufacture reverted to the mechanism initially designed and recorded.
In the first design, helical grooves were cut inside the body to propel the nib outside the pen; in the second design, the grooves were replaced by a helical movable piece. In the first design, the nib revolved as it extended, and in the second, it did not revolve. In the first design, models were fitted with a threaded propulsion extension onto which the cap was screwed; in the second design, the propulsion extension was smooth.

After World War II, the success of the retractable nib encouraged many French manufacturers to increase production of these models, further enhancing its success. That success was perhaps due to the fact that the nib was continually immersed in ink and thus always ready to write. The attraction of the hidden mechanism and the elegance of the manipulation required to extend the nib may also have contributed greatly to its lasting success. In 1930, Stylomine, in order to eliminate the inconvenience of eye-dropper fillers, introduced a retractable nib model called the Self-Filler.

JiF fountain pens manufactured during World War II. These models were equipped with a glass cartridge and lever. Since the use of gold was prohibited, the nibs were made of steel.

— By direct pressure of the fingers (Fig. f), either on the sac or on an extension (Stylomine).

— Through a compression bar (Fig. g). There are numerous systems utilizing this principle: a pin, long or short levers, jointed, off-centered, push-buttons, rotating buttons, etc. Some models vary the systems just for the fun of it, but the composition of the fountain pen is very different from that of human beings. It is useless and precludes a good understanding of the subject when one creates variations without good reason.

Capillary attraction, essential to the filling system, was not very successful when variations of the basic system were employed. Major manufacturers, such as Parker and Waterman, tried them. When filled with stacked disks, a mesh fabric or other absorbent material to increase the surface area, the resulting reservoir had a very small capacity. In Fig. h, the filling is provided by a spongy wick. Whether it is the nib that is immersed in the ink (Waterman **X Pen**) or the other end of the reservoir (Parker **61**), filling is a slow operation and takes an average of 20 seconds. Conversely, if the cap is not replaced, the ink dries very quickly. However, because these systems were unaffected by variations in pressure, they were relatively popular among aircraft crews and airline passengers.

Cartridges

The cartridge is the simplest and most practical filling system. It is also the most recently developed system and the one most innovative in design (Fig. i). Due to its simplicity and sturdiness, the cartridge succeeded in place of systems that were cumbersome, fragile, and sometimes incredibly complex. Not re-

Boxes of glass nibs

In 1936, when JiF introduced its cartridge models, which had been adopted from American pens, their shape did not conform to that of a cartridge and required modification so that the body could be opened to allow the insertion of the cartridge. The lever and its opening were eliminated, the feed was adapted so that it served as a plunger, and a rubber ring was added to make the cartridge leakproof. The assembly was all accomplished in France. World War II put an end to production. Models shown are deluxe examples having gold and silver trim (1936-1939).

quiring any mechanism, the cartridge frees the body of the pen and increases its capacity.

The average cartridge contains more ink than an equivalent rubber sac that can never be completely filled. It permits better control of ink flow and eliminates the need for ink wells and bottles (which are missed by some, who have obviously forgotten the rags and blotters required to wipe out stains, spills, and various other mishaps). With the cartridge, ink regains its positive characteristics and flows in the right direction, downward towards the tip, with sufficient consistency to permit writing, but never in sudden bursts as is the case with other systems. (It might be claimed that the forced passage of air and the strong flow of the ink theoretically "cleans" the channel. In fact, since the ink becomes more viscous because of the inevitable evaporation, the opposite might be more true.) Additionally, the cartridge prevents the writer from running out of fuel. Carried in a pocket or in a travel case, one or two spare cartridges are much less cumbersome and messy than a bottle of ink. Pens with large ink reservoirs are now obsolete. The first cartridges, made of glass, were rather fragile, but with the advent of plastic this particular weakness has been completely eliminated. The history of the cartridge really began in 1936. Cartridges existed before then, but they were fleeting fads, without any longstanding influence on the future of the fountain pen.

The oldest attempt at a cartridge filler was that of the Eagle Pencil Company. It first offered glass ink vials in 1890. Unfortunately, they were very fragile; the tip of the vial had to be sectioned off carefully and mishaps were frequent, even more so when people insisted on using eyedroppers to fill them. Ten years later, Eagle abandoned the concept and adopted the rubber sac like everyone else.

The pen that Blair patented in 1898 and

303
"STYL"
Agrafe de sureté
Visibilité totale.
Remplissa e en 6 pulsations
Aucun mé anisme intérieur.
Ecoulem nt régulier.
4
FOIS PLUS
D'ENCRE

THE DADDY OF THEM ALL
Sees That All Fountain Pens Are Not IDEAL
HIS MARK
NOT A WATERMAN'S
NOT A WATERMAN'S
NOT A WATERMAN'S
NOT A WATERMAN'S
NOT A WATERMAN'S
GO 'WAY—DON'T STOP ME!
A WATERMAN'S!
Waterman's Ideal Fountain Pen

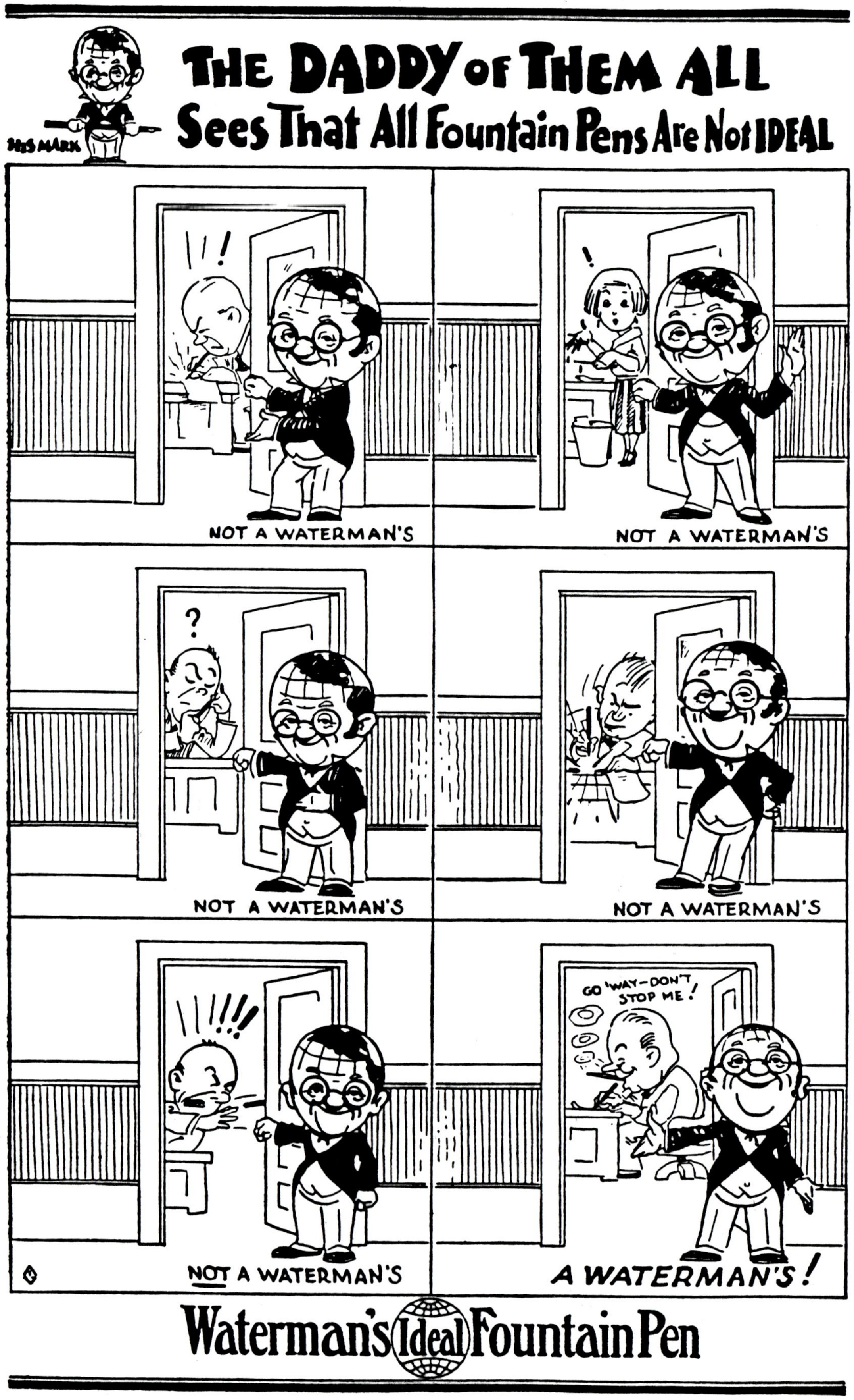

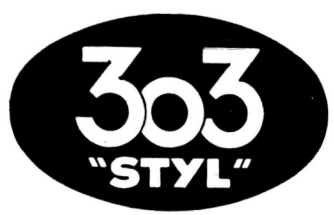

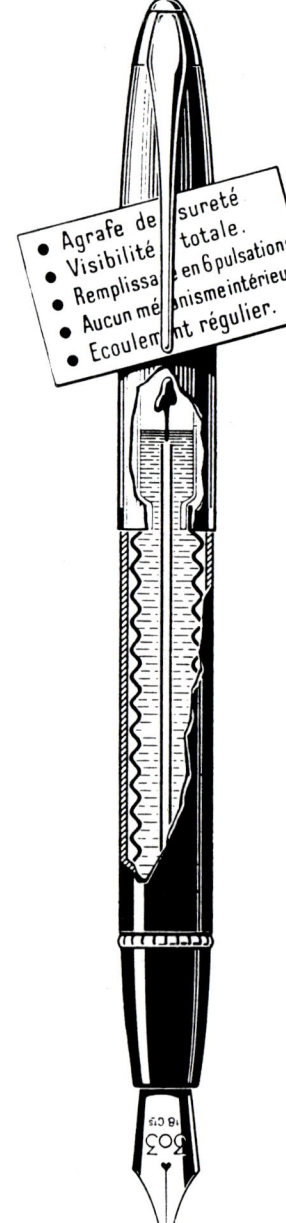

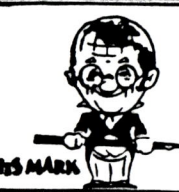

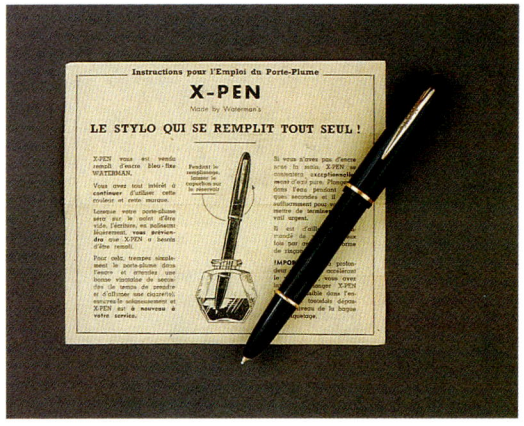

Waterman X Pen, capillary filling (stack of absorbent disks), 1951 patent issued in 1953. Manufactured in France.

Stylomine 303 with retractable nib and automatic filling. The mechanism, which extends the nib, can be unscrewed to release a small rubber bulb (1930).

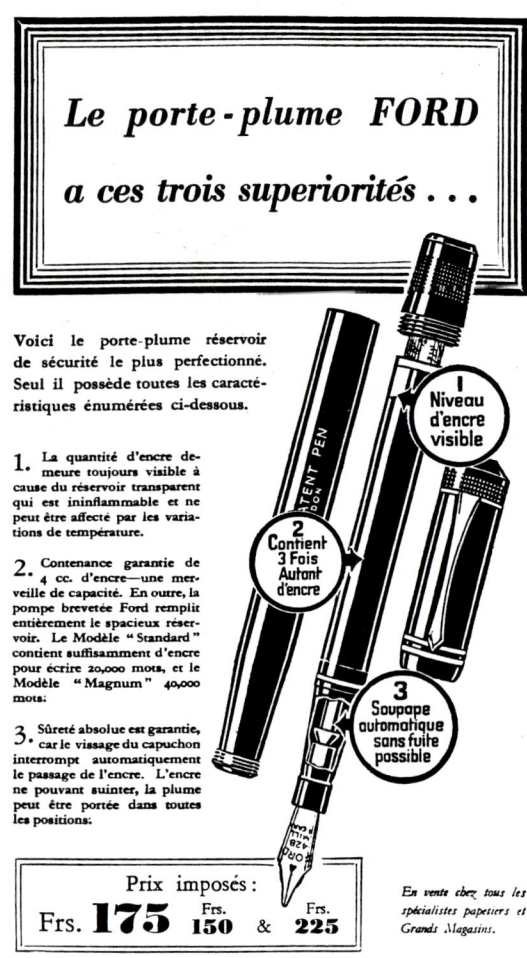

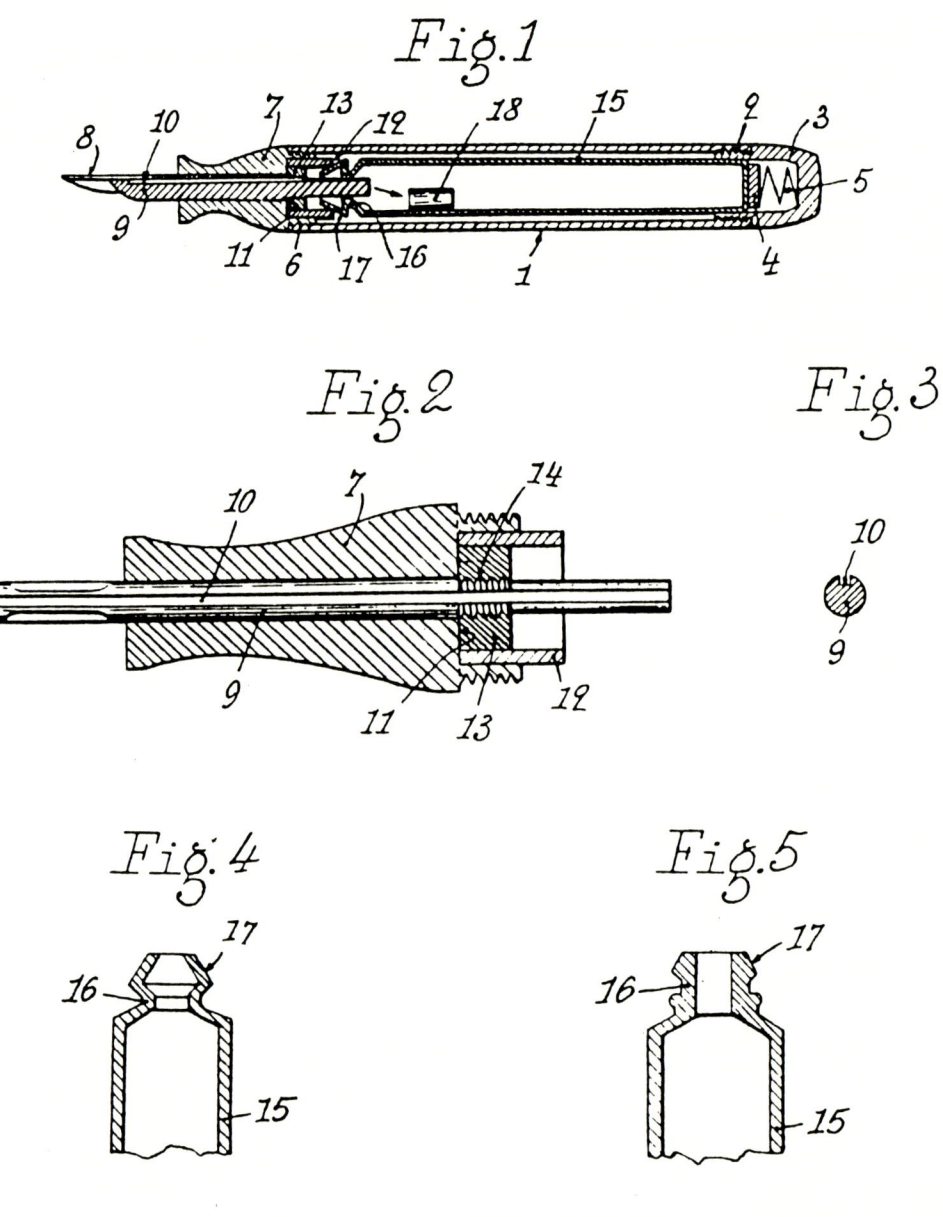

JiF-Waterman patent for the glass cartridge with truncated neck, the rubber ring holding the cartridge in the section, and the feed extension (located at the end of the capillary passage), which was designed to dislodge the cap.

Mac Niven & Cameron (c. 1918).

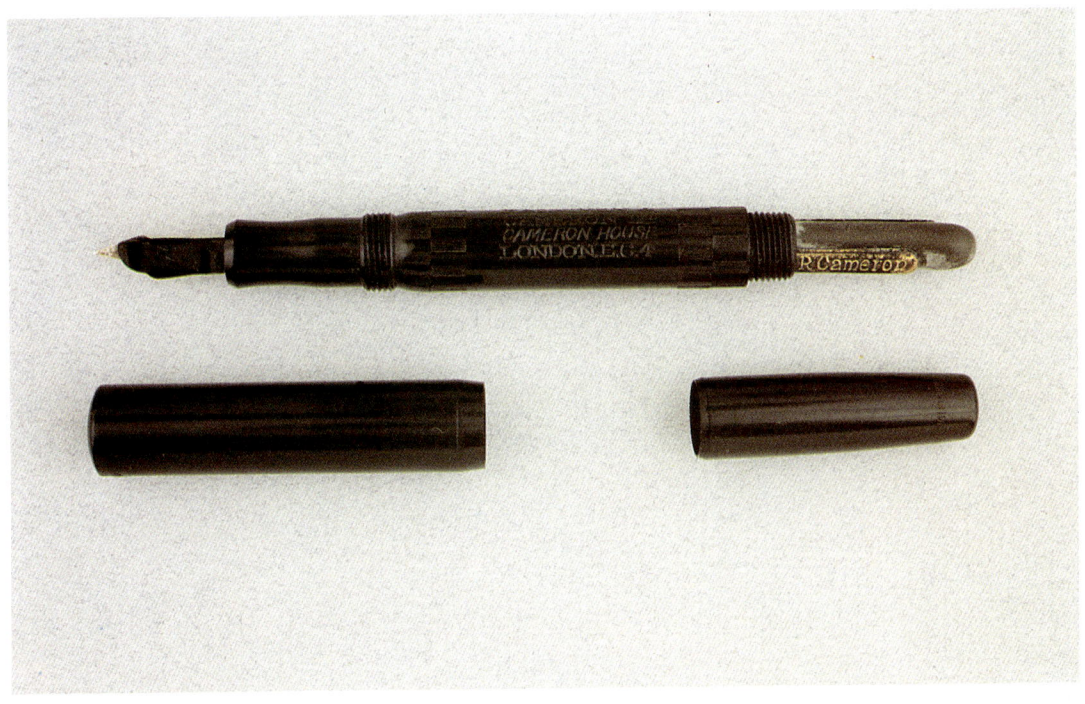

Aurora fountain pen (circa 1932): the filling is done (possibly with one hand) through a small lever located at the end of the body (shown partially extended in this photograph). First distributed in France in 1931 by Edacoto. The Aurora fountain pen and the Edacoto mechanical pencil were called the "duo moderne".

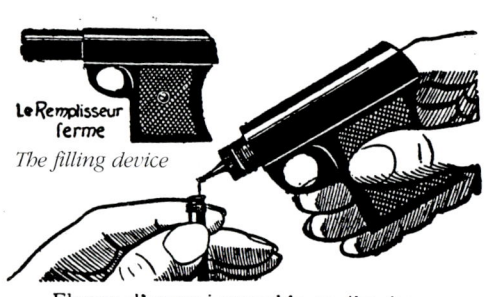

Flacon d'encre incassable en ébonite, pour remplir les Porte-Plume Réservoir

Small unbreakable bottle of ink in black hard rubber with which the reservoir of the pen holder could be refilled.

Monograph *pen holder*, patented on May 17, 1889. The revolutionary capillary filling system obviously had not yet become popular.

Waterman standard model with lever and cartridge (1953-1955). The Super Cartouche *(with a gold dot on the cap) is equipped with a large capacity cartridge and visible ink level.*

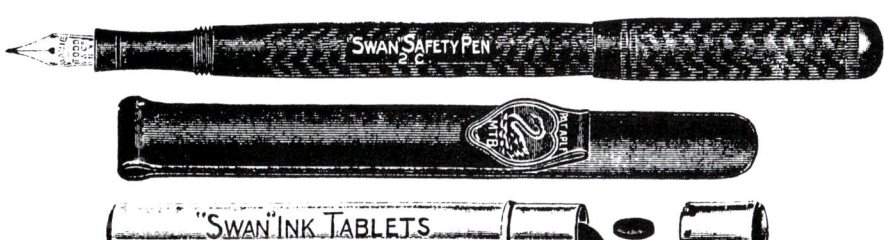

Flash Fill *by Waterman, with cartridge. Manufactured in France (1950).*

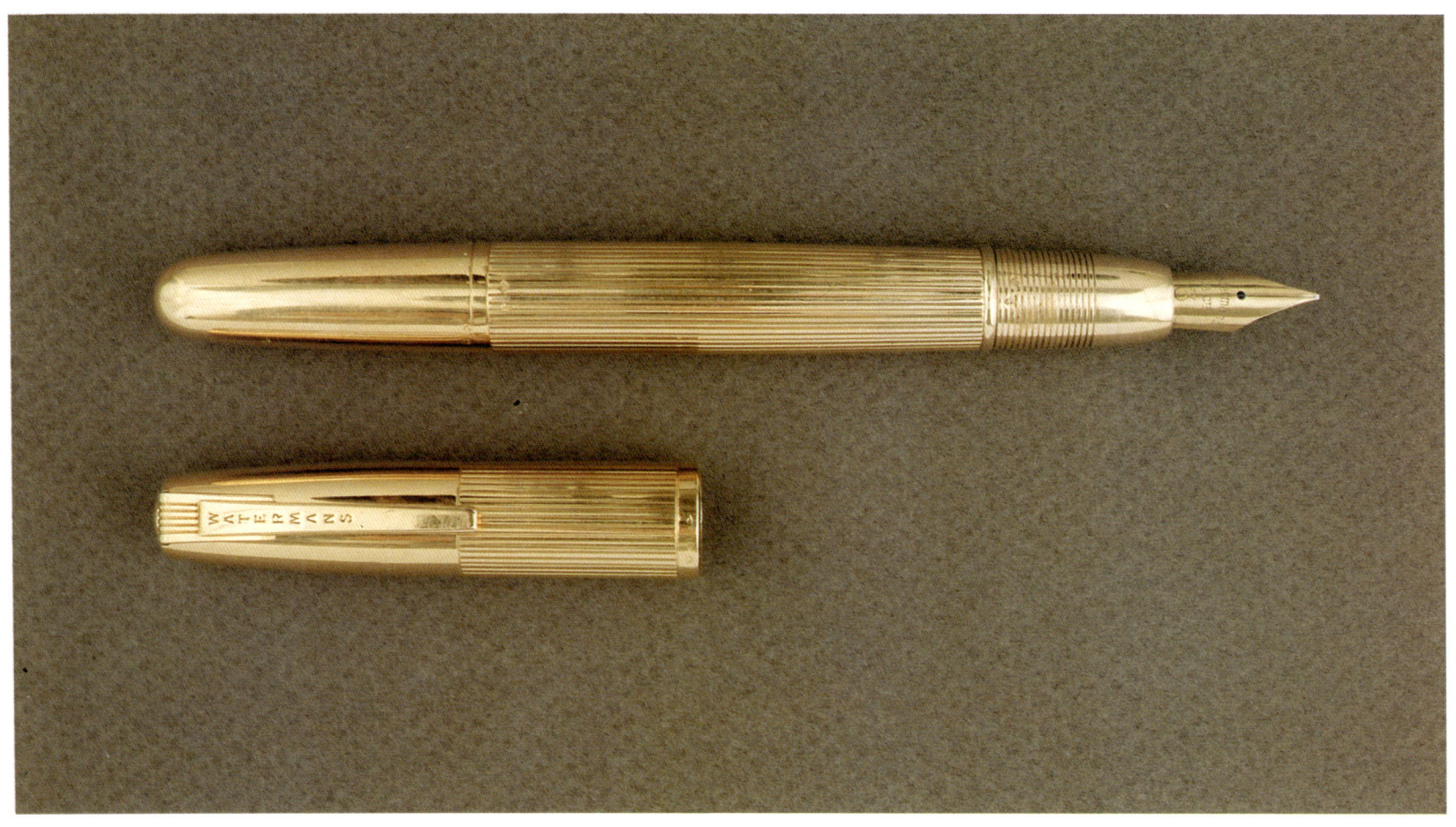

manufactured for many years is often cited as the first fountain pen to use a cartridge. Indeed, there was a cartridge used with this pen, but it was a container with solid ink. After it was inserted into the pen, it had to be filled with water. Thus, in reality, the filling system was not all that different. That model, as well as all others using water, was especially popular with people constantly on the move, such as explorers and the military. Even in the middle of nowhere, running out of ink was no longer a risk. A single solid ink cartridge was enough to refill the pen several times. Other than that, the Blair pen did not offer any of the advantages of the modern ink cartridge.

In 1921, the Frenchman Jean-Baptiste Salmon invented a pen with an independent and rechargeable ink reservoir, which he called the *Stylo Tube*. The name was registered in 1923. The reservoir consisted of a glass tube with the open end threaded and capped. The empty cartridge was unscrewed and discarded, and a full one was inserted in its place. In 1927, Salmon improved his tube. He replaced the glass with black rubber, a good idea but, unfortunately, a sales failure.

In 1936, Jif patented a fountain pen with a capsule, designed 7 years earlier by a Mr. Perraud, then technical manager of the company. The patent describes three compo-

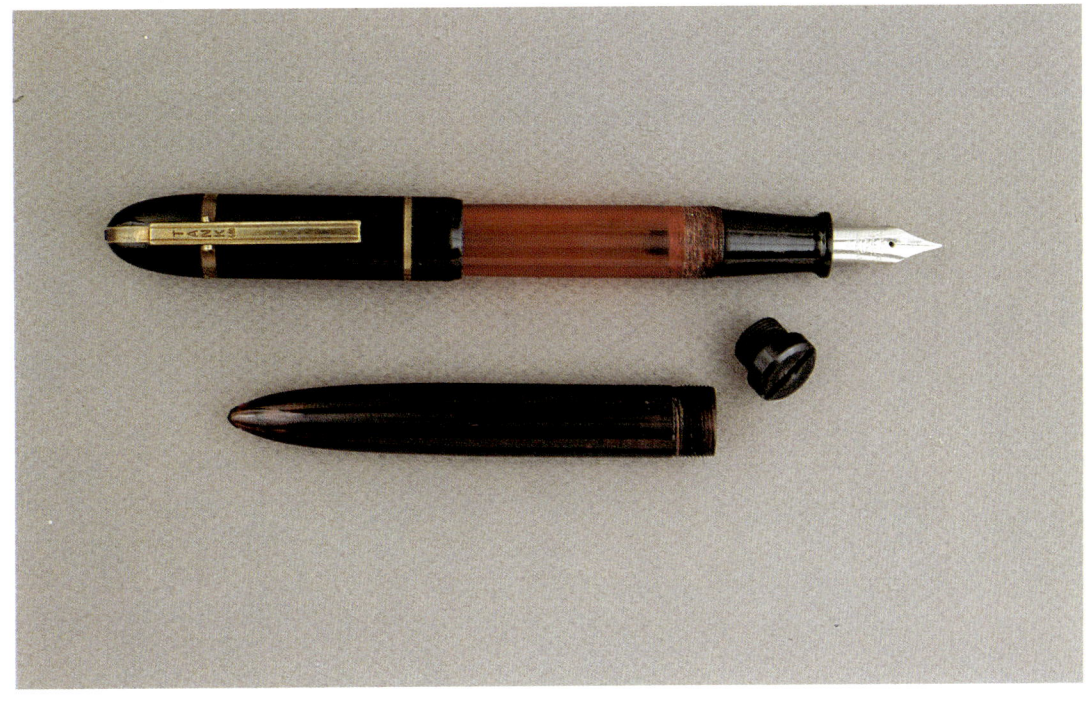

Tank 400 *by Pierre Bagnol, with a spare ink reservoir (1947)*.

nents: a glass capsule with a narrowed neck, a body with an O ring that keeps the capsule in place, and a perforator that pierces the cap. Success was instantaneous. It proved to be the birth of the first practical cartridge, and thanks to this patent Waterman had exclusive production rights for the ink cartridge.

In 1947, the Pierre Bagnol Company launched the **Tank 400**, which was very aptly named. The whole body of the pen was a cartridge. Once empty, it was discarded. Its capacity was enormous; in a publicity release, Bagnol described his model as being the equivalent of 10 fountain pens. The **Tank 400** was a moderate success.

Nibs

Together with the feed, the nib is the most essential part of a fountain pen. Obvious? Not necessarily so! Many a fountain pen is purchased today on the strength of its good looks. The appearence of the written word is of lesser importance. When it comes down to making a choice, a few customers will try the feel of the nib, but more often than not the compatibility of the nib with the individual's handwriting will be secondary to the attractiveness of a lacquered finish.

The fountain pen nib must have several qualities. First and foremost, it must glide on

Glass nibs were used mostly for mulitiple copy writing because of their hardness. Despite their attractive shapes and colors, they were never widely used because of the unpredictability of the ink flow.

Colors
Lever filling Waterman #5 and #7 pens in marbled red hard rubber (1926-1930). Originally, the #7 was sold for $7.00 and the #5 for $5.00. The type of nib was identified by the color of the band on the cap. The color was engraved on the nib and the eye was in the shape of a keyhole.

the paper surface with ease and flexibility. Since it cannot be sharpened, nor is it practical to discard it after a few pages, it must also resist the acidity of the ink and the friction of the paper. Glass is impervious to acid, but it lacks flexibility. Laminated and shaped metals are flexible enough, but steel corrodes easily and gold wears out quickly.

Glass Nibs

Although glass is inflexible and breaks easily, this did not prevent it from being used as a writing instrument. In 1837, a man named Boulard registered a patent for a glass nib, a real nib with a split end and two points. According to him, its fragility was largely compensated for by its stainless and acid resistant

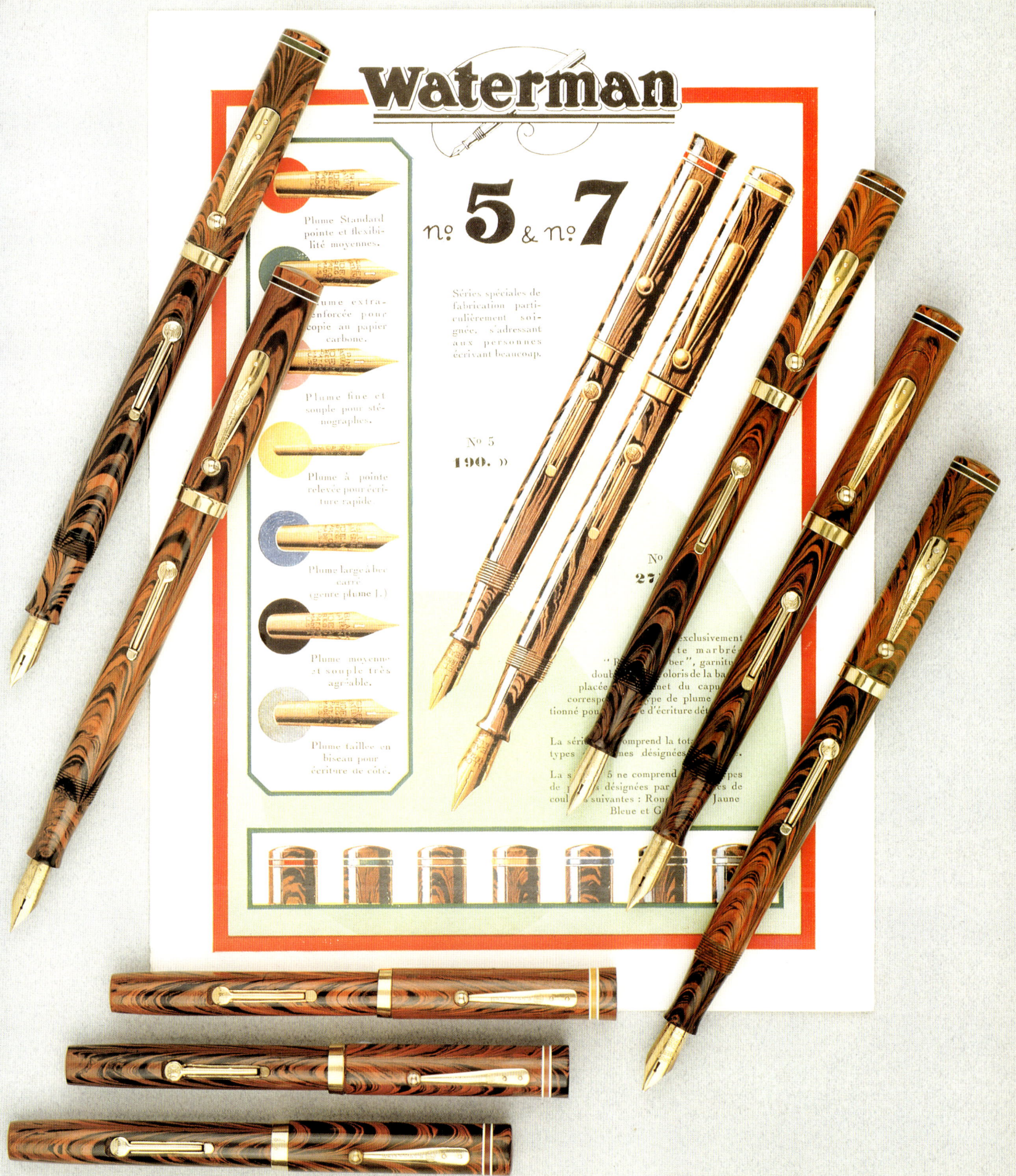

The thickness of the nib decreases from the tip to the back-end.

The width of the slit decreases towards the point.

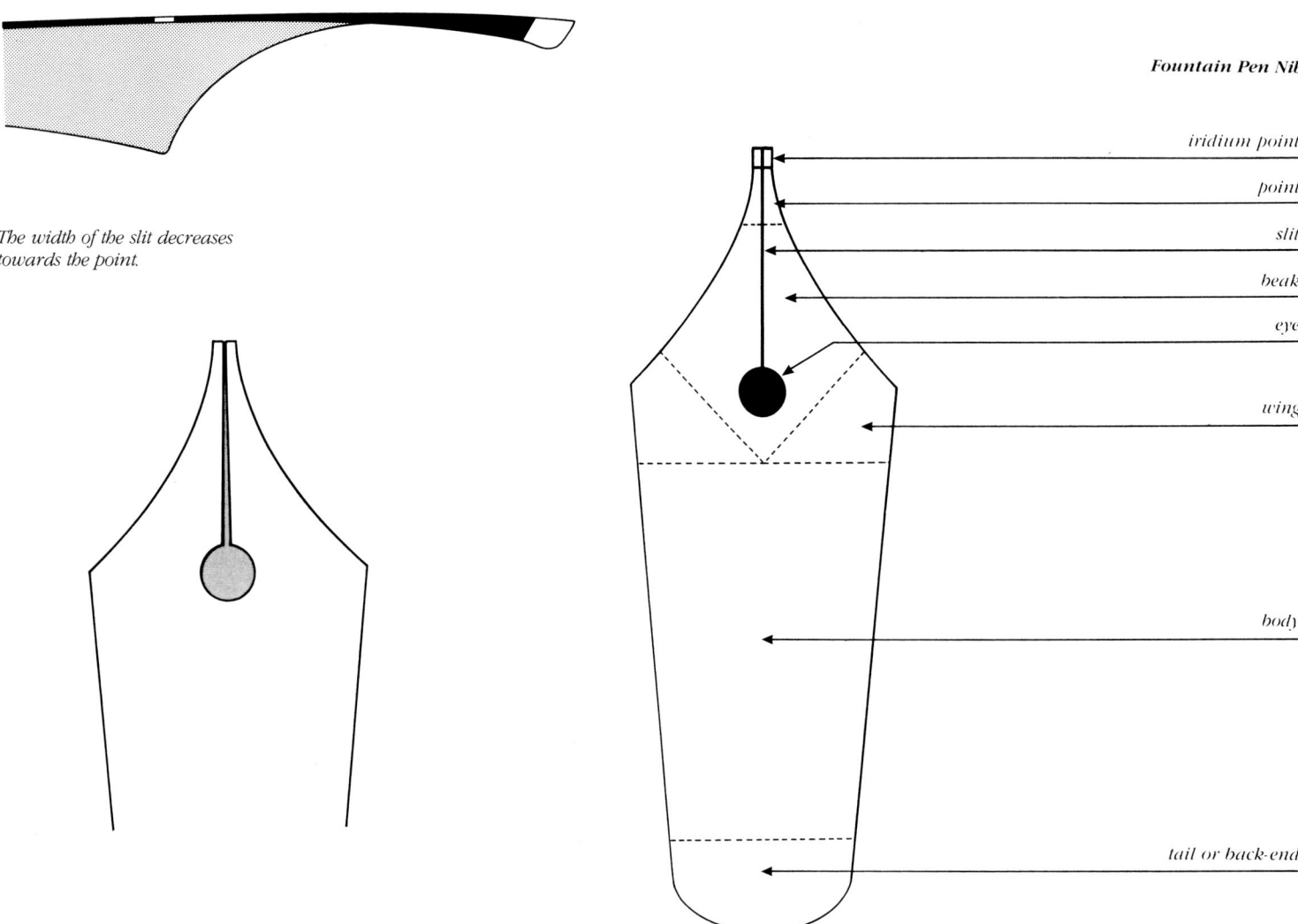

Fountain Pen Nib

iridium point
point
slit
beak
eye
wing
body
tail or back-end

qualities; it was also durable and economical.

Although the patent was granted to Gay-Lussac, production remained dormant. Successful glass nibs were not actually nibs, they lacked sharp points and had no split end; they were simply pieces of twisted or corrugated pointed glass mounted on some type of holder. Other glass nibs, more beautiful, were mounted onto a glass handle or were drawn until they were a one-piece nib and holder, at the end of which a glass bird was formed.

The period following the end of the Victorian era was the golden age of glass nibs. Traditional glass blowing techniques created charming colored and decorated objects that graced the desks and writing tables of many refined young ladies.

At the beginning of the 20th century, particularly in the 1920s, some fountain pens were equipped with glass nibs. The hardness of their points made them ideal instruments for multiple copy writing or for writing with carbon paper. They competed favorably with stylographic and manifold nib fountain pens. During World War II, when steel was requisitioned for more pressing needs, glass nibs found patrons among accountants and some businessmen, but there were few other users.

Glass nibs were used largely on inexpensive fountain pens and no major manufacturer except Montblanc ever used them, and then only to a very limited extent. Their resistance to acids and corrosion and even their low price were not enough to offset their significant shortcomings. The hardness that made them attractive to carbon paper users was not really a plus to the majority of writers, since, in addition to being hard, they were very fragile. Glass nibs that were fitted onto fountain pens were either fixed or retractable. In the case of the former, the ridges or fluting that fed the tip dried up very quickly and it was then difficult to re-establish the flow of ink. With retractable nibs, the point was better protected and did not dry as easily, but the slight disadvantage of all

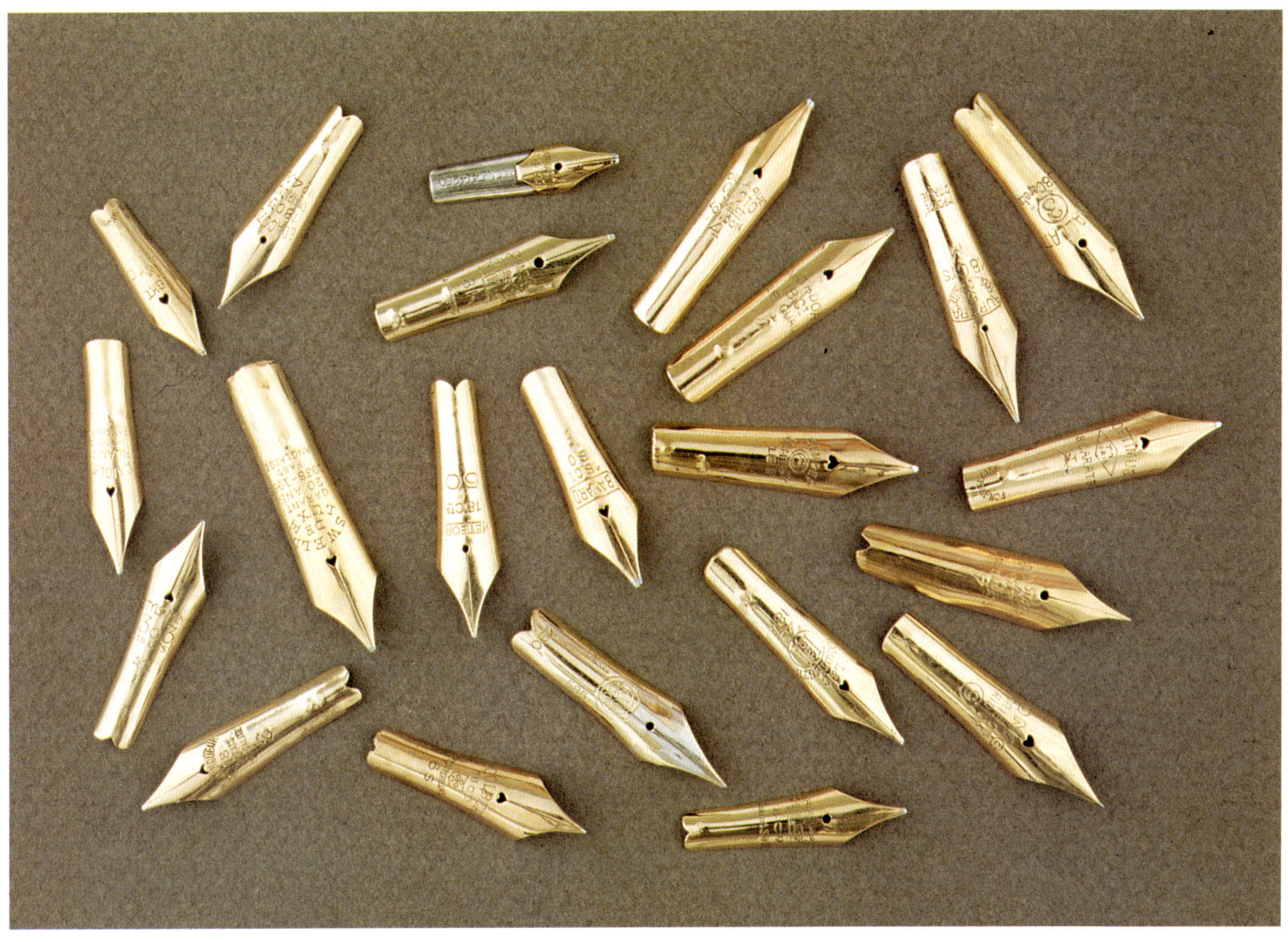

Nibs made by various manufacturers.

retractable nib pens became disastrous in this case. When the nib was extended, if the propulsion device were not fully extended, the ink would empty onto the paper as soon as the nib came into contact with the surface.

Manufacturing techniques being simple, glass nibs were made everywhere. One company that was able to make a name for itself in this domain was the German company, Haro. In France, most glass nibs were imported from Czechoslovakia.

Gold Nibs

Impervious to corrosion, gold is the ideal material for nibs once a way has been devised for dealing with the wear resulting from friction with the paper. The solution was to fix a hard element at the tip. Feather quill nibs were equipped with a similar component by the addition of bits of precious stones (diamonds or rubies). The hardness of these materials was unquestionable, but they also needed a round surface without which they inflicted irreparable damage to the paper. Around 1827, an Englishman named Doughty used ruby-tipped gold nibs. Although very expensive, his nibs were moderately successful.

Platinum has been known since 1735, but it was not until 1804 that two British chemists identified other metals in platinum ore.

Smithson Tennant discovered osmium and iridium, and William Hyde Wollaston, palladium and rhodium.

It is generally acknowledged that the first individual to obtain satisfactory results with any of these metals was John Isaac Hawkins, who, in 1823, had started experiments with feather quills, using tortoise shell and horn nibs set with ruby tips. He understood that by capitalizing on the hardness of diamonds, rubies, and other precious stones, he could affix them to a grinding wheel and, thus, cut the iridium. At the same time, he was able to make improvements in gold nibs, rendering them more durable.

In 1843, after trying ruby tips, the Frenchman Jean-Benoît Mallat initiated the

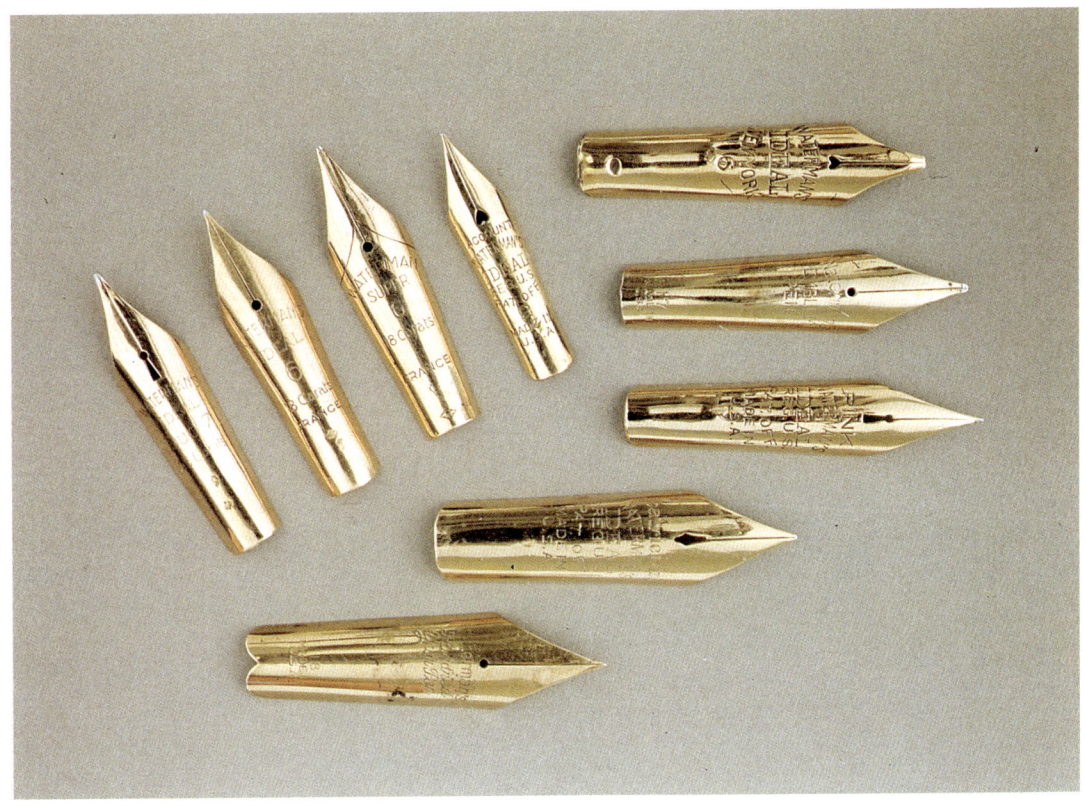

Assortment of Waterman nibs:
Patrician, Hundred Year, Duo 7...

production of gold nibs with iridium points. However, it was not until the end of the 19th century in the United States that the industry was actually born. New York rapidly became the fountain pen capital of the world, and for several decades it supplied the world market. Sitting on innumerable steel nibs, manufacturers in Birmingham, England had missed the boat. Despite its expertise in the production of fountain pens, Great Britain played only a secondary role in the gold nib industry. It was the principal European producer, but lagged far behind the United States. As for France, it was only after World War I that an interest in gold nibs developed, and even then only the major companies began production (Bayard, Gold Starry, Paillard, Stylomine, Unic). A few specialized companies supplied nibs under their own name brands or under the trade names of small pen manufacturers.

The gold used in making nibs is so soft that it is necessary to reinforce it with a small cap of iridium or osmium. Even 14 or 18 karat alloys are too soft. A karat is defined as the quantity of pure gold contained in a gold alloy, expressed as a ratio of 1:24. Pure gold is 24 karat, 18 karat gold has a gold/gold alloy ratio of 750/1,000, and 14 karat gold has a ratio of 585/1,000 (the numbers 750 and 585 often appear on nibs).

Most fountain pen nibs are made of 14 karat gold, with the exception of those manufactured for export to a few markets that require 18 karat gold. In France, gold of less than 18 karats may not be sold. The other two elements in the alloy are copper and silver.

14 karat: 14 Au + 6 Cu + 4 Ag
18 karat: 18 Au + 3 Cu + 3 Ag

These alloys are subjected to very careful analysis. They must not contain any impurity that might alter their properties and make them more fragile.

The manufacture of gold nibs involves a large number of steps, the most important of which are as follows:

1. Rolling of an ingot, 10 to 20 mm in

Sabon workshop (Bordeaux) circa 1930.

thickness. Several passes and annealing (necessary to ensure flexibility) are required before obtaining a thickness of 0.3 to 0.6 mm. Annealing is performed after each rolling, except for the last one, when the gold is hardened. Further rolling and other operations on bench presses are performed on the cold gold; the hardness thus obtained is necessary to ensure the flexibility and resilience of the nib.

2. Cutting of the first dies.

3. Soldering of the iridium point with a blow torch.

4. Sanding the area around the iridium where the soldering has melted the gold.

5. Rolling of the form between two grooved cylinders. The sides of the nib are extended. This rolling phase results in a thickness between 0.3 and 0.6 mm at the tips, but reduces the body of the nib to about 0.1 mm at the tail. For obvious reasons, the tip must be thicker. The thinning of the body improves flexibility. The reason behind the thinness of the tail is not to save gold, but to ensure a tight fit of the nib between the feed and the section.

6. Cutting of the other parts of the nib on a bench press equipped with a point and a die.

7. Marking and drilling of the eye on a press.

8. Impressing the curvature of the nib.

9. Splitting of the nib tip up to the eye with a cutting disk.

10. Repeat sanding of the iridium or osmium

Real Fakes (Fabricated Copies)
During the peak era of hard rubber, fabricated copies were used by retailers for display. The fake fountain pens, replicas of the ones sold inside the store, served a dual purpose - they could not be used by would-be thieves and prevented the real pens from being damaged by prolonged exposure to the sun. After synthetic resins and plastics were used and the fear of heat damage disappeared, the fear of theft remained, so replicas were abandoned and display pens were drilled with small holes that made them useless.

point. At this stage, the nib is graded as either fine, medium or broad.

11. Polishing of the top of the nib and dulling of the underside (to regulate ink flow).

These processes are followed by various quality control tests (elasticity, point calibration, line drawing, etc.), which are automated today. Formerly, soldering with a blow torch had to be done before the last rolling. Now, electric soldering of the iridium is done after the curvature of the nib is produced.

The average standard nib weighs about 0.30 grams, the smallest weighs about 0.15 grams, and the largest rarely exceeds 0.50 grams. They may be partially rhodium-plated, creating a white surface that contrasts with the yellow of the gold.

The numbers that are sometimes stamped on the body indicate the size of a particular model. Nib size, point type, and nib quality should not be confused. Size refers to the nib dimensions (small, medium, large, etc.), type describes the point characteristics (extra-fine, fine, medium, large, square or broad, oblique, etc.), and quality refers to its behavior (soft, semi-soft, hard, etc.). All these characteristics can be combined; for example, a nib may be large or small, fine or broad, and hard or soft. This is one of the reasons why there was such an incredible number of models offered by the

Real Fakes (Wood)
The first wooden replicas were used by individuals of limited means, essentially students. A simple fountain pen in colored wood or brass, with cap and clip, placed in the pocket of a student or a down-on-his-luck adult would give the illusion of wealth. Top: Among pen holders, two real wooden fountain pens made during World War II. The black lever model is made of plastic. The black and white marbled model is bolder in its design; both feed and section are made of wood.

Displays of French nibs in gilded steel (gold-plate is usually not used for nibs because of the difference in flexibility between gold and steel). These nibs were generally used on inexpensive fountain pens.

Before the Parker 51, other pens also had hidden nibs.

major manufacturers throughout the 1920s and 1930s.

After World War II, the decline in the popularity of the fountain pen resulted in a sharp decrease in the variety of nibs available. The number of models offered by major manufacturers, who had offered scores only a couple of decades before, could be counted on the fingers of one hand. Today, the resurgence in the popularity of the fountain pen has somewhat improved the situation.

Steel Nibs

Before World War II, a good nib had to be made of gold. Steel is harder than gold, but it is not resistant to wear. Iridium and osmium are very expensive, even more so than gold. It would have been absurd to use them as points on steel nibs since the latter were so corrosive. Steel was used for inexpensive models, low-end pens for students with limited incomes or would-be writers. When iridium was not used, the nib tips were doubled or thickened in order to ensure a smooth flow onto paper and resistance to wear, thus adapting techniques that had been used on nibs for dip pens for decades.

Today, things have changed a lot. Gold is still used for higher priced models and it retains its prestige, but the majority of contemporary fountain pens are fitted with steel nibs. Many factors contributed to this trend. During and immediately after World War II, there was a gold shortage in Europe, which, although insignificant when compared with other tragedies at the time, had serious repercussions. In France, it was forbidden to manufacture gold nibs until 1949. Beginning in 1971, gold prices, which

Lady's fountain pen, gold trim by an anonymous jeweler. The cap is embellished with an amethyst cabochon and a circle of rose-cut diamonds.

previously were stable, started to climb sharply. Finally, and most importantly, advances in stainless steel technology enabled the manufacture of steel nibs of a quality almost equal to that of gold nibs.

The resistance to corrosion and the mechanical properties of stainless steel nibs are close to those of 14 karat gold nibs. In these alloys, chromium plays an essential role in resisting corrosion, while nickel and molybdenum ensure thermal stability, and the carbon content is kept as low as possible. The processing of this type of steel requires a technology much more advanced than that required to produce 14 or 18 karat gold.

Steel nibs are sold in either their natural color or gold-filled. They even look like gold, but obviously are not. Contemporary stainless steel nibs are of excellent quality, which could not be said of steel nibs made in the past.

Nibs have been made of other alloys too, such as palladium and silver. Although these materials are slightly more subject to corrosion, their performance and price are close to that of gold nibs. The problem with them is their color; the market is not inclined to pay the price of gold for something that looks like steel.

Shapes

Whether gold or steel, the nib does not offer many options for its shape. This is unfortunate since there was an incredible variety of pens. They numbered in the thousands and were decorated with engraved figures or imaginative filigrees.

The ancestor of the fountain pen nib was nothing more than a simple tip mounted onto a hollow pen-holder, generally fed by a short tube. At the beginning of the 19th century, they were generally made of goose feather quills. The first real fountain pen nibs combined the curved line of the reed pen with the softness of the feather quill, which, together, ensured the strength and flexibility of the tip.

With its proven track record, the curved shape remained constant throughout the centuries and did not change, even with the advent of the fountain pen. For ink to feed properly, three parts of the pen must be fitted as closely as possible: the nib, the feed, and the section that keeps the parts together. For nearly 40 years, feeds and sections were bored from black rubber tubes or rods. The era of plastics and injection molding did not alter the roundness of the body. It did, however, permit some innovations; for example, a built-in nib in an interchangeable section.

Adjustable Wahl Eversharp nibs, which allowed a reduction or increase in tip flexibility for thin or broad writing (1932-1939).

Some of the "innovations" do not quite deserve the title. One example was the adjustable nib on some Wahl-Eversharp pens. Billed as revolutionary at the beginning of the 1930s, it was actually a copy of a gimmick introduced 75 years earlier. Tubular nibs, however tailored or aerodynamic, are basically of the same shape as the reeds, feather quills, and first metal nibs made by the Romans.

With its covered nib, the Parker **51** was a very good pen, a best seller in its day, and one that was copied by many; moreover, it radically (and some assert negatively) changed the world's perception of the fountain pen. The covered nib may have its defenders, who argue that it permitted the use of a faster drying ink, that the nib was protected, and that it had a design worthy of the Bauhaus architectural style. Nevertheless, it was the first fountain pen to conceal its nib. By hiding the nib, you could forget that it existed, and the fountain pen had a hard time recovering from such an assault on its image. Fortunately, the covered nib was just a fad. The current renaissance of the fountain pen is eloquent testimony; nibs have once again emerged into the light of day.

Materials

A nib must be flexible. Metal provides that property. As for the rest of the fountain pen (feed, body, cap, reservoir), its history is intimately linked to that of plastics. This poorly defined term in current usage equates plastics with synthetic resins. Following mechanical stress, an elastic distortion is temporary; however, any distortion of a plastic material is permanent. Therefore, any material that consists of macromolecules exhibiting this property is a "plastic" substance. Hence, there are natural plastics (amber, horn, tortoise shell, etc.) and artificial plastics made by man from natural substances (cellulose acetate, celluloid, black hard rubber, etc.) as well as synthetic resins that are completely man-made, especially from petroleum and coal by-products (bakelite, plexiglas, rilsan, etc.).

In 1839, after 8 years of unsuccessful attempts, Charles Goodyear inadvertently placed a mixture of rubber and sulfur on a stove. He could never have imagined that this incident would have such a huge impact on the evolution of writing instruments. Although it

Black hard rubber and red marbled rubber rods ready to be cut and drilled.

took 5 more years and bore the name of Thomas Hancock, vulcanized rubber had been invented, and sulfur had created the bond between its macro-molecular chains.

As long as ink was stored within its body, the fountain pen had to be composed of non-porous, acid-resistant, non-heat conducting materials that would also be inexpensive. Vulcanized rubber has all of these properties and, therefore, hard rubber was a critical factor in the birth and early expansion of the fountain pen industry. For more than half a century, it was the most widely used material. It was delivered to factories in the form of rods, generally 1 meter long, and of varying diameter corresponding to the parts to be shaped on the lathes: bodies, caps, sections, and feeds. Excellent pens were created from this material and the body was seamless. Hard rubber is easy to identify; if you have any doubts, just rub the pen with a woolen cloth. If it is composed of hard rubber, the static electricity created will attract small pieces of paper.

After rubber came cotton or wood cellulose. Celluloid was developed in 1869 by John Wesley Hyatt in response to a competition designed to identify a material to replace ivory for use in billiard balls. Celluloid is obtained by

"Cardinal" red hard rubber Waterman pen (1910-1929).

Engraved Gold and Silver
From left to right, three standard and one safety fountain pens: a British Swan, silver, circa 1910; a French CM, silver, circa 1922; an American A.A. Waterman, gold-filled, circa 1914; and a French pen, circa 1916.

heating nitrocellulose with camphor, which gives it its characteristic smell. The manufacturing process was perfected in Germany during the last two decades of the 19th century and, as a result, it was used in the fountain pen industry in Germany much more so than in any other country. It was supplied in rods or tubes and, like hard rubber, was shaped by a lathe. Its plastic qualities were remarkable, but it had one major flaw that prevented it from outperforming hard rubber: it was highly flammable. Many a collector, trying to disassemble the section of a pen by heating it, has found himself with a small heap of ashes on his hands. Any fountain pen that emits a faint odor of camphor after being rubbed a few times should be kept away from an open flame. Unfortunately, in the case of very old models, the smell is almost undetectable.

After 1920, celluloid was competing with cellulose acetate and, later on, with acetobutyrate of cellulose, both of which were much less flammable. Materials developed by DuPont de Nemours in the 1920s were derived from nitrocellulose, regardless of their name (i.e., radite [Sheaffer] and permanite [Parker]).

Two nameless models among the many manufactured by Fernand Laureau (circa 1930).

Ivory fountain pen with Japanese design (circa 1930).

Patricians
Faithful for a long time to hard rubber, Waterman moved into plastics in 1929, 6 years after Sheaffer. Although slow in joining the bandwagon, Waterman produced some of the most beautiful pens of that era - especially the Patrician. *It was offered in six colors: black, emerald, onyx (red and cream), turquoise (blue and gold), nacre (black and pearl), and moss agate (green, brown, and black). Its clip was no longer mounted with rivets but inset in the cap. The globe in the logo also disappeared from the lever. One year later, the ladies' model, the* Lady Patricia, *appeared (1930-1938).*

However designated, all these materials had a quality that black hard rubber lacked: they could be produced in bright colors. Lighter and less fragile, they also allowed greater freedom in the shape of the material and thus played a pivotal role in the evolution of the fountain pen. In 1923, Sheaffer was the first major manufacturer to use plastics, followed by Parker in 1926, and Wahl in 1927. Slower to react, Waterman did not follow suit until 1929, and this delay was one of the reasons for Waterman's decline in the 1930s. That same year, Sheaffer introduced designs with more fluid lines that could be injection-molded. Although Hyatt patented injection molding in 1878, it was not widely used in the fountain pen industry until the 1940s.

After cotton came cows...In 1899, the Austrians Kritsche and Spitteler developed a remarkable plastic from milk. Clotted, purified, and dried, casein treated with formaldehyde would become galalith (gala: milk; lithe: stone [Greek]). Parker first used it as early as 1904 (ivorine), but it was not produced in large numbers until much later. Its weakness was that in time it would invariably become discolored by ink. Fortunately, the introduction of the rubber sac helped correct that problem.

After the era of plant- and animal-derived products came the age of synthetic resins. The first of these was bakelite (1907). It owes its

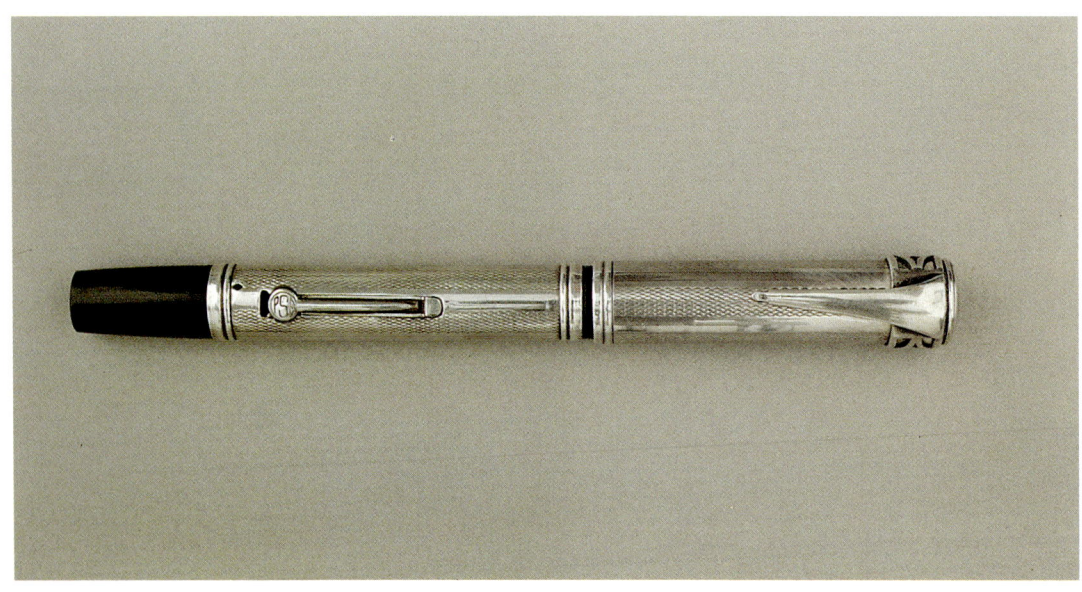

Grand Aigle Mercier *with lever, black hard rubber, silver trim (circa 1934).*

Retractable nib pens. Toledo style trim, possibly manufactured by Charcellay (circa 1925).

Waterman standard pens, gold and silver trim (France) on black hard rubber (circa 1910).

Lever filling Waterman pens in "two-tone" hard rubber (1928-1930).

"Continental" Dress
Retractable nib Waterman pens, black hard rubber, so-called "continental" trim in gold-plate (1920-1925). These gold-filled overlays and filigrees cover the entire body; the original clip matches the motif.

name to the Belgian chemist, Leo Hendrik Baekeland, who created it by condensing phenol and formaldehyde. This was followed by the development of a long list of synthetic materials that were known to the public by their trademarks, such as plexiglas (which is, in fact, methyl polymethacrylate). Since the 1940s, synthetic resins have been widely used in the fountain pen industry.

We have seen that in the first part of the l9th century, most early fountain pens were designed or manufactured in metal, for lack of better materials. Highly valued as trim and indispensable for the nibs, levers, and other filling mechanisms, metals are not suitable for the body of the fountain pen. Most metals oxidize, are corroded by ink, and, in most cases, are good heat conductors. In addition, shaping them is more difficult than is the case with plastics; in short, they are full of flaws. Yet, a few manufacturers offered inexpensive fountain pens in molded metal (brass, nickel, silver, etc.) The aluminum pen was moderately successful during World War II because it made use of whatever scraps of metal the war machine did not use. In the 1940s, there were a few good examples made completely of metal.

The points of these nibs were made of rare metals, especially iridium and osmium. Pure iridium was obtained for the first time in 1885. The first nibs were equipped with points made of iridium-osmium, a natural alloy obtained from platinum ore during the first phase of processing. Later on, synthetic alloys of iridium and osmium, and sometimes ruthenium, were used. Eventually, various alloys, such as tungsten, cobalt, rhenium, titanium, and, in low-end products, nickel were used.

Methods of Transportation

A fountain pen is essentially an ink well in a pen and not a pen in an ink well. The fountain pen owed its early success to the fact that, in theory, it offered considerable advantages over

Case with French clips (1918-1935).

The "Transportable" Pen
Top, left to right: Knapsack by Waterman (1903) with two compartments (one for the pen and one for the eyedropper); French leather case for Mont d'Or*; metal cases (*Swan*, 1908) that fit in the pocket to hold a fountain pen; leather case by Waterman to be attached to a belt; leather case with two compartments. Bottom: English clip for three instruments; Parker case for the* Vacumatic *and its matching mechanical pencil designed for the Royal Canadian Air Force aviators (World War II); two English pen keychains, one in silver-plated brass and the other in sterling; small leather case for the* Viala Lilliput*; two Telescope Cap Waterman pens with an additional tasseled cap mounted over the regular cap (1914-1927).*

the pen AND the ink well: it was easily transportable and always ready and available. In practice, a few wrinkles had to be ironed out.

The cap of the first fountain pens was simply placed over the nib. Thus, leaks were unavoidable. It was recommended that a fountain pen be carried straight up with the nib in a vertical position. To achieve this, a number of devices were invented. A few were truly simplistic, such as vertical seams in pockets that would hold pens (or cigars) straight up - a little silly. Systems with clips or bands were preferable. Since it first entered the fountain pen market, Waterman has offered the **Ideal Pocket**, a sort of metal case that fitted into the pocket and could hold several writing instruments. Soon, however, manufacturers realized that it would be smarter to place an accessory on the pen itself rather than in the pocket. This led to the development of the band around the cap that holds a clip. Such a system offered even more advantages. It prevented the pen from falling out of the pocket or rolling off a flat surface. The clip could also be adjusted to fit the depth of the pocket. For ladies, a ring affixed to the top of the cap permitted the pen to be hung by a chain or ribbon and worn as a pendant.

The screw cap and the retractable nib of the safeties improved the fit and reduced leakage. Keeping a pen in a vertical position was no longer a necessity. The manner in which pens were carried by men did not change much, since they were still positioned vertically in suit pockets. However, women now became less

Une nouveauté pour votre sac Madame,

Display case of clips in sterling, silver plate, and copper (1918-1935). It was an easy and inexpensive way to personalize a pen. They were very popular in France.

hesitant to carry pens in their purses and men could use their vest pockets. A leather case was sometimes used to separate fountain pens from handkerchiefs, notebooks, and other accessories. Many ladies' pens were trimmed with a ring and a ribbon or tassel attached for no other purpose than to assist their owners in retrieving them from the bottom of a purse.

Major manufacturers soon adopted the fixed clip. It was first offered as an option; later, its use became widespread. Waterman's **Clip-Cap** was introduced in 1905. Other than its obvious usefulness, the clip offered another advantage. Even when the pen is placed inside a pocket, the clip remains visible, allowing the model to be identified. Manufacturers engraved their names on the clip, which, in most cases, proved to be small and difficult to read. For ease of recognition, they gave their clips readily identifiable shapes. Parker's arrow is perhaps one of the best known and most successful signatures. Created in 1932 for the **Vacumatic**, it survived many changes.

Clips were used more in France than in any other country. For years, Waterman exported fountain pens without clips to France, and until World War II, many French manufacturers distributed clipless fountain pens, allowing the retailer to sell clips selected by the buyer, who was then afforded an opportunity to personalize his or her writing instrument.

Demand was strong and a large number of models were produced. Some were strictly utilitarian, including a few with imaginative anti-theft devices. Others offered a more elaborate design, so that all tastes could be satisfied. The clips were sometimes made of gold, often of silver. In the less expensive models, the

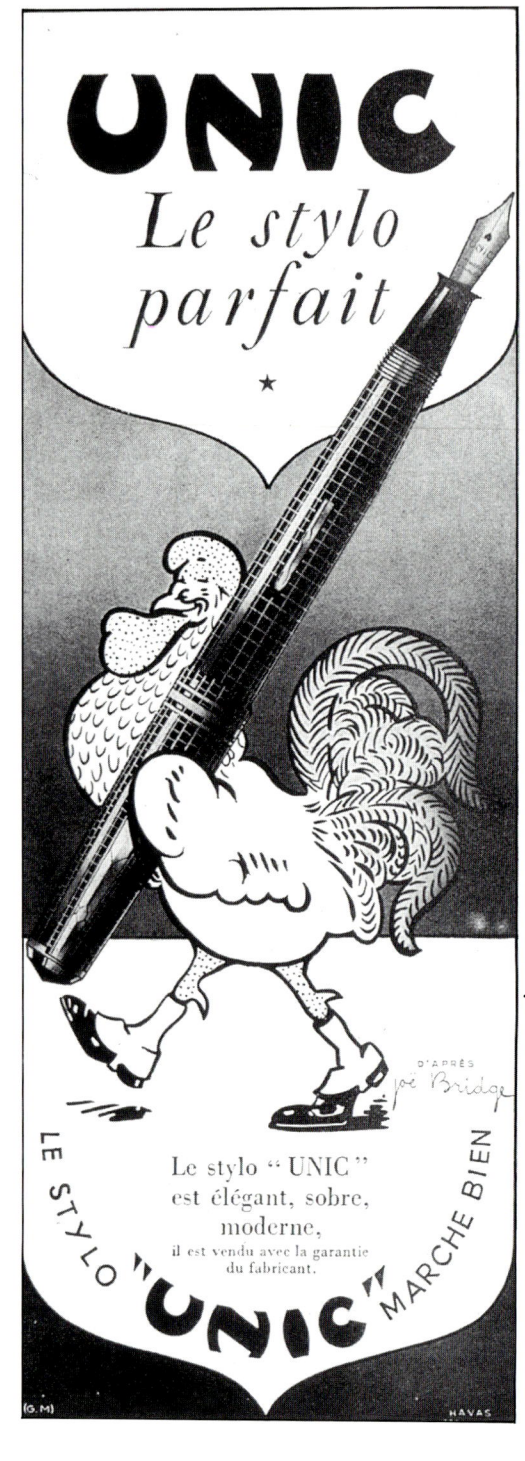

Gold-plated French clips shown flat to fully illustrate the design. Augis (1922).

clips were gold- or silver-plated, chromed or enameled. They were very fashionable during the 1920s and 1930s. Their demise is perhaps regrettable, as this was one of the rare instances in which the individuality and fancy of the user could be expressed at little cost.

State of Affairs

From 1880 until the end of World War II, there were hundreds of pen manufacturers. The Great Depression of 1929 caused the demise of quite a few in the United States. An abrupt drop

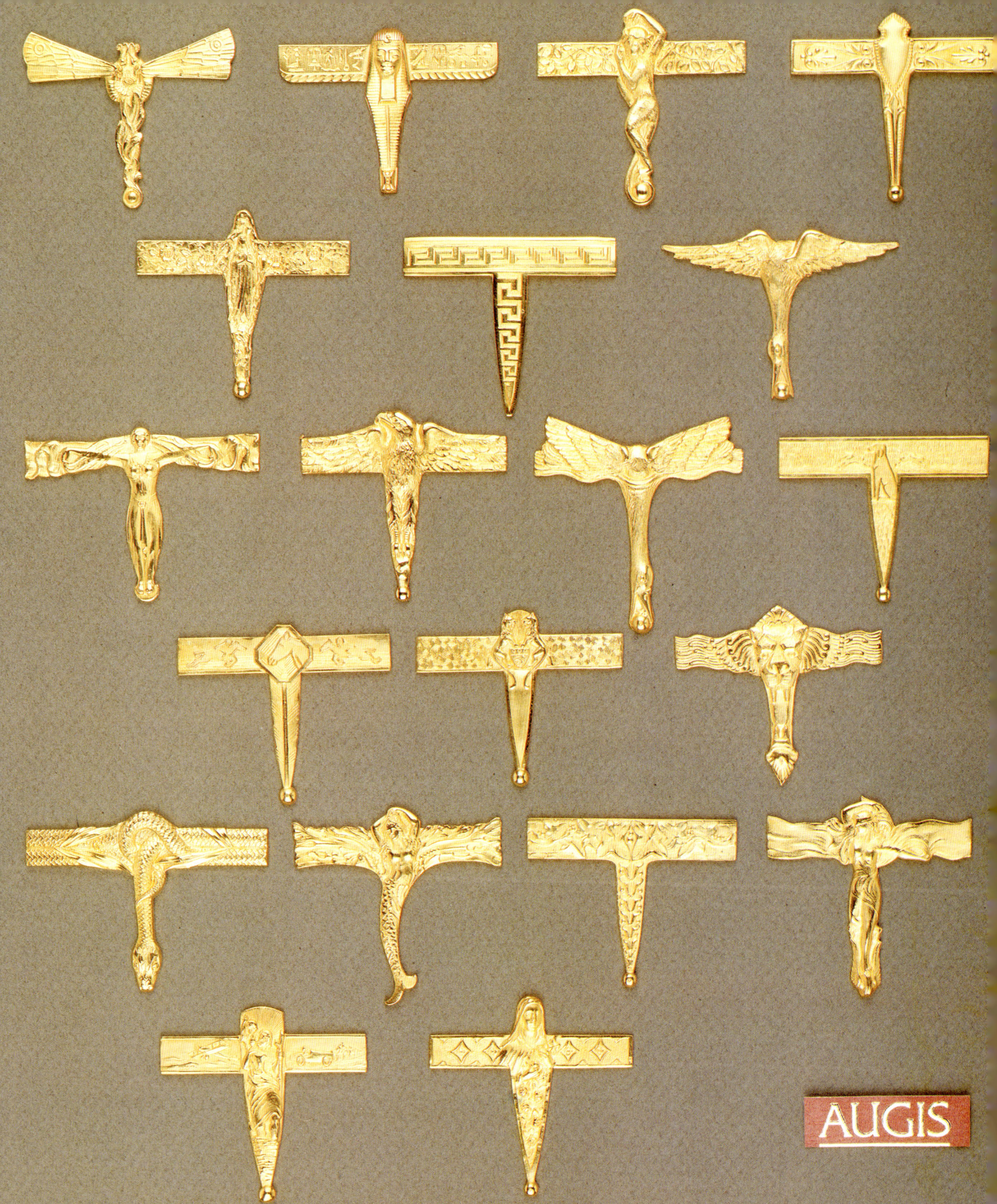

at least equal to that of the pens produced by large manufacturers. In the 1940s, injection molding upset the apple cart. Within a short period of time, only large manufacturers were able to afford the investment required for larger and more complex machinery. Increasing numbers of small craftsmen failed to remain competitive. After World War II, only the major manufacturers could survive and even then, only with difficulty.

The shortage of gold was a concern, but that situation was relatively short-lived. The real problems appeared with the introduction of a new instrument, the ball-point pen. Since it played such an important role in the chronology of writing, we should devote a few lines to its history.

Ball Versus Nib

The idea of using a ball as a tracing point was not really new. In 1888, the American J.J. Loud obtained a patent for a ball-point pen to be used on rough surfaces, such as packing crates. The primary ball turned freely against two small auxiliary balls maintained by a piston. In 1891, E. Lambert offered a model even more ingenious since it had only one ball. In 1899, Varley registered a comparable patent, and in 1910, the German Michael Baums obtained another patent for the ball-point pen. Why did it take half a century for this type of writing instrument to secure a foothold in the industry? The explanation lies in the fact that it could operate with neither the fluid ink of the fountain pen (which flowed without forming a film around the ball) nor the viscous inks of the time, which were too thick, clogged the mechanism, and were too slow to dry.

From 1935 to 1939, Frank Klimes and Paul Eisner manufactured and sold a few thousand **Rolpens**. However, they were not well designed. To force the ink down to the ball, a piston had to be twisted while a button was pushed at the same time.

The true father of the ball-point pen was the Hungarian Laszlo Jozsef (Ladislao José) Biro. He was born in Budapest in 1899 and died in Buenos Aires in 1985. At the age of 17, he

in production occurred in the 1950s. Until then, mass production and individual craftsmanship had co-existed peacefully; this explains the literally hundreds of trade names on the market at that time. The machinery required for a small-scale fountain pen factory was inexpensive and readily available. As a result, the level of quality of pens produced in small shops was

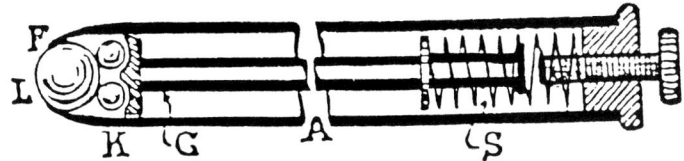

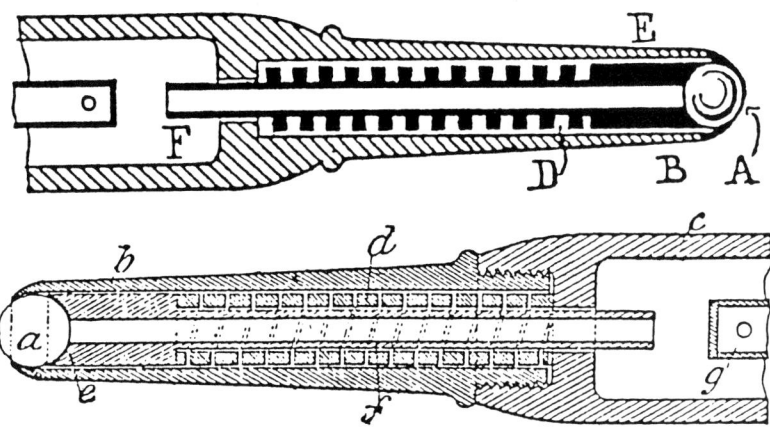

invented a washing machine and later designed an automatic gear box. In 1938, he obtained a patent for his ball-point pen in Hungary. It was a good design, but the environment for further development was not conducive. He took refuge in France and when war broke out, he fled to Argentina where he manufactured a few models and went into partnership with a British financier named Henry George Martin, with whom he founded the Eterpen S.A. Company. A contract with the Allied Armed Forces led to a series of licensing agreements in the United States. The first was with the Eberhard Faber Pencil Co., which, in turn, sold the license to Eversharp. These new instruments appealed to the Air Force because they were less sensitive than fountain pens to variation in atmospheric pressure. At the time, a reliable ink had not yet been developed and the rapid wear of the ball resulted in leaks.

In 1945, Milton Reynolds appeared on the scene. An entrepreneur and opportunist, he realized very quickly that the new instrument could be very successful. Without too much concern for Biro's patents, he commissioned his engineers to design a ball-point pen. His factories churned them out, with his models wholesaling for 70 cents and retailing for $12.50 at that time. On October 29, 1945, the first pens were placed on the market. In the first week, Reynolds sold 25,000 pens. In February 1946, production reached 30,000 pens per day. However, these instruments were disaster prone; they leaked and their thick ink resisted all attempts at cleaning.

All the major producers jumped on the band wagon, including those companies, like Eberhard Faber and Eversharp, who purchased the distribution rights, as well as dozens of smaller manufacturers. Prices fell precipitously to under a dollar. In 1948, having amassed a fortune, Reynolds retired. Faber withdrew from the market in 1947. Although Eversharp had sold millions of ball-point pens, it lost $10 million and would never recover from this setback.

In the early 1950s, there was an upswing in the economy. The major manufacturers took the time to test their models before releasing them. Seech, also a Hungarian, developed an ink better adapted to the ball-point pen. Improved technology and a better selection of materials finally led to the production of reliable instruments that wrote as well as a small metal ball would allow.

The idea of using a ball to write is much older than the two patents obtained by Biro and Reynolds. Top: Loud patent (1888). Bottom: Lambert patent (1891).

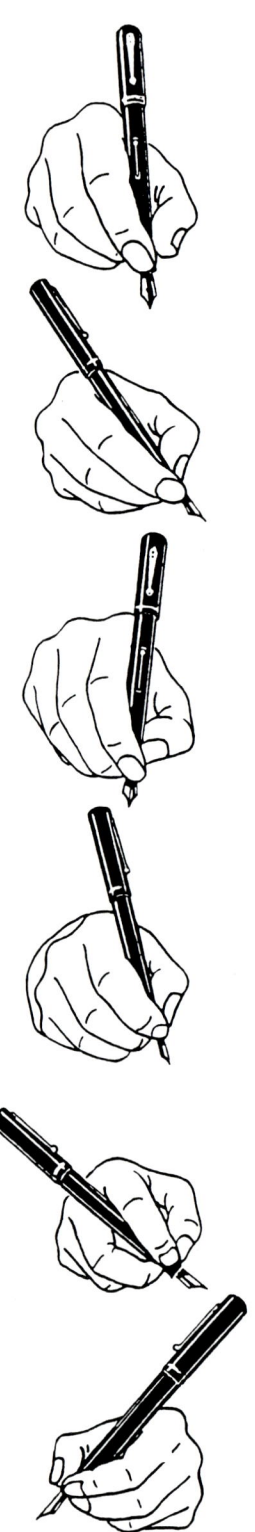

Today I
From top to bottom: Gentleman *in silver by Waterman,*
men's Opera *by Waterman, men's* Centenaire de la
Révolution Française *by Waterman,* Lady Patricia *by*
Waterman, Premier Athens *(lacquer and gold-plate) by*
Parker, Parker 75 *in sterling, green* Duofold *by Parker*
(limited edition), Parker 180 *in sterling, latticed*
Sheaffer Slim, *Sheaffer* Nostalgia *in gold-filled sterling,*
Connaisseur Prestige *by Sheaffer, gold-filled* Targa *by*
Sheaffer, (P.K.).

Waterman Hundred Year Pen *(1940).*

Writing

One has only to compare personal diaries from the 1920s with more recent examples to realize that the art of writing has been lost. Of course, children today are not less intelligent than their grandparents, but their terrible penmanship is appalling! Some may blame the fountain pen for having rendered its predecessor, the metal dip pen, obsolete. But this is not entirely true; during the first half of this century, they co-existed in perfect harmony. Of course, little by little, the fountain pen replaced the dip pen in business and personal use. This evolution was appreciated by everyone, except drycleaners. Nevertheless, it was not the fountain pen, but rather the ball-point pen that usurped the traditional metal nib in the schoolroom. A traditional gift for important milestones, birthdays, and graduations, the fountain pen was ineptly wielded by young hands that had become used to ball-point pens.

Today, some elementary school students are once again using fountain pens, one reason being that manufacturers have begun to produce inexpensive models for these young users. However, these fountain pens come equipped with large spherical nibs that react very slightly or not at all to variations in handwriting. The ball eliminates the need for careful positioning of the hand. Hence, the difference between these fountain pens and the ball-point pen is minimal. Additionally, the ink used in the fountain pen is less greasy and dries more quickly than that used in the ball-point pen.

In the first section of this book, we stressed

Silver Cartier (1938).

Today II
From top to bottom: Pelikan M 800, *Pelikan* Toledo, *gold-filled* Cross, Lamy 2000, Montparnasse *by Dupont, retractable Pilot, lacquered* Vendome *by Cartier,* Parapen *by Omas, a briar-colored Omas, gold-filled Montblanc 146, sterling Montblanc 146, black Montblanc 146, (P.K.).*

the indispensible character of the fountain pen as the writing instrument of our time. Will it remain so into the next century? We can only hope so. Of course, all sorts of innovations in materials, inks, and shapes will appear. But if the writing instrument of the future is a nib with a reservoir, whatever its facade, it will still be a fountain pen, or at least the descendant of the fountain pen. Some say that tomorrow's pen will utilize a laser beam. So be it. We will then have a remarkable instrument, reliable and with considerable autonomy. But we will have to give it another name, because it will no longer be a fountain pen. It will replace the ball-point pen, but not the nib. The nib is timeless; the feather quill nib was used for more than a thousand years and the steel nib for more than a century and a half. The fountain pen has already celebrated its centennial. Those who predict its demise are the same individuals who predicted the death of the theater with the birth of the movies. Although it is true that the ball-point pen posed a serious challenge to the fountain pen, it did not replace it. What it did was eliminate a few manufacturers who were unable to compete on the open market.

Today, the fountain pen has largely regained its place, although the new market has a slight disadvantage. The trend is toward nostalgia; the shapes of the 1920s are back with a vengeance, which is understandable since the shapes and designs of that era have never been equaled. All the major manufacturers are offering fountain pens of unparalled quality, only with new and improved technology and filling mechanisms. Although the market is demanding fountain pens "like before", we will never again see the variety and range of models offered in bygone eras. A good example is the 34 different nibs offered by Waterman in 1909. When all the different permutations of the various sizes and variations of the three base models are taken into account, there were 12,444 possible combinations. In 1933, Waterman alone offered 302 nibs and 375 models.

Fountain pen nostalgia does have its limits, which can only be satisfied through the collection of period pens. If this desire to imitate the past is kept under control, it will be a positive influence, but if it is carried to an extreme, then it will freeze the fountain pen as an image of the past and will entice manufacturers to copy from old catalogs rather than develop new models and technologies. The latter will certainly be more expensive. We can only hope that once the clientele has been regained, thanks to the nostalgia for the old pens, innovations will continue.

Lever filling Waterman,

black hard rubber,

American made

silver trim

(1915-1929).

SECTION III

A Pen and More...

To own two or three hundred beautiful fountain pens

and to write with a ball-point pen; doubtless, is collecting.

To own ten times more or one hundred times fewer,

but to occasionally dip the nibs in fresh ink

and let them glide across paper,

this is certainly the way to become a "collector".

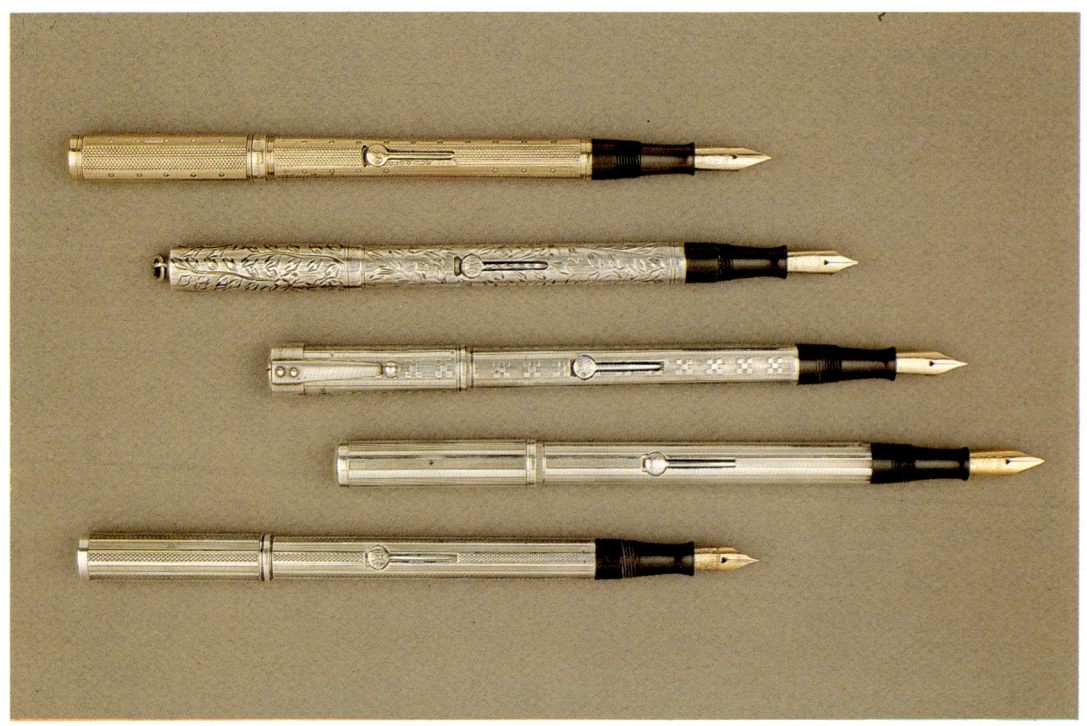

Elegant trim: barley grain, floral, and checkered. Waterman E.C. (end covered) slim models (52 1/2). From bottom to top: Three French models and two English models.

Waterman "Filigree"
These 12 pens are good examples of the "filigree" motif prevalent between 1905 and 1930. From bottom to top: Three standard models, four safeties, five P.S.F.s (pocket self-fillers). These pens, with American style trim, did not sell well in France.

Collecting Pens

It is not surprising that more and more people are becoming fountain pen collectors. Many different things are collected, whether or not they are generally considered to be valuable, all the way from Frank Sinatra autographs to cigar bands. Some collections are slightly bizarre, others are enviable, and almost all are useful, at least to the collector, who finds in them pleasure or a distraction from an increasingly demanding world. Whether they represent a source of profit or of nonprofitable knowledge, what motivates individuals is as diverse as the collections themselves, and is irrelevant; what matters is that we all benefit from the impulse. The contribution of major collectors, and especially of pioneers, is that at some point in time they are able to salvage a number of objects and knowledge that might otherwise have been, if not lost, then at least threatened with obscurity. Collectors often remedy the shortcomings of cultural institutions. When a given object or category of work does not meet the market demand or the priorities for preservation, the bold amateur remains its only chance for survival and restoration. Antique shops and museums would be empty were they not supplied by collectors estates, since personal collections frequently end up in one or the other of these after the demise of the collector.

The fountain pen is no exception to the rule. It is again in vogue, and the dark years are only a bad memory. You will hear no complaints, in spite of the corollary accompanying this resurgence in popularity: the prices of the older model fountain pens have increased exponentially. This is one more reason why the amateur collector should be careful and observe a few basic rules.

During World War II and the post-war years, the scarcity of gold had serious repercussions. Not only were gold nibs no longer manufactured, but existing gold pens and nibs were melted down. In the 1950s, the decline in popularity of the fountain pen and its very small number of collectors exacerbated the problem. Since its value had dropped to almost nothing, a few shrewd individuals started hunting for old fountain pens just to obtain their gold nibs.

Top of the line pens had beautiful cases, but the rest had to make do with modest cardboard boxes, which nevertheless have their own charm.

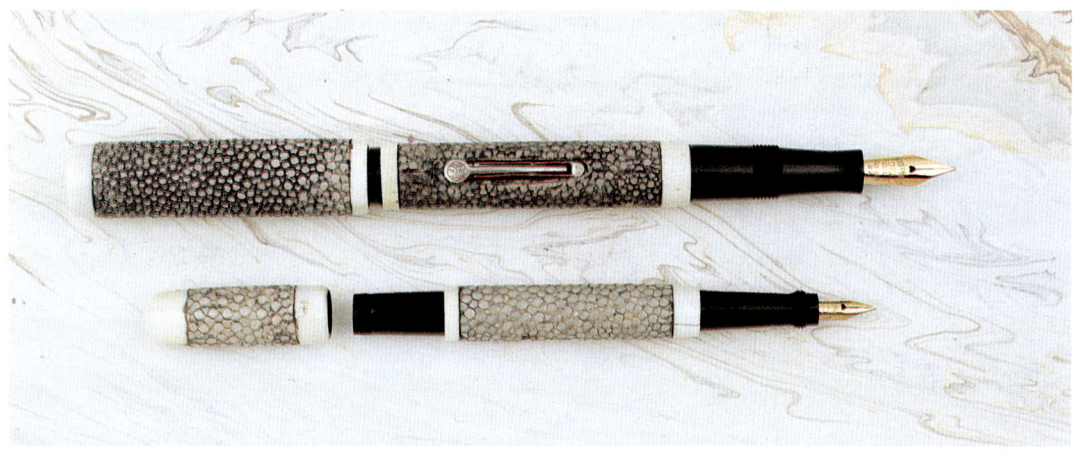

In the 1920s, the fountain pen did not escape the sharkskin fad. Two French models in ivory and sharkskin on black hard rubber.

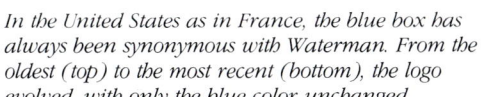

In the United States as in France, the blue box has always been synonymous with Waterman. From the oldest (top) to the most recent (bottom), the logo evolved, with only the blue color unchanged.

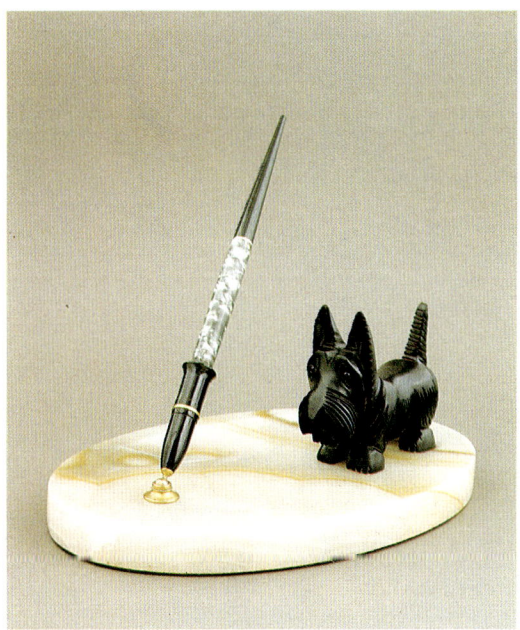

Waterman desk fountain pen with cartridge (1936-1939).

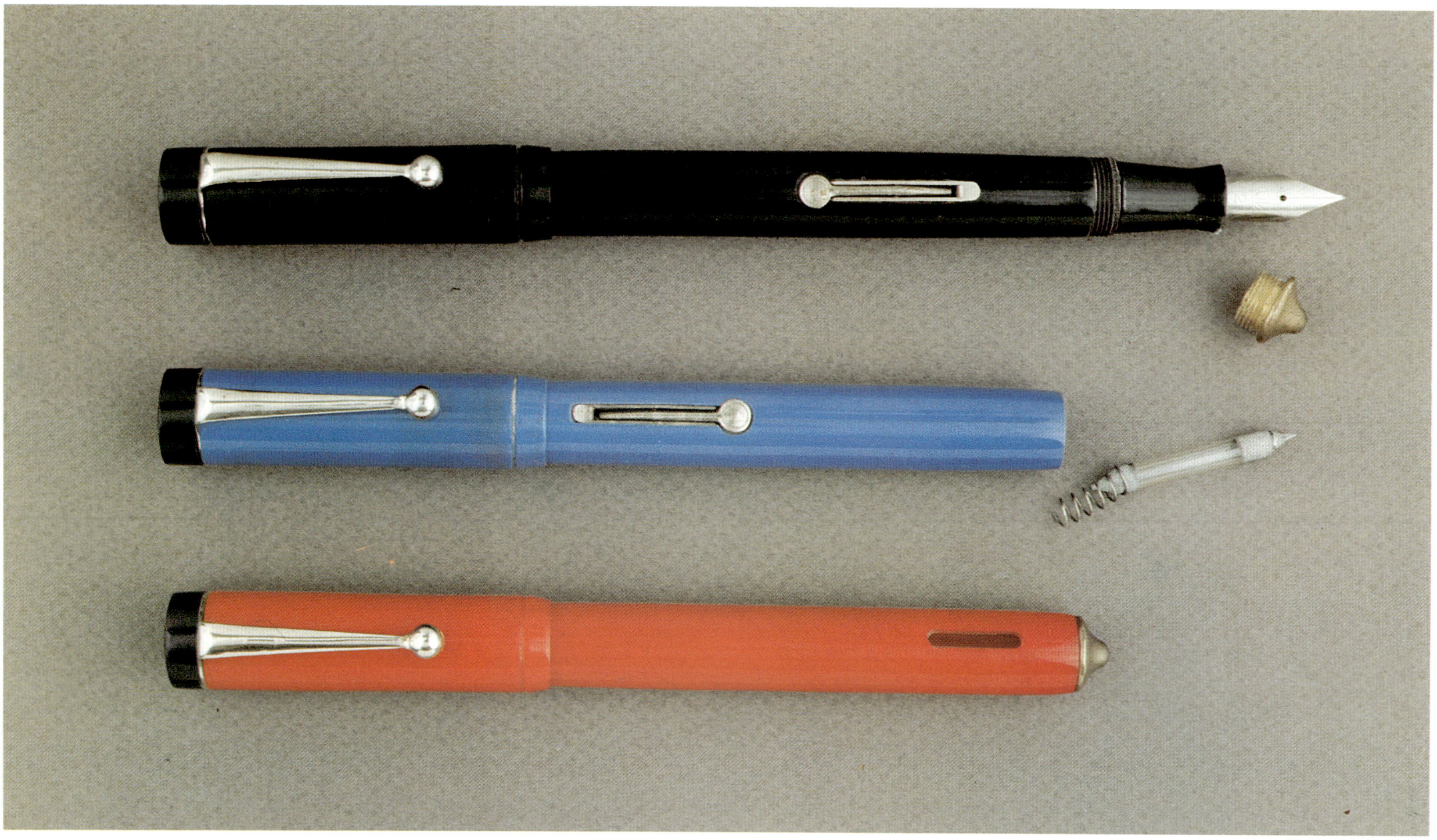

For the driver.
These instruments were made by Bayle and Griffon after the patent was obtained (1928). A neon vial linked to a brass terminal is used to check spark plugs; a small red indicator lights up when the spark plugs are operational.

The Collector
It is a well-known phenomenon: the fountain pen collector browsing in flea markets or antique shops falls in love with a model new to him or her. When the cap is removed, a razor, magnifying glass, thermometer, or even a weapon may be revealed. The link with a writing instrument is rather tenuous...even if it is not a pen, it looks very much like one...it has to be taken home to be added to the collection.

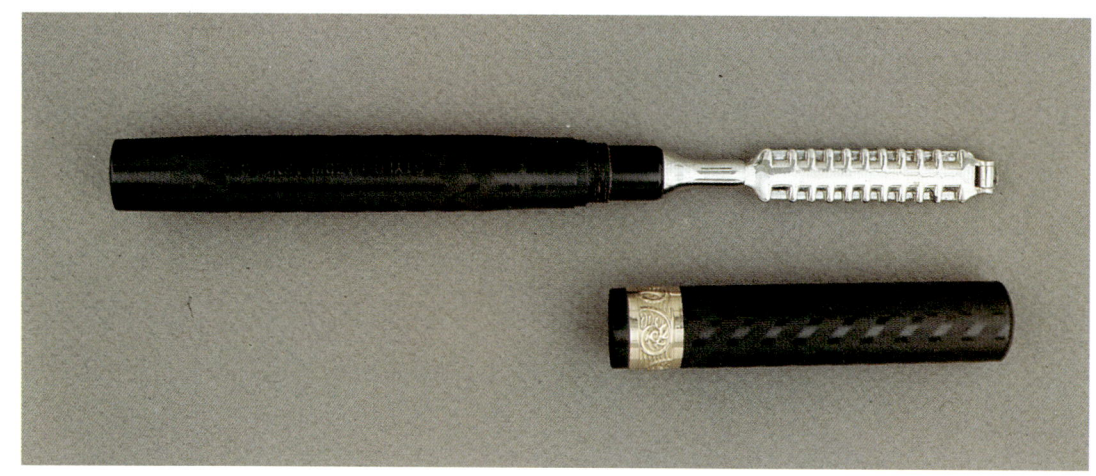

Arnold razor (1913). Of course, it does not write, but it does not shave too badly.

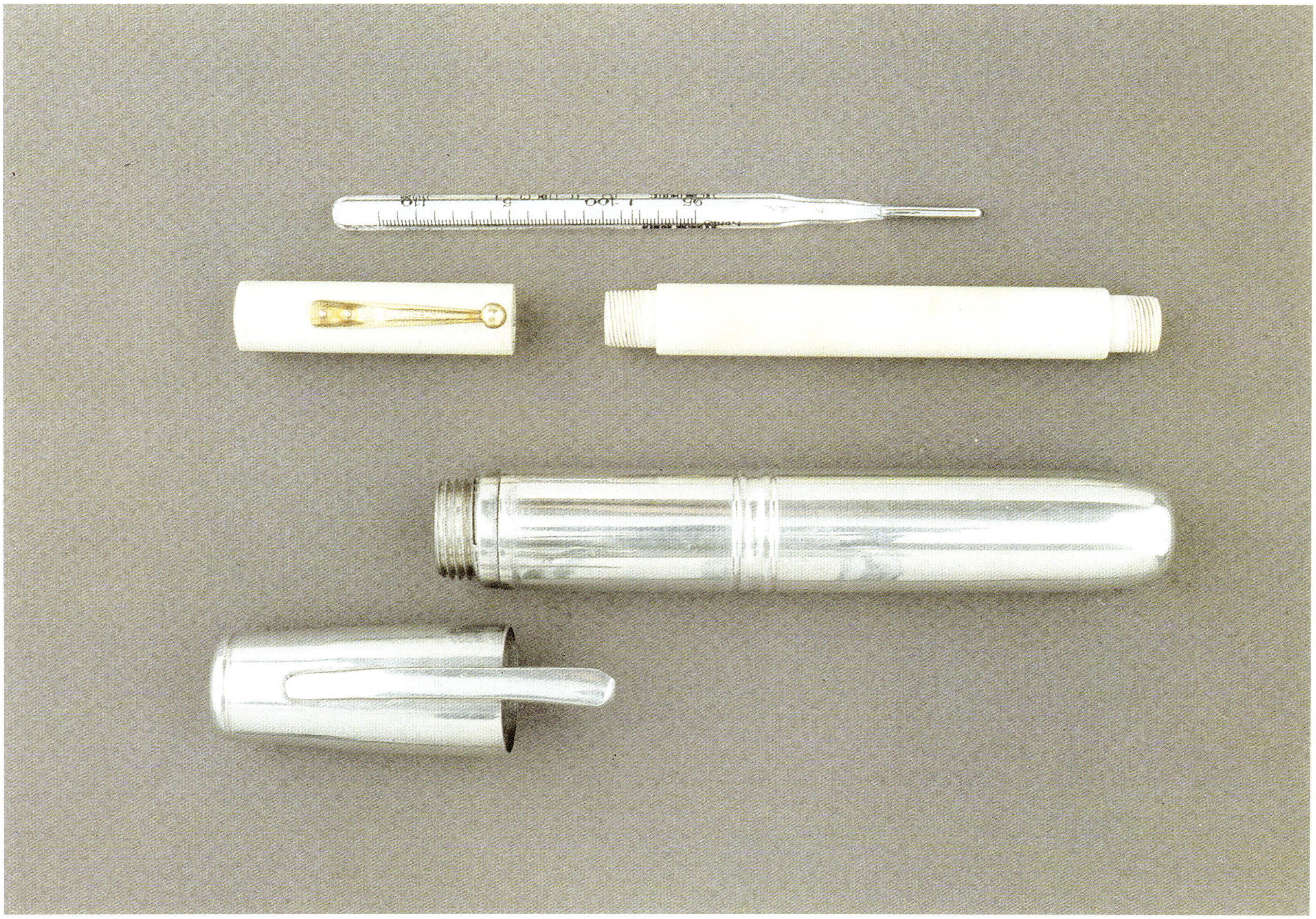

Two pens and two ways to take care of one's health. One contains a thermometer (Waterman, 1927). The other, once the cap is removed, emits a powerful odor of gin (P.H.-L.V.).

Contrary to their harmless appearance, these two "pens" are dangerous weapons. The English model is a pistol with real bullets. The other is an American "pen" that fires tear-gas cartridges.

Gold Starry case in which the history of the company's production is traced. From left to right: Standard 8 Bis (1912), 1 (1920), safety 86 (1929), safety (1933), 35 (1935), Staroid (1935), 27 (1933), silver (1935), (1938), Luxe (1938), 13 (1938), Luxe (1945).

A Stack of Safeties
Waterman pens with retractable nibs, black hard rubber and red marbled rubber, in various sizes (from 42 1/2 to 46) that were offered in the early 1920s. French bodies, clips, and bands.

The less luxurious models fitted with steel nibs were thrown away without much thought. The resumption of production and renewed interest in collecting has put an end to this wanton decimation.

Today, collectors of old fountain pens are numerous and ecclectic. Some are guided by a spontaneous love at first sight. A Parker, a Bayard, an Aurora, a Sailor, no matter what: "I like it, I want to buy it." Others are more selective. There are some Japanese aficionados who only collect Japanese pens, Italians who only have eyes for Eversharps, and Belgians for whom only a Sheaffer will do. They all have valid reasons for collecting specific models. All collectors are confronted with the pitfalls of buying and trading. Old fountain pens were once new pens and, thus, utilitarian objects that, for most part, have been both used and abused. Their owners carried them everywhere...and who could blame them? As a result, a number of surviving pens show their scars.

To the inevitable wear and tear, there are additional problems. Since a fountain pen was frequently an object to which its owner was emotionally attached, each mishap led to repair. During World War II, repairs may not have been made with the proper spare parts, and not all repairmen were equally dexterous. Today, many models cannot be restored because the original spare parts are not to be found.

With luck, a thorough knowledge of the subject, and careful searching, you can find marvelous "old timers" for less money than you would expect to spend on mediocre newcomers. The opposite can also be true. Every day somewhere in the world, and especially during weekend excursions to flea markets,

Lever filling Stellor in plastic (1932). Gold-filled (1938).

Pens Galore
French ladies fountain pens, from inexpensive celluloid to solid gold (Cartier). All date from the 1930s.

irreparable old fountain pens are bought for much more money than excellent new models.

any purchase, a thorough examination should be undertaken.

Careful Acquisition

Except when received through inheritance or as a gift, a collection is built up through buying or trading, and in both cases a certain number of precautions should be taken.

Scratches, slight discoloration of the black hard rubber, a worn inscription, and a dried up rubber sac are all marks of wear that are totally acceptable. Conversely, some defects greatly depreciate the value of a fountain pen, especially if they are not reparable. Before making

The Cap

You should insert the tip of your thumb into the inside of the cap to feel around for cracks or dents not visible from the outside. A bright light inserted into the cap will often illuminate a hairline crack not otherwise visible.

Note that the band around the rim of the cap may not always be the original, and may have been placed there to hide a dent or a crack. It is advisable to familiarize yourself with the characteristics of any particular pen. This is

Two Conklin Endura *pens, with lever, in red and blue sapphire plastic (1927-1932).*

To Twist or Not to Twist
Gold-plated Waterman pens with retractable nibs, sizes 42 and 42 1/2, with French trim, produced from 1907 to the 1930s. The grooved models are the oldest (1907 and 1912). The propulsion mechanism is threaded with a spiral groove; the nib extends in a rotating movement. In the other models, the nib extends without revolving.

especially true of fountain pens in marbled black hard rubber or with engraved metal trim: the cap may have been replaced and no longer exactly matches the body of the pen. You should evaluate the condition of the cap threads by fully unscrewing the cap (in the case of retractable nibs, you should also check the retraction mechanism). If the threads are worn or the cap is not the original one belonging to that particular pen, it will not fit snugly. You should also check to see that the cap fits snugly onto the other end of the pen. If it does not fit well, this might indicate that it has been replaced, or that the body of the pen has been vigorously polished to hide, for example, tooth marks made by a nervous writer. Some manufacturers used to mount the pen clip on the cap with small rivets, which were not very resistant to abuse. If there is no fixed clip, the mounting holes may have been hidden under a mobile clip or may have been filled with wax or another material.

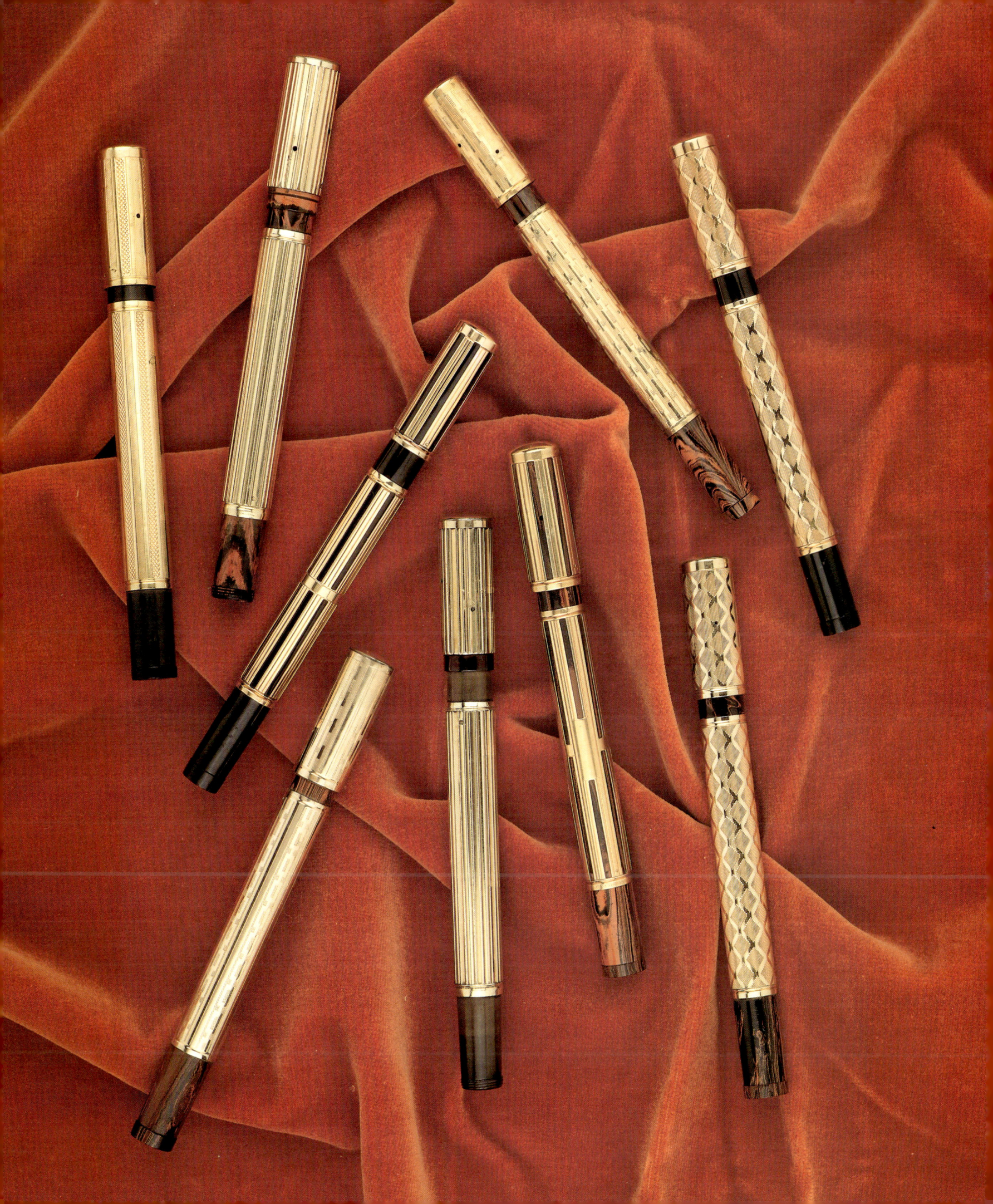

Waterman pen with European trim. Model with ring for hanging (circa 1920) and model with clip (circa 1930).

Dunhill-Namiki
Lacquered pens by Namiki (Pilot) for the Paris Dunhill store. From left to right: 1930, 1927, 1938, 1932, 1932, (K.T.).

The Body

With the cap screwed on, you should rotate the body between your thumb and forefinger in order to spot warping or other distortions. You should check to see that there are no cracks, especially between the joints and other parts, such as the section, lever, or propulsion mechanism.

The Nib

It goes without saying that a fountain pen without a nib is not worthy of its name, unless an identical replacement part is available. The nib is a very important factor in the potential depreciation of the pen. You should check to see that it is not cracked and does not show any superfluous "slits".

Ink-Vue by Waterman with visible ink level. Filling is accomplished using a jointed lever (1935-1940).

L.E.C.s for Ladies
French and American trim on E.C. or L.E.C. (lower end covered) Waterman ladies' pens (52 1/2). Black hard rubber, lever filling. American made trim is recognizable by examining the cover of the cap, which (unlike the French cap) is made of two pieces with visible joints. (United States, 1917-1929; France, 1924-1939).

The condition of the iridium tips should also be examined. Reasonable wear and tear is not harmful, but the complete erosion of one of the two tips renders the nib unusable and significantly reduces the value of less expensive pens. New iridium can be applied.

The Mechanism

With retractable nib models, you should try to retract the nib several times. If it retracts too easily, this may mean that the cork seal is loose. If the propulsion mechanism rotates without the nib appearing, this may mean that the nib is missing or, more seriously, that an internal part is missing or broken. If you acquire a retractable nib fountain pen and the propulsion mechanism does not rotate, you should not apply force because the mechanism could be damaged. Rather, you should fill the pen with cold water to soften up any dried ink, the probable cause of the malfunction. Do not be impatient, this process can be lengthy. When you are dealing with levered fountain pens, gently lift the lever while at the same time pressing a finger on the area where the axis pin is located. You will avoid breaking the lever if the compression bar encounters a dried, rigid rubber sac.

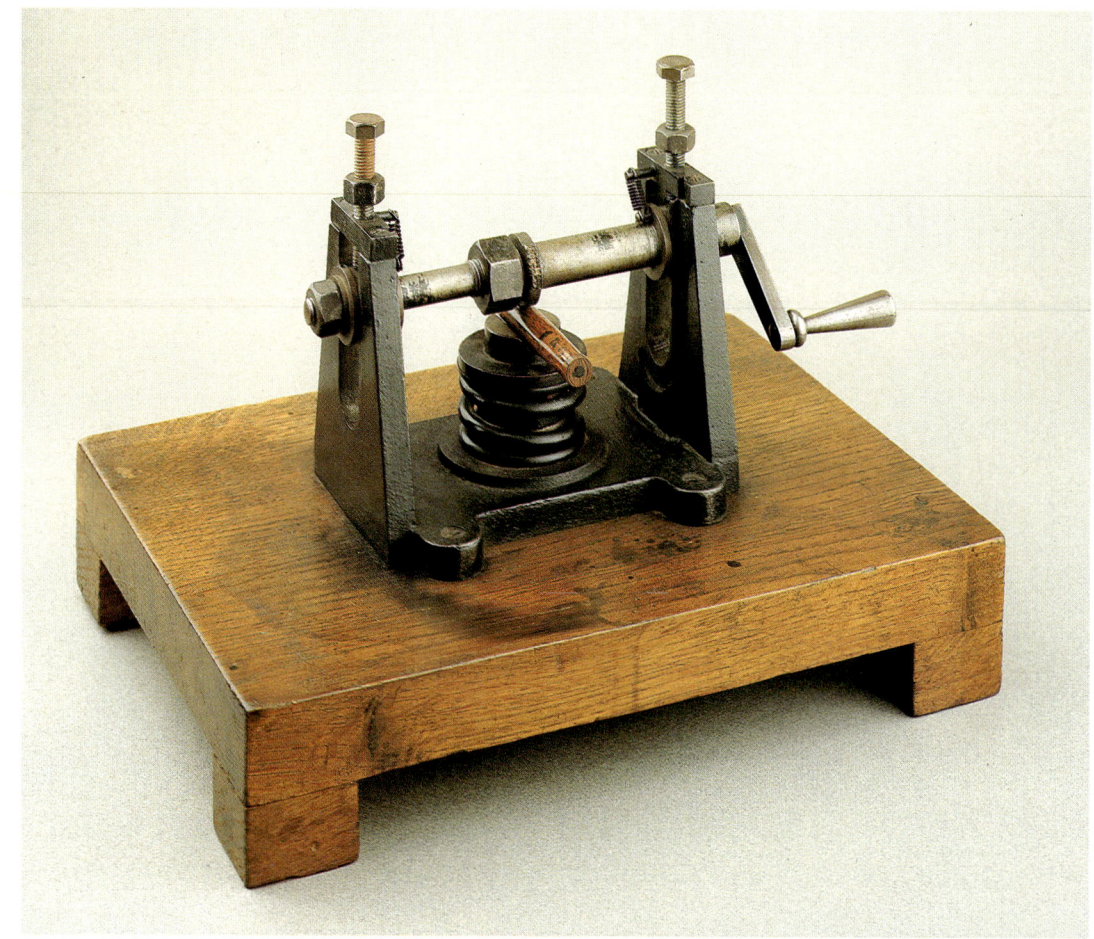

Personalized Pens
A stamping machine allowing the inscription of two to three lines of text on the body of the fountain pen. This text could be the name of the client or the name of the retailer. The process permitted the creation of countless inexpensive brand names. Large department stores or mail order companies often offered articles under their own names rather than under the name of a little known manufacturer. With a few exceptions, such as Kirby Bird for Waterman, the name of the retailer does not appear on the major brands, but the name of the owner often does.

Tools
There was a time when all reputable stores selling fountain pens had a workshop with several employees. Small repairs and modifications were performed. A few manufacturers, such as Atlet, offered complete tool kits, such as the ones pictured here.

If the pen has a push button, activate it. If it does not respond, this may indicate that the rubber sac has dried up. If it remains retracted, this may mean that the sac is missing, or even worse, that the compression bar is missing or damaged.

Piston models must be handled with great care because they are very difficult to repair. Try to manipulate the piston, but be aware that it is often adhered to the wall by dried ink. For more sophisticated models, such as Parker's **Vacumatic** or certain Stylomines, it is very difficult to judge their state of repair when you buy them at a flea market.

The Wise Buyer

Whether it is acquired through a purchase or a trade, an antique fountain pen without any major flaw will still need cleaning. This operation is not without potential pitfalls. It is always advisable to read the user's guide, when one is available.

You should also know what comprises the material of the fountain pen. Most materials are sensitive to heat, especially hard rubber. Celluloid is extremely flammable, and other nitrocellulose plastics are only a little less so.

Repair Kit
The major fountain pen brands gave their distributors tool kits especially designed for their models. The one pictured above is for Parker. This kit allowed a full range of repairs for the Vacumatic. *After World War II, the temporary decline of the fountain pen and the tendency to throw away the pen rather than repair it rendered these kits obsolete.*

Therefore, considerable care should be used when applying heat. The part should be moved through the heat source.

To dissolve dried ink, you should use only cold water. This should be done with care because certain materials become discolored or distorted when they remain immersed in water for any length of time. You should be extremely wary of bleach; it is very useful in removing ink stains but, even when heavily diluted, it can damage gold nibs. Do not use rubbing alcohol because it damages plastics. Therefore, use cold water and only cold water. In most cases, a simple bath is not going to be enough to restore the appearance of old fountain pens. You will need to tinker with them...but only on

(Preceding page)
Art Deco
Laureau's fountain pens went down in history when they were awarded a silver medal at the 1925 International Exposition of Decorative Arts. That medal saved Fernand Laureau from oblivion, since he produced very little under his own name and worked mostly as a subcontractor for Parker, among others. These retractable nib pens are made of hard rubber with a chiseled silver trim.

Illustration for the cover of a 1912 Kaweco catalog. Two years later, the call to arms was in significant contrast.

Silver Levers
Lever filling Waterman pens, 14 P.S.F.; sizes 52 1/2, 52, and 54; black and red hard rubber; French bodies with silver overlays (1915-1932).

the condition that you feel the pen is worth the effort involved.

A word of caution...if the repair required exceeds your competence or if you do not have the proper tools, it is far better to go to a professional; there are still a few very competent individuals left.

Those who have the knowledge, tools, and skill necessary also will need the spare parts. Today, one of the best ways of obtaining these is to buy fountain pens in poor condition. From these rejects, there is always at least one part you can salvage that can be used to restore a perfect body. A good repair job should not only produce a fountain pen with its former appearance, but also should ensure its proper operation, namely, its ability to write well.

If the nib and the section have been removed, they must be replaced very carefully. Frequently, the nib slightly distorts the inside of the section by creating its own imprint. The position of the nib relative to the feed is easy to determine. The tip of the feed is generally located approximately half-way between the point and the eye. If the two ends of the nib are spread apart, they should be bent slightly towards the point. If they are overlapping, they should be bent slightly away from the point. If the point is rough and does not glide easily across the paper, a few figure eights scribed onto super fine sanding paper should restore its smoothness.

In principle, all parts should be original spare parts, with the exception of rubber sacs. Whether dried up, cracked or missing, they have to be replaced. The source of the new rubber sac is of little importance.

Once the fountain pen has been cleaned and repaired, it can be filled with ink if you wish to write with it. It is important to use the correct

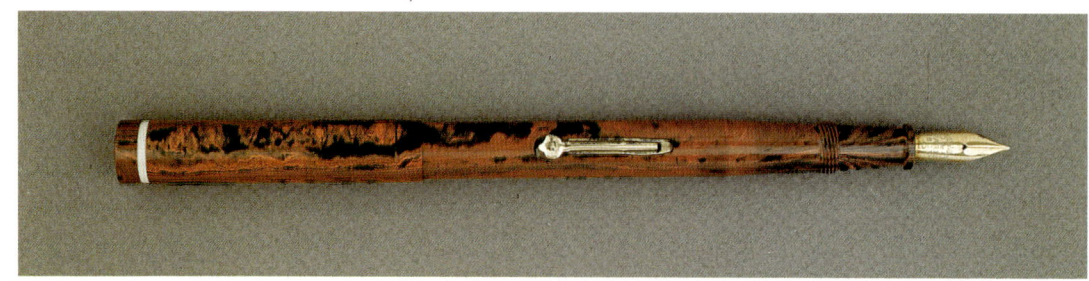

Lever filling Meteore, *marbled red hard rubber (1926).*

Lever filling Semper *by Paillard, black hard rubber (circa 1935).*

ink (e.g., blue or blue-black) for any given fountain pen. Once your prized pen is ready to join its fellows in your collection, do not forget to empty it and rinse it out several times in cold water. Never leave ink in an unused pen. It will not only clog the capillary grooves (which is not serious for a pen not used for writing), but the acidity of the ink will also increase with evaporation. In that case, there can be considerable damage, since plastics may be discolored and the metal parts may corrode.

Black rubber is sensitive to heat and also to light, which discolors it over a period of time. This means that a collection of black hard rubber fountain pens should not be left in a case exposed to direct sunlight. However, this does not mean that, like vampires, black hard rubber fountain pens cannot withstand a moderate amount of light.

If you wish to use an antique fountain pen, a few precautions are in order. Use only fresh ink. Any bottle of ink that has been open for more than 1 year should be discarded. In case of an interruption in use, even for a few weeks, you should empty and rinse your fountain pen. Do not expect that a nib shaped by another hand will adjust easily to yours. On the contrary, compatibility will take some time, but will eventually occur. So, without being rude about it, try not to lend your fountain pen to another person for any extended period of time; however, a few words written by another hand will not alter the shape of the point.

French Dressings in the 1920s
Lever filling Waterman pens, sizes 52 1/2 and 52, in black hard rubber or red marbled, French bodies (1917-1933). Four in solid gold and four gold-filled.

Avec l'ONOTO
on peut écrire en route

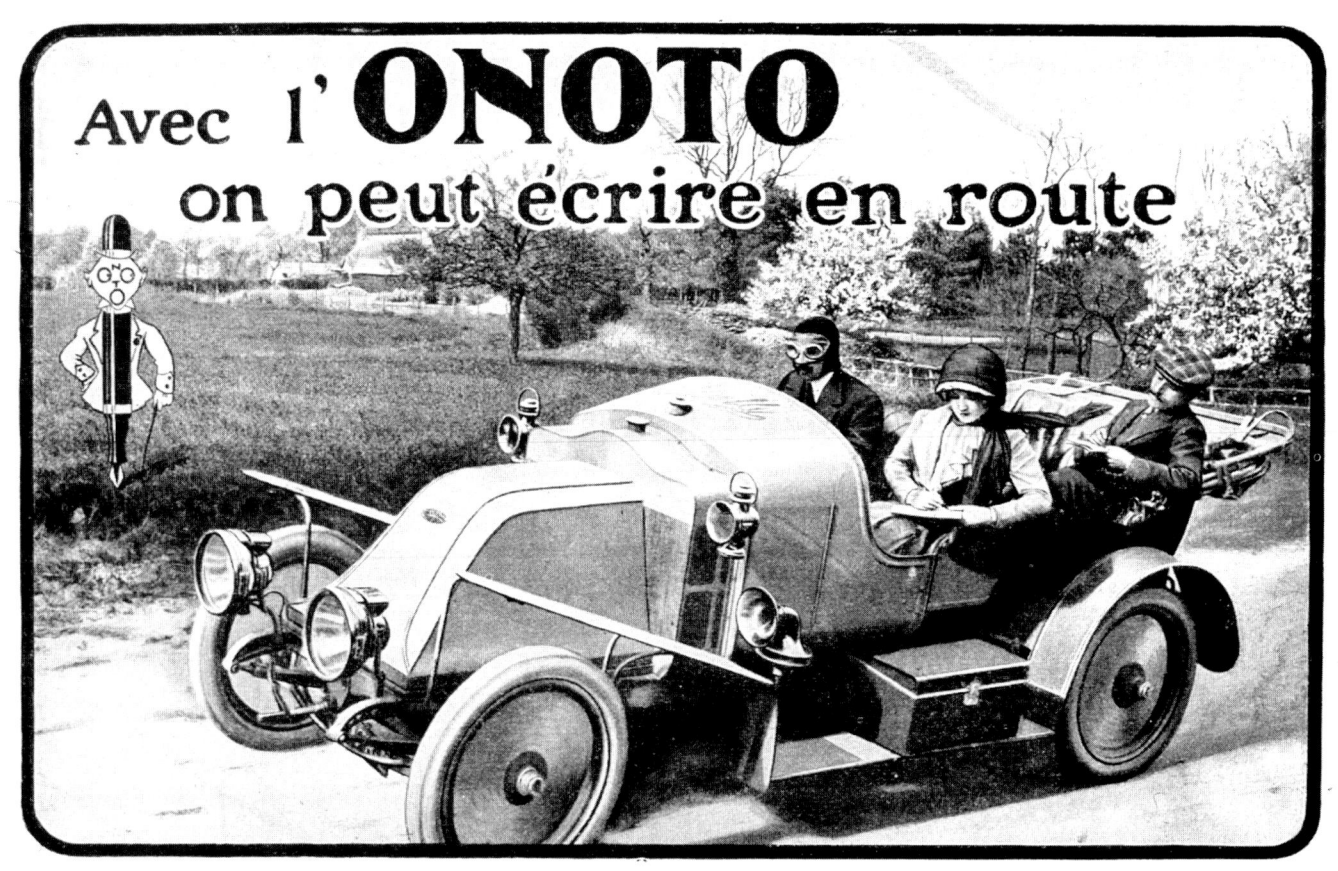

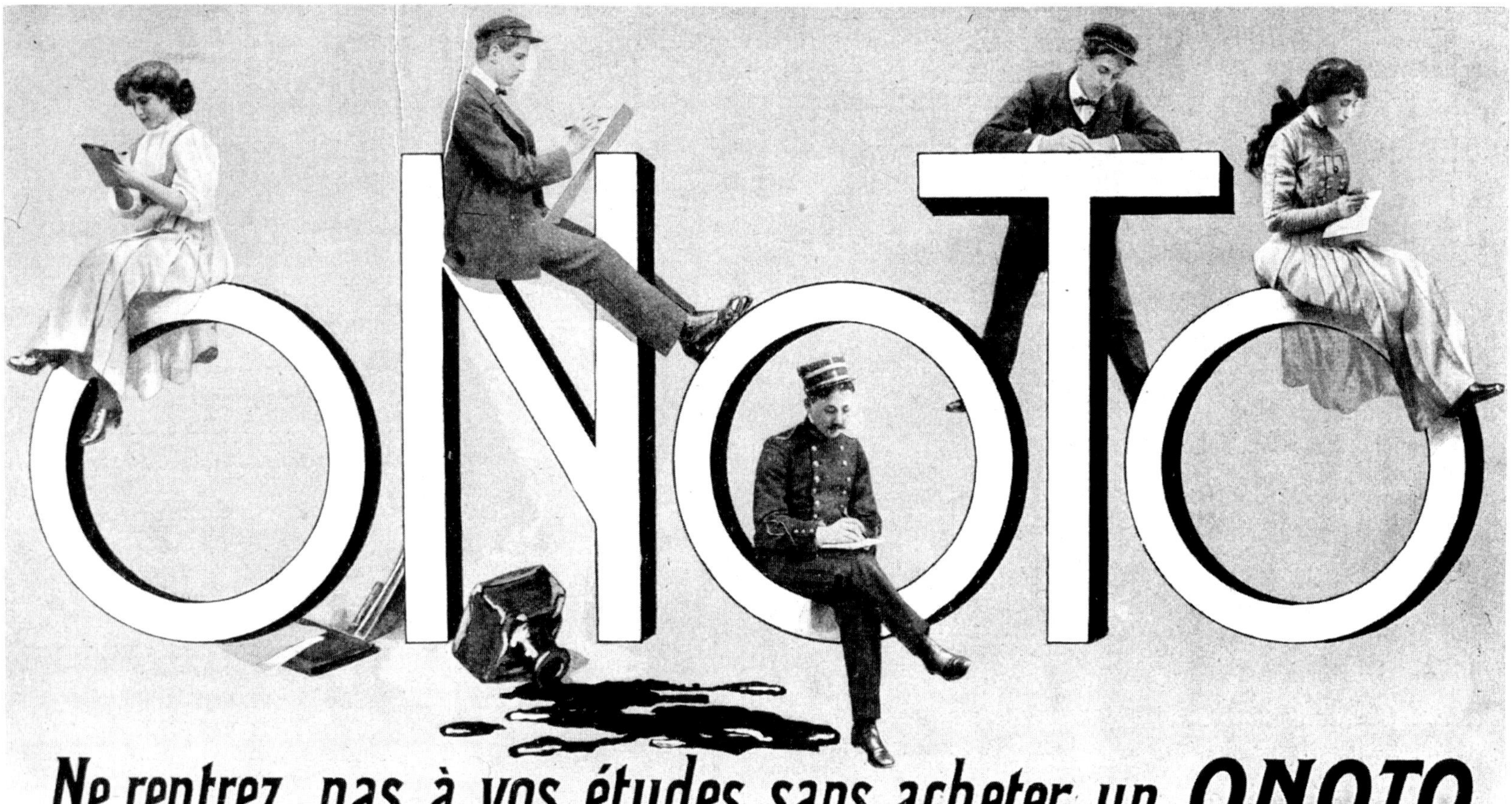
ONOTO
Ne rentrez pas à vos études sans acheter un ONOTO

Waterman ladies fountain pens, L.E.C. casings, silver, gold, gold-filled (1924-1930).

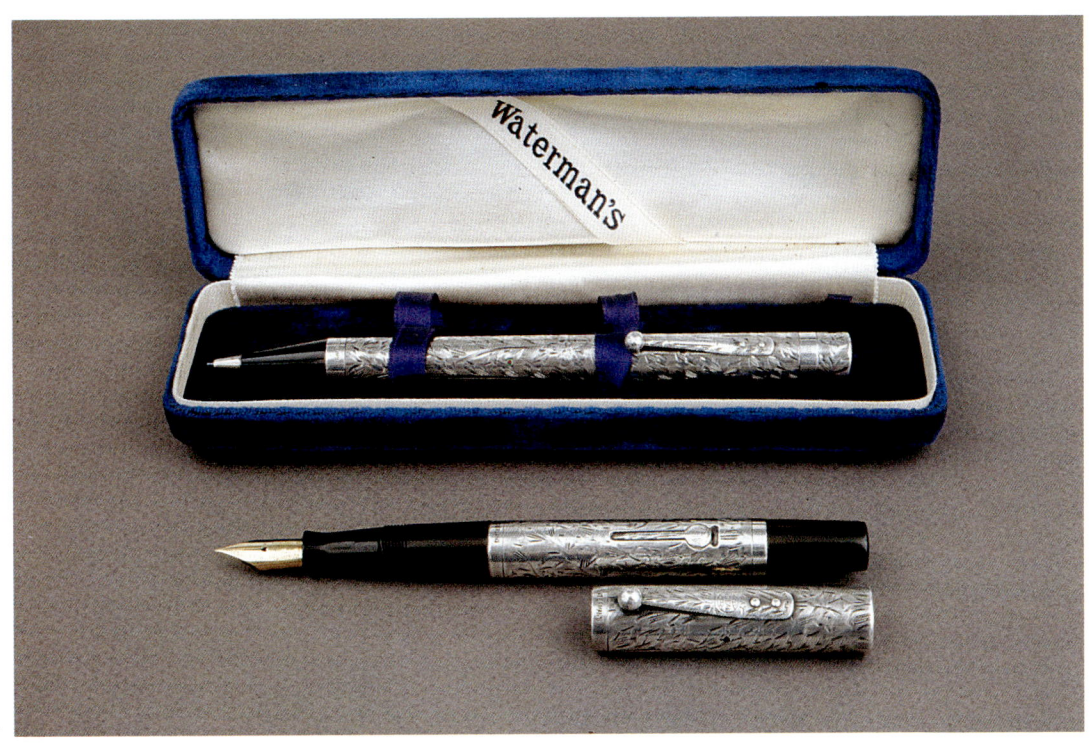

Waterman fountain pen and mechanical pencil set, silver body, floral trim (1915-1929).

Lever filling Waterman pens (52 v.), French L.E.C. casings over black or red marbled hard rubber (1924-1939).

At the beginning of this century, Hardtmuth represented Waterman in France. Soon after the opening of the Paris office (1904), the company published these promotional postcards that were given away to customers.

Waterman promotional cards (1884).

Glastnost

The ability to see the ink level while writing contributes greatly to the user's sense of security. Despite a few short-lived attempts (Waterman in 1903 and Swan 10 years later), that luxury remained unavailable for a long time. Hard rubber is not transparent and neither is the rubber of the ink sac. With the introduction of plastics, a see-through ink level was a very important sales pitch. In the 1930s, all the manufacturers joined the band wagon. The Nozac model by Conklin had an exclusive "word gauge"; its transparent body was graduated according to a unique measuring unit: 1,000 words. Determining the ink level was possible since all or part of the body was transparent. In most models, it was necessary to unscrew the body to view the reservoir or the cartridge, or to view the transparent plunger located in the barrel.

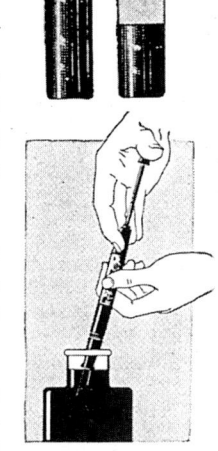

aucun obstacle n'est insurmontable...

TANK-400

Le stylo à grande contenance garanti pour l'existence.

C'est un bon stylo qu'IL désire..!

— ALORS, c'est le **TANK-400** qui LUI fera le plus plaisir...

car c'est la grande nouveauté. Ignorant la panne, supprimant l'encrier, c'est le stylo de l'homme de lettres, du journaliste, de l'homme d'affaires, du comptable, de l'étudiant. Un instrument de travail moderne et chic, dont le corps formant cartouche interchangeable à niveau visible, permet d'écrire *à la plume et à l'encre* sans arrêt.

Le TANK 400 en écrin de luxe avec ses quatre cartouches de rechange remplies d'encre
BLEU RADIO ou BLEU NOIR
Stephens' extra fluide

GARANTIE A VIE — *Où que vous soyez si le TANK 400 ne vous donne pas satisfaction entière, présentez votre bon de garantie au papetier de la ville; il vous sera échangé immédiatement et sans frais*

DESCRIPTION

1. le **CAPUCHON** avec son clip de sûreté, véritable pièce de mécanique de précision.
2. la **CARTOUCHE** interchangeable formant le corps du stylo à niveau d'encre visible.
3. la **SECTION PLUME**, qui avec ses perfectionnements, constitue l'âme du stylo le plus moderne.

Autres avantages
Entièrement en PLEXIGLAS, donc INCASSABLE, Clip, joncs et plume en métal doré à l'or fin. INALTÉRABLES.

Ets Pierre BAIGNOL & Cᵒ

USINES & BUREAUX : 19, rue de SARTORIS
LA GARENNE - COLOMBES (SEINE)

APPENDIX

Significant Names in the Fountain Pen Industry

Charles Abel (Paris, France). *Clebs, Million* (1934), *Oldchap* (1921) fountain pens.

Eugene Abel (Paris, France). *Atlet* (1924, tools for repair), *Babel* and *The Traveller* fountain pens.

Agap (Paris, France). *Analpen* and *Club* fountain pens.

Aikin Lambert (New York). Company founded in 1864. Manufacturer of gold nibs. Began manufacturing its own fountain pens circa 1890. Victim of the Great Depression of 1929, it was acquired by L.E. Waterman in 1932.

Akira Seishindo Seisakusho (Tokyo, Japan).

Alkovitszky (Paris, France). Nibs and trim.

Amalgamated Gold Pen Makers (London/Liverpool, Great Britain). Nib manufacturers.

The American Fountain Pen Co. (Boston, Massachusetts). Founded in 1899. One of the first to manufacture retractable nib fountain pens, the *Moore Non-Leakable*, without a rotary mechanism, but with a slide (patented by Morris W. Moore). Became the Moore Pen Co. in 1917. Adapted the innovations of major manufacturers in later years. Went out of business in 1956.

Ancora (Arona, Italy). Founded in 1909 by Giuseppe Zanini. After his death in 1929, his son Alfredo took over the business. The trademark is symbolized by an anchor (ancora). Went out of business in 1975.

Anglo American Co. (Paris, France). 28, boulevard Poissonnière. Importer since 1887. First exclusive distributor for L.E. Waterman.

Artus (Heidelberg, Germany). Artus Fullhaltergesellschaft Kaufman & Co. See Lamy.

Ateliers Francais (Issy-les-Moulineaux, France). Jacques Bonhomme (1920).

Augis (Champagne-au-Mont-Dore, France). Jeweler and medal manufacturer. Established in 1830. Produced a large range of high quality clips.

Aurora (Turin, Italy). Fabbrica Italiana di Penne a Serbatoio. Founded in 1919 by Isaia Levi. Products sold in France by Edacoto (1931). Still in operation.

Badois (Paris, France). Trademark registered in 1921 by Louis Badois, one of the partners of La Plume d'Or.

Pierre Baignol (Paris, France). *Tank 400* fountain pen (circa 1947).

Baruzzi-Ostal (Nice, Alpes-Maritimes, France). 1910. *Champion*, *Mignon*, and *Scribo* fountain pens.

Baudinière (Paris, France). *Luxor* fountain pens (circa 1930).

Bayard (Paris, France). In 1903, Étienne Forbin was a distributor for Nauheim & Co. (New York). In 1912, he registered many trademarks of the fountain pens that he distributed (*Bayard*, *Excelsior*, *Alpin*, etc.). Some were manufactured at his factory in Arbois (Jura). In 1922, his nephews took over the business and the trade name was changed to Panici Frères et Cie. The initials P.F. are inscribed on the nibs. The workshops on the rue des Cordelières in Paris gave the company a new distinction. It was one of the most prominent fountain pen companies until it went bankrupt, like many others, in the 1960s.

Bayle (Lyon, Rhône, France). *Mon Plaisir* fountain pens (circa 1929).

Beaufils (Paris, France). See Stellor.

A. Bendoni (Paris, France). Nib manufacturer.

B & D (Brussels, Belgium). Beirlaen and Delen. See Bermond.

Bermond (Brussels, Belgium). Trademark registered in 1919 by Beirlaen and Delen (B & D). Bermond is the contraction of BEirlaen and RayMOND. Another trademark: *The Scout Pen*. Manufactured in Germany by Mertz and Krell. After World War II, Bermond imported Biro ballpoint pens. The company still operates under the management of Penn & Co.

Blair (New York). Solid ink pens with a water reservoir.

Blanzy-Poure (Boulogne-sur-Mer, Pas-de-Calais, France). Founded in 1846 by Pierre Blanzy and Eugene Poure. One of the largest steel nib manufacturers. Produced nibs for the *Chromix* fountain pens.

The Boston Pen Co. (Boston, Massachusetts). Founded in 1894. Acquired by the Wahl Adding Machine Company in 1917.

Bourbon (Saint-Lupicin, Jura, France). Manufacturer of pen bodies.

Brause & Co. (Iserlohn, Germany). Founded in 1850 (steel nibs).

Camel Pen Co. (Orange, New Jersey). Founded in 1935 by Joseph V. Wustman. Solid ink pens filled with water. Good quality but poor sales. The company went bankrupt after 3 years.

Carey Pen Co. (New York). Founded in 1890. Went out of business in 1915.

Carpentras et Donarier (Nantes, Loire-Atlantique, France). *Solveig* fountain pens (1945).

*****Carter** (Cambridge, Massachusetts). The Carter Ink Co. was founded in 1858 by John W. Carter. It quickly became one of the major ink manufacturers in the United States. From 1926 to 1931, it manufactured high quality fountain pens in unusual colors, but of classic design.

Cartier (Paris, France). Established in 1847. Jewelers. Began manufacturing gold dip pens in 1860. Early in the 20th century, Cartier began designing trim for fountain pens. In the 1930s, the company designed a numbered series of small fountain pens. Since 1968, it has been involved in the development of high technology production. The *Must* of Cartier. The Compagnie des Technologies de luxe was founded in 1988.

Castela (Toulouse, Hautes-Pyrénées, France). *Wildaup* fountain pens circa 1922.

Caw's Pen & Ink Co. (New York). Manufacturer of ink and fountain pens (1890-1915). Received a silver medal at the International Exposition of Decorative Arts in 1900.

Charcellay (Paris, France). In 1910, Charcellay established the Franco-British Manufacturing Pen Cie. After World War I, the company expanded rapidly and became one of the most important in the French market. In the 1930s, it took the name of the Compagnie française pour la fabrication de P.P.R. Despite the quality of its products, Charcellay is not very well known because, like Laureau and Sabon, its models were marketed under other trademarks.

Chemol (Budapest, Hungary).

Chilton (Boston, Massachusetts). Founded in 1923 by Seth Chilton Crocker. The company moved to Long Island City, New York in 1932, later to Summit, New Jersey, and went out of business in 1941. The fountain pen was filled through two tubes and pressure was applied by a column of air on the sac. See Crocker.

Christian (Paris, France). Christian Boursier. See Evergood.

Cisea (Turin, Italy). *Radius* fountain pens.

*Conklin (Toledo, Ohio). Founded in 1898 by Roy Conklin, it was the first manufacturer to offer a satisfactory filling system utilizing the compression of a rubber sac (1901). Roy Conklin sold his shares in the company in 1903. Conklin remained one of the top four United States manufacturers until the early 1920s when it was surpassed by Wahl. Entered the French market in 1920. Went out of business in 1947.

Conté (Paris, France). Pencil manufacturer. *Monte Cristo* fountain pens.

Conway Stewart (Enfield, Great Britain). Founded in 1905 by Thomas H. Garner and Frank Jarvis. Manufacturer of the first Gold Starry pens. Went out of business in 1975.

Crocker Pen Co. (Boston, Massachusetts). Founded in 1902 by Seth Crocker. His son, Seth Chilton Crocker, took over the business in 1907 and in 1923, founded the Chilton Pen Co. In 1931, he sold Crocker Pen to N. Zaino. See Chilton.

A.T. Cross (Boston, Massachusetts/Lincoln, Rhode Island). The date when the company was established is difficult to determine, because the frequently quoted year 1846 refers to the birth date of Alonzo Townsend Cross (1846-1922), who succeeded his father, Richard Cross, and his great uncle, Edward Bradbury, who had made writing instruments since 1843. The elder Crosses were English and had only recently settled in the United States. Alonzo was born in Birmingham. Until 1881, the company was called Richard Cross & Son. The initial products were largely mechanical pencils and pencil caps. Beginning in 1877, Alonzo Townsend Cross manufactured the needle pen and was the first to market an instrument called the *stylographic pen*. A few years later, he became the first manufacturer of that type of instrument, while his production of nib fountain pens remained insignificant. In 1916, the company was acquired by Walter Russell Boss. In 1935, the distinctive black half cone trademark appeared. Since 1946, the superb *Century* has been produced without major modifications. In 1953, the company launched the production of ballpoint pens. In 1963, the word "Pencil" was dropped from the company's name and it became A.T. Cross Co. Still in business, it is the dean of all United States writing instrument manufacturers.

Daussy (Paris, France). *Le Pneu* fountain pens.

De La Rue (London, Great Britain). Founded in Guernsey in 1813 by Thomas de la Rue, who moved to London in 1816 and made a fortune by printing playing cards. Throughout the 19th century the company developed and diversified. It produced stamps, bank notes, stationery, and writing instruments. After manufacturing a few unsuccessful models (*Swift*), real fountain pens were first produced in 1895 (*Pelican*) and appeared on the French market in 1898. The first *Onoto* appeared in 1905 and its success lasted for many years. It is said that the name was chosen because it is pronounced the same way in most languages. The *Onoto* was produced for more than half a century. In 1958, De La Rue gave up manufacturing fountain pens and the rights were acquired by its Australian distributor.

Delolme (Saint-Claude, Jura, France). *Jemco* fountain pens.

Demilly & Degan (Paris, France). See La Plume d'Or.

Devarson (London, Great Britain). See Lincoln.

Dhome (Paris, France). DOM nibs.

Dolfina (Doorn, The Netherlands).

Dunhill (London, Great Britain). Founded in 1907 by Alfred Dunhill, who was originally a purveyor of fine tobaccos. In 1930, Dunhill sent its fountain pens to be lacquered by Namiki (Pilot). These pens are manufactured today by Montblanc (which belongs to the Dunhill group).

Dunn (New York). Founded in 1921. Went out of business in 1924.

Dupont (Paris, France). Founded in 1872 by Simon Tissot-Dupont. Manufacturer of luxury leather goods and of lighters since 1939. Manufacturer of fountain pens since 1973. Production facilities in Faverges (Haute Savoie).

Eagle Pencil Co. (New York). In 1856, Henry Berolzheimer founded a company in Furth, Bavaria. He moved to New York in 1860 and shortened his name to Berol. He became one of the most important pencil and nib manufacturers in the United States. The company produced the first fountain pens with glass cartridges in 1890 and, in subsequent years, produced low-end, affordable fountain pens.

Edac (Paris, France). Originally, the company was a mechanical pencil company. However, it produced a few fountain pens, one of which was metal with an automatic filling system. In 1931, the company distributed Aurora pens in France and sold them as sets, together with automatic pencils, under the name, *Duo Moderne*. In 1959, it became one of the manufacturers of the *Visor Pen*.

Edacoto (Paris, France). See Edac.

E.J.S. (Paris, France). See Laureau.

Eskesen (Copenhagen, Denmark).

Elmo-Montegrappa (Bassano, Italy). Founded in 1921 by Alessandro Marzotto.

Evergood (Paris, France). Facilities in the Conflans-Sainte-Honorine. Manufacturer of mechanical pencils. Christian Boursier, its founder, began producing fountain pens during World War II.

Faber-Castell (Nuremberg, Germany). Faber's history dates back to 1761. Kaspar Faber founded a pencil factory, but the initials of his successor, Anton Wilhelm Faber (1758-1819), appear in the trade name. In the second half of the 19th century, three brothers headed the company. Eberhard moved to New York in 1849 and founded his pencil company. Johann remained in Nuremberg and in 1878 established a company bearing his name (on the French market: *Le Métro* [1904], *Neptune*, *Cardinal*). The eldest brother, Lothar, managed A.W. Faber until his death in 1896. In 1898, Count Castell-Rudenhausen, who had married a Faber, became head of the company and added his own name to the trade name (on the French market: *Flamingo* [1904], *Novellus* [1905]). A.W. Faber-Castell strengthened its position in the fountain pen market by buying the Johann Faber company in 1932 and buying into Osmia in 1935. Faber-Castell is today a very important company in the area of writing and drafting instruments, but it gave up manufacturing fountain pens in 1975.

Forbin (Paris, France). See Bayard.

Fortin & Cie (Paris, France). 59, rue des Petits-Champs. Stationery manufacturer since 1802, specialized in steel nibs. Began manufacturing fountain pens circa 1894.

The Franco-British Manufacturing Co. (Paris, France). See Charcellay.

I. Frank (Paris, France). 15, rue des Petits Carreaux. Stationers under Napoleon III and distributors of John Mitchell (metal nibs). I. Frank began selling fountain pens circa 1894. In 1934, the company changed its name to Penlex.

Frazer & Geyer (New York). See Lincoln and A.A. Waterman.

Fukunaka Mannenpitsu Seisakusho (Tokyo, Japan).

Geha (Hanover, Germany). Founded in 1818 by Heinrich and Conrad Hartmann.

Globe d'Or (Paris, France). Manufacturers of *Gold*, *Black* (1945), *Dominator* (1946).

Goldirca (Hanover, Germany). Derived its name from the words Gold-IRidium-CAoutchouc (rubber). Founded in 1919.

Gold Star (Paris, France). See Gold Starry.

Gold Starry (Paris/ Saint-Leu-la-Forêt, France). In 1909, Maurice Jandelle, at the time a salesman for Éditions Delagrave, concluded an agreement with the Conway Stewart Company for the distribution of its fountain pens in France under the trademark "Gold Star". Since the name was already registered, it was changed to "Gold Starry" (1912). In 1919, two men, a technician, Paul Janvrin, and a merchant, André Petit, began producing high quality fountain pens as a small business in Saint-Leu-la-Forêt, with direct distribution but inadequate financial backing. Jandelle wished to become independent and distribute French products. In order to establish these goals, in 1921, the Gold Starry company was created from a merger of the houses of Jandelle and Petit & Janvrin, with an initial capital outlay of 750,000 francs. Factories were built in Saint-Leu-La-Forêt. Gold Starry was the first French manufacturer to offer cellulose acetate pens in bright colors. The company remained very successful until 1933, but thereafter declined. Mr. Jandelle retired and management was passed to a Mr. Pérouse, who managed a small company that produced the "Viala

Lilliput". After World War II, the situation worsened with the advent of the ball-point pen. Gold Starry changed direction and became a producer of gold and silver trim for fountain pens. The company launched the production of high-end fountain pens. It remained successful for the next 20 years. In 1973, skyrocketing gold prices and a price freeze combined to push Gold Starry into bankruptcy. It went out of business in 1980.

Grafex (Brussels, Belgium). Trademark registered in 1926 by Louis Pauwels. Belgian manufacturing facilities. Other trademarks include *Perfex*, *Pratex*, and *Victory's Pen*. Went out of business in 1950s.

Grand Aigle (Toulon, Var/Toulouse, Hautes-Pyrénées, France). See Mercier.

Grand Perret (Saint-Claude, Jura, France). *Universal Fountain Pen*, 1930.

Gravade & Sachet (Paris, France). 29, boulevard Saint-Michel. Retailers circa 1894.

Grieshaber (Chicago, Illinois). Founded in 1842. Important fountain pen and nib manufacturer.

Guillement et Servet (Paris, France). *Omo* fountain pens (circa 1920).

Haro (Regensburg, Germany). Founded in 1926 in Fromsdorf (Silesia) by HAnns ROggenbuck. Specialized in glass nib fountain pens. Factories in Frankenstein and Weisswasser. In 1946, following the post-World War II partition, the Frankenstein factory found itself in Poland and the Weisswasser factory found itself in East Germany. Roggensbuck moved to Bad Mergentheim, then to Regensburg (West Germany). The company continued manufacturing glass nibs, then moved into the production of paper goods and stationery.

H. Hebborn (Heidelberg, Germany). Founded in 1925. Acquired by Parker in 1970.

Heintze & Blanckertz (Berlin/Frankfurt, Germany). Founded in Berlin in 1849 by Heinrich Siegmund Blanckertz and Rudolph Heintze. One of the largest German steel nib manufacturers and, before 1914, one of the fountain pen pioneers. Nationalized in East Germany after World War II and took the name of Schreibfedernfabrik-Berlin-Oranienburg.

Highley Pen Co. (Highley, Great Britain). Founded in 1946 by A.A.S. Charles (son of T.H. Charles, owner of T. Hessin & Co., manufacturer of steel nibs). Since 1949, it has had controlling interest in D. Leonardt & Co. and Hessin & Co. Took over the nib production department of Brandauer in 1960.

John Holland (Cincinnati, Ohio). Founded in 1841 (George Shepard). Originally a manufacturer of gold nibs, from the 1860s on John Holland was a true fountain pen pioneer. Some think that he was the first to market a more or less operative fountain pen in 1869. The company discontinued manufacturing fountain pens in 1950 but remained in business as stationers until 1980.

Hosonumma K.K. (Tokyo, Japan).

Houston Pen Co. (Sioux City, Iowa). Founded in 1911 by William A. Houston. Went out of business in 1924.

Inubushi & Co. (Tokyo, Japan).

Jewel Pen Co. (London, Great Britain).

JiF (Paris, France). Initials of Jules Fagard. See Waterman S.A.

Kaweco (Heidelberg, Germany). Heidelberger Federhalter Fabrik KOch, WEber & CO. Founded in 1892 by Otto Koch and Rudolph Weber, it was one of the first companies to offer *safeties*. Entered the French market in 1911. It was acquired by Friedrich Grube in 1929 and went out of business in 1970.

Kintz (Antwerp/Brussels, Belgium). See Le Tigre.

Kunz (Vienna, Austria).

Lamy (Heidelberg, Germany). The "Orthos Fullhalterfabrik C.J. Lamy" was founded in 1930 by Joseph Lamy, then director of the German subsidiary of Parker. Acquired Artus after World War II and is still in business.

Lancaster (Baltimore, Maryland). Founded in 1879 by Warren N. Lancaster. One of the true fountain pen pioneers.

Lang Pen Co. (Liverpool, Great Britain).

Laughlin (Detroit, Michigan). Founded in 1880. Went out of business in 1925.

*Laureau (Paris, France). In 1922, Fernand Laureau registered the trademark *Loro*. He manufactured fountain pens bearing the trade names of his customers, with or without nibs, and subcontracted to large manufacturers, such as Parker, for the production of beautiful pen trim. He received a silver medal in 1925 at the International Exposition of Decorative Arts in Paris. In 1929, he went into partnership with J. de Soultrait to establish E.J.S. Because he did not put his name on his products, Laureau's name is unjustly forgotten today.

*LeBoeuf (Springfield, Massachusetts). During the 1920s and 1930s, the company marketed exceptionally attractive fountain pens, utilizing many different plastics not employed by other manufacturers. The company survived until the late 1930s.

Lecocq (Paris, France). Manufacturer of clips.

LeFranc (Marseille, Bouches-du-Rhône, France).

Legorrec & Billaud Founded in 1923. Nib manufacturer, still in business.

Lieber (Lyon, Rhône, France). *Fontencre* trademark.

Lincoln (New York). Founded in 1894. Sold in 1900 to Frazer & Geyer Company. Two years later, after going into partnership with A.A. Waterman, Frazer and Geyer sold Lincoln to Perry, the largest nib manufacturer in the world. Nibs were manufactured in Great Britain under the trademark Devarson. Frazer and Geyer founded the New Lincoln Pen Co., which manufactured A.A. Waterman's low-end products.

C.W. Little (New York/Whitewater, Wisconsin). Manufactured *Century* fountain pens for approximately 30 years, beginning in 1900.

Lombard (Neuilly-sur-Seine, France). *Orego*, *Argeco*, and *Eco* fountain pens.

Lopy (Paris, France). Founded in 1934 by J. Lopy. *Jerzon* fountain pens.

Loro (Paris, France). See Laureau.

Luschi (Paris, France). Nib manufacturer. Significant contributor to the development of nib production in France through the creation of nib production divisions at the facilities of several fountain pen manufacturers.

Luxor (Heidelberg, Germany). See Hebborn.

Mabie Todd (New York, United States/London, Great Britain). In 1873, Mabie Todd & Bard was created from the association of John H. Mabie and Henry H. Todd, on the one hand, and of Jonathan Sprague Bard (Bard Bros. gold nib manufacturers since 1843), on the other. In 1884, a subsidiary was opened in London and the French market was entered in 1898. *Swan* fountain pens were distributed in most European countries. In 1909, fountain pens began to be manufactured in Great Britain. The *Swan* became *The Pen of the British Empire*. The parent company foundered, but the subsidiary flourished. In 1915, the British firm bought all rights for Europe and the British colonies. In 1921, the *Merle Blanc* was launched in France. The United States subsidiary went out of business in 1938. In 1952, Mabie Todd was absorbed by Biro and became Biro Swan, which was sold to BiC (France) in 1957.

MacKinnon (New York). Duncan MacKinnon is said to be the first to have offered a good needle point pen (1875). His stylographs were manufactured by John Holland.

MacNiven & Cameron (Edinburgh, Great Britain). Founded in 1770, the company was a leader for nearly two centuries in the area of stationery items. Its role was far more modest in the area of fountain pens. *Cameron* and *Waverley* fountain pens.

Maillocheau (Tours, Indre-et-Loire, France). *Magna*, *Speed-Point*, and *Walk-Over* fountain pens.

Mallat (Paris, France). Founded in 1842 by Jean-Benoît Mallat, an eclectic inventor who perfected the steel nib and manufactured the famous *Syphoïde* (1864). The company began manufacturing pens in 1916. *Integral* (1936), *Plexigraph* (1943). Still in business (writing and drafting instruments).

Manufacture Française de P.P.R. (Paris, France). See Richard.

Manufacture Lorraine de P.P.R. (Metz, Moselle, France). *The Kid*, *Griff*, and *Le Lorrain* fountain pens.

Manufacture Parisienne de P.P.R. (Paris/Nanterre, France). See La Plume d'Or.

Maroncini (Florence, Italy).

Le Matador (Paris, France). See Richard.

Matador (Wuppertal, Germany). See Siebert & Lowen.

Mauram (Saint-Claude, Jura, France.) M & A Benoit Frères.

Menard & Deluzy (Paris, France). 108, rue de Rennes. Distributors of Hicks fountain pens, circa 1894.

Mercier (Toulon, Var/Toulouse, Hautes-Pyrénées, France). Founded by Joseph Mercier in 1915 in Toulon. Moved to Toulouse in 1932. *Grand Aigle* trademark.

Mercury (Brussels, Belgium). Trademark of Dammaerts (circa 1947), an importer of Montblancs.

Météore (Paris/Nanterre, France). See La Plume d'Or.

Molinier (Marseille, Bouches-du-Rhône, France). *Roberry's* fountain pens.

Montblanc (Hamburg, Germany). Trademark registered in 1911 by Simplo-Fillerpen GmbH, which had been founded in 1908 by August Eberstein, Max Koch, Alfred Nehemias, and Klauss Voss (C.W. Lausen and W. Dziambor acquired shares in 1909). *Rouge et Noir*, *Simplo*, and *Diplomat* trademarks. In 1924, the high-end pens were given the generic name of *Meisterstuck*. In 1934, the trade name was changed to Montblanc Simplo GmbH. The white star is a symbol for Mont Blanc, and its altitude of 4,810 meters is inscribed on all nibs of the *Meisterstuck* pens. Still in business today, it is owned by the Dunhill group.

Mooney Pen (Chicago, Illinois). Founded in 1875 by Frank H. Mooney, gold nib manufacturer. Went out of business in 1917.

Moore (Boston, Massachusetts). See American Fountain Pen Co.

E. L. Moreau (Paris, France), 1898. Business office in Paris, 17, rue de Lancry, factories in Mouy, in Oise. In 1904, the company was acquired by and changed its name to J.M. Paillard. See Paillard.

Namiki (Tokyo, Japan). See Pilot and Dunhill.

Ohmi Yoko (Osaka, Japan).

Omas (Bologna, Italy). Officina Meccanicha Armando Simoni. Founded in 1919 by Armando Simoni (1891-1958). Still in business.

Onoto (London, Great Britain). See De La Rue.

Orthos Fullhalter-Fabrik (Heidelburg, Germany). See Lamy.

Osmia (Dossenheim, Germany). Founded in 1919 by Georg Bohler. Sold to Parker in 1928, which soon sold it back. Faber-Castell bought into the company in 1935.

Pagliero (Turin, Italy).

Paillard (Paris, France). Trade name used since 1904 by E.L. Moreau. Factory in Mouy (Oise). Specialized in colors and drafting instruments. Distributors in France of the *Simple* (Montblanc). In 1919, many of the partners of the J.M. Paillard Company founded the SAPR (Société anonyme pour la fabrication des P.P.R.). *Semper* and *Scriptor* fountain pens. Still in business as a manufacturer of drafting instruments.

Panici Frères (Paris, France). See Bayard.

*****Parker** (Janesville, Wisconsin). Founded in 1892 by George Stafford Parker. The success of the *Vacumatic* fountain pen, introduced in 1933, allowed Parker to exceed Sheaffer's sales. Today, it is the largest fountain pen manufacturer in the world. Recently acquired by the Gillette Company.

Pauwels (Lons-le-Saunier, Jura/Paris, France). Manufacturer of nibs, supplies, and pen trim.

Louis Pauwels (Brussels, Belgium). See Grafex.

Pedersen (Copenhagen, Denmark).

Pelican (London, Great Britain). See De La Rue.

Pelikan (Hanover, Germany). Founded in 1871 by Gunther Wagner following his purchase of the chemist Caro Horneman's business (created in 1832), where Wagner had worked since 1863. A manufacturer of stationery and drafting supplies (dyes and inks), Pelikan entered the fountain pen market in the 1920s. Known in France for its nibs and *Graphos* pen (industrial drafting). Still in business, but since 1984, it has belonged to Condorpart AG, Switzerland (today, the Pelikan Holding AG).

Pellet (Paris, France). *Plexipell*, *Stylopell*, *Montcalm*, and *Writter* fountain pens.

Pelletier (Brussels, Belgium). Ink and stationery manufacturer. *Pelletier* and *Imperial Pelletier* fountain pens.

Penlex (Paris, France). Formerly L. Frank Company. Manufacturer of fountain pen nibs.

Penola (Helsinki, Finland).

Pentel (Tokyo, Japan). Founded in 1946.

Perry & Co. (Birmingham/London, Great Britain). Founded in 1824 by James Perry. It was the largest manufacturer of steel nibs in the world and also produced fountain pens, but its success in that area was far less spectacular. See Devarson.

E.S. Perry, Ltd. (London/Gosport, Great Britain). Founded in 1921 by Edmund S. Perry, the director of Perry & Co., Ltd. in Birmingham, first manufacturer of steel nibs in the world. *Osmiroid* nibs and fountain pens.

Pilot (Tokyo, Japan). Trade name since 1938 of the Namiki Manufacturing Co., founded in 1918 by Ryosuke Namiki and Masao Wada.

Platinum (Tokyo, Japan). Trade name since 1942 of the Nakaya Manufacturing Co., founded in 1919 by Shyunichi Nakada.

La Plume d'Or (Paris/Nanterre, France). Name taken in 1921 by a company founded in 1916 as Manufacture Parisienne de P.P.R. Trademark of *La Météore*. Other trademarks are *Zodiac*, *Prompto* (1922), and *Pullman* (1932). It is also one of the most important European manufacturers of gold nibs. First in Paris, 63, rue des Archives, then 48, rue des Vinaigriers, the factories were finally located in Nanterre, 26-30, rue des Amandiers. Nibs are marked with the initials D&D (the founders' [Demilly and Degen] initials). During World War II, gold nibs were replaced by steel nibs stamped with the word *Vaedium*. Went out of business in 1956.

A. Poulain (Paris, France). 1892. Distributors of Caw's *Dashaway*, 7, avenue de l'Opéra. The former stationery company specialized in the distribution of English steel nibs, having been founded by Roux and agents of E.W. Cuthbert (Birmingham).

Pratex (Brussels, Belgium). See Grafex.

Regnault (La Ferté-Millon, Aisne/Valence, Drôme, France). Factories were established in 1927 in La Ferté-Millon. *Ludo* fountain pens. In 1945, the company moved to Valence. In 1948, it began manufacturing Reynolds ball-point pens (United States) under license. In 1954, it acquired the patent for that trademark. Still in business.

Reynolds (France). See Regnault.

Richard (Paris, France). Before World War I, F.J. Richard opened a factory in Paris for the manufacture of fountain pens bearing the trademark *Le Matador*. Other trademarks include *Imperial*, *Royal*, *Le Gracieux*, *Le Piou-Piou*, and *Le Parigot*. In 1920, the company changed its name to Manufacture Française de P.P.R.

J.G. Rider Pen Co. (Rockford, Illinois). Founded in 1905, went out of business in 1925.

Roggenbuck (Regensburg, Germany). See Haro.

Rotring (Hamburg, Germany). Founded in 1928. Manufacturer of tubular fountain pens.

Roubeaud (Paris, France). *Pousse-Pouce* (1897), *Star* (1899) fountain pens.

Ruyter (Chicago, Illinois).

Sabatier (Paris, France). Manufacturer of *Mors*, *Rip*, *Execo*, and *Wattman* fountain pens.

Sabon (Bordeaux, France). Manufacturer of heavy equipment until 1920. Began producing fountain pens during the 1920s, manufacturing its own supply of ebonite. In 1932, it switched to nitrocellulose plastics. With the exception of two fountain pens, *Sprell* and *Yale*, most of its products were sold under their clients' trade names. Went out of business in 1950.

Sakata Seisakusho (Japan). Founded in 1911 by Kyugorou Sakata. *Sailor* fountain pens.

Salz (New York). Founded in 1907.

Sertic (Paris, France). Manufacturer of *Inoxstyl* (1928).

Sheaffer (Fort Madison, Iowa). Founded in 1913 by Walter A. Sheaffer, a jeweler and inventor of the lever filling system. In 1923, Sheaffer was the first to use a material called pyroxylin and to manufacture colored pens. In 1929, he also was the first to capitalize upon the potential of pyroxylin, introducing a line called *Balance*, which would give him an advantage in the market of the 1930s. Sheaffer, who had refused to produce pens with 18K gold nibs, did not export his products to France until the 1950s.

A. Shipman's Sons (New York).

Siebert & Lowen (Wuppertal, Germany). Founded in 1895. Manufacturer of *Matador* fountain pens.

Simplo (Hanover, Germany). See Montblanc.

Soennecken (Bonn, Germany). Founded in 1875 by Friedrich Soennecken (1848-1919), who was a calligrapher and the author of penmanship manuals. The company is the oldest German manufacturer of steel nibs (1890) and remained one of the most influential during the next 50 years. Upon Friedrich's death, his son, Alfred, succeeded him. The factories were destroyed during World War II. The 1950s were a difficult period. Despite a partnership with the Bayard Company, the company went out of business in 1967.

Stabil (Brussels, Belgium). Trademark used by Chaim Jakubowicz, a distributor for Sheaffer. Sold fountain pens bearing the trademarks *Stabil-Drake*, *Majestic* (1946). Went out of business in 1974.

Staedler (Nuremburg, Germany). Pencil factory founded in 1662 by Friderich Staedler. While retaining pencils as its major product, the company launched the manufacture of fountain pens after World War I. It is today the largest manufacturer of drafting and writing instruments in the world.

Stellor (Paris/Pavillons-sous-Bois, France). Founded in 1918 by Joseph Beaufils. 39, rue Doudeauville in Paris, then 288, avenue Aristide-Briand in Pavillons-sous-Bois. Factories in Nurieux (Ain, France) and La Ferté-Millon (Aisne, France). Closed from 1942 to 1945, the company went out of business in 1956.

Stephens (Levallois-Perret, France). Founded in the 1930s by an influential English ink manufacturer. *Autograph* and *Royal* fountain pens.

Stortz (Graz, Austria).

Stroesser (Paris, France). *Matcher Pen*, *Matcher Colombes*, and *Jeep* fountain pens.

Stylomine (Paris, France). Originally the Y.E. Zuber Company, specialized in the cutting and stamping of metals. Produced, among other items, nibs, clips, and mechanical pencils. The latter were sold under the trademark *Stylomine* (1921). Fountain pen manufacturing began in 1925, and in 1930, the launching of the *303* was an enormous success. It was filled through a system using air compression of a rubber sac. This pen with a retractable nib is one of the rarest examples of automatic filling. In 1934, the company changed its name to S.A. des Établissements Stylomine. In later years, the company introduced the tube and accordion filling system, which was copied by most French manufacturers after the original patent expired.

StyloPneu (Neuilly-sur-Seine, France). Filling was accomplished through an external rubber bulb, which compressed the reservoir, or by blowing into the pen.

Swan (Great Britain). See Mabie Todd.

Tabo (Bologna, Italy). Founded by Tantini. (TAntini BOlogna).

Tardy (Paris, France). *Optimus* (1919) and *Le Parisien* fountain pens.

Tibaldi (Florence, Italy). Founded in 1916.

Le Tigre (Antwerp/Brussels, Belgium). Registered in 1918 by Rene Kintz, it is the best known trademark in Belgium. Other Kintz trademarks are: *Le Lion*, *Le Loup*, *Le Régional*, *Lyceum Pen*, *Régent*, *Jubilé*, and *Regency*. In the 1960s, Kintz sold low-end fountain pens, such as the student model called *Tintin-Kuifje*, which was made in Germany. The company went out of business in 1985.

Tonnelier (Cholet, Maine-et-Loire, France). *Eric Pen* (1924), *Gallia*, *Reservo*, and *Lido* fountain pens.

Tourneries du Rhône (Lyon, Rhône, France). *Stylox* is the trademark.

Tourteau (Paris, France). Clips; *L.T.* is the trademark.

Tropen (Ludenscheid, Germany). Founded by Gustav Schroeder. Injection molds for plastics. Began manufacturing fountain pens in 1925. Still in business today.

J. Ullrich & Co. (New York). Founded in 1884. *Independent* and *Elk* fountain pens.

Unic (Paris, France). Founded in 1919 by Kothe and Vannier. 51, rue Rochechouart.

Viala Lilliput (Paris, France). See Gold Starry.

Visconti (Florence, Italy). Founded in 1988 by Dante del Vecchio and Luigi Poli. Specializes in fountain pens and other writing instruments made from celluloid following the original process for making celluloid as developed by the Hyatt brothers in 1869.

Gunther Wagner (Hanover, Germany). See Pelikan.

Wahl-Eversharp (Chicago, Illinois). Founded in 1905, the Wahl Adding Machine Company took over the Japanese company Ever-Sharp Pencil in 1914 and the Boston Pen Company in 1917, becoming the Wahl Company. Entered the French market in 1922 (Frazar France). The best known pen trademarks are *Personal Point* (1929) and *Doric* (1931). Wahl and its affiliate Eversharp merged in 1940. The *Skyline* model was introduced in 1941. In 1957, Parker acquired the writing instrument division of Eversharp and liquidated it 5 years later.

A.A. Waterman (New York). Founded in 1897 by Arthur Waterman (no relation to Lewis Edson Waterman). In 1902, he launched the *Modern Middle Joint Fountain Pen*, one section of which extends all the way to the middle of the body to avoid leaks and stains on the fingers. Introduced the *Automatic Self-Filling Modern Fountain Pen*, which was filled by twisting the rubber sac. The company went into business with Frazer & Geyer in 1902, which proved to be a bad move. In 1905, Frazer & Geyer took over control of the company and the name. Lewis Edson's heirs objected to the use of the name and went to court in 1912. It was decided by the court that from then on, catalogs and advertising should mention that there was no connection between A.A. Waterman and L.E. Waterman. Four years later, A.A. Waterman stopped producing fountain pens.

L.E. Waterman Co. (New York). Founded in 1888 by Lewis Edson Waterman 4 years after he registered a patent on a feed design. Circa 1910, the company controlled three quarters of the U.S. market, and until 1925, it was the largest pen company in the U.S. A reluctance in accepting technological innovations, such as the replacement of black hard rubber by plastics with a nitrocellulose base, led to its decline in the 1930s. Production stopped in 1954. The U.S. trademark was sold to Baron Bich, who sold it back to Waterman S.A. (France) in 1971. In 1972, the English trademark was also acquired by Waterman S.A.

Waterman S.A. (Paris, France). In 1914, Jules Fagard was the distributor for L.E. Waterman Co. in France and Belgium. In 1926, he founded Jif-Waterman. His widow and daughter managed the company, which became Waterman S.A.. His granddaughter, Francine Gomez, made Waterman S.A. the premier European fountain pen manufacturer and the second worldwide. The company was sold to the U.S. company Gillette in 1986. From 1967 on, all Waterman fountain pens have been manufactured in France in the factory of Saint-Herblain located near Nantes.

Wearever (North Bergen, New Jersey). David Kahn Inc. Specialized in the low-end fountain pen with steel nibs. For years the company produced more pens than Waterman, Parker, Sheaffer, and Wahl-Eversharp combined.

Weidlich Simpson (Cincinnati, Ohio). Founded in 1895 (Wright Pen Co.). Went out of business in 1921.

Wirt (Bloomsburg, Pennsylvania). The most important manufacturer of the last century. In 1878, Paul E. Wirt registered a patent for a simple fountain pen. A few were manufactured by hand, and in 1885, he designed his own machinery and launched mass production. He was the first to mechanize the production of fountain pens and claimed that one million had been made in his factories over a 10-year period; a number four times that made by Waterman. In 15 years, he claimed to have manufactured more fountain pens than all of his competitors combined. Beginning in 1900, competition would inexorably drive Wirt out of business, which finally occurred in 1937.

Wyvern (Leicester, England). Founded in 1896 by David, Alec, and Alfred Finburgh. Stayed in business until 1955.

Y.E. Zuber (Paris, France). See Stylomine.

INDEX

This index does not include the names found in the preceding appendix, "Significant Names in the Fountain Pen Industry".

Aluminum: 74, 93, 126
American: 35, 40, 44, 46, 62, 64, 68, 70, 72, 74, 80, 99, 120, 141, 144, 149, 158
Anglo-American Co.: 68
Argentina: 135
Atmospheric pressure: 38, 90, 92, 135
Aurora: 103

Babel: 73
Baekeland, L.H.: 126
Baignol & Farjon: 29
Baignol, P.: 107
Bakelite: 50, 118, 126
Band: 14, 36, 40, 50, 108, 128, 144, 150
Barnes: 50, 61
Bauhaus: 56, 118
Baums, M.: 134
Bayard: 68, 72, 74, 113
Bayle: 148
Beaufils, J.: 72
Belgium: 72
Bibax: 93
Bion, N.: 29, 30
Birmingham: 28, 29, 35, 113

Biro, L.J.: 135
Black hard rubber: 38, 40, 42, 44, 46, 50, 52, 58, 68, 74, 82, 84, 93, 106, 117, 118, 119, 120, 122, 124, 125, 126, 141, 146, 150, 152, 154, 162, 166, 175
Blair: 70, 106
Blanzy-Pour: 29
Bonhomme, J.: 72
Borrowdale: 35
Boston Fountain Pen Co.: 62
Boulard: 108
Boulogne-sur-Mer: 28, 29
Brandauer: 29
Brass: 27, 30, 115, 128, 148
Britain: 35, 64, 68, 70, 72, 80, 113, 120
Bronze: 26, 68, 70

Camphor: 120
Capillary attraction: 38, 52, 89, 90, 92, 93, 94, 96, 98, 116
Capillary passage: 38, 90, 102, 117
Cap: 14, 15, 40, 44, 46, 50, 52, 54, 64, 66, 72, 95, 98, 102, 105, 107, 113, 115, 117, 118, 122, 126, 128, 130, 148, 149, 152, 154, 156, 158, 178
Carbon: 110, 117
Cartier: 138, 152
Cartridge: 98, 99, 102, 105, 106, 107, 147, 178

Casein: 122
Caw's Nib & Pen Co.: 68, 70
Celluloid: 68, 84, 118, 122, 152
Cellulose acetate: 74, 118, 122
Cellulose acetobutyrate: 122
Chalk: 11, 22
Charcellay: 66, 70, 125
Chicago: 38, 62
Chilton: 97
China: 35
Chrome: 117, 132
Clarke, C.F.: 28
Clip: 15, 40, 50, 62, 66, 100, 115, 122, 126, 128, 130, 150, 156
Clip ring: 40, 128, 130, 132
Coal: 118
Cobalt: 126
Collecting: 78, 80, 143, 144, 148, 150
Compression bar: 36, 52, 61, 66, 72, 74, 98, 158, 160
Conklin: 58, 61, 154, 178
Conte, N.: 35
Conway Stewart: 70
Copper: 27, 113, 130
Corrosion: 110, 111, 117
Cotton: 120, 122
Coulson, B.: 61
Czechoslovakia: 111

De La Rue: 68, 82
Decatur: 40
Degen: 72
Demilly: 72
Diamond: 11, 111, 112, 117
Doughty: 111
Dunhill: 156
DuPont de Nemours: 61, 120

Eagle Pencil Co.: 99
Eberhard Faber Pencil Co.: 135
Edac: 68, 72, 74

Edacoto: 68, 103
Egypt: 22, 26, 29
Eisner, P.: 135
Eterpen: 135
Europe: 26, 28, 32, 35, 46, 70, 113, 116, 156
Ever-Sharp Pencil Co.: 62
Evergood: 74
Eversharp: 35, 62, 64, 118, 135
    Doric: 62, 64
Eyedropper: 58, 95, 96, 99, 128

Fagard, J.: 72
Feed: 15, 30, 35, 38, 40, 50, 52, 74, 90, 92, 93, 96, 98, 107, 117, 118, 119, 164
Ferris, W.J.: 38, 70
Felt: 11, 13, 22
Filigree: 17, 42, 50, 117, 144
Filling: 30, 32, 36, 38, 52, 56, 58, 61, 62, 68, 72, 73, 82, 87, 90, 93, 95, 96, 97, 98, 101, 103, 104, 106, 108, 126, 138, 141, 152, 158, 164, 166, 175
Forbin: 72
Formaldehyde: 122, 126
Fort Madison: 58
France: 14, 29, 32, 38, 40, 46, 50, 54, 64, 66, 68, 72, 74, 93, 99, 101, 103, 106, 111, 113, 116, 120, 125, 128, 130, 132, 135, 144, 146, 147, 150, 152, 154, 158, 164, 166, 175, 176
Franco-British Pen Manufacturing: 70

Galalith: 122
Germany: 28, 30, 32, 68, 74, 90, 120
Gillott, J.: 28, 29
Glass: 98, 99, 102, 106, 107, 108, 110, 111
Gold: 11, 26, 27, 34, 42, 44, 46, 58, 74, 82, 93, 98, 99, 105, 108, 111, 112, 113, 114, 115, 116, 117, 125, 132, 134, 144, 152, 166, 174
Gold-filled: 42, 44, 46, 50, 58, 62, 68, 74, 84, 116, 117, 120, 126, 132, 136, 138, 166, 174
Gold Star: 70
Gold Starry: 70, 72, 74, 113, 150
Goodyear, C.: 119
Grand Aigle: 72, 74, 124
Graphite: 35, 36
Gravity: 90, 92
Griffon: 74, 148

Hancock, T.: 119
Hardtmuth, L.& C.: 70, 176
Haro: 111
Harrison, S.: 28
Hawkins, J.I.: 112
Hayakawa, T.: 62
Heintze & Blanckertz: 29
Holland, J.: 35, 50
Horn: 112, 118
Howard, E.T.: 42
Hungary: 135
Hyatt, J. W.: 120, 122

Ink: 11, 14, 22, 27, 29, 30, 32, 34, 35, 36, 38, 40, 42, 50, 52, 66, 74, 89, 90, 92, 93, 94, 96, 97, 98, 99, 100, 105, 106, 107, 108, 110, 115, 117, 118, 119, 122, 126, 134, 135, 136, 138, 143, 158, 160, 166, 178
Ink well: 29, 34, 36, 61, 93, 95, 99, 128
Iridium: 93, 111, 112, 113, 114, 115, 116, 126, 158
Ivory: 120, 146

Jandelle, M.: 70, 72
Janesville: 50, 52
Japan: 22, 62, 122
Janvrin, P.: 72

JiF: 72, 74, 98, 99, 102
Jura: 66

Kansas City: 61
Karat: 113, 117
Klimes, F.: 135
Kollisch: 97
Kothe: 72
Kraker Pen Co.: 61
Kraker, G.: 62
Kritsche: 122

Lacquer: 11, 46, 107, 136, 138
La Plume d'Or: 72, 74, 93
Lambert, E.: 134, 135
Lancaster: 35
La Palice: 72
Laser: 138
Laureau, F.: 54, 66, 122, 164
Le Météore: 72, 74, 166
Lever: 36, 42, 50, 56, 58, 61, 62, 64, 66, 74, 82, 84, 87, 90, 98, 99, 103, 105, 115, 122, 124, 126, 154, 156, 158
Loud, J.J.: 134, 135

Mabie Todd: 80
Maginnis, J.P.: 32
Mallat: 30, 32, 66, 68, 72, 74, 113
    Plexigraf: 66
    Syphoïde: 30, 32, 68, 72
Manufacture Parisienne de P.P.R.: 72
Martin, H.G.: 135
Mason, J.: 28
Matador: 74, 90
Mercier, J.: 72, 124
Mesopotamia: 72
Methyl polymethacrylate: 126
Mitchell, J.: 28, 29
Mitchell, W.: 28, 29
Molybdenum: 117

Monograph: 104
Montblanc: 84, 110, 138
Moreau, E.L.: 68
Moseley: 97
Muslim: 22

Namiki: 156
New York: 36, 40, 58, 113
Nib: 11, 13, 14, 15, 22, 27, 28, 29, 30, 32, 34, 35, 40, 50, 52, 56, 62, 66, 74, 90, 92, 93, 94, 95, 96, 98, 107, 108, 110, 111, 112, 113, 114, 115, 116, 117, 118, 126, 128, 130, 138, 143, 144, 156, 158, 164, 166
Nib, feather quill: 111, 112, 117, 118, 138
Nib, steel: 13, 28, 29, 34, 35, 36, 98, 113, 116, 117, 138, 150
Nib, gold: 11, 30, 111, 112, 113, 114, 116, 117, 144, 162
Nib, glass: 108, 110
Nib, metal: 26, 27, 28, 30, 35, 68, 118, 136
Nib, retractable: 46, 95, 98, 101, 110, 118, 125, 126, 130, 150, 154, 158, 164
Nickel: 117, 126
Nitrocellulose: 54, 61, 90, 120, 122, 162

Onoto: 82
Osmium: 111, 113, 115, 116, 126

P.S.F.: 42, 50, 144, 164
Paillard: 68, 74, 113, 166
Palladium: 111, 117
Palmer, W.F.: 52
Panici: 72
Parker, G.: 50, 51, 52
Parker, K.: 52, 56
Parker, R.: 52

Parker: 35, 36, 50, 52, 54, 56, 62, 64, 93, 98, 122, 128, 130, 136, 162, 164
    Arrowhead: 52
    Jack Knife: 50, 52
    Jointless Pen: 52
    51: 52, 56, 116, 118
    61: 98
    75: 136
    180: 136
    Duofold: 50, 52, 54, 136
    Lucky Curve: 50, 52
    Parkette: 50
    Royal Challenger: 50
    Televisor: 50
    Vacumatic: 50, 52, 56, 97, 130, 160, 162
    Victory: 50
Pen, ball-point: 11, 13, 15, 22, 56, 64, 74, 134, 135, 136, 138, 143
Pen, eternal: 19, 29, 30
Pen, pneumatic: 73
Pencil: 11, 22, 35, 58, 62, 64, 72, 92, 95, 103, 128, 175
Permanite: 54, 122
Perraud: 107
Perry, J.: 28, 29, 35
Petit, A.: 72
Petroleum: 118
Phenol: 126
Piston: 84, 90, 96, 97, 134, 135, 160
Platinum: 111, 126
Plexiglas: 74, 118, 126
Plexor: 54
Plunger: 82, 97, 99, 178
Plumoto: 73
Pneu: 97
Point: 22, 90, 93, 108, 110, 114, 115, 116, 126
Pullman: 74
Pulsa Pen: 74
Pyroxylin: 61

Radite: 58, 122
Reed: 22, 29, 90, 92, 118
Reed pen: 11, 13, 22, 24, 26, 29, 30, 35, 117
Regular: 15, 96, 128
Remington: 64
Reynolds, M.: 135
Rhenium: 126
Ring: 30, 99, 102, 107, 128, 130, 154, 156
Rods: 38, 52, 93, 117, 119, 120
Rolpen: 135
Rome: 22, 26, 118
Rool's: 72
Rubber: 36, 42, 50, 54, 56, 58, 61, 72, 78, 93, 95, 97, 99, 101, 102, 108, 115, 119, 120, 122, 126, 152, 158, 160, 164, 166, 178
Ruby: 111, 112, 113
Ruthenium: 126

Sabon: 66, 90, 114
Safety: 15, 66, 84, 95, 96, 120, 130, 144, 150
Salmon, J.-B.: 106
Seech: 135
Sénéchal, G.: 32
Sharkskin: 146
Sheaffer, C.: 61
Sheaffer, W.A.: 58, 61, 62
Sheaffer: 35, 50, 58, 61, 97, 122
    Balance: 56, 58, 61, 64
    Connaisseur: 136
    Lifetime: 58
    Military Clip: 58
    Nostalgia: 136
    Slim: 136
    Snorkel: 58, 61
    Targa: 136
    Triumph 1250: 58

Sheffield: 28
Silver: 26, 27, 30, 42, 44, 46, 50, 62, 66, 73, 74, 93, 99, 113, 117, 120, 124, 125, 126, 128, 130, 132, 136, 138, 141, 150, 164, 174, 175
Skinner, J.: 28
Soennecken, F.: 29
Solid ink: 106
Spitteler: 122
Spittle: 28
Steel: 27, 28, 98, 108, 110, 116, 117
Stellor: 72, 74, 152
Straus, M.: 64
Stylo: 14, 15, 74
Stylograph: 14, 15
Stylomine: 70, 72, 74, 95, 97, 98, 113, 160
    303: 101
Styloplum: 73, 74
Swan: 74, 80, 95, 120, 128, 178
Synthetic resins: 118, 126

Tank 400: 107
Tennant, S.: 111
Titanium: 126
Tortoise shell: 112, 118
Tungsten: 126

Unic: 66, 72, 74, 113
    Duocolor: 66
United States: 13, 14, 34, 35, 42, 46, 62, 68, 74, 95, 113, 134, 135, 147, 158

Vacuum tube: 97
Vannier: 72
Varley: 134
Visor Pen: 74

Wahl Adding Machine Co.: 62
Wahl: 62, 64, 87, 118, 122

Personal Point: 64, 87
Wahl-Eversharp: 35, 62, 64
    Skyline: 62, 64
Waterman, A.A.: 97
Waterman, L.E.: 32, 35, 36, 38, 40, 42, 50, 52, 70, 90, 93, 94
Waterman Co.: 17, 36, 38, 40, 42, 44, 46, 50, 61, 62, 64, 68, 70 72, 74, 95, 98, 105, 106, 107, 108, 113, 120, 122, 125, 126, 128, 130, 136, 138, 141, 144, 147, 149, 150, 154, 156, 158, 160
    Centenaire de la Révolution Française: 136
    Clip-Cap: 38, 130
    Cone Cap: 44
    Duo: 113
    Flash Fill: 106
    Gentleman: 136
    Hundred Year: 113
    Ideal Pocket: 128
    Ink Vue: 158
    Knapsack: 128
    Lady Patricia: 122, 136
    Opéra: 136
    Patrician: 113, 122
    Secretary: 44
    Sleeve Self-Filler: 36
    Straight Cap: 36
    Super Cartouche: 105
    Taper Cap: 40
    Telescope Cap: 128
    X Pen: 98, 101
    58: 50
Waterman, S.A.: 74
Wirt: 35
Wollaston, W.H.: 111
Wood: 20, 22, 35, 38, 42, 74, 115, 120

Zuber, Y.E.: 72

Bibliography

Bowen Glen, *Collectible Fountain Pens*, Glenview, IL, 1982.

Castruccio Enrico, *La Penna*, IdeaLibri, Milan, 1985.

Columbain Marcel, *l'Aventure multiple des outils de l'écriture*, 1963.

Desechaliers, *Annuaires de la papeterie et de l'imprimerie.*

Edgar G., "The Story of the Fountain Pen", *London Magazine*, 1904.

Fischler George and **Schneider** Stuart, *Fountain Pens and Pencils*, Shiffer Publishing, West Chester, PA, 1990.

Fournier Lucien, "Le stylo, c'est la vie", *la Science et la Vie*, December 1925.

Fritz Paul, "La Plume: partie essentielle du stylo", *Papetier de France*, March 1962.

Gandillet Etienne, "Avec le porte-plume à réservoir, nous sommes loin de la plume d'oie", *la Science et la Vie*, March 1920.

Germano Stefano, *Signori, la Penna*, La Stilographica, Bologne.

Habert L., "Fountain Pen, les plumes d'or à réservoir", *la Nature* (no. 1 574), 1903.

Huber Jurg-Peter, *Griffel - Feder - Bildschirmstift, Eine Kulturgesicichte der Schreibgerate*, AT Verlag, Aarau, 1985.

Jackson Donald, *The Story of Writing*, Studio Vista, London. French translation, *Histoire de l'Écriture*, Denoel, Paris, 1982.

Lacroux Jean-Pierre and **VanCleem** Lionel, *la Mémoire des Sergent-Major*, Ramsay-Quintette, Paris, 1988. Italian translation, *Il Pennino*, Ulissediziona, Turin, 1988.

Lambrou Andreas, *Fountain Pen, Vintage and Modern*, Sotheby's Publications, London, 1989.

Lawrence Cliff, *Fountain Pens*, Paducah, Kentucky, Collector Books, 1977.

Lawrence Cliff, *Official P.F.C. Pen Guide*, Pen Fancier's Club, Dunedin, Florida, 1982.

Lawrence Cliff and **Lawrence** Judy, *An Illustrated Fountain Pen History*, Pen Fancier's Club, Dunedin, Florida, 1986.

Maginnis James P., "Reservoir, Fountain and Stylographic Pens", London, *Journal of the Society of Arts*, Nos. 2 761 to 2 764, Vol. LIII, October and November, 1905.

Nakasono Hiroshi, *les Porte-Plume réservoir dans le monde*, Tokyo, Kodanska, 1985.

Schneider Stuart L. and **Etter** Roberta B., *Collecting and Valuing Early Fountain Pens*, Hudson Valley Graphics, Teaneck, NJ, 1980.

Schneider Stewart and **Fischler** George, *The Book of Fountain Pens and Pencils*, Westchester, PA, 1992.

Sénéchal Georges, *l'Industrie des porte-plume réservoirs en France*, Éditions de la Revue dactylographique, Paris.

Tavanti Sergio, *La Penna silografica, origini, funzionamento e collezione*, Alberti & C., Arezzo, 1987.

Whalley Joyce Irene, *Writing Implements and Accessories*, Newton Abbot, David & Charles, 1975.

Collectors Clubs

(Between parentheses: year of founding and frequency of bulletin)

Belgium. Belgian Pen Collectors' Association (1986), *La Plume d'Oie/De Ganzepen* (x4)

Canada. Collectors' Club of Canada (1984), *Writing Instrument* (x4)

United States. American Pencil Club Collectors (1957), *Pencil Collector* (x12)
Pen Fancier's Club (1977),
The Pen Fancier's Magazine (x6)
Pen Collectors of America, Inc.
The Pennant (x4)
Pen World (non-affiliated)

France. Club des collectionneurs de plumes, porte-plume réservoir et objets d'écriture (1979)
Au fil de la plume (x4)

Great Britain. Writing Equipment Society (1980), *Journal of the W.E.S.* (x3)

Germany. Internationales Forum Historische Burowe (1981)
Historische Burowelt (x4), *HBW Aktuell* (x8)

Italy. Accademia Italiana della Penna Stilografica *Stilomania* (x4)

Acknowledgments

Roland Bergue, Roland Germain,
Bernard Jouve, Patrick Kuperfis, Kimiyasu Tatsuno,
Patrick van Hoof, Lionel Van Cleem.

Printed in Italy

by G. Canale & C. S.p.A.

Borgaro T.se - Turin